Advances in Human Factors/Ergonomics, 21B

Design of Computing Systems:

Social and Ergonomic Considerations

Advances in Human Factors/Ergonomics

Series Editor: Gavriel Salvendy, Purdue University, West Lafayette, IN 47907, USA

Vol. 1. Human-Computer Interaction (G. Salvendy, Editor)
Vol. 2. Human-Computer Dialogue Design (R.W. Ehrich and R.C. Williges, Editors)
Vol. 3. Expertise Transfer for Export System Design (J.H. Boose)
Vol. 4. Engineering Physiology: Physiologic Bases of Human Factors/Ergonomics (K.H.E. Kroemer. H.J. Kroemer and K.E. Kroemer-Elbert)
Vol. 5. Human Factors Testing and Evaluation (D. Meister)
Vol. 6. Applications of Fuzzy Set Theory in Human Factors (W. Karwowski and A. Mital, Editors)
Vol. 7 Human Reliability: Analysis, Prediction and Prevention of Human Errors (K.S. Park)
Vol. 8. Human Aspects of Occupational Vibration (D.E. Wasserman)
Vol. 9 Human Factors Research: Methods and Applications for Architects and Interior Designers (J.E. Harrigan)
Vol. 10A. Social Ergonomic and Stress Aspects of Work with Computers (G. Salvendy, S.L. Sauter and J.J. Hurrell, Jr., Editors)
Vol. 10B. Cognitive Engineering in the Design of Human-Computer Interaction and Expert Systems (G. Salvendy, Editor)
Vol. 11. Occupational Safety and Accident Prevention: Behavioral Strategies and Methods (C.G. Hyos and B. Zimolong)
Vol. 12A. Work with Computers: Organizational, Management, Stress and Health Aspects (M.J. Smith and G. Salvendy, Editors)
Vol. 12B. Designing and Using Human-Computer Interfaces and Knowledge Based Systems (G. Salvendy and M.J. Smith, Editors)
Vol. 13 Designing User Interfaces for International Use (J. Nielsen, Editor)
Vol. 14. Human Factors in Product Design (W.H. Cushman and D.J. Rosenberg)
Vol. 15. Work Space, Equipment and Tool Design (A. Mital and W Karwowski, Editors)
Vol. 16. Connective Networks in Ergonomics: General Methodological Considerations (E.A. Franus)
Vol. 17. Psychology of System Design (D. Meister)
Vol. 18A. Human Aspects in Computing: Design and Use of Interactive Systems and Work with Terminals (H.-J. Bullinger, Editor)
Vol. 18B. Human Aspects in Computing: Design and Use of Interactive Systems and Information Management (H.-J. Bullinger, Editor)
Vol. 19A. Human-Computer Interaction: Applications and Case Studies (M.J. Smith and G. Salvendy, Editors)
Vol. 19B. Human-Computer Interaction: Software and Hardware Interfaces (G. Salvendy and M.J. Smith, Editors)
Vol. 20A. Symbiosis of Human and Artifact: Future Computing and Design of Human-Computer Interaction (Y. Anzai, K. Ogawa and H. Mori, Editors)
Vol. 20B. Symbiosis of Human and Artifact: Human and Social Aspects of Human-Computer Interaction (Y. Anzai, K. Ogawa and H. Mori, Editors)
Vol. 21A. Design of Computing Systems: Cognitive Considerations (G. Salvendy, M.J. Smith and R.J Koubek, Editors)
Vol. 21B. Design of Computing Systems: Social and Ergonomic Considerations (M.J. Smith, G. Salvendy and R.J Koubek, Editors)

Advances in Human Factors/Ergonomics, 21B

Design of Computing Systems:

Social and Ergonomic Considerations

Proceedings of the Seventh International Conference on Human-Computer Interaction, (HCI International '97), San Francisco, California, USA August 24-29, 1997
Volume 2

Edited by

Michael J. Smith
University of Wisconsin, Madison, WI 53706, USA

Gavriel Salvendy
Purdue University, West Lafayette, IN 47907, USA

Richard J. Koubek
Purdue University, West Lafayette, IN 47907, USA

1997
ELSEVIER
Amsterdam - Lausanne - New York - Oxford - Shannon - Tokyo

ELSEVIER SCIENCE B.V.
Sara Burgerhartstraat 25
P.O. Box 211, 1000 AE Amsterdam, The Netherlands

ISSN: 0921-2647
ISBN: 0 444 82183 X

This book is printed on acid-free paper.

PREFACE

A total of 2,183 individuals from industry, academia, research institutes, and governmental agencies from 43 countries submitted their work for presentation at the Seventh International Conference on Human-Computer Interaction held in San Francisco, California, U.S.A., 24-29 August 1997. Only those submittals which were judged to be of high scientific quality were included in the program. These papers address the latest research and application in the human aspects of design and use of computing systems. The papers accepted for presentation thoroughly cover the entire field of human-computer interaction, including the cognitive, social, ergonomic, and health aspects of work with computers. These papers address major advances in knowledge and effective use of computers in a variety of diversified application areas, including offices, financial institutions, manufacturing, electronic publishing, construction, and health care.

We are most grateful to the following cooperating organizations:

Chinese Academy of Sciences
EEC-European Strategic Programme for Research and Development in Information Technology-ESPRIT
Human Factors and Ergonomics Society
IEEE Systems, Man and Cybernetics Society
Japan Ergonomics Society
Japan Management Association

The 251 papers contributing to this book cover the following areas:

Cognitive Modelling
Interface Design and Evaluation
Multimedia, Virtual Reality and World-Wide Web

The select papers on a complementary part of human-computer interaction are presented in a companion volume edited by M. J. Smith, G. Salvendy and R.J. Koubek, titled *Design of Computing Systems: Social and Ergonomic Considerations.*

We wish to thank the following Board members who so diligently contributed to the success of the conference and to the direction of the content of this book. The conference Board members include:

Organizational Board
Lajos Balint, *Hungary*
Hans-Jorg Bullinger, *Germany*
Satoshi Goto, *Japan*
Yoshio Hayashi, *Japan*
Bengt Knave, *Sweden*
Vladimir M. Munipov, *Russia*
Takao Ohkubo, *Japan*
Susumu Saito, *Japan*
Ben Shneiderman, *USA*
Jean-Claude Sperandio, *France*
Constantine Stephanidis, *Greece*
Hiroshi Tamura, *Japan*
Thomas J. Triggs, *Australia*
Gunnela Westlander, *Sweden*
Kong-shi Xu, *P.R. China*
Lian Cang Xu, *P.R. China*

Ergonomics and Health Aspects of Work with Computers
Arne Aaras, *Norway*
Marvin J. Dainoff, *USA*
Biman Das, *Canada*
Issachar Gilad, *Israel*
Martin Helander, *Sweden*
Kitti Intaranont, *Canada*
Henry S. R. Kao, *Hong Kong*
Waldemar Karwowski, *USA*
Peter Kern, *Germany*
Helmut Krueger, *Switzerland*
Thomas Laubli, *Switzerland*
Holger Luczak, *Germany*
Choon-Nam Ong, *Singapore*
Michael Patkin, *Australia*
Vern Putz-Anderson, *USA*

Steven L. Sauter, *USA*
Valery F. Venda, *Canada*
Gunnella Westlander, *Sweden*
Lian Cang Xu, *P. R. China*
Andrew B. Whinston, *USA*
John R. Wilson, *UK*
Hidekazu Yoshikawa, *Japan*
Bernhard Zimolong, *Germany*

Human Interface and the Management of Information
Yuichiro Anzai, *Japan*
Ram R. Bishu, *USA*
Gunilla Bradley, *Sweden*
Pascale Carayon, *Framce*
Barbara G. F. Cohen, *USA*
Johnson A. Edosomwan, *USA*
Frida Marina Fischer, *Brazil*
Michitaka Hirose, *Japan*
Fumio Kishino, *Japan*
Danuta Koradecka, *Poland*
Yasufumi Kume, *Japan*
Mark R. Lehto, *USA*
Ann Majchrzak, *USA*
Hirohiko Mori, *Japan*
Shinji Moriya, *Japan*
Nobuto Nakamura, *Japan*
Shogo Nishida, *Japan*
Ahmet F. Ozok, *Turkey*
Leszek Pacholski, *Poland*
Kyung S. Park, *Korea*
Robert W. Proctor, *USA*
Shen Rongfang, *P.R. China*
Lawrence M. Schleifer, *USA*
Joseph Sharit, *USA*
Thomas B. Sheridan, *USA*
Tsutomu Tabe, *Japan*
Mitchell M. Tseng, *Hong Kong*
Eberhard Ulich, *Switzerland*
Tjerk W. van der Schaaf, *Netherlands*
Bernardo Villarreal, *Mexico*

Human-Computer Interaction
Beth Adelson, *USA*
Sebastiano Bagnara, *Italy*
Valentina M. Bondarovskaya, *Ukraine*
George J. Boggs, *USA*
Martha E. Crosby, *USA*
Takao Enkawa, *Japan*
Klaus-Peter Faehnrich, *Germany*
Pierre Falzon, *France*
Ephraim Glinnert, *USA*
Pekka Huuhtanen, *Finland*
Sheue-Ling Hwang, *Taiwan*
Julie A. Jacko, *USA*
John Karat, *USA*
Richard J. Koubek, *USA*
Kari Lindstrom, *Finland*
Yili Liu, *USA*
John Long, *UK*
Anthony A. Maciejewski, *USA*
Nicolas Marmaras, *Greece*
David L. Morrison, *Australia*
Masaru Naniwada, *Japan*
Celestine A. Ntuen, *USA*
Katsuhiko Ogawa, *Japan*
Jose A. Pino, *Chile*
Michelle Robertson, *USA*
Dominique L. Scapin, *France*
Pentti K. Seppala, *Finland*
Kay Stanney, *USA*
Kim Vicente, *Canada*
Nong Ye, *USA*

This conference could not have been held without the diligent work of Kim Gilbert, the conference administrator, Myrna Kasdorf, the program administrator, and Jennifer Gruver, the registration assistant, who were all invaluable in the completion of this book. Also, a special thanks goes to Dr. Kay Stanney, the student liaison, for all her outstanding efforts.

Michael J. Smith
University of Wisconsin-Madison
Madison, Wisconsin 53706

Gavriel Salvendy
Purdue University
West Lafayette, Indiana 47907

Richard J. Koubek
Purdue University
West Lafayette, Indiana 479

June 1997

HCI International '99

The Eighth International Conference on Human-Computer Interaction, HCI International '99, will take place in Munich, Germany, August 22-27 1999, at the Hilton Park Hotel. The conference will cover a broad spectrum of HCI-related themes, including theoretrical issues, methods, tools and processes for HCI design, new interface techniques and applications. The conference will offer a pre-conference program with tutorials and workshops, parallel paper sessions and panels and post-conference industrial tours. The main conference reception will take place in the famous Hobräuhaus.

Prof. D.-Ing. Hans-Jorg Bullinger
Program Chair
Fraunhofer IAO
HCI International '99
Nobelstr. 12
D-70569 Stuttgart
Germany
PHONE: +49-711-970-2188
FAX: +49-711-970-2299
EMAIL: HCI99-Info@iao.fhg.de

The proceedings will be published by Elsevier Science Publishers.

CONTENTS

Cognitive Modelling

Cockpit Systems Design in Future Military Aircraft

Dr. Axel Schulte

ESG Elektroniksystem- und Logistik-GmbH, Experimental Avionics Systems,
P.O. Box 80 05 69, 81605 München, Germany. E-mail: aschulte@esg-gmbh.de

This paper describes an approach of how to support the design of the crew interface of future air transport/weapon systems. Due to increasing demands put on crews of military aircraft, effective cockpit systems will be required in order to reduce workload and to improve crew performance. This paper presents an approach to crew assistance in tactical flight missions. The underlying tasks are tactical decision making, low-level flight planning and flight guidance. The integration of the Tactical Situation System as part of a knowledge based crew assistant and a flight guidance display system incorporating sensor and synthetic vision components offer a promising solution to improve the situational awareness of the crew. Respective prototypes have been successfully tested and evaluated in a simulated environment.

1. INTRODUCTION

Human operators might be overtaxed due to the various tasks resulting from military air transport missions in hostile environments. Guiding the aircraft through adverse weather conditions in ground proximity puts further demands on crew performance [1]. Most accidents can be at least partly attributed to human erroneous actions due to a lack of situational awareness [2]. Human-centered automation [3] offers a promising approach to the solution of the obvious problems of those design philosophies in flight deck automation just utilizing the available technology regardless of human needs. The main scope of present programmes on on-board pilot assistance is the enhancement of the crew's situational awareness. Situational awareness is guaranteed if the pilot has all relevant information for the present flight situation at his disposal and is therefore able to cope with the posed tasks. In order to improve the situation assessment capabilities it must be ensured that the attention of the cockpit crew is guided towards the objectively most urgent task of the situation and, if necessary, the workload reduced to a normal degree which can be handled by the crew. [2]

2. SYSTEM DESIGN

In order to meet the above design principles of the human-centered approach an appropriate crew assistant system has to incorporate the capabilities of situation assessment and planning in the field of tactical air-operations. It should carry out respective tasks in parallel to the crew. Additionally the approach has to consider the visual perception aspects of human performance. Enhanced and synthetic vision systems are today's solution in order to improve the crew's situational awareness in the context of low-level flight guidance. While classical flight director systems avoid or at least decrease the involvement of the pilot, enhanced/synthetic vision systems keep the pilot active in the flight guidance loop. This is achieved by depicting information suitable for each human performance level [4] instead of just appealing to the skill-based level as required for aircraft stabilization and control.The following sections give a brief overview over the approach and the functional implementation of the system design.

2.1. Approach

While performing a tactical low-level flight mission, deviations from a preplanned trajectory might be induced by the crew while reacting to a suddenly changing mission scenario. Under adverse weather conditions this creates a high crew workload and a loss of situational awareness concerning the aircraft's position and attitude relative to the terrain and the desired flight trajectory. The suggested crew assistant system is the *Tactical Situation System*. It yields the capability of taking workload off the crew by giving decision aids while keeping up the situational awareness. This is achieved by the integration of the following functional capabilities: Situation interpretation and assessment through terrain and threat analysis; Planning through on-board mission management and optimal trajectory generation; Situation visualization through enhanced and synthetic vision. The following section provides a closer view to the functions and architecture of theTactical Situation System.

2.2. Functions

The Tactical Situation System (TSS) consists of the functional modules *Tactical Situation Interpreter, Low-altitude Flight Planner, Tactical Map, Synthetic Vision* and *Enhanced Sensor Vision* as described as follows.

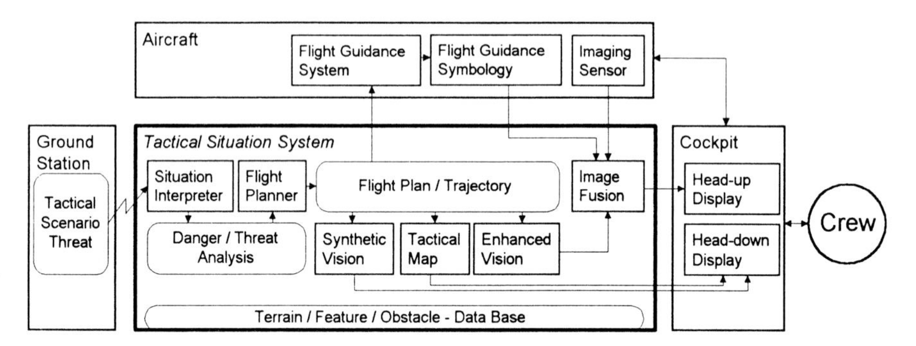

Figure 1. Modules and Integration of Tactical Situation System

Figure 1 depicts the top-level architecture of the TSS. It is designed as an on-board cockpit system with interfaces to the crew via cockpit displays and controls, to the aircraft systems, and to ground stations via data link. Based on the current tactical scenario the TSS performs a situation assessment resulting in/a danger and threat analysis. The latter provides the input for the flight planning, by computing an optimal trajectory in terms of survival probability according to the crew-given mission constraints. The trajectory is visualized on various cockpit displays such as *Synthetic Vision* and *Tactical Map*. Additionally the planned flight path can be issued to an aircraft-hosted flight guidance system in order to generate flight director commands or perform automatic flight control. Respective flight guidance symbology is fused with a sensor image and the latter enhanced by the *Enhanced Vision* overlay for display on a cockpit head-up equipment. The following sections give some more details on the various submodules.

The *Tactical Situation Interpreter* is a knowledge-based module for situation interpretation. Its main contribution is the computation of a *threat map* which gives a kill probability over the considered operation area. The calculation is based upon digital terrain elevation data (DTED) and the military threat models. Furthermore, the danger/threat analysis incorporates aspects of visibility, collision probability, and considers the terrain elevation structure and computes the corresponding radar shadow. Additional details are given in [1].

The *Low-altitude Flight Planner* calculates a three-dimensional route between mission-given waypoints with a maximum probability of survival. This is achieved by avoiding threatened areas if possible, minimizing the exposure to unknown threats and keeping clear of the terrain. Therefore, the mission constraints, the precalculated danger map, the terrain elevation data and the aircraft performance data are taken into consideration. The output of the planner is a detailed trajectory and a waypoint/flightleg-oriented representation. The actual numerical optimization is based on dynamic programming [5].

The *Tactical Map* provides the primary crew interface to the Tactical Situation System. Basically, it is an interactive electronic moving map display for navigational and operational purposes. The aim of the interface is the improvement of the pilot's situational awareness. The display utilizes digital terrain elevation data and cultural feature data (DFAD), tactical and threat information as well as a variety of navigational elements in order to create a topographical map with task-specific overlays. The various feature classes can be displayed selectively providing a very efficient decluttering of the screen contents. Additionally the system provides an interface to the Low-altitude Flight Planner by allowing the pilot to enter waypoints interactively by marking them on the map.

Synthetic Vision and *Enhanced Sensor Vision* represent the primary flight guidance instruments on the head-down and head-up displays respectively. A graphical three-dimensional representation of the terrain structure constitutes the basis of both systems. The enhanced sensor vision utilizes the benefits of an imaging sensor (such as surface resolution) in combination with a synthetic overlay (viewing distance, depth preception) and flight guidance symbology. Thereby, the visual enhancement of poor sensor images can be achieved. Deficiencies of the applied databases are recovered by taking advantage of the imaging sensor.

3. SYSTEM EVALUATION

The *Tactical Situation System* has been designed and prototyped in a generic cockpit simulator at ESG and then integrated and evaluated in the Daimler-Benz Aerospace DASA Airbus development flight simulator for Future Large Aircraft at Hamburg. The tests were carried out with respect to technical feasibility, human-machine-interaction considerations, and real-world conditions. The main objective of the experiment was to investigate whether the crew performing a tactical flight mission under adverse weather conditions could be effectively assisted by the prototype system in terms of situational awareness improvement.

3.1. Apparatus

The experimental system is a full scale three-seat fixed-base flight simulator equiped with a collimated wide FOV visual simulation system. The cockpit hosts two 10 inch high resolution CRT displays for each crew member and a collimated head-up display for the pilot flying. Flight control is provided by Airbus cockpit controls including sidestick control. The crew's control actions are passed to an Airbus flight control system. Flight director signals are provided by a low-level flight guidance system.

3.2. Subjects and scenario

The subjects were seven German Air Force pilots: four of them tactical transport instructor pilots, two test pilots with fighter experience, and one civil transport pilot. Each of them had to perform a tactical low-level transport mission of about 45 minutes. The mission contained portions of transit and tactical flight. Low-level flight was performed at 250 ft AGL utilizing terrain masking in mountain valleys. Additional features of the mission were a tactical drop procedure, an intended deviation from the preplanned track with an unguided recovery, and the performance of an unguided go-around pattern at the destination airfield. The whole mission had to be performed under night and low visibility (400 meters) visual conditions.

3.3. Evaluation results

One of the main objectives of the described experiment was the knowledge elicitation for future developments in the field of cockpit systems design. Therefore, a continuous assessment of the prototypes was performed during the simulated flight by applying the method of observation during task performance. Additionally, debriefings with questionnaires were conducted. The observations showed that the preferred configuration of the Tactical Display was a height-colour-coded terrain relief with a collision-warning overlay. This supported terrain aviodance and the low-level flight guidance task extremely well. Crew coordination aspects were promoted significantly by the integrated display concept. Concerning the head-up display, a strong enhancement could be achieved by adding a (simulated) FLIR image whereas the ability of ego-motion estimation in the Synthetic Vision was regarded as insufficient. Overall, a total of seven one-hour low-level flight missions were successfully conducted under visual conditions prohibiting unaided terrain masking. An unguided go-around manoeuvre was performed successfully three times. Two touch-and-go procedures were conducted utilizing the Synthetic Vision navigation capabilities.

4. CONCLUSIONS

The aim of the investigation was to create flexible system prototypes for cockpit avionics in order to elaborate and evaluate user requirements with operational personnel under human-machine-interaction considerations. The proposed prototype system facilitates the demonstration of advanced mission management technologies. The experimental results show that the proposed approach to visual flight guidance assistance is extremely powerful and yields a high potential for further developments in the field of human-centered cockpit design.

REFERENCES

[1] A. Schulte and W. Klöckner. Perspectives of Crew Assistance in Military Aircraft through Visualizing, Planning and Decision Aiding Functions. In AGARD MSP, 6th Symposium on Advanced Architectures for Aerospace Mission Systems, Istanbul, 1996.

[2] R. Onken. Knowledge-Based Cockpit Assistance. In *The Role of Intelligent Systems in Defence*, St Hugh's College, Oxford, 1995.

[3] C.E. Billings. *Human-Centered Aircraft Automation*. Technical memorandum 103885, NASA Ames Research Center, Moffett Field CA, 1991.

[4] Jens Rasmussen. *Information Processing and Human-Machine Interaction. An Approach to Cognitive Engineering*. North-Holland, 1986.

[5] Ulrich Leuthäusser and Friedhelm Raupp. An efficient method for three-dimensional route planning with different strategies and constraints. In AGARD CP 504, Amsterdam, 1991.

The Crew Assistant Military Aircraft (CAMA)

M. Strohal, R. Onken

Universität der Bundeswehr München, Werner Heisenberg Weg 39, D-85577 Neubiberg, Germany, Fax: 00498960042082, E-mail: Michael.Strohal@unibw-muenchen.de

This paper describes the concept of the knowledge-based Cockpit Assistant Military Aircraft (CAMA) and its functions as an example of human-centered automation. A general survey of *CAMA* with its structure, functions and interfaces will be given, added by a brief description of the individual system modules.

1. INTRODUCTION

In future combat transport aircraft, constraints created by low level flying in a high risk theater, the high rate of change of information and short reaction times will produce physiological and cognitive problems for pilots. From the cognitive point of view, low level flying over rapidly changing terrain elevation coupled with complex and dynamic tactical environment will result primarily in difficulties to maintain situation awareness. It still seems impossible to ensure the pilot's situation awareness as the dominating requirement for high level mission performance and safety.
However, with CAMA a novel approach breaks new ground to effectively enhance situation awareness in future combat aircraft. This knowledge-based aiding system is being developed and tested in close cooperation between the DASA (Daimler-Benz Aerospace), DLR (German Aerospace Research Department), ESG (Elektronik- und Logistiksysteme GmbH) and the University of the German Armed Forces, Munich. The central idea for the development of *CAMA* is, to ensure that the crew will have all necessary and useful information without overloading, according to human-centered automation [1]. Design criteria were established, which aim at a cooperative function distribution between man and machine like that of two partners [2]. Both man and machine are active in parallel by assessing the situation and looking for conflict solutions at the same time. In contrast with current man-machine interaction, both assist each other while heading for the same goals. Consequently Billings [1, Page 84] demands: „Each element of the system must have knowledge of the others' intent. Cross monitoring (of machine by human, of human by machine and ultimately of human by human) can only be effective if the agent monitoring understands what the monitored agent is trying to accomplish, and in some cases, why." Hence, the level of understanding what each element of the system is doing should be as high as possible. Derived from the demands on automation a knowledge-based aiding system should comply with two basic requirements [3]:

Requirement (1): As part of the presentation of entire flight situation the system must ensure to guide the attention of the cockpit crew towards the objective most urgent task or subtask.

Requirement (2): If requirement (1) is met, and if there (still) occurs a situation of over-demanding cockpit crew resources, the situation has to be transformed - by use of technical means - into a situation which can be handled normally by the cockpit crew.

Basic requirement (1) is to ensure situation awareness of the crew. In part, it can be transferred into the functional requirement for the assistant system of being capable to assess the situation on its own.
Pilot's workload has become a critical issue as the mission complexity has grown. It is particularly desirable to reduce the need to compose the relevant information from numerous separately displayed data. The ability of the assistant system to detect conflicts, to initiate and to carry out its own conflict-solving process and to recommend and explain this solution to the pilot, gives the pilot sufficient time to cope with unanticipated events and to act reasonably (requirement 2.). This appears to be a flexible situation-dependent, and cooperative share in situation assessment and conflict resolution between the electronic and the human crew member.[3].
If the above mentioned design-criteria and requirements are perfectly fulfilled, this will result in an electronic crew member which is capable:

- to understand the abstract goals of a mission,
- to assess mission, environment and system information the crew needs,
- to detect the pilot's intent and possible errors as part of situation analysis,
- to support during planning and decision making by recommendations of the conflict solver and
- to know, how to present it to the crew effectively by the dialogue manager

2. THE CREW ASSISTANT MILITARY AIRCRAFT *(CAMA)*

The CAMA-program was planned for four phases including a pre-contract feasibility study, a module development phase in a limited scenario for each module, and an integration phase with several testing steps. The actual integration phase will end in June 1998 with a man-in-the-loop full mission simulation campaign. After simulator tests the system will be demonstrated in flight experiments which are scheduled for winter 1999. It is planned, that CAMA will be integrated in the experimental cockpit of the ATTAS test aircraft of the German Aerospace Research Department (DLR).
CAMA assists the crew during a tactical mission to enhance situation awareness with an *interpretation of*:

- the altering tactical situation
- the actual weather situation
- the flight trajectory ahead to avoid safety critical ground proximity
- other safety relevant events

and through mission execution *services like*:

- an optimized 3D/4D trajectory flight plan
- time-management with regard to Time Over Targets (TOT)
- landing guidance without ground infrastructure
- evaluation and recommendation of alternates

Necessary *communication* with ground facilities like Command and Control Centers or Air Traffic Control (ATC) are provided by data-link.
The overall *information flow* from *CAMA* to the crew and vice versa is controlled by the dialogue management.

A *general survey of CAMA* with its structure, functions and interfaces will be given, added by a brief description of the individual system modules.
The module **Tactical Situation Interpreter** (ESG) monitors tactical events and threat characteristics to analyze the transport mission situation. Threat data are assessed based upon digital terrain and elevation data (DTED) as well as the threat's models. The algorithm allows to calculate a position-dependent threat value taking terrain masking against the opponents radar into account. An internal *threat map* contains a complete representation of the tactical situation. [4].
The **Flight Situation and Threat Interpreter** module (UniBw) combines stored mission data with current or proposed plans and the results of the situation interpretation modules. Its main contribution is to find any plan-conflicts and to initiate a conflict-solving process.
The **Mission Planner** (UniBw) creates and maintains a take-off-to-landing mission flight plan, including routes, profiles, time- and fuel-planning based on knowledge about the mission plan, gaming area, destination, ATC instruction, a/c status, environmental data, etc.. Events like failures of A/C systems, weather or threat changes and ATC or C&C instruction and information are taken into consideration. The mission planer covers the flight under Instrument Flight Rules (IFR) as well as tactical routing. Time management, especially with regard to a TOT (time over target), fuel calculations and routes/profiles calculations will assist the crew. The calculated trajectory is presented as proposal to be accepted or modified and serves as knowledge source for other function blocks.
The **Low Altitude Planner** (ESG) calculates the trajectory based on knowledge about weapon and system capabilities. Minimum risk routes are chosen to bypass hostile defenses.
Based on the aforementioned *threat map* the LAP generates an optimized low level flight plan by calculation of a minimum risk route through the gaming area. In generating plans, account is taken to the current situation and available resources, such as fuel or time, while complying with waypoint restrictions and other mission constraints [4].
The module **Terrain Interpreter** (DASA) contains a digital terrain data base to warn the cockpit crew if the projected aircraft path is getting too close to the ground or an obstacle. This eliminates several traps such as controlled flight into terrain or descend into ground short of the runway.
The aircraft may need to be updated with fresh information during the mission. The **External Communication Interface** (DASA) will provide the crew and assistant system with external data, like weather forecasts, the intention of external war-fighting units or changed tactical situations that might effect the planned mission.
The module **System Interpreter** (DLR) monitors and analyses onboard systems to determine the current state of the aircraft systems. Any detected malfunction is evaluated to determine the degree of degradation of the overall system capability.
The module **Computer Vision External** (UniBw) will assist the crew by computer vision during the approach phase to avoid collisions in high density air traffic and to ensure a quasi ILS/MLS landing at any unequipped landing field. To improve the aircraft state estimation, a camera-system will be used to determine the relative position to the runway. Two cameras with different focal lengths are used in parallel for bifocal vision. A wide-angle lens is used for initialization and stabilization and the tele-lens for object tracking. The system has been tested in real-time with a hardware-in-the-loop simulation. Image processing combined with the current inertial sensors are able to perform precise landing guidance. [5]
To improve the reasoning capability of the pilot model, the eye movement of the pilots will be evaluated. With the module **Computer Vision Internal** (UniBw) a camera system similar to

the hardware configuration of the module Computer Vision External is used to register head and eye movements. This information, for instance the point of gaze to a control surface or to a special indicator, could be used to confirm the need for a warning or a hint.
The **Pilot Behavior Reference** (UniBw) module describes a rule-based model of expected pilot-behavior concerning the actual flight plan and the module **Pilot Intent and Error Recognition** (UniBw) evaluates the pilot's activities and mission events in order to interpret and understand the pilot's actions [6].
The information flow from the machine to the crew and vice versa is controlled exclusively by the module **Dialogue Manager** (UniBw). The many different kinds of messages require a processing in order to use an appropriate display device and to present the message at the right time. As output devices both, a graphic/alphanumeric color display and a speech synthesizer are used. Brief warnings and hints are used to make the crew aware of a necessary and expected action and are transmitted verbally using the speech synthesizer. More complex information, e.g. the current flight plan, are depicted on a Horizontal Situation Display.
The Horizontal Situation Display is an interactive touch-sensitive map display organized in a number of layers which allows the crew to select optional map-presentations in any combination. It allows to depict tactical and threat information as well as a variety of navigational elements and a topographical map similar to the currently used low flying charts paper-maps. A second alpha-numerical display contains the flight-log and is used for in-flight departure-, approach- or missed-approach-briefings and a drop-briefing to assist with a combined linguistic and graphical briefing, describing the characteristics and any dangers associated with the current flight phase. The input information flow is established by use of speech recognition in addition to conventional input mechanisms. Intuitive direct voice input relieves the pilot of a lengthy and tedious alpha-numerical input task. Voice control seems to be the best solution to deal with mass data. The total on-board vocabulary will be very large and is broken down into sub-sets according to context. In order to improve speech recognition performance, almost the complete knowledge of *CAMA* is used for contextual decoding to provide situation dependent syntaxes. Thus, the complexity of the overall language model is reduced significantly such that the system can achieve high recognition rates.
The use of speech input and output devices also reflect the idea of human-centered development with respect to efficient communication.

REFERENCES

1. C. Billings, Human-Centered Aircraft Automation: A Concept and Guidlines, NASA Tech Memorandum 103885, Moffet Field, 1991
2. J.M. Reising, Automation in Military Aircraft, Proc. of the HCI, Orlando, 19A, ‘93
3. R. Onken, Basic Requirements Concerning Man-Machine Interaction in Combat Aircraft, Workshop on Human Factors/Future Combat Aircraft, Ottobunn, Germany, ‘94
4. A. Schulte, W. Klöckner , Perspectives of Crew Assistance in Military Aircraft through Visualizing, Planning and Decision Aiding Functions, AGARD Mission Panel, 6th Symposium on Advanced Architectures for Aerospace Mission Systems, Oct. 1996
5. S. Fürst, S. Werner, D. Dickmanns, E. Dickmanns, Landmark navigation and autonomous landing approach with obstacle detection for aircraft, AeroSense ´97, Orlando
6. Stütz, P., Onken R., Adaptive Pilot Modeling Within Cockpit Crew Assistance, HCI ‘97

ELICO a platform for dynamic adaptation of user-system interaction

F-R. Monclar [a and b], I. Jacob [a], J-P. Krivine [a]

[a] Electricité de France (R&D), 1 ave. Général de Gaulle, 92141 Clamart, France

[b] LIRMM, 161 rue ADA, 34090 Montpellier, France

In this paper, we claim that in certain situations *cooperative systems* should be able to dynamically adapt their interaction with the user while solving a problem. ELICO is a framework for the development of such systems. It is applied to fault restoration of power systems.

1. A NEED FOR DYNAMIC CHANGE OF USER-SYSTEM INTERACTION

Our objectives are to develop *cooperative systems* able to behave in different interactive modes ranging from "highly automated" to "strongly interactive". We distinguish two important classes of application where cooperation is needed. The first one is illustrated by planning tasks, computer-aided design or data analysis systems as described in [1-4]. In this class of application, user intervention is necessary for chaining calculations, building and adjusting hypotheses, etc. Solving is a *"backtrack-on-error"* process where the user can put to the test his hypotheses in automatic or interactive mode.

The second class of applications is illustrated by traffic control or supervision of industrial processes. In these situations, the environment is dynamic and only partially controlled. Due to the real-time constraints, it is impossible to backtrack on each decision. Furthermore, we will show that some critical or unexpected situations require a dynamic change in the interaction with the operator. This is the case of the AUSTRAL system, currently in the industrialization phase at EDF which is our source of experiments [5]. It provides functions for the supervision of power distribution networks.

2. A CONCEPTUAL BACKGROUND FOR COOPERATIVE SYSTEMS

Here, we present the conceptual background of our work. We show that concepts derived from Distributed Artificial Intelligence can characterize the needs brought by cooperative systems. This led to the implementation of a development shell called ELICO described in the next section.

2.1 Problems and Agents

A **problem** is the basic notion in our approach, it represents tasks to be solved in terms of *states* to be reached. Problems will be solved by **agents** which are entities that can participate in the solving process. In our particular case, only two agents can be distinguished : the system and its user.

2.2 Roles and Cooperation modes

The notion of **role** characterizes the responsibilities that will be given to the agents, e.g. the problems they will be in charge of in the organization. We define a role as a set of problems. The **cooperation mode** represents the role distribution to the agents associated with a certain force (obligation, preference, forbidden). Note that in the case of "preference", the agent specified might not be the one who eventually plays the role.

2.3 The cooperation context

During the solving process, agents will be notified of **events** occurring either on the domain (alarms, etc.), on the solving process (failures, errors) or at the cooperation level (mode revision, etc.). The **cooperation context** records all events, providing a picture of the situation. It will help to characterize situations where the mode has to be adapted.

2.4 Cooperative problem solving behavior

Cooperative problem solving behavior will determine how the above elements are used to solve given problems. For each *problem* to be solved, the *role* it belongs to must be determined. The *cooperation mode* determines which *agent* should be called to solve it. Since the system is placed in a dynamic environment, *events* can occur at any time. They are stored in the *cooperation context* and might trigger a change in role distribution.

3. IMPLEMENTATION OF ELICO

The result of the conceptualization is a general framework called ELICO for the development of cooperative systems. It has been implemented as an extension of the LISA language described in [6] which handles structures representing ***goals*** and ***methods*** together with a ***control engine*** that uses these structures to solve concrete problems.

As shown in figure 1, the implementation of *problems* and *methods* respectively correspond to that of *goals* and *methods*. *Roles* are sets of *problems* while *agents* refer to sets of *methods*. To be more precise, *agents* are actually the system's point of view on the *agents*' (user and system) capacities. The *control engine* has been completely re-written to implement *cooperative problem solving behavior* : to solve each *problem*, its role should be determined. The *cooperation mode* will point to the agent that should play this role. Then, a *method* of the chosen *agent* is executed.

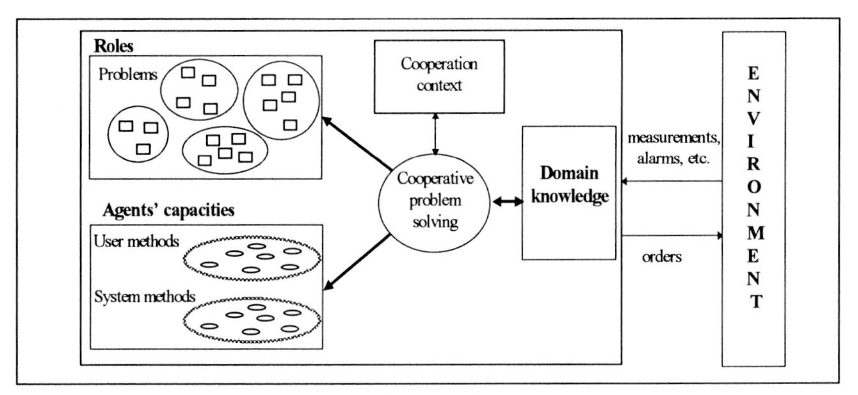

Figure 1 : Architecture of Elico

Remember that if an agent is "preferred" and no method exists belonging to him, another agent might be called. In that case an event is also generated to report that the cooperation mode was not respected. Each time an *event* occurs, it is stored in the *cooperation context*. Modification of the cooperation context might trigger a change in the *cooperation mode*.

4. APPLICATION TO RESTORATION OF DISTRIBUTION NETWORKS

ELICO was used to develop a prototype called AUSTRALIA for restoration of power distribution networks. We identified a number of *roles* and defined *cooperation modes* to assign the *roles* to the *agents* (user and system).

4.1 Roles

The roles defined in AUSTRALIA are :

- **domain-decision** : decisions to be taken in the solving process such as "*choice of location hypothesis*" or "*choice of a restoration plan*",
- **domain-operation** : problems requiring action on the environment e.g. that modify the state of the network,
- **domain-perception** : problems related to acquiring the state of the network,
- **reasoning** : problems dealing with the different calculations performed by the system, of which "*construct possible location hypotheses*" is an example,
- **control-decision** : these problems are related to strategic decisions at the control level. An example is the"*choice of the method to be executed*".

These will later be assigned to the agents by means of *cooperation mode* which we describe in the next section.

4.2 Cooperation modes used

Cooperation modes assign roles to agents. At the beginning of each session, the user chooses the cooperation mode he wants to use. With AUSTRALIA, it is possible to choose from an **automatic mode** (assigning all the roles to the system), a **user oriented mode** (where the user plays the decision-domain role), a **control decision mode** (where the user plays both decision-domain and decision control roles) and a **manual mode** (assigning all the roles to the user).

4.3 A short scenario

In a simple but realistic scenario, the system is used in *automatic mode* when a fault occurs. After collecting fault information from fault detectors, the system solves the problem "*choice of location hypothesis*" and isolates the fault by opening switching devices neighbouring the faulted area. The next step is to restore a part of the downstream area, by closing the open feeder. After closing a feeder, the protection device is activated again. This either means that the fault location hypothesis was wrong, or a second fault occurred. An event is generated and the *domain-decision* role is assigned to the user. The cooperation mode changes to reflect the fact that the situation is too critical for an automated mode. This way, the user will be responsible for any further decisions to be made on the domain. He will, for instance, be called upon to solve the problem "*choice of the location hypothesis*" for a second time.

5. CONCLUSION AND FUTURE WORK

A conceptual background and an environment called ELICO for the development of cooperative systems was used to develop a prototype for fault restoration of power distribution networks. This application raises the need for a dynamic change of cooperation mode between user and system.

The dynamic adaptation of role distribution brings the operator into the decision process when the first problems arise. He is then better able to understand what has happened and quicker to react.

Today, ELICO is used to develop a version of the industrial fault restoration function. After a validation phase with the collaboration of the operators, we expect it to be put into real operation in the medium term.

REFERENCES

1. Kant, E. "*Interactive problem-solving using task configuration and control*", IEEE Expert Vol. 3, n° 4, pp 36-49, 1988.
2. Willamowski, J., Chevenet, F. and Jean-Marie, F. "*A development shell for cooperative problem-solving environments*", 3rd Int. Conf. ESNC, 1993.
3. Clarke, A. and Smyth, M. G. "*A co-operative computer based on the principles of human co-operation*", Int. J. Man-Machine Studies n ° 38, pp 3-22, 1993.
4. Fischer, G. "*Communication requirements for cooperative problem solving systems*", Information System, Vol 15, pp 21-36, 1990.
5. Krivine, J-P. and Jehl, O. "*The Austral System for diagnosis and power restoration : an overview*", ISAP, 1996.
6. Krivine, J-P. and Delouis, I. "*Cooperative expert systems : collaboration among artificial and human experts for power systems planning* ", ESAP, 1993.

A decision support system for production scheduling and control

M.-B. Chen[a] and S.-L. Hwang[b]

[a]Front End Manufaturing Division, MOSEL VITELIC INC. Science-Based Industrial Park, Hsinchu, Taiwan ROC

[b]Department of Industrial Engineering, National Tsing-Hua University, Hsinchu, Taiwan ROC

The purpose of this study is to develop a decision support system (DSS) to help production engineers in dealing with scheduling control and subcontracting problems. The Knowledge of the system was acquired through deep interviews in a real plant. After constructing the data base and user interface, an experiment was conducted to verify the effect of the DSS. The results indicated that the outcomes of the scheduling with DSS were significantly better than those without DSS. Furthermore, the DSS was evaluated by the shedulers in a real fartory and was obtained positive responses.

1. INTRODUCTION

Production scheduling and control are the main tasks of production management in a system. Currently, there are some approaches to deal with production scheduling and each has advantages and disadvantages. Mathematical programming approach [1-2] may find an optimal solution, but contains too many assumptions and constraints which may lose reality of system. Heuristics oriented approach [3-4] is easier to use and flexible, but may not find an optimal solution. Simulation based approach [5] needs complicated assign priniples and artificial intelligent based approach [6-7] is very difficult to formulate knowledge.

Some problems of production scheduling and control may not be quantified, especially those problems dealing with decision making. On the other hand, subcontracting problems need a highly experienced manager to solve them. Therefore, the purpose of this study is to develop a decision support system (DSS) to help production engineers in dealing with scheduling control and subcontracting problems. By the DSS, the information of the products from materials to finished parts can be saved and ommunicated among different departments of the plant, which provide decision makers on-line system situations.

The DSS of this study was designed based on a pneumatic company which represented a typical mid-scale industry with subontrating problems. Through deep interviews with managers and engineers in the related field, the knowledge was acquired. Then the system analysis was applied to identify the necessary information to be displayed and the requests of the users. The goal was to make sure the knowledge in the DSS would match the one in decision makers. Finally, the data base and user interface were constructed.

2. METHODOLOGY

2.1 The Struture of the DSS

The system design of the DSS is according to the concept of DDM [8] which includes 3 systems: 1) Dialogue Generation and Management System (DGMS), 2) Data Base Management System (DBMS), and 3) Model Base Management System (MBMS). Therefore, the system in this study includes the integration of data bases of all departments, different decision models, and the user interface (Fig. 1). In procedural oriented, the structure of the DSS in this study consists of four parts:

1) production control --- confirm order, schedule plan, part purchase and assembly,
2) data base,
3) tooling base,
and 4) output interface.

As to the user interface, the interface was designed as an adaptive one to fit either the new or the skilled users. In other words, the software was designed to meet the users' property and knowledge level.

2.2 Experiment

An experiment was conducted to verify and evaluate the effect of the DSS. The students who majored in industrial engineering and had taken the course of production management participated the experiment. To prevent the order effect, half of the subjects run the production scheduling and control by manual first and then by the help of DSS. The other half of the subjects run the scheduling and control by DSS before by manual. The dependent variables were the finish time, outcome of the scheduling and subjective evaluation.

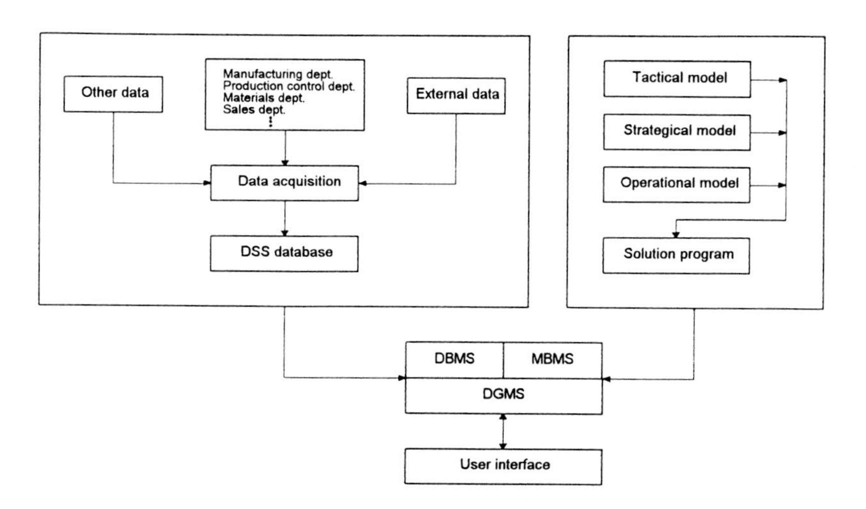

Fig.1 Functional Structure of the DSS

3. RESULTS

The results indicated that the subjects when using DSS saved 61.7% of finish time than using manual scheduling and control (Fig.2). Moreover, the outcomes of the scheduling were evaluated by experts to judge if the outcomes satisfied some requirements, such as due date, the amount of the materials needed, the least cost, and the reasonable inventory level. The results revealed that the outcomes of the scheduling with DSS were significantly better than those without DSS ($p<.05$). Furthermore, the subjective evaluation of the DSS interfae by the subjects showed positive results on user control, consistency, clarity, feedback, and the error tolerance of the system.

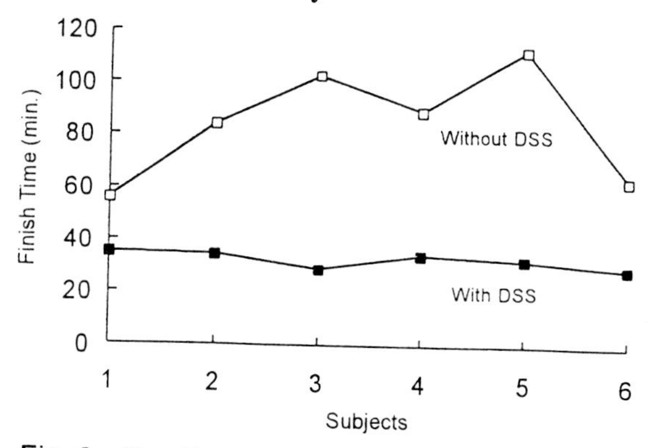

Fig. 2 The Comparison of Finish Time with/out a DSS

Finally, the DSS was evaluated by the schedulers in a real factory and was obtained positive responses. The DSS can solve the problems in production control, and also help the schedulers in data searching and data acquisition. The heuristic rules generated by the experts in this DSS can guide the new schedulers to understand operating procedures and the severity of scheduling problems.

4. CONCLUSIONS

Most of the previous Decision Support Systems provide alternatives for the users to select. The DSS developed in this study generates the alternatives through some heuristical rules and support the users in making fast and high quality decisions. The other advantages of this system are as follows,

1) By the DSS, the new schedulers can understand their tasks faster and also decrease the information operating time.
2) The DSS was designed in procedural oriented, and provided "menu popup" for the skilled schedulers and "experienced rules" for the new schedulers.
3) The system also provides an on-line help system.

REFERENCES

1. Y.G. Huang, L.N. Kanal and S.K. Triphi, "Reactive Scheduling for a Single Machine Problem definition, analysis, and heuristic solution", Int. J. computer intergrated manufacturing (1989), VOL.2, No.1,2-14.
2. P.C. Chang, "ISCES: An Integrated Scheduling and Control Expert System", Proceeding of The First International Conference On Automation Technology, National Chiao Tung University, Taipei, (1990), 575-585.
3. P.M. Sanderson and N. Moray, "The Human Factors of Scheduling Behavior", Ergonomics of Hybrid Automated Systems II pp.399-406 (1990).
4. An Collinot, C. Le Pape, "Adapting the behavior of a job-shop scheduling system". Decision support system 7(1991) pp.241-353.
5. Y.K.P. Chau, "Decision Support using Traditional Simulation and Visual Interactive Simulation", Information and Decision Technology (1993), pp.63-76.
6. O. Charalambus and K.S. Hindi, "A review of artifical intelligence-based job-shop scheduling systems", Information and Decision Technology, 17(1991), p.189-202.
7. Willian J. Stevenson "Production /Operation Management", IRWIN, (1986).
8. M.S. Scott Morton and Andrew M. McCosh, "Management Decision Support System", John Wiley & Sons (1978).

A Time-Based Interface for Electronic Mail and Task Management

Kelvin S. Yiu[a], Ronald Baecker[b], Nancy Silver[b] and Byron Long[b]

[a]Department of Electrical and Computer Engineering, University of Toronto, 10 King's College Road, Toronto Ontario, M5S 3G4, Canada

[b]Department of Computer Science, University of Toronto, 10 King's College Road, Toronto Ontario, M5S 3G4, Canada

Email overload is a growing problem for many users in the workplace [1]. Users often have trouble in retrieving messages for later use or in remembering to reply or to act on a particular message. There are two causes. The first is due to the problems associated with maintenance and retrieval in a semantic hierarchical structure. The second is due to the fact that current email systems are designed around the assumption that messages are informational and are read upon arrival, and that important messages are filed.

The use of a semantic hierarchy for filing presents many problems for dealing with a large volume of data. Filing and maintenance is very time consuming and cognitively intensive. Since there can be hundreds of new messages arriving each day, it is difficult to file and maintain a reasonable hierarchy that facilitates efficient retrieval. Moreover, categories can become obsolete over time, and messages in different categories may become semantically related. Therefore, users must spend time periodically to reorganize their mail hierarchies.

People use email for more than communication. On the surface, email is a form of asynchronous communication. In reality, email are actually used for purposes such as document delivery and archiving, work task delegation, storing personal names and addresses, and scheduling appointments [1]. Also people need better tools to remind them of their tasks [2, 3] and current email systems lack such support.

1. THE TIMESTORE PROJECT

TimeStore is an email system that uses the time of arrival as the principal arrangement to display electronic mail. We designed and built TimeStore around the philosophy that the user should not have to do any filing. Time-based visualization can complement or replace the traditional semantic based email by using an aspect of human memory that most existing email systems ignore: temporal organization in autobiographical memory [4].

Our project started with a study on how users organize their computer environment [5]. The 14 subjects (Mac, DOS, UNIX, VMS) all used a semantic organization for their filing. File organization was strongly influenced by the visual display of the system. Subjects did use time stamps and date notations were used in naming files and folders. This finding led to the first TimeStore prototype.

Long [6] implemented this prototype to study time-based visualization as an alternative to folders for organizing and retrieving email messages. The original version was an add-on for the Macintosh version of Eudora. Time was represented along the x-axis of a two-dimensional graph and the message senders were listed along the y-axis and sorted in various ways.

Subsequently an Eudora user study and a TimeStore user study [7] found that the addition of a time-based system did aid in retrieval. Users were able to see patterns of correspondence activity and the lack of dots acted as a reminder to contact a specific. Since users were reluctant to give up semantic hierarchies, support for Eudora mail folders was added.

2. THE CURRENT TIMESTORE PROTOTYPE

The new version of TimeStore manages email but also has an additional objective: to provide integral support for task management as well as other personal information where time is the primary method for access. [8]

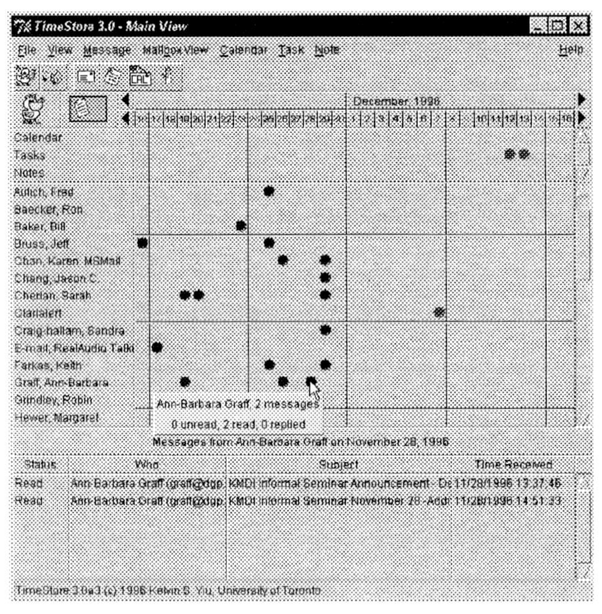

Figure 1. TimeStore's main window.

TimeStore plots information as dots on a two-dimensional graph (Figure 1) where the x-axis displays time and the y-axis displays a list of names. The display is divided into three sections. The top section displays a calendar, a task list organized by due date, and a note list organized by date of creation. The purpose of this section is to remind the user of tasks and to provide a context for the messages displayed below. The middle section plots messages as dots with the sender names listed along the y-axis. Messages are plotted using the time that the message was downloaded. The bottom section is used to display detailed information when the user clicks on a dot.

In place of the traditional folders, TimeStore provides mailbox views for creating dynamic mail folders. Similar to the substream concept found in the Lifestreams project [9], users can specify keywords to search for in certain fields in the message. The result is displayed in the same form as the main view. New messages that satisfy the specified criteria will be included automatically. Moreover, messages can appear in multiple views, which eliminate the problem that a message can only appear in one folder. The inbox is one view into the mailbox where only unread and recent messages are listed. This provides a single location for the user to get new and unread messages.

TimeStore also provides integrated task management by allowing the user to create tasks from within the message window. Tasks created this manner are associated with a message and the user can record additional notes about the task that were not in the message itself. The User can click on the "View Mail" button to read the associated message.

The interaction is simple and consistent for all data. The user moves the pointer over a dot to see the number of messages (or tasks, etc) and a breakdown showing the number of messages that have been unread, read and replied. A single click will cause TimeStore to list the message at the bottom section of the main window, and a double click will open the messages in a new window.

3. USER EVALUATION

Our objective in the usability testing was to understand whether time-based visualization is useful. Moreover, we also wanted to know if the integration of other personal information into an email system is useful. For a full description of the methodology and results, see Yiu [8].

3.1 Methodology

Usability data was gathered using a combination of interviews and think-aloud sessions with screen and audio capture by the user's computer using Microsoft Camcorder [10]. Five users were recruited and asked to use TimeStore as their primary email system while keeping their original email system as a backup and for features not implemented. During a three-week test period, users used Camcorder to record their TimeStore sessions. A microphone was provided to record the user's thoughts expressed aloud. The resulting AVI movie files can be played back on any Windows PC using the standard MediaPlayer.

3.2 Results and Discussions

User responses were positive. Four out of five users liked the time-based visualization for email messages. They liked the ability to see trends in correspondences with their friends and associates. However, TimeStore used the receive time in the display rather than the send time and this caused some confusion along the users. A very important discovery was that users were often unable to remember exactly when a message arrived. They often had to click on a succession dots in order the find the desired message.

Most users thought that the ability to associate a message with a task was useful. A problem with the tasks is that TimeStore requires them to be associated with due dates. However, users had are tasks that "they just have to do" but did not have a particular due date.

Only one user used mailbox views consistently because TimeStore already categorize messages by name. Mailbox views were used to help him find messages from mailing lists that interests him. He also used mailbox views to help track a conversation with his friends.

The inbox was a source of confusion for one user. The user expected the inbox to only contain new messages. One user did not use the inbox because he did not read all of his incoming messages. This problem became more apparent when a user subscribed to mailing lists because TimeStore's mailbox view was able to let him filter out messages of interest.

An interesting insight occurred during a final interview session. The user commented that TimeStore has made him view his email more like a database than messages in folders. Therefore, the user's expectations of TimeStore grew and TimeStore's inability to provide other statistics to answer questions such as "Why cannot I view by how much time I spent on a person?" became apparent.

4. CONCLUSION

The majority of the users found the time-based concept useful. They especially like the fact that messages are automatically arranged by sender. They also like the ability to see patterns and trends, which can remind them to contact people. Users found the ability to associate messages with tasks useful, but they also wanted to have tasks that do not have due dates.

Users cannot remember the exact date for a number of messages and therefore future versions of TimeStore must provide a better compensation for such. A possible solution can be to show messages from a range of dates. Users also disliked the fact that TimeStore provided no way to organize names. This can be resolved by providing a hierarchical structure to manage names. Other recommendations can be found in Yiu [8].

REFERENCES

1. Whittaker, S. and Sidner, C. "Email Overload: Exploring Personal Information Management of Email." *CHI'96 Conference Proceedings*, 1996: 276-283.
2. Nardi, B., Anderson, K., and Erickson, T., "Filing and Finding Computer Files." Apple Computer Inc. Advanced Technology Group Technical Report 118, July 1995.
3. Malone, T. "How Do People Organize Their Desks? Implications for the Design of Office Information Systems." *ACM Transactions of Office Information Systems, Vol. 1, No. 1,* January 1983: 99-112.
4. Conway, M. A., *Autobiographical Memory: An Introduction.* Open University Press, Milton Keynes, 1990.
5. Fitzmaurice, G., Baecker, R., and Moore, G. "How Do People Organize their Computer Desktops?" University of Toronto, unpublished work, 1994.
6. Long, B., "TimeStore: Exploring Time-Based Filing" University of Toronto, unpublished work, 1994.
7. Silver, N., "Time-Based Visualizations of Electronic Mail." University of Toronto, Department of Computer Science, M. Sc. Thesis, 1996.
8. Yiu, K. S., "Time-Based Management and Visualization of Personal Electronic Information." University of Toronto, Department of Electrical and Computer Engineering, M.A.Sc. Thesis, 1997.
9. Freeman, E. and Fertig, S., "Lifestreams: Organizing You Electronic Life." *AAAI Fall Symposium: AI Applications in Knowledge Navigation and Retrieval*, November 1995.
10. Microsoft Corp, Camcorder, 1997. URL: http://www.microsoft.com/msoffice/office97/camcorder/default.htm.

User Modeling by Graph-Based Induction

Kenichi Yoshida
Advanced Research Laboratory, Hitachi Ltd.
Hatoyama, Saitama, 350-03 Japan

The analysis of user behavior is one important function of the user-adaptive interface. The acquisition of the user behavior model is crucial. In a previous study, we presented a framework which uses task dependency information to construct the user behavior model. In this paper, we extend our previous framework and realize a method which also speeds up file access in computer systems.

1. INTRODUCTION

The analysis of user behavior is one important function of the user-adaptive interface. Such analysis enables understanding of the user's intention and releases the user from tedious tasks often required when using a nonadaptive interface. The acquisition of the user behavior model is crucial. Most studies meant to develop a user-adaptive interface system, such as [1], and [4], only analyze the sequence of user behaviors, *i.e.* command sequence, from which to automate the repetitions. Since the command sequence sometimes does not typify the user's behavior, the user model constructed only from the command sequence does not adequately reproduce the user behavior. In [5], we present a framework that also analyzes the computational processes activated by the user commands to build the user behavior model. An important feature of the framework is the analysis of data dependency between the user commands. The experiments have shown evidence that this additional information gives a better user behavior model and improves the command selection accuracy of the user-adaptive interface.

In this paper, we extend our previous framework, and realize a method to speed up file access in computer systems. In conventional computer systems, Least Recently Used (LRU) based caching techniques are used to speed up file access. Kroeger and Long [3] propose a method to speed up file access by predicting files to be prefetched. They use trie data structure [2] to memorize previous I/O sequences and predict files to be used in the future. Although their predicting cache system has a higher cache hit rate than the conventional LRU cache system in a simulated environment, the modeling method they use has a theoretical limitation. Their method lacks a framework to analyze a complex I/O sequence in the multi-task environment. This decreases the performance of their prefetch cache system in the multi-task environment.

2. OVERVIEW

Figure 1 shows our extended framework. The operating system reports process and I/O information, and we use a directed graph as the representation language of this information. The GBI program constructs the user model from this information and the

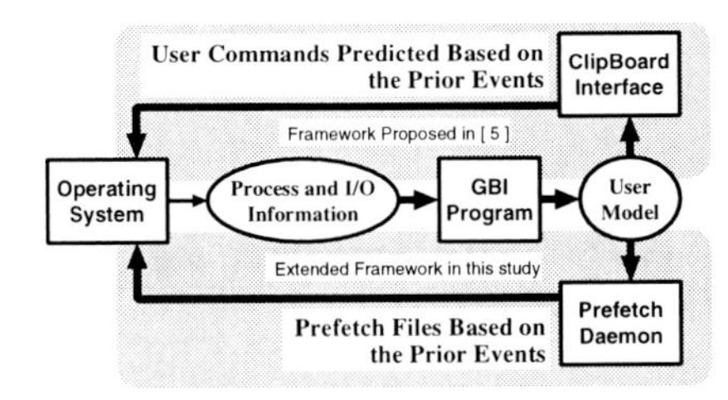

Figure 1. Using User Model in Computer System

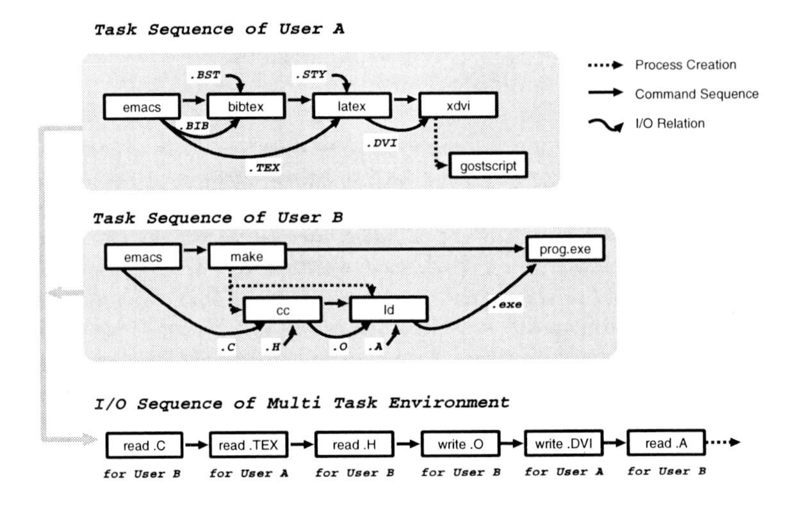

Figure 2. Process and I/O Information

system uses this user model 1) to predict the next user command, and 2) to prefetch files.

Figure 2 shows a situation where two users use the same computer for a different purpose: user A uses the computer for document processing and user B uses the same computer for programming. To be accurate, the operating system records 1) all process creations, and 2) all I/O operations. Although the I/O operation sequence of each task has some regularity, the overall I/O operation sequence in the computer is affected by the subtle timing of each task in progress. A simple analysis of this overall I/O operation sequence does not result in adequate user models.

The graph structure can store the information in the multi task environments. In our study, the information acquired by the operating system is analyzed by the GBI program. The GBI program extracts typical subgraphs from the input graph so that the extracted subgraphs represent typical events in the computer system. These extracted events repre-

to determine the user's membership in the given groups. In DOPPELGÄNGER [11], not only group membership is learned, but group models themselves are learned from user models and dynamically revised. Learning group models is an example of discovering information that is implicit in user models. This task is traditionally tackled with deductive reasoning; inductive learning algorithms will be a valuable complement.

2.4. User Modeling as Open Learning Process

The benefits that machine learning offers to user modeling will be limited, if the user modeling process remains focussed on single applications. Assumptions about the user can be acquired more reliably based on observations of several applications, and they can be of use for more than one application.

Therefore, user modeling should be an open learning process. *Open* means that a user modeling system constructs and represents *all kinds of assumptions* (behavior- and mentally-oriented) about *many users*, and communicates with *several providers and consumers* of user information. User modeling becomes a *learning process*, when assumptions about the user are incrementally constructed from observed user behavior and existing user model contents.

3. LaboUr – an Architecture for Learning about the User

Based on the notion of user modeling as open learning process, we developed the LaboUr architecture for user modeling systems. The central entities within this architecture are the *learning components*, each of which incorporates a learning method for user model construction or evolution. The learning method considers its own history or context information and perhaps also domain knowledge and current user model contents when forming assumptions about users. Observations about users are the main input for model construction. They are reported via a *communication interface* that can also be used between LaboUr components. A learning component possesses a filter mechanism that recognizes appropriate observation data and, if necessary, transforms them into an input format suitable for the learning method. A second filter refrains generated assumptions from being entered into a *user model* if strength of or confidence in an assumption is not high enough. Model evolution components obtain input from user models and, in return, manipulate user model contents or update group models. Finally, assumptions are communicated to applications, mainly query-driven, but also proactively.

At first sight, a LaboUr system is designed as a centralized user modeling server (cf. [11]), which maintains models of a multitude of users, and processes observations and queries (perhaps sent across a network) of a multitude of application systems. However, smaller LaboUr systems can exist on personal, perhaps mobile computers and cooperate with bigger servers. Also, the components of one LaboUr system can be distributed across a network. So, the LaboUr architecture provides enough flexibility for a wide range of user modeling applications.

REFERENCES

1. D. W. Aha, D. Kibler, and M. K. Albert. Instance-based learning algorithms. *Machine Learning*, 6(1):37–66, 1991.
2. R. Armstrong, D. Freitag, Th. Joachims, and T. Mitchell. Webwatcher: A learning apprentice for the world wide web. *Proc. of the 1995 AAAI Spring Symposium on Information Gathering from Heterogeneous, Distributed Environments*, March 1995.

3. M. Bauer. Machine learning for user modeling and plan recognition. In V. Moustakis J. Herrmann, editor, *Proc. ICML'96 Workshop "Machine Learning meets Human-Computer Interaction"*, pages 5–16, 1996.
4. J. Finlay. Machine learning: a tool to support improved usability? In V. Moustakis J. Herrmann, editor, *Proc. ICML'96 Workshop "Machine Learning meets Human-Computer Interaction"*, pages 17–28, 1996.
5. R. Kozierok and P. Maes. A learning interface agent for scheduling meetings. In W. D. Gray, W. E. Hefley, and D. Murray, editors, *Proc. of the International Workshop on Intelligent User Interfaces, Orlando FL*, pages 81–88, New York, 1993. ACM Press.
6. M. Krogsæter, R. Oppermann, and C. G. Thomas. A user interface integrating adaptability and adaptivity. In R. Oppermann, editor, *Adaptive User Support*. Lawrence Erlbaum Associates, 1994.
7. N. Lavrac and S. Dzeroski. *Inductive Logic Programming – Techniques and Applications*. Ellis Horwood, New York, 1994.
8. P. Maes. Agents that reduce work and information overload. *Communications of the ACM*, 37(7):31–40, July 1994.
9. M. F. McTear. User modelling for adaptive computer systems: a survey. *Artificial Intelligence Review*, 7(3-4):157–184, August 1993.
10. T. Mitchell, R. Caruana, D. Freitag, J. McDermott, and D. Zabowski. Experience with a learning personal assistant. *Communications of the ACM*, 37(7):81–91, July 1994.
11. J. Orwant. Heterogeneous learning in the Doppelgänger user modeling system. *User Modeling and User-Adapted Interaction*, 4(2):107–130, 1995.
12. J. R. Quinlan. *C4.5: Programs for Machine Learning*. The Morgan Kaufmann Series in Machine Learning. Morgan Kaufmann, San Mateo, CA, 1993.
13. C. G. Thomas and G. Fischer. Using agents to improve the usability and the usefulness of the world-wide web. In S. Carberry, D. Chin, and I. Zukerman, editors, *Fifth International Conference on User Modeling*, pages 5–12. User Modeling, Inc., 1996.
14. P. E. Utgoff. Incremental induction of decision trees. *Machine Learning*, 4(2):161–186, 1989.
15. W. Wahlster and A. Kobsa. User models in dialog systems. In A. Kobsa and W. Wahlster, editors, *User Models in Dialog Systems*, pages 4–34. Springer, Berlin, Heidelberg, 1989.
16. Geoffrey I. Webb and Mark Kuzmycz. Feature based modelling: A methodology for producing coherent, consistent, dynamically changing models of agent's competencies. *User Modeling and User-Adapted Interaction*, 5(2):117–150, 1996.
17. D. Wettschereck. *A Study of Distance-Based Machine Learning Algorithms*. PhD thesis, Oregon State University, June 1994.
18. S. Wrobel. *Concept Formation and Knowledge Revision*. Kluwer Academic Publishers, 1994.
19. K. Yoshida and H. Motoda. Automated user modeling for intelligent interface. *International Journal of Human-Computer Interaction*, 8(3):237–258, 1996.

An interactive environment for dynamic control of machine learning systems[*]

Engelbert Mephu Nguifo

CRIL, Université d'Artois - IUT de Lens, Rue de l'Université SP 16, 62307 Lens Cedex, France. E.mail: mephu@cril.univ-artois.fr

ABSTRACT

This paper describes an interactive model that allows a dynamic control of machine learning (ML) systems. The model is based on the design of ML systems. It integrates different tools with various purposes in an open environment in order to facilitate the interaction between the ML system and another agent which is often an expert or a user. This integration is done in such a way that every change of data in one of the tools is automatically taken into account and interpreted by the other tools. Such dynamic process improves the interaction with the ML system, and consequently the interpretation of ML results. A prototype of such model is described.

1- INTRODUCTION

Machine learning (ML) has given rise to many systems that are implemented in different ways, and have been applied to different artificial and real domains. ML systems have proved to be valuable tools. Such systems learn new knowledge by examining new data (examples and their description) provided by an expert. They need to control the learning process in order to validate the learned knowledge. There are two kinds of control: internal and external. The former consists in using heuristics inside learning algorithms to guide the knowledge search. The latter is based on the interaction with the expert to validate the learned knowledge. In this paper, we address only the second kind of control.

Here, we assume as a general matter that, artificial intelligent systems are intended to shed light on solutions for a set of tasks, and not to solve any complex task [1]. The best solution comes out from interactions between the system and the expert. A few attempts have been done around the SOAR system which is a general cognitive architecture [2]. Thus dealing with the issues of interaction is the challenge for artificial intelligence, especially for ML [3].

ML systems often interact with the expert-user by means of initial data, results obtained, and also different parameters required by the system during the learning process. An example of such ML environment is the WEKA workbench [4] where different ML systems are integrated for learning purpose. This simple way of interaction allows learning to be a rather static and cyclic process where first of all, the expert provides initial data, then the system

[*] Financial support for this work has been partially made available by the Ganymede project of the Contrat de Plan Etat/Nord-Pas-de-Calais.

learns new knowledge from these data, or derives new results, and finally the expert evaluates these results or knowledge. If the evaluation fails, then the expert changes initial data, and restarts the process. This kind of evaluation is an external and static control done by the expert-user through his own knowledge or an experimentation with an automatic and appropriate tool. As the evaluation and the learning tools often use different representation in practice, the whole process may be time-consuming for the expert due to data translation. Consequently, the machine learning system may be hard to use in real domain applications.

Our goal is to help the expert during the interpretation of the learning results. Such help is done by adding an indirect dialogue between the expert and the system. In practice, learning is not a single task, but should be consider in a multi-tasking environment since it has links with other tasks such as explanation, teaching, problem solving, etc. We propose here a model for learning system, by integrating various tools with different purposes in order to dynamically control the learning process. The dynamic control does not require too much time for its implementation, since this should be done initially when adapting the learning system to the real domain application. As this is done once at all, the evaluation time is considerably reduced, and the expert can dynamically interact directly with the learning system, or indirectly by using his everyday favorite tools. The communication between different tools within our environment is done through the procedures and protocol of the computer operating system.

Section 2 gives a general overview of static control of ML. Section 3 describes our dynamic model of control. In section 4, a prototype of our model is introduced.

2- MACHINE LEARNING SYSTEMS AND CONTROL

Various ML systems [4] have been developped among which ID3, C4.5, LEGAL [5]. These ML systems have given rise to different paradigms with internal and external control. Internal control of ML systems is not sufficient to validate the learned knowledge. Additional control needs to be done by interacting with an external user (expert). This is already done by ML systems which often interact with the expert-user by means of initial data, results provided, and also different parameters required by the system during the learning process.

This simple way of interaction is achieved in a rather static and cyclic process (figure1) where first of all, the expert provides initial data, then the system learns new knowledge from these data, or derives new results, and finally the expert evaluates these results or knowledge. If the evaluation fails, then the expert changes initial data, and restarts the process. In the WEKA workbench for example, the expert can choose one of the different ML systems provided. If he had few programming capabilities, then he could also integrate any familiar ML systems.

This kind of evaluation is an external and static control done by the expert-user through his own knowledge or an experimentation with an automatic and appropriate tool. For real applications, the expert can use different external tools for control such as data editors, statistical computer systems or experimental tools.

As the evaluation of the learning tools often use different representation in practice, the whole process may be time-consuming for the user due to data translation. The user should translate the learning results in an appropriate manner for the control tools. Consequently, the

ML system may be hard to use in real domain applications, as the data translation is done several times.

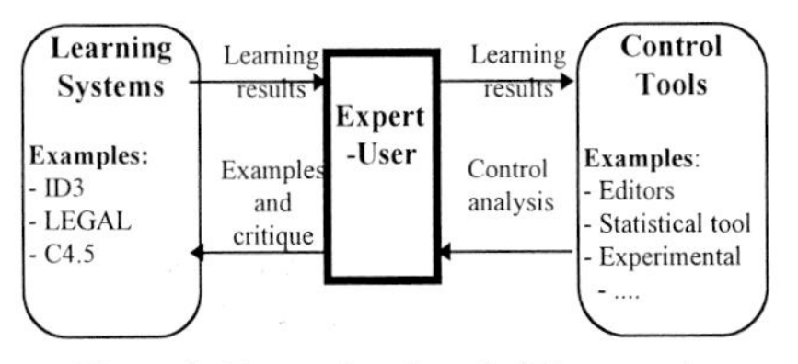

Figure 1: External and static ML control

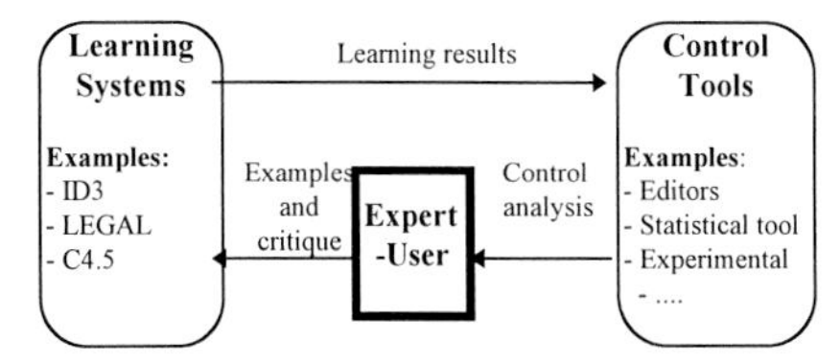

Figure 2: Dynamic control for ML systems

3- THE DYNAMIC MODEL OF CONTROL FOR ML SYSTEMS

This section describes a dynamic model of ML control that allows to reduce limitations encountered with static control. Our goal is to help the expert during the interpretation of the learning results. Such help is done by adding an indirect dialogue between the expert and the system. In practice, learning is not a single task, but should be consider in a multi-tasking environment since it has links with other tasks such as explanation, teaching, problem solving, etc. We propose here an interactive environment for learning system, by integrating various tools with different purposes in order to dynamically control the learning process.

In our model, learning system and control tools are linked together by means of a communication channel. Learning results are forwarded instantaneously to the control tools. The expert could then rapidly interpreted the learning results with the control tools. The control tools could also be linked together. For example, an editor tool can be connected to a statistical tool in order to achieve a task.

Such interactive environment has various capabilities for:

- viewing the results produced through different manners. This allows the expert-user to choose the best tool for interpreting results.
- filtering and performing various calculations on the results provided. The expert can add new information before the interpretation of the results.
- directly manipulating the results in a flexible and user-friendly manner. The expert can keep usage of his favorite tool to interact with the system.
- keeping track of the different learning processes, by managing and storing sets of data. The expert can backtrack to previous data and results already obtained.

Learning is context-dependent since any general learning system need to be adapted to application data, and to take in account initial knowledge over the application. The dynamic control does not require too much time for its implementation, since this should be done initially when adapting the learning system to the real domain application. As this is done once at all, the evaluation time is considerably reduced, and the expert can dynamically interact directly with the learning system, or indirectly by using his everyday favorite tools.

The communication between different tools within our environment is done through the procedures and protocol of the computer operating system, as for example the dynamic data exchange (DDE) in the OS/2 operating system, or the communication protocol of the UNIX operating system.

4- A SIMPLE PROTOTYPE OF DYNAMIC CONTROL

A simple prototype[+] of this environment was implemented on a particular application in Biology which consists in predicting the complete macromolecular conformation of a molecule. We have used the ML system, LEGAL [5], the standard spreadsheet EXCEL (Microsoft product), and a three-dimension molecular editor PCDRA [6], with the capabilities to dynamically share data. This prototype runs on an IBM PC compatible workstation using OS/2 operating system. The communications are performed through the use of standard tools (DDE procedures and protocol) that are independent of our methods. Such approach avoids many integration problems.

LEGAL calculates the structural motif prediction scores which are incorporated as new column within EXCEL. PCDRA uses these results to display the protein structure. Any modification of the settings of LEGAL that affects the prediction score, are dynamically reflected by EXCEL and PCDRA. It is thus possible to fine tune the learning parameters and validate predictions through successive cycles and comparative study of the results.

The EXCEL spreadsheet functionalities can also be easily used over the structural motif prediction scores. Our prototype has shown to be a useful tool for biologist when examining protein structure.

5- CONCLUSION

In this paper, we have argued that control of ML systems need to be done dynamically in order to increase their usability. A model of ML control is described and an implementation of a prototype is given. This model is a preliminary translation from ML systems to multi-agent systems where different agents cooperate during the learning task.

Further research [7] deals with the interconnexion between ML operators and dialogue operators in order to minimise the increase of communication flow between the various tools of the environment.

REFERENCES

1. Schank R.C., 1991, Where's the AI?, *AI Magazine*, 12(4):38-49. Winter.
2. Rosenbloom., Laird, & Newell, A. 1993. *The Soar Papers: Readings on Integrated Intelligence*, MIT Press.
3. Herrman J., & Moustakis V. 1996. ICML Workshop proceedings on ML meets HCI, Bari (Italy).
4. Holmes G., Donkin A., & Witten I.H. 1994. WEKA: A Machine Learning Workbench. In Proc. of 2nd Australia and New Zealand Conf on Intelligent Information Systems, Brisbane (Australia).
5. Mephu Nguifo E. 1994. Galois Lattice: a Framework for Concept Learning. Design, Evaluation and Refinement. In Proc. of 6th Intl. Conf. on Tools with Artificial Intelligence, New Orleans (LA), IEEE Press.
6. Afshar M., & al. 1990. PCDRA: PC interactive molecular representation and modelling system; *J. Molecular Graphics*, 8:39-44.
7. Mephu Nguifo E., & al. 1996, ML operators meet Dialogue operators. In ICML Workshop Proc. on ML meets HCI, Bari (Italy).

[+] This prototype was implemented when the author was at the Laboratoire d'Informatique, de Robotique et de Microélectronique de Montpellier (LIRMM).

From Informative Patterns to Information Architecture

Kurt Englmeier and Eva Mang

Ifo Institute for Economic Research, Munich, Poschinger Str. 5, D-81679 Munich, Germany, Email: {eva, kurt}@ifosysv.ifo.de

1. KNOWLEDGE AGENCY

Modern information technology gives rise to the chance for Information Retrieval Systems (IRS) to be established in a wider application area, far beyond the scope of laboratory experiments and prototypes. When it comes to large databases exploring information becomes difficult and causes information overload and even anxiety. Databases in the economic domain, for example, often contain millions of time series. Users without a profound expertise both in the domain and handling the retrieval system are at loss to browse and to locate the desired information. The strategies, however, to focus and narrow are well understood by information-search specialists, e.g. economic researchers.

1.1. Navigating Information and Expertise

Right now an information system like the Web means information framed on a two-dimensional hypertext page. It means users navigating via blind clickable links and search-engine requests, drilling down to try to find what they want. And it also means content displayed within an application on a computer screen which is steered and controlled by its users. Such systems support the users to find and get information, in most cases in the form of reports (this is the active side of information retrieval).

On the other hand, these reports, as well as their counterparts from the print media, inform the users. They are in a rather passive role while perceiving the content. Whenever they start to read, they invite information, but once they dive into the content (if it is useful and good), the content's author has the active role and pushes them along, and the content is steering. In this situation content is pushed to you, in contrast to the invitational pull the users make when they navigate along the hypertext links.

Meanwhile it is unfortunately also the truth that there is a host of documents available which one simply cannot browse. Navigating through the information space drains too much effort in the overall task of finding a solution for an information problem. This kind of information overload could be avoided, if expertise in the information space can support the users steering their system along the navigational links, in the way an author steers a reader through the context of his/her report in order to clarify its content. This expertise aims at a closer linkage of the active and passive roles in information retrieval.

The image to hold in mind is *navigating information and expertise*. This means there are human agents behind the scenes, working to hold this trait alive. They design rules for the appropriate navigation links in a way reminiscent to what they do in writing their reports. These structures lead to meta-data, that become the steering components in the retrieval.

Meta-data in the economic domain orchestrate the arrangement of time series and texts, the application of statistical methods, the composition of appropriate visualizations and the providing of navigation possibilities.

From this target we derived the important trait for retrieval systems: It enables the users to move seamlessly between a medium they steer (interaction while active navigation) and a medium that steers them (rather passive perception of information clusters).

1.2. Cooperative design of an Information Architecture

In general, there are *different actors* involved in this *role of agency*. And likewise, the design of the IRS is the result of a corporate process with a variety of skills involved. The appropriate allocation of responsibility amongst all the actors has a decisive impact on the success of the design solution [1].

The new quest is to take complex information and to convey it to an audience as simply as possible - in other words, to communicate information efficiently. This however leads to a shift in the rationale of Information Exploration Tools, which already took place in many areas of information agency. It challenges a design of human-computer interaction (HCI) which enables knowledge communication.

Herein, the design of the HCI borrows its conceptual model from an Information Architecture that ensues from the perception of the tasks in information agency. It aims at implementing the strategies of information-search specialists into the HCI of information systems to guarantee a widespread use of the underlying data. The architecture derived from the strategies and fixed in meta-information is the result of creating systemic, structural, and orderly principles to make information captured in databases understandable and clear [6]. From another point of view, the architecture is a common sense relationship between the different actors which manifests itself in the concrete IRS as their artifact [4].

2. HUMAN-CENTERED INFORMATION RETRIEVAL

Incorporating prior knowledge and interpreting results are decisive for an extensive IR process. The design of key technologies for *Information Exploration* has as well to take place within a framework determined by a careful elucidation of the steps along a realistic retrieval process. Knowledge is conveyed by providing *patterns related to* the *knowledge* of the information-search specialists *by analogy*. [5] This ushers in a correct arrangement, processing and visualization of data.

2.1. Tacit Aspects

Information exploration can be heavily supported by visualization techniques. Visualization will not be a *substitute* for mathematical and statistical methods of IR, it definitely is a useful addition if not another *instrument* for knowledge discovery. To distill information from numerical raw data it may be helpful to just visualize dependencies and regularities between data in different ways. This may happen without applying any mathematical methods or whilst applying them.

Every task or information problem has explicit and *tacit aspects*. This means that there is one part of the problem that is easy to describe exactly. But it is not completely described if the tacit or intuitive view is ignored. There are creative parts in any approach to whatever task or

the same, and even if we could do it to some limited extent we should not. One reason, if nothing else, is that this would defy the purpose of making the system intelligent - or it would narrow the definition of intelligence to suit the specific conditions. Intelligence is rather the capability to maintain an equilibrium, or reach a new equilibrium, for many diverse conditions.

In order to achieve such a capability for a joint cognitive system it is necessary to have a thorough understanding of the problem. Thus rather than starting from the available solutions, tools, and techniques, we should try to look at the problem with fresh eyes. In terms of the analogy used above, we should consider the design of a joint system that would amplify the human ability to move from one place to another. Various such systems exist, ranging from horses to cars (assuming we stay on the ground). In this analogy, the problem is not how to build a better car, but how to improve the human ability to move intentionally, efficiently, safely and quickly on the ground. Building a better car is an engineering problem, but building a better transportation function is rather a problem for cognitive systems engineering.

In the building of intelligent systems, the available tools have more often been a disadvantage than an advantage. The disadvantage is due to the power of the tools, i.e., they make it very easy to achieve something that appears to have the desired properties. A good example is the various tools for building expert systems or knowledge based systems, such as the KBS shells that flourished in the late 1980s. These made it possible quickly to build artefacts that showed a modicum of intelligence, but they so constrained the solutions that alternative approaches were rarely considered. People were enticed to focus on **how** artefacts could be built, rather than **why** they should be built or **what** they should do. The result was a proliferation of structural solutions that relied on the aggregation of "essential" information processing components, greatly supported by the "philosophy" of physical symbol systems (Newell, 1980). The alternative, which is slowly being rediscovered, is to consider instead a functional solution, as exemplified by cognitive systems engineering.

3. INTELLIGENCE OF INTERACTION

What, then, should we aim at in the building of really intelligent joint systems? One solution is to amplify using the tools that we have available, such as processing speed, number of knowledge units, etc. This is unfortunately not the right way to go because the tools are likely to be arbitrary relative to the goal - simply because we do not have a good definition of what intelligence is. In any specific situation or domain, it is possible to define relevant goals for the joint system. Examples can be given for driving a car, flying a plane, performing surgery, pumping oil, producing a device or gadget, controlling a power plant, etc. It would, however, be more useful if we can find a generic approach by characterising the performance of a joint system independently of the specific domain. From a cybernetic point of view the basic characteristic is that the joint system is able to maintain an equilibrium or to reach a new equilibrium despite the occurrence of disturbances. By definition, an equilibrium is a state in which the system can remain for an extended period of time, hence survive. Conversely, a non-equilibrium state will only be transitory or temporary, which means that the system has little chance of surviving in these states. Cybernetics shows us how it is possible to build machines that have a purpose, which is different from machines that have intelligence. The same distinction, of course, applies to joint cognitive systems.

The ability to maintain an equilibrium depends on the ability to control the situation. Control is effective if the actual events have been anticipated, since this means that they can be reacted to in a quick and efficient manner. The ability to predict can be amplified in several ways and making a prediction itself involves several distinct functions. One of these is the generation of plausible alternatives (postulating plausible disturbances and appropriate response - thinking ahead). A second is the capability to go through the alternatives and relate them to the current objectives. A third is the ability to be able to evaluate the possible developments, so that the optimal path can be chosen. The reason why chess-playing machines, such as Deep Blue, have been so successful is not because they have become more intelligent in the true meaning of the word, but only because they have become faster at evaluating possible future paths of action.

Given the current technology, the best ways in which we can amplify human intelligence, hence improve the intelligence of the joint system, is by using the tenacity and speed of computers. The tenacity can be used to produce the comprehensive and complete set of alternatives and guard against the human tendency to make conceptual shortcuts or premature conclusions. The speed can be used to traverse the paths (given that enough data is available to make this possible). The final step, the evaluation, is precisely the one that involves intelligence, or it is at least so complex in its multi-attribute nature that it is not feasible to consider amplifying this as a primary goal. It is rather something that can be achieved by an effective co-operation between human and machines.

Altogether these steps will make it possible to amplify the control of the joint system hence make it more intelligent in every sense of the word. Note, however, that this is not achieved by endowing the machine with some magical replication of human capabilities, but rather by identifying precisely what the machine can do that will contribute to the amplification of the performance of the joint system as a whole. The solution is in the joint cognitive system, not in disembodied machine intelligence.

REFERENCES

Armer, P. (1963) Attitudes toward intelligent machines. In E. A. Feigenbaum & J. Feldman (Eds.), (1963). *Computers and thought.* New York: McGraw-Hill.

Ashby, W. R. (1952). *Design for a brain.* New York: Wiley.

Feigenbaum, E. A. & Feldman, J. (1963). *Computers and thought.* New York: McGraw-Hill.

Hollnagel, E. & Woods, D. D. (1983). Cognitive systems engineering: New wine in new bottles. *International Journal of Man-Machine Studies, 18*, 583-600.

Newell, A. (1980). Physical symbol systems. *Cognitive Science, 4,* 135-183.

Newell, A. & Simon, H. A. (1956). The logic theory machine. *IRE Transactions on Information Theory*, September.

Turing, A. M. (1950). Computing machinery and intelligence. *Mind*, October, *59*, 433-460.

von Neumann, J. (1958). *The computer and the brain.* Yale University Press.

Communication and shared knowledge in human-computer systems

David Benyon

Dept of Computer Studies, Napier University, 219 Colinton road, Edinburgh EH14 1JD, UK

1. AGENT-DEVICE SYSTEMS

When we sit in front of a VDU it is easy to focus attention on the human-computer interaction. Yet when we take a picture with a modern, automatic camera or we engage in a telephone conference it is equally easy to overlook the role of information technology (and increasingly intelligent information technology) in these activities. Humans no longer interact with computers. They become part of a network of multiple interacting systems. Some of these systems will be humans, others will be 'intelligent' software agents others are information processing devices and still others will be 'dumb' devices.

A useful abstraction of a network of interacting systems is to consider it as the interaction of just two types of entity - agents and devices. Both agents and devices are themselves systems (relatively complex 'wholes'). Agents are intentional, autonomous systems in that they have beliefs and desires and can formulate their own goals [14]. People are agents and we are beginning to create artificial agents e.g. [11], [12]. Devices are either syntactic systems (such as hammers, buttons or switches) or they are information processing systems (such as computers).

Interaction between agents and devices is effected through the exchange of messages - expressed as signals travelling through a communication channel. For syntactic systems these messages do not mean anything. Syntactic systems simply react to an input by producing an output [2]. Information processing systems operate at a symbolic processing level; they interpret the messages which they receive in terms of some semantic system which they possess. Agents are able to evaluate the meaning of messages by interpreting them in terms of their existing knowledge and their goals [2].

Agents undertake goal-directed activities by making use of devices. A goal is as a state of the environment or of the agent which is preferred by that agent to the current state. An agent formulates goals based on its desires and knowledge of the devices and agents which it perceives as existing in the environment. Goals are thus constrained by the agent's world view; it's *weltanschauung*.

Given that the agent has formed a goal, the agent then selects a device which will enable it to achieve that goal. The process of selecting a suitable device depends on how the agent conceptualises the device. The agent does this at one or more of three levels (following Dennett [6]). The agent may consider the 'intentional' stance - whether the perceived purpose of the device is appropriate for the current goal. Alternatively, the agent can consider the 'design' stance - does the device possess the necessary, logical capabilities to accomplish the goal. The agent may also conceptualise the actual interaction with a specific device (the 'physical' stance).

The process of selecting a device is based on the agent's conceptualisation of the device: the agent's (mental) model of some possible future interaction. Once a device is selected, the tasks necessary to accomplish the goal are prescribed by the logical structure and functioning of the device i.e. by the way that it has been designed, or has evolved.

At some point, the agent physically interacts with a device by performing an action. That is to say, the agent transmits some signals which the device is capable of receiving (e.g. the agent types a command on a keyboard). The agent then has to evaluate any signals transmitted by the device (or some linked device) and decide to what extent the agent's goals have been met. Conceptual interaction is based on the agent's model(s) of the devices. Physical interaction concerns phenomena observed and signals transmitted by the agent. In terms of the familiar 'gulfs' of execution and evaluation [13], physical interaction is equivalent to the articulatory gulf and conceptual interaction is equivalent to the semantic gulf.

One important result of this analysis is that tasks are dictated by the design of the device; tasks are device-dependent. Of course an agent may misunderstand how the device works and may therefore engage in some activities which do not satisfy its goal. Alternatively the agent may undertake unnecessary activities. It is exactly these problems which should be the focus of agent-device system designers.

2. COMMUNICATION AND SHARED KNOWLEDGE

Agents make use of devices in order to achieve goals within the actual and perceived constraints of an environment. The environment is itself a system which is interacting with other systems. Thus there are multiple levels of agent-device systems. The tasks which an agent actually has to undertake are dictated by the logical structure of the device. The tasks which an agent thinks it has to undertake are dictated by the agent's conceptualisation of the device.

From the perspective of a designer who is intent upon creating an agent-device system, the important thing is to decide how the knowledge possessed by the agents and devices is to be abstracted and represented and to decide how that structure is to be communicated to the other agents and devices. Impoverished agent-device interaction will result if there is a mismatch between the sender and the receiver's abilities in either of these respects.

2.1. Communication

For example, if I am using a computer and the computer issues the message 'dictionary facility is not installed' I may think that the dictionary system is not installed, when in fact the dictionary facility is installed in a different directory. Poor interaction results because we take these messages to mean different things; we are using different semantic systems. To the computer 'not installed' means that the utility is not located where the computer expected it to be. To me 'not installed' means that the utility is nowhere in the system. Of course any particular problem such as this can be 'fixed' by changing the content of the message displayed or by increasing the system's ability to search elsewhere for the required facility. Such an *ad hoc* approach characterises most software products. A more satisfactory solution is to make the knowledge and communicative capabilities of the component systems explicit.

2.2. Sharing knowledge

Consider another example. If a change was made to the Glossary in earlier incarnations of Word, the computer would ask if the user wanted to save the changes ('save changes to Glossary'). If the user responded 'yes' the system would offer a default option that the glossary should be saved as a file called 'Standard Glossary' in the Word folder. So the system has the knowledge of where to store its glossary and what to call it. It also knew that the user wished to save changes to it. However, if the user responded 'yes' to this option, the system would ask if it should replace the existing Glossary, this time offering the default option of 'do not replace'. The reason that Word behaved in this inconsistent manner is because the 'Offer Default' function (or device) did not reveal its knowledge to the standard 'save as' device. The designer of an agent-device system must capture a conceptual representation of the whole domain and consider how different physical representations can reveal this structure to the other agents and devices in the system.

2.3. Modelling the domain

Although seeing human-computer systems as shared-knowledge, communicating, interacting systems consisting of a number of agents and devices is useful for HCI design, the implications of this view go far beyond current interaction problems. The construction of interface and other agents, [11], [12], adaptive systems [3], [15], critiquing systems [9] and the like demands that the knowledge possessed by agents, their abilities to communicate and their strategies for control are made open, explicit and are understandable by other agents and devices in the whole system.

The challenge which these developments offer is to find an appropriate formalism for abstracting, or modelling the domain. A job of the human-computer system designer is to describe domains in such a way that all agents and devices can be constructed or instructed so that knowledge and control over the interaction can be shared. It has previously been argued that using data as the building block of models is particularly appropriate for agent-device systems [1], [4]. The object of the model will be the domain; data is the material from which the model is constructed.

Several well-developed techniques such as entity-relationship (ER) models to represent structure and dataflow diagrams (DFDs) to represent functioning are available to help the designer of a human-computer system to understand the distribution of knowledge and the communication required. These have been applied to human-computer systems [5], [10].

Data models are appropriate conceptual models for modelling domains of agent-device systems since they provide a common medium for describing all agents and devices. Since information is derived from data, the human as an information processor implies the human as data processor. Computers are also data processors. In general all communication takes place through the exchange of signals [7], [8] which are data. Other domain modelling techniques such as objects or tasks are less suitable for conceptual models because both objects and tasks are dependent on the device employed to accomplish the goal.

3. A NEW AGENDA FOR HCI

The conceptualisation of HCI as a network of interacting agents and devices has some important repercussions for design. Rather than designing systems which support existing human tasks, we are entering an era in which we develop networks of interacting systems which support domain-oriented activities. The agent-device view of human-computer systems emphasises that all interaction is mediated by devices. It also focuses attention of the distribution of information throughout the whole work system.

The repercussions of this view are significant for the whole research agenda of HCI. We must shift attention from humans, computers and tasks to communication, control and the distribution of domain knowledge between the component agents and devices. We need to consider the transparency, visibility and comprehensibility of agents and devices, the distribution of trust, authority and responsibility in the whole system and issues of control, problem-solving and the pragmatics of communication. Users are empowered, not by having task-centred software at their disposal, but rather by having domain-oriented configurable agents and devices with which they communicate and share their knowledge.

4. REFERENCES

1. Benyon, D. R. Task Analysis and Systems Design: The Discipline of Data. (1992) *Interacting with Computers,* 4(2) 246 - 259
2. Benyon, D. R. (1993) A Functional Model of Interacting Systems; A Semiotic approach in Connolly, J. H. and Edmonds, E.A (eds.) *CSCW and AI* Lawrence Erlbaum, London,
3. Benyon, D. R. and Murray, D. M. (1993) Applying user modelling to human-computer interaction design *Artificial Intelligence Review* (6) pp 43 - 69,
4. Benyon, D. R. (1996) *Information and Data Modelling*, Second edition. McGraw-Hill, Maidenhead
5. Benyon, D. R. (1995) A Data-Centred Approach to User Centred design In Nordby, K., Helmersen, P. H., Gilmore, D. J, and Arnesen, S. A. (Eds.) *Human-Computer Interaction: INTERACT-95.* London: Chapman and Hall.
6. Dennett, D. C. (1987) *The Intentional Stance*. Bradford Books, MIT press
7. Eco, U. (1976) A Theory of Semiotics Indiana University Press, Bloomington.
8. Eco, U. (1984) Semiotics and the Philosophy of Language Indiana University Press, Bloomington,
9. Fisher, G. (1989) Human-Computer Interaction Software: Lessons Learned, Challenges Ahead *IEEE Software,* (January) pp 44-52
10. Green, T. R. G and Benyon, D. R. (1996) The skull beneath the skin; Entity-Relationship Modelling of Information Artefacts *International Journal of Human-Computer Studies* 44(6) 801-828
11. Kay, A.(1990) User Interface. A personal View in B. Laurel (ed.) *The Art of Human-Computer Interface Design* Addison-Wesley
12. Maes, P. and Kozierok, R. (1993) *Learning Interface Agents* AAAI '93 Conference on Artificial Intelligence, Washington,
13. Norman, D. A. (1986) Cognitive Engineering. in Norman, D. A. and Draper, S. W. (eds.) User Centred System Design. Lawrence Erlbaum
14. Storrs, G. (1989) Towards a Theory of HCI In *Behaviour and Information Technology*, 8(5) pp 323-334
15. Totterdell, P. Browne, D. and Norman, M. (1990) *Adaptive User Interfaces*, Academic press

Problems in Human-Machine Interaction and Communication

L. J. Bannon

Interaction Design Centre, University of Limerick, Limerick, Ireland*

> *'treat computers as nothing more than fancy conduits to bring people together;*
> *never treat information as being real on its own; its only meaning is in its use by people;*
> *never believe that software models can represent people'*
> Jaron Lanier, *Agents of Alienation*, 1995

1. INTRODUCTION

In recent years, with the rapid expansion of the World Wide Web and the concomitant explosion of information available via the Internet, there has been increasing interest in mechanisms to provide forms of filtering and various forms of automated "intelligent assistants" or agents, in order to allow us to sort through this information jungle. In this paper, I wish to make some remarks as to the potential efficacy of a variety of mechanisms that are usually referred to as intelligent agents in helping us with this task. I claim that the general discourse as to the potential of such intelligent agents is problematic, as it assumes that we are in a position to embed within artifacts a level of "intelligence" that is far from being realisable. It appears that just when much of the earlier rhetoric about the capabilities of artificial intelligence (AI) systems have been shown to be overstated, we are in danger again of assuming that it is possible to program software to behave in a way that mimics human intentions and approximates human behaviour. I argue that the general problems of AI that have been discussed in an earlier period still remain in our discussion around intelligent agents. Certainly we can develop software that can perform certain kinds of services for us in the way of determining from keywords etc. the relevance of certain articles, and in doing simple matching against a user's profile of interests, but I argue that this is a far cry from the general kinds of intelligence imputed to machine agents in much of the current debate.

In order to illustrate my general argument, I will refer to one particular form of agent here, namely the idea of an intelligent interface agent with whom we can converse. While some may feel that this particular kind of agent is not representative of the kinds of agents that are being discussed currently on the Web, I believe that fundamentally the arguments are the same, and since I wish to refer to earlier discussions on this theme, I will stay with this particular example of agents. While some may dismiss this discussion as mere hand-waving and of little interest, my argument is that, on the contrary, if we take seriously the critique offered here, it could have a fundamental impact on the kinds of research we carry out, and the scenarios we develop as we envision how people interact with and through computer artifacts.

2. EARLY VIEWS ON AGENTS

In the early days (1980's) of the field of human-computer interaction (HCI), if one asked people in the field what kind of interaction we should be aiming for with machines, the

* This work has been supported by grants from the Irish Forbairt agency and the EU Training and Mobility of Researchers Programme - COTCOS.

overwhelming majority of people would argue that some form of conversational metaphor, with the computer acting in the guise of another human-like agent, would be the most natural and beneficial. The utility of such a metaphor was rarely examined in detail. Yet even back then, there were people who argued that this might not necessarily be the most appropriate kind of interaction metaphor. The work of Pelle Ehn and other Scandinavians was arguing for a less glitzy, but perhaps more useful notion – the tool metaphor. In this scenario, we posit the computer user as a skilled craftsperson who has access to appropriate tools. Tools are built to fit the capabilities of the user on the one hand, and the nature of the work activity on the other. Here the focus is on working through the computer on a task, rather than focusing on the interface *per se*. While this approach was inspired by a host of cultural, historical, and political factors that I cannot delve into here (but see Ehn, 1988, for further background), the tool idea was one which had appeal to many designers, as one which made the human user the locus of intelligence, and the controller of the tools and media. There were others in the early eighties that also questioned the conversation metaphor as being the ideal form of human-computer interaction. Ben Shneiderman's discussion of successful interfaces pointed out that many had a quality of direct manipulation about them, where the user felt in control and directly manipulating the objects of interest, rather than requesting an intermediary to accomplish tasks (Shneiderman, 1983) . In the influential edited collection by Don Norman and Steve Draper, *User Centered System Design* , Brenda Laurel (1986) argued for the need for first-personness in the interface, rather than third-personness, again alluding to the feeling of engagement and control that the user has in such situations. Interestingly, she has since become a proponent of the agent idea, working at Apple initially and later at Interval. In another paper in this volume, Don Norman seemed to go both ways, arguing on the one hand for powerful yet transparent tools, and for the need to engage the user directly, while at the same time arguing for intelligence in other situations.

In one of the most powerful presentations at CHI'83, John Seely Brown gave a talk with a sub-title:; "When artificial ignorance is better than artificial intelligence", which made the forceful point that we should beware of building systems that try to guess what it is that the person is trying to accomplish. He illustrated the problems that this can lead to, when the machine interpretation of events may become different to the human's interpretation of events, yet where this discrepancy may not be evident in the interface. The examples he was using were from the important studies of Suchman on the "intelligent" interface for Xerox copiers described in more detail in Suchman (1987). Finally, in a conceptually difficult but important paper on interface metaphors, Ed Hutchins (1987) provides an account of a series of different metaphors and expresses reservations about the conversation metaphor. He notes: "..taking the problem of human-computer interaction to be a communicational problem assumes that the computer will be another intelligent agent rather than a tool or a structured medium that the user can manipulate ... communication should not be the only organizing metaphor for human-computer interaction."

Despite this long and distinguished list of commentators and researchers who have pointed to problems with the intelligent agent approach to interface design, there has always been an irresistible pull for many people towards the idea of being able to interact with your computer as if it were a human being. In this view, an "ideal" communication would be one where the computer behaved as an "intelligent interlocutor", in Steve Draper's terms (ms.). Many within the AI field have presented this view as the goal of their work, an ideal to which we should aspire. Advocates of this approach over the years include Nicholas Negroponte, head of the MIT Media Lab, who has consistently argued the case for having intelligent agents available to do our bidding.

3. THE RETURN OF AGENTS

In the past few years, while much of the general agenda of AI research has been drastically modified, there has been a resurgence of interest in the idea of agent-like interfaces and software that would act apparently "just like" human agents and make our lives significantly easier as we worked with and through computers and networks. This kind of discussion has become a commonplace on the Internet, so much so that much of the discourse on agents is presented as unremarkable and mundane. On closer examination however, I would argue that there is a lot of glossing between usages of the term where it is being used in a very innocent and un-rhetorical way simply to describe a piece of software that can perform certain well-defined tasks, and those uses where there is an implication that the software agent has a variety of capabilities which allow it to be viewed for many purposes as a substitute for a human agent with similar capabilities. It is this latter viewpoint which I believe to be seriously misguided. My concerns are twofold. One, I believe that, as a general design frame for the field of HCI, at a practical level, it is problematic and liable to lead to a number of blind alleys. Two, I believe that it is yet another attempt to resuscitate the dying embers of the AI fire, where these researchers still believe that, with just a little more work, and money, we will have machines that finally perform "just like humans".

My own position, articulated in several papers over the years, is that we should focus our attention on augmenting the capabilities of people with our technology, and not continue to try to substitute their abilities through machines (Bannon, 1990). The rationale for this view is both moral, ethical, conceptual and pragmatic. It is not possible in these few remarks to exhaustively delineate the problems with the alternate approach, but I will focus on a few issues here. On the one hand, if the terms agent is being used in a very neutral and simple fashion for a piece of pre-programmed software, then I do not see the point in it being labelled as an "agent", especially as this term is being used by others to represent capabilities that are supposedly "human-like". But it is to this latter usage that I wish to particularly orient. In the past few years, we have seen Apple try to flog us primitive hand-held portable computers (the Newton) with limited capabilities under the guise of "personal digital assistants", and we have seen them attempt to provide a computerized human agent in their Knowledge Navigator video. What I find somewhat insidious with this video is that it leads one to believe that the kind of behaviour of our personable agent interface is just around the corner. While some aspects of the agent's performance may indeed be achievable in the near future, the video elides a series of quite fundamental issues about language understanding and disambiguation of context which have no real chance of being resolved in either the near, or I would say, far, future. The result is an amusing but not necessarily informative view on human-computer interaction. Microsoft has also muscled in on the act with their problematic "Bob" character, though it has been done in such a ham-fisted way that the general public reaction seems to have been uniformly negative, and of course Negroponte has championed these computational agents continually for many years, more recently within his Wired magazine columns.

Criticisms of this approach have been a commonplace among various HCI "invisible colleges" for some time – indeed back in the early eighties the UCSD HCI group engaged in an extensive discussion of the topic, although most of the notes produced at that time remained internal and unpublished. Again, around 1990, there was another flurry of interest and debate about the topic among certain circles, on email, but again this material was not made publicly available. Since that time I personally have not become involved in the issue, until my interest was re-awakened by the increasingly strenuous claims being made on the Web about these agents, and then the refreshing, if unexpected, counter-attack by none other than Jaron Lanier, the virtual reality guru. Whether or not one wishes to accept Lanier's accusation of the "evil" of this approach, I certainly support his strong criticisms of the concept, and the way it is being presented. In particular, I agree with the 3 major statements he makes, which I reproduce at the head of this essay. The first argues that we should focus on computers as media that support human-human communication, a point I, (and I am sure many others!) made explicitly back in 1986 (Bannon, 1986). In this perspective, the computer is not seen as the entity that one is interacting with, rather one wishes to operate *through* the interface, in Susanne Bødker's words, on the object of interest, or engage in communication with another person (Bannon & Bødker, 1991). The second point made by Lanier is that we should not reify "information" as some discrete de-contextualized entity , rather it only takes on meaning in its use by people for particular purposes in particular settings. This again is a view that I subscribe to, and in the paper with Susanne Bødker, we critique much of the extant HCI field for ignoring this point, arguing for the importance of understanding the use situation (Bannon & Bødker, 1991)

Finally, Lanier's third point - that we should not attempt to model people in our software systems - is one that has been a cornerstone of much Scandinavian – and some other groups –

work on software development for many years. It calls into question much of the work on "user modelling" that has existed in HCI over the years, as this approach is necessarily incomplete. While within narrowly circumscribed domains, it may be possible to develop a user model that has some predictive capacity, this model is unable to evolve as the domain is enlarged. It is for precisely such reasons that the early work on Intelligent Computer Assisted Instruction (ICAI) systems - which seemed to promise much in the way of diagnosing student problems - has been discontinued. Of course, there are a large number of other issues relating to user modelling, concerning who has access to these models, who can change them, etc. but they would take us too far afield.

4. CONCLUSION

Recent discussions within the HCI and AI communities have focused on the potential of interface "agents" to provide useful services both in selecting appropriate information for users based on their embedded user model, as well as hinting that the communication between user and computer can itself be handled most appropriately with an interface agent. This note has attempted to surface some initial objections to such an approach.

REFERENCES

Norman, D. A. (1986). Cognitive engineering. In D. A. Norman & S. W. Draper (Eds.), User centered system design (pp. 31-62). Hillsdale, NJ: Lawrence Erlbaum Associates, Inc.

Laurel, B. (1986) Interface as Mimesis. In D. A. Norman & S. W. Draper (Eds.), User centered system design (pp.67-85) Hillsdale, NJ: Lawrence Erlbaum Associates, Inc.

Bannon, L. (1990) A Pilgrim's Progress: From Cognitive Science to Cooperative Design. AI & Society, 4,4, Fall Issue, 1990, 259-275.

Bannon, L. (1986) Computer-Mediated Communication. In D. A. Norman & S. W. Draper (Eds.) User Centered System Design: New Perspectives on Human-Computer Interaction. (Hillsdale, N.J.: Lawrence Erlbaum Associates).

Bannon, L. & Bødker, S. (1991) Beyond the Interface: Encountering Artifacts in Use. Book Chapter in J.M. Carroll (Ed.) (1991) Designing Interaction: Psychology at the Human-Computer Interface, pp.227-253. (New York: Cambridge University Press)

Draper, S. (ms.) Prosthesis v. Intelligent Interlocutor: Two modes of interaction (private ms.)

Ehn, P. (1988). Work-oriented design of computer artifacts. Falköping, Sweden: Arbetslivscentrum/Almqvist & Wiksell International.

Ehn, P., & Kyng, M. (1984). A tool perspective on design of interactive computers for skilled workers. In M. Sääksjärvi (Ed.) Proceedings from the Seventh Scandinavian Research Seminar on Systemeering, (pp. 211-42). Helsinki: Helsinki Business School.

Hutchins, E. (1987) Metaphors for interface design. UCSD ICS Report 8703, La Jolla, California. Presented at NATO workshop on Multimodal Dialogues including voice, Venaco, Corsica, Sept. 1986.

Lanier, J. (1995) Agents of Alienation. ACM Interactions, 2,3, 66-72.

Seely Brown, J. (1993) When user hits machine, or, when is artificial ignorance better than artificial intelligence? Invited presentation, ACM CHI'83 Conference, Boston.

Shneiderman, B. (1983) Direct manipulation: A step beyond programming languages. IEEE Computer, 16, 8, 57-69.

Suchman, L. (1987). Plans and situated actions: The problem of human-machine communication. Cambridge: Cambridge University Press.

Machine intelligence in HCI revisited: from intelligent agents to intelligent systems

L.Bálint

Department of Natural Sciences, Hungarian Academy of Sciences
Nádor-u.7., Budapest, H-1051, Hungary*

This paper is introducing a Special Session on "Machine Intelligence" at HCI International '97. The session is devoted to exploring some crucial questions (and possibly answers) about machine intelligence in man-machine systems.

1. BACKGROUND

Machine intelligence is widely claimed as one of the key features of complex systems based on human-computer interaction. However, while there is a common agreement regarding the role of intelligent machine agents in these systems, a considerable debate has also been evolving about the many details related to the properties, development, operation and applications of such intelligent machines, since the early introduction of human-computer cooperation in solving complex tasks. Although some straightforward disciplines and principles like artificial intelligence, knowledge based systems, cognitive ergonomics, intelligence amplification, etc. did get general recognition, if specific questions about machine intelligence are arising, the most frequent approach is characterized by attempts of simply copying human intelligence elements by information processing machine agents (computers). Neither the models, nor the construction and application of interactive human-computer complexes did arrive until now at a solid state by which even the most important questions could be possible to answer appropriately.

2. OBJECTIVE

The paper below is an introduction to a Special Session on "Machine Intelligence" at HCI International '97. The Session has been organized by the author with the aim of providing a critical (although far not complete) overview about past results, present problems, and potential future evolution related to the topic.

---

*Thanks are due for the kind support of the Hungarian Scientific Research Fund (OTKA T-022213).
The author is indebted to all contributors to the "Machine Intelligence" Session at HCI International '97.

3. MOTIVATION

Since many years, intensive work has been devoted worldwide to investigate how human intelligence should or could be copied/substituted/assisted by (intelligent) machine operation. However, a crucial question is still unanswered: should and could intelligent machines really substitute intelligent humans or they are just to complement human cognition and behavior in intelligent human-machine systems? And what machine functions should we call intelligent in the sense of intelligent human-machine systems?

A definite answer to these questions would not only help specifying and developing intelligent machines for man-machine system applications but could also provide a basis for deciding how machine components should be constructed for sake of elevating overall intelligence of such man-machine complexes on the system level.

The above questions are investigated, from different viewpoints, by the contributors of the "Machine Intelligence" Session at HCI International '97. A number of specific aspects about the very basic question are covered by the papers, from cognitive ergonomics to human behaviour, from interaction modeling to interface design, from adaptive operation to HCI applications. But all of them are concentrating on the key aspect of what kind of assistive intelligence should be involved by the machine components (in our specific case by the HC interfaces) in order to achieve highest amount and highest complexity of intelligence at the man-machine system level (specifically, at the level of the entire human-computer complex).

4. SOME KEY ASPECTS AND RELATED OPEN QUESTIONS

It is widely understood that in case of intelligent systems (i.e. intelligent human-machine complexes), intelligence amplification as a result of applying intelligent machine agents is a key aspect. However, intelligent systems are always built in order to perform a certain task. Thus, we need task-independent intelligence on system level.

This means that the assistive intelligence brought in by the machine agent should complement human intelligence in order to amplify that specific system capability, related to the very task. The machine, being companion to the human, has to be intelligent in this sense and should be defined, selected/developed, and also evaluated in view of the task which is also complex enough as far as its definition/specification is concerned.

But can we at all speak about companion intelligence, complementing intelligence, intelligence amplification, and intelligent machine agents in general? Can we aim at a task-independent, well established way of defining, characterizing, and developing this kind of general machine intelligence? Or should we be satisfied, at least today, by task-specific solutions to the widely acknowledged intelligence amplification problem?

Predictably the above questions cannot be answered in their entirety. However, some specific partial questions may be derived and investigated separately, as listed below.

(a) How intelligent systems are related to cognitive systems? How system intelligence is related to the intelligence of its components?

(b) How HCI development is related to system functions/tasks? How HCI intelligence is to be specified? How system models relate to cognitive ergonomics aspects?

(c) How to handle shared knowledge in HCI? How domain-oriented activities/functions are to be taken into consideration? How domain-knowledge and cognitive processing is to be encountered in HCI?

(d) How intelligent machine support eliminates the obstacles in human communication? How to derive guidelines for applying intelligent machine support?
(e) How AI is related to the HCI problem? How human cognition can be taken into consideration when developing intelligent solutions?
(f) How specific (extreme) task requirements can be taken into account when developing intelligent support systems? How to achieve intelligent coordination support in case of many collaborating humans in the system?

The next Section briefly summarizes those answers provided by the Session contributors to the above questions.

5. SUGGESTED ANSWERS - A LIBERAL VIEW OF MACHINE INTELLIGENCE

By trying to answer questions in (a), Erik Hollnagel's contribution "Building joint cognitive systems: a case of horses for courses" emphasizes among others that intelligent systems are to be considered as joint cognitive systems characterized by internal as well as external interactions. Intelligence amplification can be achieved by using intelligent system components so that the resulting system intelligence is not a simple union of the intelligence fractions brought in by the components, but an additional increase of intelligence, characterizing the entire system, is resulting.

As a result of looking for the answers to questions in (b), Chris Stary states in his paper "The role of interaction modeling in future cognitive ergonomics: Do interaction models lead to formal specification of involved machine intelligence?" that HCI development requires task-based design methods utilizing work-flow based execution of the functions. Formal specification of the intelligence in HCI can be achieved by using a task-oriented approach where well established models and controlled separation/integration techniques are applied. The models themselves should be based on complex cognitive ergonomics aspects and should represent not only modes and media of the interaction but the problem domain and the user, as well, in their entirety.

David Benyon's talk on "Communication and shared knowledge in HCI" emphasizes the importance of sharing knowledge in HCI, when making an attempt to answer those questions in (c). HCI itself is considered as an interactive multi-agent system (characterized by distributed cognition, cooperation and communication). Supporting the humans is nothing else in this approach than supporting domain-oriented activities. Furthermore, human-computer communication is taken as the basis for these domain-oriented activities, while human-computer interactions are considered as distribution of domain knowledge and cognitive processing between the agents. Since user tasks are in close relation with problem domains, task support reguires domain-oriented configurable intelligent agents.

Another more or less specific aspect is investigated in view of questions in (d) by George Weir's contribution "Strategies for intelligent support in second-language-interaction". Here, the specific obstacle of human communication is stemming from using/understanding different languages by the users. The suggested solution is based on intelligent support for second-language interactions. Development strategies for such an intelligent support can be well specified and also guidelines can be appropriately formulated for the introduction of these kinds of intelligent support solutions.

Artificial intelligence aspects are handled in his talk "Problems in human-machine interaction and communication" by Liam Bannon when trying to answer the questions in (e). While the present state of AI theory and practice is critically evaluated in view of the earlier hopes, besides the unsolved problems some successful attempts and the promising future are also mentioned. The importance of combining conceptual/philosophical approaches and an empirical establishment is emphasized. The key role of understanding human cognition is also stressed because lack of knowledge about human intelligence is a major obstacle of getting ahead in intelligent HCI development.

Finally, P.H.Carstensen and M.Nielsen go into details about supporting coordination of diverse human activities in a multi-user human-computer system. Answering questions in (f), their paper "Towards computer support for cooperation in time-critical work settings" results, for a specific application example, in such corollaries as how to approach, conceptualize, and support (by intelligent machines) the coordination of time-critical human activities. The importance of special tools for transforming difficult tasks into simple ones is emphasized, together with the fact that, for achieving efficient coordination, special organization of the work and appropriate sharing of the tasks among the participants is absolutely necessary.

6. CONCLUSION

The contributions to the Special Session on "Machine Intelligence" at HCI International '97 are devoted to exploring some crucial questions (and possibly answers) about machine intelligence in man-machine systems. Although each contribution deals with separate, well distinguishable aspects of how machines can serve as companions to humans in their activities, the basic goal is common in all papers/presentations: how the machines can help in achieving most intelligent solutions on the system level.

By summarizing the suggestions provided by the Session contributors (i.e. their answers to the questions listed above), an interesting and surprisingly straightforward combination of ideas and approaches can be derived, as it is briefly formulated below.

Intelligent man-machine systems (eg. in HCI) are joint cognitive systems (with internal and external interactions) where intelligent machine agents should provide intelligence amplification. Careful combination/integration of the intelligent agents in order to perform well structured system functions should provide intelligence surplus above the union of agent intelligence fractions. By utilizing the background knowledge about human cognition together with a task and work-flow based approach, domain-oriented activities can well be supported. Structured execution of the system functions, sharing tasks and domain knowledge among the system agents, distributing cognitive processing, and involving well controlled and supported cooperation and communication may lead to domain-oriented configurable/adaptive intelligent agents which are able to fulfil the basic requirements. Formal specification based system development can well be applied even in case of multi-user human-computer systems and time-critical system functions, by using theoretically and empirically established task and user models and controlled separation/integration of the system functions/agents.

Obviously, the contributions are not able to answer all the open questions but they provide an interesting and stimulating selection from some crucial topics related to machine intelligence in HCI. The papers probably inspire further work in the theory and practice of intelligent machines for human-computer interaction, or at least will provoke some exciting ideas (maybe contradictory ones) regarding the messages stemming from the carefully selected and prepared presentations.

Are interactive media as large as life and half as natural? *

L. E. Merkle †
merkle@csd.uwo.ca

R. E. Mercer
mercer@csd.uwo.ca

Cognitive Engineering Laboratory, Computer Science Department
University of Western Ontario, N6A 5B7 London Ontario Canada

Cultural, social, and cognitive issues are being recognized as fundamental to Human-Computer Interaction (HCI). Interactive media, the products of HCI, drive and are driven by the intricate connections of environments, organizations and individuals: environments, whose notions of space are no longer physically bounded such as in the World Wide Web or in Virtual Realities; organizations, in which technology can be used both to empower workers and enforce procedures, as in Information Technology or in Computer Supported Cooperative Work; individuals, whose omnipresent paraphernalia (such as the ubiquitous computer apparatus) can determine their acceptance or exclusion within a group, society or market.

Questions that should be addressed in HCI are: To what extent can we characterize interactive media in order to understand better how they affect and how they are affected by people, computers, organizations, and the associated involved context? How can interactive media design help us to consider biological, sociological, ethical, political, technological, and other necessary issues concomitantly? Which foundations should we choose for interactive media studies in order to encompass all the above mentioned points?

The scaffold we choose to address these questions has a dialogical constructivist flavor. We have labeled it *Architectonics of Interacting.* We have been using the following in its construction: Cybernetics and Cognitive Science to understand organisms and their co-evolution within environments; Science and Technology Studies to distinguish degrees of artificiality in organisms and artifacts under a common framework; and finally, Media Studies to focus on interactive phenomena as mediated activity. In this scaffold, it is fundamental to include both the commonalities and the dissimilarities of the involved parts or participants of an interactive system, which includes people, computers, and their associated context.

Keywords: Autopoiesis, constructivism, cybernetics, design, dialogism, groupware, mediated learning, pragmatics, self-organization, semiotics.
Key authors: Bakhtin, Latour, Peirce, Maturana, Varela, Vygotskii, Winograd.

*This work is supported by: the Brazilian Government through the Council for Scientific and Technological Development (CNPq Process 201437 – 94.5 GDE), and through the Paraná Federal Center of Education and Technology (CEFET-PR Process 23064.000238/95 – 67); the Canadian Government through the Natural Sciences and Engineering Research Council (NSERC Grant 0036853).

†On leave from the Academic Department of Informatics, CEFET-PR Paraná Federal Center for Education and Technology, 80230-901 Curitiba Paraná Brazil

1 FOUNDATIONS – LARGE AS LIFE

Although Cognitive Science has played an important role in HCI, new foundations are required to address social, ethical, and aesthetical issues within HCI. More than one common denominator has guided our choice of foundations within Cognitive Science, Science and Technology Studies, and Media Studies to scaffold our work. Firstly, although using diverse methodologies and traditionally belonging to different areas of the sciences and the arts, all of these fields strongly suggest that action is an essential factor in cognition, intelligence, communication, etc. Meaning is not given or transmitted, it is actively constructed. Secondly, action and cognition are always situated. Environment and organism, musician and instrument, science and technology, human and computer are temporally and spatially coupled as wholes that cannot be broken apart without losses. Thirdly, recurrent loops are always present. Doing and knowing, developing and using, expressing and experiencing, interacting, and communicating (in the original sense of sharing, participating) coalesce through multiple levels and loops. Fourthly, doing language is crucial for human development and evolution. Lastly, these approaches resonate with our constructivist epistemological commitments.

Our view of interactive media is indebted to several authors and studies within the above mentioned areas. In Cognitive Science the work of Humberto Maturana and Francisco Varela [12, 13] is fundamental. Grounded in theoretical biology, their broad view of cognition encompasses the analysis of a large spectra of cognitive phenomena, ranging from cell dynamics, to organizations, but passing on language issues. This scope is, for us, essential in an understanding of situated cognition. Lev Vygotskii's work, sometimes referred to as sociohistorical or social constructivist, complements our analysis by focusing its studies on the mediation of conceptual or material tools in education[18, 19].

Our link to Science and Technology Studies is Bruno Latour's work, [8, 16] sometimes labeled as heterogeneous constructivism. Its complementary sociological, cultural and semiotical perspectives on technology development gives us the necessary framework for the problematic cohabitation of different theories, with diverse objectives, in our project.

In media studies, Mikhail Bakhtin's work on philosophy and media criticism [1, 2] enables us to conceptualize media not only by its structure, but also within the dialogical and unfinalizable system where interaction occurs, key points in our understanding of the relationships between humans and technology. Peirce's works [15], with pragmatism and semiotics, complete the scene by giving a more structured analysis of sign communication.

The coalescence of these influences requires a study of the possible synergy among these theories [14]. Our intention is not to produce a single unified theory of media. We do not believe that this is even possible. Our goal is to show that distinct but complementary perspectives are required within a theory for interactive media.

2 DEVELOPMENT – HALF AS NATURAL

From a theoretical perspective the Human-Computer Interaction field is passing through a transformation, if not a paradigm shift. Its theoretical foundations and its practices are changing from a machine-centered view toward a human-centered view. We suggest that a human-centered view is not sufficient. An interaction-centered view, with the contribution not only from the human sciences but also from the arts is required. For that, we are proposing an *architectonics* (the systematic arrangement of knowledge). This term goes beyond architecture but still relates to design and could encompass ontology. It has a more constructive flavor and a less established meaning within HCI.

Terms and theories in HCI Studies like distributed cognition [6], ubiquitous computing [21], situated action [17], graspable interfaces [7], theory of activity [4, 3, 10], language-action [23], and hermeneutic computer science [20] illustrate the need to disperse the natural within the artificial and vice versa. It is clear that we as individuals are both indebted to our history and culture as well as to our own actions. Our practices and theories in HCI have to change accordingly.

Following Latour[8], we conceive humans and artifacts (nonhumans in Latour's terminology) as part of the same medium. The way a medium is partitioned depends on the researchers' perspectives and theories used and developed to mediate their studies and interests. For example, computer scientists probably emphasize different issues than psychologists in interactive system design. They use a different language and refer to different objects when talking about the same medium.

For methodological and historical reasons the creation of artifacts, for example tools, has been considered part of humankind's capabilities. Because we are interested in the creation of tools for human activities, we are also interested in the counterpart of this process: how the artifact is essential for human development. The theory from Maturana and Varela [12], which connects cognition to the concept of self-producing systems, understands language and cognition from a biological perspective instead of from an informational one, which uses a computational metaphor.

The borders between the human and the artifact are understood as constructed, not given. The artificial and the natural are seen as interdependent. Therefore a framework to understand what happens in this new interconnected medium, where multiple interdependent loops inbetween and within human and artifact are present. Bakhtin's theory suits our needs affording an analysis based on multiple voices where boundaries are all constructed through action.

3 CONCLUSIONS – BUT TWICE AS ALIVE

This study is intended to shed light on our relation with life, language and technology, in particular, on the design of artifacts that consider the artificial and the natural at the same time, on the importance of acting in our daily interactions with and through technology, and on the importance of action as a concept on our design metholodogies.

The next steps in our work include a study on how Cognitive Science, Human-Computer Interaction, and Media Studies complement each other in media design [14]. The cited architectonics and its mapping onto the understanding of interactive media design processes [22], are in progress. To ground and validate our emerging architectonics, we are applying our ideas to computer supported non-verbal interaction to enrich emotional awareness in groupware.

We invite the reader to reflect on interactive media based on what the Unicorn says when it describes Alice, : "We found it today. It's large as life, and twice as natural." in "Through the Looking Glass"[5]. Are we "half" as natural and are our computers "twice" as alive? We thank Lila Kari for her comments during the writing of this text.

and Causal Analysis. In the first part, the analyst detects the errors performed by the operator(s) involved in the accident. These actions are the manifestations of erroneous behaviors, the final result of an erroneous cognitive process. In the second part of the method, the analyst investigates causes of human error using a classification to be shortly described below.

2.1. Classification of Human Errors

The classification, the core of HERMES, was originally proposed by Hollnagel and it is described here with reference to the latest updated version (Hollnagel and Marsden, 1996) that we modified and adapted to our needs.

According to Hollnagel, there are two fundamentally different ways to consider erroneous actions. One is with regard to their manifestations or **phenotype**, i.e., how they can be observed; the other is with regard to their causes or **genotype**. In the analysis of an event involving human errors, we can observe phenotypes but we can only infer genotypes. A phenotype is, to some extent, what Reason calls an active failure. The classification distinguishes between phenotypes and genotypes; these are further classified into three major groups to support a more detailed analysis, they are: (1) Person-related causes, Prc; (2) System-related causes, Src; (3) Environment-related causes, Erc. While factors related to cognition and emotional states are classified as Prc, those that can be attributed to the technological system and to the environment are included in Src and Erc. Phenotypes are grouped separately; they are the result of an interaction between genotypes and the context.

A well defined method allows to use the classification tables to analyze errors and investigate their causes. In the analysis, we start from the phenotype and we reconstruct the causal chain through the tables. It is possible to sketch it as a fault tree, where the active failure is represented by the root, and causes by its leaves.

3. DAVID: THE ANALYSIS TOOL

We developed DAVID (Dynamic Analysis of Video in Incident stuDies), a prototype of a software tool implementing the main principles of HERMES. The idea is to support the expert in the organization of data concerning errors, and in the investigation of errors causes. DAVID is therefore composed of two modules: a data Organizer and an Analyzer.

With this tool, we aim at exploiting the potentialities offered by the multimedia and by HERMES for the video analysis of human error causes.

3.1. Data Organizer and Analyzer

The expert examines the video recording of an event and detects errors. However, all the data concerning these errors need to be arranged conveniently in order to make appear the information that can be useful for the causal analysis. For this purpose, the Organizer interface provides a ten columns table where the analyst inputs errors characteristics (Table 1); for example, row no. 7 is devoted to error description.

By means of the Analyzer the analyst can graphically trace back the erroneous cognitive process. The interface is basically structured in two parts: while the left one is devoted to the identification of genotypes, the right one displays the reconstruction of the causal chain as a fault tree. Since the architecture of the classification is totally transparent to the expert, the use of the Analyzer is rather simple. At the beginning of each error analysis, the expert classifies the active failure according to a list of phenotypes. The selected category becomes the *root element* of the fault tree and it is placed in the 'account of events'. Causes are proposed on the left-hand side of the interface, the selected categories are added below the root element, and they become the *leaves* of the fault tree. The expert can comment the chosen categories and demand for their explanation (Figure 1).

We integrated DAVID in a multimedia environment, whose architecture was conceived together with the tool. This environment relies on a Selector for the realization of video scenarios, and on a relational Database for the storage of the analysis results. The Selector is an off-the-shelf desktop for digital video editing (Video Machine©) by which the expert digitizes and decomposes in sub-sequences (clips) the video concerning the accident/incident. The objective is the realization of scenarios that are subsequently analyzed (Pedrali and Bastide, 1996).

1	Time
2	Available information or stimulus
3	Event signaled
4	Knowledge and/or belief state component
5	Intention
6	Expectation
7	Decision/Action
8	Source for Decision/Action
9	Immediate feedback
10	Comments

Table 1 - Organizer table.

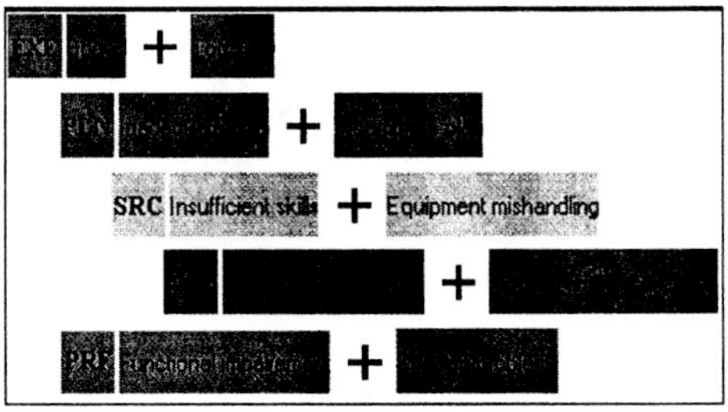

Figure 1 Graphic representation of the fault tree.

4. CASE STUDY

We applied DAVID in the analysis of video recordings concerning a series of sessions, carried out in an AIRBUS A340 full flight simulator, that took place at Airbus Training Centre. The simulation concerned non-precision approaches, the two pilots crew (captain and copilot) made seven approaches over two different airports (New York-JFK and Toulouse-Blagnac). After the sessions, the copilot played the role of the domain expert: he examined the simulation, detected the errors made and analyzed them by means of DAVID. What it is of interest in the results we obtained is the double validation of our approach, from the following points of view: (1) the expert analyzing the errors made by the operator; (2) the operator analyzing his/her own errors.

5. CONCLUSION

Though usability tests are necessary to completely validate DAVID -- from the point of view of its utilization and the accuracy of the results we can obtain -- this study gave us a first feedback. The acknowledged advantage in this type of analysis shows the fundamental interest to discern the set of causes responsible of an active failure, and the reconstruction of the causal chain in the cognitive process. It is extremely useful for the improvement of human-machine interactions discovering that some kind of causes exert their influence in a particular phase of the cognition. Moreover, by ascertaining that some kind of active failures are more frequent in certain working conditions can be extremely important in accidents/incidents prevention. For these reasons, applications of our approach in the domain of operators training is envisaged.

By the introduction of a video tool, we overcame also the problem of static analyses, that is always present in a 'paper pen' study; DAVID returns the dynamism demanded by errors analyses.

6. ACKNOWLEDGEMENTS

The work described in this paper is sponsored by the EEC as part of the European Strategic Programme for Research and Development in Information Technology (ESPRIT), DG III, Directorate General for Industry (Project No. 23917 - SafetyNet).

7. REFERENCES

Hollnagel, E., Marsden, P. (1996). Further development of the phenotype-genotype classification scheme for the analysis of human erroneous actions. EUR 16463 EN, CEC-JRC.

Pedrali, M., Cojazzi, G. (1994). A methodological framework for root cause analysis of human errors. In N. Johnston, N. McDonald, R. Fuller (Ed.), *Human Factors in Aviation Operations: Proceedings of the 21st Conference of the European Association for Aviation Psychology (EAAP)*, Vol. 3, Avebury Aviation, pp. 143-148.

Pedrali, M., Bastide, R. (1996). DAVID: a multimedia tool for accident investigation. In M.A. Sasse, R.J. Cunningham, R.L. Winder (Ed.), *People and Computer XI, Proceedings of HCI '96*, Springer, pp. 349-368.

Integrating System's Functional Model and its Structural Model toward Denotational Dialogue Specification

K. Matsubayashi[a], Y. Tsujino[b] and N. Tokura[b]

[a] Department of Applied Mathematics and Informatics, Ryukoku University,
Ohtsu, Shiga 520-21 JAPAN
E-mail: *shaolin@math.ryukoku.ac.jp*

[b] Department of Informatics and Mathematical Science
Graduate School of Engineering Science, Osaka University,
Toyonaka, Osaka 560 JAPAN

This paper presents an extension onto our previous researches for integrating system-side description and that of user-side. Higher abstraction level (task level) of dialogue description is introduced, and the relationship between this and command level are specified. The structures of the dialogue are given in the syntax representation, and semantics functions are introduced to designate the "meaning" of the dialogue, based on the notions of Denotational Semantics. With our framework, we can have possiblities of having a way to clarify problems between a system and a user.

1. INTRODUCTION

The development of formal foundation and specification of various aspects of interaction between computer systems and their users have still been a big challenge. Giving formal description of the interaction is expected to be one of the good solutions for helping clarifying the problems and ensuring the quality of application programs. Well-known researches such as [1], [2], [3] and [5] shows that formal approaches can give several good solutions for identifying some kinds of problems and charasteristics in user-interface or in the interaction.

Our alternative standpoint is summerised as follows: we should like (1) to give fundamental and formal framework to define the "meaning" of the dialogue, (2) to deal with both the user and the system as on equal terms as possible, and (3) to get the specification which reflects various charasteristics of the interaction. In our previous papers and reports, we have showed that denotational approach for formal description of human-computer dialogue have much possibilities in order to describe various aspects of the interaction, such as hierarchical structures of the dialogue and abstraction levels of the interaction, both of that should be helpful for providing good understandings of human-computer dialogue in multiple views.

We capture the interaction as a triple of a user, a computer system, and the dialogue between them (Figure 1). Dialogue is a symmetric process of exchanging information between a user and a computer, both of which have their own "internal entities", that presents the state of the subject, and that are referred and get changed as the dialogue proceeds. Dialogue structures are defined as BNF-like syntax representation, and the meaning of the dialogue are given as semantics functions over the syntax. That is, the meaning of user's input symbol are described as a mapping from a current state of the computer (i.e. computer's internal entities) to the next state and the system's output. Similarly, the meaning of system's output symbol are described as a mapping from concrete representations of system's output symbols (actual output on the display) and a user's knowledge (user's internal entities) to the next state and the user's next output to the system.

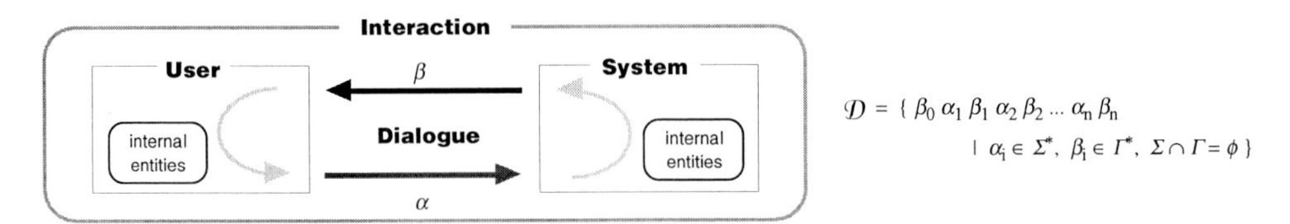

Figure 1. Interaction between a user and a system.

We have applied the technique of specification to some sample application programs and successfully obtained brief and concise specifications of the syntax and semantics of user's input symbols to the system. This also captures such charasteristics of the dialogue as hierarchical structures and the abstract levels [4].

Our next step is to extend the framework toward (1) handling higher abstraction levels, as well as (2) the specification of the syntax and semantics of system's output symbols with user's entities. Following sections describe the outline of these topics.

2. SPECIFICATION OF TASK LEVEL DIALOGUE

This section deals with the extension of our framework to handle higher task level. With this extension, we can get another viewpoint of human-computer dialogue as more user-centered perspective.

2.1. Definition of task level

In our previous papers, we only dealed with two abstraction levels: command level and lower articulation level. For example in command level, a set of user's input symbols to the system is defined as a set of commands as much as actually being prepared in a target application program. But in task level, we face with a problem how to determine what is exchanged between two subjects: in other words, to determine what is "task" for a system. Actually, few application programs explicitly implement "tasks" in the same way as commands.

We capture the task level dialogue as a series of a user's requirements to achieve some goal and a response from a system. These requests are essentially defined external to the computer system. Hence a set of tasks is infinite in essence - actually, this is dynamically created and modified by the user -, but we regard a finite set of tasks that is given *a priori* so as that developers of a target application program assume (or select as a result of task analysis). So, in the task level, we could prepare tasks and actions only enough to describe user's activities. For example, they may include a parameterized input symbol DuplicateDocument (doc), that corresponds to a sequence of SelectFile (*f*) and CopySelection in the command level.

2.2. Specifying syntax and semantics of task-level dialogue from system's side of view

As widely accepted, a ***task*** is "a structured set of activities in which actions are undertaken in some sequence" [6], so we specify the syntax of task-level dialogue so as to reflect the task structures. And the syntax thus represents canditates of sequences of sub-tasks and actions to accomplish the task. For example, suppose the task to replace all words a appeared in a certain document with b, represented as a non-terminal symbol **ReplaceAll** (a, b), can be decomposed as

ReplaceAll (a, b) ::= doReplaceAll (a, b) Resp
| doFindAndReplace (a, b) Resp ***FR***
| LookForString (a) Resp Replace (a, b) Resp ***FR*** | ....

FR ::= ε | doFindAndReplace (a, b) Resp ***FR***
| LookForString (a) Resp Replace (a, b) Resp ***FR***

Bold-faced symbols such as **ReplaceAll** are non-terminal symbols that represent tasks, while plain symbols such as doReplaceAll are minimal actions in task level. Underlined symbols represent system's output, and Bold and italic symbols like ***FR*** are non-terminal symbols in the syntax.
This syntax is also actually infinite in essence, for there are many "verbose" representations available: for example, **ReplaceAll** (a, b) can be decomposed as **ReplaceAll** (a, c) **ReplaceAll** (c, b). To avoid such verbosity, we set some restrictions on task structures such as (1) in the top-level syntax specification, non-terminal task symbol on the left-side must not be re-appeared in the right side, (2) verbose actions (a repetetion of typing 'A' and pressing delete key, for example) cannot be included, etc.
Semantic specification from system's side of view are given as defining semantic functions that describes how a user's input symbols (actions in the task level) changes the system's internal entities and outputs. For example, we prepare a semantic function doTask to designate the meaning of user's input symbols in the task level. Below is a semantic specification over the first canditate of **ReplaceAll** shown above. replace-all-string is a support function to increase the readability of the specification.

doTask[doReplaceAll] $a\ b\ f\ m\ buf$ = **let** $(copy, find) = buf$ **in**
let f' = replace-all-string (f, a, b) **in** $(f', m, (copy, a))$

Note that the result of excecuting every canditate of the task will be exactly the same. It means that the value returned by the semantic function will be the same.

3. SPECIFICATION OF THE DIALOGUE FROM USER'S SIDE OF VIEW

This section deals with another extension of our model to give the meaning of human-computer dialogue from user's side of view. Below we will show an example of the specification in the task level, but all other lower levels can be specified in the same manner.

3.1. User's Internal Entities

In order to specify the meaning of the dialogue viewed in the light that the user receives the output from the system, gets the user's entities changed, and sends next input to the system, we must determine what is suitable for user's internal entities.
We assume that the user has the following two entities: **T**' (knowledge of the task and the pointer which the user is currently executing) and E_T' (knowledge of the system). The former is a subset of the syntax of task-level dialogue, and the latter be a subset of the system's internal entities. The extent how much these entities is complete represents an user's level of knowledge.

3.2. Semantic Specification

With these entities, we have to define two new semantic functions think and decide to designate how system's output symbols change the user's internal entities and determine his/her next input symbols to the system. Note that we are currently dealing with the case that only the pointer among the user's internal entities gets changed.

think[doFindAndReplace Resp] *d t e*
decide[doFindAndReplace Resp] *d t e*

For example, These two semantic functions must be specified so as that the user observes the display (*d*), compare his mental model of the system (*e*) and finally deside which will be the next input in the current task execution (*t*).

4. DISCUSSION

We have showed the first step to specify whole stages of interaction cycle between a system and a user. Actually, there are some difficult questions when specifying details. Currently we set some important limitations to simplify the model : (1) we define user's knowledge simply as a subset of system's internal entities, (2) the knowledge of the user is fixed during the dialogue, and (3) the sterategy that the user select among canditates of action sequences to accomplish a certain task is capsulized in a form of support functions. Some of these limitations will be removed, but some may need more studies.

At the time of this writing this paper, we are specifying all levels of the dialogue of some simple application program as an example. An extract from this specification, that has been omitted here due mainly to the lack of the spece, will be showed in our oral presentation.

5. CONCLUSION

We have shown that our extension to task level and user-side specification is expected to have much possibility considering whole states of human-computer interaction between a computer system and a user. We are currently investigating further extension of our framework, including taking away the limitation of fixed user's entities, as well as development of emulation of human-computer dialogue based on the specification.

REFERENCES

[1] S.K. Card, T.P. Moran and A. Newell : *The Psychology of Human-Computer Interaction.* Laurence Erlbaum Associates, Hillsdale, NJ. (1983)

[2] D. Kieras and P.G. Polson : *An Approach to the Formal Analysis of User Complexity.* International Journal of Man-Machine Studies, 22, pp.365-394. (1986)

[3] T.P. Moran : The Command Language Grammer : A Representation for the User Interface of Interactive Computer Systems. International Journal of Man-Machine Studies, 15, pp.3-50. (1981)

[4] K. Matsubayashi, Y.Tsujino and N. Tokura : *A Denotational Approach for Formal Specification of Human-Computer Dialogue.* Symbiosis of Human and Artifact Vol.B (Proceedings of HCI International '95). Ed. Y. Anzai, K. Ogawa and H. Mori. North-Holland, Elsevier Science, pp.71-76. (1995)

[5] S.J. Payne and T.R.G. Green : *Task-Action Grammers - A Model of the Mental Representation of Task Languages.* Human-Computer Interaction, Vol.2, pp. 93-133. (1986)

[6] J. Preece et al.: Task Analysis. *Human-Computer Interaction*, pp. 409-429, Addison-Wesley. (1994)

[7] D. Scott and C. Strachey : *Towards a Mathematical Semantics for Computer languages.* Proceedings of Symposium on Computers and Automata. Polytechnic Institute of Brooklyn Press, New York, U.S.A. pp.19-46. (1971)

Modeling the Sources and Consequences of Errors And Delays in Complex Systems*

Chaya K. Garg, Victor Riley, and Jonathan W. Krueger

Honeywell Technology Center, 3660 Technology Drive, Minneapolis, MN 55418, U.S.A.
E-mail: garg_chaya@htc.honeywell.com, riley@htc.honeywell.com, and krueger@htc.honeywell.com.

1. INTRODUCTION

Many incidents and accidents involving new systems occur due to human errors that were not foreseen during the development of the system. Many of these errors, in turn, are due to confusion, loss of situation awareness, high workload, or other difficulties induced by the design of the system. Aspects of the system that raise these problems range from the design of the user interface to the roles and responsibilities of the automation and human operator; accordingly, they span a range from concrete, visible causes to abstract, invisible ones. Often, the more abstract and less visible causes are the ones that cause the greatest difficulty for system designers because they are the hardest to anticipate.

The next generation air transportation management system is likely to raise a very broad range of issues arising from the extensive reliance on automation, the potential for differences between automation and operator solutions and strategies, and the increased complexity of the operational environment. In particular, the free flight (RTCA, 1995) concept will reduce the ability of the controller to anticipate the future trajectories of aircraft and thereby anticipate and resolve conflicts. This may make the system vulnerable to a large range of human error.

Human performance modeling tools have typically focused on modeling human performance or human-system interaction at the task level. Models developed using these tools yield task timeline, task sequence, and workload profile data. In turn, these data are used to study the effects of new automation concepts on crew and system performance, evaluate alternate design concepts, perform crew size studies, and effect an appropriate allocation of functions.

We have approached modeling at a level of abstraction that is higher than the task level. This is because the primary objective of our modeling effort was to investigate the sources of errors, delays, and goal mismatches between humans and automation in air transport management transactions and to assess their impacts on overall system performance. A secondary objective was to define a methodology that not only attempts to overcome the problem of scenario dependence, but also facilitates scenario reuse.

The Mixed-Initiative Model-Domain Modeling Environment (MIM-DoME) methodology focuses on characterizing a domain's agents (human, automation, and the external environment) using a critical but sufficient set of characteristics, and identifying key interactions among these characteristics. These characteristics and interactions are implemented using a development environment that allows an analyst to set up scenarios of interest, manipulate the types of errors and conflicts that may occur between operator and automation, and examine their effects on overall system performance. The level of abstraction used in this approach is at the level of activities that are represented as perceptions, inferences, plans, decisions, and actions. This level of abstraction, which is higher than the task level,

* This work was supported by NASA contract NAS2-14288 as part of the Advanced Air Transport Technologies Program. Dr. Mike Shafto was the Contracting Office Technical Representative.

allows us to use a single model to address a variety of questions regarding the sources and consequences of delays and errors in air transportation management transactions. Currently, the tool outputs an event path from a given start state to a hypothetical end state.

Below we provide the steps involved in developing models using the MIM-DoME methodology. The methodology is described with respect to the model that we developed as a proof of concept to investigate concerns relating to the next generation of air transport management systems.

2. MODEL DEVELOPMENT

Step 1: Outline scenario. While this step is self-explanatory, care should be taken that the scenario selected for use will allow the analyst to represent errors, delays, and goal mismatches. In our study, we focused on the pilot and Flight Guidance System part of air transport management transactions. The scenario was defined based on a preliminary Function Allocation Issues and Tradeoffs (FAIT) analysis (Riley, 1993), which indicated the possibility of numerous conflict areas between agents in the scenario.

Step 2: Determine the system's level of automation. The Mixed-Initiative Model (MIM) (Riley, 1993) provides a framework that allows the analyst to identify, in a scenario-independent, systematic, and almost comprehensive manner, the characteristics that influence problem-solving and performance in a domain. Because the MIM contains parallel representations of machine and human operator information processing, conflicts of knowledge, strategies, and goals between the two can be readily represented. It also provides a framework within which to organize the activities that are performed by the domain's agents. For a detailed description of the nodes represented in this model, see Riley (1993).

Step 3: Identify characteristics. In this step, characteristics and their associated legal values are identified for the environment in which the system will operate, the machine components of the system, and the human operator. Characteristics are significant attributes associated with a domain or with the agents in the domain. Characteristics act individually or in groups to influence system and individual agent performance. Example characteristics identified for the machine - world sensors node included data link delay and weather radar delay and accuracy. A complete list of characteristics and their legal values is available in Garg and Riley (1996).

Step 4: Outline behaviors. Behaviors represent interactions between and among characteristics within a node or across nodes. Behaviors in conjunction with delays determine the information that flows through the model. We did a preliminary FAIT analysis to ensure that many of these interactions are identified and represented as behaviors. The behaviors that we defined for the purpose of this study are documented as if-then rules in Garg and Riley, 1996. We also used behaviors to represent the possibility of several types of error on the part of the pilot and automation due to variables such as workload, sensor inaccuracies, fatigue, trust in automation, and automation related complacency.

Step 5: Specify delays. The sources of delay are identified and listed as characteristics in Step 3. Step 5 consists of coming up with reasonable estimates for delay periods. Delays determine when the values of variables are updated during model execution. It is conceivable that, due to delays, the world, automation, and pilot use different values for the same variable. For example, a delay by the pilot in reading the displays causes the pilot to not use the currently displayed trajectory for an aircraft close to it, while the collision avoidance system uses the currently displayed trajectory for the aircraft to determine if there is potential for conflict.

The Mixed Initiative Nets (MI-Nets) notation (Krueger, 1996) used to implement this model seems almost tailor-made for implementing and modeling the effects of these delays. This is because MI-Nets are basically augmented high level timed mark Petri nets that allow an analyst to associate maturation and expiration times with delay values. These maturation and expiration times determine when the value is available for propagation through the model.

Step 6: Specify control flow and order of evaluation. Control flow defines the path that the model takes or determines the order in which the nodes in the MIM are traversed in each cycle. A node has control or is defined as enabled once all input edges to it are enabled or convey control. More than one node can have control concurrently. Additionally, logical statements and functions can be used to direct the control flow of the model. Once a node is enabled or has control, the behaviors associated with the node are fired in the order specified by the analyst.

The MI-Net (Krueger, 1996) notation provides an interface that allows the analyst to implement the control flow and to specify for each node the behaviors that belong to it and the sequence in which the behaviors should be evaluated.

Step 7: Implement model. This step entails representing the nodes in the MIM, the characteristics, behaviors, control flow, and the order of evaluation. The MI-Nets notation was used to instantiate the MIM in DoME.

DoME (Krueger, Engstrom, Ward, Peterson, Maloney, and Hoplin, 1996) is an application framework for developing and using formal, domain-specific, graphical languages. DoME is implemented in Smalltalk and runs on most computing platforms. New notations are defined using DoME's Tool Specification Language, and additional functions can be added with DoME's interoperable extension languages: Projector and Alter.

The MI-Net notation and tool were built using DoME. This notation, which is still evolving, is a high level, timed mark Petri-net augmented with extensions to guide the firing of transitions during execution. It is supported by a tool that includes a simulation engine which allows us to execute models developed using this notation. Both static and dynamic analytic methods can be applied to this formalism. A formal description of MI-Nets is available in Krueger (1996).

In the MI-Net notation nodes are implemented as classes internally and represented as labeled rectangles with rounded edges. Characteristics are implemented internally as places within the classes. Characteristics are represented as ovals within the appropriate classes. Next, behaviors are implemented as transitions with input and output edges to relevant characteristics. For example, the behavior to determine if there is a conflict situation is implemented as shown in Figure 1. The rectangular box is called a transition. It has input edges from characteristics whose values are used in the guard function contained in the transition box or from characteristics whose values are passed to other characteristics through output edges. Output edges pass a single number or string to the output characteristic or evaluate a function or logical statement to output a value.

The control flow is implemented by drawing labeled edges between classes. For each behavior, two property assignments are made. The "belongs to" property assignment is made to indicate the class that the behavior belongs to, and the "firing order" property assignment is made to specify the priority with which the behavior should be fired relative to other behaviors that belong to the same class.

Step 8: Simulation set-up. This step consists of performing routine simulation set-up tasks such as initializing variables, inputting delay periods, outlining events in the events file, specifying stop states, and listing variables for data collection.

Step 9: Run model and collect data. This step consists of specifying standard run time parameters such as turn trace on or off and turn debugging on or off. Analysts can also specify the frequency with which the model is traversed.

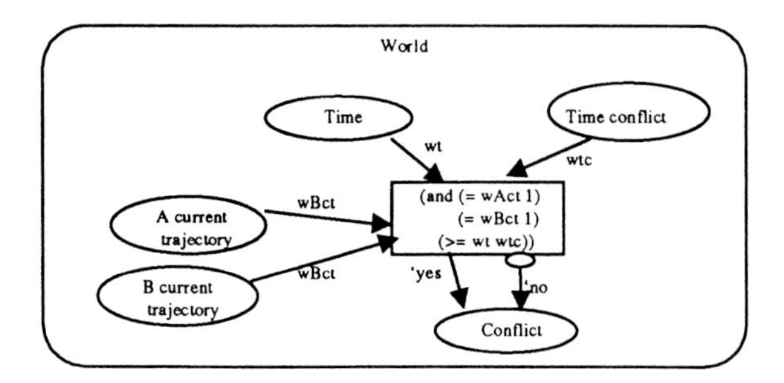

Figure 1. A sample transition implemented using the MI-Net Notation

Step 10: Data Analysis. The primary outputs of interest are the event paths followed by the model based on the errors and delays encountered, the associated event timelines, and the variable states. Data collected during the model run can be imported to standard statistical packages and Excel for analysis and graphing.

Once the model is implemented and used to run a baseline scenario, it can be modified to address a wide variety of "what-if" questions (Garg and Riley, 1996).

3. CONCLUSIONS

The MIM-DoME modeling methodology provides an effective and useful means of investigating a range of errors that may occur in dynamic systems. The MIM in conjunction with the MI-Nets development environment and the "activity" level of abstraction used to implement models provides a unique capability for investigating error propagation and its associated event paths across a wide range of what-if conditions.

REFERENCES

Garg, C., and Riley, V. (1996). Analyzing the Dynamics of a Next Generation Air Transport Management System: The MIM-Dome Modeling Tool. Honeywell Technology Center, Minneapolis, MN.

Krueger, J. (1996). "The MI-Net Development Environment." Honeywell Technology Center, Minneapolis, MN.

Krueger, J., Engstrom, E., Ward, J., Peterson, T., Maloney, J., and Hoplin, D. (1996). "DoME Users manual. Software Version: 4.14." Honeywell Technology Center, Minneapolis, MN.

Riley, V. (1993). "Function Allocation Issues and Tradeoffs: User's Manual." DOT subcontract number 19800(2014)-1981, Honeywell Technology Center, Minneapolis, MN.

RTCA. (1995). Final report of the RTCA Task Force 3 Free Flight Implementation. (Washington, D.C.: RTCA).

Task Network Modeling of Human Workload Coping Strategies

J. F. Lockett, III

U.S. Army Research Laboratory, Human Research and Engineering Directorate, ATTN: AMSRL-HR-MB (Lockett), Aberdeen Proving Ground, Maryland, 21005-5425, United States

Models designed to predict human performance should account for changes in task execution attributable to human response to workload. This paper documents the use of a workload modeling tool called WinCrew, simulator trial data, and extensive rules for resolving dual task conflict to develop WinCrew inputs for modeling workload coping strategies.

1. INTRODUCTION

One promising method for predicting the performance of man-machine systems is stochastic task network simulation. Over the past 10 years, the U.S. Army has developed several computer-based modeling tools using this approach, many of which have application outside the Department of Defense. Among other features, the tools enable performance evaluation of trade-offs in system design options such as the number of operators, level of automation, task allocation, and user interface design. The tools use mental workload as one indicator of the acceptability of a system design to human operators. In general, the higher the predicted workload, the worse the predicted human performance and resultant system performance. There is a problem, however, with this basic approach. Mental workload researchers have reported an inconsistent relationship between workload and performance and suggest that models that fail to account for this may be overestimating human performance (Hart, 1989, Hart & Wickens, 1990). Models should consider human response to various workload levels. When the Army developed its workload modeling tools such as Crewcut and later WinCrew and IMPRINT version 3.0, it included a workload coping strategy modeling feature. In keeping with the goal to maintain a balance between the amount of input, effort, and expertise required to build a model and the diagnosticity and validity of the output, the tools have built-in defaults, algorithms, or pre-canned effects to aid in model development. Six basic predefined coping strategies were developed, based on likely behaviors and those suggested by Hart (1989) until anticipated research efforts quantitatively revealed what the strategies were and when they are employed. An advanced capability was also developed to build more complex strategies in the event that research results revealed the strategies to be more complex than the six original choices (Little, Dahl, Plott, Wickens, Powers, Tillman, Davilla, & Hutchins, 1993). Unfortunately, and largely because of funding cuts and reorganizations in the late 1980s and early 1990s, research about strategies was not completed or was canceled. Tool validation studies in the early 1990s that may also have revealed generalizable strategies were also curtailed or canceled.

Fortunately, the United Kingdom Ministry of Defense, Defense and Engineering Research Agency (DERA) has been conducting a series of simulator-based trials which may yield information about coping strategies employed by fighter pilots during low level flight (Fennel, O'Connor, Cowling, Murphy, Sayers, Enterkin, Fearnside, & Hodgson, 1995). Under auspices of The Technical Cooperation Program, subgroup U, Technical Panel 7 (TTCP UTP-7), Human Factors in Aviation Systems the U.S. Army and the UK DERA began a cooperative effort in 1996 to model workload coping strategies. The remainder of this paper documents the use of the WinCrew tool in representing information from UK efforts.

2. APPROACH

The general approach was to use the UK data and subject matter expert (SME)-developed rules for task conflicts in low level flight and determine WinCrew input parameters for coping strategies.

As part of a future offensive aircraft (FOA) development program, researchers in the UK collected subjective workload rating data (among other data) during trials simulating a low level flight mission segment. The pilots were fast jet pilots, and the FOA simulated was a single seat concept. During post-trial interviews, the pilots were asked to identify what their task strategy was for high workload periods as reported by them subjectively during the trials (Fennel, et al., 1995). Using this information, the UK researchers (including fast jet pilot SMEs) developed an extensive set of rules for characterizing terrain following task performance at various levels of workload. The rules varied according to terrain type (e.g., rough or flat), flying in visual or instrument mode, overall workload (e.g., low, medium, or high), degree of visual or cognitive conflict (e.g., acceptable, marginal), criticality and interruptability of tasks, time since completion of last copy of a task, and workload when the task started. The rules were incorporated into a series of MicroSaint (Micro Analysis and Design, 1993) models developed for the terrain-following aspect of the simulated mission (Cowling, 1995).

Based on the sources of the UK-developed set of rules, a review of the post-trial interview data, and a re-review of the original qualitative strategies suggested by Hart (1989), a proposed paradigm for modeling coping behavior during high workload in WinCrew was developed. The primary factors across the sources appeared to be workload level, task importance and urgency as compared to other ongoing tasks, and mental resource conflicts. Aims were to limit the paradigm to existing WinCrew input variables and to minimize model complexity (i.e., input burden) for the model builder.

The WinCrew software includes six basic strategies that can be triggered when an operator's overall workload exceeds a modeler defined threshold value. These strategies are
A -- No effect. All tasks are performed regardless of workload value. This is the default strategy.
B -- Do not begin the next task. The next task is not started by any other operator. This is sometimes referred to as "task shedding"
C -- Perform tasks sequentially beginning with ongoing tasks and then performing the next task.

D -- Interrupt ongoing tasks in favor of starting the next task. Restart ongoing tasks in "windows of opportunity" (i.e., when the operator's workload drops below the threshold value).
E -- Reallocate the next task to the contingency (backup) operator if a backup is present in the design.
F -- Reallocate ongoing tasks to the contingency operator.

More complex strategies may be created by building "if-then-else" type expressions with variables, the basic six strategies, and mathematical and logical operators (= =, : =, >, <, +, -, /, *, |, &). Variables that are available for use in the expressions are
P = priority of the next task (Each task in WinCrew can be assigned a priority from 1 to 5)
H = highest priority of any one ongoing task
T = total workload for the operator after beginning the next task
S = operator's workload threshold value (defined in terms of W_T: the total aggregate workload at a given moment)

The most important aspect of the WinCrew strategies is that they represent dynamic task scheduling based on operator workload. In other words, the model builder does not specify exactly which task the operator will do next in performing a mission. The next task or tasks will be determined in part, based on the momentary workload of the operator.

3. PROPOSED PARADIGM

The paradigm proposed for modeling coping strategies using WinCrew represents many of the variables from the UK set of rules and uses only existing WinCrew features. The paradigm involves filling in the WinCrew "task priority" variable to classify each operator task for a mission as
5 = critical and uninterruptable
4 = critical but interruptable
3 = noncritical
Task priority values of 1 and 2 are not used in the paradigm.

Next, an advanced workload strategy of "if P>H then D else C" is entered for the operator. This advanced strategy translates to this: if the priority of the new task is higher than that of any ongoing task, then perform the new task and suspend ongoing tasks until workload goes below the threshold. The "else C" portion of the advanced strategy can be adjusted, depending on the availability of another operator or automation (else F) in the crew station concept. This classification of tasks and strategy for determining which one is attended to accounts for the task importance and urgency factor found across the information sources used to develop the paradigm.

Finally, the threshold value at which the strategy is triggered is set. The threshold can be determined by an SME review of a WinCrew model run without a coping strategy. Predicted workload peaks are decomposed into their component tasks. The SMEs are asked to review ongoing tasks during the workload peaks and to identify which combinations represent the margin of their ability to perform. The workload level at the peaks corresponding to these combinations can be used to determine a general threshold value. Otherwise, a value of "60" is the WinCrew default for experienced operators (Little et al., 1993). Because the workload algorithm in WinCrew considers mental resource conflicts, this accounts for the workload level and resource conflicts determined to be factors across the information sources used to develop the paradigm.

4. DISCUSSION

Despite a strong theoretical grounding, any proposed method for predicting performance as a function of human coping behaviors must be validated against data from trials collected with a man-in-the-loop. Sensitivity analyses must also be performed to determine if the addition of more inputs, features, or variables produces better prediction of performance. Is the additional burden of building models with these refinements worth resultant refinement or added robustness in the output produced? The next step in modeling human workload coping strategies is to develop both a complete WinCrew model and MicroSaint model of the UK FOA low level flight mission and then compare the output of each model to pilot and aircraft performance data. Model outputs with and without a coping strategy in effect will be compared to see how well either correlates with task performance observed in trials. If both models prove to be equally valid and if the addition of coping strategies yields better prediction, the conclusion will be that the WinCrew tool provides the model builder with the best capability for modeling human performance with a minimum of input.

REFERENCES

Hart, S.G. (1989). Crew workload management strategies: A critical factor in system performance. In Proceedings of the Fifth International Symposium on Aviation Psychology, 22-27. Columbus, OH: Ohio State University.

Hart, S.G. and Wickens, C.D. (1990). Workload assessment and prediction. In H.R. Booher (Ed), MANPRINT: An Emerging Technology. Advanced Concepts for Integrating People, Machines, and Organizations. New York: Van Nostrand and Reinhold.

Little, R., Dahl, R., Plott, B., Wickens, C., Powers, J., Tillman, B., Davilla, D., Hutchins, C. (1993). Crew reduction in armored vehicles ergonomic study (CRAVES) (ARL-CR-80). Aberdeen Proving Ground, MD: U.S. Army Research Laboratory.

Fennel, J., O'Connor, S., Cowling, G., Murphy, P., Sayers, S., Enterkin, P., Fearnside, P., & Hodgson, S. (1995). FOA workload prediction: Initial simulation results and workload modeling comparisons (DRA Customer Report Ref: DRA/AS/MMI/CR95222/1). Farnborough, United Kingdom: Defence Evaluation and Research Agency.

Micro Analysis and Design. (1993). Micro Saint for Windows user's guide. Boulder, Colorado: Author.

Cowling, G. P. (1995). An aircrew workload model for a future offensive aircraft (DRA Customer Report Ref: DRA/AS/MMI/CR95332/1). Farnborough, United Kingdom: Defence Evaluation and Research Agency.

Predicting Human Performance in Complex Systems - A Method and Case Study

Ron Laughery and Beth Plott[a]

[a]Micro Analysis and Design, Inc., 4900 Pearl East Circle, Suite 201E,
Boulder, Colorado 80301, USA rlaughery@madboulder.com

1. BACKGROUND

Years of research have been devoted to evaluating general issues in human-computer interface design. As a result, guidelines for evaluating the human interface have been produced. However, the determination of which design is better is often a function of operational factors (i.e., which menus are likely to be accessed from where and how often). General guidelines may offer limited advice on this, but what is needed to supplement the guidelines are tools for evaluating the performance of different design options *as a function of the possible scenarios of operation and use.* Experimentation with human operators is the ideal way, but is often too expensive and time consuming.

Computer modeling of the human-computer interface is a technology that can be used for evaluating aspects of human-computer interface designs. This approach is called *task network* modeling. The general concepts and techniques of task network modeling are discussed in detail in Laughery and Corker (1996). Essentially, task network modeling involves the extension of a task analysis into a network that defines task sequence. Then, task timing, decision logic, and interactions with other tasks are added to the model to accurately depict the human interacting with the system. This model can be used to collect data and run parametric studies in manner conceptually the same as with human-based research. In this paper, we will present an introduction to the technology and then a case study of its use in bank teller system design.

2. TASK NETWORK MODELING FUNDAMENTALS

Task network modeling involves the extension of a task analysis into a predictive model based around a network representation of the human's activity. This concept is illustrated in Figure 1 which presents a sample task network for accessing a computer system menu item.

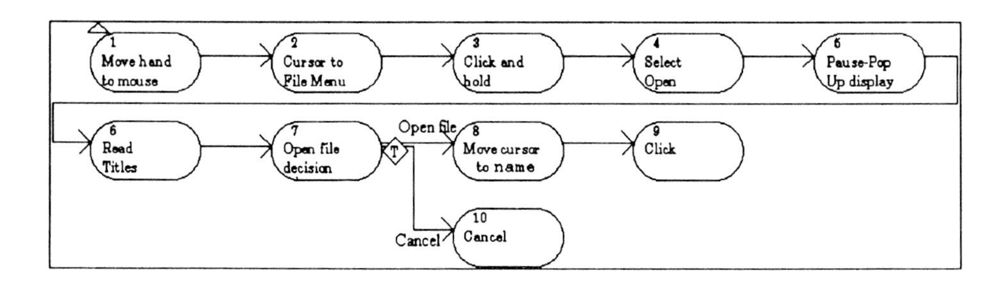

Figure 1. An Example of a Task Network Model Representing a Human Using a Menu System

Task network models of human performance have been subjected to validation studies with favorable results (e.g., Lawless, Laughery, and Persensky, 1995; Allender, *et al* 1995).

A modeling software tool that supports task network modeling is *Micro Saint.* The basic ingredient of a Micro Saint model is the task analysis as represented by a network or series of networks. The performance of the tasks can be interrelated through shared variables. The relationships among different components of the system (which are represented by different segments of the network) can then communicate through changes in these shared variables. For example, when an operator enters a command on a keyboard, this may initiate change in computer state or the information that is presented on an operator display. This task network is built in Micro Saint via a point and click drawing palette. Through this environment, the user creates a network as shown in Figure 2.

To reflect complex task behavior and interrelationships, more detailed characteristics of the tasks can be defined. By pointing and double clicking on a task, the user opens up the Task Description Window as shown in Figure 3. whereby information describing task behavior and linkage with other system elements can be defined.

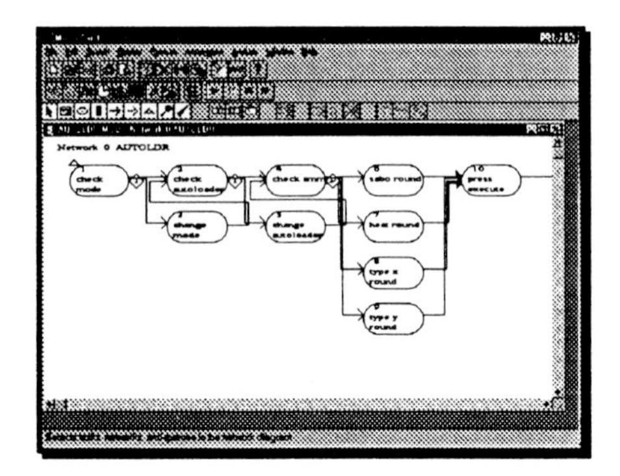

Figure 2. The Main Window in Saint for Task Network and Viewing

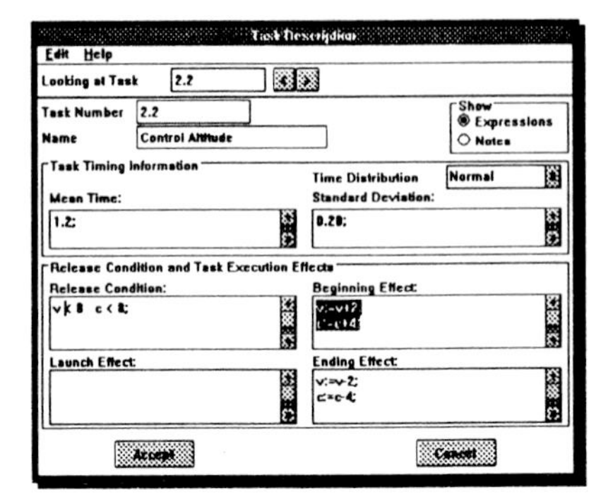

Figure 3. The User Interface in Micro Saint for Providing Input on a Task

Another notable aspect of the Task Network Diagram Window shown in Figure 2 is the diamond-shaped icons that follow some tasks. These are present every time more than one path out of a task is defined. Implicitly, this means that a decision must be made by the human to select which of the following potential courses of action to be followed. By opening a window into the decision, decision logic and algorithms can be developed to any level of complexity. Micro Saint also offers many other features that facilitate the construction and use of task network models such as automatic error trapping, scenario development features, data gathering and analysis, and model animation features that allow the user to view the dynamic activities of the humans and system he or she is using. From these basic building blocks, a task network model can be built to describe human and system behaviors of any size or level of complexity.

3. A CASE STUDY OF USING TASK NETWORK MODELING TO EVALUATE BANK TELLER SYSTEM DESIGN

The transaction completion process in the Japanese banking industry has many areas that could be improved to better meet customer and bank needs. The current process includes several tasks that are time-intensive, inefficient, and may prevent tellers from keeping up with customer demand. Potential methods that could be used to decrease the time required for each transaction and improve accuracy include hardware improvements (e.g., printers, data entry systems), software enhancements (e.g., the design of the software), and process redesign. A provider of equipment to the Japanese banking industry determined that a cost-effective and efficient method to investigate the Japanese banking industry problems and evaluate potential solutions was simulation modeling.

One purpose of the Japanese banking project was to identify the effects of changing various parameters on the overall transaction process. In particular, the goal was to obtain results from the effects of changing system, personnel, timing, and/or resources.

The basic sequence of events for most transaction types is as presented in Figure 4. More detailed models of each of these transactions were developed in the model.

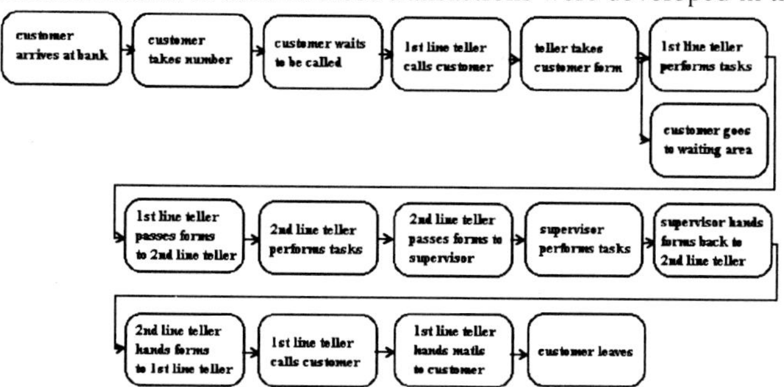

Figure 4. The basic sequence of events for most transaction types

Each transaction had its own unique set of steps that were required.

For the base Japanese banking model, a graphical animation was constructed along with the simulation to show the general layout of the bank along with various output parameters. While the model executes, the animation is updated to represent the current state of the system. The animation background is shown in Figure 5.

The model was constructed so that certain system input parameters could be easily manipulated. For example, the user can easily modify customer arrival rates based on branch and type of day, transaction type ratios, number of tellers, and probability of transaction types.

Model_Output - The output data collected from the model includes time per transaction, average transaction time for each transaction type, total customer time in the bank, customer wait times, teller utilization, transaction queue times at the teller stations, and the maximum queue lengths.

Proposed transaction process changes were evaluated and analyzed in order to understand, identify, and test opportunities for process improvement or reengineering. Some potential improvements identified include cross training of tellers and the introduction of bank automation. In sum, simulation provided a cost effective tool to evaluate the payoff of human factors improvements on employee productivity and customer service.

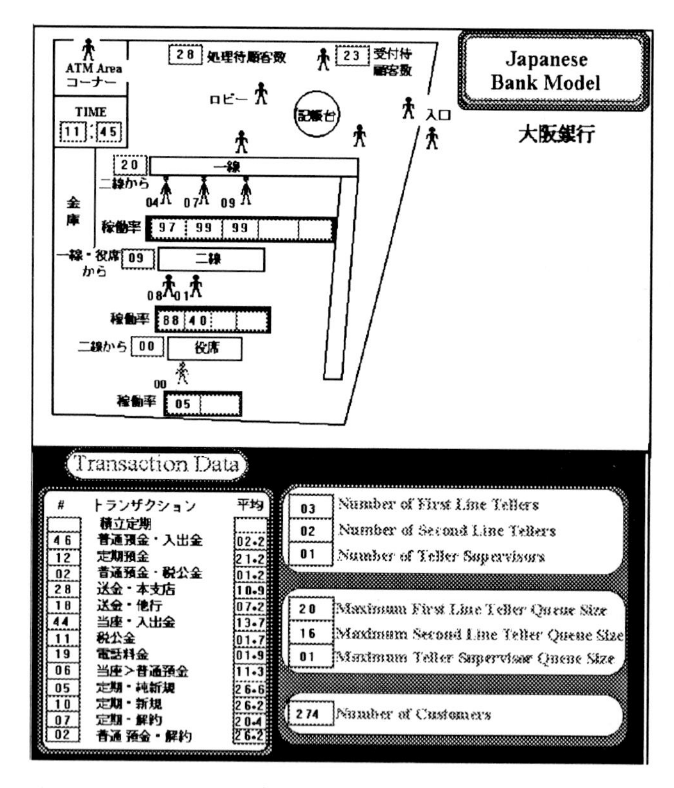

Figure 5. Graphical Animation Background

SUMMARY

Task network modeling provides one way to assess the value of human-computer interface designs in the operating environment. Together with usability analysis to ensure model accuracy and systems analysis to define how the system will be used, a better assessment of the value of the human interface can be gained.

REFERENCES

1. Lawless, M.L., Laughery, K.R., and Persensky, J.J., *Micro Saint to Predict Performance in a Nuclear Power Plant Control Room: A Test of Validity and Feasibility.* NUREG/CR-6159, Nuclear Regulatory Commission, Washington, D.C., August 1995.
2. Allender, L., Kelley, T., Salvi, L., Headley, D.B., Promisel, D., Mitchell, D., Richer, C., and Feng, T., "Verification, Validation, and Accreditation of a Soldier-System Modeling Tool. In the Proceedings of the 39th Human Factors and Ergonomics Society Meeting, October 9-13, San Diego, CA, 1995.
3. Laughery, K.R. and Corker, K., Computer Modeling and Simulation of Human/System Performance in the *Handbook of Human Factors*, G. Salvendy, Ed. Wiley and Sons, 1997.

Multi-agent human performance modeling in OMAR*

Stephen Deutsch
BBN Corporation, 10 Moulton Street, Cambridge, MA 02138, USA
sdeutsch@bbn.com

1. INTRODUCTION

Simulation has been used for many years as a tool to evaluate system performance, but it is only recently that attempts have been made to include realistic models of human operators in these evaluations. The Operator Model Architecture (OMAR) is a simulation system that addresses the problem of modeling the human operator. Its development focused first on the elaboration of a psychological framework that was to be the basis for the human performance models, and then on the design of a suite of software tools to support the development of these models. In using OMAR to build human performance models, particular attention has been paid to the representation of the multi-tasking capabilities of human operators and their role in supporting the teamwork activities of the operators.

The ability to model human operators and their interactions with other team players and with their target systems has opened up several new areas for investigation through simulation. Procedure development is an important one. Simulation can now have an impact on both operator-procedure and maintenance-procedure development. It will now be possible to pursue procedure development by evaluating procedures far earlier in the design cycle of a system than was previously possible. The evaluation of both operating and maintenance procedures as part of the design process can lead to significant improvements in system operability and reduce downstream maintenance costs.

2. TEAMWORK AND THE HUMAN PERFORMANCE MODEL

The modern workplace is seldom the province of a single person. A person's work is most frequently part of a larger effort, linked more or less closely to the work of others, either at a nearby workplace or at a remote site. It is becoming more common for people to work with others at remote sites as the capabilities of networked systems improve to support this mode of operation. On the one hand, team members can be viewed as resources to assist in the accomplishment of a task, while on the other hand, they can be the source of interruptions to ongoing tasks as they seek assistance in meeting their own objectives. To address the goal of modeling the teamwork of human players, human performance models must be capable of accurately reflecting human *multi-task behaviors* and the accompanying shifts in *attention* that these behaviors demand.

Proactive activities require that attention be focused to support a given task. Similarly, reactive demands are made on attention by interruptions that may be auditory or visual. To improve the behaviors of the human performance models, it was important to examine the nature of teamwork and particularly its impact on attention, to improve those aspects of the model that represent multi-tasking, and to enable the human performance model to exhibit reasonable behaviors in the teamwork aspects of the tasks being executed.

2.1. An architecture for modeling human multi-task behaviors

The basic architectural components of the OMAR human performance model (Deutsch & Adams, 1995) are implemented at the symbolic level. They are patterned after the large-scale structure of the brain as outlined by Edelman (1987, 1989) and Damasio (1989a, 1989b). The

* The research reported on here was conducted under USAF Armstrong Laboratory Contract No. F33615-91-D-0009.

performance of a particular capability is typically implemented through the participation of a small number of functional centers. Subsets of these centers, operating concurrently, typically have links between them operating in both directions—*reentrant signals* as described by Edelman (1987). The operation of a set of "lower" level centers is coordinated by a "higher" center, or *convergence zone* in Damasio's (1989a, 1989b) terminology. Damasio's convergence zones are much like Minsky's (1986) hierarchy of agents in *The Society of Mind.*

The Simulation Core (SCORE) language, the OMAR procedural language, provides the basis for representing functional capabilities. A SCORE procedure is used to represent a simple capability. The procedure is typically in a *wait state,* pending activation based on a match to a particular stimulus pattern. Several such procedures may represent the components of a particular functionality. The procedures form a network along whose links pass the signals that each of them may generate. For any signal generated, one or more related procedures may be enqueued on it. Depending on the response of their respective pattern matchers, some procedures may be activated by a stimulus, while others may ignore it. The pattern of activation in a complex of procedures differs due to variations in the stimulus patterns.

Within any given complex, several procedures may be running concurrently, some representing automatic processing, others representing components of attended processing. In the several layers of concurrent processing, a sensory input may have initiated the "lowest" processing level, with each subsequent "higher" processing layer starting up at the behest of the initial "output" from the next lower level. The behaviors of the concurrent processes are based on those discussed by Jackendoff (1987) in *Consciousness and the Computational Mind.*

2.2 Attention

Within this architectural framework, attention is not simply one component or one complex of procedures. Following Neumann (1987), attention is a "generic term for a number of phenomena each of which is related to a different selection mechanism." In building the model of attention, a selected subset of these phenomena related to air traffic control and aircrew tasks were implemented. The focus was on auditory and visual processing, since they are the most important forms of attention in managing the air traffic control and flight deck workplaces.

The work on auditory attention addressed the verbal communications of air traffic controllers and aircrew members, either in person, via telephone, or over the party-line radio. The work on visual attention focused on the air traffic controller's use of the synthetic radar screen and visual support for, and coordination of, manual workplace tasks. Both visual and auditory attention have proactive and reactive components. Verbal communication is the basis for proactive coordination of flight-deck activities, while party-line radio communications are frequent interruptions to ongoing activities. In the visual domain, the appearance of a new aircraft icon on the radar screen may interrupt an ongoing activity, while proactive visual attention is required to accomplish simple manual actions such as setting mode control panel values.

The processing of auditory messages is modeled by a set of concurrent processes. At the lowest level is a "hearing" process that is initiated in response to the onset of the auditory input. Shortly after the hearing process is initiated, it in turn triggers a cognitive "message-understanding" process. Lastly, an "attended" cognitive process represents the hearer's attending and reacting to the spoken message. The nature of the communication over the party-line radio made it necessary to factor in another layer of complexity. From an aircrew's perspective, many air traffic controller messages are directed to other aircraft in the airspace and can be ignored at some level. Tanenhaus, Spivey-Knowlton, Eberhard, & Sedivy (1995) suggest that this is determined as soon as a verbal discriminator appears. The air traffic controllers identify the target aircraft for each message as the first utterance in a message, but it is clearly not the case that both the "understanding" and "attended" processing stop at this early point in message processing. An aircrew member will not initiate a verbal communication with another crew member while the "ignored" message is still in process. The speculation represented in the model is that the "understanding" process continues processing the incoming message, and it is this process that "flags" the end of the message, so that another verbal message resuming the interrupted person-to-person conversation among crew members may be initiated.

The presence of the party-line radio means that auditory interruptions are an expected occurrence. In particular, it must be possible to stop an intra-crew conversation at the onset of a radio message and resume the conversation at the completion of the interruption. SCORE priorities assigned to in-person and radio conversations are used to implement the processing of the interruption by a new radio message. The intra-crew verbal transactions are typically a statement-response pair. An interruption anywhere in the exchange will be resumed, not at the point of the interruption, but by the initial statement of the exchange being repeated.

Visual attention, as modeled in OMAR, also has proactive and reactive components. The proactive processes are executed primarily in support of related cognitive and manual processes. The reactive processes are concerned with the response to visual events. The air traffic controller's synthetic radar workplace is a visually rich environment. The visual events modeled include the appearance of a new aircraft icon on the radar screen as the aircraft approaches the air traffic controller's airspace, the movement of the aircraft icon across the screen as its position is updated, and a flashing light on the telephone to announce an incoming call from a neighboring controller. The initial response to a visual event is the simple act of identifying the event, followed by a sequence of signals that trigger the appropriate network of procedures to respond to the event. The events that occur are not unexpected and there is typically a goal that has set up and is governing the response to each event. The execution of the response at each stage is mediated by the priority associated with the response procedure and that of the other ongoing procedures.

Visual attention in support of cognitive procedures takes several forms. Probably the most complicated activities take place as the air traffic controller "sits down" at the radar console to take over control of the airspace from the previous controller at the start of a shift. There are several aircraft in the sector and in neighboring sectors. Flight strips, arrayed at the side of the radar screen, provide flight-plan information on active and pending aircraft. The air traffic controller's initial acts are to "read" the active flight strips, associate each with the appropriate aircraft icon on the radar screen, and initiate a procedure for managing that aircraft's transit through the airspace. Implicit in the procedure for managing the aircraft is the memory of where the aircraft icon appears on the radar screen. The expert air traffic controller, like the expert chess player, maintains the knowledge of where the pieces are on the board by constantly revisiting them. In the human performance model, this memory is local to the procedure for managing the particular aircraft, rather than a slot in a global short-term memory resource.

3. AIR TRAFFIC CONTROL STUDIES

The human players modeled in the air traffic control environment include the en route and approach controllers and the aircrews of the aircraft in their sectors. Recently developed scenarios have examined the ATC hand-off of aircraft from one controller to a neighboring controller in high traffic density conditions. Earlier studies examined aircrew and ATC procedures for approach and landing sequences, including top-of-descent negotiation and procedures in which radio-based voice communication was replaced by data-link for aircrew/ATC communication. The studies, typically based on high-density air traffic conditions, have required the simulation of large numbers of human players. The efficiency of OMAR in human performance modeling has made this possible.

The aircrew processing of an ATC directive is depicted in Figure 1. Following company-based policies, the pilot-not-flying (PNF) responds to the ATC directive, while the pilot-flying (PF) also attends to the communication and has responsibility for acting on the directive. The onset of the ATC communication may well have interrupted verbal communication between crew members on the flight deck that will have to be resumed. Radio communications to other aircraft in the sector are attended in a closely related manner and may similarly be the cause of interruptions to verbal communication to coordinate flight deck activities.

Expectations also play a critical role in the coordination and accomplishment of team tasks and their modeling is essential to the evaluation of operating procedures. In addition to the basic processing of the auditory message, the expectations that support cross-checking are important components of OMAR human performance modeling capabilities. Based on the expectation generated by the ATC directive, the pilot-not-flying will check that the pilot-flying has properly set the mode control panel (MCP) in compliance with the directive and intervene if the directive is not accurately executed in a timely manner.

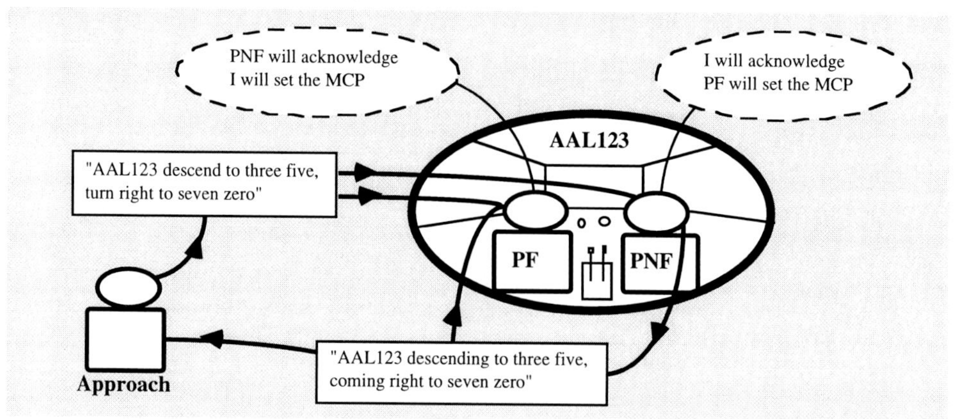

Figure 1. Aircrew processing an ATC directive.

4. CONCLUSIONS

Teamwork, an essential component in the successful operation of many systems today, makes significant demands on the multi-tasking capabilities of individual team members. Additional demands on attention are made by the mix of proactive and reactive activities that force the interruption and resumption of tasks in the complex flow of activities to accomplish individual and team goals. By addressing the nature of teamwork and carefully modeling its impact on multi-tasking and attention, the behaviors of the OMAR human performance models were improved to better represent this elaborate set of human capabilities. These advances in human performance modeling have made it possible to develop and evaluate new operating procedures involving large numbers of team players in complex time-critical environments.

REFERENCES

Damasio, A. R. (1989a). The brain binds entities and events by multiregional activation from convergence zones. *Neural Computation, 1,* 123-132.

Damasio, A. R. (1989b). Time-locked multiregional retroactivation: A systems-level proposal for the neural substrates of recall and recognition. *Cognition, 33,* 25-62.

Deutsch, S. E., & Adams, M. J. (1995). The operator-model architecture and its psychological framework. *6th IFAC Symposium on Man-Machine Systems.* MIT, Cambridge, MA.

Edelman, G. M. (1987). *Neural Darwinism: The theory of neuronal group selection.* New York: Basic Books.

Edelman, G. M. (1989). *The remembered present: A biological theory of consciousness.* New York: Basic Books.

Jackendoff, R. (1987). *Consciousness and the computational mind.* Cambridge, MA: The MIT Press.

Minsky, M. (1986). *The society of mind.* New York: Simon and Schuster.

Neumann, O. (1987). Beyond capacity: A functional view of attention. In H. Heuer & A. F. Sanders (Eds.), *Perspectives on perception and action* (pp. 361-394). London: Lawrence Erlbaum Associates.

Tanenhaus, M. K., Spivey-Knowlton, M. J., Eberhard, K. M., & Sedivy, J. C. (1995) Integration of visual and linguistic information in spoken language comprehension. *Science, 268,* 1632-1634.

Mapping Instructions Onto Actions: A Comprehension-Based Model of Display-Based Human–Computer Interaction

Muneo Kitajima[a] and Peter G. Polson[b]

[a]National Institute of Bioscience and Human–Technology, 1-1 Higashi Tsukuba Ibaraki 305, Japan

[b]Institute of Cognitive Science, University of Colorado, Boulder, CO 80309-0345, USA

1. COMPREHENSION-BASED MODELS

This paper describes a cognitive model, LICAI, **LI**nked model of **C**omprehension-based **A**ction planing and **I**nstruction taking, that simulates the cognitive processes involved in comprehending and following hints, and successfully performing steps by exploration [5][6]. The cognitive processes specified in LICAI are implemented using the *construction–integration (CI) architecture* developed by Kintsch [1]. The CI architecture is symbolic-connectionist and has been applied successfully to model cognitive processes such as text comprehension, word problem solving, and action planning. In the *construction* phase, a CI model generates a connectionist network that includes alternative meanings of the current text or alternative actions that can be performed on the interface display. In the *integration* phase, a CI model spreads activation among the constructed networks and selects a contextually appropriate alternative. Links in the network are established by common symbols; when two nodes share symbols, they are connected.

The CI architecture has evolved from a detail model of skilled, text comprehension — a highly automated collection of cognitive processes that make use of massive amounts of knowledge stored in long-term memory [1][2]. This is a very different foundation from the other cognitive architectures. For example, one of the primary foundations of Soar [7] was the General Problem Solver, a model of deliberate cognition (i.e., problem solving and action planning) in situations where the problem solver has limited background knowledge.

2. THE LICAI MODEL

LICAI, shown in Figure 1, simulates the processes involved in reading instructions and performing a task on the interface by exploration with three different processes. These processes are expressed by various CI cycles.

a) E-mail: kitajima@nibh.go.jp; related papers available from WWW: http://www.aist.go.jp/NIBH/~b0544/

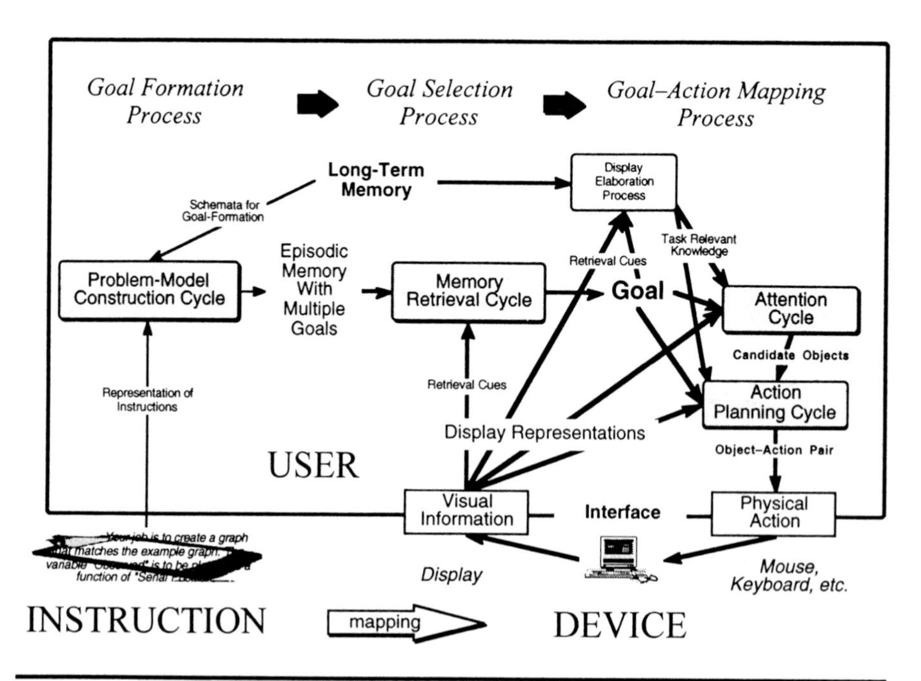

Figure 1. An illustration of LICAI [6], adapted from [3].

2.1. Goal Formation Process

When reading instructions, a user attempts to extract goals that should be accomplished on an interface. LICAI assumes that this *goal formation process* is analogous to solving word problems, and models it as a problem-model construction CI cycle, which is a strategic form of the basic text-comprehension process that generates representations specialized for interacting with devices; that is, the goals that control solution of a task described in instructions. Instructions are processed by executing a single CI cycle for each sentence. In the construction phase, LICAI generates a network that includes semantic representations of a sentence as well as elaborations that translate the semantic representation into goals. In the integration phase, LICAI selects a single meaning for the sentence and links this representation with the memory representation of earlier parts of the text. Thus, after reading the entire text, the memory contents represent the result of instruction comprehension. If the text contains descriptions of multiple goals, LICAI stores them in episodic memory.

2.2. Goal Selection Process

After reading instructions, the user tries to select a goal from the episodic memory. LICAI assumes that this *goal selection process* is done by a single CI cycle using the current application display as retrieval cues. In the construction phase, the episodic memory and the current interface display constitute a network. In the integration phase, a goal consistent with the current interface display is selected; that is, the most highly activated goal is selected. Since the sources of activation are the nodes representing the display and the pattern of links in the network is largely determined by the argument overlap mechanism, a goal that overlaps the currently visible screen objects is likely to be selected. For example, if the representation of the goal includes matching labels on any screen objects, it will be selected.

2.3. Goal–Action Mapping Process

After selecting a goal, the user tries to generate a sequence of one or more actions that will accomplish the selected goal. This *goal–action mapping process* involves two CI cycles, *the attention cycle* and *the action-planning cycle*, which is a generalization of the model of skilled, display-based, action planning developed by Kitajima and Polson [4].

2.3.1. Display Elaboration Process

The initial display representations contain only limited information about the identity of each screen object and its appearance, including visual attributes (e.g., color, highlighting). The poor display representations are augmented by retrieving relevant knowledge from long-term memory. This display elaboration process is simulated by a random memory sampling process: the retrieval cues are the current goal and the propositions representing the current display. The elaboration process is stochastic. The probability that each cue retrieves particular information in a single memory retrieval process is proportional to the strength of the link between them. LICAI carries out multiple memory retrieval in a single elaboration process. A parameter controls the number of times each argument in the display and goal representations is used as retrieval cues. The predictions and implications that follow from this stochastic elaboration process is discussed in detail by Kitajima and Polson [4].

2.3.2. Attention Cycle

The elaborated display representation is the model's evaluation of the current display in the context defined by the current goal. In the goal–action mapping process, the model first limits its attention to three screen objects out of ~100 objects displayed on the screen by applying an attention CI cycle. These screen objects are candidates for the next action to be operated upon. During the construction phase, a network is generated that consists of nodes representing the goals, the screen objects and their elaborations, and candidate object nodes of the form *Screen-Object-X is-attended.* Any screen objects are potential candidates. During the integration phase, the conflict is to be resolved. The sources of activation are the goals and the screen objects. The targets are the candidate object nodes. When the spreading activation process stabilizes, the model selects the three most highly activated candidate object nodes. These nodes represent screen objects to be attended to during the next action-planning cycle.

2.3.3. Action-Planning Cycle

The second CI cycle is an action-planning cycle. As preparation for constructing a network, the candidate objects carried over from the preceding cycle are combined with any possible actions to form object–action pairs of alternatives. The model considers all possible actions on each candidate object. Examples would include *Single-click Object23*, *Double-click* Object23, *Move Object23*, *Release Object23*, and the like, where *Object23* represents one of the candidates.

During the construction phase, the model generates a network that includes the goals, the screen objects and their elaborations, and representations of all possible actions on each candidate object. During the integration phase, the sources of activation are the goals and the screen objects, and the targets are the nodes representing the combinations of object–actions. At the end of the integration phase, the model selects the most highly activated object–action pair whose preconditions are satisfied as the next action to be executed. The action representations include conditions to be satisfied for their execution. The conditions are matched against the elaborated display representations. Some conditions are satisfied by the current screen, others by information that was retrieved from long-term memory in the display elaboration process. For example, the model cannot select an action *double-click a document icon* unless the icon is currently pointed at by the mouse cursor, and the information "the icon can be double clicked" is available. Observe that if information about a necessary condition is missing from an elaborated display representation, the model *cannot* perform that action on the *incorrectly* described object.

3. NATURE OF INSTRUCTION–ACTION MAPPINGS

LICAI describes the underlying mechanism that controls users' instruction mapping onto interface actions. A series of CI cycles depicts the various component processes executed. Mapping from instructions to an action is successful if the goal formation processes generate the correct goal, if the goal selection processes select that goal, and if the goal–action mapping processes generate the correct action sequence. LICAI predicts that the instruction–action mapping processes will become more difficult with longer instructions, more screen objects, and/or an increase in possible actions. LICAI suggests that interacting with a novel application by following instructions will become difficult unless the instructions are worded carefully and the interface is designed to facilitate exploration (see [3] for more detailed discussion).

REFERENCES

1. Kintsch, W. (1988). The role of knowledge in discourse comprehension: A construction–integration model. *Psychological Review*, **95**, 163-182.
2. Kintsch, W. (in press). *Comprehension: A paradigm for cognition.* Cambridge University Press.
3. Kitajima, M. (in press). Successful technology must enable people to utilize existing cognitive skills. In *Cognitive Technology: Methods and Practice*. Elsevier.
4. Kitajima, M., & Polson, P.G. (1995). A comprehension-based model of correct performance and errors in skilled, display-based human–computer interaction. *International Journal of Human–Computer Systems*, **43**, 65-99.
5. Kitajima, M., & Polson, P.G. (1996). A comprehension-based model of exploration. In *Proceedings of CHI'96 Conference on Human Factors in Computing Systems,* 324-331.
6. Kitajima, M., & Polson, P.G. (in press). A comprehension-based model of exploration. *Human–Computer Interaction.*
7. Newell, A. (1990). *Unified theories of cognition.* Cambridge, MA: Harvard University Press.

The human scheduler's mental models and decision aids of the interactive scheduling system

Nobuto Nakamura[a], Joji Takahara[a], and Tamotsu Kamigaki[b]

[a] Dept. of Industrial and Systems Engineering, Hiroshima University,
4-1, Kagamiyama 1 chome, Higashi-Hiroshima 739, Japan
nakamura@pel.sys.hiroshima-u.ac.jp

[b] Dept. of Computer Science, Hiroshima Denki Institute of Technology,
20-1, Nakano 6 chome, Aki-ku, Hiroshima 739-03, Japan
kamigaki@c.hiroshima-dit.ac.jp

This paper describes the scheduler functions of the chemical process, where for each function the corresponding mental models and computer-based decision aids are analyzed. Based on the proposed framework for the decision process, the interactive scheduling system is designed for a practical example: bulk loading and flexible container packing scheduling for chemical products.

1. INTRODUCTION

The scheduling tasks are almost routine works, but there are not the same routines everyday because real-life problems such as batch splitting, process breakdown and maintenance, material availability etc., always happen [1]. This paper recognizes these complexities, no by attempting to design a totally automated scheduling system, but by including the human scheduler interaction as an integral part of the solution strategy.

In an interactive scheduling system, generally, the human scheduler makes his/her own mental model instantly in each stage of decision process. Therefore, the computer aids provided to the human scheduler must be desired to correspond with this mental model in order to that each decision can be done promptly and correctly.

In this paper, considering the bulk loading and flexible container packing scheduling as a practical example, we describe the decision process of the human scheduler, and determine the associated mental models and computer aids.

2. FUNCTIONS OF THE HUMAN SCHEDULER: CORRESPONDING MENTAL MODELS AND DECISION AIDS

Table 1 lists the categories of human scheduler functioning in order in which they normally would occur in performing a scheduling task in chemical process and shows for each function the corresponding mental model and the potential computerized decision aids. (This section referred to Sheridan's supervisory control model [2].)

Table 1. Human scheduler functions with assumed mental models and computer aids

Decision Process	Associated Mental Model	Associated Computer Aid
1. PLAN		
a) understand total demands (filling values)	filling values; maintenance	total demands understanding aid
b) recognize goal, constraints	aspiration level; constraints	aspiration level and constraints recognizing aid
c) determine strategy, procedure	normal procedure; guidelines	procedures training aid
2. CALCULATION		
set target values and calculate mass balance	data input; alternative mass balance equation	calculation procedure aid
3. CHECK		
a) check the schedule	goal; feasibility	simulation and visualization aid
b) decide change or not	criteria for change or not	judgment aid
4. ORDER		
print out and order	cumulative memory; experience	cumulative record and analysis aid

1. PLAN

1a. The first scheduler function under PLAN is to understand the total demands of the products in the planning horizon (i.e., a week). From this understanding the human scheduler decides to accept the normal procedure or the other one. The computer aids then must display the total demand and its trend for the short future period.

1b. The second scheduler function under PLAN is to set a goal and constraints, which usually have been given to the human scheduler. He/She recognizes whether the new constraints are added or not and decides the scheduling policy.

1c. The third scheduler function under PLAN is to set general strategy and procedure. In normal condition, the human scheduler uses the given strategy and knowledge in his/her brain. In the case that the normal strategy cannot be used, the human scheduler considers the several strategies. The computer aid here is that of the general operating procedure and guidelines which must be committed to memory.

2. CALCULATION

First, the human scheduler teaches to computer the initial values of the requirement at the first time-duration. After that, the computer calculates the requirement at second time-duration and so on, successively, from filling values per time, tank capacity, remainder values at the ending time by the computing procedure already stocked in the computer. The computer aid in this case is calculation aid, that is, the computer conducts the calculation, instead of the human scheduler.

3. CHECK

3a. The schedule which has been made in cooperation with the human scheduler and the computer is checked by them, whether it satisfies the goal or not, and whether it is feasible or not. The computer aid displays the human scheduler the infeasible parts of the schedule, using the visual technique (for example, Figure 1).

3b. The human scheduler decides whether the schedule must be changed or not, based on his/her knowledge and experience. If it must be changed, the decision is feedback to 2 or 1. The mental model includes a relative small number of options to resolve an error or a conflict. The computer aid can advise which to use and how to use it.

4. ORDER

The last step of the decision process is ORDER. This is output of the final decided schedule which is ordered to the shop floor. The mental model is visual check for the schedule totally. The computer aid should provide some cumulative record and analysis in a form that can be accessed and used later during the PLAN phase.

3. EXAMPLES AND CONCLUSIONS

As a practical example, we treated a scheduling problem of the chemical plant as shown in Figure 2. This plant sends a chemical product produced in the plant to the day silos (D1 $\sim$ D3), after that sends it from the day silos to the storage silos (BL1 $\sim$ BL4 and FC1 $\sim$ FC3) using pipe line for pressure sending, and last releases it from the storage silos.

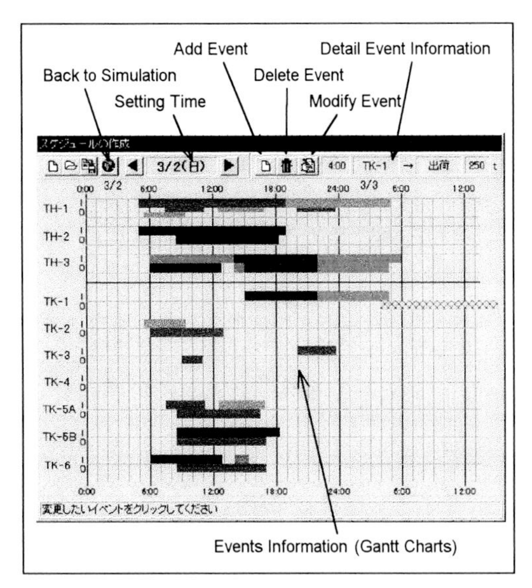

Figure 1. Gantt chart

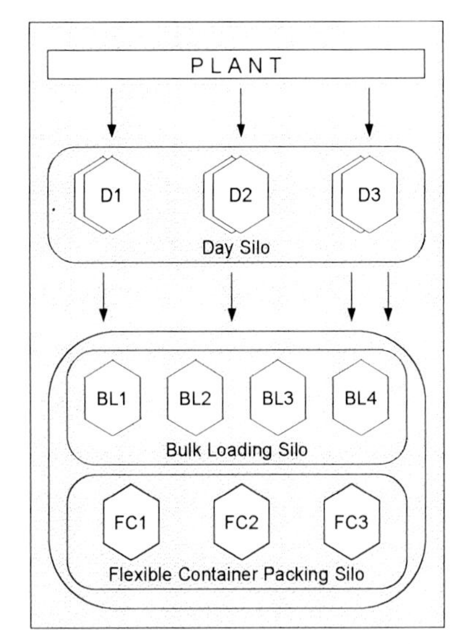

Figure 2. Chemical plant model

The product is shipped out to the customers from the storage silos, according to customers' orders. Then, the scheduling problem here is to make a bulk loading and flexible container packing schedule for shipment.

Based on the above investigation, we have designed a human-computer interactive model for scheduling the bulk loading and flexible container packing. From the results of the field test, this model showed good quality performance more than a totally automatic scheduling system.

REFERENCES

1. A. Bauer, R. Bowden, J. Browne, J. Duggan and G. Lyons: Shop Floor Control Systems, Chapman & Hall, 1991.

2. T. B. Sheridan: Telerobotics, Automation, and Human Supervisory Control, The MIT Press, 1992.

The Role of Interaction Modeling in Future Cognitive Ergonomics: Do Interaction Models Lead to Formal Specification of Involved Machine Intelligence ?

Chris Stary

Department for Business Information Systems, Communications Engineering, University of Linz, Freistädter Straße 315, 4040 Linz / Austria, email stary@ce.uni-linz.ac.at

The role of representations for upcoming user interface generations is reviewed and exemplified for modeling the organizational and cognitive dimension of human-computer interaction.

1. INTRODUCTION

Design principles or measurements in Cognitive Ergonomics aim to improve the usability of user interfaces in a way that the cognitive and intellectual attributes of individuals working in an organizational environment can directly be addressed, in order to ease the use of artifacts. Key principles to achieve this addressing, such as task conformance and adaptability of interactive software, although they focus on the elimination of hindrances through interaction devices or styles for the effective and efficient accomplishment of tasks, involve several dimensions of interaction [1]: user-related social and cognitive issues, the organization of work, and the technical components of the user interface.

These dimensions have to be reflected by development techniques. Up to now, interaction modeling has been primarily based on interaction media and modalities, such as graphical user interfaces utilizing menus, windows, icons, and a pointing device. Few approaches exist that openly discuss the required openness of interaction models towards the representation of the problem domain and the users. In the following, we will review some techniques for representing cognitive processes as well as organizational issues.

2. FORMAL REPRESENTATION OF HUMAN KNOWLEDGE AND COMMUNICATION

Considering the development of interactive systems several representations have to be distinguished [2]: (i) The *mental models* of the end users of the target system: these are those internal models that are tempted to be specified explicitly by means of knowledge-elicitation methods; (ii) The *conceptual model* of the target system, i.e. the external model that has to be developed by the designer; (iii) The *developer's conceptual model* of the *mental model* of the *end users* of the target system: this model the developer has in mind in the course of knowledge elicitation and the specification of the conceptual model of the target system.

Formal specifications of machine intelligence therefore focus on (ii) the externalized model of the designer.

In addition, it has turned out that only physically observed and performed actions through end users (motor actions and activities of perception performed through human senses) can directly be mapped to formal representations, e.g. [3]. Only parts of internal processes, such as the performance of memory at the syntax layer, have been investigated thoroughly, e.g. [4]. As a consequence, developers cannot trace the external and internal processes related to cognition completely and in a transparent way.

Opacity, complexity, ambiguity, and a lack of proper metaphors are considered to be the major bottlenecks for the acquisition of knowledge for human-oriented interfaces to computer systems, although several issues are under investigation:

- the *nature of problem solving*, since problem solving is often a dialectic process of argumentation. If artificial problem solvers have to manage a complex problem, they face the same problems like humans. The 'cognitive' difficulty to maintain all sub-arguments and their entailments through interaction or to induce them while they receive information, is frequently overwhelming. Due to the consequent decomposition of a complex argument into linear summarization's, results or intermediate products of this dialectic process are suppressed [5].

- the *accuracy of acquisition and representation*: There are too many hindrances between the users and the machine to accomplish their tasks accurately. Users miss shared 'understanding' and argumentation about activities in their context, e.g. [6].

- *pragmatic aspects of language*: Techniques from ethno-methodology, such as conversation analysis, are based on the assumption that communication usually occurs without any external disturbances, since most of the time only one speaker is talking. In addition, listeners often concentrate on what the talker tries to communicate. This situation corresponds to the situation designers assume when they create user interfaces. The knowledge about how to behave in the course of human-to-human communication is often not conscious, but influences our communication essentially. Up to now, there is no model or representation to capture these features of communication for human-computer interaction. If we enable machines to accept these inputs from humans in the intended context, the support will become more effective.

The addressed deficiencies are related tightly to the conceptual deficiency of user interfaces to provide knowledge about situations when interaction and problem solving occur. User interfaces should develop some kind of 'awareness', whether the user who is currently supported has an understanding about what is actually going on at the interface from a semantic or pragmatic point of view. The provision of log-files is only a way to acquire syntactic 'sugar'. If an interactive system is not enabled to check whether the user has a semantic understanding of the occurring interaction, it is unlikely that any type of traditional help function or user model will help him/her in case of troubles.

3. FORMAL REPRESENTATIONS OF HUMAN WORK AND TASKS

Although conceptual modeling of work organization and the process of task accomplishment is considered to be a major source of knowledge for human-oriented user interface development, e.g. [7], most of the design representations lack consistency and task conformance. Tasks are either identified to be a linked set of goals that, after decomposition, can be handled through elementary operations, such as proposed in [8], or designers are assumed to identify a set of interaction styles from data flows in a straightforward way, as for instance proposed in [9]. Following traditional design approaches user interface design representations are usually sets of (composed) objects that do not allow to trace the design process, and thus, to evaluate the conformity to tasks and user needs. First approaches in considering software engineering concepts to structure user-interface development have been restricted to prototyping and scenario-based object-oriented programming, e.g. [7].

Neither engineering approaches for the development of general purpose software nor UIMS-specification schemes can be utilized as they are for the required migration of workflows into user-interface design representations. They all lack at least one of the required features for task-oriented system development [10]:

1. Both do not provide techniques for integrating interaction and data handling, although the UIMS approach provides a structured architecture for interactive systems. However, recognizing the need for control knowledge does not imply necessarily to know how to represent this knowledge in the design phase. Hence, a strategy to achieve a non-reductive representation has to be provided.
2. Both are incomplete, since the essential context of any interactive system, namely tasks (including workflows) and user roles is lacking.
3. They are not seamless in the sense as they do not allow smooth transitions between the phases of development. As a consequence, complicated transitions and transformations from analysis to design and implementation may cause a loss of information and misconceptions of the resulting behavior.
4. Both lack balanced semantic expressiveness for representing the structure and behavior of a system. The consequence is an unbalanced emphasis of either dynamic specifications or data models in the design phase.

The migration of workflow representations into design representations for the development of interactive systems has to support the integration of workflows into the design process in an unifying way. In the TADEUS approach [11] (Task Analysis, Design, End-User Systems) the integrated design of workflows, problem domain data, user profiles, and interaction modalities is provided. Several aspects of workflows have been migrated into user interface design representations:

- business goals and rules through the identification of workflows and their relationships to tasks and roles of people involved in task accomplishment.
- purposeful activities through refining tasks into activities that have to be performed in a certain sequence for task accomplishment.

- the consistent and complete refinement to interactive behavior and data manipulations.
- profiles of users according to their functional roles in an organization, their perception of tasks, and preferences concerning interaction styles.

In TADEUS the flow of data and control can be made transparent through different relationships between activities. Designers are provided with a notation that allows static and dynamic specification and mutual adaptation of the models at several layers of abstraction. Finally, designers are enabled to keep the relationships between the models as loose or tight as they feel adequate, without loosing the different perspectives, and the consistency of the initially separated specification of four different models. Using TADEUS, consistent and context-sensitive user interface development as it occurs in reality of software development is supported.

Concluding, the answer to the question raised initially ('Do interaction models lead to formal specification of involved machine intelligence?') is definitively 'yes', otherwise individuals will still be forced to adapt towards the behavior of machines and not vice versa.

REFERENCES

[1] Stary, Ch. & Totter, A. (1996). Cognitive and Organizational Dimensions of Task Appropriateness, in: Proceedings ECCE'96, European Conference on Cognitive Ergonomics, EACE, Granada, September 1996.

[2] Norman, D.A. (1983), Some Observations on Mental Models, in: Mental Models, eds.: Stevens, A.L.; Gentner, D., pp. 7-14, Lawrence Erlbaum, Hillsdale, New Jersey.

[3] Olson, J.R.; Olson, G.M. (1990), The Growth of Cognitive Modeling in Human-Computer Interaction Since GOMS, in: Human-Computer Interaction, Vol. 5, No 2&3, pp. 221-266.

[4] Lim, K.H.; Benbasat, I.; Todd, P.A. (1996), An Experimental Investigation of the Interactive Effects of Interface Style, Instructions, and Task Familiarity on User Performance, in: Transactions on Computer-Human Interaction (1996), ACM, Vol. 3, No. 1, pp. 1-37.

[5] Brown, J.S., Newman, M. (1985), Issues in Cognitive and Social Ergonomics, in: Human-Computer Interaction, Vol. 1, No. 4, pp. 359-392.

[6] Preece, J. (ed.) (1994), Human-Computer Interaction, Addison-Wesley, Wokingham.

[7] Rosson, M.B; Carroll, J.M. (1995), Integrating Task and Software Development for Object-Oriented Applications, in: Proceedings CHI'95, ACM, pp. 377-384.

[8] Johnson, P. (1992), Human-Computer Interaction, McGraw Hill, London.

[9] Janssen, Ch., Weisbecker, A., Ziegler, J. (1993), Generating User Interfaces from Data Models and Dialogue Net Specifications, in: Proceedings INTERCHI'93, ACM/IFIP, pp. 418-423.

[10] Stary, Ch. (1996), Integrating Workflow Representations into User Interface Design Representations, in: Software - Concepts and Tools, Vol. 17, pp. 173-187, December.

[11] Stary, Ch. (1997): Workflow-Oriented Prototyping for the Development of Interactive Software, in: Proceedings IEEE COMPSAC'97, to appear.

Do people in HCI use Machine Learning?

Vassilis S. Moustakis[a,b]

[a] Department of Production and Management Engineering, Technical University of Crete, Chania 73100, Hellas.

[b] Institute of Computer Science, Foundation for Research and Technology – Hellas, Science and Technology Park of Crete, P.O. Box 1385, Heraklion 71110, Hellas.

Implementation of Machine Learning (ML) in Human Computer Interaction (HCI) work is not trivial. The article reports on a survey of 112 professional and academics specializing in HCI who were asked to state level of ML use in HCI work. Analysis showed that about one third of those who participated in the survey has used ML in conjunction with a variety of different HCI tasks. Neural networks, rule induction and statistical learning emerged as the most popular ML paradigms across HCI workers although intensive learning, such as inductive logic programming are gaining popularity among application developers*.

1. INTRODUCTION

Machine Learning, or ML for short, represents one of the fastest growing technologies today with an abundance of prototype and fielded industrial applications. Robotics, computer vision, manufacturing, medicine, knowledge acquisition, execution and control, design, planning and scheduling, among others, are areas, which have uncovered the potential of the technology. Work over new media and networks has also identified a niche for ML in navigation and retrieval of information.

This article specializes on the use of ML in the HCI community; it complements earlier research related to the application of ML in a wide variety of tasks reported by Moustakis et al. (1996). Moustakis et al. surveyed experts in ML to capture perception regarding appropriateness of generic ML methods to a set of generic intelligent tasks. Results reported herein derive from a survey which was distributed to HCI workers, both in industry and academia to assess degree of use, or non-use, of ML in HCI work. Using World Wide Web resources 185 persons working in HCI were contacted and asked to fill in a research protocol.

In the sections that follow, we overview the protocol we used in this study, present the method and data analysis procedures and discuss results of analysis.

---

* Research reported in this paper was partially supported via the Knowledge Extraction for Statistical Offices, KESO (ESPRIT) project. I would like to thank survey participants. Needless to mention that opinions, views and results reported herein represent the author and do not actually correspond to the opinion of participants. Protocol and raw survey data may be retrieved via ftp from the following address: *ftp.ics.forth.gr/pub/machine_learning*. File *survey96.txt* includes the protocol and file *survey96.xls* includes raw survey response data. For further information please contact the author at: **moustaki@ics.forth.gr**

The text that follows is a brief presentation of the subject matter. Detailed presentation of protocol, statistical data analysis and results may be found in Moustakis and Herrmann (1997).

2. PROTOCOL AND DATA ANALYSIS RESULTS

To elicit information about the use of ML in HCI work, we conducted a survey by means of electronic mail. The survey was based on a structured protocol which is divided into five parts, namely:

1. *Area of HCI involvement.* Eight different areas of HCI involvement were includes in the protocol. Responders were asked responders to indicate the ones in which they are mostly involved. *User interface design/evaluation*, *product or services design and evaluation*, *computer supported cooperative work*, *education and training*, and *virtual reality* emerged as the most popular areas on HCI involvement. On the other hand *human robot interaction* received the fewer votes. Contingency 2x2 analysis between areas revealed significant interaction between top running areas.

2. *HCI task involvement.* The second part of the questionnaire included indicative HCI tasks. Responders were asked to select tasks with which they are mostly involved in professional or academic work and advised to select as many areas as they considered necessary. Average task selection was 4.90 per responder. *Adapting, customizing, or optimizing systems according to user need*, *usability engineering*, *user modeling*, *multimedia system design and evaluation*, and *modeling of cognitive behavior* followed by *information retrieval* emerged as the top running tasks. Collectively they account for more than 60% of all responses. Contingency analysis across responses revealed few significant interactions between tasks.

3. *HCI application involvement.* The third part of the questionnaire captured application demographics of responders. Participants could select more than one area of HCI application involvement. *Education and training*, followed by *public sector*, *media or entertainment* and *aviation/aerospace* lead the list of HCI application domain involvement. Average involvement was 1.88 application domains per responder.

4. *Machine Learning expertise and use (or not use) in HCI work.* The fourth part of the questionnaire captured: (a) level of awareness with respect to alternative ML paradigms; (b) use of ML in HCI work; and, (c) satisfaction with ML by those who have used it in HCI work. ML paradigms were motivated from earlier work on ML research and application classification – for instance, refer to Kodratoff et al. (1994), Langley and Simon (1995), or to Moustakis et al. (1996). Out of 112 participants 41 indicated that they are either have used or using ML in their work. However, ML users do not usually confine themselves to a single ML paradigm. Detailed frequency of ML use analysis revealed an average of 2.3 paradigms per responder who has used ML. *Neural networks*, *statistical learning methods*, *rule induction*, and *case based learning* emerged as the most popular paradigms among participants. Average ML knowledge scores were not that high; they ranged between 3.6/7.0 maximum for *neural networks* to 2.3/7.0 minimum with respect to *reinforcement leaning*, *conceptual clustering* and

inductive logic programming. Finally, average satisfaction across all ML users proved significant, i.e., 3.3/5.0.

5. *Why ML has not been used in HCI work.* Finally, responders were asked to indicate the reason(s) for not having used ML in their work. Results indicate that between those who have not used ML in HCI work they did so because they either perceived ML as not being necessary to what they were doing, or, because they were not aware that ML could be useful to their work, or, because t hey have not seen enough success stories of using ML in work similar to theirs.

3. DO USERS OF ML DIFFER FROM NON-USERS?

Focused on HCI task involvement and clustered responses to examine differences between users and non users of ML. AutoClass (Cheeseman and Stutz, 1996) generated three classes out of 112 vectors of binary data extending over 21 attributes. Use or no use of ML was not included to avoid biasing result. The process converged to three classes, namely class A, B and confirmed that significant differences between those who have used ML and those who have not used could *not* be identified.

4. DISCUSSION

Work reported herein was based on a structured survey aiming towards identifying and modeling use of machine learning, ML, in HCI work. The survey was conducted in May 1996. Analysis of responses contributed to the formulation of some critical initial conclusions, namely:

- A significant number of HCI workers (both from industry and academia) is well aware and knowledgeable about ML.

- More than a third of HCI workers has actually reported that they have, or are, using ML in HCI work, and doing so they are satisfied with the result. Certainly average satisfaction may be improved and this represents a challenge for the future.

- Implementation of ML in HCI tasks is not trivial. HCI work is synthetic and so has to be application of ML into it. However, this poses a great challenge to ML, which has to mature enough before it is able to support complex real world tasks. More than that it will require effective process models leading to an agenda for action appropriate for the situation at hand.

- A high percentage, e.g., slightly less than two thirds, of those who participated in the survey does not use ML in HCI work. A sample of reasons for doing is so is reported in the text – see Table 9. Leading reasons include *unawareness about the potential*, *perception that ML is not necessary* and *lack of concrete case studies*. In fact these reasons mix with each other to form leading contingencies between *lack of concrete case studies* and *misperception* on the one hand, and *unawareness and misperception* on the other. To overcome these barriers people should try to promote their work and researchers in both fields should cooperate to advance pedagogy about ML. Several years ago Churchman and Shainblatt (1965) pondered that effective implementation of management science in the workplace would require a cooperative dialectic between the researcher and the manager. Likewise ML and HCI colleagues should join forces and try to understand each other's needs, ideas, biases and perceptions. Both ML and HCI are cognitive sciences. Both focus on humans and draw from human centered needs.

The rendezvous between HCI and ML awakens several interesting bearings and one unique characteristic. The characteristic draws from the history of both fields; one has lived longer than the other. Bearing on the meeting itself one may wonder about the role of the other upon itself; *what can, or should, HCI offer to ML*, or conversely, *what can, or should ML offer to HCI*?

To probe further on the interaction between HCI and ML the reader may refer, among other sources, to two special issues (Moustakis, 1996; Herrmann and Moustakis, 1997), or to Notes of a workshop organized during the 13th International Conference on Machine Learning, on "ML meets HCI" (Herrmann and Moustakis, 1996).

REFERENCES

Cheeseman, P., and J. Stutz. 1996. Bayesian Classification (AutoClass): Theory and Results. *In Advances in Knowledge Discovery and Data Mining*, pp. 153–180, ed. U. M. Fayyad, G. Piatetsky-Shapiro, P. Smyth, and R. Uthurusamy. Menlo Park, California: AAAI Press/MIT Press.

Churchman, C. W., and A. H. Schainblatt. 1965. The researcher and the manager: A dialectic of implementation.*Management Science*, 11(4):B69--87, 1965.

Herrmann, J. and V.S. Moustakis. 1996 (editors). Workshop notes: ML meets HCI. International Conference on Machine Learning, Bari, Italy, July 1996. Papers may be retrieved on the WWW at: http: //www.ics.forth.gr /~moustaki/ public_html/ ICML96_HCI_ML/hci_ml.html

Herrmann, J. and V.S. Moustakis. 1997. Special issue: Machine Learning meets Human Computer Interaction. *Applied Artificial Intelligence: An International Journal* (to appear).

Kodratoff, Y., V. S Moustakis, and N. Graner. 1994. Can Machine Learning Solve my Problem?*Applied Artificial Intelligence: An International Journal*, 8(1):1-36.

Langley, P., and H. Simon. 1995. Applications of machine learning and rule induction. Communications of the ACM, 38(11): 54-64.

Moustakis, V.S. (guest editor) 1996. Special issue: "Machine Learning meets HCI", International Journal on Human of Human Computer Interaction. 8(3): 217-360.

Moustakis, V.S. and Herrmann, J. 1997. Where do machine learning and human computer interaction meet? *Applied Artificial Intelligence: An International Journal.* (to appear).

Moustakis, V.S., M. Lehto, and G. Salvendy. 1996. Survey of expert opinion: which machine learning method may be used for which task? *International Journal of Human Computer Interaction*, 8(3): 221-236.

Modeling the User's Problem-Solving Expertise for Effective Decision Support

C. Chiu
Department of Information Management, Yuan Ze Institute of Technology,
Neili, Taoyuan, Taiwan 320

ABSTRACT

Aiding a decision maker's problem-solving capability is one of the potential applications of Decision Support Systems (DSSs). To be better as a decision-making aid, a DSS should be broadened to help user in exploring and processing each problem-solving stage. This research proposes a methodology that is based upon the use of AI (artificial intelligence) techniques to support modeling the user's problem-solving behavior dynamically. A experimental prototype is constructed and the results based on this proposed method indicates the promising future environment that is able to provide more effective decision support.

Keywords: fuzzy logic, DSS, user model, problem-solving expertise.

1. INTRODUCTION

Problem-solving is all the activities that lead to the solution of a problem. In general, problem-solving process includes several stages: intelligence, design, choice, implementation, and monitoring (Huber, 1983). Furthermore, a DSS should augment its concept to consist of more support to Type II work in addition to traditional intelligence-design-choice view of decision making (Sprague, 1987). One of the typical functions of Type II work is problem-solving (Panko and Sprague, 1984).

According to Sprague and Carlson, the DSS, in its broad sense, is a an adaptive system that provides the decision maker the capabilities and flexibilities to search, explore, and experiment with problem area (Sprague and Carlson, 1982). Also Angehrn and Luthi advocate that a DSS should aid a decision maker in a manner compatible with their mental representation rather than simply delivering information system technology or problem-solving techniques (Angehrn and Luthi, 1990). Liang points out that model management systems should consider human cognitive limitations during decision-making process (Liang, 1988).

Dynamic reasoning about a user's problem-solving expertise during user-system interaction offers an effective approach in determining user requirements and for enhancing the relationship between a user's cognitive abilities and software sophistication. To model users more accurately, however, a better reasoning methodology is required to accomplish this goal by gaining more insights about user. One of the most important insights is about a user's problem-solving expertise. This expertise can be derived by careful inspection on the actual user-computer dialog behavior.

This research proposes a methodology that is based upon the use of AI (artificial intelligence) techniques to support modeling the user's problem-solving behavior dynamically. The proposed research architecture helps to produce a user's problem-solving model that can more effectively reflect the user's performance progression along the novice-expert continuum. This environment could be used to aid determining what should adapt to the user once the user's problem-solving expertise is captured. Therefore, a DSS with such a reasoning component can be used to aid each stage of decision making process; thereby improving the process effectiveness and decision quality.

2. PPROPOSED METHODOLOGY

The primary objective of this research is to develop an innovative methodology that can analyze the user's dialog behavior, evaluate the user's performance, and infer the user's level of expertise in a given problem domain.

Figure 1 presents the overall system framework. The system is composed of several modules including Knowledge Association Map (KAM), Performance Stability Recognizer (PSR), Knowledge Index Analyzer (KIA), and Fuzzy_Classfier (FC). These components are described in detail below.

- KAM is used to store the causal relationship of pairwise interaction states;
- KIA is to analyze the user's knowledge index;
- PSR is used to detect a user's dialog performance pattern during a specific interactive session
- FC is aimed for resolving inconsistent and uncertainty inferences during the user's interaction.

3. IMPLEMENTATION

The implementation prototype is constructed based on the decision problem of financial statement analysis. Each user is expected to work on a puzzle like environment to tackle problem scenarios. The purpose of this design is not only for a user to solve the problems encountered, but also to help learning the contents itself during interaction. In order to successfully finish the puzzle (as shown in Figure 2), a user needs to be proficient and knowledgeable when adopting appropriate financial ratio(s) to evaluate a corporate financial performance. There are four Groups (from A to D) that a user may alternatively enter to provide answers system initiated or to select suitable items to proceed a puzzle. For Group A, five categories of financial ratio analysis are grouped together.

According to the above five financial categories, twenty two major financial ratios are stored in group B. Group C introduces income statement and balance sheet related concepts that with the latter one consisting of assets, liabilities, and owner's equity information. A user is required to appropriately select both the denominator and numerator of financial index to form a meaningful financial ratio. Group D includes several performance evaluations options which is used as the measuring mechanism of corporate operation financial performance. These options are classified into three types: (1) the higher ratio the better performance; (2) the smaller ration the better performance; and (3) the more extreme ration the worse performance. In order to decently accomplish a whole round puzzle, a user is required to finish each group with a positive correlation value. That is, the subsequent activity in one group must be conceptually associated with the one in preceding group.

The calculation of both PSI and KI indexes are based on the tracks navigated. Further details can be found in (Chiu, 1996). This paper mainly reports how to synthesize the uncertainty information (i.e., both PSI and KI) using fuzzy approach that is similar to the human judgment.

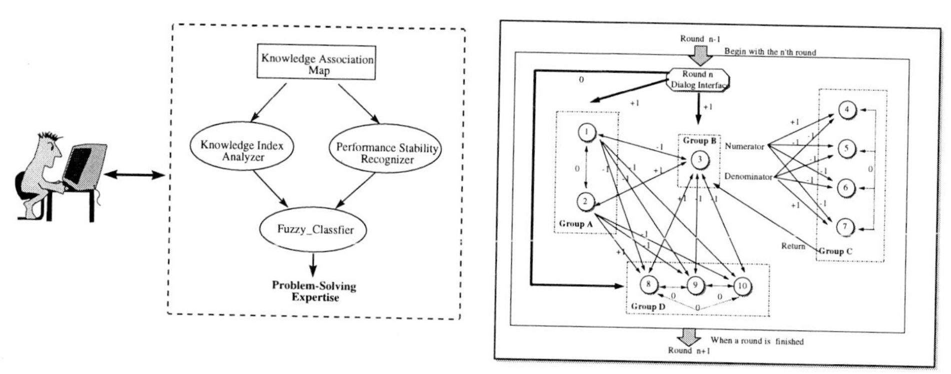

Figure 1. System Architecture

Figure 2. A Portion of Puzzle in Support of Financial Statement Analysis

3.1 The Design of Fuzzy Classifier

Typically the internal reasoning method for fuzzy production rules adopts the compositional rule of inference proposed by Zadeh (Zadeh, 1973). However there exist many other methods that are available to resolve such problems. In this research, a simpler fuzzy based method is used to tackle the problem that is without explicit linguistic production rule sets to describe the relationship between the information from FC v.s. the information from PSR and KIA (Lee and Lee, 1996). Assume that the original fuzzy membership and other related classification categories are given. Besides, assume P is the factor influencing the PSE (output from FC) value most. The internal reasoning method in this research is defined as:

$$\mathbf{P = \max_i \, [PSI_i \, W_i \times (1 + \max_j \{KI_j \, W_j\}/2)]}$$

where W_i and W_j are the weighting imposed upon each attribute of PSI_i (output from PSR) and KI_j (output from KIA) in order to provide an emphasis of impact from each attribute. To simplify the discussion here, both W_i and W_j are treated with identity vectors; that is all the impact from each attribute is equally weighted.

When P is large, its impact toward PSE is large. Therefore the complement value of P, that is, 1-P can reflect positively toward membership grade for PSE. Thus a transformed membership function can be derived, and a defuzzified value M can be calculated by:

M = 0/Least Proficient + 0/Less Proficient + 0/Normal + Y/Proficient + X/Very Proficient

Thus PSE = a*Y + b*X, where $\mu_{Proficient}(a)=1$, $\mu_{Very\ Proficient}(b)=1$.
An example and its detail explanation can be found in (Chiu, 1996).

4. CONCLUSION

This research has intensively investigated the nature of problem-solving behavior from the decision support view point. Furthermore, this research has proposed a method to better understand a user's problem-solving expertise from directly observe and analyze the user's computer dialog behavior. Currently a prototype system used to support financial analysis activities is under development, and the fuzzy logic based approach is employed to concurrently transform these derived dialog events information into problem-solving expertise. This inferred understanding about the user can be further utilized as feedback information to aid the system's adaptation to the following interaction as well as to improve a user's decision-making capability. Further research will proceed an empirical study to test and evaluate human subjects within the proposed financial statement analysis platform.

ACKNOWLEDGEMENTS

This research was supported by National Science Council, Taiwan, Republic of China, under the contract number NSC85-2416-H-155-002.

REFERENCES

Angehrn, A. A. and Luthi, H. J., "Intelligent Decision Support Systems: A Visual Interactive Approach," *INTERFACES*, Vol. 20, No. 6, 1990, pp. 17-28.

Chiu, C., "Modeling User's Problem-Solving Expertise for Effective Decision Support," Technical Report, NSC85-2416-H-155-002, 1996, National Science Council, Taiwan.

Huber, G. P., "Cognitive Style as a Basis for MIS and DSS Design: Much Ado about Nothing?" *Management Science*, Vol. 29, No. 5, 1983, pp. 567-679.

Lee, H. and Lee, C., "Inexact Strategy and Planning – The Implementation of Route Planning in Taipei City," *The Proceedings of the 7th International Conference on Information Management*, 1996, pp. 340-347.

Liang, T. P., "Meta-Design Consideration in Development Model Management Systems", *Decision Sciences*, Vol. 19, 1988, pp. 72-92.

Norcio, A. F. and Stanley, J., "Adaptive Human-Computer Interfaces: A Literature Survey and Perspective," *IEEE Transactions of Systems, Man, and Cybernetics*, Vol. 19, No. 2, 1989, pp. 399-408.

Panko, R. and Sprague, R. H., "DP Needs New Approach to Office Automation, *Data Management*, 1984.

Sprague, R. H., "DSS in Context," *Decision Support Systems*, Vol. 3, No.3, 1987, pp. 197-202.

Zadeh, L. A., "Outline of a New Approach to the Analysis of a Complex Systems and Decision Process," *IEEE Transactions on Systems, Man, and Cybernetics*, Vol. 3. No. 1, 1973, pp. 28-44.

Technology needs for Web-based technical training

T. A. Plocher[a] and S. L. Hillman[b]

[a]Honeywell Technology Center, 3600 Technology Drive, Minneapolis, Minnesota, 55418

[b]Honeywell Industrial Automation and Control, 2820 West Kelton Lane, Phoenix, Arizona 85023

1. BACKGROUND

Honeywell annually invests millions of dollars in training for its sales, engineering, and field service personnel. The three major Honeywell learning centers—the Industrial Automation and Control (IAC) Automation College, the Home and Building Control (HBC) University, and the Commercial Flight Systems Group (CFSG) Customer Support/Training Center—also have responsibility for training the customers and users of Honeywell products. The revenue generated by customer training, per se, is significant. However, the importance of our training services to customer satisfaction with Honeywell products, while difficult to quantify, also has an important impact on revenues. Given this flow of training-related dollars, enhancements to Honeywell's training services resulting in more efficient course production and delivery, and increased training effectiveness, have the potential for major financial benefits.

Advanced learning technologies hold a great deal of promise as one means of streamlining course development and delivery, increasing the effectiveness of training, and enhancing the resulting on-the-job performance. An internal research and development project was undertaken by the Honeywell Technology Center, in collaboration with the Honeywell Automation College and several of its major industrial processing customers. The goal of the study was to determine which of these new learning technologies would have the most significant impact on the Honeywell training business in terms of meeting the future needs of both our internal and external customers.

1.1 A Major Paradigm Shift

The most consistent and common need described by organizations participating in the study was the need for a major paradigm shift for a significant portion of their training services from the current "schoolhouse" model to one of interactive distance learning.

The majority of technical training courses offered by the Honeywell training centers are still conducted using a combination of stand-up lecture and hands-on laboratory. Courses are taught either at the customer's site or at one of the

Honeywell training facilities. Customer personnel are typically either sent on travel to Honeywell for the course or are pulled out of normal work rotations for courses taught on-site. The study concluded that significant portions of this current curriculum, particularly those involving electronics or process fundamentals, procedural learning, and troubleshooting, were candidates for an alternative instructional delivery model. That new model had to be responsive to a series of problems and constraints voiced by the participants in this study.

- Cost in time and travel to participate in classes taught at centralized locations.
- Desire to pay for instruction as it is "consumed," rather than bear large up-front cost.
- Disruption of work schedules and rotations; loss of productive work during employee training leave.
- Emerging trend in industry to encourage individual employee responsibility for education; *flexibility* in learning environment needed to facilitate "after hours" use or use during "slow periods" on the job.
- Need for *focused*, refresher training *on demand*, as the job dictates.
- Desire for a more effective and *interactive* alternative to lecture.

The common solution proposed by study participants was to evolve toward a learning paradigm based on individualized, interactive multimedia instruction delivered on demand via distributed network.

2. A WEB-BASED INTERACTIVE LEARNING ENVIRONMENT (ILE)

In our concept for a Web-based Interactive Learning Environment, the Honeywell learning centers will serve as host sites for instruction delivered to internal customers via corporate intranet and to external customers via the Internet. Courses in the form of intelligent tutoring packages will be hosted on a learning center server. Each tutor will contain simulations and other multimedia productions, lessonware that includes instructional plans, presentations, explanations, drills, and tests, and a student model which prescribes conditions and pathways for advancement and remediation.

Customers with authorized access to the host learning environment will select, at their convenience, a tutor on their topic of interest, and pay as they progress. With the bulk of the tutoring "intelligence" residing in the courseware on the host site processor, only minimal intelligence is required on the learner's platform. The tutor will present instructional materials to a Distance Learning Interface for delivery to the learner. The Interface will monitor the learner's responses and passes them to the tutor for analysis by the student model. The next item for instruction will be selected by the tutor, and so on until the learner signs off. Results of the session will be recorded and sent to a Learner Record and Assessment System, where they are available for review and analysis by a Honeywell instructional manager. The learner also will interact with other learners and with the instructional manager on the selected topic via a newsgroup, electronic mail, or desktop video conferencing.

3. CHALLENGES TO ACHIEVING THE ILE

The Integrated Learning Environment is only a vision right now for industries like Honeywell. In order for it to be realized, numerous challenges, some technical, some organizational, will need to be met.

3.1 The Economics of Courseware Authoring

A dramatic reduction in multimedia courseware authoring time is required. Popular authoring languages, such as Macromedia Authorware and Asymetrix Toolbook, generally consume 100 to 500 hours of authoring time in order to produce one hour of computer-based instruction. This is extreme compared to the 20 to 40 hours of preparation generally experienced per hour of classroom instruction. Transitioning an entire curriculum over to computer-based training becomes economically infeasible under these conditions.

How can this challenge be met? Perhaps a whole new authoring paradigm is needed. Until that comes along, however, a number of things can improve the current onerous development-to-class time ratios. These include course planning aids, libraries to facilitate re-use of courseware, particularly simulations, wizards and agents to help retrieve information, and template-based metaphors for learning strategies, individualized testing and navigation throughout the course. The RIDES tools developed at USC [1] are one promising attempt to automate some of the more routine courseware authoring tasks.

3.2 Making Adaptivity a Reality

A significant shift envisioned by the ILE is an increase in the adaptiveness of instruction to individual skills and needs. An intelligent tutor has a major advantage over a human instructor in being able to observe and measure each and every action taken by the student, as well as the amount of time the student takes to perform an action or complete a lesson. An instructional pathway through the material is selected which best fits the student's needs. The learner's time is used optimally and the instruction is focused on areas of need. A few commercial products for intelligent tutoring systems exist [2], but none has been developed with the purpose of hosting instruction on a distributed network. Research, such as Brusilovsky's [3], and Johnson's [4] in the area of server-side intelligence for Web-based tutoring systems holds a great deal of promise for meeting this challenge in the near future.

3.3 Interfaces With Legacy Courseware

Many corporate training centers have already invested significant amounts of money in computer-based instructional materials. At Honeywell, for example, dozens of these so-called "legacy" materials exist on CD ROM, authored with multiple versions of several different tools, including Asymetrix's Multimedia Toolbook, Macromedia's Authorware, and

PowerPoint. Corporate learning centers cannot afford to re-invent these materials in a new format. Rather, the challenge is to conceive an ILE architecture that allows maximum re-use of legacy materials, either in part or in their entirety.

3.4 Retaining the "Human" Element

In shifting from an instructor-intensive approach to learning (i.e., lecture/lab) to a distance learning environment, great care must be taken to provide, in some way, for all the functions performed by the human instructor. A conventional computer-based training (CBT) package, for example, does not provide for individual student attention nor for dynamic interaction between an instructor and student. Intelligent tutoring systems, such as RIDES [1], succeed in capturing more of the functions of the human tutor than do CBTs. However, even if the state-of-the-art tutoring systems are the foundation of the integrated learning environment, a fundamental question remains, "how much interaction with a human instructor is still required and how best do we satisfy that requirement?". Further, do we need to provide also for interaction between students in the same course? Electronic mail, newsgroups, bulletin boards, video conferences, and, perhaps even occasional in-person visits by an instructor are all potential mechanisms for supporting this requirement.

3.5 Changing Roles for Instructors

In the ILE, the traditional instructor becomes an *instructional manager*. This represents a cultural change to training organizations that most likely will take place with a certain amount of turmoil. Instead of presenting stand-up lectures and conducting laboratories, he/she serves as chief author and maintainer for a set of instructional programs, i.e., tutors. The instructional manager also serves as a monitor of the students enrolled in his/her instructional programs at any given time. Among his/her tasks is that of reviewing learner progress and certifying the completion of instruction and any related job qualification. They also provide the human element in the ILE, interacting and intervening with remote learners as they deem necessary via newsgroups, e-mail, voice, or video conferencing.

REFERENCES

1. J. McCarthy, Introduction to Sonalysts' Intelligent Training Technology, Sonalysts, Inc. Waterford, CT, 1996.
2. A. Munro, Q.A. Pizzini, D.M. Towne, J.L. Wogulis, and L.D. Coller, Authoring Procedural Training by Direct Manipulation, USC Behavioral Technology Laboratories Working Paper WP94-3, University of Southern California, Redondo Beach, CA, 1994.
3. P. Brusilovsky, E. Schwarz, G. Weber, ELM-ART: An Intelligent Tutoring System on World Wide Web, in Intelligent Tutoring Systems, C. Frasson, G. Gauthier, and A. Lesgold (Eds.), Springer, 1996, 261.
4. W.L. Johnson, Automated Management and Delivery of Distance Courseware, in Proceedings of the WebNet96 Conference, 1996.

Using Multiple Technologies for Distance Learning

Lisa Neal

EDS, Three Valley Road, Lexington, MA 02173 USA
Phone: +1 617-861-7373
Email: lisaneal@media.mit.edu

ABSTRACT

This paper describes how collaborative technologies, including a corporate intranet, email, videoconferencing, audioconferencing, Internet Relay Chat (IRC), NetMeeting, Virtual Places, WorldsAway, and other Internet-based conferencing tools, are used to teach classes to geographically-dispersed participants. The paper covers the motivation for distance learning, the selection and use of delivery technologies, deployment strategies and issues, and participant feedback.

We found that the use of multiple technologies provided an effective and cost-saving alternative to face-to-face instruction and allowed us to reach more people in more locations than we would have without distance learning. The use of multiple technologies provided richer communication than any one technology alone. However, the use of each technology was far more complex than we had imagined at the outset. There were also striking differences between the distance learning and face-to-face classes including instructor preparation and lecture style, student interactions, the types of assignments that worked well, and the importance of establishing protocols.

1. INTRODUCTION

Businesses and universities are searching for ways to reduce the cost and increase the availability of education. This includes innovative university programs where a residency is not required, continuing education, and corporate education. The benefits of distance learning are location flexibility for both instructor and student, fewer disruptions to work and family life, and lower cost (although there may be expenses arising from the delivery mechanisms). These benefits only accrue if education is effective. The challenge is to have a location-independent class which is as effective as a face-to-face class or where the other benefits achieved are worth some small sacrifice in effectiveness.

This is difficult to accomplish because part of the educational process is the interaction with the instructor and other students. Students and instructors are used to classrooms, and typically need to adjust their learning and teaching styles, respectively. An instructor has to plan lessons and assignments differently, and delivery can be subject to far more technical difficulties than occurs in a traditional classroom.

At EDS, we taught classes on Emerging Technologies in Human-Computer Interaction (HCI) and Emerging Technologies in Collaborative Environments using a technology-based solution to replace face-to-face classroom instruction. The goal was to make the class as rich and compelling a learning experience for the students as the more traditional face-to-face delivery. The challenge was to balance the constraints of the technologies with the needs of the subject matter, course structure, and participants.

Delivery technologies were selected based on cost, availability, and perceived or hoped-for (at least in the first delivery) effectiveness, and were matched to objectives in order to select and deploy the most appropriate technology for each class meeting. Feedback was used to make adjustments, including attempted blendings of technologies. In the process of planning, delivering, and evaluating the classes, we learned how to match tasks to technologies, the subtleties of the technologies employed, and how distance learning differs from face-to-face instruction. In planning subsequent classes we incorporated many of these lessons.

2. CLASS STRUCTURE, TOPICS, AND STUDENTS

The courses were structured as an instructor-led seminar with guest speakers and student research projects. Organizing each course as a four-week non-contiguous class was deemed the best way to give the students time to select, reflect upon, and research a topic for their class research projects. Distance learning was well-suited to the course structure. Guest speakers were able to lecture or lead discussions without being constrained by location. Students juggled work and class commitments, but, as one student observed, "having time between classes gave me a chance to digest and explore" and to relate what they learned to their jobs. Class sessions were typically three times a week for 2 or 2 ½ hours. This seemed to be an appropriate length of time for the students to fit into their work schedules. We found that it was more difficult to have a longer technology-based class session than a face-to-face session due to the greater difficulty for the instructor to detect students' attention level and to pace a session Students occasionally experienced problems with interruptions from their coworkers, who did not realize they were participating "in a class" at their desk. Students individually or in teams produced impressive class projects, many choosing topics related to their jobs.

Emerging Technologies in HCI and Emerging Technologies in Collaborative Environments seemed appropriate topics for interactive delivery technology. Not only was it beneficial for students to sample unfamiliar technologies, but also the participants could evaluate the effectiveness of the technologies, an appropriate exercise for an HCI or Collaborative Environments seminar. The technology was both delivery mechanism and course content. Most students had little previous exposure to technologies such as videoconferencing and IRC. It would have been hard to replicate their first hand and observed technology experiences in a face-to-face classroom or laboratory.

Four Emerging Technologies in HCI and Emerging Technologies in Collaborative Environments distance learning classes were taught by the author for a total of almost 80 students. These classes were interspersed with on-site face-to-face classes, which offered the author a reminder of the many differences between the delivery techniques. Students in the distance learning classes were in many countries; the only problem we experienced delivering a class to international students was the narrower window in which classes could meet due to different time zones. Many of the enrolled students in the distance learning classes would not have been able to participate in a multiple-day class due to work commitments or to travel costs.

3. DISTANCE LEARNING TECHNOLOGIES

A variety of technologies are currently being used to replace or supplement the face-to-face classroom [1]. EDS was attracted to a hybrid approach based on the Open University's success [2]. We chose not to strictly imitate the Open University's approach since we had more distance learning methodologies available at EDS than the Open University instructors had at their disposal.

Of the technologies available at EDS, videoconferencing, audioconferencing, and a variety of Internet-based conferencing tools were selected based on cost, availability to participants, appropriateness for the course structure, and their potential to provide an interactive virtual classroom. A Web site was built that included the class schedule, syllabus, evaluation forms, RealAudio and text transcripts from class meetings, student projects, and student biographies. (The initial Web site led to the development of a Virtual University, which was the repository of the materials classes and provided information not only to enrolled students but to other employees as well.) Finally, email and the telephone were chosen for one-on-one instructor-student communication.

The technologies we selected varied in how easy they were to use, how comfortable their use was for the instructor and students, and the learning situation which they best facilitated. We discovered that protocols had to be established for each technology, and that the instructor played a key role in both establishing and enforcing its use. Audioconferences, for example, were convenient and familiar but the lack of visual information limited their usefulness. IRC was too multi-threaded and participants' typing speed was too slow for it to be effective as a sole delivery mechanism for a class session.

This led to the combining of audioconferences with IRC and other Internet-based technologies; Virtual Places, a hybrid IRC/virtual world, was useful for the expressiveness of the avatars, for the web-based information that could be shown, and for the "tour buses" that students could "ride" to visit Web sites. Microsoft NetMeeting was added to share applications, have a shared whiteboard, and to include audio and video. CompuServe's WorldsAway, along with other virtual worlds, was used for class parties. Most students had no prior experience with these technologies, and welcomed the opportunity to explore new ways to collaborate. Students gained further experience by collaborating with other students for class projects, and many found that they were able to use the technologies introduced during class as part of their work collaborations, thus increasing the value of the class to them.

4. CONCLUSIONS

We found that the use of multiple technologies provided richer communication than any one technology alone. The use of the technologies proved, in itself, to be a valuable pedagogical experience. Although the rationale for the media selection was largely accurate, we did not fully enough weigh the subtleties of the selected technologies for teaching and learning. This oversight was mitigated somewhat by combining technologies, where the strength of one addressed the weakness of another, and adding new ones such as NetMeeting. Table 1 summarizes, for each technology, how it was most effectively used and the communication benefits and difficulties.

The strength of the relationships students formed with the other students and the instructor is largely attributed to the mix of technologies, each of which fostered varied personal communication styles and allowed people's personalities and senses of humor to be apparent. (Some instructors even incorporated some of these technologies into face-to-face classes to enrich the students' relationships.) The humor in some students' home pages and in their virtual world visits greatly increased their familiarity with each other. This led to better team projects and to the sense of community and networking that is typically a by-product of education. Networking, which is especially important in a large organization, was further enhanced by the wide geographic distribution of each class (typically a mix of the U.S., Europe, and Asia/Pacific).

Students were better able to fit classes into their work schedules when the classes required no travel and met over a longer duration with shorter sessions for each class meeting. Students reported that this schedule increased their retention of material and facilitated deeper exploration into materials since students were using their own open-ended time rather than being given a fixed block of time in the classroom for exercises and readings.

From the instructor perspective, the class preparations were much more time-consuming and the matching of class meeting to delivery technology much more complex than anticipated at the outset or than required for classroom delivery. An instructor needs to plan beforehand what kind of interactions he or she wants to accommodate and why. Primarily due to the greater difficulty of "managing" the interactions, we expect that this type of class does not scale up well, i.e. would not work for a much larger class than 20 students. It is challenging for the instructor to convey information without the visual cues that are so abundant in face-to-face delivery. It is also challenging for the student to absorb that information with no visual cues from an instructor. Both learning and teaching require practice, protocols, and patience.

ACKNOWLEDGMENTS
Many thanks to Tom Gillespie, Gail McLaughlin, and Jenny Preece for their comments on this paper.

REFERENCES

1. Neal, L., Ramsay, J., and Preece, J. (1997) Report on the CHI 97 Special Interest Group on Distance Learning, SIGCHI Bulletin, in press.
2. Preece, J. and Keller, L. (1991) Teaching the practitioners: developing a distance learning postgraduate HCI course, Interacting with Computers, 3, 1, 92-118.

Table 1: Summary of technology use

technology	situation best for	communication benefits	communication difficulties
videoconferencing	one-to-many information broadcasts such as lectures and student reports; also instructor-controlled discussion	visual information from speaker; created sense of community for class; students got to know others at shared videoconference sites	speaker could only see the last site where someone had previously spoken; transmission delay made interaction awkward
audioconferencing	full group interaction such as discussion	comfortable and familiar technology; highly interactive; easiest availability	lack of visual cues to know when to jump in with a question or comment; no displayed lecture materials
IRC	supported group interaction but also allowed for private chats between participants; class evaluations where prepared questions are asked	everyone could contribute simultaneously; contributors automatically identified; highly interactive and often humorous; transcript of session	slowness of typing speed; following the typical multi-threaded dialogue
audioconference/ IRC	audio lecture with questions or comments by IRC; supporting visual information for lecture	could question speaker or make comments without interrupting speaker; easier to follow lecture with supporting visual information	multitasking difficulties for speaker; needed 2 phone lines if out of office
audioconference/ NetMeeting	lecture with visual information through application sharing or shared whiteboard; benefits of audioconferencing with the visual point of reference and chat	easier to follow lecture or presentation; could have audio or video as well in NetMeeting when microphones or cameras available	multitasking difficulties for speaker; needed 2 phone lines if out of office; slow speed over modem
audioconference/ Virtual Places	lecture with visual information at Web site or "tours" of Web sites, so that entire class viewed the same information; benefits of audioconferencing with the visual point of reference and chat	easier to follow lecture or presentation, ability to view Web sites when relevant to class discussion, use of avatars allowed greater communication freedom and more humor	multitasking difficulties for speaker; needed 2 phone lines if out of office
virtual worlds	informal encounters, such as a "field trip" and "end-of-class party"	use of avatars allowed more humor and creativity of expression	occasional difficulty connecting because of traffic or firewall

PEBBLES: Providing Education By Bringing Learning Environments to Students

Laurel A. Williams[†*], Deborah I. Fels[†], Graham Smith[‡], Jutta Treviranus[§] and Roy Eagleson[*]

[†]Ryerson Polytechnic University, 350 Victoria St., Toronto, ON, CANADA, M5B 2K3
[*]University of Western Ontario, 1151 Richmond St., London, ON, CANADA, N2L 3G1
[‡]Telbotics Inc., 317 Adelaide St. W., Suite 202, Toronto, ON, CANADA, M5V 1P9
[§]ATRC, University of Toronto, 130 St. George St., Toronto, ON, CANADA M4L 3P3

1. INTRODUCTION

When a child is away from school for extended time due to illness, s/he is isolated from normal classroom experiences. A remote controlled communication system, PEBBLES, was developed to allow a student to communicate and learn with his/her regular class.

This research addresses issues in remote control, interface design and video conferencing. Providing effective control of remote systems is complicated and many approaches have been employed most often for directing remote manipulators [3], [4], [5]. The remote controlled system described here must move as directed by a child. Less research has been done in designing interfaces for children rather than adults [6]. It has been suggested that input device design should be based on children's preferences thus children are included in this project's design process [1]. Research in video conferencing indicates that interaction is improved through a realistic sense of presence and mutual awareness [2], [7]. Video-mediated presence has been investigated in a structured, adult environment. This project attempts to provide video-mediated presence in a less structured environment. Presence is provided through physical, audio/visual, and control representations.

This paper describes PEBBLES and reports the results of two pilot studies. We examine the effectiveness of PEBBLES in allowing a remote student to participate in classroom activities and to have a sense of presence in the classroom. We focus on the control of PEBBLES to participate in activities, gaining the attention of the teacher, and attitudes.

2. SYSTEM DESCRIPTION

PEBBLES consists of a remote and classroom component (Figure 1). A modified video conferencing system provides two-way audio and video via ISDN. The remote end provides the child with control of the system in the classroom. Preferred control methods were gathered in an informal study with children. A game pad was specified as an input device.

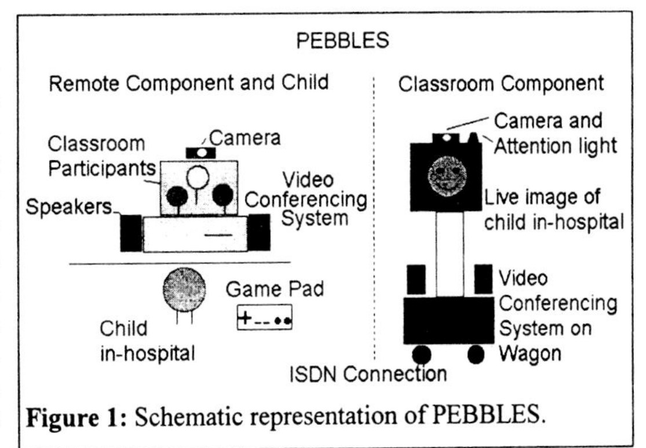

Figure 1: Schematic representation of PEBBLES.

This interface provides seven control actions (left, right, up, down, zoom in, zoom out, and attention). The monitor on the classroom end is mounted on a pedestal such that it and the camera pan left and right together. The camera tilts and zooms in response to up, down, zoom-in, and zoom-out controls. The attention control activates a light to gain attention in the classroom.

3. METHODOLOGY

The first of two pilot studies involved five participants (cub scouts), attending a two-hour computer badge session. One cub participated from a remote location using PEBBLES. The second study was held during a two-week multimedia workshop. Five to ten participants (ages 12+, with and without disabilities) were present at the camp on any one day. The remote participant, a senior female student with a neuromuscular disability, used PEBBLES from home.

Each study included training sessions of 1-2 hours to familiarize the remote student with PEBBLES and allow him/her to practice controlling the system. The trials consisted of briefing and debriefing sessions and the badge/workshop activities. Video cameras located in the classroom and remote locations were used to gather data and analyse the pilot sessions. Between the studies modifications were made to PEBBLES to solve some technical problems and to change the attention mechanism to a flashing rather than solid light.

Use of PEBBLES to participate in activities is characterized by identifying the control tasks employed and then evaluating: errors by different control tasks, successful completion of the intended action, and attention light usage. Errors are classified as: overshoot - visually bypassing the intended target; undershoot - not going far enough; zoom-off-target - zooming in on the incorrect target; and wrong button - pressing the incorrect control. Success gaining attention in the classroom is dependent on acknowledgment of an attention request. Subjective attitudes were gathered through discussions and video taped comments.

4. RESULTS

In the first pilot study, three tasks were analysed: 1 - reading/finding the blackboard (10 occurrences), 2 - reading the computer screen (11), and 3 - finding a person (15). A one-way ANOVA performed on occurrence level for control errors between tasks indicated a significant difference between control tasks for: overshoot ($F[2,33]=7.0$, $p<0.05$); undershoot ($F[2,33]=5.5$, $p<0.05$); zoom-off-target ($F[2,33]=9.1$, $p<0.05$). A one-way ANOVA performed on elapsed time for each task indicated a significant difference between tasks ($F[2,33]=5.1$, $p<0.05$). In the second study, three tasks were also analysed: 1 - following a group discussion (5 occurrences); 2 - reading the computer screen (6); and 3 - finding a person (10). One-way ANOVA tests as above indicated no significant differences. Error levels for each study are shown in Figures 2 and 3.

In the first study, 29 of 36 total task attempts were successfully completed while in study two 19 of 21 tasks attempts were successful (see Figures 4 and 5).

In both studies, the attention light was used to ask a question, answer a question, or contribute to a discussion. In study one, five of nine attempts to gain the attention of the instructor were unsuccessful and in study two, two of three attempts were also unsuccessful.

In study one, the cubs said that their friend was able to participate fully and complete his computer badge remotely. They thought that the game pad control was “really cool”. In study two, the remote participant and the instructors expressed an overall positive experience with PEBBLES. The remote student specified that she was included and not

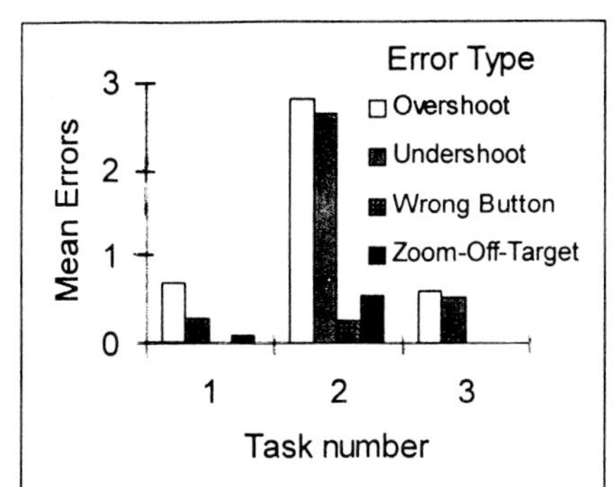

Figure 2. Mean number of errors for each error category by task for study 1. Note. Task 1 is reading the blackboard, task 2 is reading the computer screen and task 3 is finding a person.

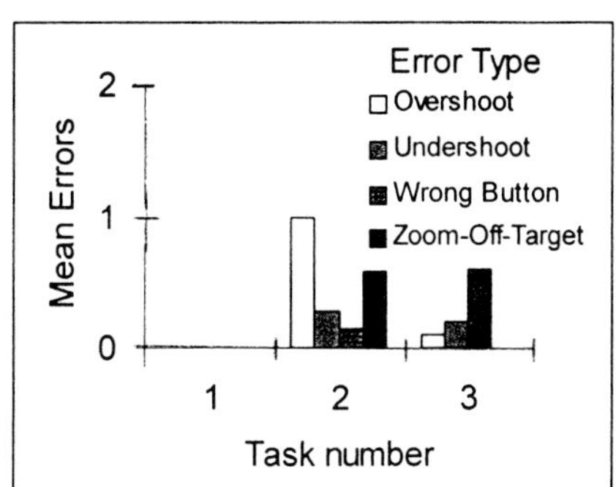

Figure 3. Mean number of errors for each error category by task for study 2. Note. Task 1 is participating in a discussion, task 2 is reading the computer screen and task 3 is finding a person.

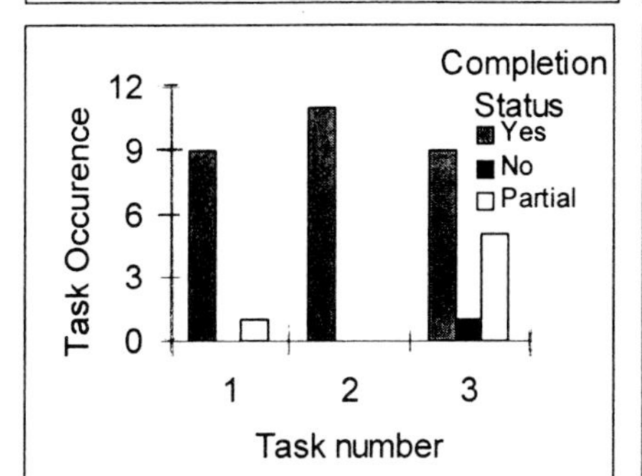

Figure 4. Number of successful, failed and partially successful occurrences for each task category for study one. Note that the total number of occurrences in each task is different (10, 11 and 15 respectively for tasks 1, 2 and 3).

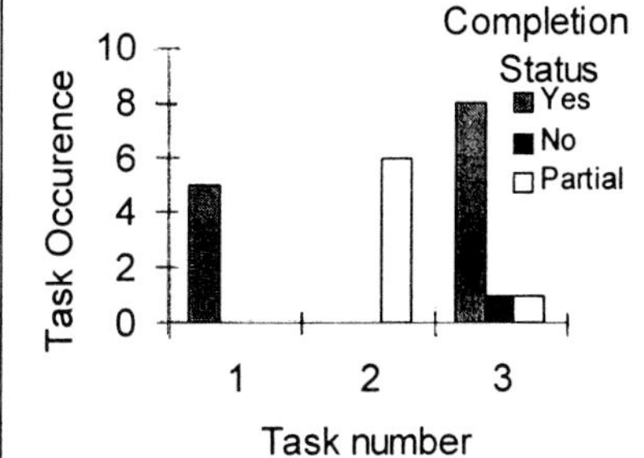

Figure 5: Number of successful, failed and partially successful occurrences for each task category for study two. Note that the total number of occurrences in each task is different (5, 6 and 10 respectively for tasks 1, 2 and 3).

isolated. The two camp instructors indicated that they experienced difficulties in monitoring the remote student's work and giving assignments to the entire class. However, they stated that a majority of th teaching activities were not affected by PEBBLES.

Comments made during the sessions indicate a sense of presence in the classroom. When the remote user visually located a classmate s/he stated that "I am going over to see (name of person)." Classroom participants referred to the remote user as if s/he was in the room.

5. DISCUSSION

While a statistical comparison between the two pilot studies cannot be made because of the lack of subjects, some descriptive comparisons and observations can be made.

Finding a person seemed to be the most difficult task to complete in both studies. This task was performed to locate a person in the classroom, but not necessarily to interact with the person directly. The remote participant seem to be satisfied with a partial view of the person in many cases. This may indicate that audio feedback during communication is sufficient in some cases.

The use of the attention light was largely ineffective despite modifications. The attention light was successful only when teachers were looking directly at PEBBLES. Note that this may also be true for children who are in the classroom.

In the first study, the response to the game pad controller was positive. It was easily recognized and afforded a natural interface for remote control. In the second study, its use seemed more awkward. The participant in this case experienced difficulty with computer interfaces in general whereas the first participant was more comfortable with computer input devices. Experience with computers may have a significant impact on the use of PEBBLES.

Participants reported having a positive experience with PEBBLES. Remote participants reported that they did not feel isolated from their peers and they were included the classroom activities and discussions (indicating classmate's awareness of presence).

There remain a number technical and logistical issues, however, initial results point to a successful implementation of a system that allows a remote user to participate in regular class(es).

ACKNOWLEDGEMENTS

The authors would like to thank Cinematronics, Telbotics, The Bloorview MacMillan Centre, PicTech, the 44th Toronto Cubs and the UT Multimedia Workshop. Funding: Ryerson Polytechnic, Bell Canada, and Canadian Engineering Memorial Foundation.

REFERENCES

1. Alloway, N. (1994). Young children's preferred option and efficiency of use of input devices. *Journal of Research on Computing in Education.* 24(1). 104-109.
2. Buxton. W. (1992). Telepresence: Integrating Shared Task and Person Spaces. *Proceedings of Graphics Interface '92*, 123-129.
3. Hackenberg, R.G., (1986). Using natural language and voice to control high level tasks in a robotic environment. Intelligent Robots and Computer Vision: Fifth in a Series, *SPIE, 726*, 524-529.
4. Kameyama, K., Ohtomi, K. (1993). A shape modeling system with a volume scanning display and multisensory input device. *Presence, 2(2).* 104-111.
5. Masanic, C., Milner, M., Goldenberg, A.A., Apkarian, J. (1990). Task Command Language Development for the UT/HMMC Robotic Aid. *Proc. of RESNA 13th Annual Conference*, Washington D.C., 301-302.
6. Rimalovski, I. (1996). The Children's Market. *Interactivity.* June, 30-39.
7. Tang, J.C., Rua, M. (1994). Montage: Providing teleproximity for distributed groups. *Proc. of Human Factors in Computing Systems - CHI'94.* Boston, MA, 459-464.

Computer Based Learning: GroupSystems® in the Wireless Classroom

Constance Charlier Keating, Ph.D.*

Bogota 7697, (5503) Carrodilla, Provincia de Mendoza, Argentina

The Center for the Management of Information, Netwave Technologies, Inc., and the Ventana Corporation are participating in an ongoing stream of computer based educational research. The focus of this section of the overall project is on the application of GroupSystems® technology in the wireless classroom.

1. INTRODUCTION

GroupSystems® (GS) and GroupSystems® for the World Wide Web (GS_{Web}) software are information technologies that support transformation of the classroom environment from a traditional lecture-based model to an interactive, collaborative learning environment. A Wireless Local Area Network (WLAN) creating an on-premise data communication system without wired connections makes new uses of this technology possible. This flexibility in networking allows users the mobility to access information and network resources as they attend classes, collaborate with other users, or move to other locations. It also allows the network itself to be readily moveable.

The objective in this research stream is to transform Management Information Systems design expertise and the experiences and knowledge gained from numerous GS classroom field studies into a set of successful strategies for using this software in a wireless computing classroom environment. The goal of this paper is to share experience in an innovative integration of technology and pedagogy with cross-curriculum applications in a completely portable and wireless environment. The results indicate strong support for the notion that this strategy provides a framework for collaborative learning resulting in increased efficiency and mobility.

---

* Research conducted by the University of Arizona, Center for the Management of Information, Management Information Systems Department, P.O. Box 210108, Tucson, AZ 85721-0108.

2. WIRELESS LOCAL AREA NETWORKS

A Wireless Local Area Network (WLAN) is an on-premise data communication system that reduces the need for wired connections and makes new applications possible, thereby adding new flexibility to networking. It is a flexible data communication system implemented as an extension to, or as an alternative for, a wired LAN within a building or campus. Mobile WLAN users can access information and network resources as they attend meetings or classes, collaborate with other users, or move to other campus locations. The benefits of WLANs extend beyond user mobility and productivity to enable portable LANs; with WLANs, the network itself is movable. The Center for the Management of Information wireless classroom uses Toshiba Pentium laptops with Netwave Technologies, Inc. CreditCard Netwave Adapters allowing the cordless connection of PCMCIA-capable laptops to a Netwave local area network (LAN). The system uses both fully cordless networking and cordless access to a cable-based network using the Netwave Access Point.

3. EDUCATIONAL APPLICATIONS OF GROUPSYSTEMS® SOFTWARE

GroupSystems® (GS) are general-purpose collaborative problem-solving tools that have been shown to increase project group productivity. This software has repeatedly demonstrated its effectiveness as a set of group problem-solving tools in the business world, substantially improving the productivity of face-to-face work groups. Problem-solving teams in the field have reduced their labor costs by an average of 50%, and have shortened the elapse time for their projects an average of 90% (Nunamaker et al., 1991; 1995). Wired classroom pilot results include increased student interest, participation, and learning; as well as improved efficiency in the use of collaborative learning techniques by teachers. The challenge is, as with the use of any new technology, how to best take advantage of its potentials. Through this research program, teachers-in-practice and researchers are discovering means to improve the use of collaborative learning strategies and increase both teacher efficiency and student learning. Data collection is accomplished through a variety of benchmarking software, time tests, user feedback, and observation.

GroupSystems® is typically based on a network of personal computers, usually one for each participant. Individuals interact through this network of PCs whenever and wherever they need to work together and collaborate: in class sessions, study group meetings, between classes, across the country, or around the world. Simultaneous input leverages student times and creates better quality ideas. Anonymous participation encourages insightful comments without fear of reprisals. Instant documentation automatically records every idea for distribution and future work. Process support facilitates movement from Point A to Point B in a group process. Access to information lets the group use outside data while working in GS programs.

4. RESEARCH RESULTS

Through the use of initial time tests, GroupSystems® was eliminated as a viable alternative, instead, the Internet-based program GroupSystems® Web (GS_{Web}) was evaluated. Internet programs need to be able to transfer compact data quickly and efficiently. Most Internet programs are designed to run over low bandwidths, and usually rely on modems with speeds ranging from 9600 bytes per second (bps) to 28,800 bps to communicate the data. These specifications proved much more viable in the wireless environment. GS_{Web} was specifically designed to run over the Internet via the World Wide Web, and requires use of an Internet Web Server to run. Web Servers are used to take requests from clients, run them through the program, and send the results back to the clients. The Java programming used in developing GS_{Web} requires the use of Netscape Navigator.

4.1. Wireless Use of GS_{Web}: Technical Results

Researchers at the Center for Management of Information (CMI) are directly involved in the development of GroupSystems® Web (GS_{Web}). This allowed an increased amount of freedom in testing the software. Programming problems could be reported and immediately corrected. Different configurations and strategies could be employed and adjusted with great speed. In addition, because the software was developed in-house, the program could be tested both hardwired into the CMI backbone and as a standalone network.

Initial testing was conducted with four wireless clients tied into the CMI backbone. The clients were connected using a wireless access point attached through an 8-port hub. The clients logged into "Circus," one of the two CMI web servers. Circus is a Gateway 2000, 100 Megahertz Pentium Computer with 32 MB of RAM running Windows NT 4.0 Server and the Netscape Internet Fast Track Web Server. Through Circus, the GS_{Web} program transferred data to the clients using the Netscape Navigator browser. All data was transferred via the TCP/IP protocol.

Four students were involved in pre-testing through the use of a series of GS_{Web} classroom exercises to see if the program would work as intended over the wireless network (i.e. identically to the way it works in a wired environment). Initial results were poor due to a constant refresh that occurred while running Netscape. A refresh in this scenario refers to the action taken by the program to constantly send messages to the other members of the network to check to see if any of the users have altered or added information. This is the element which allows for true simultaneous interaction. This constant refresh made it appear that GS_{Web} was running very slowly. This process puts a very large amount of traffic on the network causing congestion and thereby slowing transmission rates as well as using additional bandwidth. Through discussions with the developer, it was discovered that the polling rate could be reconfigured to a 25 second interval without affecting the clients; the problem was solved.

In addition to running the test tied to a backbone, Netscape Fast Track Internet Server and GS_{Web} were set up on a portable Windows NT 4.0 server so that testing could be done on a standalone network. Adequate memory to run Netscape Fast Track Internet Server (a minimum of 24 MB of RAM; 40 MB of RAM were used) is necessary for use on the wireless local area network (WLAN). In the laboratory user experiments, GS_{Web} operated inconsistently. Once a variety of technical issues were resolved, researchers were able to use all of the GS_{Web} tools with success. A variety of classroom exercises were completed by the student team including: discussion forums, voting, categorization, and sharing of voting results. Once these experiments were consistently successful, standalone WLAN classroom testing was possible.

4.2. Wireless Use of GS_{Web}: Classroom Results

Classroom testing was conducted using GS_{Web} with a graduate level Education class to elicit their impressions of the Software and Hardware experience and to gather standard course evaluation feedback. Although in prior tests a combination of several GS_{Web} tools were employed, to remain purposeful and of maximum benefit to the educational requirements of classroom use, these activities were more focused. The *Topic Commenter* tool was employed for this purpose.

Consistent with the results of previous testing of GroupSystems® software in a wired environment, the overall impressions of the students (most of whom are currently employed teachers) were very favorable. They reported a great level of enthusiasm and excitement concerning potential classroom applications, both as users and as instructors using the system to teach others. This was the first test of GS_{Web} in a standalone network environment of any kind (wired or wireless) as well as the first test of its use in a classroom situation.

5. CONCLUSIONS

This research provides a glimpse at the classroom of the future where advanced technology assists collaborative learning strategies resulting in important pedagogical advances. The wireless collaborative computing classroom can exist with positive results. Teachers can use WLANs and the Web as fully portable teaching tools even more easily than a flip chart and a marker. The technological components for the virtual office and classroom exist and the challenge for today is to logically integrate them. The keys to success lie in research and development of the core components of this workspace and the interfaces needed for their integration and coordination. Systems integration of the various components can provide an environment for total collaboration. Wireless technology can take this one large step beyond.

REFERENCES

1. Nunamaker, J.F., Dennis, A.R., Valacich, J.S., Vogel, D.R. and George, J.F. (1991). Electronic Meetings to Support Group Work. *Communications of the ACM*, 34(7):40-61.
2. Nunamaker, J.F., Briggs, R.O. and Mittleman, D. (1995). Electronic Meeting Systems: Ten Years of Lessons Learned. In: Coleman and Khanna (Eds.) *Groupware: Technology and Applications*. Prentice-Hall.

Evaluating effects of self-directed ergonomics training

Barbara G. F. Cohen
Leader, National Office Safety and Health Team
Internal Revenue Service
Department of Treasury
Washington, D.C., U.S.A.

An evaluation of a written ergonomics training course, The Practical Guide to Ergonomics*, was conducted to assess the effects of self-directed learning in 1) comprehension and retention of material and 2) daily application of ergonomic principles as a result of the course book. A pilot group comprised of 23 data clerks and first line managers completed a written questionnaire immediately before taking the course. Three months later the questionnaire was again administered to 61% of the pilot group. Pre and post ergonomic observations were also conducted in addition to group interviews. Findings indicate that trainees acquired and retained considerable ergonomics information through the self-directed course. Moreover, a trend of reduction of fatigue and of the visual system complaints was observed. The evaluation concluded that this self-directed course is a viable, cost effective effective, and an economic training instrument.

1. INTRODUCTION

Ergonomics training as an intervention to prevent or mitigate musculoskeletal, visual, and psychosocial strain has been demonstrated in workshop and classroom settings (Dainoff, 1995; Dainoff et al, 1995; Cohen et al, 1996). An effective ergonomic training intervention was conducted in classroom/hands-on workshop style at a field site of a large agency as part of a larger international study (Dainoff, 1995). Success of this intervention was attributed to focusing on achieving healthy work postures and workstation adjustments, and motivating trainees to routinely apply that learning. It was determined that all employees would benefit by similar training.

---

* The course entitled, The Practical Guide to Ergonomics was developed for IRS by Marilyn Dainoff, CPE, and Marvin J. Dainoff, Ph.D., CPE.

As budget constraints necessitated a wider-reaching and less costly ergonomic training vehicle to address the organization's many sizable groups throughout the country, The Practical Guide to Ergonomics was developed as a stand-alone, self-directed instructionmodule. This manual was based on the aforementioned interactive training. The aims were to provide the basis for achievement of healthy working postures, application of sound ergonomic principles, and the motivation to alter one's "non-ergonomic" working habits.

The purpose of the present evaluation study was to measure the effectiveness of the ergonomics course book for practicality of widespread use for an agency of over 100,000 employees. Although individual styles of learning, e.g., visual, auditory, kinesthetic, etc., may be better served via multi-method training techniques, most computer operators are acclimated to visual learning. Thus it was expected that a well-written, illustrated manual would benefit most trainees.

2. METHOD

2.1 Subjects

Twenty-three volunteers, comprised of data clerks and their first-line managers, participated in the study. Daily tasks included intensive data input via computers, particularly for the clerks. All operators and managers were required to be knowledgeable about yearly-changing tax regulations. Tenure at their current jobs ranged from 8 1/2 to 29 1/2 years. All had completed high school and almost half the group had taken some college work but did not have a degree. None of the participants had ever participated in any previous ergonomics training.

Sixty-one percent of the original group completed the post test and participated in all the post training evaluation measures. The remaining 39% were either in mandatory training for the next filing season or absent at the scheduled post-test evaluation time. Closely matching the original group in all aspects, the post-test group was comprised of both first-line managers and employees. Their job tenure ranged from 8 1/2 to 23 years, and their educational level was also equivalent.

2.2 Procedures

Assessment was conducted via the following four methods:

(A) Anonymous written survey questionnaires were administered before the training and repeated three months afterwards. They were designed to elicit demographic information, self-assessment of knowledge of ergonomic information, and self reports of respondents' well-being during the workday (e.g., fatigue, musculoskeletal and visual status).

(B) Group discussions were conducted regarding comprehensibility after reading each chapter. Participants were prompted to respond to questions about their understanding of the chapter contents and encouraged to present any specific difficulties or comments that may have occurred to them during the readings.

(C) Observations of work conditions by an ergonomist via a walk-through of the work

Effects of Ergonomic Training as an Ergonomic Intervention

Marilyn Hecht Dainoff[a] and Marvin J. Dainoff[b]

[a]Marvin Dainoff, Associates, Inc., Cincinnati, OH 45231

[b]Center for Ergonomic Research, Department of Psychology, Miami University, Oxford, OH 45056

1. INTRODUCTION

The MEPS project "Musculoskeletal, Eyestrain, Psychosocial Stress" represents an unprecedented example of international multidisciplinary cooperation and coordination, the objective of which is to examine the effects of various kinds of ergonomic interventions, including corrective lenses, on a combination of musculoskeletal, postural, and psychosocial outcomes. These studies have been conducted in several different countries. Each country has utilized the same standardized research protocol, but ergonomic interventions are individually designed.

Preliminary results of this research have been presented at several international conferences. In this paper, the focus will be restricted to the training portion of the U.S. component of this study.

2. METHOD

2.1. Design

The basic research design, repeated in each country, consisted of four components: Pre-Test, Intervention, One Month Post-Test, and One Year Post-Test. A research protocol was developed which was repeated for each test component. The protocol consisted of a series of standardized measures administered to each of the participating subjects.

2.2. Research Protocol

The protocol was composed of standardized measurements of musculoskeletal load, demographic and psychosocial factors, medical and optometric status, and ergonomic factors. Musculoskeletal load was assessed using a computerized field-portable apparatus (Physiometer) which allowed simultaneous measurement of EMG of right and left trapezius, and postural angles of head, neck, and trunk. These measurements were carried out during a

40 minute period of data entry at the worksite. Ergonomic analyses of the workstation included workstation dimensions and luminance/illuminance measures. Demographic and psychosocial questions—including attitudes toward work, family and economic situation, and symptoms of psychological stress—were obtained by a combination of interview and questionnaire. A similar approach was used to obtain subjective ergonomic assessments of subjects' workstations. Finally, each subject received professional medical and optometric examinations.

2.3. Intervention

The U.S. intervention which was carried out at the Cincinnati Service Center of the Internal Revenue Service, consisted of three components. A group of 26 female data entry employees participated in the study. Their average age was 41.02; the mean number of years working at IRS was 15.71.

2.3.1 Corrective lense component

An optometric assessment of the visual function of each participant with respect to the particular visual demands of the workplace as carried out. When needed, appropriate corrective lenses were prescribed according to specific optometric criteria.

2.3.2 Workstation redesign component

A complete redesign of each workstation was carried out so as to provide an optimum ergonomic workplace. The redesign included the following elements: (1) ergonomic chairs which all had height, seatpan angle (including forward tilt), and backrest angle adjustment; (2) motorized adjustable worksurfaces which allowed operators push-button control of the worksurface height from a sitting to a standing posture and provided ample storage areas; (3) fully adjustable divided keyboards with interchangeable sections; (4) height and angle-adjustable copyholders specially designed to accommodate a stack of documents 3 inches thick; (5) customized monitor supports; and (6) adjustable footstools.

2.3.3 Training component

Training for the volunteers took place in several stages. First, an intensive, participative classroom course on ergonomics was presented to introduce the concept of ergonomics and how it could be of benefit at work, at home, and in other aspects of life. Topics covered included anthropometry, seated posture, including physiology of the spine, how to sit and why; light and glare; position of monitor and copy; arm/wrist anatomy and proper posture; proper keying and writing; analyzing and modifying the workspace, including link analysis; dealing with stress; interrelationship of health on the job and at home. The training course was interactive and involved exercises, examples both from work and outside work, and discussions. Before arrival of the new equipment, participants were encouraged to make as many ergonomic changes as possible using their old equipment.

A part of the training involved empowerment, as the participants had had very few possible choices in the arrangement and use of their old equipment—and no training in those that were possible (such as adjustment of their chairs).

The second stage involved hands-on training with their new equipment as it arrived. Most of this process was straightforward, with the exception of the alternative keyboards. The adjustable keyboards had a "standard" key layout, and the keyboard company was willing to reprogram the keys to be compatible with the older, specialized, non-standard keyboard in use at the IRS. This involved not only different physical placement of the keys, but also special codes. An iterative process was used for mapping the keys, with the users literally trying out the keyboards and then discussing and voting; and it took several iterations until a satisfactory setup was achieved. The keyboards also presented a problem of touch, as they required only a small amount of pressure to function, in contrast to the old keyboard which the workers pounded on, partly to get auditory feedback that the key had registered.

As the volunteers became used to their equipment, the training ergonomist became a "coach," periodically walking from workstation to workstation, reminding the volunteers of the principles they had learned and providing feedback as to how they were doing. The coaching continued intermittently throughout the project, with the coach asking questions about how people felt and giving suggestions about how they could be more comfortable.

A third method of training involved a newsletter written for volunteers, which included special instructions, updates, humor, and discussion of the ergonomic issues which were common to many volunteers. These and other written materials were kept by each volunteer in a special notebook distributed for the purpose.

3. RESULTS

The three components of the intervention were carried out simultaneously and were functionally interdependent. That is, effectiveness of the ergonomic equipment and, in fact, motivation to use new eyeglasses, are necessarily related to the extent to which training was or was not effective. Accordingly, the analyses will consist of a set of univariate tests on a set of critical outcome variables across the time periods corresponding to the point at which measurements were taken. The null hypothesis is that there is no difference across time periods. The alternative hypothesis predicts a positive impact from the intervention over the period from the pre-intervention baseline assessment (Commencement) to the 30 day post-test, but no differences between the 30 day and one year post-test.

3.1. Confirmation of hypothesis

Results indicate that the prediction of improvement from commencement to 30 days post-test, but no change from 30 days post-test to one year post-test was statistically confirmed for each of the following:

Physical Examination. Critical items from the physical examination reflecting objective determination of specific signs or precursors of a musculoskeletal disorder included: (1) the number of painful pressure or trigger points, (2) number of positive signs of disorder from clinical examination of the shoulder joint, (3) number of positive signs of disorder from clinical examination of neck mobility.

Musculoskeletal Pain. Participants reported their average intensity of pain or discomfort experienced in the following regions of the body: neck, shoulder, forearm/hand, back, and legs.

Static Load Assessment. Direct observations were made of the extent to which each participant's working posture were likely to result in high static loads.

Visual Problems. Participants reported the extent of the following visual problems: fatigue, burning/itching, red eyes, or double/hazy vision.

Ergonomic Evaluation. Participants were asked for their subjective evaluations of chair comfort and height adjustability of the keyboard support surface.

3.2. Exceptions

Postural Angles. During the work samples, angle sensors were used to determine head, arm, and trunk flexion. Results indicated that pattern of improvements in average flexion angle for head and trunk conformed to the predictions described above. On the other hand, the results are opposite to expectation for shoulder flexion. In this case, there was an increase in average flexion angle from commencement to the 30 day post-test. We attributed the increased flexion to an appropriate and reasonable postural readjustment to the alternative keyboard. This interpretation is based on the positive objective and subjective findings regarding upper arm pain during this period.

Psychosocial Variables. There was no change in any of the psychosocial variables across any of the time periods throughout the study.

4. CONCLUSION

These results indicate a consistent pattern of improvements in musculoskeletal and visual outcome indicators resulting from an integrated ergonomic-visual intervention of which training was a key component. Moreover, these levels of improvement were maintained over a prolonged period of time (peak tax season) during which there was no contact between the trainers and the participants and during which task demands were intense. The lack of change in psychosocial variables would seem to indicate that the intervention had its effect primarily on musculoskeletal and visual indicators. The results support the notion that visual and musculoskeletal interventions should be combined, and that effective training is an essential element.

A model for ergonomics training evaluation

M. Robertson

Institute of Safety and Systems Management
University of Southern California
Los Angeles, California USA 90089-0021

1. INTRODUCTION

A systematic evaluation process should always be incorporated into the design of training programs. Linking the evaluation process to the objectives of the training program, as well as to the identified organizational needs can establish the benefits of the program on the individual and organizational levels. At the individual level, these benefits may include an increase in workers' knowledge and skills in understanding the application of ergonomic principles to workstation set-up, and use of proper postures. Benefits to the organization can include: reduced worker compensation cases, decreased musculoskeletal disorders and reduction in lost work days.

Since research has shown the associated negative health effects, specifically work related musculoskeletal disorders (WMSDs), with the introduction of VDTs into the workplace (e.g., Sauter, Dainoff, & Smith, 1990) several ergonomic intervention strategies have been proposed and implemented. These strategies include: ergonomic training, re-design of workspaces, environmental re-design, bio-mechanics and physiological interventions (work-rest cycles); enhancement of user control, job redesign, and organizational re-design. All of these intervention tactics are critical components that must be applied within a systems perspective in order to effectively minimize the negative health effects arising from the intensive use of VDTs (e.g., Robertson & Rahimi, 1990; Robertson and O'Neill, 1995; Bammer, 1993; Hendrick, 1994).

This paper focuses on the importance of ergonomics training programs for reducing the negative health effects from VDT work as well as present a model for evaluating the effectiveness of an ergonomic training program. Several organizational examples of the application of the training evaluation model are presented.

1.1. Importance of ergonomic training programs

The importance of ergonomic training programs has been demonstrated by several researchers and practitioners (e.g., Verbeek, 1991; Dortch & Trombly, 1990; Green and Briggs, 1989). To have a successful ergonomic training program, it should be incorporated into the overall organizational strategic plan for health and stress reduction. This establishes the importance of such an effort, company wide. It also is important to establish an evaluation plan to track the effectiveness of the ergonomic training program that is linked to organizational performance measures.

While engineering controls such as workstation re-design or the use of adjustable furniture are frequently suggested (Verbeek, 1991), administrative controls such as training must also be used so that employees and management understand the need to change work habits when using office technology. In addition, training can assist in ensuring that both managers and employees fully understand and participate in the ergonomics program. Training also helps employees to understand workstation set-up and the use of proper postures to avoid discomfort and WMSDs (Gross & Fuchs, 1990; Kukkonen, et al., 1983; Robinson, 1991).

Without training these concepts, the presence of other administrative and engineering controls will result in only limited success. For instance, Green & Briggs (1989) found the availability of adjustable furniture alone did not prevent the onset of overuse injury in some users. They found workers who suffered injuries, worked with the same equipment as non-sufferers, but that sufferers expressed greater dissatisfaction with the new furniture as well as greater discomfort. Workplace and anthropometric measurements revealed that sufferers were adopting more awkward postures than non-sufferers, which the researcher attributed to inadequate knowledge regarding the proper adjustment of their workstations. This ignorance existed despite an abundance of printed literature, leading the researchers to suggest the need for verbal training on furniture adjustment as the solution. Verbeek (1991) implemented a training program for use of adjustable furniture, and found that it significantly improved user posture when compared with an untrained group. Kukkonen et al. (1983) also were able to reduce neck and shoulder tension through education of workers who were found to be unfamiliar with the proper adjustment of their furniture and, as a consequence, were using poor working positions. Their lectures also included information on the foundation of ergonomics, covering physiology, work habits and relaxation methods.

Training of managers and supervisors also is necessary in order to provide a responsive environment in which employees are encouraged to utilize their training through reinforcement and reward. Since supervisors have more influence on the daily performance of individual employees, their participation in the training process is essential for the success of the ergonomic training program. Smith & Smith (1984) found that supervisors responded best to training that emphasized situations over which they had some measure of control, and that such training made them more cooperative and supportive of change.

2. ERGONOMIC TRAINING EVALUATION MODEL

In this section, we present a multiple measure, four level training evaluation model. This model is based on a training evaluation model proposed by Kirkpatrick (1975) and Goldstein (1986). The success measurements used at each level of the evaluation should be developed during the design phase of the training program. This is necessary so that the training objectives, and the measures of success for the program, are closely linked.

2.1. Four level training evaluation model

The model includes four levels (see Table 1.)

Table 1.

Level I	Reaction to the training program.
Level II	The Learning of principles, facts, techniques, and attitudes.
Level III	Behavior relevant to job performance.
Level IV	Results of the training program related to organizational objectives.

Measurements that may be taken at each of these training evaluation levels are as follows: Level I: post-training questionnaire asking the trainee to evaluate the value and usefulness of the training; Level II: pre-post questionnaires/tests assessing how well the trainee learned the information taught as well as observations/interviews with the trainee; Level III: assessment of the trainees' behavior on the job--how well was the trainee able to transfer the knowledge and skills to the job--this may be completed by observations and/or interviews; and Level IV: results and impacts of the training program on organizational performance measures, benchmarking and tracking of organizational performance measures.

To evaluate an ergonomic training program, we suggest incorporating some of the following measurements at each of the evaluation levels. Level I measurements may include questions on the post-training questionnaire regarding the usefulness and value of the training as well as the relevancy of the training to the workplace. For Level II, various measurements may be taken prior to the training such as an observational analysis of posture and work habits, postural discomfort surveys, and/or a paper and pencil pre-knowledge VDT ergonomic training test. Other measures that can be used in Level II include: post-training discomfort surveys identical to the pre-training survey measures, and/or a post-training ergonomic knowledge test. Level I and II may also serve as a type of formative evaluation of the instructional materials indicating if the training program was well received and the training materials were clear and understandable and the training objectives were met. For Level III, various measurements may be taken after the ergonomic training in order to evaluate the transfer of the training by the trainee to the workplace. These may include: an observational ergonomic analysis and interviews of what the trainee changed in their workplace as a result of the training. Additionally, within the post-training ergonomic knowledge test, open ended questions may be asked of the trainee regarding how they are going to use the training when they return to work. For Level IV, various organizational performance measurements may be taken related to the results of intensive VDT work, such, such as reported CTDs and WMSDs, time off work, worker's compensation rates, and health and stress related costs. Pre- and post-training organizational health and safety performance measurements should be tracked over time. Tracking these pre- and post- health and safety performance measures, as well as other training costs, determines the basic variables for calculating a Return on Investment (ROI) for the training program.

3. APPLICATION OF THE TRAINING EVALUATION MODEL FOR VDT ERGONOMIC TRAINING

In this section, we present examples of how two companies successfully applied this four level training evaluation model and used various measurements of each level.

3.1. Background of the VDT ergonomic training program

Two companies designed, developed, implemented and evaluated a VDT ergonomic training program to reduce adverse health effects from VDT work. The first case involved a large telecommunication company where telephone information operators, customer service representatives and supervisors were trained. The second case involved engineers working on computer aided design systems in a large aerospace company.

3.2. VDT ergonomic course content

The content area of these VDT ergonomic training programs included: definition of ergonomics, basic physiology of the upper extremities, causes of discomforts and injuries, ergonomic principles regarding workstation layout, recommendations for analyzing the employees workstation, what to do when they feel uncomfortable (management's policies on who to contact), and relaxation and exercise techniques to relieve VDT stress.

3.3 VDT ergonomic training evaluation results

Overall, positive effects of the ergonomic VDT training for both companies were demonstrated. Since these companies conducted all four levels of training evaluation, clearly documented results of the training program was accomplished. These results demonstrated the success of these VDT training programs.

For Level I evaluation, the trainees from both companies rated the training highly favorable, useful and informative, and relevant to the job. For Level II evaluation, there were significant changes in the amount of knowledge gained by the trainee's concerning VDT ergonomics as measured by self-reporting knowledge gained scores for company 1, and by pre and post training scores for company two.

For Level III evaluation, significant positive behavioral changes of the trainees' as measured by self-reported behavioral changes and observed changes were found and reported for both companies. For company 1, over 80% of the trainees reported that they had applied the ergonomic knowledge to their jobs. Areas to which they had applied the ergonomic knowledge were on the correct placement of the VDT screen, the position of their wrists at the workstation, and their sitting posture. Follow-up observations and interviews by the corporate ergonomist confirmed these self-reported behavioral changes. For company 2 it was found that the experimental group (trained CAD engineers) was able to significantly reduce time spent in awkward postures. In addition, a significant decrease in overall discomfort was found in the experimental group. The control group did not change significantly during this period. Follow-up interviews and observations by the corporate ergonomist supported these results, as over 80% of the trainees said that they were able to apply many of the principles taught in class to their workplace. Of the changes to the workplace reported, most were adjustments to the chair or the placement of the monitor. Many of the participants reported that the awareness that came from training led to changes in posture and an increase in the number of breaks for exercises or movement. Of the reasons given for not making desired changes, the constraints imposed to the work environment was the most frequent response. In both companies, there was some ability to change the workstation configuration, within some defined constraints, which provided some means of user control at the individual level. Additionally, for Company 1 ergonomically designed furniture and workstations were provided at all locations.

For Level IV training evaluation, a Return on Investment (ROI) analysis for both of the VDT training programs revealed that the programs resulted in an acceptable payback for the individuals trained and their companies. A significant decrease in work related musculoskeletal disorders and overall loss work days for company 1 was demonstrated. As for company 2, two potential RSI cases were curtailed after the individuals adopted the VDT ergonomic principles to their workstations and job routines.

Other successful components of the training were the commitment by top management to the ergonomic program itself, active involvement of the employees, positive response by management to employee's requests regarding VDT workstation redesign or reconfiguration, and continuous support of management in applying the VDT ergonomic principles to the work environment. Furthermore, an essential part of sustaining the results of a successful ergonomic training program is training the supervisors so that they know what employee behaviors to reinforce. This was effectively completed in company 1 as part of the overall strategic plan to address health issues associated with VDT work.

4. CONCLUSION

Applying a systematic evaluation process to demonstrate the effectiveness, benefits and positive successes of an ergonomic training program is essential. This paper presented a systematic, four level, training evaluation model which was used by two organizations in evaluating their VDT ergonomic training program. Positive effects of the VDT ergonomic training programs were demonstrated at each evaluation level. Overall, the value of ergonomics training programs at the individual and organizational level in preventing WMSDs and adverse health effects from VDTs have been shown. Furthermore, coupling an effective training program with other work organizational factors in a systematic approach can alleviate WMSDs associated with VDT work.

REFERENCES

1. Bammer, G. (1993). Work-related neck and upper limb disorders - social, organisational, biomechanical and medical aspects. In L.A. Gontijo, and J. de Souza (Eds.), Segundo Congresso Latino Americano e Sexto Seminario Brasileiro do Ergonomia (pp. 23-38). Florianopolis: Ministerio do Trabalho Fundacentro/SC.

2. Bell, J.D. & Kerr, D.L. (1987). Measuring training results: Key to managerial commitment. Training and Development Journal, 41, (1), 70-73.

3. Dortch, H.L., III & Trombly, C.A., (1990). The effects of education on hand use with industrial workers in repetitive jobs. American Journal of Occupational Therapy, 44(9), 777-782.

4. Gagne, R.M., Briggs, L.J., & Wager, W.W. (1988). Principles of instructional design (3rd ed.). New York: Holt, Rinehart and Winston.

5. Goggins, R. & Robertson, M. (1994). A systematic evaluation of a VDT ergonomic training program. Proceedings of the 38th Annual Human Factors Meeting, 192, Santa Moncia, California.
6.Green, R.A. & Briggs, C.A., (1989). Effect of overuse injury and the importance of training on the use of adjustable workstations by keyboard operators. Journal of Occupational Medicine, 31(6), 557-562.
7.Gross, C.M. & Fuchs, A., (1990). Reduce musculoskeletal injuries with corporate ergonomics program. Occupational Health and Safety (1), 29-33.
8. Hendrick, H.W. (1994). Macroergonomics as a preventative strategy in occupational health. In G.E. Bradley and H.W. Hendrick (Eds.). Human Factors in Organizational Design and Management-IV. Amsterdam: North-Holland, pp. 713-718.
9.Kirkpatrick, D. (1975). Techniques for evaluating training programs. In D.L. Kirkpatrick (Ed.), Evaluating Training Programs (pp. 1-17). Madison, WI: American Society for Training and Development.
10.Kukkonen, R., Luopajarvi, T., & Riihimaki, V., (1983). Prevention of fatigue amongst data entry operators. In T.O. Kvalseth (Ed.) Ergonomics of Workstation Design. Butterworths.
11. Kroner, W., Stark-Martin, J.A., and Willemain, T. (1992). Using advanced office technology to increase productivity. The Center for Architectural Research, Rensselaer Polytechnic Institute, Troy, NY.
12. Luopajarvi, T., 91987). Workers' education. Ergonomics, 30(2), 305-311.
13. Robertson, M.M. & O'Neill, M.J. (1994). A systems analysis for integrating macroergonomic research into office and organizational planning. In A. Grieco, G. Molteni, E. Occhipinti, & B. Piccoli. (Eds.), *Work With Display Units, Selected proceedings book for WWDU'94,* (pp. A6-A7), Elsevier, North Holland.
14. Robertson, M. & Rahimi, M. (1990). A systems analysis for implementing video display terminals. IEEE Transactions Engineering Management, 37 (1), 55-61.
15. Robinson, M. A. (1991). Ergonomic modernization of an office environment and its effect on physical discomforts, job satisfaction, work environment satisfaction, and job performance. Unpublished thesis, University of Southern California. LA, CA.
16. Sauter, S.L., Dainoff, M.J., & Smith, M.J., (1990) Promoting Health and Productivity in the Computerized Office, London: Taylor & Francis
17. Verbeek, J., (1991). The use of adjustable furniture: Evaluation of an instruction program for office workers. Applied Ergonomics, 22(3), 179-184.

Ergonomics Training in Industrially Developing Countries : Case Studies From "Roving Seminars"

P A Scott[a] and H Shahnavaz[b]

[a]Department of Human Movement Studies, Rhodes University, P.O. Box 94, Grahamstown, South Africa

[b]Department of Human Work Science, Lulea University of Technology, 97184 Lulea, Sweden

INTRODUCTION

In Industrially Developing Countries (IDC's) which are striving to acquire and implement advanced and complex technologies, it is essential to establish an awareness of the principles and benefits of Ergonomics. This is best achieved through education and training of workers operating in IDC's, which in turn should ensure efficient utilisation of the transferred technology resulting in a more cost effective production line. Further more, a knowledge of Ergonomics should serve to establish a more congenial working ambience and safeguard the workforce from the multitude of hazards which are so evident in most IDC's. However, due to a lack of awareness of the potential benefits of Ergonomic intervention, the shortages of Ergonomic resources in both trained personnel and material, plus our failure as a professional body involved in ergonomics research and application, there is very little known on the subject in many parts of the world. The result being that Ergonomics is in its infancy in most Industrially Developing Countries.[1]

IDC's which have substantial problems within their work sites are particularly in great need of Ergonomics in terms of both theoretical knowledge and practical application. Ergonomics can help to create the correct interface and harmonious interdependence between technology, human operators and the prevailing environment. It should be evident that the key issue overcoming the existing problems regarding the high rate of accidents, injuries and inefficiencies within IDC's work sites is a basic knowledge of Ergonomics adapted to the specific needs of the industry, while taking cognizance of the overall state of development within the country. In developing countries one needs to consider the specific socio-cultural characteristics of the workers and local working conditions; Ergonomic principles can then be adjusted to the particular needs, resources and technological level and infrastructure of each country.

1.1 Ergonomics Education

Formal Ergonomics educational programmes leading to a degree are essential in order to produce qualified Ergonomists with the required competence for teaching and researching at all levels. Unfortunately courses of this nature are available at very few universities within IDC's. According to the information published by the International Ergonomics Association (IEA) in 1994, out of 235 educational programmes that are offered world wide, only 10 programmes (4%) are offered in IDC's. While courses offered in Industrialized Countries (IC) are not only often inaccessible to the majority of students from IDC's, they are also not entirely suitable, because they do not consider the specific conditions and requirements of developing countries. Further more, foreign students often do not return to their home country after completing their university education in the IC's. However, due to the good work being conducted by many truly international Ergonomists who often give freely of their time, the theoretical principles of Ergonomics are being spread on the global front, and fortunately there are a growing number of universities in IDC's which are beginning to introduce Ergonomics in their educational programmes, emphasising specific IDC's Ergonomic needs.

1.2 Ergonomics Training

Besides formal educational programmes there is a need to establish an awareness and appreciation for Ergonomics at all levels of workers, within all work environments, small and large. This can best be achieved by running short, practical courses. Short-term training courses at the workplace, organised for a cross section of the work force is of particular importance. The European Community, for example emphasised training as the key element in the development of safe and proper working condition and has created legal instruments for its member states. Small group training, mainly short-term, on-site and action oriented, emphasising learning through supervised involvement in applied situations considering local problems and solutions, are prerequisite for an effective training programme.[2]

1.3 "Roving Seminars in Southern Africa"

The International Ergonomics Association (IEA) is a cosmopolitan body striving to establish a global understanding of Ergonomics and a particular thrust has been to reach under developed countries. In an attempt to fulfil this goal, a sub-committee of Ergonomics in Industrially Developing Countries has been established and over the last few years several seminars and workshops have been conducted in various developing countries. In 1996 the focus was on Southern Africa.

1.4 Objectives

The main objective of the Roving Seminars was to promote an awareness of the theoretical principles and practical applications of Ergonomics; the aim being to enhance the safety and well-being of the worker and to improve the quality and quantity of productivity. However, in order to improve working conditions, which in many situations are exceptionally poor, one has to teach employers and employees how to continually evaluate the general situation. This must be followed with a specific work site analysis where specific problems need be identified and prioritised in order to establish appropriate intervention strategies.

1.5 Personnel

As information about the seminars was distributed to the various centers throughout Southern Africa, there were several queries as to who should attend the courses. It was necessary to convey the message that improvement of work conditions, be it environmentally orientated, physical demands, task specific or workstation design/redesign, is the responsibility of **all** involved in the particular work site. It is essential that a cross section of all workers, at all levels, involved in a particular firm, gain insights into the benefits of running an industry on sound Ergonomics principles. Attendance at the five seminars conducted in June/July 1996 included managers, floor workers, occupational health nurses and physiotherapists, doctors, engineers, supervisors, industrial psychologists and academics. Courses ranged from one to four days, with 16 to 26 people attending each seminar.

2. METHOD

The main material used was the recently published "Ergonomic Checkpoints" (ILO, 1996)[3], with all participants having their own copy to work with during the course, and then to take back to their own work setting to use *in situ* on an on-going basis.

After a general introductory talk on Ergonomics, plus an outline of the principles and format of the course, the basic concepts of the Checkpoints were explained to the group as a whole. All participants were then divided into smaller groups and each group was given specific sections of the Checkpoints to discuss amongst themselves. The course leaders moved round to various groups encouraging full participation from all delegates and sorting out any queries or problems which may have arisen from the group discussions.

Having worked through the Checkpoints (often with lively discussions), factory visits were arranged for the delegates to experience the practical application of both the guidelines within the Checkpoints plus personal suggestions. Still in groups of four to five people the participants walked through various factories (motor, paint and food industries, and mining sites were visited). With the Checkpoints to guide them, the delegates were required to identify problem areas and to offer possible solutions. The time spent within the different factories was from two to four hours.

Once back at the seminar venue, the various groups not only presented their findings of the specific site visited, but were encouraged to bring in examples from their own working experience. The emphasis was very much on delegate participation with course leaders only involved when there was a problem, or in the summing up of the findings of the groups. In this way a thorough working knowledge of all 128 Checkpoints was established with most people enthusiastically participating.

2.1 Evaluation

In order to assess the success of the Roving Seminars and to identify possible strengths and weaknesses as perceived by the participants, an evaluation form was handed out at the completion of the workshop. The immediate evaluation was based on five categories; materials, lectures, procedures, participation and overall assessment. Responses were indicated on a five point scale: 1 very poor; 2 poor; 3 average; 4 good; 5 very good. A follow-up questionnaire was sent out six months after the completion of the course; this comprised of 20 questions on the benefits of the course, need for further workshops, examples of specific application, responses of managers and co-workers at their work site, plus open suggestions. It was considered necessary to evaluate the success/failure of the training programme from two perspectives i.e. of participants themselves as well as their perception of the receptiveness of management and colleagues within their organisations.

3. RESULTS

While over 100 delegates attended the five seminars, questionnaires were only given to the 52 people who participated in the three day workshops. All 52 delegate completed the initial questionnaire, while 29 (56%) responded to the follow-up questions. In both questionnaires the responses are presented as a percentage of the total number of responses.

Table 1
Initial Evaluation : percentage of responses "very good" or "good".

Area	Responses (%)
Materials Presented	94
Lectures	96
Procedures	76
Participation	95
Overall Picture	95

Table 2
Follow-up Questionnaire: percentage of responses in top two categories.

Area	Responses (%)
Personal Response	
Workshop beneficial	100
Need for follow-up workshop	100
Improved Awareness	100
Felt more knowledgeable	100
Felt more capable	79
Perception of Colleagues' Response	
Co-operation of co-workers	98
No resentment	100
Management interest	68
Management concern	41
Management more receptive	71

It is evident from the above tables in which the responses of delegates have been summarised, that the Ergonomics workshops conducted in Southern Africa were a great success. Not only did they serve to establish a general awareness of Ergonomics, but equipped with something tangible to work from viz the Ergonomics Checkpoints, the delegates felt more confident in being able to identify problem areas and to offer solutions, so implementing intervention strategies to improve working conditions. Several respondents reported that they had organised an Ergonomics Committee within their industry and this had resulted in a team effort to resolve problems. Others, although no formal group had been formed, commented on more open communication and greater co-operation across hierarchial levels.

Two specific adjustments to work-station design were identified: i) the simple raising of a working area to reduce the forward lean of workers thus reducing the stresses on the back of six operators working at that station; ii) the other was to reposition a machine which resulted in less physical effort from two workers and an overall improvement in workflow.

However, while floor workers and even supervisors appear to have learnt a great deal at the workshops and have succeeded in implementing immediate and effective minor Ergonomic interventions, it is evident that more effort needs to be made to get managerial involvement, or at least to advise workers as to how to get more support and commitment from management. It is clear from the responses indicated in Table 2, as well as from personal experience, that a greater effort needs to be made to convey the benefits of Ergonomics to management and to establish a greater involvement and commitment from them. A further request from the participants of the workshops was for more specific practical examples of common problems and possible solutions.

4. CONCLUSIONS

Ergonomics training should be part of regular job training. The best method of providing training is in the form of conceptual information and practically implementation, describing the required procedural "know- how" for intervention strategies.

Through regular Ergonomics discussions, groups are able to increase employee's ability to work more safely and to improve their own working condition through personal involvement and responsibility. This is achieved through the exchange of experiences and ideas with fellow employees, thus establishing an ambience of "co-operative co-responsibility"[4]. In so doing one will establish an on-going continuous improvement process to be more efficient and reliable, while at the same time being less stressed and less likely to be injured, creating an optimal working environment. This in turn should improve productivity, in all probability with less effort. In 1996 Scott stated that any industry run on sound Ergonomic principles will not only improve working conditions, but should also result in an increase in productivity (most likely with less effort); the result being that each worker, industry and the Nation as a whole must benefit[5].

5. REFERENCES

1. H. Shahnavaz, The Ergonomics Society Lecture (1995): Making ergonomics a world-wide concept. Ergonomics, Vol.39 (12) 1391-1403 (1996).

2. H. Shahnavaz, What can we do to help ergonomics practice in the industrially developing countries? Advances in Applied Ergonomics, USA Publishing, 1006-1010 (1996).

3. Ergonomics Checkpoints, International Labour Office (ILO) Geneva, Switzerland (1996).

4. P.A. Scott, Compatibility and cooperative coresponsibility. In: Human Factors in Organizational Design and Management - V (ed.) O. Brown Jr and H.W. Hendrick. New York, North-Holland, (1996).

5. P.A. Scott, Editorial: Ergonomics SA, Vol.8 (1) 1 (1996).

Comparative Hypertext Approaches to Ergonomic Training

Wei Xu and Marvin J. Dainoff

Center for Ergonomic Research, Department of Psychology, Miami University, Oxford, OH, 45056, USA

Training users to utilize ergonomic information effectively in workplace design requires the application and integration of multiple sources of information (e.g., interdependent criteria for furniture adjustability and keyboard/display location). To aid users in coping with such a complex task, the use of hypertext representation of complex documents is increasing. However, when searching in a complex hypertext application, users may experience navigation disorientation and cognitive overload. This study examined the relative effectiveness of a semantic relationship for a hypertext interface, and the relationship between semantic representations and complexity of search task. The goal was to design a more effective hypertext interface for complex documents in support of users' problem solving.

1. BACKGROUND

1.1 Navigation in hypertext

Hypertext has widely been used to facilitate retrieval of information in computer systems. However, navigation disorientation in searching hypertext documents is a common difficulty. In attempting to aid hypertext users to navigate more effectively, various linking structures have been employed. Mohageg (1992) divided hypertext links into two categories: organizational and relational. Organizational links are based on more conventional schemes such as hierarchical structures (i.e., tables of contents), whereas relational links tend to reflect the underlying semantic relationships among hypertext elements (nodes). Some researchers have advocated emphasizing semantic navigation and shifting from "making navigable" to "making hypertext understandable" (McKnight, Dillon, and Richardson, 1991).

However, researchers have also appreciated the difficulty of designing a semantic representation for a hypertext interface. There are two challenges to be addressed. First, we need to find a more understandable and appropriate semantic representation to support users' information searching. Second, we also need to consider the relationship between users' search tasks and semantic representations, because the effectiveness of a semantic representation might depend on the search task. Search tasks in hypertext may vary with different complexity levels. Different search tasks may factor different types of semantic representations. Little attention has been paid to the relationship between users' search tasks and the semantic representations for hypertext interfaces.

1.2 Ecological Interface Design.

The current study examines the applications of an alternative strategy, an ecological interface design (EID) approach, in hypertext interface design. The EID approach has been proposed as a framework for the design of interfaces for complex human-machine systems (Vicente and Rasmussen, 1990).

EID is mainly based on Rasmussen's abstraction hierarchy (AH) approach as a multilevel knowledge representation framework or a work domain. (See Rasmussen, Pejtersen, and Goodson, 1994). It consists of two dimensions: abstraction levels and part-whole decomposition. The abstraction levels—typically five in number—combine physical and functional components of the work domain. The part-whole decomposition allows the viewing of behavioral elements at different scales of analysis.

The five levels from top to bottom are: (1) *functional purpose* [system goals]; (2) *abstract function* [underlying measures of priority and flow within the system such as value (money), information, mass, energy]; (3) *generalized function* [basic system functions in general terms]; (4) *physical function* [system functions in general physical terms]; (5) *physical form* [actual physical components satisfying the above levels]. Each lower level is the *means* by which the higher level *ends* are achieved.

2. "HYPERERGO": TWO HYPERTEXT APPROACHES TO REPRESENTING ERGONOMIC INFORMATION

This study consists of a comparison of two different hypertext representations of complex ergonomic information. The first approach uses a more traditional hypertext structure based on tables of contents. This is labeled the Category Hierarchy (CH). The second approach attempts to represent the underlying semantic structure of the data bases using an abstraction hierarchy approach (AH) derived from ecological interface design principles.

2.1 Common technical content of Hyperergo

For both hypertext approaches, the technical content was identical. The primary source of technical information was ANSI-HFES 100 (1988), The American National Standard for VDT Workplaces. The knowledge domain for Hyperergo included the following categories: (1) *physical components* [chairs, desks, keyboards, monitors]; (2) *fundamental ergonomic principles* [physiological efficiency, visibility, reach]; (3) *interactions*[interdependencies between pairs of physical components—such as seatpan height and worksurface height]; (4) *system integration* [interdependencies among several components]; (5) *physical/ physiological interactions* [physical discomfort in body areas potentially attributable to physical components].

2.2 Alternative versions of Hyperergo

The alternative versions of Hyperergo consisted of different semantic structures. In the CH version, the above five elements of the domain knowledge were represented at the first layer of the hypertext representation (chapter headings), with second and subsequent layers representing decomposition in subcomponents. In the AH version, the first level of the hypertext representation consisted of the three basic ergonomic principles: efficiency, visibility, reach (item 2 above). The second layer consisted of the relational functions: interactions, integration, body discomfort (items 3, 4, 5). The third layer consisted of individual components (item 1) which could be further decomposed at lower levels. Activating a higher level node (such as visibility) would constrain lower level node availability to only information which was relevant.

2.3 Predictions

The focus of the study was the training of designers to use ergonomic information. For simple design problems (e.g., what is the optimal height of a chair?) there was no reason to expect any advantage of the AH over CH. However, as the problems become more complex—

involving interdependencies between or among components—it was predicted that AH would be more effective. Hence, a statistical interaction was predicted.

3. METHOD

3.1 Procedures

Hyperergo was implemented on a Macintosh II CX using HyperTalk, a programming language for HyperCard (v. 2.3). Subjects were 24 students at Miami University majoring in engineering, systems analysis, management information systems, and architecture. Subjects were paid for their participation. All subjects had minimal ergonomic experience prior to the study, as assessed by a pre-experimental questionnaire.

Subjects were randomly assigned to either the CH or AH conditions. Each subject participated in a training session and an experimental session. Each session lasted 90 minutes and were conducted on consecutive days. The training programs were designed as modular self-guided tutorials—using a combination of hypertext and audio tapes. Subjects were required to successfully complete all modules.

During the experimental session, subjects were given 16 search questions. Questions varied in complexity: the first four were simple search questions, the next five were complex search tasks, the next three were global problem solving tasks, and the final four questions were simple search tasks. All questions were given in the same order. On each task, subjects were instructed to press the "pause" button when the desired information was located, and to then type the information into a second computer. Thus, only search times were recorded by the primary computer system.

3.2 Design and measures

The study was a 2 x 3 mixed factorial design with one between-subjects factor (interface type), and one within-subjects factor (search tasks). Four dependent measures were recorded: (1) *search time* [total time interacting with Hyperergo exclusive of time spent answering the questions]; (2) *time per node* [time spent interacting with Hyperergo independent of number of nodes visited]; (3) *navigation disorientation—objective measures* [extent to which subjects became lost. Measured by the deviation ratio between number of nodes visited relative to number of nodes visited using a predetermined optimum path]; (4) *navigation disorientation—subjective measures* [rating scales measures of the extent to which subjects experienced disorientation.]

4. RESULTS

Analysis of variance of the first three dependent measures indicated that, in each case, there was a significant difference between the two interfaces, and a significant interaction between type of interface and type of search task. Thus, the results support the predictions. For the simple tasks, the differences between CH and AH were not significant, however, as task complexity increased, the increasing benefit of the AH became evident. Means and standard errors for task and conditions are shown below.

	Average Search Times		Deviation Ratios		Time per Node	
	AH	CH	AH	CH	AH	CH
Task 1	7.41 (0.39)	8.35 (0.56)	1.10 (0.008)	1.10 (0.010)	1.80 (0.034)	1.89 (0.033)
Task 2	13.40 (1.45)	28.72 (1.65)	1.36 (0.017)	1.69 (0.024)	1.92 (0.032)	2.21 (0.034)
Task 3	20.86 (1.81)	76.23 (4.12)	1.57 (0.031)	2.22 (0.046)	2.08 (0.041)	2.51 (0.038)

In addition, the subjective ratings showed a significantly higher perception of being lost when using the CH system. The results support the interpretation that the AH hypertext implementation of Hyperergo provides a better semantic representation of a complex problem domain than does CH.

5. REFERENCES

McKnight, C., Dillon, A., and Richardson, J. (1991). *Hypertext in Context.* Cambridge: Cambridge University Press.

Mohageg, M. (1992). The influence of hypertext linking structures on the efficiency of information retrieval systems. *Human Factors*, 34, 351-367.

Rasmussen, J., Pejtersen, M., and Goodstein, L. (1995). *Cognitive Systems Analysis.* New York: Wiley.

Vicente, K.J. and Rasmussen, J. (1990). The Ecology of Human-Machine Systems II: Mediating "Direct Perception" in Complex Work Domains. *Ecological Psychology*, 2, 207-249.

ENHANCING THE INTERFACE TO PROVIDE INTELLIGENT COMPUTER AIDED LANGUAGE LEARNING

M. Murphy[a], A. Krüger[b], and A. Grieszl[c]

[a] School of Information and Software Engineering; University of Ulster; Newtownabbey, Co. Antrim BT37 OQB, N. Ireland

[b]Institut für semantische Informationsverarbeitung, Universität Osnabrück, Sedanstr. 4, D-49069 Osnabrück, Germany

[c]Institut für Logik und Linguistik, IBM Wissenschaftliches Zentrum, Vangerowstr. 18, D-69115 Heidelberg, Germany

This paper describes the design of a knowledge-based approach to Computer Assisted Language Learning which provides enhanced user interface capabilities.

1. INTRODUCTION

The demand for software for Computer-Assisted Language Learning (CALL) is increasing considerably. This demand is driven by a growing need for foreign language skills in Europe and the Far East and by the move towards equipping the independent learner with facilities for distance learning. However a drawback of most, if not all, of the currently available CALL software is that it cannot provide very helpful feedback to the learner. There are little or no facilities for individualised tuition and as a result CALL systems are often perceived by learners, as well as tutors, as being dumb and inflexible, which is demotivating to the learner and restricts the independent use of CALL systems considerably [1].

The aim of the RECALL project* was to work towards providing a more adequate and user-oriented interface for CALL. The project focused on the design of an application called CASTLE which takes into account the strengths and weaknesses, preferences, level of proficiency, and preferred learning strategies of each individual student. This was accomplished by extending an existing IBM CALL program with a module that provides detailed linguistic analysis of the learner's response to the exercises of the program (Diagnosis Module); a module that creates a dynamic model of the learner, representing his or her learning history (Learner Module) and a module that controls the system's reactions to the learner's input and the structure of the materials offered to the learner on the basis of the Diagnosis and Learner Modules (Tutoring Module).

*RECALL Repairing Errors in Computer Assisted Language Learning. This work has been funded under the EU's Telematics Applications of Common Interest - Language Engineering LE1- 1615.

2. FORMULATING THE REQUIREMENTS

The requirements and functional specification for RECALL was formulated from market studies, a usability and effectiveness study based on the source IBM CALL program, empirical error analysis, and a compilation and analysis of didactic information. The sources revealed that learners who expressed reservations against CALL software complained about inadequate user interfaces and the general lack of user friendliness, as well as the lack of adaptivity and individualisation.

The RECALL design sought to overcome these reservations by providing an intelligent multimedia interface which encourages active language learning by using a communicative role-playing scenario, allowing the input of freely formulated sentences, in combination with an underlying set of grammatically-based remedial exercises. In storing information about the individual student, and in particular through getting hold of her grammatical weaknesses, the system is able to offer those exercises that can meet the learner's language needs.

3. THE THREE CORE COMPONENTS

3.1. Diagnosis Module

The task of the Diagnosis Module is to check the learner input for correctness and to provide a detailed linguistic error description in the case of erroneous input. In the remedial exercises, where the interaction mode is more restricted and correct answers can be anticipated, error detection can be performed by simple pattern matching with the anticipated correct and incorrect solutions. When dealing with more complex exercise types such as the communicative role-playing scenario, where the learner is allowed to enter *almost* free input, simple pattern matching is inadequate for assessing the learner's errors; an analysis based on linguistic methods is called for. In this case the Diagnosis Module starts an analysis of the learners' input by means of a parser operating on a grammar, error grammar and lexicon. The error rules [2] [3] in an error grammar represent incorrect constructions and annotations in these rules activate further error handling. On the basis of our research on error classification, error rules and error entries have been constructed which allow a number of different error types to be handled. Consequently, the output of the Diagnosis Module is an error description which is sent to the Tutoring Module where it triggers the appropriate error feedback strategy and sent to the Learner Module where it is used to update the Learner Model.

3.2. Learner Module

The task of the Learner Module is to create and maintain a model of the learner's proficiency in the language. The Learner Module has three main components: a Learner Model database; a Stereotype Library and a series of Implicit Acquisition Rules. The Learner Model contains the model of the individual learner which is to be regularly updated in the course of her interactions with the system. Based on a functional architecture for user modelling systems proposed by Benyon & Murray [4] three different groups of information have been identified for the Learner Model: *Profile information:* personal learner details; *Student Model information:* the system's estimate of the learner's (grammatical) proficiency in the language and *Cognitive Model:* relatively stable characteristics of the learner. Most of the items under Profile Information as well as those found within the Cognitive Model are stable and unlikely to require updating, whereas items under Student Model are dynamic and will require updating as the learner interacts with the system.

When the Learner first interacts with CASTLE a stereotype model is instantiated which infers default information about the learner's proficiency in the domain. This default information is updated by the implicit acquisition rules as the learner progresses through the exercises in the role-playing scenario and remedial exercises. The Learner Model information is used by the Diagnosis Module when selecting the most probable error hypothesis and by the Tutoring Module when selecting an appropriate feedback strategy and selection of exercises to present to the Learner.

3.3. Tutoring Module

The central agent in the RECALL system is the Tutoring Module. The main task for the Tutoring Module is to formulate adequate feedback to the learner when errors are made and to select exercises from the exercise library which are suited to the students' individual strengths and weaknesses. The Tutoring Modules offers an adventure-like communicative role playing scenario where the learner is required to enter (almost) free text in the context of a communication situation such as asking for directions or buying items from a shop. If an error occurs a number of times within the role-playing scenario, e.g., mixing up adjectives and adverbs, the Tutoring Module selects a remedial exercise which concentrates on repairing this error. Even though the learner's primary goal is to create syntactically correct sentences, the user is also asked to pay attention to the game she is playing. Thus, the user is tutored in a two-fold way: by the language tutor paying attention to language errors and by a game tutor which decides whether the goal of a particular scene in the story has been fulfilled.

When an error occurs, the language tutor communicates feedback which is triggered by the error description provided by the Diagnosis Module. In order to do this the Tutoring Module has access to a template-based feedback library where suitable feedback text for a number of error types is stored. From the language tutor's perspective, CASTLE distinguishes between three possible cases: (i) the input is correct; (ii) the input is correct, however an error of minor importance occurred (e.g. a typing error) and (iii) the input is not correct with respect to the learning goals *and* there are further errors of minor importance.

In the first case the student will be notified of the correctness of her input. In the second case the minor error will be acknowledged but not tutored, the system might correct a typing error automatically, but will acknowledge that the input is correct. The third situation leads to the first step of a multilevel strategy which shall tutor the learner to correct a more important error. The multilevel strategy consists of four levels. For example, the learner is asked to offer something in order to get a newspaper from a man sitting on a bench:

Learner Input: **Want you mony?*

CASTLE Response Level 0:	notification of error: *That is not correct.*
CASTLE Response Level 1:	hint: *Consider the use of 'to do' in your input.*
CASTLE Response Level 2:	explanation: *A question without interrogative starts with a form of 'to do'.*
CASTLE Response Level 3:	answer: *Do you want money?*

If the student were to enter: **Do you want mony?* committing the typing error again, the system would acknowledge the typo in this stage of tutoring, as Case 2 applies.

If the Learner makes the same type of error 3 times, the Tutoring Module recommends that the learner undertakes a remedial exercise(s) which can tutor her on this error. Thus, rather than employing the traditional CALL response strategy that the learner has entered erroneous input, the CASTLE Tutoring Module is able to provide a richer response strategy to the learner, based on the type of error which has occurred and the level of help which has been sought by the learner when answering questions. The CASTLE interface is further enhanced by the game tutor which focuses the Learner on tasks which have to be accomplished in order for her to progress in the game. This provides an innovative level of interaction with the CALL program which serves to motivate the Learner and enhance the communicative aspects of CASTLE.

4. CASTLE DEMONSTRATOR

The CASTLE demonstrator concentrates on teaching English as a second language for German students. It runs under Windows™ 3.1 but can also be run as 16-bit application from Windows™ 95. The GUI has been written using HTML, running under Netscape Navigator™. The Diagnosis Module incorporates a C-based chart-parser with a feature based grammar and lexicon making use of an attribute-value-formalism of the PATR-II family. The Learner Module and the Tutoring Module are both written in PROLOG. Initial user evaluation sessions have been carried out in Germany and Ireland. Tutors and learners have indicated that the level of functionality provided by the adaptive interface is beyond that encountered in CALL applications currently on the market.

5. CONCLUSION

The RECALL project has concentrated on developing an enhanced user interface for contemporary CALL products. This has been realised through the inclusion of a sophisticated error diagnosis module, a computationally feasible learner modelling component and a multi-faceted tutoring module which tutors at the linguistic level and at the game level. Future work will concentrate on providing a robust and generic shell which allows tutors to develop their own CALL products which exhibit enhanced user interface capabilities.

REFERENCES

1. Gust, H., Ludewig, P., Reuer, V., Unsöld, R.F. (1994): *Evaluation auf dem Markt verfügbarer Software zum computergestützten Spracherwerb*. Study on behalf of the IBM Germany Information Systems. Universität Osnabrück.
2. Krüger-Thielmann, K. (1992): *Wissensbasierte Sprachlernsysteme: Neue Möglichkeiten für den computerunterstützten Sprachunterricht*. Tübingen: Narr.
3. Schwind, C. (1994): Error analysis and explanation in knowledge based language tutoring, in: Appelo, L., de Jong, F.M.G. (eds.): *Proceedings of the 7th Twente Workshop on Language Technology (TWLT 7)*. Enschede: University of Twente, 77-91.
4. Benyon, D., Murray, D. (1993): Applying user modeling to human-computer interaction design. *Artificial Intelligence Review* 6, pp. 43-69.

The Role of Case-Based Reasoning in Instructional Design: Theory and Practice

B. Asiu[a] and M. D. McNeese[b]

[a]Cognition and Performance Division, Armstrong Laboratory, Brooks Air Force Base, Texas 78235-5352

[b]Human Engineering Division, Armstrong Laboratory, Wright-Patterson Air Force Base, Ohio 45433-6573

1. INTRODUCTION

Instructional system development is a set of procedures based on general systems theory to guide instructional designers in the analysis, design, development, implementation and evaluation of learning environments. However, the implementation of this systematic approach to instruction has been difficult for several reasons. First, the process requires high levels of expertise in subject matter content, instructional design processes and learning theory (Rowland, 1993). Second, many models of instructional design are highly proceduralized and linear with a weak or nonexistent theory base (Andrews & Goodson, 1980). And third, the nature of instructional design as a complex cognitive skill is not well understood (Perez, Johnson, & Emery, 1995).

A partial solution to this problem is the implementation of instructional design automation and performance support tools. The expert system approach to lesson design has been demonstrated in several projects but none are as yet in wide use. The reasons for this may lie in the nature of knowledge acquisition and expert systems, and in the complexity of instructional design expertise. Duchastel (1990) argues that instructional design knowledge as a body of expertise is far from refined enough to represent in symbolic form. The task of implementing expert cognitive behavior via software is further hampered by the lack of supported models of instructional design expertise (McNeese & Asiu, 1995). Another complicating factor is that the reasoning process needed to solve instructional design problems requires an iterative combination of analysis and synthesis (Boose, 1989).

Case-based reasoning offers an alternative to the expert system approach that promises to be more robust in addressing the complexity of instructional design, can better recognize the richness of learning context and is more consistent with the way designers solve instructional problems. The elements within case-based reasoning that advance the multidimensional aspects of instructional design are: 1) the *case* itself (context-specific contents through a variety of knowledge structures, e.g., Schank's (1990) stories as content form, 2) the ability to *change/adapt the case* to match a new situation occurring in the context of design tasks, and 3) the ability *to index any case* (or set of cases) to other material (general principles, cognitive knowledge, rules, other cases, concept maps, etc.) using similarity measures as access/retrieval cues. Two key points regarding these elements are: a) the type of cases utilized and b) the way cases are indexed to other objects. The extent to which one may use (or reason) from a case is related to the similarity of the user's task at hand, the given familiarity of the system's cases as viewed by the user, and the user's own context of cases representative of their experience. Further considerations include whether the system has built-in indexing designed for the novice user (a *closed architecture*), or whether the cases may be adaptively reconstructed (made similar yet different from pre-existent cases) by an expert user (*an open-adaptable architecture*). Novices given an open architecture without some form of scaffolding, or experts given a closed architecture, may prove disastrous from a human computer interface standpoint.

This paper reports on one study in a series of formative evaluations of a case-based lesson planning software tool called GUIDE-- Guide to Understanding Instructional Design Expertise. The software was developed in Asymetrics ToolBook® by the Armstrong Laboratory. GUIDE consists of two specific features to help instructional developers create pedagogically sound instruction: 1) elaboration of Robert Gagné's Nine Events of Instruction (a general sequence of learning events to support assimilation and retention of learning) and 2) a range of carefully selected example lessons elaborating the application of the nine events. The GUIDE approach substitutes two essential components of expertise absent in novices: 1) an organized knowledge base containing the declarative information of instructional design via elaboration of the Nine Events and 2) extensive past experience in the successful application of instructional design via the sample lessons within the case base. The purpose of this study is to better understand expert and novice instructional design problem solving using case-based reasoning as implemented via the GUIDE software.

2. METHOD

Previous research has focused primarily on an *information processing* approach to study instructional design expertise via quantification of designer behaviors and thinking protocols under contrived tasks. However, there has been discussion in the literature that traditional approaches do not adequately account for the complexity and constraints of design problem solving in naturalistic contexts (Banathy, 1987; Duchastel, 1990; Rowland, 1993). The more recent perspectives of ecological psychology and situated cognition are consistent with these views and suggest that instructional design problem solving should be viewed as a complex activity involving key dimensions such as the *designer* (expertise, schemas, personal experience), *instructional design* (instructional models, knowledge base) and *problem solving artifacts* (information, design tools, design products), and the complex interactions between these dimensions, as critical to decision making and performance (McNeese & Asiu, 1995). As a user-centered approach, the group interview technique is perhaps better suited to capture various transactions between the designer, technology, and context from the perspective of the instructional designer as an actor within this complex system. Six novice and two experienced instructional designers were questioned in a group interview format to elicit information on instructional design problem solving, as well as designer perceptions and experience, using the GUIDE case-based lesson design support tool.

3. RESULTS

Interview analysis provided several observations regarding instructional design problem solving, the instructional design environment, and user response to the GUIDE case-based reasoning approach. First, novice developers appear to implement instructional design in a mechanistic manner, without understanding or applying the robustness of the process as recognized by the more experienced developers. Second, users responded that instructional design is frequently a team effort while the case-based approach in collaborative settings remains undefined. Third, experienced designers expressed a preference for maximum flexibility, minimum constraint from case-based support while novice developers preferred the system to be more directive in nature, even at the expense of flexibility and comprehensiveness. Finally, novice developers expressed difficulty in matching the support within the GUIDE case base to their specific design task. This last finding challenged the assumptions regarding the design of GUIDE as a case-based lesson design support tool. Thus a pilot effort was undertaken to confirm this emerging hypothesis.

One experienced and three novice instructional developers (these subjects were not involved in the previous study) participated in a concept mapping activity to explore user conceptualizations of the design task and GUIDE case base. Convenient heuristics for analyzing concept maps are provided by Zaff, McNeese, and Snyder (1993) and were applied in this study. The single expert concept map showed strong integration of GUIDE within the lesson design process, and explicit recognition of case-based lesson support features. The novice maps showed weak contextual relationships between GUIDE and the lesson design task, as well as confusion between the case base and elaboration of Gagné's Nine Events of Instruction. There were also no clear decision criteria regarding case selection or any indication of mapping across the cases. In follow up discussions with participants, it was clear that the experienced developer was able to more clearly and easily articulate herself in the concept mapping activity while the novice developers struggled in the mapping process, perhaps showing that they were less able to clearly describe their own thinking process in applying GUIDE to instructional design problems.

4. DISCUSSION

That the novice developers did not see the GUIDE case base as familiar and representative may be because the cases were selected by experts as the best-available exemplars for quality instruction, not because they were seen as most relevant to novice designers. Furthermore, experienced researchers provided the elaboration between individual cases and the Nine Events of Instruction, and in doing so perhaps organized the elaboration around deep structural traits of the case base not salient to the novice. A novice may be extremely knowledgeable in domain specific areas (e.g., knowledge of mechanics) but have very shallow knowledge in the target area (e.g., instructional design). If they have not developed particularly good metacognitive strategies that allow them to generate active questions or structure new information beyond just a shallow level, there is a need to bridge the gap between what they know and what is given in the case through the use of built-in scaffolds to support assimilation. These scaffolds would need to connect what is given to the proper context of what they already know possibly allowing them to take an area they are already knowledgeable in and anchor/map that analogously into the new area required in their task. Thus the quest for the case-based reasoning system is to help the user acquire, assimilate, and possibly adapt the relevancy of a set of cases to their own task environment.

Lessons from this implementation of case-based reasoning reinforce theoretical considerations that cases need to be perceived as rich with domain specific knowledge and relevant to the problem at hand. What remains is to create a second implementation of GUIDE with expanded cases, relevant problem anchors, and structured queries to guide the novice developer given their mechanistic implementation of problem solving processes, limited metacognitive strategies and inability to abstract beyond their immediate context.

REFERENCES

Andrews, D. H., & Goodson, L. A. (1980). A comparative analysis of models of instructional design. Journal of Instructional Development, 3(4), 2-16.

Banathy, B. H. (1987). Instructional systems design. In R. Gagné (Ed.), Instructional technology: Foundations (pp. 85-112). Hillsdale, New Jersey: Lawrence Erlbaum Assoc.

Boose, J. H. (1989). A survey of knowledge acquisition techniques and tools. Knowledge Acquisition, 1, 3-73.

Duchastel, P. C. (1990). Cognitive design for instructional design. Instructional Science, 19, 437-444.

McNeese, M. D., & Asiu, B. (1995). User-centered knowledge acquisition to model instructional design expertise. In H. Halff (Chair), Human factors in generative computer-based training. Colloquium conducted at the 1995 meeting of the Human Factors and Ergonomics Society, San Diego, CA.

Perez, R. S., Johnson, J. F., & Emory, C. D. (1995). Instructional design expertise: A cognitive model of design. Instructional Science, 23, 321-349.

Rowland, G. (1993). Designing and instructional design. Educational Technology Research and Development, 41(1), 79-91.

Schank, R. (1990). Tell me a story: A new look at real and artificial memory. New York: Charles Scribner's Sons.

Zaff, B. S., McNeese, M. D., & Snyder, D. E. (1993). Capturing multiple perspectives: A user centered approach to knowledge design and acquisition. Knowledge Acquisition, 5(1), 79-116.

Designing Interactions for Guided Inquiry Learning Environments

Noel Enyedy (enyedy@socrates.berkeley.edu), Phil Vahey, and Bernard Gifford
University of California at Berkeley: 4533 Tolman Hall
Education in Mathematics, Science and Technology
Berkeley, CA 94720

1. Introduction

Cognitive science perspectives on learning encourage researchers interested in computer-mediated instruction (CMI) to revisit the theories of instruction that inform their efforts to design environments that actively engage students in learning. We advocate an approach to CMI that emphasizes student appropriation of inquiry skills by making available student-controlled interactive simulations, dynamic-representations, and contextualized learning activities. These activities guide students toward a deeper understanding of the salient aspects of formal domains—those aspects whose mastery are considered by experts as being essential to understanding the larger domain. In this paper, we first contrast the inquiry approach with three other methods of instruction: *transmission*, *intelligent tutoring,* and *discovery*. Next we discuss the important aspects of guided inquiry learning environments, using examples from the Probability Inquiry Environment (PIE), a computer-mediated inquiry environment designed to teach middle school students elementary concepts in probability.

2. Three Approaches to Instruction

The *transmission* approach is exemplified by the whole-class lecture, in which the instructor presents the same lesson material to the assembled students. This approach hinges on the assumption that it is possible to displace non-normative ideas held by students with well-articulated normative ideas presented by the instructor. Unfortunately, the transmission approach can lead to inert knowledge that students find hard to apply to problematic situations. The *intelligent tutoring* approach, developed in part to overcome this weakness, is strongly influenced by the assumption that it is possible to improve problem solving skills by using adaptive feedback within carefully sequenced problem sets and curriculum materials. Anderson's research [1] demonstrates the effectiveness of this approach in domains where the problem space is well-specified, such as basic proofs in geometry.

However, recent research on learning has shown that students can retain resilient and robust misconceptions even after such formal instruction, and even into their adult years [2,3]. Some researchers have interpreted this as evidence that conceptual change requires a *discovery* approach to instruction, proposing that learning is best advanced when students are situated in open-ended instructional settings where they are more or less free to work on domain-related problems and activities of their own choosing. While such an approach ideally allows students to discover the important connections and interrelations underlying formal domains, the approach is also inefficient and students may be sidetracked by activities and concepts that are peripheral to the core knowledge of the domain.

3. An alternative: The Probability Inquiry Environment (PIE)

Guided inquiry shares some of core assumptions of the discovery approach, but differs in the degree to which the learning process should be supported. Both approaches start from the assumption that students have deep-rooted and resilient intuitions that cannot simply be replaced, nor overcome by learning how to reproduce problem solving methods, but that intuitions should instead be used as building blocks for knowledge in formal domains [4,5]. However, unlike discovery learning, the carefully crafted activities and student-controlled simulations of guided inquiry are designed to focus students on particular intuitions, and to guide students towards a normative conception of the domain. Such an approach builds on the insights of distributed cognition [6, 7], which redefines intelligence as a system mutually constituted by the student, the symbolic and physical environment, and other social actors. The instructional implication of distributed cognition is that productive learning environments must support activities that encourage students to engage in the social and representational practices of formal disciplines [8], but need not require students to re-invent the discipline.

We advocate an inquiry cycle consisting of *predict*, *experiment*, *conclude*, and *principle generation*. Because students come to instructional activities with a naive understanding of the domain, and attempt to apply their understandings to the activities, interactivity must be designed so that the strategic points of a learners' trajectory through the environment are seeded with opportunities for interactions that are specifically designed to transform their understanding of the domain. Anticipating these strategic points requires a detailed understanding of how students intuitively understand and interact with the domain. Technology-mediated inquiry has been effectively used to teach middle school students Newtonian physics [9], and promises to be equally effective for mathematics.

An essential component of each step in the guided inquiry cycle is social interaction. In order to promote conceptual change, the environment must support and encourage social exchanges between students of similar and different levels of expertise such that the students become part of a community of learners [10]. Social interaction provides an opportunity for students to become aware of intuitions different from their own, helping students to articulate their own ideas. Social interaction, in the service of inquiry, helps students make their thinking visible, and supports reflection, negotiation and revision of their ideas.

There are at least two ways social interaction can be supported by the interface of the environment. First, one can build into the software seamless links to synchronous and asynchronous, location-independent, networked communication (such as E-mail and on-line discussion tools), which can support the communicative aspect of the learning process. An alternate method is to design the interface for multiple physically present students who work collaboratively at a single computer. In contrast to networked communication, which is designed to enable conversations *on-demand*, designing for collaboration around a single computer involves structuring the interface to *prompt* conversations. In PIE, which is designed for two students at a single computer, students are often asked to respond to a set of key questions. Some of these questions provide a place for each student to respond individually, whereas others provide only one place where the students must construct a joint response. In the case of the jointly constructed responses, each student is provided with an "agreement bar" where they can individually show how much faith they have in the answer. This type of support makes each student's thinking visible and leverages social norms, such as a desire for consensus, to prompt students to discuss their ideas.

4. A STUDY OF MIDDLE SCHOOL STUDENTS USING PIE

One of the main methods which PIE uses to help students to articulate their intuitions and make their thinking visible is to guide the students through the process of identifying interesting questions and making predictions. The *prediction* step of the inquiry cycle helps students to identify and explore their own ideas about a subject domain. Computer mediation can scaffold the process by posing interesting questions and by highlighting salient aspects of the phenomena. For example, using the outcome space when determining probabilities is a vital part of understanding probability theory, but the outcome space is often ignored by middle school students. When middle school students used PIE to evaluate the fairness of an unfair game of chance, they ignored the outcome space, instead relying on their belief that any game based on luck must be fair. Although the outcome space was visible on the computer screen during the predictions, the students never counted the outcomes to determine the fairness of the game. Even though the students initially ignored the outcome space, the questions predisposed them to focus on that aspect of probability later in their investigations.

After students make predictions in PIE, they use a simulation that allows them to critically evaluate their intuitions and generate and test new conjectures. To create an effective instructional experience, PIE augments its *experimentation phase* with multiple, linked dynamic representations to help students understand the mathematics in the activity. The interactive simulation begins with a concrete representation of a random event such as a coin flip. The result of each event is displayed and animated across a probability tree which spatially depicts the outcome space, and is dynamically connected to a histogram which keeps track of the frequency of events. We found that animating a token across the probability tree helped students to better understand the connection between individual events, the outcome space, and the frequency of outcomes over time—connecting and relating intuitions about probability that, prior to the simulation, were isolated and fragmented. This resulted in most students questioning their early predictions that the game was fair.

However, even such crafting of experiences is not enough to ensure conceptual change. For the inquiry process to lead to a lasting re-conceptualization of the domain, students must critically evaluate the evidence that supports or challenges their intuitions and conjectures and then draw conclusions based on this new organization of knowledge. A strength of guided inquiry is that in the *conclusions phase* students can be scaffolded by questions and prompts that helps them formalize their intuitions. For example, one pair of students, Q and T, disagreed as to the fairness of the game. While T realized the game was unfair, Q kept insisting that the game was just luck, and so must be fair (Figure 4.1). It was while considering their conclusions that T noticed the difference in the number of outcomes that scored a point for each team. She was finally able to convince Q that the game was unfair by contrasting a partitioning of points that would make the game fair with the actual partitioning.

Figure 4.1 Students negotiating the importance of the outcome space.

T: See, Team B is losing by a lot. Told you it was unfair
Q: This game is just luck, it's just a penny game, it's just luck
T: Why is it fair, Q?
Q: Because it's a game of luck, it's just throwing pennies
T: I'm not talking about who's got the penny, I'm talking about right here [pointing to the tree]. They keep losing, and I'm trying to figure out why...wait a minute! See how this is AA right here? and this is AB AB AB [referring to the outcomes]...

The last aspect of the inquiry process is possibly the most important to support. Going through a process of inquiry does not automatically protect the student from producing inert knowledge that is bound to the learning context. Students must explicitly transform the data-bound findings of their investigations into general principles, and then practice applying the principles to novel contexts. In our early versions of PIE, students were not explicitly asked to generalize their findings and draw connections to other contexts. As a result students had difficulty transferring what they had learned beyond the specifics of the immediate investigation.

5. CONCLUSION

Guided inquiry is an appropriate model for the design of future generations of CMI that can productively build on the intuitions that people spontaneously employ. Guided inquiry environments have the potential to break the tradition of self-contained, hermetically sealed learning environments which lead to knowledge that is compartmentalized and inert, while at the same time minimizing the false starts and unproductive tangents of pure discovery learning. Perhaps most importantly, guided inquiry provides a model for instruction that encourages students to actively engage in and reflect upon their own learning. Ultimately, this will help students appropriate both content knowledge and the skills necessary for life-long learning.

REFERENCES

[1] Anderson, J., Corbett, A., Koedinger, K., Pelletier, R., Cognitive tutors: Lessons learned. *The Journal of the Learning Sciences,* v4 (n2): (1995).

[2] Confrey, J., A review of the research on student conceptions in mathematics, science, and programming. In *Review of Research in Education*, C. Cazdem (Ed.). American Educational Research Association, Washington DC, (1990).

[3] Shaughnessy, J.M., Research in Probability and Statistics: Reflections and Directions. In *Handbook of Research on Mathematics Teaching and Learning,* D. Grouws (Ed), (1992).

[4] Smith, J., diSessa, A., and Roschelle, J., Misconceptions Reconceived: A constructivist analysis of knowledge in transition. *The Journal of the Learning Sciences, 3*(2), (1993).

[5] Vahey, P., Enyedy, N., and Gifford, B., *Beyond Representativeness: Productive Intuitions About Probability*. Accepted for publication in the Proceedings from the 19th Annual Conference of the Cognitive Science Society (1997).

[6] Salomon, G., No Distribution Without Individuals' Cognition: a dynamic interactional view. In G. Salomon (Ed.) *Distributed Cogntitions: Psychological and Educational considerations,* New York, NY: Cambridge University Press, (1993).

[7] Pea, R., Practices of distributed intelligence and designs for education. In G. Salomon (Ed.)*Distributed Cogntitions: Psychological and Educational considerations,* New York, NY: Cambridge University Press, (1993).

[8] Gordin, D.N., Polman, J.L., & Pea, R.D., The Climate Visualizer: sense-making through scientific visualization. *Journal of Science Education and Technology*, 3, (1994).

[9] White, B. & Frederiksen, J., The ThinkerTools Inquiry Project: Making Scientific Inquiry Accessible to Students and Teachers, *Causal Models Rep. 95-02, McDonnell Foundation,* (1995).

[10] Brown, A., and Campione, J., Psychological Theory and the Design of Innovative Learning Environments. In *Innovations in Learning: new environments for education,* L. Schauble and R. Glaser, (Eds.), Mahway, NJ: Erlbaum, (1996).

Granular interface design: decomposing learning tasks and enhancing tutoring interaction

A. Patel[a] and Kinshuk[b]

[a]CAL Research & Software Engineering Centre, Bosworth House, De Montfort University, Leicester LE1 9BH England Email: apatel@dmu.ac.uk Phone/Fax: +44 116 257 7193

[b]GMD FIT, Schloss Birlinghoven, D-53754 Sankt Augustin, Germany
Email: kinshuk@gmd.de Phone: +49 2241 14 2144 Fax: +49 2241 14 2065

The concept of granularity has been applied in Intelligent Tutoring Systems (ITS) for the purpose of diagnosis, plan recognition, reasoning and belief revision. These implementations have been related with the knowledge management within the system. This paper considers granularity from an interface design viewpoint. The learning tasks are decomposed into smaller components at varying levels of granularity with the perspective shift enabled through the user interface. They remove the need for the system to engage in complex inferencing about the user knowledge as the system can provide a status feedback (e.g. correct/incorrect) at a coarser grain size and require the student to use a fine grained interface for more detailed interaction. The students also find it easier to focus on and grasp these smaller components.

1. INTRODUCTION

The process of education involves traversing the granularity of various disciplines to varying extents, from detailed to abstract and from intrinsically simple to complex representations of knowledge - the complexity arising from implicit knowledge, implied context and inferred semantic. The students also traverse the aggregation granularity in the process of an educational model progression along the part-whole dimension. At an introductory level, the students generally learn the details to be able to apply them at an advanced level. As they progress in a discipline, they learn to combine the *parts* into *whole*, within appropriate environmental constraints such as behavioural factors. The very nature of the educational process therefore favours a granular approach towards the design of tutoring systems where students can easily move between different grain sizes. As Hobbs (1985) observed, " We look at the world under various grain sizes and abstract from it only those things that serve our present interests".

2. BACKGROUND

This paper describes a Computer Integrated Learning Environments (CILE) approach employing Intelligent Tutoring Tools (ITTs) for the teaching and learning of numeric disciplines at an introductory level. It discusses the adoption of granularity in interface design to decompose the tutoring tasks with a view to reduce both the complexity of the system-student interaction and the cognitive load on a student at an introductory level. The ITTs are *mixed-initiative* systems with an *overlay* type of student model. The structural details of the ITTs are discussed in Patel & Kinshuk (1996a). They may be mixed and matched with other technologies (e.g. video) as well as human teachers, in various configurations to suit classroom based, open and distance learning (Kinshuk & Patel, 1996).

3. GRANULARITY WITHIN AN ITT

The ITT provides fine-grained tutoring through a short and simple feedback regime. Instead of attempting to infer complex steps and build up complicated feedback messages, the ITT advises a student to move to an appropriate level of detail - either to an intermediate step within the current interface or through calling up a fine-grained interface (Patel & Kinshuk, 1996a, 1996b). The learning tasks within an ITT are decomposed into smaller components.

This is illustrated with reference to the Capital Investment Appraisal ITT. An investment project may be viewed from various perspectives. Each of this perspective is an abstraction of a project's details to enable comparison between different projects or evaluation against some policy norm. 'Payback' (figure 1) is an abstraction, along the dimension of time but implicitly along the dimension of uncertainty and risk as longer duration means greater uncertainty. Similarly 'ARR' or the Accounting Rate of Return (figure 2) is an abstraction in terms of an average percentage return on the average investment. The 'NPV' (figure 3) or the Net Present Value is an abstract representation of a project's surplus cash flow in absolute terms but discounted on the basis of time and a rate representing interest, cost of capital or opportunity cost.

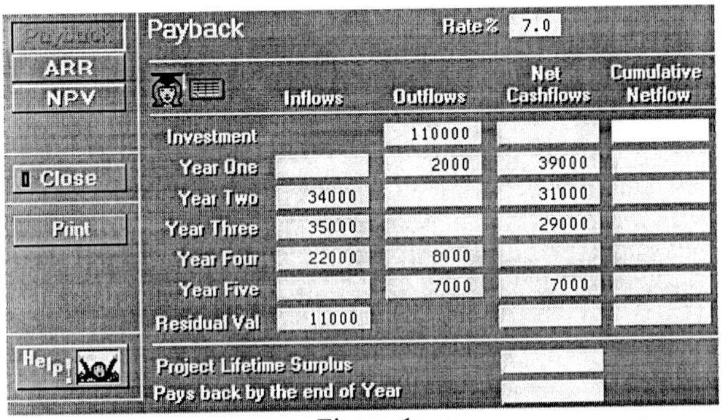

Payback — Rate% 7.0

Payback | ARR | NPV | Close | Print | Help!

	Inflows	Outflows	Net Cashflows	Cumulative Netflow
Investment		110000		
Year One		2000	39000	
Year Two	34000		31000	
Year Three	35000		29000	
Year Four	22000	8000		
Year Five		7000	7000	
Residual Val	11000			

Project Lifetime Surplus

Pays back by the end of Year

Figure 1

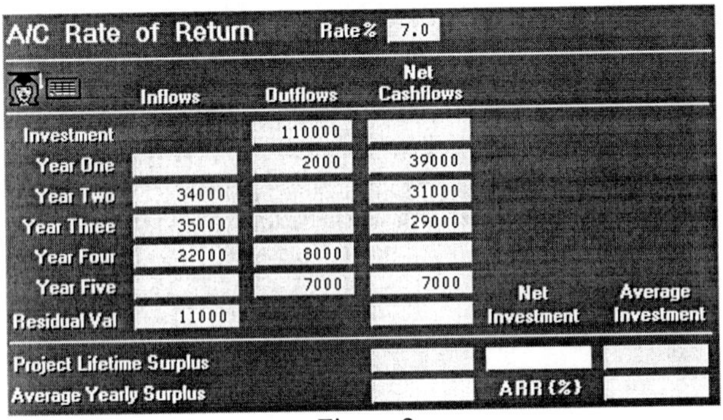

A/C Rate of Return — Rate% 7.0

	Inflows	Outflows	Net Cashflows
Investment		110000	
Year One		2000	39000
Year Two	34000		31000
Year Three	35000		29000
Year Four	22000	8000	
Year Five		7000	7000
Residual Val	11000		

Net Investment | Average Investment

Project Lifetime Surplus

Average Yearly Surplus | ARR (%)

Figure 2

The ITT represents these perspectives on different screens, keeping the project details on the same area of the screen, enabling a student to appreciate that the different techniques evaluate a given project from different perspectives. A student can move between these perspectives through the buttons on the control panel. Within these screens, each node represents an instance of a basic or a derived concept and the nodes are connected to each other through mutual relationships based on the underlying concepts. Dependencies are created by the sequence in which the nodes are instanciated, so that given a relationship between three nodes, if two nodes are already instanciated, a legal instance of the third node must satisfy the relationship. Such granular representation of the concepts allows the system ease in inferring missing conceptions and misconceptions in a student's knowledge boundaries, and enables learning of the concepts at various level of understanding (Lelouche & Morin, 1996).

Net Present Value Rate % 7.0

	Inflows	Outflows	Net Cashflows	Discount Factors	Present Value
Investment		110000			
Year One		2000	39000		
Year Two	34000		31000		
Year Three	35000		29000		
Year Four	22000	8000			
Year Five		7000	7000		
Residual Val	11000				
Project Lifetime Surplus				Net P. V.	

Figure 3.

On the 'NPV' screen, a student can choose to apply the discount factor as an abstract notion and let the system provide the values or choose to calculate it. If a student chooses to calculate it and enters an incorrect value, the system informs that it is incorrect and advises the student to use the fine-grained interface shown in figure 4. The interface can be called up using the 'Formula' button on the control panel (not shown in Figure 3). If the student still has difficulty in understanding the concept of discount factor, the system provides a more descriptive explanation with an example as shown in Figure 5, accessible through the 'Explain' button. The fine-grained interface is also laid-out along the dimension of aggregation granularity, to show that the discount factor is a reciprocal of the compound factor - a function of time duration and the rate per time duration.

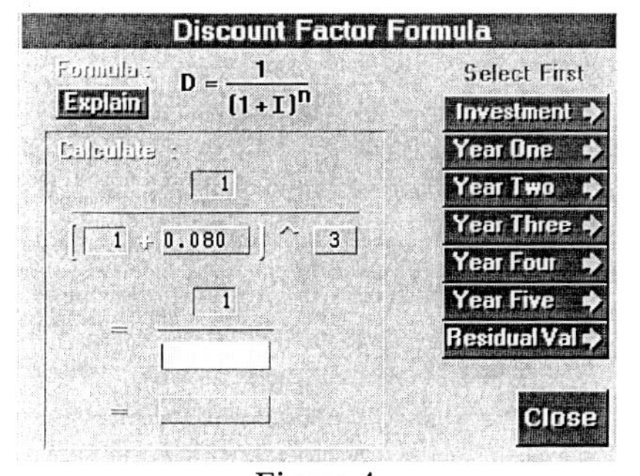

Figure 4

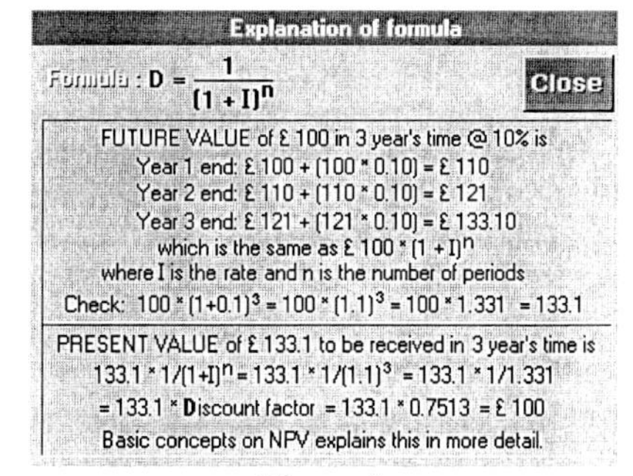

Figure 5

The tutoring on the finer details is necessary until the students internalise these concepts and are able to process them more efficiently in their minds. Beyond this point, it causes boredom if they are forced to use the fine-grained interface. An ITT, therefore, does not force the use of the fine-grained interface as a routine. It provides a status feedback (e.g. correct/incorrect) at a coarser grain size and advises the student to use the fine-grained interface for more detailed interaction. Similarly, the ITT does not force a rigid path to the solution and does not require that all the intermediate steps on the interface be complete, enabling a student to manipulate the finer details mentally and enter the aggregate outcome. It accommodates a *serial* approach of systematically working through the details as well as a *holistic* approach of rapidly assimilating details and focusing on the goal.

4. GRANULARITY IN THE DOMAIN KNOWLEDGE

The domain knowledge can be viewed as consisting of three levels. At the **introductory application level**, a student learns how to use the basic tools of a subject discipline and the *Basic* ITT is designed to suit this level. At the **advanced application level** two types of integration are possible. A *Ranking* ITT provides a suitable interface for holding and comparing the results of multiple instances of an ITT while the *Linking* ITT combines other ITTs to extend the size and complexity of the problems covered. At the **actual application approximation level**, the students learn how to account for behavioural and environmental factors seen in the real world. This viewpoint enables creation of larger tutoring systems where a student can drop to a Basic ITT for more detailed interaction but otherwise operate at a relatively coarser grain size.

5. CONCLUSION

McCalla and Greer (1991) noted that students seem to reason at many grain sizes and appear to have both deep and shallow knowledge at the same time, especially in case of problem solving abilities. They observed that the relationship of partial knowledge to more complete knowledge is also a granularity relationship and as students refine their understanding, they are, in Hobbs' (1985) terms, articulating their knowledge to finer grain size, apparently along at least three dimensions: aggregation, abstraction and goals. The students could also move in the opposite direction - from fine grained knowledge of particular situations to an understanding of inclusive, generic, coarse-grained knowledge.

The granular interface design for tutoring systems, as described above, is therefore in harmony with a student's state of knowledge and the ongoing process of knowledge acquisition at any given time. It enables easy shift in perspective and facilitates both bottom-up and top-down approaches to learning. While it provides all the necessary details without cluttering the screen, it also prevents fragmentation of concepts by adopting a granular interface design - a well-balanced interplay of the parts and the whole.

REFERENCES

Hobbs J. (1985). Granularity. Proceedings of the 9th International Conference on Artificial Intelligence, pp432-435.

Kinshuk & Patel A. (1996). Intelligent Tutoring Tools - Redesigning ITSs for adequate knowledge transfer emphasis. Proceedings of 1996 International Conference on Intelligent and Cognitive Systems (Ed. C. Lucas), pp221-226.

Lelouche & Morin, (1996).The Formula : A Relation? Yes, but a Concept too!! Lecture Notes in Computer Science, 1108, pp176-185.

McCalla G.I. & Greer J.E. (1991) Granularity-Based Reasoning and Belief Revision in Student Models. Student Modelling:The Key to Individualized Knowledge-Based Instruction (Eds. Greer J.E & McCalla G.I.), Springer-Verlag, pp 39-62.

Patel A. & Kinshuk (1996a). Applied Artificial Intelligence for Teaching Numeric Topics in Engineering Disciplines. Lecture Notes in Computer Science, 1108, pp132-140.

Patel A. & Kinshuk (1996b). Knowledge Characteristics: Reconsidering the Design of Intelligent Tutoring Systems. Knowledge Transfer (Ed. A. Behrooz), pp190-197.

Discount usability engineering

Paulo J. Santos[a] and Albert N. Badre[b]

[a]Bell Laboratories, Lucent Technologies
200 Laurel Avenue, Middletown, New Jersey 07748-4801, USA

[b]College of Computing, Georgia Institute of Technology
Atlanta, Georgia 30332-0280, USA

Learnability evaluation has traditionally required expensive and time-consuming techniques. Practitioners have refrained from performing extended learnability evaluation due to its prohibitive costs. We propose a discount method for evaluation of an interactive system's learnability. Our method is based on automated logging of user actions, detection of user mental chunks, and observation of chunk size as it grows over time with experience. We introduce a model for chunk detection, and present experimental results validating the use of chunk size as an indicator of learnability.

1. INTRODUCTION

One important component of an interactive system's usability is its learnability. Learnability, also known as "ease of learning", can be defined as a measure of the effort required for a typical user to be able to perform a set of tasks using an interactive system with a predefined level of proficiency. Most authors of usability and interface design publications (e.g., [1–4]) recognize the importance of ease of learning in interfaces. Nielsen [1] even claims that "learnability is in some sense the most fundamental usability attribute".

Despite the consensus that learnability is an important issue in usability, few authors discuss at length the issue of learnability evaluation. Some stress the importance of evaluating learnability for the first-time user, which may be important but is by no means a complete study of a system's learnability. Measuring intuitiveness, or initial ease of learning, does not provide the analyst with information about the learnability over an extend period of use.

The most significant obstacle to extended learnability studies is probably cost. Observing users over an extended period of time would consume huge resources, particularly observer time. Often, usability evaluators can not afford to spend the time or money they would like to make a complete usability study. Discount usability techniques [5] have become quite popular, because they are more likely to result in a better return on the usability expenditures. The proponents and supporters of discount usability contend that most of the problems with an interactive system are found in the early phases of testing. It is common to hear in this industry that 90% of the problems are found with 10% of the effort, but the remaining 10% of the problems take 90% of the effort. Although we know of no scientific backing for these specific figures, we agree that usability evaluation has diminishing returns. In these days of tightening budgets and competitive schedules, corporate usability engineers are often faced with the need to apply discount usability in order to stay within their resource limitations.

2. A METHOD FOR STUDYING LEARNABILITY

Our proposed method is based on automatically performing three steps: logging user actions, detecting user mental chunks, and observing chunk size as it grows over time. In the past, these steps have been traditionally done manually, making the study of learnability prohibitively expensive. By automating these steps, the cost of studying interface learnability is dramatically reduced and the time to perform those studies is also compressed.

The method can be implemented by means of monitoring and analysis software that runs in the background of the user's environment, simultaneously as the user uses the system. The software needs little or no configuration to setup, preserves confidentiality of the users data and interaction, and produces numerical results that can be easily compared and plotted.

2.1 Logging user actions

The first step in our method consists of automatically monitoring user activity. This can be achieved by using a logging tool, of which many are available in the research and commercial domains (e.g., [6,7]). A logging tool runs in the background, does not intrude upon the user's interaction, and produces a detailed interaction log with precise timing of every user action.

2.2 Chunking and chunk detection

A chunk is a coherent cognitive unit that the user holds in short term memory while performing a cognitive operation. A body of literature, generally referred to as the chunking literature, establishes relationships between cognitive stages and observable behavior. User behaviors have been shown to be intimately related with their cognitive skills. Independent results [8–10] show that experts organize the components of their mental process in larger chunks than novices.

2.2.1 Automatic chunking in human-computer interaction

Chunk boundaries can be identified by pauses in the observed behavior. The feasibility of chunk identification by pauses has been demonstrated in several domains, including games [10, 11], tactical decision scenarios [8], and human–computer interaction [12].

The general operation of the chunk detection algorithm discussed in [12] is as follows:

- For each sequence of events in the interaction log, apply the predictive model to predict execution time.
- If a pause in the interaction log cannot be justified by the predictive performance model, assume that the pause was caused by a shift of the user to the acquisition phase, and assess a chunk boundary at that point.

Figure 1 — An example of chunking

An illustration of chunking mechanism is provided in Figure1. The top timeline represents a sequence of user events. By applying the chunking algorithm, we assess chunk boundaries where pauses are too long to be justifiable by a predictive model for execution time, and aggregate the units between boundaries into groups, which we call chunks.

In [12] we report the results of an experiment in which we matched chunks detected as described above with user verbal protocols. In that study, we found that the chunks correspond, for the most part, to the execution of user goals.

2.3 Chunks as indicators of learnability

One important observed characteristic of chunks is that they grow as users become more experienced. In many other domains, expert subjects have been shown to form larger chunks and more regular chunks than inexperienced users. In the human–computer interaction domain, one would expect this to also be verifiable. More experience users can formulate higher level goals and execute more complex execution plans, which generally require more events (keystrokes or mouse clicks).

We will therefore use chunk size as an indicator of expertise. The variation of chunk size would then be an indicator of learning. We hypothesize that chunk size grows with experience, and that the variation of chunk size is an indicator of learning by the users.

2.3.1 Expirical validation of chunk size as an indicator of learning

To validate our hypothesis, we conducted an experiment using an application that involves a high degree of strategy and significant mouse-clicking activity: a computer game. The game selected provided many opportunities to learn strategies. When a strategy involves clicking in multiple cells, the system should be able to detect that a new larger chunk was formed. Users unfamiliar with a particular strategy will reason on cells individually, and probe them one at a time, with significant pauses between them. On the other hand, users that recognize a pattern and have proceduralized a strategy to solve it will probe multiple cells in quick succession. Distinguishing between both types of activity (single or multiple cell probing) is fairly straightforward, and can be done by inter-chunk pauses analysis. We then need to assert that the larger and more complex chunks are observed in the interaction of the more "experienced" user. A full report of this experiment was published in [13].

The results of this experiment strongly support the claim that the groups of events identified by our chunk detection algorithm share characteristics with chunks. Specifically, the size of events recorded grow with experience and skill. This is the same characteristics that has been shown of chunks.

3. CONCLUSIONS AND FUTURE WORK

An analysis of chunk size over time provides an indication of user expertise. Since expertise is a direct result of learning, the variation of chunk size over time is an indicator of learnability.

Future reserach plans include instrumenting beta versions of software to record keystroke information from the usage of beta testers. We will then be able to analyze chunk sizes from the logs, and have some indicators of learnability available for the software producers before the final version of the product is released.

REFERENCES

1. Nielsen, J., *Usability engineering*, Academic Press, Boston, 1993.
2. Preece, J., Rogers, Y., Sharp, H., Benyon, D., Holland, S., Carey, T., *Human–Computer Interaction*, Addison-Wesley, Wokingham, England, 1994.
3. Rubin, J., *Handbook of usability testing: How to plan, design, and conduct effective tests*, John Wiley & Sons, New York, 1994.
4. Shneiderman, B., *Designing the user interface: strategies for effective human–computer interaction*, 2nd edition, Addison-Wesley, Reading, Massachusetts, 1992
5. Nielsen, J., *Usability engineering at a discount*, in Salvendy, G. and Smith, M.J. (eds.), *Designing and Using Human-Computer Interfaces and Knowledge Based Systems*, Elsevier Science Publishers, Amsterdam, 1989, pp. 394-401.
6. Badre, A.N., Guzdial, M., Hudson, S.E., and Santos, P.J., *A user interface evaluation environment using synchronized video, visualizations and event trace data*, Journal of Software Quality, 4, pp. 101-113, 1995.
7. Badre, A.N. and Santos, P.J., *CHIME: a knowledge-based computer–human interaction monitoring engine*, Technical report GIT/GVU-91-06, Graphics, Visualization and Usability Center, Georgia Institute of Technology, Atlanta, Georgia, 1991.
8. Badre, A.N., Selecting and representing information structures for visual presentation. *IEEE Transactions on Systems, Man, and Cybernetics 12*, 4 (Jul/Aug 1982), pp. 495-504.
9. Barfield, W., Expert-novice differences for software: implications for problem-solving and knowledge acquisition. *Behaviour and Information Technology 5*, 1 (1986), pp. 15-29.
10. Chase, W.G. and Simon, H.A., Perception in chess. *Cognitive Psychology* 4 (1973), 55-81.
11. Reitman, J.S., Skilled perception in Go: Deducing memory structures from inter-response times. *Cognitive Psychology 8*(1976), pp. 336-356.
12. Santos, P.J. and Badre, A.N., *Automatic chunk detection in human–computer interaction*, in *Proceedings of the Workshop on Advanced Visual Interfaces AVI'94* (Bari, Italy, June 1–4, 1994), ACM Press, New York, 69–77.
13. Santos, P.J., *Automatic detection of user transitionality by analysis of interaction*, doctoral dissertation, Georgia Institute of Technology, Atlanta, Georgia, 1995.

Age Differences Between Elderly and Young Workers in Effectiveness of Computer Skill Training and Task Cognition

Hiroyuki Umemuro

Department of Industrial Engineering and Management, Tokyo Institute of Technology, 2-12-1, O-okayama, Meguro-ku, Tokyo, JAPAN

1. INTRODUCTION

Recently information technology is being rapidly introduced into office work environments. At the same time, percentage of older people in the population is growing in Japan and will affect the structure of labor force. In this vein, training of aged workers in computer operating skills is becoming more important.

This research aims to propose a guideline to design training of aged persons in computer skills. Mental aspects of aged persons, especially negative cognition of computer tasks, are thought to have more influence on the efficiency of learning new skills. For the purpose, the former part of this paper investigates the relationship between cognition of trainees on the task they are doing or learning ("task cognition") and effectiveness of training through an experiment. In the latter part, the effectiveness of two training methods, exploratory and instructional training, is compared through another experiment.

2. TASK COGNITION AND TRAINING EFFECTIVENESS

2.1. Task Cognition and Learning of Aged Persons

Doing a task such as computer operation can be considered as a cognitive process. At the same time, there exists another cognitive process whose object is the cognitive process doing the task. It will be called *task cognition* in the rest of this paper.

Being a kind of meta cognition, task cognition is thought to affect choices of strategies and/or knowledge while doing a task, and organization of new skills into one's knowledge while learning a new task.

There are quantitative and qualitative differences between younger and older persons' knowledge and intelligence. It implies that the learning processes of younger and older persons are assumed to be different in their characteristics. Generally it is considered that older people tend to suffer more from metal aspects. So task cognition is also expected to have more influence on learning new skills of older people.

2.2. Experiment 1

An experiment was held to investigate the age difference in the relationship between task cognition and training effectiveness of new computer skills. The task is to input statistic data into input forms on Apple Macintosh personal computers (PCs). Subjects use Apple's "Macintosh Basics" CAI program to learn basic skills to use computers,

Table 1. Averages of task cognition points and correlation with performances

		Schedule 1				Schedule 2			
		avg.	*LS* (whole)	*LS* (early)	OT_1	avg.	*LS* (whole)	*LS* (early)	OT_1
Aged									
difficult	(early)	2.17	-0.45	-0.41	-0.68	1.80	0.75	0.85×	0.41
	(last)	1.50	-0.21	-0.27	-0.36	1.80	0.07	0.33	-0.44
tiresome	(early)	2.67	-0.69	-0.64	-0.84*	2.20	-0.08	0.17	-0.37
	(last)	2.67	-0.67	-0.67	-0.73×	2.80	-0.91*	-0.54	-0.80
fun	(early)	3.67	0.48	0.64	0.28	3.60	0.69	0.03	0.60
	(last)	3.50	0.44	0.65	0.13	3.60	0.69	0.03	0.60
helped	(last)	3.83	0.54	0.45	0.51	3.80	0.84×	0.41	0.55
Young									
difficult	(early)	2.00	0.23	-0.00	0.42	2.38	0.16	0.41	0.21
	(last)	1.75	0.26	0.44	0.57	2.50	0.53	0.85**	0.72*
tiresome	(early)	3.88	-0.15	-0.47	-0.35	4.13	0.36	0.14	0.31
	(last)	4.00	0.47	0.32	0.48	4.13	0.34	0.28	0.33
fun	(early)	2.75	0.36	0.61	0.37	1.88	-0.44	-0.13	-0.31
	(last)	2.13	-0.29	0.01	0.04	2.25	-0.50	-0.11	-0.42
helped	(last)	2.25	0.13	0.11	-0.17	3.50	0.45	-0.25	0.24

LS: learning speed, OT_1: operation time on the first form
$** : p < 0.01$, $* : p < 0.05$, $\times : p < 0.10$

including operations of mouse and menus, according to two different schedules. In schedule 1, subjects learn CAI little by little in parallel with the input tasks. In schedule 2, subjects learn throughout CAI before they start the tasks.

Subjects are 11 adults aged among 50 to 62 years old, and 16 university students aged among 19 to 21. All subjects are checked with interviews as having no or very little experiences with PCs. Each subject is required to process 25 input forms. Operation time (OT) for each form is recorded. Learning speeds (LSs) are calculated using log linear model of learning for the whole learning process (using all 25 OTs) and for the early stage of learning (using first 5 OTs). Subjects are also asked to answer the questionnaire to measure their task cognition when first 5 forms are finished (at "early" stage) and when all 25 forms are finished (at the "last" stage). Questionnaire consists of four questions: "For you, is this task i) difficult? ii) tiresome? iii) fun?" and iv) "Did CAI helped your learning the task?" Subjects answer in points scaled from 1 (least) to 5 (most).

2.3. Results and Discussion

Averages of task cognition points and correlation with performances (LSs and OT) are shown in Table 1. Averages of "tiresome"($p < 0.001$) and "fun"($p < 0.001$) are significantly different between ages. The aged subjects tend to show more positive cognition to the task.

Some of the signs of correlation differ between ages. Correlation between "difficult"(last) and LS ($p < 0.05$), "tiresome"(last) and LS ($p < 0.05$), "tiresome"(last) and LS at early stage($p < 0.01$), "tiresome"(last) and OT of 1st form($p < 0.01$) are significantly different between ages. Correlation between "difficult"(last) and LS ($p < 0.05$), "tiresome"(last) and LS at early stage($p < 0.05$) are significantly different between schedules. The results show that the relations between cognition and performances are different between ages and it may depend on training schedules.

Some interactions of age and schedules are also observed. In aged group in schedule 2, "difficult (early)" has positive correlation with performances. In schedule 1, however, it has negative correlation with them. These tendencies cannot be seen in younger group. On the other hand, "difficult (last)" have very little correlation with performances in aged group. It derives that, for aged people, the cognition "difficult" is related with performances in early stage of learning and the relation may depend on training schedules.

3. EFFECTIVENESS OF EXPLORATORY TRAINING OF THE AGED

3.1. Exploratory Training and Aged Persons

In the results of Experiment 1, aged subjects in parallel schedule whose learning speeds are fast and operation time of 1st task is long show the tendencies not to have negative cognition "difficult" and "tiresome." Aged subjects in schedule 2 showed reverse tendencies. In other words, aged persons do not become negative in "try and learn" process even if it takes long time in earlier stage. It derives the hypothesis that exploratory training[1] is effective for aged persons. As exploratory training is a process of organizing new skill into knowledge which trainee already has, the trainee's profiles such as occupation is also supposed to have influence on the effectiveness of the training.

3.2. Experiment 2

The second experiment was held to compare the effectiveness of exploratory training and traditional instructional training for aged people. The tasks are to search information and to input messages using an information system for a university's department. The system is a hypertext system with tree structure written in HTML. Subjects first attend to one of the two training courses described below, then asked to complete four sample tasks. One training course is exploratory training; subjects are encouraged to explore in the system freely during the training session. Another course is instructional; an instructor guides throughout the system and all subjects follow the instruction. Sample tasks consist of two search tasks and two input tasks. Among each set of two tasks, one is "known" task which trainees can practice during training while another is quite new to them.

Subjects are 29 adults aged among 42 to 65 years old and 16 university students aged among 19 to 21. 16 of adult subjects are currently employed as office workers and 13 are not. All subjects are checked with interviews as having no or very little experiences with PCs. Three groups, employed, not-employed and young, are divided into two groups each corresponding to two training methods. Operation time and accuracy of sample tasks are recorded. Subjects are also asked to answer the questionnaire to measure their task cognition during the training and when sample tasks are finished. Questionnaire consists of six questions: "For you, are these tasks i) difficult? ii) tiresome? iii) fun? iv) suitable? v) interesting?" and vi) "Did the training session helped your learning the tasks?" Subjects answer in points scaled from 1 (least) to 5 (most).

3.3. Results and Discussion

Averages of operation time and task accuracy of sample tasks are shown in Table 2. Operation time is significantly different between ages($p < 0.001$). Among aged groups,

Table 2. Averages of operation time and accuracy of sample tasks

task		operation time(sec)				task accuracy			
		search		input		search		input	
known/new		known (task1)	new (task2)	known (task3)	new (task4)	known (task1)	new (task2)	known (task3)	new (task4)
aged	ex.	107.9	257.4	137.9	332.3	1.000	0.875	0.250	0.750
(employed)	in.	97.9	393.6	281.6	298.6	0.875	0.875	1.000	1.000
aged (not-	ex.	67.3	796.2	228.0	439.7	0.833	0.500	0.667	0.667
employed)	in.	96.9	403.7	182.0	201.3	1.000	0.286	0.857	0.857
young	ex.	37.8	98.3	98.5	118.8	1.000	0.875	0.375	0.750
	in.	56.6	69.8	69.3	97.9	1.000	1.000	0.625	0.875

ex.: exploratory training, in.: instructional training

Table 3. Averages of task cognition points of aged subjects

	"difficult"		"tiresome"		"fun"		"suitable"		"interest"		"helped"	
	ex.	in.	ex.	in.	ex.	in.	ex.	in.	ex.	in.	ex.	in.
employed	2.06	2.94	2.13	2.56	4.13	4.38	4.00	3.56	2.88	4.25	4.63	4.38
not-employed	3.83	2.29	3.33	2.21	3.58	4.57	2.92	3.86	2.83	3.43	3.67	4.43

ex.: exploratory training, in.: instructional training

there is a significant interaction of training conditions and occupation. For subjects currently not employed, exploratory training results in longer operation time than instructional training. For subjects currently employed, exploratory training results in shorter operation time.

Task accuracy is not significantly different between ages and between occupations, while the main effect of training conditions is significant ($p < 0.05$). For both young and aged subjects, instructional training results in more accurate task performance.

Table 3 shows the task cognition points of aged subjects. Interactions of training conditions and occupations are significant for the points of "difficult" ($p < 0.001$), "tiresome" ($p < 0.01$), "fun" ($p < 0.05$), "suitable" ($p < 0.01$) and "helped" ($p < 0.05$). For subjects who are currently employed, exploratory training tends to result in more positive cognition. Besides, the instructional training results significantly higher cognition points for "fun"($p < 0.001$) and "interesting"($p < 0.01$).

None of the adult subjects have any experience in university-related jobs similar to the tasks used in this experiment. The results suggest that effectiveness of exploratory training is influenced by the trainee's knowledge even if it is not directly related to the task to learn. Subsequently, the trainee's occupation and background knowledge need to be considered in relation with the skill to be acquired when a training method is selected.

REFERENCES

1. J.M. Carroll, R.L. Mack, C.H. Lewis, N.L. Grischkowsky and S.R. Robertson, "Exploring Exploring a Word Processor", Hum.-Comput. Interaction, 1 (1985) 283

Design principles for a training software agent: issues concerning agent-user interaction

A. Farias[a] and T. N. Arvanitis[b]

[a]The Virtual University, ITESM, E. Garza Sada 2501, Sur, Mty. Mexico

[b]School of Cognitive and Computing Sciences, University of Sussex, Falmer, Brighton, BN1 9QH, United Kingdom

1. INTRODUCTION

We are currently experiencing a paradigm shift in the way we use technology. The actual trend is moving away from using computers as calculation engines, to a culture where computation serves as a communication and simulation medium [1]. Computers of today are tools which let us build intricate models of complex aspects in our world. Still, there is no doubt that computers will get more complex with time [2]. For these reasons, it is feasible to think that we need to build computational aids to augment our learning, memory and problem solving abilities. These tools can be cognitive artifacts which provide us with the necessary skills to cope with the complexities of our technological societies.

In this paper, we present an interaction framework to develop computer systems that can help people learn dealing deal with complex knowledge domains. This framework supports asynchronous collaboration between computer users in distributed environments like the World Wide Web. It is common that users of the Web have different levels of expertise, and at the same time, share a common information need: be it an interest in a topic, a need to learn to use a system, or an answer to a situation of concern. We based our design in these differences, and defined a situated communication structure, modelling the interaction between experts and novices in real world situations. Also, we allowed redundancy and mobility to provide multiple representation styles and support information migration. Thus, facilitating different learning practices and providing "just-in-time" assistance and guidance.

2. THEORY: SOFTWARE AGENTS FOR "CALM TECHNOLOGY"

An unavoidable fact about computers is that they are becoming "ubiquitous". In the near future, computers will be present in everyday artifacts, creating a vast network of small, interconnected, and communicating objects. In order to cope with this technological overflow, we need to shift our attention away from computation. This requires leaving computers in a proximal periphery and focusing on "calm technology" [3]. In other words, building technology that gets our attention when we need to use it, but that relegates itself to the background, and stays aside, actively waiting to assist us. Calm technology includes systems that operate and do things in our behalf, empowering and not enslaving us. Systems that can learn about our needs, preferences and interaction styles, but still leave us in control of the situation. The goal is to build consistent, correct and situated models of technology, so that we can trust and understand its operation, and delegate it important tasks.

In here, we propose a general approach to develop this kind of technology. We look at methods to use software agents in information delivery systems and computer learning environments. The aim is to build agents which serve as simple "communicators and containers of knowledge" but not creators of it. Our software agent framework is based on the premise that information itself - not any particular application - is becoming the focus of work. This creates the opportunity to reduce the visibility of user interfaces as we know them today, presenting the user with fragments of the information they need, where and when they need it [4].

3. APPROACH: THE CONSULTANT AGENT

The kind of software agent we present here is conceptualized as a medium which supports the collaborative process between people. Therefore, these agents always maintain users' smartness and autonomy [5] . They achieve this by gaining experience and knowledge from certain users and by communicating this knowledge to other users. Thus, these software agents, rather from being distrusted because they do things the user would prefer to do, are useful precisely because they perform actions s/he would rather not do.

The design methodology we adopt concentrates on including software agents in Computer Based Training systems. We have conducted some preliminary research to establish a set of requirements and defined an agent to support training in real working environments [6]. This agent facilitates the transfer of skills and practice between people with different levels of expertise, both novice and expert practitioners of a knowledge domain. We called this agent a "Consultant". The name derives from the dual role it takes in a learning environment. Firstly, it collects knowledge and experience from expert users and, secondly it communicates with novice users and makes this knowledge available to them. In other words, the agent works both as a personal "assistant" for expert users and as a "demonstrator" for novice users.

The Consultant agent is an independent module within a computational environment and has control over its interface. It is personalised, long-lived and cooperative. It perceives the user's interactions within the environment and provides help by using a knowledge and media (knowledge-media) base and by collaborating with other agents. Also, the agent keeps a record of the history of the user's actions and acts depending on a set of high level themes which are application dependent.Thus the Consultant agent is both a knowledge-base [7] and a learning agent [8]. It acquires knowledge by building a context-dependent knowledge-media base, and gets experience by interacting with the user, and negotiating advise from other Consultant agents.

4. DESIGN: INTERACTION PRINCIPLES

In order to define the communication between a Consultant agent and different kinds of users (both experts and novices) we devised a set of five main interaction principles:

1. Consult or Receive
2. Evaluate or Volunteer
3. Instruct or Train
4. Inspect or Memorize
5. Control or Customize

These principles have dual names because they can be analyzed from two perspectives: a representational or structural view, and a functional or operational view. From a structural point of view, they provide the usability of the model, and convey a working metaphor based in the actual dynamics when training and learning takes place (see Section 4.1). From an operational perspective, we suggest a set of current methodologies to give functionality to our Consultant agent architecture (see Section 4.2).

4.1. Structural Dimensions:

CONSULT : The user (whether expert or novice) asks the Consultant agent for more information on a particular learning situation.
EVALUATE: The Consultant agent evaluates the novice user's performance, or the expert user evaluates the Consultant agent's proficiency in the knowledge domain.
INSTRUCT: The expert user "trains" the Consultant agent about his needs and preferences in the particular knowledge domain.
INSPECT: The user (whether expert or novice) "inspects" his agent's memory, or asks for more help, compiled/analysed by the Consultant agents of other users.
CONTROL: The user (whether expert or novice) sets the Consultant agent's external appearance (interface) and examines/modifies its "brain" or internal mechanisms.

4.2. Functional dimensions:

RECEIVE (or push knowledge): The Consultant agent makes use of a compiled knowledge-media base to provide non-intrusive advise. The user can request just-in-time assistance in a particular learning context. The agent will decide, according to a set of customizable interaction primitives, how and where to present this declarative knowledge.
VOLUNTEER (or pull knowledge): The user actively seeks for a particular piece of information in the knowledge-media base, thus enriching the agent's proactive assistance. He can rate the effectiveness of the agent's support, and quality of knowledge. In the same manner, the agent can use the "user profile" to evaluate the user's proficiency.
TRAIN: The agent sets the "user profile". This profile stores the user's knowledge preferences and interaction behaviours. There are a number of different techniques that can be employed to train the agent. They are classified into two categories: "explicit" and "implicit" training. During implicit training, the system uses techniques such as programming by demonstration [9], so that the agent can record the user's actions without direct intervention (i.e. machine learning system). In explicit training, the user sets his profile by intentionally "telling" the agent what are the relevant pieces of information in a particular situation. In this method the agent actively queries the user in order to know more about his preferences (i.e. rating-based system).
MEMORIZE: The agent keeps track of the user's actions by recording a set of situation-action pairs in a given context. The criteria to record the user behaviour is defined by his profile. This procedural knowledge is available to the user for inspection. He can "down-load" procedures into similar contexts in which they were recorded, thus recalling his past actions [10]. If the agent lacks experience in a given situation, it can "ask" other agents whose users have similar profiles, then "borrow" their knowledge and make it available to user. These methods provide

both a "content-based", and "collaboration-based" aggregation and filtering of knowledge [11]. CUSTOMIZE: The user can define some of the agent's interface primitives. For instance, the use of facial expressions and a voice to provide feedback. Also, the agent internal model is available to the user in the form of dialogue boxes, for inspection and modification. This could provide a way to "fine-tune" the agent's behaviour and affect the frequency, quality and amount of support.

5. DISCUSSION

In brief, the main objective of our research is to establish a framework to design software agents that can assist users while learning a complex knowledge domain. We use a learner-centred approach where the aim of the Consultant agent is to be a communicator of knowledge between users with different levels of expertise.

In addition, we devised a set of principles for the interaction between the user and the Consultant agent. By maintaining a number of functions, we provide flexibility and modularity in the design and allow for mobility of information. We prefer to concentrate on defining an open interaction and information metaphor instead of constraining our design to a particular interface. In other words, a Consultant agent can easily fit into different contexts; for example the World Wide Web in a training scenario, or a Personal Digital Assistant or embedded system in professional or entertainment scenarios.

REFERENCES

1. P.J. Denning and R. M. Metcalfe RM (eds.), Beyond Calculation: The Next Fifty Years of Computing, Springer-Verlag, New York, 1997.
2. D. Norman, Why It's Good That Computers Don't Work Like the Brain, in P.J. Denning and R. M. Metcalfe RM (eds.), Beyond Calculation: The Next Fifty Years of Computing, Springer-Verlang, New York, 1997.
3. M. Weiser and J. S. Brown, The Coming Age of Calm Technology, in P.J. Denning and R. M. Metcalfe RM (eds.), Beyond Calculation: The Next Fifty Years of Computing, Springer-Verlag, New York, 1997.
4. T. Fernandes, Computing, 1 (1997) 47.
5. J. Lanier, Wired Magazine, 4 (1996) 157.
6. A. Farias and T.N. Arvanitis, Building Software Agents for Training Systems: A Case Study on Radiotherapy Treatment Planning, in K.S.R. Anjaneyulu, M. Sasikumar and S Ramani (eds.), Knowledge Based Computer Systems: Research and Applications, Narosa, New Delhi, 1996.
7. S. Russell and P. Norvig, Artificial Intelligence: A Modern Approach, Prentice Hall, New Jersey, 1995.
8. P. Maes, Communications of the ACM, 37 (1994) 31.
9. A. Cypher, Eager: Programming Repetitive Tasks by Demonstration, in A. Cypher (ed.), Watch What I Do: Programming by Demonstration, MIT Press. Cambridge, Mass. 1993.
10. B. J. Rhodes and T. Starner, Proceedings of The First International Conference on The Practical Application of Intelligent Agents and Multi Agent Technology (PAAM'96), London (1996) 487.
11. M. Balabanovic and Y. Shoham, Communications of the ACM, 40 (1997) 66.

Action Workflow Loops in Distance Education - Design Principles for Integrating CAL and CMC

Paul Johannesson
Department of Computer and Systems Sciences
Stockholm University and the Royal Institute of Technology
Electrum 230, S-164 40 Kista, Sweden
email: pajo@dsv.su.se

1 Background

Distance education is a form of education that offers many advantages, in particular that students can freely choose a time and place for their studies. However, distance education also gives rise to many problems. The most severe problem is that distance education puts a heavy burden on the individual student, who has to work with limited contact and support from teachers and fellow students. It has been argued that this problem can be overcome by learning environments that utilize modern information technology. Simplifying somewhat, information technology can be used for distance education in two different forms: Computer Assisted Learning (CAL) and Computer Mediated Communication (CMC). CAL systems may range from information resource databases over drill and practice systems to simulation models and full-fledged tutorials. CMC refers to a wide range of asynchronous, text or multi-media based computer conferencing systems.

There are indications that CMC may have a profound impact on the educational system and dramatically alter the relationships between students, teachers, and educational institutions, [Mason89], [Berge95]. However, CMC is not without its problems. In particular, the medium may alienate students lacking computer or writing skills. One approach to addres these problems is to combine CAL and CMC. In this paper, we suggest a set of novel design principles for incorporating computer mediated communication in CAL systems. These design principles build on the action workflow approach to workflow management.

2 The ActionWorkflow Approach

The ActionWorkflow approach to organisational communication and action is based on a theory of work structure as language action, [Medina-Mora92]. The workflow in an organisation is viewed as a set of interleaved "loops" of actions in which people interact. Figure 1 shows the basic sequence of actions in the action workflow loop. The loop always involves two agents, a performer and a customer, and it describes a particular action that the performer agrees to carry out to the satisfaction of the customer. The loop consists of four phases:

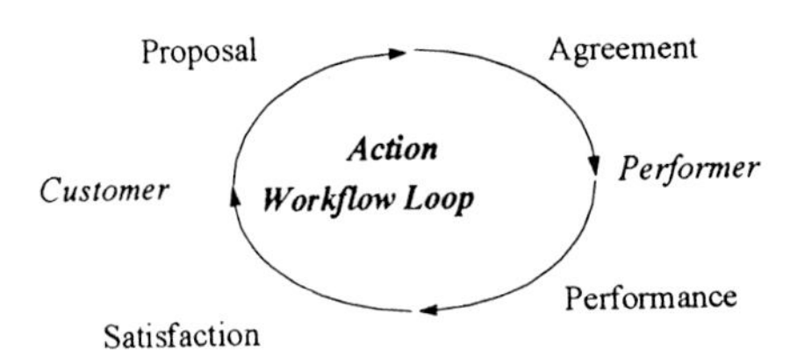

Fig. 1: The structure of an action workflow loop

1. *Proposal*
The customer requests the performer to carry out an action.
2. *Agreement*
The two agents come to mutual agreement on the conditions of satisfaction. This phase is often implicit and rests on a shared background of standard practices.
3. *Performance*
The performer declares to the customer that the action is complete.
4. *Satisfaction*
The customer declares to the performer that the action has been carried out satisfactorily.

During the execution of an action workflow loop, new loops may be initiated, as shown in Figure 2. Optional loops are indicated by dotted arrows and may execute asynchronously with the loop that initiated them.

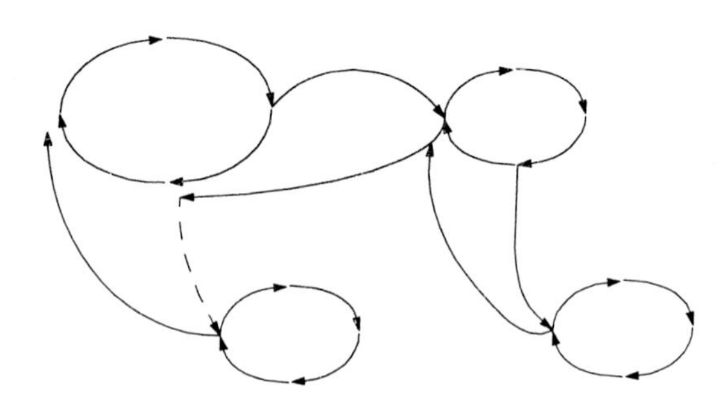

Fig. 2: Interrelated action workflow loops

3 Design Principles

In order to combine CAL and CMC, the following three design principles are suggested:

> 1. All communication between the CAL system and the student shall be structured in the form of action workflow loops.
> 2. Whenever the student is to take an action in a workflow loop, she should be allowed to initiate new loops, which may involve communication with teachers and/or fellow students.
> 3. When the CAL system closes a loop, it should give detailed information about the student's actions in the loop.

An example of applying these design rules is shown in Figure 3 below. The example shows the interactions between a student and a simple "drill and practice" CAL system, which presents questions to the student and corrects them. The main loop (TASK) starts with the student asking for a task, which the system unconditionally accepts. In order to carry out the request, the system starts another loop (QUESTION) and presents a question to the student. The student may, before accepting the question, start another loop (CLARIFICATION) and ask for complementary information and help. Before submitting the answer, the student gets the opportunity to initiate yet another loop (CRITIQUE OF QUESTION) and engage in a computer mediated communication with the teacher about the question. The next action of the QUESTION loop consists of the system assessing the student's answer. According to design principle 3, this assessment should be detailed and informative. Thereafter, the main loop (TASK) continues, and before acknowledging, the student may initiate another dialogue (CRITIQUE OF SOLUTION) with the teacher about the solution of the question.

Adhering to the design principles above provides a number of advantages. First, the principles assist the designer of a CAL system to identify those steps in the dialogue, where a student should be allowed, or even encouraged, to initiate new dialogues with other people. In this way, CMC is naturally integrated into the CAL system. Furthermore, the principles encourage the designer to think carefully about which agents should be involved in each loop; in the example above, the sub-loops initiated by the student involved either the CAL system or the teacher, but they might well have involved fellow students, newsgroups readers, or others. Finally, the principles support the designer in providing an adequate level of feedback from the system by insisting that it be verbose when closing workflow loops.

The design principles discussed in this paper are being implemented and evaluated within a project, "Hypermedia and Communication for Active Learning", supported by the Swedish Council for the Renewal of Undergraduate Education, see URL http://www.dsv.su.se/~pajo/GRUB.html.

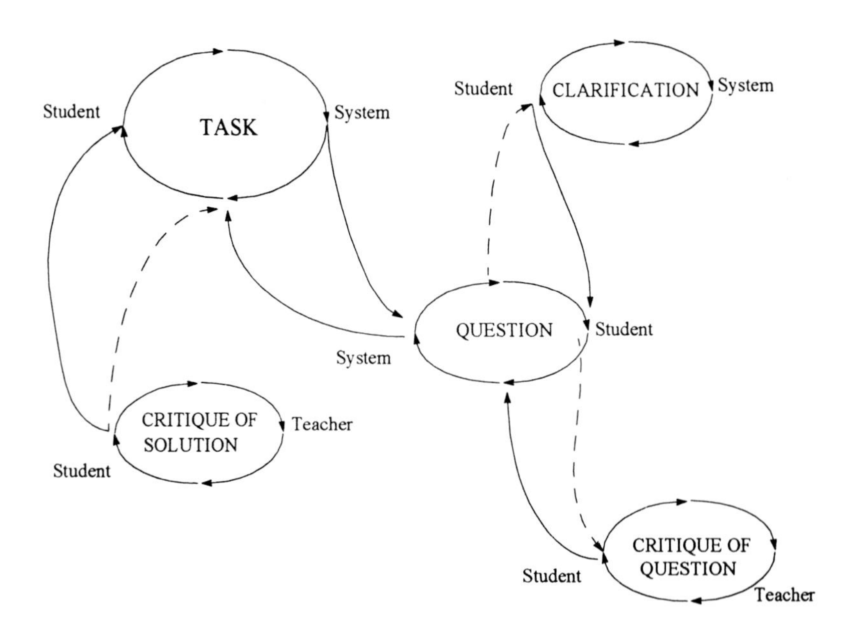

Fig. 3: Workflow loops in a simple CAL system

References

[Berge95] Z. Berge and M. Collins: *Computer Mediated Communication and the Online Classroom,* Hampton Press, 1995.

[Mason89] R. Mason and A. Kaye: *Mindweave: Communication, Computers and Distance Education,* Pergamon Press, 1989.

[Medina-Mora92] R. Medina-Mora et.al.: The Action Workflow Approach to Workflow Management Technology, *Proceedings of the Fifth International Conference on CSCW,* 1992.

Reducing Restriction of Activity in Media Communication with *Demand Driven Viewer*

Naohiko Kohtake[1] Yoshinobu Yamamoto[2] Yuichiro Anzai[1]

[1]Department of Computer Science, Keio University
3-14-1, Hiyoshi, Kohoku-ku, Yokohama 223, JAPAN
E-mail: {kohtake, anzai}@aa.cs.keio.ac.jp

[2]Electro-Technical Laboratory
1-1-4, Umezono, Tsukuba, Ibaragi 305, JAPAN
E-mail: yoshinov@etl.go.jp

Abstract

The design and implementation of a new media communication system - *Demand Driven Viewer* (*DDV*)- that reduces restricted activity of the user is described. DDV offers a panorama view, by translating a part of this view to picture frames where each user wants to see. To accomplish this, the user's head movement changes his field of view. The underlying concept is based on the fact that every person's view changes as the head moves. Our results demonstrate that a number of users can benefit from substantial speed-up to translate by *DDV*.

1. Introduction

In our experience, we feel something strange in a media communication compared to a direct communication due to the nature of the medium. We think a restriction of activity is one of the reason of this feeling. In today's media communication, even if remote camera control is possible, range of exploration is limited and logistics of control are slow. Moreover it is hard to transmit data in response to each host simultaneously. The problem with this approach are well understood. The Multiple Target Video(MTV) system[2] proposed the use of multiple cameras as a means of providing more flexible access to remote working environment Users were offered sequential access to several different views of a remote site. However, the configuration of camera will never be suitable. Furthermore, switching between views introduces confusing spatial discontinuities. Another approach involved the Virtual Window concept[3], which uses the video image of a person's head to navigate a motorized camera in a remote site. Unfortunately, this system provides a restricted field of view on a remote site. In this paper, we point out three main restrictions and going to address these problems.

2. Required Conditions

First, according to our observations of the previous media communication system, even if several users communicate with the same remote site, the required view by each user varies. It is impossible to provide reliable and efficient distribution of view when each user is free to choose the view he wants. To realize greater freedom, we fix several cameras spokenwise with each view interleaved in a continuous fashion. This is known as the panorama view. This will give the user choice of view in real-time. An important point to realize is that all users do not have to control the camera and to transmit the entire panorama view(Fig.1). In this way, it is possible to transmit the data to each user simultaneously using less bandwidth and thus obtain greater efficiency.

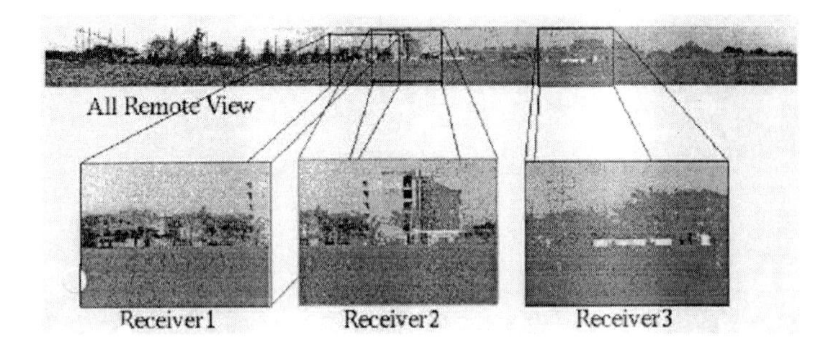

Figure 1: Picture Frames where each user want to see

Second, the previous system provided a restricted field of view on a remote site. There were discontinuities on the edge of scenes and between views from different cameras. When the person rotates his vision, they effectively increase their field of view. Movement might allow people to compensate for the discontinuities of current media communication. The second approach supports a continuous field of view, and active exploration rather than depending on passivepresentation.

Third, in the real world, a person's view naturally pans simultaneously with head movement in all three dimensions. However, in many current interfaces the user must perform panning and zooming using a mouse or a joystick. This forces the user to integrated two separate tasks. The third approach is the navigation of direct manipulation.

These analyses combine to suggest that it may help reduce the restriction of activity in a media communication system configuration. As a first attempt to provide this support, a prototype system, *DDV*was developed.

3. Implementation

Pentium PCs with a capture board as the control and several video cameras on the remote site were used. The machines are one a lightly-loaded Ethernet connected to several other LANs. Each user wears a lightweight helmet with a three- dimensional tracker mounted on the top and a head mounted display in front of his face(Fig.2). As the user moves from side to side, the display smoothly pans the panorama view. *DDV*'s software operates in two steps. The first step computes the position of user's head by the tracker on each user's site (Receiving Site). In order to reduce susceptibility to notice when the tracking, we apply a low-pass filter. The second step translates this parameter to the server (Remote Site)to get the view that the users want to see.

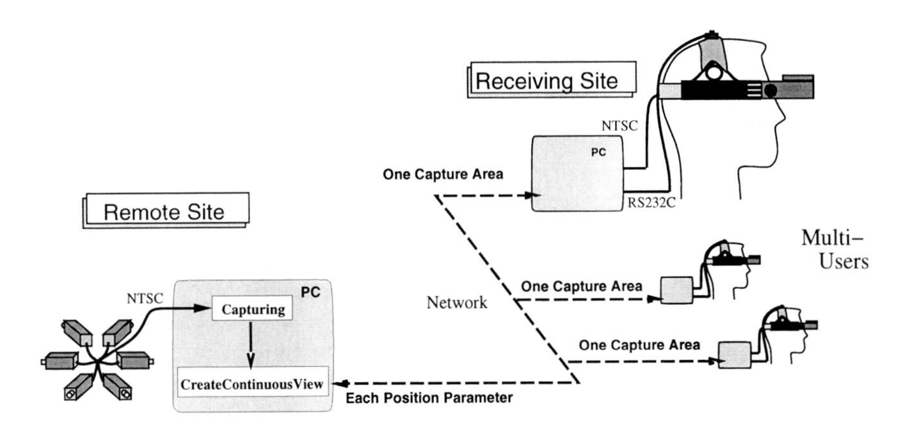

Figure 2: System Configuration

4. Performance

To obtain results using *DDV*, measurement of frame rates (Frame/Sec) were recorded with respect to the number of users at receiving sites. The system was tested over 20 repetitions per condition. Two normal pentium PCs were used, one situated at a remote site and the other at the receiving site. The remote site PC was used to process the pictures captured by the six cameras. It then transferred these frames through an Ethernet. At the receiving site, the PC would distribute the frames to the connected users. The result was that frame transmission was maintained at a constant fps value for up to 8 users, after which, degradation of fps fluctuated negatively.

To point out one advantage of this configuration, only one pentinum PC was used to distribute frames to connected users. A preferred configuration would be one PC for one user, so that frame transmission would hold at a constant rate resulting in a quality environment. One advantage of *DDV*is that it only sends one capture area of the panorama

Users	Max.	Min.	Ave.
1	6.12	5.78	6.02
2	6.07	5.68	5.97
3	6.10	5.64	5.99
4	6.06	5.67	5.88
5	5.90	5.56	5.83
6	5.98	5.62	5.87
7	5.89	5.54	5.76
8	5.73	5.50	5.61
9	5.21	5.01	4.90

Figure 3: FPS to Number of Users

Figure 4: User at Receiving Site

view. This means that the modern local area network such as Ethernet and FDDI would handle this bandwidth. To compare one capture area with the panorama view, the latter would result in slow transmission. In fact it would be impossible because the bandwidth required for panorama view was more than 40Mbyte/sec compared to a bandwidth of 40Kbyte/sec with one capture area. Ethernet can handle up to 1.25Mbyte/sec. These are just an evaluation of the results retrieved from the frame rate.

5. Conclusion

We have implemented a new media communication system - *Demand Driven Viewer (DDV)*- that reduces restricted activity of the user and recorded measurement of frame rates with respect to the number of users at receiving sites.Based on these analyses, an important contribution in using *DDV*was that it was able to provide less bandwidth when a user is seeing the view simultaneously. These abilities of *DDV*greatly enhance the user's sense of presence in media communication.

References

[1] Cooperstock,J.R., Tanikoshi,K. and Buxton,W.,"Turning Your Video Monitor into a Virtual Window". In *Proc. IEEE PACRIM, Visualization and Signal Proceedings,* pages 308–311,1995.

[2] Gave,W., Sellen,A., Heath,C. and Luff,P., "One is not Enough: Multiple Views in a Media Space". In *Proc. INTERCHI'93 Conference Proceedings,* , pages 335–341,1993.

[3] Gaver,W., Smets,G. and Overbeeke,K., "A Virtual Window On Media Space". In *Proc. CHI'95 Conference Proceedings,* pages 257–264, 1995.

[4] Hix,D., Templeman,J.N. and Jacob,R., "Pre-Screen Projection: From Concept to Testing of a New Interaction". In *Proc. CHI'95 Conference Proceedings,* pages 226–234,1995.

[5]Yamaashi,K., Cooperstock,J.R., Narine,T and Buxton,W., "Beating the Limitations of Camera-Monitor Mediated Telepresence with Extra Eyes ". In *Proc. CHI'96 Conference Proceedings,* pages 50–57,1996.

Lock-On Pointer: A Foundation for Human-Object Interaction

Nobuyuki Matsushita Mineo Morohashi Yuichiro Anzai

Department of Computer Science, Keio University
3-14-1, Hiyoshi, Kohoku-ku, Yokohama 223, JAPAN
E-mail: {matsu, moro, anzai}@aa.cs.keio.ac.jp

Abstract

We design and implement a pointing device called Lock-On Pointer as the instinctive User Interfaces. Direct object-pointing is useful for direct manipulation in real word. However, we feel annoyance when we point objects with pointing devices, such as a laser pointer, since we can't control subtle movement of our hands. Also, object-pointing becomes more difficult with a distance from us to the object. "Lock-On Pointer" is a new pointing device with *Lock-On mechanism* that has 2 modes, Straight mode and Lock-On mode. We can choose right mode whenever we want. Thus Lock-On Pointer is able to identify the objects easily, and will give an agreeable feeling for the way of pointing objects.

1. Introduction

Computers are becoming increasingly portable and ubiquitous, as recent progress in hardware technology has produced computers that are small enough to spread a lot around the environment. Weiser proposed the idea known as *ubiquitous computing*[1], that tries to make the world where a large number of computers embedded all around us so that they become invisible part of our surroundings. In addition, there are some examples[2][3] in which computers are embedded in all appliances, and connected each other. Using traditional computer network, we need not to take each computer's location into consideration owing to location transparency. In contrast, in case of appliance network as mentioned above, such case could occur that we would like to point and control objects directly. There will be many situations in which we would like to interact with the objects surrounding us. As a consequence, there arise a possiblity to specify them directly so that we should be able to control them. *Augmented Reality*[4][5] or *Augmented Interaction*[6]can make these possible by presenting a virtual world that enriches the real world, that endeavors to integrate information displays into the everyday physical world. Instead of blocking out the real world, this approach annotates reality to provide valuable information, such as descriptions of important features or menus for controlling. From these point of view, we can assume that a large number of objects with computers will be embedded all around us.

However the essential interface, with which we control the objects or about which we get information, is not only pointing secondarily with seeing Head Mounted Display(HMD) or See-through Display, but also pointing the objects directly in real world. Pointing the objects in real world would be the most instinctive way to select and control the objects. This is exactly what we say as instinctive User Interface. To do so, we have to identify an object on which we would like to focus. We are particularly interested in the role of pointing devices that point the objects.

It's possible to point objects with pointing devices, such as mouse, easier through display. Even if the objects are hard to determine due to the size, it can be done by zooming the window of the display. However, on the other hand, to interact with objects existing in real world, things are not so easy. We are confronted by following two difficulties, when we use traditional pointing devices, such as laser pointer. The first is that when we would like to point objects located far ahead, it's difficult to focus on them with high accuracy. Even if amplitude of pointer's movement is little, the movement of the position grow larger with distance in which we point, compared with close distance. The second is that we can't avoid movement of the hands with which we manage pointing devices. This causes laxity for pointing objects, particularly in case of pointing long-distance objects.

Thus, considering these situations, it would be worthwhile to research how effective pointing device should be for instinctive User Interface. Lock-On Pointer that we propose will be one solution for the way of effective pointing.

This paper is organized as follows. Section 2 states an overview of Lock-On Pointer with the mechanism. Next, the prototype system we implemented to investigate the effect is described in Section 3. Finally, we note conclusion in Section 4.

2. Lock-On Pointer

Lock-On Pointer allows pointing at an object directly, like a laser pointer. Moreover, to change a pointing position in a narrow pointing scope, Lock-On Pointer can **"lock on"** the object. Hence Lock-On Pointer makes it possible to point in which is independent of distance wherever the user wants.

To obtain a position in which user want to point, Lock-On Pointer equips 3D Tracker that senses an angle of direction of Lock-On Pointer (Fig.1).

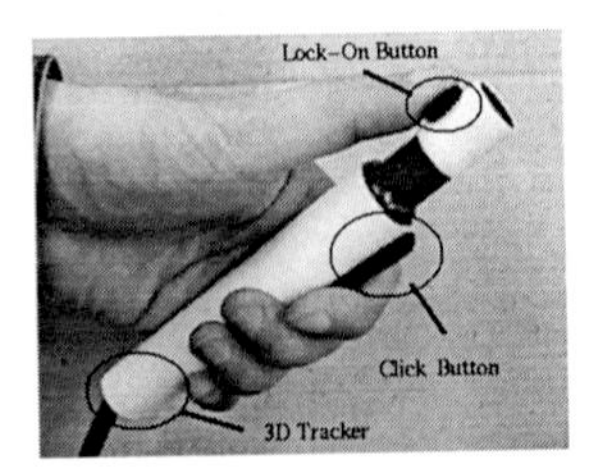

Figure 1: Overview of Lock-On Pointer

For controlling, Lock-On Pointer offers two pointing modes, *Straight mode* and *Lock-On mode.* In Straight mode, Lock-On Pointer points directly to the object in alignment with the pointing line. In Lock-On mode, the position of the pointer cursor moves slowly compared to the real movement of the pointer, and doesn't select an object in a straight line. The user can switch over these two pointing modes whenever the user want, so that the user would be able to control the pointer cursor in detail, even in long distance, by "locking on" the object with the Lock-On Pointer.

Lock-On Pointer has two buttons, Lock-On button and Click button (Fig.1). The user can switch over the pointing modes by using former one. While the user is pushing Lock-On button, it is in Lock-On mode. Releasing the button means, it is in Straight mode. If Lock-On Pointer directs toward an object that the user wants to point, he can click the Click button to select the object.

"Lock-On" mechanism functions as follows: First, the user points an object directly with Lock-On Pointer, in Straight mode. If the selected object is what the user want to select, click the Click button. If not, the user can change the mode to Lock-On mode. Pushing the Lock-On button enables fine adjustment of the pointer cursor position. When the user pushes it, a Virtual Screen appears between the user and the object(Fig.2). The image of the pointing line on the Virtual Screen is projected onto the Real Screen. This spot becomes the new pointing position of the Lock-On Pointer. Using this system, the user is able to control the pointer cursor more accurately and therefor can point to a certain object no matter how far it is. This system can be said to be a reduction of zooming technique into the real world.

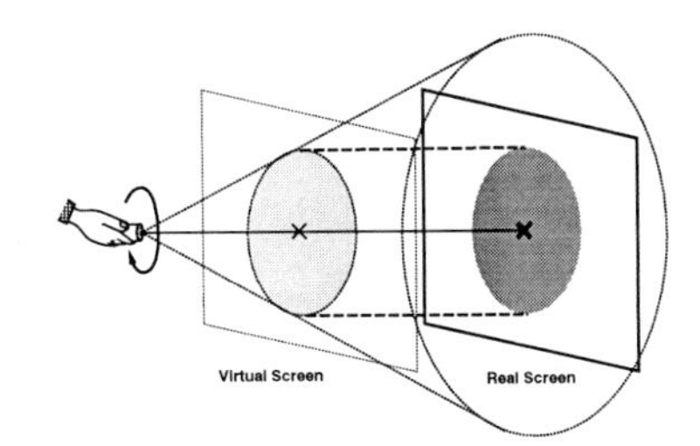

Figure 2: Key Map of Lock-On Pointer

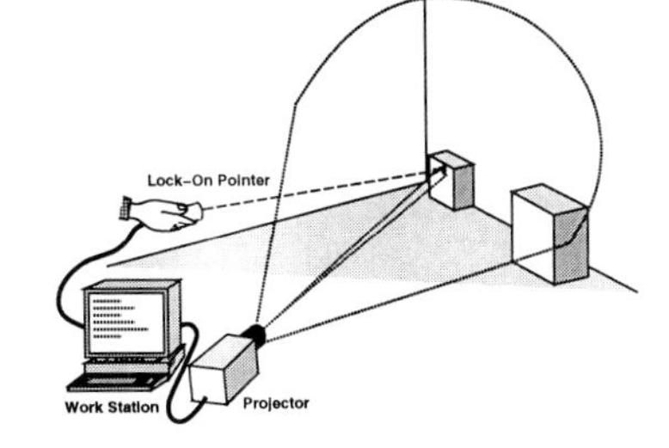

Figure 3: Prototype System

3. Prototype System for Lock-On Pointer

We build a Lock-On Pointer prototype to further investigate the effect of instinctive User Interface. The prototype consists of the Lock-On Pointer, an SGI Indigo2 and a projector (Fig.3).

Since Lock-On Pointer has a 3D Tracker, Lock-On Pointer detects the location where the user wants to point by the position and direction of Lock-On Pointer. For that, we have to measure the real position of all target objects. In this implementation, we structures a simple allocation in which there exist several objects that is independent of distance in front of a wall.

For feedback of a pointing position, we project the pointer cursor onto a real world, wall and objects. The reason why we choose the way is to show the importance of direct pointing in real world. This is only prototype, so we choose the easiest way to implement it. In near future, the projector will be more smaller, and some of objects might have displays with them. Thus we will be able to make the system more smartly.

Pointing sequential procedure is as follows:

1. the user points in which he wants to select.
2. the pointing position of Lock-On Pointer is calculated by the "Lock-On" mechanism.
3. the pointer cursor is projected as feedback of pointing.

To show feedback of the pointing position, the system projects the pointer cursor onto real objects by a projector. Since we have to make this system in real world, not in virtual world, we use projector and show the feedback in real world. If we make this system in virtual world, it makes no sense to point object directly. Furthermore, we can calibrate the rate of distance between Real Screen and Virtual Screen.

To take off a swing of Lock-On Pointer as a reflection of movement of hands, we strike an average of latest one second.

4. Conclusion

We have designed and implemented a new pointing device that allows us to point a spot, without subtle movement of hands, on which we would like to focus. Lock-On Pointer, with angle toleration, provided smooth and accurate pointing function. We may go on from this to the conclusion that Lock-On Pointer is a new pointing device that is easy to manage for pointing.

References

[1] M.Weiser: The computer for the twenty-first century. *Scientific American*, pp.94–104, 1991

[2] Y.Fujita, S.Lam: Menu-Driven User Interface for Home System. *IEEE Transaction on Consumer Electronics*, Vol.40, No.3, pp.587-597, 1994

[3] G.Leeb: A User Interface for Home-Net. *IEEE Transaction on Consumer Electronics*, Vol.40, No.4, pp.897-902, 1994

[4] S.Feiner, B.Macintyre, and D.Seligmann: Knowledge-based augmented reality. *Communication of the ACM*, Vol.36, No.7, pp.53–62, 1993

[5] M.Bajura: Merging virtual objects with the real world: Seeing ultrasoundimagery within the patient. *SIGGRAPH'92*, pp.203–211, 1992

[6] J.Rekimoto, K.Nagao: The World through the Computer: Computer Augmented Interaction with Real World Environments. *UIST'95*, pp.29–36, 1995

A Sensor Network Management System
- An Overview and Example -

Mitsuhiko Ohta Yuichiro Anzai

Department of Computer Science, Keio University
3-14-1, Hiyoshi, Kohoku-ku, Yokohama 223, JAPAN
E-mail: {ohta, anzai}@aa.cs.keio.ac.jp

Abstract

This paper describes an overview of a Sensor Network Management System for Intelligent Sensor Actuator Network (ISAN). This system is effective for a Human-Computer Interaction using many sensors and actuators. And multiple applications share many sensors embedded into environments. This paper introduces the overview of a Sensor Network Management System and suggests some instances of using this system.

1. Introduction

We suggest Intelligent Sensor Actuator Network (ISAN) which supports a safety and an amenity life. ISAN is a network that consists of a lot of sensors and actuators connected each other. These sensors and actuators are embedded into real world environments, for example, room, house, building, road and so on. They are sensing environments, and ISAN gives various information to people.

When construct a system such as a multimodal operational system and a teleoperation system, so far there has been necessary to make networks that connects necessary sensors and actuators, and these applications to manage these sensors, actuators and networks. And these sensors will be used by the system and another systems can't use.

We introduce a Sensor Network Management System for ISAN, which manages ISANs and allows applications accessing various sensors. And it also offers interfaces that can operate a robot and use sensors on the robot. This system will be helpful to construct Human-Computer Interaction systems.

In the following, we describe the overview of a Sensor Network Management System and ISAN. And we suggest some instances of using this system.

2. Intelligent Sensor Actuator Network (ISAN)

Our destination of ISAN is making circumstances of infrastructure, so that we would be in safe from potential danger in environments in human life. And ISAN is available to make applications need some sensors and actuators such as Human-Computer Interaction.

Each sensor devices are connected to sensor modules that have a processor, memories and I/Os. The sensor module can perform some control functions like polling sensors locally rather than relying on a central computer. Sensor modules are connected to a PC or a work station (call "host" in this paper) through some network media like serial-bus and ether net.

3. An Overview Of A Sensor Network Management System

This system manages ISAN and gives applications using ISAN (call "clients" in this paper) interfaces that operate sensors and actuators of ISAN. And it also offers interfaces that can operate a robot and use sensors on the robot. The system and an overview of ISAN is given in Figure 1.

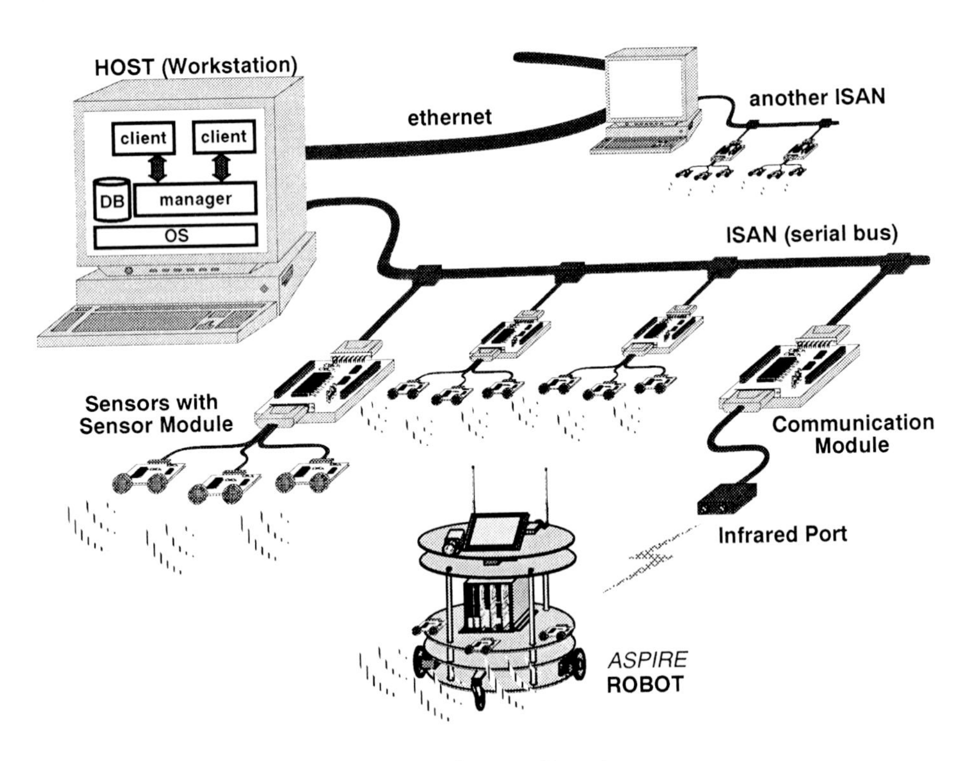

Figure 1: System Overview

This system can make clients use sensors and actuators connected with various media without concern with the difference of a medium. And this system can also make clients use sensors not connected to ISAN and sensors connected to the other network. Clients can use these sensors with the same interface as if they are connected to the ISAN.

Now, serial-bus, infra-red, and ethernet are available for network media. Clients can be programmed using class libraries offered by the system. Many systems for Human-Computer Interaction need a lot of sensors and actuators. To construct these applications, this system will be very useful.

This system makes one sensor module perform different operations requested by many clients at the same time. In ISAN, there will be a sensor module that can not perform complicated operation and multiple operations because of its simple hardware. Such operations will be performed in the manager's side. Clients send requests which operate in the sensor modules to the manager. These operations are given by a script. The manager investigates the job list of the sensor module, decides where to operate, and sends the script to the sensor module or creates process to perform the script in the host. So clients can use sensors without concern where the script will be performed and share many sensors embedded into environments.

For example, one client sends a request to the manager which asks to send a signal when the value of the sensor is over "100". If the sensor is not operating other requests, the manager sends this request to the sensor module. The sensor module receives the request, checks the sensor data, and sends a message to the manager when the value of the sensor became over "100". When the other client sends a different request to the same sensor module, the manager makes a message to send the sensor module and a process in the host. When the sensor module receives this message, it stops its operation and do the new operation that send a sensor data to the process. The process receives sensor data and manage two operations.

This system also allows clients use many sensors of robots as they are connected to this network. We implemented a server on *μ-PULSER* [1] which is an operating system for a personal robot and used *ASPIRE* robot [2]. This server sends data from sensors to the manager at the request of clients. The robot can communicate through the Communication Module connected on ISAN with IrDA.

4. Examples of using this system

For instance of this system, we introduce our laboratory guiding system which can switch user interface depend on a location of a user. This guiding system guides our laboratory using the *ASPIRE* robot and workstations with Identification Pendant (ID Pendant) [3] put on the human user. The ID pendant is small box which is capable to send user ID using infrared signals every second, which is received by a ID receiver module on the robot and connected to the ISAN. Depending on a user's location, the guiding interface can be switched from the robot to the workstation (See Figure 2).

This guiding system gets the location of user using ID receivers and sensors connected to the ISAN. Then it tells the user to follow the robot though the nearest workstation and calls the robot to move there. The robot guides the user to the demonstration place, then it checks whether the user is following using ID receivers on the robot. If the user leaves from the robot, it searches the user using ID receivers and sensors of the ISAN, and guides using the nearest terminal of a workstation. This guiding system can dynamically change a user interface depend on the user's location.

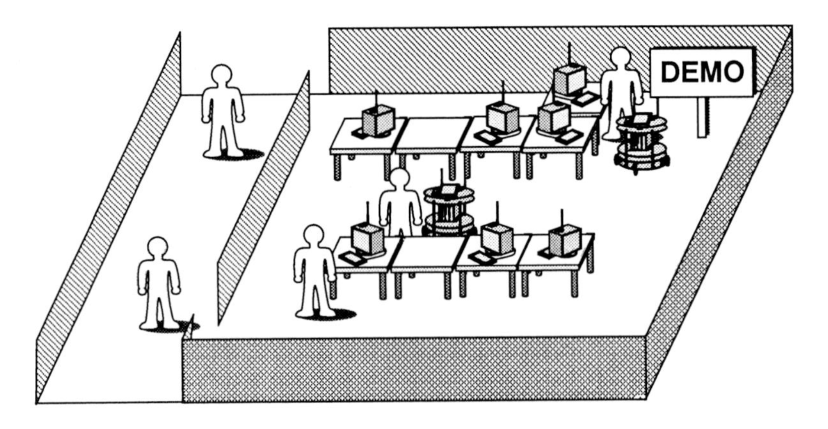

Figure 2: The guiding system

5. Conclusion

We describe the overview of a Sensor Network Management System for Intelligent Sensor Actuator Network (ISAN) in this paper. And we introduce a instance of using this system which guide a laboratory. This guide system can switch its user interface depend on a location of the user. Using this management system, such application for Human-Computer Interaction can be made easily.

References

[1] T.Yakoh, T.Sugawara, T.Akiba, T.Iwasawa, and Y.Anzai. "*PULSER*: A Sensitive Operating System for Open and Distributed Human-Robot-Computer Interactive Systems". In *Proc. IEEE Int. Workshop on Robot and Human Communication*, pages 404–409, 9 1992.

[2] Nobuyuki Yamasaki and Yuichiro Anzai. "The design and implementation of the personal robot hardware architecture *ASPIRE*". In *Proceedings of the 48th Annual Conference of Information Processing Society of Japan*, volume Vol.6, pages 91–92, 1994.

[3] Kaoru Hiramatsu and Yuichiro Anzai. "User Identification in Human Robot Interaction Using Identification Pendant". In *Proceedings of the 6th International Conference on Human-Computer Interaction*, Vol.2, pages 237–242.

A Fundamental Mechanism of Intelligent Sensor-Actuator Networks

Soutaro Shimada[1] Yoshinobu Yamamoto[2] Yuichiro Anzai[1]
[1]Department of Computer Science, Keio University
3-14-1, Hiyoshi, Kohoku-ku, Yokohama 223, Japan
[2]Electro-Technical Laboratory
1-1-4, Umezono, Tsukuba, Ibaragi 305, Japan
E-mail: shimada@aa.cs.keio.ac.jp, yoshinov@etl.go.jp, anzai@aa.cs.keio.ac.jp

Abstract

This paper describes a learning architecture for Intelligent Sensor-Actuator Networks (ISANs). An ISAN is a robotic environment to stave off dangers and to guarantee amenities in everyday human life. It is a network of sensors, actuators and computers equipped in the real world environment, such as a room, an office, a house, etc., towards *a new type of Human-Artifact Interaction.* Because the real world changes continuously, the ISAN must adapt to the changes all the time. Thus, it is necessary that the ISAN has a learning capacity so that it can automatically modify configurations of sensors and actuators. We propose an architecture for the ISAN based on distributed objects and a classifier system to adapt to the environmental changes dynamically.

1. Introduction

Our daily life is becoming convenient and comfortable by recent scientific development. We can buy anything at convenience stores, and we can know anything from our desktop computers connected to the WWW. However, dangers in daily life are still remaining or even growing, for example, traffic accidents, accidents and injuries caused by a mistake on a machine operation, and so on. Since artifacts are created aiming at comfortable life of human beings, it is unacceptable that human beings are suffering from accidents created by the artifacts. We should improve this undesirable situation and start designing the artifacts to become more responsible for their own action.

Intelligent Sensor-Actuator Networks are a technology that makes artifacts have responsibilities for their activities and are towards a new type of Human-Artifact Interaction. An ISAN is a network that consists of a lot of sensors and actuators designed to stave off dangers and guarantee amenities in everyday human life. It watches our daily life through sensors, and if we are faced with a danger, it acts to stave off the danger by using actuators. Our life with the ISANs will become safer and more comfortable.

Because the ISANs are faced with the real world environment that changes continuously, they must adapt to the environmental changes all the time. Thus, the ISANs must have a mechanism for adaptation so that they can handle a novel situation without stopping their services. In this paper, we propose a learning architecture to accomplish this.

This paper is organized as follows. Section 2 describes an overview of the ISAN and goes to difficulties of machine learning techniques in the ISAN. Then, a learning architecture for the ISAN is proposed in Section 3. Finally, we note future directions in Section 4.

2. Computational Basis of the ISANs

An ISAN is a network that consists of a lot of sensors and actuators equipped in the real world environment and some computers that manage the data created by them. From the computational viewpoint, the ISAN is based on the distributed objects environment(DOE)[1] so that all sensors and actuators can be handled similarly to software objects by the computers. The DOE has a capacity that manages data of sensor/actuator location, sort, state, and so on. Also there are sensor-fusion objects that gather some sensor inputs and create some kind of meta information and controller objects that control actuators for some actions. These objects are used by the ISAN itself or external applications to obtain some information or stave off dangers.

We prepare four abstract classes for ISAN object, that is Sensor, SensorFusion, Controller and Actuator. Each kind of concrete classes is derived from the abstract classes, and each object is instantiated from the concrete classes. An object combination generates actions of the ISAN(Figure 1). Namely, organizational actions are emerged from local object bindings: which Sensor objects are used by SensorFusion objects, which SensorFusion objects are used by Controller objects, and which Actuator objects are controlled by Controller objects. So, it is essential to know appropriate combinations of objects for the environment.

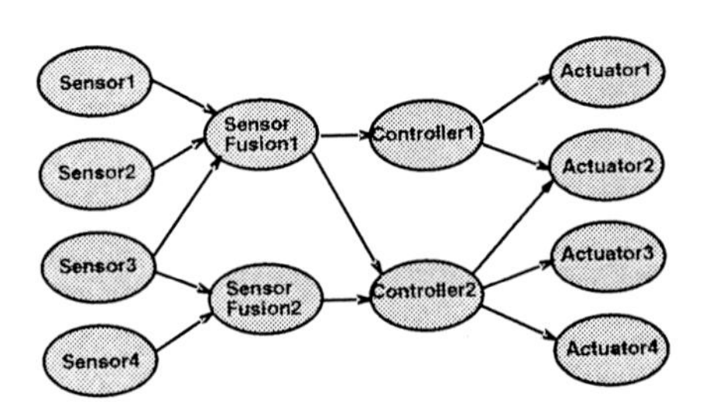

Figure 1: ISAN Object Combination

Useful combination of the objects depends on a situation that may change all the time. If the situation changes, the ISAN composed of the objects must recompose itself to act effectively on a new situation. However, it is difficult to know appropriate object combinations for every situation, so that a learning mechanism that makes ISANs to adapt to the changing situation is needed. The next section describes a learning architecture for the ISANs.

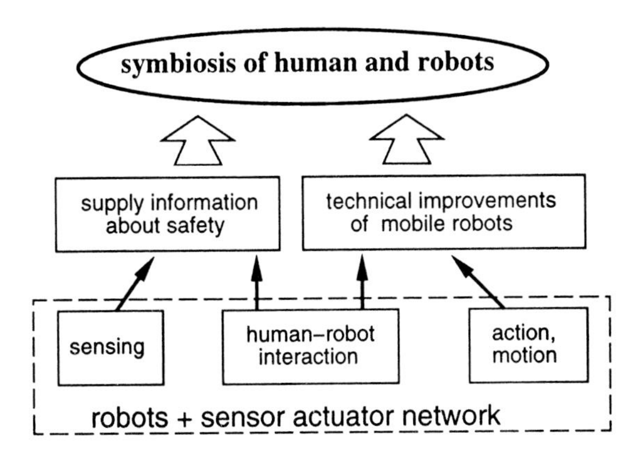

Figure 1: Relation between ISAN and our goal

spatial posision information of all sensors, the database has to be updated all the time because any sensor has possibility to be broken or moved away.

Finally, the mapping problem between spatial address and logical (computational) address arises. It must be noted that any logical address must be dealt with ordinal computer network protocols such as TCP/IP to provide network transparency. Therefore one or more routing service nodes must be available, and they serve address translation requests.

4. PROTOTYPE IMPLEMENTATION

A prototype we have created has the following specifications: each module has one or more kinds of sensors, one 10MHz Z80 chip as CPU, and one RS–485 250Kbps communication interface. They are bus-connected with twisted-pair cables and can be attached or detached at any time. There is an IBM–PC as a master controller and a gateway to traditional computer networks. An overview of the prototype is shown in Figure 2. To adapt the dynamic network changes such as node destruction or node addition, a polling method is used to collect sensor information on the gateway. One of advantages of polling over interrupt is that collisions that spoil networks never occur, therefore all nodes are assured to get a token. Since one of disadvantages is that the mean response time is very long, a scheduling mechanism of polling list has been implemented. It has improved wait time significantly when the sensing period of nodes are not balanced.

5. CONCLUSION

A new concept to change the world into safe and comfortable was described. We believe that ROIS must be widely applied to artifacts, and ISAN technologies that is suitable to realize the world have to be more developed. We have pointed out some problems in constructing ISAN, and have shown a prototype of ISAN node.

Figure 2: A prototype of ISAN node

REFERENCES

[1] J.Jarvinen and W.Karwowski. Analysis of self-reported accidents attributed to advanced manufacturing systems. *International Journal of Human Factors in Manufacturing*, Vol. 5, No. 3, pp. 251–266, 1995.

[2] E.Segantini. The safety of robots. *Tecnologie Elettriche*, Vol. 17, No. 5, pp. 132–137, 1990.

[3] Mark Weiser. The computer for the twenty-first century. *Scientific American*, Vol. 263, No. 3, pp. 94–104, 1991.

[4] Augstin A. Araya. Questioning ubiquitous computing. In *23rd Annual ACM Computer Science Conference*, pp. 230–237, 1995.

[5] Ken Sakamura. TRON project, 1984.

[6] XEROX PARC. ISAT study of distributed information systems for MEMS, 1995.

[7] T.Sato, Y.Nishida, and H.Mizoguchi. Robotic room: Symbiosis with human through behavior media. In *Robotics and Autonomous Systems*, pp. 185–194, 1996.

[8] Joost van Lawick van Pabst and Paul F.C. Krekel. Multi sensor data fusion of points, line segments and surface segments in 3D space. In *Proceedings of the SPIE Sensor Fusion IV*, Vol. 2059, pp. 190–201, 1993.

Information Presentation and Control in a Modern Air Traffic Control Tower Simulator

Richard F. Haines[a] Sharon Doubek[a] Boris Rabin[b] Stanton Harke[b]

[a]RECOM Technologies, Inc., Computational Sciences Div., NASA-Ames Research Center, Moffett Field, Calif., 94035.

[b]Surface Development and Test Facility project, NASA-Ames Research Center, Moffett Field, Calif., 94035

The proper presentation and management of information in America's largest and busiest (Level V) air traffic control towers calls for an in-depth understanding of many different human-computer considerations: user interface design for graphical, radar, and text; manual and automated data input hardware; information/display output technology; reconfigurable workstations; workload assessment; and a comprehensive understanding of air traffic control procedures that are in use today. This paper discusses these subjects in the context of the *Surface Development and Test Facility* (SDTF) currently under construction at NASA's Ames Research Center, a full scale, multi-manned, air traffic control research simulator which will provide the "look and feel" of an actual airport tower cab.

1.0 The Surface Development and Test Facility

The SDTF consists of a two story building of approximately 2560 sq. feet on each floor. The first floor contains specially designed rooms for Ramp Controllers, Pseudo Pilots (who will recreate actual ground and air traffic movement), Test Engineers, Briefing, and a Computer Room (cf. Fig. 1). The second floor consists of a completely glass-enclosed, symmetrical, twelve-sided tower cab area approximately thirty four feet in diameter within which ten or more people will work during typical day - night shifts (cf. Fig. 2). A variety of graphical, radar, and alpha-numeric text displays and airport sub-system controls are located around the perimeter of the room in specially human factor-designed consoles for standing and seated operators. These displays provide touch screen control of aircraft Terminal Radar Approach Control facility (TRACON) arrival and departure information, and two kinds of graphs (airfield activity histograms, runway balancing). This information is presented on up to twenty-seven 16" liquid crystal, high-brightness color, touch-screen displays; each possesses 1280 by 1024 pixel resolution. Special consoles are located in the center

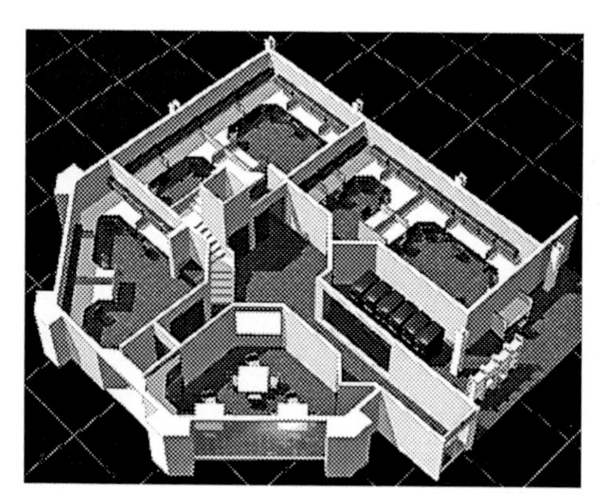

Figure 1. Artist rendering of first floor of SDTF

Figure 2. Artist rendering of second floor cab of SDTF

of the room for use by the supervisor and other staff. They can be reconfigured to simulate the floor layout "footprint" of any Level V U.S. tower. Other personnel (ramp controllers and pseudo-pilots) will also use high resolution color monitors to display two-dimensional airport plan views and other information.

The outside world is simulated in 3-D by two Silicon Graphics Inc. 121-Onyx- "Infinite Reality" engines (each with 12 parallel 250 MHz processors; 4-way interleaved 512 MBytes main memory; and other features). These engines drive twelve high brightness, CRT projectors whose images are back-projected onto adjacent, seamless optical screens (each 90" h x 120" w [150" diagonal]) in size. The initial outside scene library of the SDTF includes six of America's largest and busiest airports with up to two hundred different moving vehicles displayable at any one time. Image resolution of 1.5 minutes arc per pixel makes it possible to portray airplane registration numbers and fuselage markings at relatively large visual distances. Day and night, weather, and realistic runway lighting effects also can be displayed. In addition, realistic (vehicle movement-following) background sound effects are generated along with real-time audio communications among all personnel.

2.0 Facility Applications

This unique, flexible, engineering and simulation facility will be used to develop and evaluate new and innovative technology, procedures, software, and hardware for America's air transport system within a non-critical environment. It will support FAA operations in a stand-alone mode as well as function as an integral part of the National Simulation Capability. Future remote access to the facility is planned from the Federal Aviation Administration's Tech Center in Atlantic City, New Jersey via a secure IntraNet link. Operational users for the SDTF may include air traffic controllers, airline operators, airframe manufacturers, airport operators, pilots, scientists, test engineers, and others.

3.0 Information Classes

Specific classes of historic, real-time and future (predicted) information to be presented will include *airplane:* identification, taxi, routing, departure control, approach and landing control, and *simulation system:* maintenance, calibration, test logistics, and current status. Two dimensional realistic air and ground radar displays will present on-screen "tagged" airplane identification integrated with actual airplane operational data. Each type of data requires graphical user interfaces (GUI) that are initially familiar to the end user(s), readily reconfigurable to meet new research requirements, maintainable, and highly reliable (as new software is added or deleted). One such development tool used is UIMX (Visual Edge). Current 'Surface Movement Advisor' (SMA) interfaces employ a layered, three-level access window approach; the "Departure Splits" screen, for example, permits rapid manual selection of various second-level screens related to airfield activity, departures, arrivals, and inquiry.

In addition, a variety of other information will be available in the SDTF including: *Outside-in Information:*: realistic external, 360 degree view of surrounding environment; synthetic reconstructions of airport environment from different vantage points, traditional radar displays. *Inside-out Information:* runway lighting status displays/controls, weather displays, etc.

4.0 Human Factors Research

Planned human factors and related studies will use a facility-wide audio-visual record and playback capability, highly accurate out-the-window visual simulation, GUIs which are easy to interpret at a glance under a wide range of lighting conditions, real time data-feed, repeatable "incident" conditions, and algorithms to monitor and model low probability-of-occurrence operator errors, time-and-motion studies, etc. Such studies will help to enable user-validation and acceptance of new technology and operating procedures. Investigations to determine the best mix of automation and human interaction in the air traffic control environment will be carried out.

Some of the facility computer displays (e.g., communication controls) are high brightness (liquid crystal) displays with touch-screen capability. Their use will make direct comparisons possible with other more traditional means of data entry and equipment manipulation. Special emphasis will be given to the design and evaluation of human-computer interfaces needed to present the many different kinds of information displayed at the various controller and supervisory positions. Computer-aided design (CAD) and other analytic, real-time computer-based tools will be used to develop and evaluate the facility.

5.0 Summary

NASA's Surface Development and Test Facility at Ames Research Center will support a wide variety of theoretical and applied research. Its various performance monitoring systems will permit research to be carried out on various procedural and related issues. Its flexible physical layout coupled with its powerful computational support environment will also make it possible to realistically recreate existing Level V U.S. airports and future airport facilities still in the planning stage.

Information Engineering: Creating an Integrated Interface

Michael J. Albers
Department of English. Texas Tech University.
Lubbock TX 79406

An effective software interface must focus on the user's complex problem-solving strategies. A complex task is a complete piece of the big action which involves a question posed in terms of the user's job and normally requires the use of several menu options. Supporting problem-solving strategies requires information to be presented in a way which matches the user's cognitive processes, assumptions, implications, and expectations. We must go beyond seeing the interface as a collection of independent functions and begin to think in terms of what Bowie calls "information engineering," where, at each point, we support the problem-solving strategies that address the user's complex real-world tasks.

1. LITERATURE REVIEW

Providing answers to the complex, open-ended questions asked by the user means understanding the user's cognitive processes and what factors influence the mental processing of information. Providing users with the means of manipulating a program while minimizing cognitive workload means gaining a much deeper understanding of the user's cognitive processes. Bereiter and Scardamalia found that most communication acts (including operating software) occur close to the level of cognitive overload; poor design causes cognitive overload, leading to errors. Fennema and Kleinmuntz found most users attempt to minimize perceived effort at the expense of optimal solutions. In studying how people use computer programs, Cypher found most do not perform a job in the linear progression described in the task analysis; instead, they often interleave multiple jobs.

Rasmussen (1986) explains how all parts of an interface must be treated as an integrated part of the system and calls for considering the user's schema in order to create a design compatible with the user. The schema the user brings to the system profoundly affects the perception of the design's usability.

Information presentation has major effects on people's decisions; depending on the presentation, they may actually make opposite choices and believe them to be best (Tversky & Kahneman, 1981; Russo, 1977). Slovic (1972) found that users tend to change their assessment strategy to fit to the presentation method, rather than transform the information to fit a better assessment strategy. Given different routes, the user almost always takes the route of minimal cognitive effort (Johnson, Payne & Bettman, 1988). Potential problems can occur when minor, but easy to present, information occupies an excessive amount of the interface and, subsequently, diminishes the salience of important, but harder to present, information.

2. PROBLEMS WITH OUR CURRENT DESIGN METHODS

Effective design focuses on the user's real-world complex tasks. Determining these complex tasks starts with the task analysis and the audience analysis and continues throughout design and development. Unfortunately, the task analysis often gives a list of what functions the programmers think management wants the system to do, and the audience analysis often amounts to little more than "it's a group with high school education or less, so we better keep it real simple."

As a method of improving task analysis and audience analysis and of providing answers for complex questions, Norman (1986) calls for user-centered design which emphasizes that to the user, "the interface is the system" (p. 61). The interaction between user and interface must drive all design considerations and include consideration of how and, more important, why the user interacts with the system. Woods and Roth (1988) define the critical question as "how knowledge is activated and utilized in the actual problem-solving environment" (p.420). I believe the cause of many current design shortcomings are not that the answers derived from the task and audience analysis were incorrect, but rather, the result of a methodology which misplaced its emphasis and, consequently, asked the wrong questions. I propose a fundamental flaw in current design methodology is the lack of attention during early analysis to defining the user's complex problems. This lack of attention is reflected in three different ways.

1. The failure to anticipate the user's needs, rather than a lack of information, forms the basis of most problems. Often, design decisions are made with reference to other software concerns, rather than to human concerns (Lanier, 1997).
2. Users ask problem-oriented or procedure-oriented questions, and not system-oriented questions. The audience analysis must reveal how and why users ask these questions (Rosenbaum & Walters, 1986). To support complex problem-solving, we must gain an in-depth understanding of the user's mental processes and fit the interface to them.
3. Rather than address complex problems, most interface design centers around individual tasks; each menu option of the program exists in its own world, never connecting to any other option.

3. IMPROVING DESIGN METHODS

Tasks can be defined in terms of the program or in terms of the real world. With program-based tasks, audience analysis consists of defining what the user needs to do to execute the function. Program-defined tasks are based on the psychology of performance and the task becomes just the sum of its operations. On the other hand, with the design goal of defining complex tasks, audience analysis expands to include the "strategies and skills that users need to learn to adapt program operations to their conceptual models" (Mirel, 1992, p. 17). Systems designed for problem-solving are based on the psychology of problem-solving; the design assumes that tasks vary by situation, and the user has dynamic strategies (Leveson, 1997). To help solve the task, the interface provides a conversation between the program and the user. However, many interfaces do not seem designed for conversation; instead, they are an entity which people happen to use. But the central design goal must be to remove the interface from the user's attention, so they can focus solely on the task (Rakin; van Dam).

As a method of providing a consistent human-computer interface, design guidelines were created. However, design guidelines often fail to address user tasks and maintain an overly narrow focus. They tend to emphasize pieces of a single screen, but do not focus on the user task. (Ritter and Larkin, 1994) However, the usefulness of a system results from the total experience integrated across the displays and documents and guideline's usefulness diminish because "a design is highly dependent upon task context and user behavior" (Gerlach & Kuo, 1991, p. 528). Mirel (1992) found user's were very inventive in how they use the system and that the "situational demands and interactions define users' tasks and task needs more than technical ones" (p. 34). Good design must support these demands.

4. IMPROVING INTERFACE DESIGN

Helping users solve real-world problems means focusing everything the user sees on solving the user's problem. A user interface has two logical parts. The first, the physical interface, is by far the easiest: the buttons, windows, etc. provide the controls between the user and the program. The second part of the user interface is the content: the actual text of labels on the entry fields (as opposed to the placement/font), the error messages, the help text and documentation. It also includes the navigation between screens and the interrelation of components of the program. Via the interrelation of components, the user accomplishes problem-solving. A consistent user interface is not sufficient to make a program usable; effective interface layout and content must integrate with the documentation. Failure to coherently integrate every piece reduces overall program effectiveness.

The design of software interfaces must be reconceived so that the design is done from the problem-solver's point of view. Carroll et al found that people already understand the task-relevant concepts of the program. They know what they want to do, it is our responsibility to ensure they can easily accomplish it. People are faced with many complexities and obstacles in performing a job and have developed many different strategies in overcoming these obstacles. We must acknowledge these obstacles and strategies and create designs that minimize them. Design must never focus on each menu option of the program as existing in its own world, never connecting to any other option.

The differences arise from how programmers and users view software. The programmer, views the software with an inside-out perspective and sees a set of disjoint screens that interact according to the program specification. The user, on the other hand, works with a top-down problem-solving method, views the software with an outside-in perspective, and brings to the program a conceptual schema of how it works. We must strive to create designs that fit the conceptual schema, directly address user tasks in user terms, and do not limit the user's problem-solving strategies.

5. CONCLUSION

Supporting the user means interface designs that assist in answering the user's complex questions. To accomplish this goal, the designer must adhere to three principles:

- Design for the real world tasks the user wants to solve.
- Design to fit the user's conceptual schema of the real-world tasks.
- Design for different and dynamic problem-solving strategies.

An effective interface communicates with the user in terms the user understands. With effective communication, the user doesn't have to try and figure out what the program wants. Instead, the interface essentially disappears and the user focuses on accomplishing the real-world task.

REFERENCES

Bowie, J. (1996) Information engineering: Communicating with technology." *Intercom* 43(5), 6-9.

Carroll, J., Smith-Kerker, P., Ford, J. and Mazur-Rimetz, S. (1988) "The Minimal Manual." *Human-Computer Interaction,* 3.

Fennema, M., & Kleinmuntz, D. (1995). Anticipation of effort and accuracy in multi-attribute choice. *Organizational Behavior and Human Decision Processes,* 63(1), 21-32.

Gerlach, J., & Kuo, F. (1991). Understanding human-computer interaction for information system design. *MIS Quarterly,* 15(4), 527-550.

Johnson, E., Payne, J., & Bettman, J. (1988). Information displays and preference reversals. *Organizational Behavior and Human Decision Processes,* 42, 1-21.

Lanier, J. (1997) The frontier between us. *Communications of the ACM,* 40(2) 56-57.

Leveson, N. (1997) Software engineering: Stretching the limits of complexity. *Communications of the ACM,* 40(2) 129-131.

Mirel, B. (1992). Analyzing audiences for software manuals: A Survey of instructional needs for 'real world tasks'. *Technical Communication Quarterly,* 1(1), 15-35.

Norman, D. (1986). Cognitive engineering. In D. Norman & Draper, S. *User centered system design.* (pp. 31-61). New Jersey: Erlbaum.

Raskin, J. (1997) Looking for a human interface: Will computers ever become easy to use? *Communications of the ACM,* 40(2) 98-101.

Rasmussen, J. (1986). *Information processing and human-machine interaction: An approach to cognitive engineering.* New York: North-Holland.

Russo, J. (1977). The value of unit price information. *Journal of Marketing Research,* 14, 193-201.

Slovic, P. (1972). From Shakespeare to Simon: Speculations--and some evidence--about man's ability to process information. *Oregon Research Institute Research Bulletin.* April

Tversky, A., & Kahneman, D. (1981). The framing of decisions and the psychology of choice. *Science,* 211(30), 453-458.

van Dam, A. (1997) Post-WIMP user interfaces. *Communications of the ACM,* 40(2) 63-67.

Woods, D., & Roth. E. (1988). Cognitive engineering: Human problem solving with tools. *Human Factors,* 30(4), 415-430.

Designing for complexity

Erik Hollnagel

Principal Advisor, Ph.D., OECD Halden Reactor Project, P. O. Box 173, N-1751 Halden, Norway

1. INTRODUCTION

The extensive use of information technology in process control and control rooms has had two important effects: firstly, that the amount of information has significantly increased and secondly that the information can be transformed and presented in a variety of ways.

The increasing amount of information stems both from the growing complexity of the process, from improvements in measurement technology, and from the stereotypical organisational response to adverse events, which is to compensate by adding further instrumentation or barriers. Conventional instrumentation has over the years gradually been miniaturised, but there is a lower limit for that due to the perceptual and motor characteristics of humans. The real revolution happened when the VDU was introduced in the control room as a display device. The VDU was both universal and of unlimited capacity, in the sense that it displayed a subset of a potentially infinite amount of information that was held in store. It removed the physical imitations to how much information could be made available to the operators, and information presentation was no longer confined by the available "real estate" or surface area of the control panels. The use of the VDU also meant that the information could be transformed in any way that was feasible, and that it could be displayed instantaneously. It was no longer constrained by the physical instruments, such as dials, numbers (digital values), curves on chart writers, or annunciators. Both form and colour could be changed practically at will, and new combined measurement values could be defined or calculated.

After some initial mistakes and hard learned lessons, the design of man-machine interfaces for complex systems came to rely on the assumption that the interface should be as simple and easy to use as possible. Since operators were not capable in practice of handling all the information that could be thrown at them it was necessary effectively to reduce the amount of information. It was also well known from practical experience, as well as from problem solving psychology, that the proper representation could facilitate the understanding and comprehension of a situation and in many cases make a solution obvious (Duncker, 1945). This notion was reinforced both by writings in perceptual psychology (Gibson, 1979) and in understanding (Brunswick, 1956). This created an approach to interface design that can be called **information structuring** (Hollnagel, 1996). The main principles were: (1) to provide the right information, (2) presented in the right way and, (3) at the right time. Unfortunately, while this has a very attractive ring to it, this is easier said than done.

In order to provide the right information it is necessary to know what the situation is and which information the operators require. This exercise is fairly easy to do in retrospect, as discussed by Woods et al. (1994), but there is a wide step from that to being able to specify in advance which information is required. To accomplish that it is necessary to identify all the potentially important situations and describe their information needs. Quite apart from the fact that this is susceptible to the *n+1* fallacy, it creates the problem of reconciling the design of *n* specific displays with the goal of maintaining global situation awareness. The balance between the two is difficult to strike, particularly as current display technology forces the use of piecemeal or fragmented information presentation.

In order to present the information in the right way one must know both what the situation is and what the users need. The latter presents a very difficult problem, since the users' state of mind can only be predicted in a rather general way. One solution is to apply some type of user model, either based on stereotypes or on specific theories of human information processing (e.g. May et al., 1993). Another is to rely on intuitively appealing design principles, although these may be difficult to make operational and to apply to other than the chosen exemplars. The basic problem for design of information presentation is that the variation between and within users and domains is too large and that it is impossible to anticipate the specific needs of a specific user at a specific time. Finally, a third and more practical, solution is to design a display as well as possible and then train the users to understand it.

In order to present the information at the right time, it is necessary to know what the situation is so that the information can be presented neither too early nor too late. Fortunately, most dynamic processes can be described in terms of limited number of states, and knowing the transition between the states can be used to determine the right time. Similarly, the time constants of the process can be used to synchronise information presentation, at least in an ordinal sense. In most cases an approximate solution is therefore possible.

The ease by which specific situations can be analysed has made the design problems seem simple, but experience has shown that generic design principles do not emerge from the accumulation of particular solutions. Designing for simplicity via information structuring can be a powerful technique if both the target function and the interaction are well defined, i.e., if the situation can be constrained to match the premises of the design. When this is not possible in practice, the alternative of designing for complexity should be considered.

2. SIMPLICITY AND COMPLEXITY

The work situation in industrial process control is complex rather than simple and the operators' situation has rightly been characterised as "coping with complexity" (Rasmussen & Lind, 1981). In the current design tradition the emphasis has been on making this complexity seem simple. This can, however, only be achieved if the transformations from complex data to simple representations can be relied upon under all possible conditions. As argued above, this is not a principle that can be easily trusted, and many design ideas have in fact only been shown to work under simplified or restrained conditions.

The principles of designing for simplicity are illustrated by Figure 1. Here the complexity, and richness, of data from the process is reduced by mapping or projecting a subset of the data onto a specific information structure. This discretisation of the world complexity can clearly

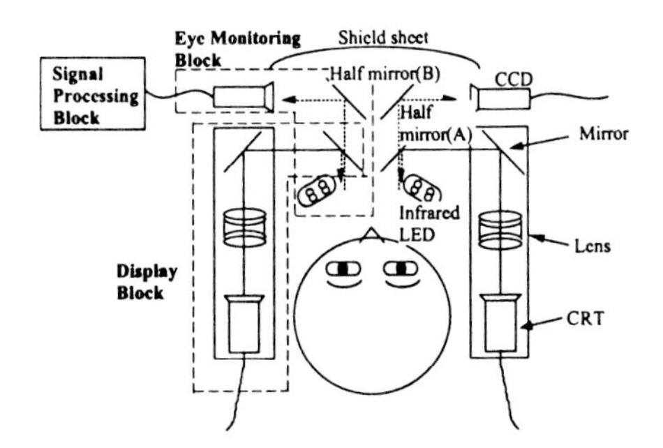

Figure 2. Block diagram of the developed Eye-Sensing HMD.

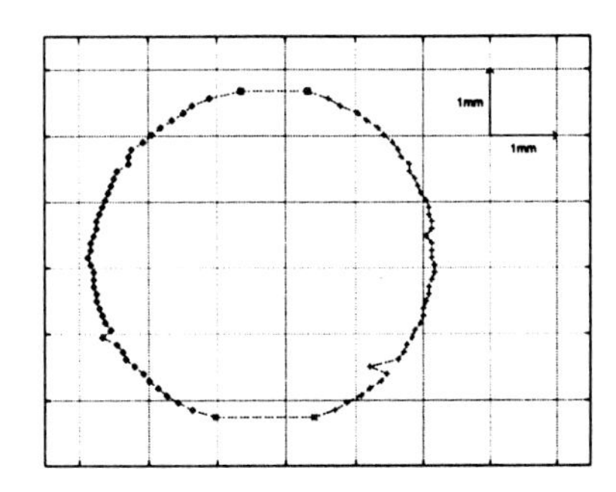

Figure 3. Both ends data of pupil reproduced by Signal Procesing Block.

2, pictures on CRTs are projected through lenses by way of mirrors and half mirrors to the user's both eyes. A user is supposed to watch picture on about 43 inches' screen at 1 meter distance from the screen. User's view through the display will change in accordance with his head motion which will be detected by the position detecting sensor on the top of the HMD.

3.2 Eye Monitoring Block

The purpose of Eye Monitoring Block is to monitor the user's eyes and to get pupil images as binary image, that is, black and white image. In order to obtain a pupil image, four infrared LEDs placed in front of each eye but outside user's field of view to emit infrared rays. The infrared rays illuminate user's each eye without his sensing. The light reflected on his eye surface comes into each CCD camera through focus adjustable lens fixed on camera by way of two half mirrors.

The resolution characteristics of the developed Eye Monitoring Block is as follows; time resolution of about 16.7[msec.] depends on NTSC video format. Concerning spatial resolution, vertical and horizontal are different. While vertical resolution is 0.109[mm], horizontal is 0.0467[mm]. It is because the former depends on NTSC video format in the same way as time resolution, whereas the latter depends on CCD resolution regardless of NTSC format.

3.3 Signal Processing Block

By processing the video signal obtained in Eye Monitoring Block, ocular information is then obtained in Signal Processing Block. They are pupil diameter, center position of pupil in the camera coordinate system and eyeblink. Center position of pupil can be easily transformed into gaze point in the display coordinate system by advance calibration.

Figure 3 shows both ends data of pupil reproduced by Signal Processing Block. It can be seen from the figure that horizontal resolution is better than vertical one. As seen in Figure 3, some parts in the figure are not reproduced correctly, because LED images are overlapped on pupil image.

4. EXPERIMENT OF EYE GAZE INTERACTION BY USING ES-HMD

A basic experiment was conducted on realizing Interaction Tool Function as seen in Figure 1. In order to evaluate the performance of eye gaze interaction, conventional mouse device was also used to compare with eye gaze interaction.

4.1 Experimental procedure

In the experiment, a subject executed the task as shown in Figure 4. First, the subject watches

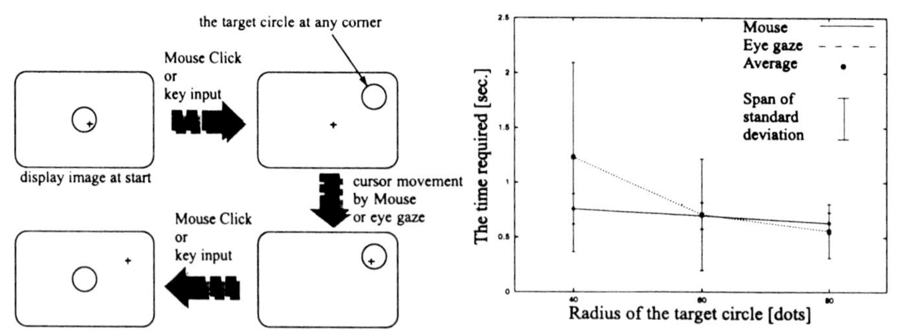

Figure 4. Procedure in the experiment.

Figure 5. Average time required for cursor pointing.

the target circle (indicated by O mark in the figure) at the center on the display. After he has fixed Mouse cursor or Eye gaze cursor (indicated by + mark) in the target displayed at center, he clicks Mouse button or hit any key on the keyboard. Then, the target circle jumps to any corner out of the four corners on the display. When he finishes moving Mouse cursor or Eye gaze cursor to the target displayed at one corner, he must push the button again. In the experiment the time required for moving Mouse cursor or Eye gaze cursor is measured with the radius of the target circle as an experimental parameter. The radius values of 40, 60, 80[dots] correspond to 20%, 30%, 40% of horizontal width of a display, respectively.

4.2 Result

Figure 5 shows the average time required for cursor pointing for a subject who is 23-year-old male student without glasses. In Figure 5, the horizontal axis indicates the radius of the target, while vertical axis, the average time required to reach the jumped target with vertical line in each point being the span of standard deviation. As seen in Figure 5, the standard deviation of Eye-gaze-cursor resulted in larger value than that of Mouse-cursor. This is because the subject is not so accustomed to using Eye-gaze cursor as Mouse-cursor. However, the target circle becomes larger, the average times in both cases become almost the same value. Therefore, eye gaze interaction has as good performance as conventional mouse device in the case of larger target.

5. CONCLUDING REMARKS

In this paper, a new framework of CAI system was proposed from the viewpoint of MADI. Then, in order to realize such CAI, a new HMD was developed, which can monitor the user's both eyes and provide ocular information such as eye-gaze, pupil dynamics and eyeblink. Finally, a basic experiment was conducted by using the developed HMD in order to show that the eye gaze interaction has as good performance as conventional mouse interaction in larger target.

REFERENCE

1. S.Fukushima, D.Morikawa, M.Takahashi and H.Yoshikawa, A Physiological Study on Mental Workload in Terms of Visual or Auditory Information Presentation (2) Analysis of Eyeblinks, Human Interface News and Report, Vol.11, 123/130 (in Japanese) (1996).
2. S.Taptagaporn and S.Saito, Analysis of Pupil Movements for Ergonomic Evaluation of Visual Environments, Proc. of the 6th Symposium on Human Interface, 617/624 (1990).

Development Strategies on an Intelligent Software System for Total Operation and Maintenance of Nuclear Power Plants

Soon Heung Chang[a], Han Gon Kim[b], and Seong Soo Choi[a]

[a]Department of Nuclear Engineering, Korea Advanced Institute of Science and Technology 373-1 Kusong-dong, Yusong-gu, Taejon 305-701, Korea

[b]Center for Advanced Reactors Development, Korea Electric Power Research Institute, 103-16 Munji-dong, Yusong-gu, Taejon 305-380, Korea

ABSTRACT

We propose an intelligent software system that has the integrated functions for total operation and maintenance of NPPs in this paper. The objectives of this system are to replace most of operators' physical tasks and to advise on their cognitive tasks for the whole spectrum of plant operation. The intelligent software system is composed of five major modules for operators and one diagnostic module for maintenance personnel. It is expected to improve operational efficiency of NPPs greatly by supporting operators' cognitive interfaces and to enhance the safety of NPPs by lengthening the operator response time when the plant is under emergency conditions.

1. INTRODUCTION

The roles of digital computers have been much restricted traditionally in the main control room (MCR) of nuclear power plants (NPPs) such as gathering plant data and displaying safety parameters. They are being extended, however, as large programs on advanced control rooms (ACRs) are underway with the strides in computer technologies, artificial intelligence (AI), and human factors engineering since mid 1980s. The control room of N4 PWR of France [1], that of ABWR of Japan [2], and Nuplex 80+ of USA [3] are the typical examples of the ACR.

Three important trends in the evolution of ACRs are 1) increased automation by adoption of soft control, 2) the development of compact, computer-based workstations based on human factors engineering, and 3) the development of intelligent operator aids using various AI techniques. Among these trends, however, the roles of intelligent operator aid systems have been limited such as alarm processing or computerization of operating procedures. Therefore, as the next step of technology development, operator aid characteristics will be enhanced in ACR design in order to replace or to assist operators' cognitive tasks. The application possibilities of AI techniques have been demonstrated in nuclear field through some of operator aid systems such as Operator Companion, IDA, DISYS, JOYCAT, and OASYS [4] and the design of some ACRs such as ISACS-1 [5], the MCR of APWR, and Toshiba's intelligent man-machine system.

In this paper, we propose the integrated software system which can be used as a core software of the modern MCR of NPPs. It will be installed on the mock-up system of ACR being developed in Korea Advanced Institute of Science and Technology (KAIST).

2. OVERALL CONCEPT OF THE SYSTEM

The operating states of NPPs can be divided into normal, abnormal, and emergency states. The operator's tasks under each state can be summarized as follows: monitoring of safety/performance related parameters, identification/prediction of plant behavior or abnormality, finding the strategy to protect unexpected reactor trip or to maintain NPPs safe conditions, and carrying out specific actions/procedures according to the strategy. That is, the operators should perform many kinds of tasks not so easy to perform by one or a few crews according to plant states. Therefore, 6~8 operational crews have participated in the operation in the conventional MCR in NPPs. In ACRs, 3~4 crews, which consist of 2 operators, 1 supervisor, and/or 1 safety advisor, operate NPPs. This reduction has been possible by the innovation of hardware and software of MCR.

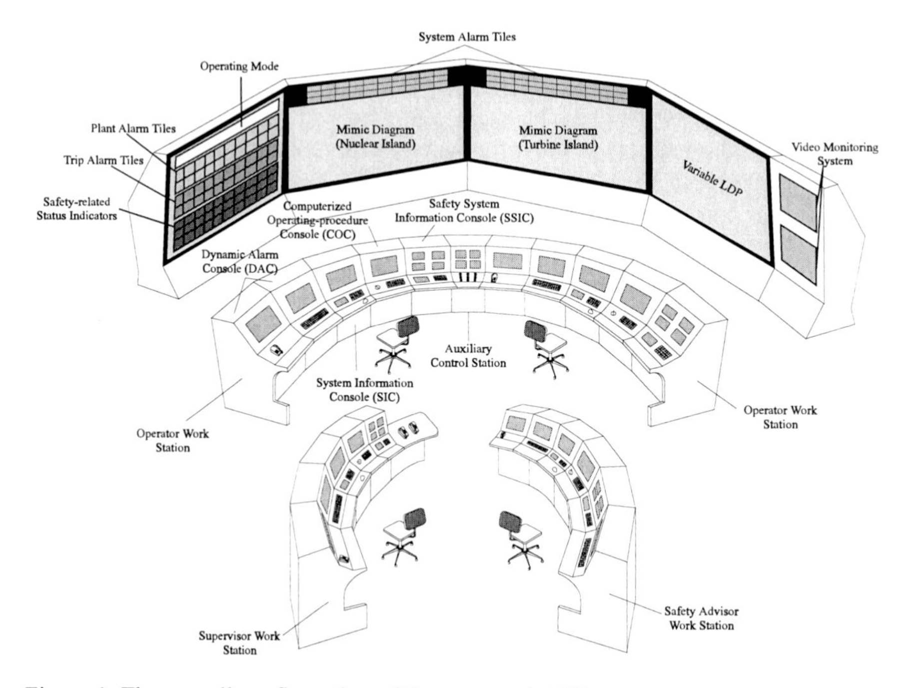

Figure 1. The overall configuration of the proposed ACR

The objectives of the intelligent software system proposed in this paper are to replace most of operators' physical tasks and to advise on their cognitive tasks for the whole spectrum of plant operation. Operators' roles would be changed to decision-maker or supervisor from performer by this system. Furthermore, it will make maintenance more effective by on-line monitoring and diagnosis of important systems/components.

The intelligent software system is composed of six major modules such as "Information Processing," "Performance Diagnosis," "Event Diagnosis," "Alarm Processing," and "Procedure Tracking" modules for operators and "Malfunction Diagnosis" module for maintenance personnel. These modules, which are under development, will be installed on the mock-up of ACR shown in Figure 1 as the operating software.

3. IMPORTANT FUNCTIONS OF THE MAIN MODULES

3.1. Monitoring and control

The main functions of "Information Processing" module are 1) providing processed data from field signals through signal validation for an ACR composed of a few compact work stations, 2) monitoring performance parameters to maintain normal conditions and safety parameters to identify threats to safety functions, 3) planning adequate procedure-based operational strategies, 4) displaying operating information within limited VDU area on the basis of the established information hierarchy of the plant, 5) providing equipment-level control means closely coupled with monitoring, and 6) coordinating the other modules in an efficient manner. It includes MCR configuration by consideration of HMI.

The main roles of "Alarm Processing" module which uses both spatially dedicated alarm tiles and variable, hierarchical CRTs are as follows; 1) eliminating meaningless alarms using time delay technique, suppressing less important ones with various suppression techniques such as mode dependency and state dependency, and dynamically assigning ranks on individual alarms according to system-oriented and mode-oriented prioritization to discern important ones, 2) improving operators' situation awareness by structuring alarm hierarchy, and 3) providing suitable procedures timely.

In NPPs, most of the operations are performed on the basis of the procedures, which are composed of the general operating procedures (GOPs), abnormal operating procedures (AOPs), and emergency operating procedures (EOPs) according to plant states. All procedures are composed of decision making and actions. Operators should identify current plant status and select appropriate procedures using various operating information. "Procedure Tracking" module is in charge of dynamic tracking of procedures through self decision-making and automatic control. For this purpose, qualitative decision making tasks in the procedures are computerized using fuzzy membership and binary control portions among actions are automated.

3.2. Diagnosis for operators

The NPP operators do not require the detailed failure causes of the systems/components. Therefore, the diagnosis modules provide the overall plant status in view of performance and safety.

"Performance Diagnosis" module is being developed which has the following functions: 1) calculating current performance indices and checking whether the performance of an NPP is degraded by mathematical analysis and heuristic rules, 2) optimizing plant control strategy, 3) diagnosing the cause of degradation using rule-based expert systems and artificial neural networks, and 4) providing guidelines to recover the degradation.

The main objective of "Event Diagnosis" module is to identify events and to diagnose the their causes using various AI techniques. First, it classifies event category into incidents, accidents within design base accidents, and severe accidents beyond design base accidents. Then, the module identifies an event itself and its possible causes by fuzzy expert system that uses alarm data and process variables as diagnostic information and it also guides proper actions or procedures to recover plant state or to mitigate event consequences. Another function of this module is the prediction of abnormality to warn operators.

3.3. Diagnosis for maintenance personnel

Failure diagnosis systems for the maintenance personnel of NPPs can be developed for various systems/components. To select the target systems for fault diagnosis, the analysis on the historic failure data in USA/Korea NPPs has been performed. As the result of the analysis, "Malfunction Diagnosis" module is being developed for rotating systems such as the reactor coolant pump, the turbine generator and so on. The major functions of this module are 1) on-line monitoring of system conditions, 2) prediction of system failure, 3) identification of the root causes of failures, and 4) advising optimal treatment.

4. CONCLUSION

The intelligent software system proposed in this paper has the integrated functions to reduce operators' physical loads and to support their cognitive interfaces over all the spectrum of plant operation including emergency conditions. In addition, the system will improve plant availability through preventive maintenance.

This system, which is sponsored by Korea Electric Power Research Institute, is being developed by KAIST and Massachusetts Institute of Technology. Some of major parts of the software system have already been developed and some of them are under development. After the development of the major modules is completed, the system will be structured and demonstrated using a simulator to validate its performance and stability.

REFERENCES

1. "N4: The 1500MWe PWR nuclear island technical description," Framatome, 1993, p. 340.1.
2. M. A. Ross, "Advanced control rooms: balancing automation and operator-responsibility," Nuclear Engineering International, pp. 37-39, March 1993.
3. F. Ridolfo, D. Harmon, and K. Scarola, "The Nuplex 80+^{TM} advanced control complex from ABB Combustion Engineering," Nuclear Safety, vol. 34, no. 1, pp. 64-75, January-March 1993.
4. S. H. Chang, K. S. Kang, S. S. Choi, H. G. Kim, H. K. Jeong, and C. U. Yi, "Development of the on-line Operator Aid SYStem OASYS using a rule-based expert system and fuzzy logic for nuclear power plants," Nuclear Technology, vol. 112, pp. 266-294, 1995.
5. K. Haugset, O. Berg, S. Bologna, N. T. Fordestrommen, J. Kvalem, W. R. Nelson, and N. Yamane, "Improving safety through an integrated approach for advanced control room development," Nuclear Engineering and Design, 134, pp. 341-354, 1992.

Simulation-Based Interface Evaluation Method of Equipment in Power Plants

T. Nakagawa[a], M. Kitamura[a], Y. Nakatani[a], N. Terashita[a] and Y. Umeda[b]

[a]Industrial Electronics & Systems Development Lab., Mitsubishi Electric Corp., 1-1 Tsukaguchi-Honmachi 8-Chome, Amagasaki, Hyogo 611, Japan

[b]Takahama Atomic Power Station, Kansai Electric Power Co., Ltd., Tanoura, Takahama-Cho, Ohi-Gun, Fukui 919-23, Japan

Although many human errors are occurring in the maintenance field of power plants, systematic approaches to the reduction of human error are not enough compared with operation fields. The authors propose a new method of evaluating and analyzing human interface design from the viewpoint of human error, and have implemented the method on the DIAS system. This system consists of maintenance personnel and equipment interface simulators, and it generates a dynamic interaction between humans and the working environment. The authors applied this system to evaluate the interface design of actual Transformer Protection Relay Panels and their layout in a room in a nuclear power plant. Our customer accepted our evaluation and proposal to modify the panel design.

1. INTRODUCTION

Many human errors are occurring in the maintenance of power plants. Most measures to combat these human errors are reviewing and modifying procedures or increased maintenance personnel training. On the other hand, modifying the human interface design is less common because it is very expensive. Thus, it is important to evaluate and analyze the human interface in the design phase. The authors propose a new method of evaluating and analyzing the human interface design of plant equipment from the viewpoint of human error, and the DIAS system has been developed based on this method. The DIAS system undertakes a dynamic analysis and quantitative evaluation of human interface design. For dynamic analysis, we use the framework of the SEAMAID system[1]. The SEAMAID system is a evaluation and analysis support system of the human interfaces in the central control room of a nuclear power plant, and it consists of three simulators; an operator simulator, a human interface simulator and a plant simulator. The target of the DIAS is the maintenance field, and it consists of a maintenance personnel simulator and a human interface simulator. This system is implemented on a workstation, and simulates the cognition and behavior of maintenance personnel. Through this simulation, DIAS evaluates the length and the course of the visual field, and so on. As a quantitative evaluation, DIAS calculates the possible human error rate by applying THERP to the maintenance procedures of the equipment. A maintenance procedure consists of a sequence of tasks including checking meters, switching off the power, and so on. DIAS selects a suitable human error rate for each task from the THERP database

2. CONFIGURATION OF DIAS SYSTEM

2.1. Distribution Simulation System

The Distribution Simulation System consists of the Maintenance Personnel Simulator and the Equipment Interface Model, and simulates the interaction between the maintenance personnel and the Equipment Interface. (Fig. 1)

The Maintenance Personnel Simulator. This uses the maintenance knowledge that represents the standard maintenance procedures by a hierarchical augmented Petri net model, and executes them in real time. The Petri net model was adopted because of its merit of describing the state of transition by the use of place and transition, where we can represent serial maintenance procedures, and visualize the structure and the dynamic process of the state of transition comprehensively on a graphic display.

We consider the standard human cognitive/behavioral features, such as the visual aspects (the focused view and the peripheral view), the operation speed, and the walking speed. Because maintenance personnel generally refer to maintenance manuals, we do not consider cognitive resources such as focal/peripheral working memory load. The maintenance knowledge can be easily edited on the graphical knowledge editor.

The Equipment Interface Simulator. This represents the interfaces of the equipment and their layout. The analyst can easily make and modify the interfaces and the layout through the graphical interface editor, which enables the analyst to evaluate alternative designs.

2.2. Interaction Analyzer

DIAS does not propose new designs, but can quantitatively evaluate design proposals and compare them from the viewpoint of human error. During simulation, the interaction analyzer observes the length of movement, the length of eye movement, the time for a task process, the equipment selection error index, the interface design data, and so on, it sums up these indices. Using this data, the interaction analyzer selects a suitable human error rate for each task from the THERP data by asking some questions about PSFs (Performance Shaping Factors), and shows the selection and an explanation to the analyst. With these process, the interaction analyzer calculates the HEP (Human Error Probability) of the whole task sequence as a quantitative evaluation. Besides the HEP, we consider the effects of human error on the plant, equipment, and maintenance personnel. Because no attempt is made to quantify this parameter, persons in authority judge the degree of the effects by 4 (large), 2 (midium), and 1 (small). Each HEP is multiplied by its effect.

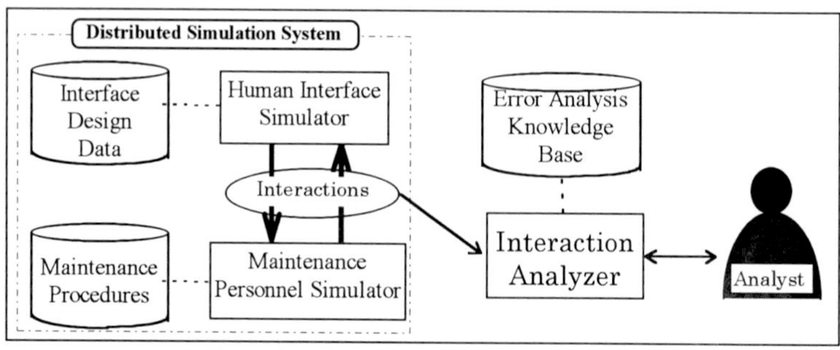

Fig. 1 Configuration of DIAS SYSTEM

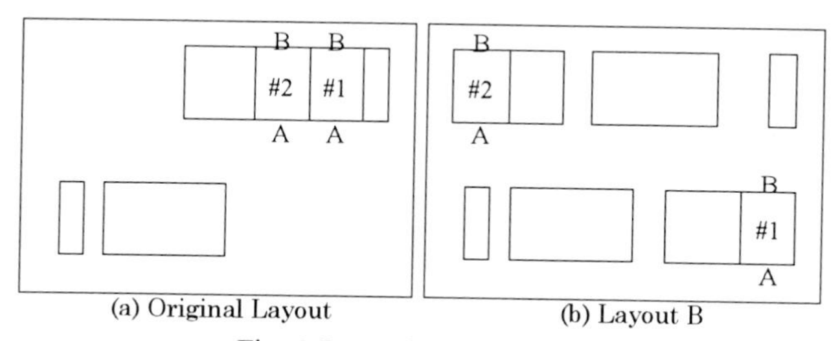

Fig. 2 Room layout patterns

The equipment selection error index represents the degree of the selection of the incorrect equipment. The maintenance personnel may select the incorrect equipment when there are several pieces of equipment with a similar interface to the target equipment. The equipment selection error index is calculated by the following procedure.

(1) For each equipment selection task, it extracts equipment with a similar interface to the target equipment.

(2) For each piece of similar equipment, it calculates the number of satisfaction of each of the following conditions.

a. within a visual field b. on the route to the target
c. close to the target d. neighbor to the target

3. CASE STUDY

This method was used to evaluate an actual transformer protection relay panel design at a nuclear power plant, in order to improve the human error rate of the panels and their layout. First two kinds of interface design and three kinds of layout were evaluated. (Cases A-C). The proposed design which was derived from the above evaluations was also evaluated (Case D).

Case A: Original layout (Fig. 2a) with original interface
Case B: Layout A with modified interface
Case C: Layout B (Fig.2b) with modified interface
Case D: Layout B with a final panel interface (Fig. 3) derived from other analysis cases.

Evaluation was done against the human error rate (standard calculation and their weighted calculations according to dangerousness, influence on the equipment, and influence on the plant of each task) and the dynamic analysis of the degree of equipment selection error, the working time, the distance of movement, and the distance of viewing point movement.
In the original layout in Fig.2a, the equipment items #1 and #2, which have the same interface, are laid out side by side, while the two panels are separated in Layout A and B (Fig. 2b). Figure 3 shows an example of the computer screen of the equipment interface simulator. There are many switches in order to isolate the protection relay system. In this figure, the behavior of the maintenance personnel simulator is also shown. The large squares denote the current peripheral view of a personnel, and the small ones with a cross denote his/her focal view. The movement of the visual field is shown by the arrows.

3.1 Consequences

Figure 4 summarizes a comparison of four kinds of design regarding the indices of the HEP, the task process time, the length of movement, the length of viewing point movement, and the

equipment selection error degree. Here, each index value of Case A is 100, which is the base of comparison. According to this figure, the best design is Case B with the modified panel, and the next best is Case C with the final panel. Based on these analyses and financial considerations, our customer selected Case C. We proposed further improvement based on the simulation, which included the addition of a partition board between panel #2 and its neighbor, because personnel can see the neighbor in their peripheral view. We also proposed the re-layout of the switches according to the operation procedure. (Case D) The customer accepted our proposals, and the new design panels are now in actual use.
In the original interface design, switches seem to be laid out without any explicit design concepts. On the other hand, in the final design, the movement of the view point is generally downward and rightward. The switches are classified into various groups, and these groups are laid out separately.

REFERENCES

1. Nakagawa, T., Nakatani, Y., Yoshikawa, H., Takahashi, M., Hasegawa, A. and Furuta, T.: Simulation Based Evaluation of Man-Machine Interface Design in Power Plants, Cognitive Systems Engineering in Process Control (CSEPC 96), 1996.
2. Swain, A. D. and Guttman, H. E.: Handbook of Human Reliability Analysis with Emphasis on Nuclear Power Plant Applications, NUREG/CR-1278, 1983.
3. Department of Defense: Human Engineering Design Criteria for Military Systems, Equipment and Facilities, Military Standard 1472 C, 1981.
4. Sanders, A. F.: Some Aspects of the Selective Process in the Functional Visual Field, 47(6), 1970.

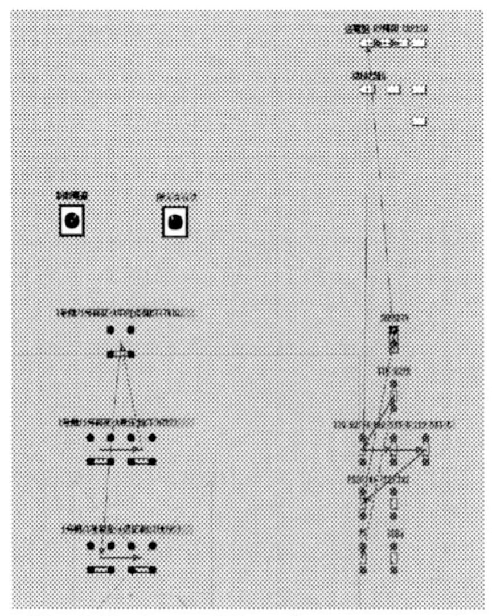

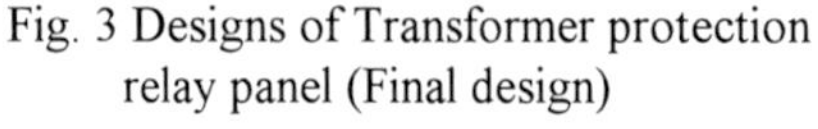

Fig. 3 Designs of Transformer protection relay panel (Final design)

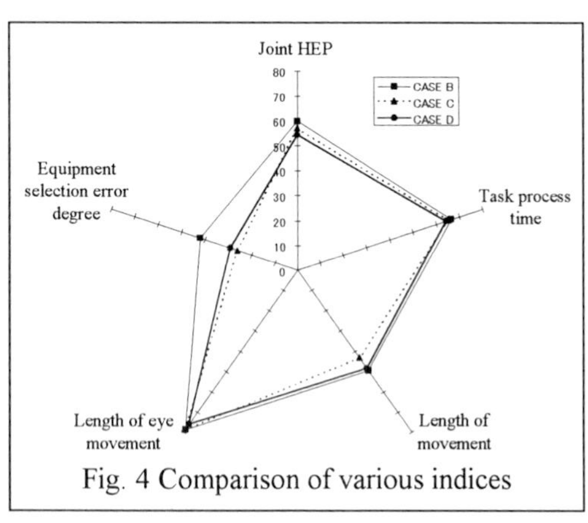

Fig. 4 Comparison of various indices

Associate Systems: A framework for human-machine cooperation

Norman D. Geddes

Applied Systems Intelligence, Inc.
10882 Crabapple Road, Suite 2, Roswell, Georgia, USA

Function allocation has been an important concept and practice in the design of human-machine systems. While function allocation has lead to improvements in system design, it also has the potential to introduce undesirable properties into human-machine systems. This paper proposes the concept and practice of function integration in place of function allocation as a means for designing effective systems. Function integration is based on cooperation and collaboration between active agents and components of the system. An example of a functionally integrated system architecture, the associate system, is provided as a brief illustration of the concept.

1. BACKGROUND

The concept of function allocation between human and machine elements of a system has been a central idea in the development of principled approaches to the design of human-operated systems [1,2]. Common approaches to the design of human-machine systems [3,4], define a sequence of steps for assigning functions to the humans operators and to the machine or computer components of the system. These steps typically consist of the following:

- Determine the processes and functions needed within the system to accomplish its purposes. This step is an analysis of the system requirements, with attention to the physical characteristics and performance of the system under design. Task analytic methods are often applied to decompose the aggregated processes of the system into more specific processes and functions, until a level of description of the functions is reached that is easily characterized in terms of potential design realizations, physical characteristics and performance. To accomplish this decomposition, information and assumptions about the future operational environment of the system and about the meaning of its requirements must often be made by the system designers.
- Evaluate the effectiveness of the alternative potential design realizations for each function. The candidate design realizations may include the human operators or users of the system, as well as software elements and hardware elements. This evaluation may be as simple as the application of "Fitt's list" types of criteria to as sophisticated as model-based evaluations that include both performance and cost.
- Allocate functions to specific system elements, selecting the specific single design realization that will be used to implement each function within the overall system. Overall design quality is based on the most effective allocation of function to single design component. From this view, humans would be allocated functions that they were most

competent to perform, while functions that would be inherently difficult for humans would be allocated to other system components.

The sequence of steps described above is common to "top down" design practices. The philosophical perspective behind the use of top-down design for human-machine systems is that the human is a component of the system whose role in the system operation is not fundamentally different from any other component. Functions that are allocated to the human components are expected by the designers to be faithfully and willingly performed by the human components, and functions not allocated to the humans are not expected to be of concern to them. Increasingly, there is evidence that this perspective on function allocation may not be appropriate for human participation in complex systems.

2. CHALLENGES TO FUNCTION ALLOCATION

The conventional view of function allocation in the context of top-down design poses some important problems in human-machine systems design, particularly with respect to systems containing advanced computer-based automation. These problems are not simply the result of humans being unreliable system elements whose behaviors have a random noise component of failure within them. The difficulties are more the result of the fundamental assumption underlying the function allocation process that treats the human as a component of the system. Among the more common challenges to conventional function allocation are *function aliasing*, *function fragmentation*, and *function rigidity*.

Function aliasing in a system occurs when identical or highly similar functions have different design realizations. The most obvious examples of function aliasing are found at the user interface, where it is not at all uncommon in current computer-based user interfaces to have similar system functions with very different interface mechanizations. Function aliasing is also an important concern deeper within the system design. An example is use of different control system approaches to control two different but highly similar elements of a complex system.

The operational consequences of function aliasing in human-machine systems are well-known. More function aliasing increases the training time and the errors in the use of the system. These effects are felt by the operators and the maintainers of the system, who find their expectations about system behavior and structure are violated by the presence of seemingly capricious differences in the system implementation. While the development and application of design and implementation standards has been influential in reducing function aliasing, standardization represents an incomplete solution.

Top down design decomposition is vulnerable to function aliasing. As the decomposition process expands, it can become increasingly difficult for the design team to be aware of the potential similarities in widely separated portions of the decomposition structure. In addition, the rationale for reducing function aliasing is not always apparent to the designers. The selection of an alternative design realization is more often a local decision that considers the effectiveness of the specific alternatives for that specific function than the more general and system-wide concern for consistency.

The need to capitalize on human strengths and to mitigate machine weaknesses is often spread throughout the operational requirements of a complex system. When human strengths lead to the allocation of small portions of many disparate tasks to the human operator, yet human limitations prevent human performance of a complete task for many of the tasks, function fragmentation exists.

Function fragmentation has several negative effects on the overall human-machine system. Participation in many partial tasks increases the cognitive demand on the user as a result of context switching between the tasks. The lack of continuity in tasks results in a breakdown of situation awareness across the system. Perhaps more important, operators can lose their sense of responsibility for the system, feeling that they are a slave to the machine rather than the agent in charge. This leads to long term dissatisfaction with the system and serious operational blunders.

The decomposition process undertaken in top down design tends to produce function fragmentation as a result of its assumption that the human is a system component. Components are not seen as having any self awareness of their roles in the system, nor any view of the overall goals and performance of the system beyond their local need for input data or signals. As a result, each function allocation decision can be made based on its local properties, without regard for how the allocation decision fragments the higher level functions across components.

Function rigidity describes the relative difficulty of performing novel behaviors within the system in response to unanticipated situations. Systems with high rigidity do not easily permit deviations from the original designed functions. For human-machine systems, low rigidity (or high flexibility) is nearly always a system goal. The most compelling reason for human participation in the system is their inherent flexibility of function. If the system exhibits function rigidity, however, the human is hindered by the system design from making appropriate responses to novel situations.

Function rigidity is also encouraged by the conventional top down function allocation process. System economy and simplicity drive the designers to allocate each low-level function to a single component of the system. The process strongly discourages allocating the same function to multiple distinct components. Even in cases in which redundancy is desired, allocation is normally to multiple copies of the same component rather than to entirely different components.

Attempts at design optimization by function allocation lead to function aliasing, fragmentation and rigidity. These phenomena are human issues, rather than machine issues, affecting the designer, the operators and the maintainers over the entire life cycle of the system. The consequences of function allocation are undesirable, producing difficulties in system operations and maintenance, long term dissatisfaction and loss of responsibility, increased cost of ownership and loss of opportunity to respond to unplanned situations.

While function allocation has lead to improvements in human-machine system design, a new approach is needed that can guide designers towards systems in which the human and machine are more integrated in their behaviors without the loss of human participation.

3. FUNCTION INTEGRATION

I propose that *function integration* replace *function allocation* in our human-machine systems design processes. The function integration approach to human-machine systems design overcomes the limitations that are an inherent part of the conventional top-down, function allocation approach. The mitigation of these limitations is the result of deliberate changes in emphasis during the design process.

Function integration is a human-centered design approach that considers the humans in the system as active agents whose presence and purpose transcend the system itself rather than strictly as system components. The function integration design method includes a decomposition of the system purposes, but emphasizes identification and preservation of alternative methods, overlap of participation in each alternative, and explicit support for limitations in performance of the alternative methods by either human or machine. The goal of function integration is the principled design of the emergent properties of the system that result from the interactions between the components and agents.

Function aliasing is avoided by the identification and deliberate design of classes of similar function alternatives that define behavior patterns for the system. Multiple alternatives for each function are supported to provide complete behavior pattern classes. The requirement to implement complete behavior pattern classes insures that the user and the system have consistent behaviors.

Function fragmentation is reduced by deliberate design of the interactions between multiple candidate participants to allow considerable overlap in participation. The interactions must permit participants to be aware of their own state and that of others who may be involved in the task. Functions are designed to be performed as collaboration, with many agents and components as participants.

Function rigidity is reduced by inclusion of many alternative methods to accomplish system goals. To the maximum extent possible, every function is supported by multiple competent components and agents, not all equal in capability. Since the interactions between participants are explicitly supported, participants have greater opportunity to adjust the designed behaviors to meet the circumstances.

The basic processes underlying function integration have been identified and reported by others. What has not been achieved is the organization of the processes into a defined design method and a broad recognition of the need to replace function allocation with function integration. As a preliminary step towards the definition of function integration, I propose the following steps:

- Analysis of purposes and alternatives. This step is a task-analytic decomposition of the system from a purposes and methods viewpoint that emphasizes the definition of a broad set of alternative methods for achieving each system goal or sub-goal. This decomposition would be similar to GOMS [5] or PGG [6] analysis.
- Synthesis of system behavior patterns. The alternative methods for accomplishing the system purposes are inspected to collect similar alternatives and identify recurring classes of alternatives. These classes become the basis for a defined set of system behavior patterns.
- Elimination of marginal alternatives. Rather than selection of only one or a few of the potential alternatives for allocation, function integration emphasizes support of many alternatives. The designer should retain all alternative methods that exceed a cost effectiveness threshold or are needed to complete a behavior pattern. Only the marginal alternatives are discarded.
- Integration of participant behaviors. Each alternative is analyzed for participation by multiple system components and agents, rather than allocated to a system component. Support for the interactions of the participants in each alternative is designed using the system behavior patterns defined earlier. Typical participant interactions include start, pick up, merge, wait, handover, drop, redo, undo, and halt for each alternative.

The result of this design process can be a system of unusual robustness and flexibility. Although function integration applies directly to human-machine systems, it is a useful view for the design of any agent-oriented system.

4. FUNCTION INTEGRATION IN AN ASSOCIATE SYSTEM

An example of a successful class of human machine systems that embody function integration as a design approach is the *associate system* [7,8]. Associate systems are computer-based aiding systems that are intended to operate as an associate to the human user. The associate metaphor is based on three behavioral capabilities:

• Mixed initiative. This capability is met when the human user and the associate each individually possess the information and the knowledge to recognize the need to take action, to determine the course of action, and to execute the course of action successfully.

• Bounded discretion. Despite the associate's capabilities to exercise the initiative in responding to the situation, the human user is in charge. The associate may only perform those activities it has been authorized to perform and that are consistent with the human user's intentions.

• Domain competency. An associate is expected to be broadly competent in the operational domain, but may have less expertise than its human counterpart. Its domain skills are less those of a narrow expert and more those of a well-integrated generalist. Its formal knowledge includes specific knowledge about its human user and about other machine functions that it and the human jointly control.

An associate system is designed to follow the human user's lead, aiding whenever necessary without the need for explicit instructions if within its bounded discretion. The human user preserves the opportunity to perform all system tasks completely manually or with aiding from the associate. In the extreme, the associate also has the capability to perform all of the system tasks autonomously, although perhaps not as well as the fully rested and alert human user, and only if authorized. An associate system provides as its goal, a completely functionally integrated system.

All of the associate systems built to date share a number of important architectural features that are the result of designed function integration with their human users. A typical architecture is shown in Figure 3.

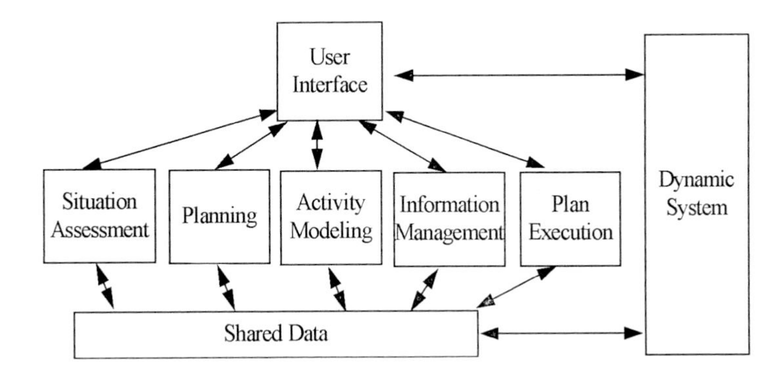

Figure 3. Associate components are designed for function integration with the user.

Map based route planning in an associate provides an example of the effects of function integration. Route planning over terrain and in the presence of hostile objects is a difficult task for both human and computer. Attempts to generate acceptable tactical routes completely by computer are rarely successful--human adjustment after the route is determined is nearly always required. In an associate system, the computer and the human normally work in collaboration to produce a route, where human interaction is supported at many levels, from general parameter setting to explicit editing of waypoints and speeds. In addition, the associate supports both extremes of full manual route definition and full automatic route definition. Although each agent is competent to produce a route alone, the integration of their efforts produces superior results.

This integration is achieved by providing both the human and the machine components with the information to perform the planning function and a set of interaction methods that follow system-wide behavior patterns. The human user of an associate can always participate in the system tasks in the same ways, knowing what effect his participation will have on the behaviors of the associate.

4. SUMMARY

Function integration is a distinctly different design approach from function allocation. While function allocation can lead to confusing system behaviors and structure, task fragmentation and loss of situation awareness, and system inflexibility, the design approach described by function integration can overcome these system limitations.

Function integration emphasizes cooperation and collaboration between system components and agents to produce the desired system behaviors as emergent behaviors of the system. It is achieved by explicitly designing shared behaviors that are consistent and supported by multiple methods and multiple participants. Although this paper has provided a preliminary description of the design process to support function integration, the maturation of this concept will require experience and refinement by the rest of the human-machine systems design community.

REFERENCES

1. Fitts, P.M & Posner, M.J. *Human Performance*. Monterey, CA: Brooks/Cole Publishing Co. (1962)

2. Chapanis, A. *Man-Machine Engineering*. Belmont, CA: Wadsworth (1965)

3. McCormick, E.J. & Sanders, M.S. *Human Factors in Engineering and Design*. New York: McGraw Hill (1982)

4. Rasmussen, J. *Information Processing and Human Machine Interaction*. New York: North-Holland (1986)

5. Card, S.K, Moran, T.P & Newell, A *The Psychology of Human Computer Interaction*. Hillsdale, NJ: Lawrence Earlbaum Associates (1983)

6. Sewell, D.R. & Geddes, N. D. A plan and goal based method for computer human system design, *Human Computer Interaction, INTERACT 90*. New York: North-Holland. (1990) 283-288.

7. Lizza, C.S. & Banks, S. B. Pilot's Associate: A cooperative, knowledge-based system application. *IEEE Expert*, June 1991.

8. Edwards, G.R. & Geddes, N.D. Deriving a domain independent architecture for Associate Systems from essential elements of associate behavior. *Associate Technology: opportunities and challenges*, Lehner, P.E (ed) Fairfax, VA: George Mason Univ. (1991)

Control versus dependence: Striking the balance in function allocation

Erik Hollnagel

Principal Advisor, Ph.D., OECD Halden Reactor Project, P. O. Box 173, N-1751 Halden, Norway

1. INTRODUCTION

Function allocation has been a central issue for human factors engineering ever since the beginnings in the late 1940s. Over the years function allocation has been based on several different principles, which all - explicitly or implicitly - have referred to a view of the nature of human action. In the traditional, engineering, type of function allocation - known as the "left-over" principle - the technological system was designed to do as much as feasible (usually from an efficiency point of view) and the rest was left for the operators to do. This implied a rather cavalier view of humans and in particular did not include any explicit assumptions about their capabilities or limitations - other than that they hopefully were capable of doing what needed to be done.

In classical human factors engineering the dominating approach to function allocation has been the compensatory principle, also known as Fitts' List (after Fitts, 1951). In this approach the capabilities (and limitations) of people and machines are compared on a number of salient dimensions, and the function allocation is made so that the respective capabilities are used optimally. The underlying assumption is that the human operator can be described as a limited capacity information processing system. This approach requires that the situation characteristics can be described adequately *a priori*, and that the capabilities will be more or less constant, i.e., that the variability will be minimal. Human action is furthermore seen as mainly **reactive** to external events, i.e., as the result of processing input information using whatever knowledge the person may have - normally described as the operator's mental model.

A third approach, which has been emerging during the 1990s, can be called the complementarity principle (e.g. Grote et al., 1995). According to this function allocation should serves to sustain and strengthen human ability to perform efficiently. It is therefore necessary to consider the work system in the long term, including how the work routines change as a consequence of learning and familiarisation. The main concern is not the ephemeral level of efficiency (or safety), but rather the ability of the joint system to sustain acceptable performance under a variety of conditions. The complementarity principle looks at human-machine co-operation rather than human-machine interaction, and emphasises the conditions provided by the overall socio-technical system. It is consistent with the view of cognitive systems engineering which emphasises the functioning of the joint cognitive system (Hollnagel & Woods, 1983). Human action is seen as **proactive** as well as reactive, driven as much by goals and intentions as by the "input" events. Information is furthermore not only passively received but actively sought, hence significantly influenced by what the person assumes and expects to happen corresponding the paradigm expressed by Neisser's perceptual circle (Neisser, 1976).

2. FEEDBACK AND FEEDFORWARD

If we accept the complementarity principle as the basis for function allocation, the overriding concern is that the joint cognitive system (JCS) retains control of the process instead of being controlled by it. The JCS must be able to keep the process in a stable equilibrium by using a judicious blend of feedforward and feedback driven control. In feedback control the output from the system is compared with the desired output, and the differences are used to produce compensatory action. In feedforward control the controlling actions (signals) are given in anticipation of changes, rather than as compensation after they have occurred, cf. Figure 1. (For the feedforward to be effective it must, of course, refer to an updated representation of the system state as shown by the dotted line.)

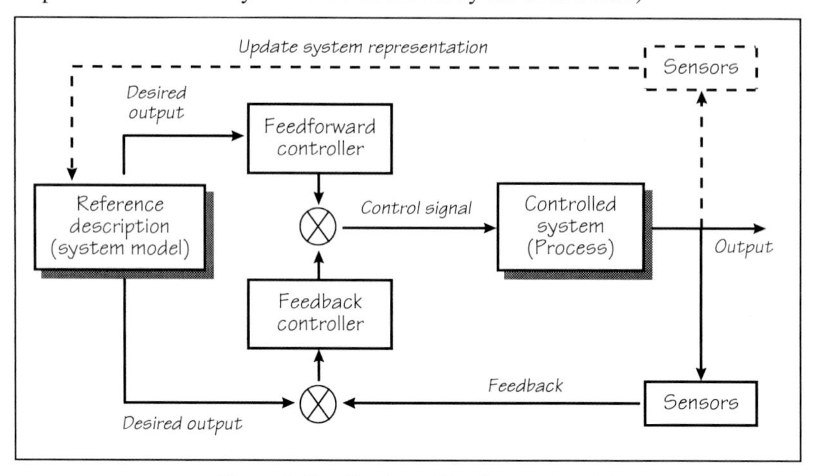

Figure 1: Feedback and feedforward control.

Control systems in general, including cognitive systems regardless of whether they are natural (people) or artificial (machines), can use feedback control only as long as the rate of change in the controlled system is slow enough for the feedback to be processed. If there is too much information to process, then the response will straggle and performance will deteriorate. In humans this may occur as the heuristic responses to input information overload described by Miller (1960). One such heuristic is to reduce the level of discrimination (categories of input) and thereby gain valuable time. The heuristics, however, mean that the feedback is less precise, which in the end may cause the JCS to lose control of the process. An alternative is to rely more on feedforward and to anticipate responses. If the JCS can anticipate correctly the output from the process, then the processing of the feedback needs less effort and time is thereby gained. Yet the anticipation can only be correct if the underlying system representation also is.

The feedback/feedforward balance can be described in terms of attention. It is a basic fact, or rather a tautology, that whatever we have to do consciously will require attention. Attention is needed to observe developments of the controlled process to detect deviations and changes in the system output. At the same time attention is needed in order to analyse (or process) the feedback. Even if sensory and cognitive resources may not be in direct competition (Wickens et al., 1983) there are clear limits on attention and a trade-off must therefore be established. If full attention is required to do something, then there is little or no capacity left to maintain an overview or to think ahead - that is, to predict or to control by feedforward - and performance becomes more imprecise. Conversely, if attention is directed at foreseeing future developments and planning interventions rather than at the outcome of actions, then the disregard of feedback will increases the likelihood that the controlled process will drift away from the target state.

3. MAINTAINING CONTROL

This relation between process demands and the capacity of the JCS controlling the process is illustrated by Figure 2. The basic assumption is that if demands are less than capacity, then control can be maintained. Since capacity clearly is limited, the demands may easily become too high. The maximum capacity will furthermore depend on whether the performance is based on feedback alone or whether it uses a mixture of feedback and feedforward. Thus the JCS can sustain a higher level of process demands if control is accomplished by a mixture of feedback and feedforward, rather than by feedback alone. In terms of processing needs feedforward driven control is probably the least demanding, but it may also be the most brittle.

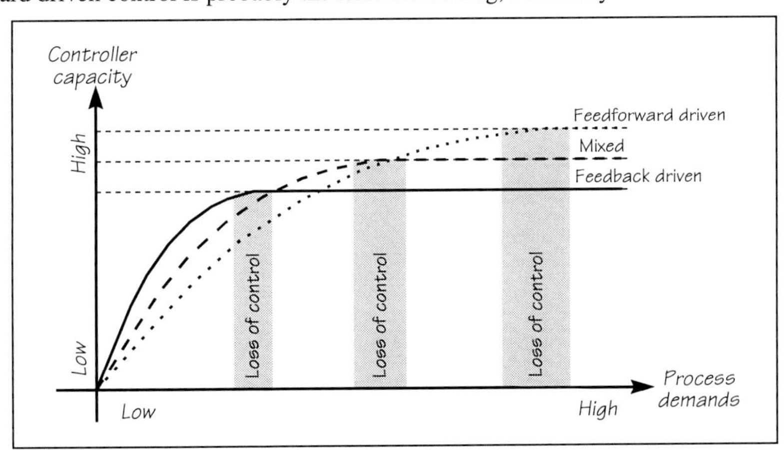

Figure 2: Relation between process demands and performance mode.

As noted above a JCS can use feedback control only as long as the rate of change in the controlled system is slow enough for the feedback to be processed. Similarly, feedforward control requires that the predictions take no more time than allowed by the system dynamics (and, of course, that they are correct). This clearly has implications for interface design and function allocation: if the interaction requires complex decisions and prediction, then sufficient time must be allowed for them. Function allocation should therefore serve to ensure that the amount of attention required to control the process is adequate. If the demands are too small, then control may be lost because the operators cannot maintain an adequate situation understanding: they become dependent of the process, such as in highly automated system.

Unfortunately this only becomes clear when the operators need to take control, e.g. when the automation fails. If the demands are too large, then operators may fall behind the process and control may be lost. (This could happen e.g. if the interface burdens them with unnecessary tasks.) Neither the demands nor the capacity can, however, be defined independently of the situation characteristics. To achieve the right balance in function allocation therefore requires that the coupling between context and control can be analysed and described, such as in the Contextual Control Model (Hollnagel, 1993 & 1997).

Without going into any details of this type of modelling, process demands can be described as a function of subjectively available time and number of goals, while controller capacity can be described in terms of the cognitive control mode. The model describes the performance characteristics of each control mode and the conditions under which control may transition from one mode to another - either being lost or regained. By emphasising the temporal characteristics of the interaction some issues become important such as how to reduce the need of navigating in the information space, how to identify possible ambiguities (mode errors) in observation and control, how to ensure that the interface matches the task demands rather than being as simple as possible, how to enable operators to understand the dynamics of the process and anticipate developments, etc. System design cannot be based on a complete structural description but must look instead at the functions, such as how the JCS maintains a dynamic equilibrium with the process. The structural approach is incomplete because the reactions of people cannot be predicted as well as the reactions of artefacts, and because situation descriptions cannot be entirely decomposed. The functional approach provides a way of describing the dynamics of the interaction, which effectively supplements the structural approach although many details still need to be worked out.

REFERENCES

Fitts, P. M. (Ed). (1951). *Human engineering for an effective air navigation and traffic-control system.* Ohio State University Research Foundation, Columbus, Ohio.

Grote, G., Weik, S., Wäfler, T. & Zölch, M. (1995). Complementary allocation of functions in automated work systems. In Y. Anzai, K. Ogawa & H. Mori (Eds.), *Symbiosis of human and artifact.* Amsterdam: Elsevier.

Hollnagel, E. (1993). Modelling of cognition: Procedural prototypes and contextual control. *Le Travail Humain, 56(1)*, 27-51.

Hollnagel, E. (1997). Context, cognition, and control. In Y. Waern (Ed.). *Co-operation in process management - Cognition and information technology.* London: Taylor & Francis (in preparation).

Hollnagel, E. & Woods, D. D. (1983). Cognitive systems engineering: New wine in new bottles. *International Journal of Man-Machine Studies, 18*, 583-600.

Miller, J. G. (1960). Information input overload and psychopathology. *American Journal of Psychiatry, 116*, 695-704.

Neisser, U. (1976). *Cognition and reality.* San Francisco: W. H. Freeman.

Wickens, C. D., Sandry, D. & Vidulich, M. (1983). Compatibility and resource competition between modalities of input, central processing, and output: testing a model of complex task performance. *Human Factors, 25*, 227-248.

Human Electronic Crew Teamwork: Cognitive Requirements for Compatibility and Control with Dynamic Function Allocation.

R.M. Taylor

DERA Centre for Human Sciences, Farnborough, Hants GU14 6TD United Kingdom.

1. INTRODUCTION

In an increasingly complex and automated aircraft environment, aircrew tasks are now more about thinking than doing: more cognitive than physical in nature. This has led to interest in the requirements for cognitive quality in aircrew systems, and a need for engineering principles to design cognitive tasks. Of particular interest are the cognitive requirements for control (strategic, opportunistic) and compatibility (usability, intuitiveness) in systems where both human and machine components influence the quality of cognitive functioning. Such joint cognitive systems require new levels of HCI, as in co-operative teamwork with dynamic allocation of functions between human and electronic crew members.

2. AIDING TECHNOLOGIES

Recent developments in real-time data acquisition, fusion, and processing, and in computer modelling and AI inferencing techniques (Expert Systems, KBS, Neural Networks), are being used to provide adaptive aiding and dynamic function allocation (DFA) in advanced aircrew systems e.g. Co-pilote Electronique, CASSY Cockpit Assistant, Pilot's Associate technologies.[1] Coupled with concepts for co-operative human-computer interaction and human-electronic crew (H-EC) teamwork, these technologies aim to enable better management of aircrew workload, and to enhance situation awareness (SA). They seek to support the pilot efficiently and unobtrusively, while allowing the pilot to remain at the top of the system control hierarchy, with the ultimate responsibility for generating and setting goals and directives i.e. staying in charge. Co-operation is achieved by incorporating a model of human decision making and control abilities into the control automation, by monitoring pilot performance and workload through behavioural and physiological indices, and by predicting pilot expectations and intentions with reference to embedded knowledge of mission plans and goals. Thus, a fuller integrated matching of human-machine capabilities is provided for enhanced performance through a synergistic H-EC relationship.

2.1 Autonomy and Control

Joint H-EC cognitive systems raise new requirements for autonomy and control of functions with DFA e.g. command, authority, trust, feedback, and error rectification. One focus of research has been the logic or rules for automatic invocation of levels of aiding, to maintain pilot skill and avoid complacency e.g. cyclical invocation.[2] Other work has sought solutions for the interface through pilot selectable levels of autonomy (LOA) for functions, with the required pilot operational relationship and interaction.[3] Discrete LOA modes are proposed (Inactive, Standby, Advisor, Assistant, Associate), based on Sheridan's taxonomy, with tailorable functional clusterings for flexible responding, to avoid too rigid automation imposed by design. In Associate mode, under full DFA, the proposed system maintains advisory functions and accepts pilot allocated tasks, but also takes over tasks as the context demands. These modes aim to provide bounded, communicable structure for delegated levels of authority, minimising mode confusion, and building trust and confidence.

2.2. Teamwork and Autonomy

Pilot's tend to view EC autonomy simply as either automatic, with or without status feedback; semi-automatic, telling what will happen and asking permission to proceed; or advisory, providing information only.[4] Our initial research examined the extent to which DFA and LOA are features of current HE-C teamwork. Aircrew provided ratings of dimensions of teamwork, comparing cockpit interfaces in RAF Harrier and Tornado aircraft.[5] DFA (real time role/task distribution), but not LOA (degrees of independent functioning), was a discriminator of teamwork quality. Function allocation is essential for the effective co-ordination of goal-oriented team activities. But autonomy threatens goal maintenance. Flexible function allocation is beneficial for ill-structured problems involving uncertainty, with good communication between team-members. In mature teamwork, with good communication, and leadership initiative turn-taking, the transitioning of authority is smooth and flexible. The locus of control is driven more by situation and context, than by the preservation of a sole source of control authority. Goal tracking and feedback is the key.

3. TRUST AND AWARENESS WHEN TEAMWORK FAILS

3.1 Un-cooperation Awareness

We have sought to identify the psychological dimensions of trust, awareness and compatibility for specification of cognitive requirements with DFA when H-EC teamwork fails. Trust is built on awareness of proven performance; in uncertainty, proven doubt is safer than blind trust. Trust and awareness of flight automation aiding were studied using the Multi Attribute Task (MAT) simulation software.[6] The MAT flight simulation comprises three task elements, namely manual Compensatory Tracking (CT), Systems Monitoring (SM), and Resources Management (RM). SM and RM tasks were operated in manual, semi-automatic, or fully automated modes. 12 non-aircrew subjects flew task profiles with increasing frequency of events requiring actions. Task aiding was invoced automatically, with on-screen caption warning and feedback on mode change and status. The invocation logic was manipulated to be either logical (cooperative, reducing workload), simulating normal aiding, or illogical (un-cooperative, increasing workload), simulating automation failure. Subjects provided subjective ratings of automation trust, reliability and SA (3-D SART). The results showed that manual compensation equated performance across the conditions. Most subjects seemed unaware of the cooperation manipulation. Ratings of SA, trust and confidence remained unchanged, similar to the dissociation reported between confidence and accuracy in computer-aided decision-making. Monitoring the status and functioning of aiding is difficult under automatic invocation without operator control. In such a system, design safeguards would be needed to prevent unacceptable unpredictability with variable function allocation, and to minimise inappropriate allocation e.g. establishing operator willingness to accept aiding.

3.2 Adaptive Iconics

The basic MAT battery provides no explicit differentiation between manual, automatic and enviromentally induced screen changes. Automation functioning has to be inferred from understanding of the control rules. To investigate more intuitive, cognitively compatible (CC) displays of automation status, a modified version of the MAT battery display screen was created. Eurofighter uses moding icons to identify system functions. It was considered that a familiar, dynamic physical instantiation of automation functioning, using icons with properties of agency, might assist in understanding automation status. In the modified MAT battery, automation activity was represented using dynamic visual icons, depicting the form and location of control actions. Droid-like, R2D2 characters emerged when the automation was invoked. These Adaptive Iconics (AIs) were re-positioned and animated beside SM parameters and RM functional elements, where the automatic action was perceived as taking place. An experiment with 20 non-aircrew subjects compared performance with the original and AIs-modified MAT task.[7] Ratings of CC using the SRK-related experiential CC-SART measure (Dimensions: Level of processing; ease of reasoning; activation of knowledge), showed enhanced CC by AIs of the RM task (and CT), with improved RM ease of reasoning. However, SM performance was poorer with the AIs. This was interpreted as over-reliance on SM automation with partial aiding, induced by the presence of AIs, without clear indication of their restricted support. AIs improved the cognitive quality of the interface for one rule-based task, but created undesirable side effects on performance and goal tracking for another.

3.3 System-Induced Deluded Control

Automation of mission planning functions anticipates predictable events and brings forward decisions on the mission time line. In theory, plan automation enhances SA by freeing limited attentional resources for unexpected events. We have sought to measure how *cognitive rigidity* or set arising from plan automation inhibits creativity and responsiveness to changed situations with knowledge-based tasks. In an experiment simulating a multi-aircraft ground attack mission, subjects (30 non-aircrew) were required to control aircraft over a target, and to exit the area safely, avoiding radar detection and enemy aircraft contact.[8] Survival was the briefed prime directive. Subjects received different levels of mission support, i.e. ad hoc preparation; automatic planning and mission rehearsal; auto-plan, rehearsal, plus automatic plan execution. In a second mission phase, enemy aircraft appeared to counter the planned attack. The results showed advantages for automatic plan generation and execution with no threat aircraft present. However, when the planned route came under attack, rather than freeing attentional resources, the automation produced over-dependence or "blind trust" in the plan, resulting in high losses from enemy contact. Subjective SA and CC ratings, and a new perceived control scale (PC-SART), indicated poor threat awareness with plan automation. Subjects exhibited *delusion of control*, rigidly interpreting experience through cognitive schema set by the plan. The expected lag in recognition of the changed situation, normally ascribed to schema refinement, appears exaggerated by plans, causing extended *awareness hysteresis*. Active involvement in plan generation and execution, provides better adherence to directives, enhanced goal awareness and better strategic control. Plan automation creates a form of *goal blindness*: failure to see the goal for the plan. So, emersive planning and rehearsal technologies, and assertive automation, should be implemented with caution. Mission support should help to prototype and critique responses to unexpected events.

4. THE COGNITIVE COCKPIT

Synthesising the lessons learned, it seems that goal control is the key to successful teamwork. We have sought to develop an approach to an intelligent H-EC *Cognitive Cockpit*, designed to be cognition sensitive, compatible, adaptive, and supportive to control of pilot goals, in accordance with cognitive engineering principles. This is done by structuring all automated support using Rasmussen's SRK framework, which ensures both invocation and representation of the automation are cognitively compatible. Using a cooperative perceptual control model, we are developing principles for supporting goal awareness (current & desired) and error awareness (diagnosis & rectification), tailored to SRK requirements, with consideration of Hollnagel's modes of control. Feedback is provided to the pilot through not only AIs, but also through schema-based *Goal Balls*, indicating action-to-goal effectiveness and risk, and *SystemCrew Balls*, representing supplied H v EC workload against required workload. Support assertiveness is tailored for goal closure in uncertainty, using tutoring, expert advisor, and critiquing techniques, intended to overcome cognitive rigidity without substitution by EC mind set. This approach could resolve the conflicting control requirements for teamwork and autonomy with DFA, by developing a view of EC as an extension of pilot cognitive functioning dealing with uncertainty, rather than as an independent cognitive agent.

REFERENCES

1. R.M. Taylor and J. Reising (eds), The Human Electronic Crew: Can We Trust the Team? WL-TR-96-3039, Wright Patterson AFB, OH., December 1995.
2. P. Hancock, S. Scallen, and J. Duley, Pilot Performance and Preference for Cycles of Automation in Adaptive Function Allocation, NAWC Tech Report, Warminster, PA, 1994.
3. R. Yadrick, C. Judge, and V. Riley, Decision Support and Adaptive Aiding for Battlefield Interdiction Mission. WRDC-TR-92-3078, p122-126, Wright Patterson AFB, OH. July 1992.
4. R. Lynch, C Arbak, B. Barnett, and J Olson, Displays and Controls for the Pilot/Electronic Crewmember Team. In, WL-TR-96-3039, p 130-138, Wright Patterson AFB, OH., Dec 1995.
5. R.M. Taylor and S.J. Selcon, Operator and Automation Capability Analysis: Picking the Right Team, AGARD-CP-520, Ch. 20, NATO AGARD, Neuilly-sur-Seine, April 1993.
6. R.M. Taylor, R. Shadrake, J. Haugh, and A. Bunting, Situational Awareness, Trust, and Compatibility, AGARD-CP-575, Ch. 6, NATO AGARD, Neuilly-sur-Seine, January 1996.
7. R.M. Taylor and S. Finnie, Adaptive Iconics: A Physical Instantiation of Automation Agency, In M Mouloua (ed), Proceedings of the 2nd Automation Technology and Human Performance Conference, Univ of Florida, 1997 (In Press).
8. R.M. Taylor, S. Finnie and C. Hoy, Cognitive Rigidity: The Effects of Mission Planning and Automation on Cognitive Control in Dynamic Situations, In Proceedings of the 9th Intlernational Symposium on Aviation Psychology, Ohio State University, 1997 (In press)

Interface Design and Evaluation

Natural User Interface (NUI): a case study of a video based interaction technique for a computer game

M. Rauterberg

Institute for Hygiene and Applied Physiology (IHA)
Swiss Federal Institute of Technology (ETH)
Clausiusstrasse 25, CH-8092 Zurich, Switzerland
http://www.ifap.bepr.ethz.ch/~rauterberg

To compare the advantages and disadvantages of a *Natural User Interface* [1] a field study was carried out. During five days of the largest computer fair in Switzerland four different computer stations with (1) a command language, (2) a mouse, (3) a touch screen, and (4) a Digital Playing Desk (DPD) interface was presented for public use. In this version of the DPD the user has to play a board game by moving a real chip on a virtual playing field against a virtual player. The task was to win the computer game "Go-bang". The reactions of the virtual player were simulated by "emoticons" as colored comic strip pictures with a corresponding sound pattern. We investigated the effects of these four different interaction techniques with an inquiry with a questionnaire. Results of the inquiry: 304 visitors rated the usability of all four stations on a bipolar scale. The touch screen station was rated as the easiest to use interaction technique, followed by the mouse and DPD interface; the "tail-light" was the command language interface. One very important result was a significant correlation between "age" and "DPD usability". This correlation means that older people prefer significantly more a graspable user interface in form of the DPD than younger people.

1. INTRODUCTION

There are two main contrary directions for new interface technology: (1) [immersive] virtual reality, and (2) augmented reality or ubiquitous computing. Enthusiasts of the virtual reality approach (VR) believe "that all constrains of the real world can be overcome in VR, and physical tools can be made obsolete by more flexible, virtual alternatives" ([2] p. 87-88). We restrict the notion of "virtual reality system"--in the context of this paper--to systems with head mounted display and data gloves or suits. In these VR applications the user has to leave his natural physical and social environment and to *immerse* in the simulated world. The following two unsolved problems are important: (1) how to simulate tactile and haptic feedback, and (2) how to overcome the social isolation for collaborative tasks. The effect, that the social nearness between real persons is of tremendous importance for collaboration, was investigated and shown in [3]. As we stated in [4], that the effects of tactile and haptic senses are important, we are looking for a realization of a user interface where the user can control the human-computer interaction by his hands dealing with real and virtual objects in the same interface space. The DigitalDesk of Wellner [2] was one of such systems.

Inspired by the ideas of Wellner [2], we were interested in a way to test empirically the advantages or disadvantages of the DigitalDesk in comparison with established interaction techniques. The DigitalDesk has the following three important features: (1) it projects electronic images (virtual objects) down onto the desk and onto real objects, (2) it responds to interaction with real objects (e.g., pens or bare fingers: hence *Digital*Desk), and (3) it can interpret the scene on an appropriate semantic level (e.g., read paper documents placed on the desk; cf. [2]).

2. SYSTEM DESCRIPTION

To run a laboratory investigation or a field study we need a fast, reliable and robust implementation of the whole system. First, we decided to minimize the task complexity and to restrict the user's action space to cognitive planning processes. For public use a simple computer game seems to be best. We implemented a version of the computer game "Go-bang". The user has to play the game by moving a *real* chip on the *virtual* playing field (see Fig. 1). To compare this interface type with the most established dialog techniques we implemented the same game algorithm on three other stations with (1) a command language, (2) a mouse, and (3) a touch screen interface.

Command interface (CI): This station run on a 386er PC with a color screen (17") in an upright position. The user has to enter the co-ordinates of the desired place of a playing field with 12 by 12 positions (e.g., A1, L12). To start a new or to cancel a game she or he has to enter the command NEW. The internal state of the algorithm is presented as text in a special output field (e.g., "Make the next move").

Mouse interface (MI): This interface run on a 386er PC with a color screen (17") in an upright position. To move the user has to click with the mouse on the desired place. To start a new or to cancel a running game she or he has to click on the button NEW. The internal state of the algorithm is presented as text in a special popup window.

Touch screen interface (TI): This station run on 386er PC with a color touch screen (21") in an inclined position of 30 degrees. To make a move the user has to touch with a finger the desired place. To start a new or to cancel a running game she or he has to touch the button NEW. The virtual player was shown on a second colored screen (17") served by a second 386er PC in a client-server architecture.

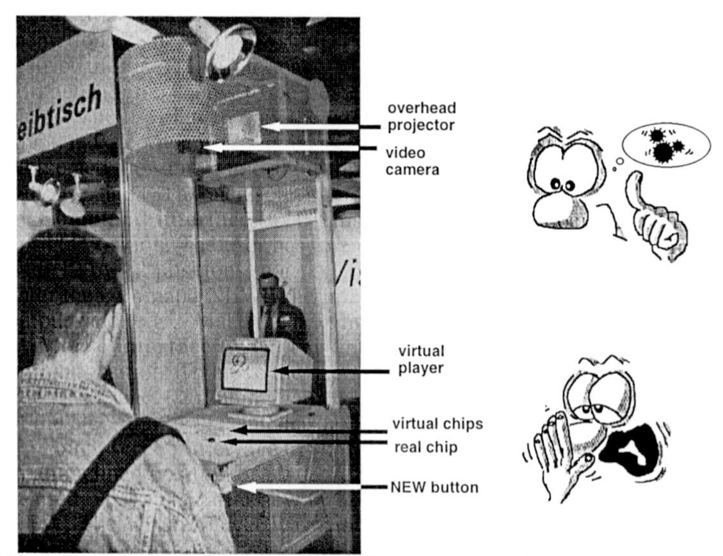

Fig. 1: The Digital Playing Desk--the front view for the users.

Fig. 2: The two emoticons "reasoning" and "yawning".

Digital Playing Desk (DPD): This station was completely realized in C++ on standard hardware components: (1) a Pentium PC, (2) an overhead projector of high luminous intensity and the projection panel GehaVision™, (3) a high resolution video camera, and (4) the video board MovieMachinePro™. For the virtual player a second 386er PC was connected in a client-server architecture. A user has to make a move by putting a *real* chip on the desired place of the *virtual* playing field (see Fig. 1). The computer's output is the projection of a *virtual* chip on the desk. If a user wanted to cancel or restart a game, then he or she had to press the real NEW button in the front of the station.

The Virtual Player: The CI and MI have their output screens in an upright position. This upright position makes it impossible to give an opponent--in the metaphor of a game--an individual representation. The classical solution is, to give feedback about the machine's internal states as text or graphics in defined areas on the screen. This solution leads always to a partition of the screen into a working area (e.g., the playing field) and the feedback area. This superimposing of qualitatively different feedback's in the same output space can be overcome, if we use an additional output device. This can be done, if we separate the working area (the playing field) from the feedback area and if we change the upright position of the working area to a horizontal position. Therefore we composed the system of a flat table to project the playing field onto and of a second screen to present the virtual player.

All--from the user's point of view--important internal states of the game algorithm were presented by the virtual player with six different facial expressions and a corresponding sound (see Fig. 2; partially animated comic strip pictures: "Reasoning" = a face with a balloon of animated turning wheels and a machine-like sound, "Waiting for the next move" = a yawning face with a corresponding sound, "Initial state" = a face with blinking eyes, "Incorrect move" = an angry facial expression with an indignant cry, "Be the winner" = a happy facial expression with an arrogant laughter, "Be the loser" = a shrinking face with a disappointed cry). We call these six comic strip pictures "emoticons" (acronym for "emotional icons"). All emoticons of the virtual player are shown on a second color screen (17").

3. VALIDATION OF THE DIGITAL PLAYING DESK

To present our four dialog techniques to a broad population of heterogeneous users, four special stations were constructed. All stations were presented in a central exhibition area during five days at the largest computer fair of Switzerland in September 1995. The official number of visitors was approximately 70'000. Most of these people passed the exhibition area and many of them came into close contact with one of the stations (e.g., playing at one of the stations or observing other people playing). With a questionnaire we got some personal information's of several users of at least one of our four stations. The stand personal was instructed to request users to answer the questionnaire. It was especially necessary to ask women, because they behaved very reserved.

Subjects: The questionnaire was answered by 304 visitors (61 females, 243 males, 5 anonymous data). The average age of the women was 31 ± 13 years, and of the men 30 ± 14 years (T-Test: $p \leq .724$). As a *control variable* the "computer experience" was measured in millimetre on the corresponding bipolar rating scale ["no experience": 0 ... 90 mm: "expert"]. *There* was a significant gender difference in computer experience; the computer experience of men was higher than the experience of women (men: 58 ± 22 mm; women: 48 ± 23 mm; T-Test: $p \leq .002$).

Dependent Measures: The questionnaire consisted of two parts: (a) personal data (age in years, gender, computer experience in form of a bipolar rating scale, and (b) usability ratings. For each of the four interfaces the following five aspects were asked with a multiple choice question: "did you play", "did you loose", "did you win", "did you play draw", "did you cancel". The *dependent measure* per interface type is the number of millimetres of the user's marking on the bipolar rating scale ["very easy to use": 0 ... 70 mm: "very difficult to use"]. We also differentiated between "real station contact" and "only observer status".

Results: All persons, who answered that they had no real contact to one of the stations, had a significant higher amount of computer experience ("no contact": 70 ± 17 mm, N = 10; "with contact": 55 ± 23 mm, N = 276; T-Test: $p \leq .04$). We can not find a significant difference between the "contact" and "no contact" group in the usability ratings.

The usability rating was best for the touch screen interface (median = 3; mean rank = 1.9), followed by the mouse interface (median = 6; mean rank = 2.4) and the Digital Playing Desk (median = 9; mean rank = 2.5). The "tail-light" was the command language interface (median = 17; mean rank = 3.3). These differences are significant (Friedman Test, df = 3, Chi^2 corrected for ties = 89.1, $p \leq .0001$). One very important result was a significant correlation between "age" and "Digital Playing Desk's usability". This correlation means that older people prefer significantly more a graspable user interface in form of the Digital Playing Desk than younger people. ($R = -0.202$, $N = 179$, $p \leq .006$). All other correlations are not significant.

4. DISCUSSION AND CONCLUSION

To carry out an inquiry with a questionnaire only for scientific purposes in the context of a commercial fair was more difficulty than we expected. The only motivation for the user to fill out such a questionnaire--in contrast to all the lotteries of the commercial issuers around us (with their very attractive prizes)--was to bring in his or her personal opinion into an scientific research process. This argumentation was the most convincing reason to participate. Overall we got more filled out questionnaires than we expected, but much less than observed visitors at one of the four stations. To increase the number of answered questionnaires the stand personal has to be active: to go to the user and to ask for participation.

We could find two main results: (1) the touch screen interface was estimated as the easiest to use, and (2) the significant correlation between age and the usability ratings for the Digital Playing Desk. If we assume, that the average age of the populations in all high industrialised countries will increase in the next two or three decades, then this result will be of tremendous importance for the development of modern computer technology for elderly people!

The general advantage and disadvantage of immersive VR are the necessity to put the user into a complete modelled virtual world. This concept of immersing the user in the computer's world ignores the on-going process of interacting with the real world. In the same interface space the mixing of real and virtual objects is not possible. But, humans are--most of their time--part of a real world and interact with real objects and other real humans.

Augmented Reality (AR) recognises that people are used to the real world and that the real world cannot be reproduced completely and accurately enough on a computer. AR builds on the real world by augmenting it with computational capabilities. AR is the general design strategy behind a "Natural User Interface" (NUI) [1].

A system with a NUI supports the mix of real and virtual objects in the same interaction space. As input it recognises *and* understands physical objects and humans acting in a natural way (e.g., object handling, hand writing, etc.). Its output is based on pattern projection such as video projection, holography, speech synthesis or 3D audio patterns. A necessary condition in our definition of a NUI is that it allows inter-referential I/O, i.e. that the same modality is used for input *and* output (see [4]). For example, a projected item can be referred directly by the user for his or her nonverbal input behavior.

REFERENCES

[1] Rauterberg M & Steiger P, Pattern recognition as a key technology for the next generation of user interfaces. In Proc. of SMC'96 (Vol. 4, IEEE Catalog Number: 96CH35929, pp. 2805-2810). Piscataway: IEEE (1996).

[2] Wellner P, Interacting with paper on the DigitalDesk. In *Com. of the ACM*. 36(7), pp. 86-96 (1993).

[3] Rauterberg M, Dätwyler M & Sperisen M, The shared social space as a basic factor for the design of group-ware. In: K. Brunnstein & P. Sint (Eds.) Intellectual Property Rights and New Technologies. (Schriftenreihe der Österreichischen Computer Gesellschaft Vol. 82, pp. 176-181), Oldenbourg (1995).

[4] Rauterberg M & Szabo K, A Design Concept for N-dimensional User Interfaces. In Proc. of 4th Intern. Conf. INTERFACE to Real & Virtual Worlds (1995) pp. 467-477.

Ecological Display Design for the Control of Unmanned Airframes

Jan B.F. van Erp and Bart Kappé

TNO Human Factors Research Institute, Kampweg 5, Soesterberg, The Netherlands.

1. INTRODUCTION

Unmanned airframes may be used for reconnaissance, target tracking, and battle damage assessment. Important operator tasks are the steering of the platform along the desired path, and the control of the on-board camera. The (sensory) information presented to the human operator is scant. No auditive, haptic, or tactile information is available, and the visual information is of degraded quality (Agard, 1995). This results in poor operator performance. Because information on airframe and camera position and heading is essential but can not be deduced from the outside world images, it is usually presented by additional pictorial displays, by walking tapers, or by points and lines on an electronic map. This paper describes an alternative approach: present the characteristics of the visual information as normally used in steering and orientation tasks. The visual system has evolved to use information contained in optic flow, and may pick-up this information without effort (Gibson, 1950). This is fundamentally different from pictorial or numerical information presentation. We call this *ecological display design*. The ecological approach contrasts sharply with the organismic approach (Vicente, 1995), which tends to ascribe skilled behaviour to elaborate mental constructs and cognitive processes (Kirlik, 1995). The ecological approach places emphasis on environmental analysis (Effken, Kim & Shaw, 1997).

Ego- versus world- frame of reference

Searching for and tracking of targets and other relevant points in 3D space calls for a different state of knowledge than steering the airframe along the desired route, viz. global awareness versus local guidance. The information for global awareness must be presented in a world referenced (north-up) display (Wickens, 1992; see also Roscoe, 1968). However, local guidance tasks predominantly need correspondence between display and control in terms of left, right, etc. This means that steering the airframe requires an *ego-referenced* display (heading-up).

2D versus 3D egocentric displays

The characteristics of an ego-referenced 3D (perspective) display are more ecological than of a 2D display (Warren & Wertheim, 1990): correspondence between display and controls, forward cone of visual space, zoomed-in, and the 3D perspective (see Figure 1). A 3D presentation has advantages for local guidance, as shown by the results of experiments in a flight simulator: with a 3D display, tracking error is reduced (Haskell & Wickens, 1993; Prevett & Wickens, 1994). The specific character of an ego-referenced 3D display is determined by the elevation angle. An elevation angle of exactly 90° (looking down) or 0° (looking to the horizon) leads to compression of one dimension, and thus to a 2D display. Ellis, Kim, Tyler, McGreevy & Stark (1985) found a U-shaped tracking error curve, with best performance at 45°. Kim, Ellis, Hannaford, Tyler & Stark (1987) replicated these results, and found little effects in the range between 30° and 60°.

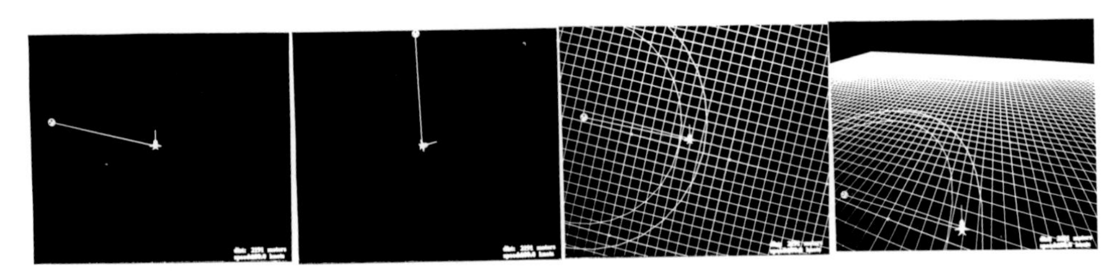

Figure 1. The four displays used: pictorial heading-up and north-up, and ecological 2D and 3D. Airframe speed and standoff distance depicted in the right corner.

One might argue that relatively large elevation angles support lateral tracking, and that relatively small elevation angles support vertical tracking. In steering the airframe, lateral tracking may be more important, because the airframe often flies at a fixed altitude. This pleas for an *elevation angle between 45° and 60°*.

Digital versus pictorial presentation of the standoff distance

Designing the ecological display, three principles for airborne displays (Roscoe, 1968) are relevant for depicting the standoff distance to a target: 1) display integration (which states that it is beneficial to integrate indicators into a single presentation), 2) pictorial realism (which states that graphically encoded information (symbols) can be readily identified with what they represent), and 3) pursuit tracking (which states that pursuit tracking is superior to compensatory tracking). This pleas for a *graphical* presentation of the standoff distance which allows *pursuit tracking*.

2. METHOD

To evaluate the ecological display design, a controlled simulator experiment was conducted. Ten paid male under graduate students participated. The experimental task was tracking a target ship with the camera, while flying the airframe in a circle (counterclockwise) around the target ship. An important instruction was to keep a standoff distance of 2250m to the target, without coming closer than 2000m.

The following dependent variables were calculated: RMS of the camera tracking error; percentage time closer than the minimum standoff distance, and standard error (s.e.) from 2250m standoff distance. A full factorial DISPLAY DESIGN (4) × AIRFRAME SPEED (3) within subjects design was used. See Figure 1 for the display designs. Airframe speed had three levels (60, 120, and 180 knots), and was introduced to vary task difficulty.

Participants always came in pairs for two consecutive days: one could rest, while the other was engaged in the experiment. After arrival, they received a written instruction, and an introduction training, during which the experimenter explained controls and images. During the experiment a recurrence training was given before every new display condition. Participants completed three 200 s scenarios in succession.

Two high resolution monitors (MITSUBISHI HL7955SBK) depicted the camera image (generated by an EVANS AND SUTHERLAND ESIG 2000 (800×600 pixels, 30 Hz, 4° × 3°), and the different display designs generated by a SILICON GRAPHICS IRIS 4D (1280×1024 pixels, 30 Hz, 120° × 96°). The pictorial designs included symbols for the airframe with heading, and camera heading and pitch. The ecological displays included an earth-fixed grid, target fixed distance circles, and a symbol for airframe and camera heading and pitch. The ecological 3D display was generated for an elevation angle of 48°. For more details on mockup and instrumentation, see Van Erp & Kappé (1996).

3. RESULTS AND DISCUSSION

Camera control (RMS tracking error) showed a main effect of airframe speed only; $F(2,18) = 3.59$, $p < .01$. This indicated performance degradation with increasing airframe speed.

The dependent variables on airframe control both showed a main effect of display design; percentage of time with standoff distance too small: $F(3,27) = 11.96$, $p < .01$ and s.e. of the standoff distance: $F(3,27) = 3.83$, $p < .025$). A post hoc test revealed that performance in the two pictorial displays is worse compared to the ecological displays for both measures.

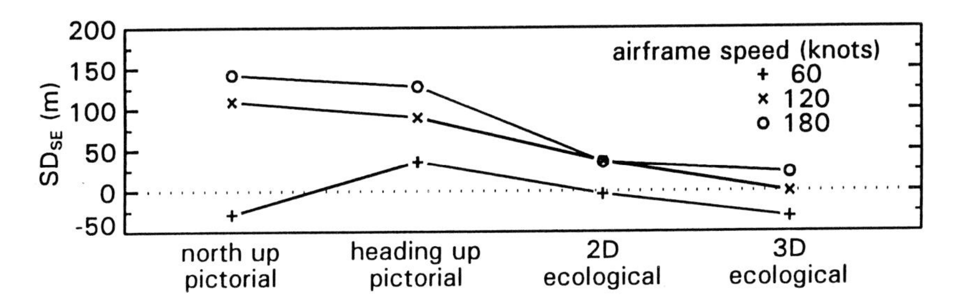

Figure 2. Interaction display design × airframe speed for the standard error of the standoff distance (SD_{SE}).

There was a significant interaction between display design and airframe speed on s.e. of the standoff distance: $F(6,54) = 3.06$, $p < .025$. A posthoc Tukey test showed that performance was independent on airframe speed with the ecological displays only, see Figure 2.

Absence of an advantage of the north-up display compared with the heading-up display may be explained by the fact that participants concentrated on the angle between airframe heading and camera heading, which should be fixated at 90° when the standoff error is correct. Performance improvement with the ecological displays is 30%. They help both in controlling the course of the airframe and preventing the airframe from coming too close. Furthermore, performance is less sensitive to task difficulty. The finding that performance with the 3D version is not significantly better is inherent to the task of lateral tracking only.

Although not all hypothesized effects were found, the results show that applying ecological display design rules (presenting a caricature of the visual information as normally used in steering and orientation tasks) improves the control of unmanned platforms.

REFERENCES

AGARD (1995), *proceedings specialists' meeting on design and operation of unmanned air vehicles*, 9-12 october 1995, Ankara Turkey.

Effken, J.A., Kim, N. & Shaw, R.E. (1997). Making the constraints visible: testing the ecological approach to interface design. *Ergonomics 40 (1)*, 1-27.

Ellis, S.R., Kim, W.S., Tyler, M., McGreevy, M.W. & Stark, L. (1985). Visual enhancements for perspective displays: Perspective parameters. *Proceedings of the 1985 International Conference on Systems, Man, and Cybernetics* (pp. 297-305). Wachtberg Werthoven, Germany: Forschungsinstitut für Antropotechnik.

Gibson, J.J. (1950). *The perception of the visual world.* Houghton Mifflin, Boston, Mass.

Haskell, I.D. & Wickens, C.D. (1993). *The ecological approach to visual perception.* Boston, MA: Houghton Mifflin.

Kim, W.S., Ellis, S.R., Hannaford, B., Tyler, M., M.W. & Stark, L. (1987). A quantitative evaluation of perspective and stereoscopic displays in three axis manual tracking tasks. *IEEE Transactions on Systems, Man, and Cybernetics, 17 (1)*, 61-71.

Kirlik, A. (1995). Requirements for psychological models to support design: toward ecological task analysis, in Flach, J. et.al. (eds.) *Global Perspectives on the Ecology of Human-Machine Systems*, (vol.I). Lawrence Erlbaum, Hillsdale NJ, 68-120.

Prevett, T.T. & Wickens, C.D. (1994). *Perspective displays and frame of reference: their independence to realize performance advantages over planar displays in a terminal area navigation task.* Technical report ARL-94-8/NASA-94-3, Aviation Research Laboratory, Institute of Aviation, Illinois.

Roscoe, S.N., 1968. Airborne displays for flight and navigation. *Human Factors*, 1968, 10 (4), 321-332.

Van Erp, J.B.F. & Kappé, B. (1996). *Computer Generated Environment for steering a simulated unmanned aerial vehicle.* Report TNO-TM 1996 A-039, TNO Human Factors Research Institute, Soesterberg, The Netherlands.

Vicente, K.J. (1995). A few implications of an ecological approach to human factors, in Flach, J. et.al. (eds.) *Global Perspectives on the Ecology of Human-Machine Systems*, (vol.I). Lawrence Erlbaum, Hillsdale NJ, 68-120.

Warren, R. & Wertheim, A.H. (1990). *Perception and control of self-motion.* Hillsdale, NJ: Lawrence Erlbaum Associates, Publishers.

Wickens, C.D. (1992). *Engineering psychology and human performance.* New York, HarperCollins.

User Interface Agents in a Public Information System

Nestor Pridun and Peter Purgathofer

Institute for Technology Assessment and Design, University of Technology, Vienna. A-1040 Vienna, Möllwaldplatz 5. nestor@iguwnext.tuwien.ac.at / purg@igw.tuwien.ac.at

1. INTRODUCTION

While agent technologies are currently widely discussed, there seem to be only a handful of applications that actually make use of such methods. In a recent project, the authors were asked to implement interface agents within the visitor information system for a recently built public facility (Ars Electronica Center, Linz, Austria). The center is dedicated to showing future technologies to its visitors, one of them being digital agents. This paper describes the context in which the system was developed, shows the decisions that were made and presents the finally implemented system with special focus on the user interface agents.

In an overview of differrent types of agents in [Nwan96], an interface agent is described as a piece of autonomous and learning software. An interface agents performs tasks on behalf of its owner/user or assists him. Interface agents are characterized best by metaphors such as "personal assistant" (coined by Pattie Maes, [Maes94]), "guide" [OrSa93] or "coach" [Selk94]. These concepts served as a starting point and inspired the design of the agents in the information system.

2. DESIGN CONSIDERATIONS

The requirements of visitors using an information system in a museum differ from the "regular" use of information processing systems.
- Museum visitors use the system between once in a lifetime and several times a year;
- The system is used by a number of people with very different age, qualifications and knowledge in computers as well as in the domain of the museum;
- The average usage time is only a few minutes;
- The user's main expectations are to quickly access further information on the exhibits, find their way around and, frankly, be entertained.

These considerations and preconditions led to a relatively simple agent design. In the system, agents perform tasks such as guiding, explaining and entertaining the user. While the system was not supposed to be a full blown intelligent agent system, several typical properties were considered to be important.

Picked from the list of agent properties described in [Fone93], the properties of agents that are relevant in this context are:
- autonomy: spontaneous execution of actions that will eventually benefit the user;
- personalizability: educatability in the task and how to do it;
- anthropomorphism: human-like behaviour;
- discourse skills: ability to cooperatively solve problems and answer questions in discourse;
- competence: ability to show what can and what can't be achieved.

We decided early that the success of our agent system is to be measured not by the stringent requirements of "hard" intelligent systems, but against softer criteria such as plausibility and suspension of disbelief [OrSa93].

Beyond classification, there are a lot of contradicting concepts and ideas in the field of agents. Prominent proponents like P. Maes [Maes94] or N. Negroponte see agents as the future technology enabling us to manage enormous amounts of information in networks. On the other side, critics like J. Larnier [Lani95] or B. Shneiderman [BrLa92] offer more pessimistic judgments ranging to the claim that agents will (or should) never be able to do any of our serious business. The discussions in this field of conflict were a major source of conceptual information for the authors to decide what could be done, what should be done and what must be done to implement user interface agents to fit the requirements of that special situation.

For a visitor information system in a museum, a separate set of problems between users and agents is created:

- Too much autonomy can come in the way of the user when she does not immediately see the benefit;
- How can a user interface agent show competence and discourse when users don't expect contact with virtual life forms?
- How can the necessary trust be achieved when the user has only a few minutes of contact with the agent?
- How anthropomorphic do we want our agents to be in this context?

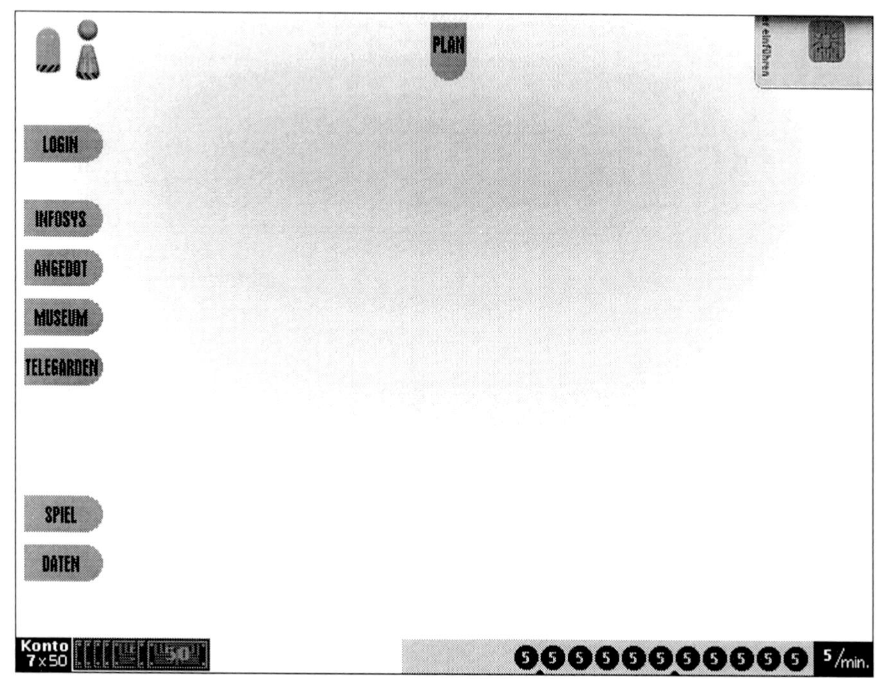

Fig. 1: Main screen of the Ars Electronca Center information system.Two agents in their "home position" can be seen in the top left corner. Note that this is a B/W reproduction of a color screen. In the original, most elements are orange on a blueish backdrop.

3. THE SYSTEM

The implemented design for the information system tries to go beyond today's metaphorical user interfaces. Just like cinema was once "filmed theater" and has now developed it's own language, multimedia systems today often imitate reality. Our design was an experiment to find a dedicated language of multimedia interaction. Fig. 1 shows the system as it presents itself to the user.

The system now includes four user interface agents that perform different tasks and try to solve separate problems of museum visitors, each in a different way. On screen, they look like pieces in a board game. This appearence was chosen as a metaphor to give the impression of servility to the user, just like pieces in a board game do only the things the user wants them to do. In the following, these agents are briefly described.

The "conferencier" welcomes the user at the system, guides him through a "cyberspace reception" where the user chooses a name, age, sex and level of knowledge (only the last of which is of further importance) that make up his "digital presence". Later on, the conferencier shows up from time to time to give feedback (e.g. when the user acts on a higher level of knowledge than declared in the beginning) and alerts (e.g. when the user has to leave the terminal), if necessary. That way, the conferencier acts as an identification figure for the system and its behaviour. Fig. 2 shows the conferencier agent, giving the user a welcome.

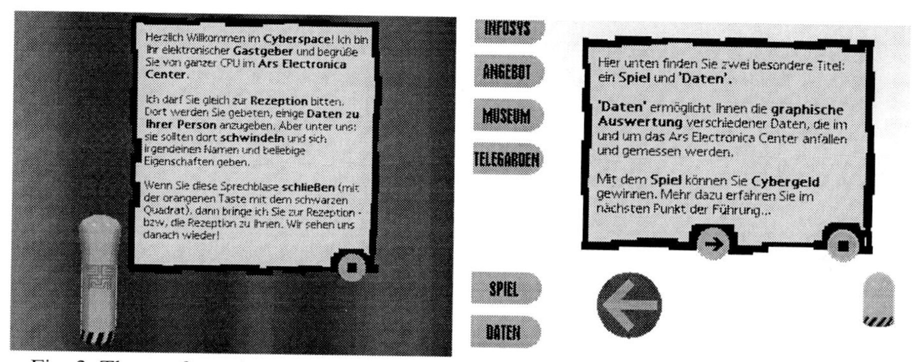

Fig. 2: The conferenciere agent welcomes a user. After morphing through a dark red curtain, he bows and introduces himself.

Fig 3: The guide agent, explaning interface elements during a guided tour through the information system.

A "guide" agent offers guided tours through the museum. Users can choose from a number of guided tours that show them different aspects of the house. In contrast to normal guided tours that usually take about an hour, are offered at fixed times and are for groups of people, the guide agent offers personalized, short tours at any time and with the possibility to interrupt. Tours include both VR and RL aspects as the visitor has to walk from infotermial to infotermial during the tour. The guide is depicted in Fig. 3, describing a user interface element during the general information system tour.

A "first aid" agent jumps in whenever he observes that the user has problems with the user interface of the system. The information system has a quite unusual, "post-wimp" user interface with much more "interactive action" (e.g dragging instead of clicking) as usual. Users often try to interact with the system in a conventional way (clicking). In these instances the first aid agent helps the user by doing the dragging action for them. This way, the agent not only resolves the situation by doing what the user inteded to do, but also shows the user how to do it herself (if she wants to) the next time. Fig. 4 shows several states of a first aid agent intervention to access off-screen information for the user.

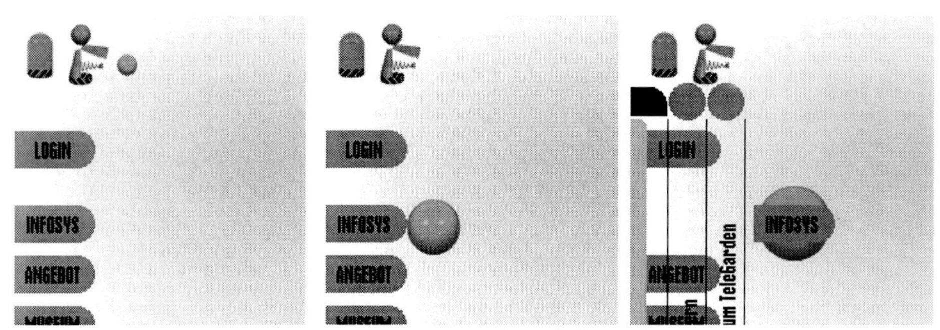

Fig. 4. The first aid agent emits a (blue) ball that drags information onto the screen. The ball does this the same way the user would do it, only from the "other side of the screen".

A "quizmaster" agent acts while the visitor plays a Q&A game about the things in the museum, inducing typical quiz elements like nervousness and restlessness. Furthermore, the quizmaster agent gives feedback on the user's answers to a question, using a positive (smile) or negative (frown) gesture. Since none of the agents have a face, the quizmaster performes these gestures with his whole body.

4. CONCLUSION

The system has been in use since September 10th, 1996. Brief evaluations of it's use were made, suggesting that the users identify the agents with the respective tasks each agent offers, once the users have seen them "in action". Additionally, the users were positive about the entertaining qualities the agents have. It looks like both plausibility and suspension of disbelief were achieved with this relatively simple, but functional agent design.

REFERENCES

BrLa92 Anthropomorphism: From Eliza to Terminator 2 Panel Session with Susan Brennan, Branda Laurel and Ben Schneiderman. In: Proceedings of CHI'92 May 92 pp. 67 ff.

Fone93 L. N. Foner: What's an Agent Anyway? - A Sociological Case Study. FTP Report - MIT Media Lab May 93.

Lani95 J. Lanier: Agents of Alienation. In: ACM Interactions July 95 Vol 2 No 3 pp. 66 ff.

Maes94 Pattie Maes: Agents that Reduce Work and Information Overload. In: Communications of the ACM July 94 Vol 37 No 7 pp. 31 ff.

Nwan96 H. S. Nwana: Software Agents: An Overview. In: Knowledge Engineering Review, Vol. 11, No 3, Sept 1996, pp. 1 ff.

OrSa93 T. Oren, G. Salomon, K. Kreitman, A. Don: Guides: Characterizing the Interface. In: B. Laurel: The Art of Human Computer Interface Design. Addison-Wesley, 1990, pp 367 ff.

Selk94 T. Selker: Coach: A Teaching Agent that Learns. In: Communications of the ACM July 94 Vol 37 No 7 pp. 92 ff.

An Automatic Document Coloring and Browsing System

Tomoyuki UCHIDA[a] and Hidehiko TANAKA[b]

[a]President, Yun Factory Corporation, 7-17-10 Koyama Shinagawa-ku, Tokyo, 142 Japan email:tomo@yun.co.jp

[b]Department of Electrical Engineering, Faculty of Engineering, The University of Tokyo, 7-3-1 Hongo Bunkyo-ku, Tokyo, 113 Japan email:tanaka@mtl.t.u-tokyo.ac.jp

1 INTRODUCTION

In recent years, the number of online documents, for example WWW, has increased. Nevertheless they are still expressed in traditional black-and-white. Color devices such as color CRT are available, and document expressions which make full use of this chromatic faculty is in great demand[1]. We will study effective coloring of documents to increase their legibility and understandability.

The difficulty of making the most of the good effects is due to its complicated side effects such as unpleasant feeling, distraction and individual variation[2]. Previous researches tend to consider only on gaudy primary colors and on fixed devices such as paper. But there are quiet colors and dynamic devices which can change coloration, and there is still much to investigate on document coloring.

Thus we measured by experiment the influence of color given to a reader, and studied the way to make the most of colored documents. And finally we developed an automatic Japanese document coloring and browsing system (named CERAS) was developed. From an estimation experiment, CERAS improved the speed and accuracy of reading.

2 COLOR EFFECT

There are various psychological effects of using colors[1]. First of all, color can express up to millions of attributes human can distinguish. Especially a phenomenon called pop-out that can discover the stimulus of a purpose in a glance from within a plural stimulus is very effective. Next, color can bring the sense of warmness, size, distance and weight. Besides that, color brings feelings such as beauty and joy, as it is now acknowledged in the advertisement field. If the effects of these colors are used effectively, it will be expected that a reader can get the outline of a document faster and understand the content of a document further and enjoy reading more.

In our experiments, coloring promotes understanding and remembering the content of a document.

The coloring rules are established taking into the advantages and side effects. But it is difficult to construct only one coloring rule that always extracts maximum effect in all styles of documents, reader's attitudes and purposes. For that reason, it is more desirable that a document browser selects best coloring rules dynamically.

3 AUTOMATIC COLORING SYSTEM 'CERAS'

We developed a colored text browsing system named CERAS. This system gives color to an input plain text and displays it on a computer CRT. Users can customize the coloring rules and tune the expression interactively by a GUI. Document processing is implemented on a UNIX workstation, and GUI on a Windows95 PC.

Morpheme analysis is carried out to sentences in the text, and coloring points; type of character, part of speech and keywords, are extracted. CERAS gives color expressions to extracted points considering the reader's customize information and feedback information from the GUI. The browser image is shown in Figure 1. It displays a colored text in a big window. Operation is carried out by mouse.

A "Pi menu" is used as a GUI to input user customize coloring information and to tune the color expression interactively in the system. Pi menu shown in Figure 2 is displayed on top of the text screen centering around the mouse pointer, when the left button of the mouse is clicked. The user moves the mouse pushing the button to the direction of a menu button and selects a function. The tuning expression is assigned to the first layer of the Pi menu. Customizing is assigned to the second and third layer. For example the bottom menu button of the first layer is assigned to weaken the expression. Therefore if this button is kept selected, the expression becomes more quiet and ends as a black-and-white expression. The top button is assigned to strengthen the expression and makes the expression gaudy.

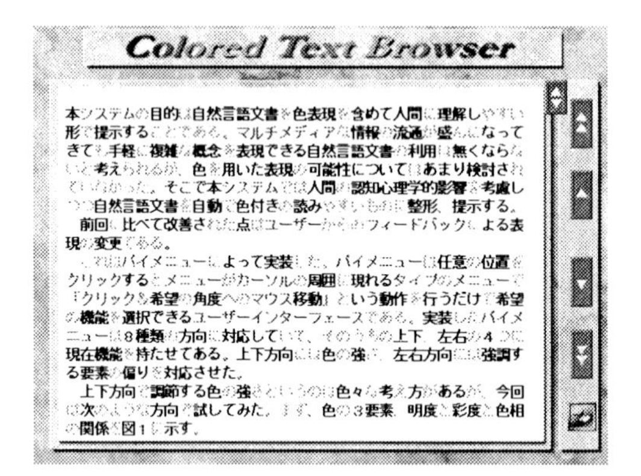

Figure 1: Appearance of the colored text browser.

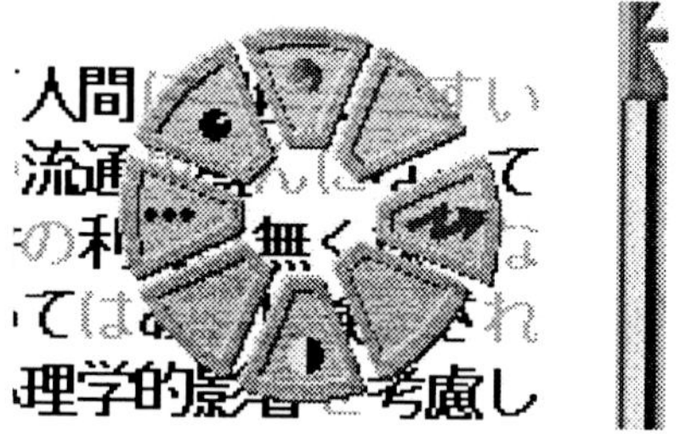

Figure 2: Feedback GUI "Pi menu".

CERAS acts in either General coloration mode or Specific coloration mode. In General coloration mode, documents are colored by general features, such as type of character and part of speech. Specific coloration mode is used when a user has a specific point to read, and the user can enter this mode anytime, to specify a keyword by clicking the right button of a mouse on it. In this mode, words related to the keyword are also colored by using a thesaurus. The related words are expressed by the color whose strength is associated with the strength of the relation. The user can quickly find out interesting parts within a document.

4 ESTIMATION OF CERAS

Keyword coloration function of CERAS was evaluated by a news article classification experiment.

A testee searches one or two designated descriptions inside a displayed document in speed-reading and classified the document into two kinds.

First, a condition sentence is presented on the CRT to a testee, and the testee chooses and clicks one word as a coloring keyword. Next, a news article of approximately 1,100 characters is presented on the CRT, and the testee classifes it and pushes either the right or the left mouse button. Two patterns of combinations of colored and black-and-white document are prepared and presented to a testee. In colored documents, keywords are colored red and colored words are displayed in deep blue.

The number of testee is 14. Time required for a testee to click the mouse button from the presentation of condition sentence is measured. It is equal to the reading time in black-and-white, because keyword is not designated in this case.

The time for choose a keyword and point it by moving the mouse is 601ms. This means that the overhead time for the utilization of interactive coloration function of CERAS is 601ms.

Table 1: Average classification time

	mono(1target)	color(1)	mono(2)	color(2)
Match	12.86	3.936	10.37	4.048
Unmatch	16.47	3.675	15.98	6.647
Total(sec)	15.47	3.745	13.18	5.347

Next, the average classification time is shown on Table 1. The required time is reduced to 24.2% by coloration in one target condition and to 40.6% in one of two targets condition. And it costs less time to classify by coloration in all cases.

The correct answer rate is raised by coloration in 3 out of 4 cases. In total, it rises 5% in one condition, 2% in two conditions.

There is a linear relationship between the location of a search goal inside an article and classification time. Therefore we can estimate the reading speed according to this relationship. It can be said that there is a correlation between distance from the top of an article to the search goal and classification time in the case of a black-and-white news article, because a testee is apt to search the target from the top to the end.

The required time to read one content word is estimated 27.71ms and 31.22ms in one condition and two conditions in the case of a black-and-white document. The time to check around a mark is 923.3ms for a keyword, and 123.6ms for a related word.

We consider that a user searches an unknown word which satisfies a semantic target in the related words.

There are n documents on the same condition. The number of keyword marks, related word marks, content words from top to target are a, b and t.

According to the fact that the time required for semantic search costs 24.4% more than lexical search[3], semantic search costs $123.6 \cdot 1.244 = 153.8$(ms) for each related word marks. The overhead of coloring is $\frac{601}{n}$(ms). The average time required for searching in the coloration documents is $\frac{601}{n} + 923.3a + 153.8b + 3073$ (ms).

In black-and-white document, we consider to 10% priming effect additional to the above. It costs $\frac{462.3 \cdot 1.244 \cdot 0.9}{14.81} = 34.95$(ms) for each content words. Therefore, it takes $34.95t + 2311$(ms) to search the target.

For that reason, in the case under the following condition, CERAS is advantageous.

$$t > \frac{17.2}{n} + 26.42a + 4.401b + 21.80$$

$$t > 0, a \geq 0, b \geq 0, n > 0$$

We tried to apply this condition to news articles using this experiment as an example. The parameters are as following: $a = 0.4500, b = 18.05, t = 355.4, n = 20$

When b is free, the condition is $b < 72.8$. This means CERAS can use 72.8 related word marks from the top to the target in average. It means 3.03 marks per line, too. In addition that the limit value is 4 times as much as that in the present system.

In a news article speed-reading case, it is presumed that related word coloration is effective for speed when the number of coloration marks are less than 3 per line on an average.

5 CONCLUSION

The number of online documents and those who read documents on color devices is increasing very much. We estimated the effect of coloring a document by experiment, and developed CERAS which generates color expression automatically and responds quickly to the user's customize and tuning request through a GUI.

In General coloration mode, CERAS presents colored expression which improves comprehension. And an user can customize its color and tune interactively while reading. In Specific coloration mode, CERAS has many advantages in fast-reading. Adding a special feature makes it easy to search a keyword, which has generally been carried out in document searching area. But, there are no systems which can change keywords and expressions interactively nor give expressions to words related to the keywords. People frequently search their interesting point when they read relatively less important documents such as news paper articles. For that reason, raising efficiency of this search is generally effective. To raise this efficiency, it is effective that a system adds a special feature to the interesting point, and raises the search speed. CERAS accomplishes this in interactive keyword specification and coloration to keywords and related words. Coloration is suitable to speedup a search utilizing its pop-out function. Moreover, interactively specification of keywords interactively is considered effective when the object of interest is changed while reading. And there is the effect of preventing the oversight of an interest point in relative word coloration. An user can find an interest point which does not include the keyword but its synonym. From an estimation experiment, CERAS is effective for fast-reading of news articles, both in time and in accuracy in Specific coloration mode. And there is enough margin till a defect appears.

Still, there is room of precise estimation and examination, but this system has important significance as the tool that estimates the possibility of color expression of a document. From now on, we will modify the system and establish a presentation method which will increase the effectiveness of colored documents.

References

[1] William Winn,"Color in Document Design," IEEE Trans. Professional Commun., vol .34, no.3, pp.180-185, 1991.

[2] R. John Brockmann,"The Unbearable Distraction of Color," IEEE Trans. Professio nal Commun., vol.34, no.3, pp.153-159, 1991.

[3] Ken Goryo,"What's reading," Tokyo Univ. Press, 1987.

Free vs. Guided Programming by Discovery

Haider Ali Ramadhan

Department of Computer Science, Sultan Qaboos University, PO Box 36 Al-Khodh 123, Sultanate of Oman, Phone (968) 515418, E-mail: haider@squ.edu

This paper discusses a synthesis-based framework for designing intelligent programming systems for novices. The framework integrates visualization features, case-based learning, microworlds and ITS, to come up with an effective environment that facilitates learning programming through discovery and tuition.

1. INTRODUCTION

Programming is a cognitively demanding task. Novice programmers face several difficulties in learning to program [1,2]. First, they need to build effective programming knowledge. This is a knowledge about the exact syntax and semantics of the available programming constructs. This includes acquiring a clear mental model of how individual programming concepts such as iterations, conditionals and procedures dynamically behave, how the computer executes programming statements and how this execution effects the internal and hidden components of the underlying machine.

Second, novices need to acquire programming skill which deals with the ability required to connect the low-level syntax and semantics of constructs to produce properly integrated higher-level plans (programs and algorithms). In other words, programming skill concerns the ability to combine different programming constructs to come up with a solution for a given problem. This requires a clear mental model of how these constructs behave and how the computer executes them, so that reasoning during problem solving can be facilitated. Empirical studies have shown that novices frequently have difficulty in putting together programming statements to solve problems [1,2].

Traditional programming tutors and systems, such as Proust [3], Bridge [2] and the Lisp Tutor [1], tend to ignore the significance of a pre problem-solving, dynamic, visible and free discovery environment. Consequently, even after using these systems, we would expect that novices still tend to exhibit some of the classical misconceptions that have been reported found in their programming knowledge, such as using conditionals instead of iterations [2].

Learning how to compose programming statements and language constructs to form higher level plans (programs) is a very important task in learning to program. However, understanding the semantics and the behavior of these constructs is not less important. Consequently, before requiring novices to put programming statements, functions and constructs together to solve a problem, we should help them first understand their dynamic behavior. Novices should be exposed to problem solving process only after building sufficiently correct underlying conceptual programming knowledge and a robust mental model of language execution and machine behavior.

These two approaches of free and guided programming support have already been demonstrated in individual systems. There is, in principle, no reason why they should not be combined to provide a single exploratory and guided programming environment. It is true that there has not yet been much clear success for intelligent programming tutoring systems in real world situations, except perhaps for the Lisp Tutor and Proust, and that in the short term, energy spent on developing programming microworlds may provide more payoff. In the long term though, it will be pleasing to see a marriage between the use of good exploratory and ITS techniques. The way forward, therefore, should be a joint endeavor between these two areas.

2. THE IMPLEMENTATION

To assess the effectiveness of the proposed design framework, we have developed a prototype system called DISCOVER. The system synthesizes free with guided programming and supports domain visualization and case-based learning. The system is designed to help novices acquire both programming knowledge and programming skill. This is accomplished in two phases:

- In the first phase, the exploratory phase, the system helps novices through visualization features to explore the dynamic behavior of programming statements and of the notional machine to build a robust mental model of language execution and machine behavior.
- In the second phase, the guided phase, novices put together program statements and language constructs, explored in the first phase, to solve problems under the intelligent guidance of DISCOVER. For each problem, the system provides the user with relevant example cases along with their model solutions. Users can use these cases to tackle their own problems.

DISCOVER's user interface represents a dynamic programming notional machine which is based on a concrete model of the notional computer proposed by Mayer [6]. However, DISCOVER presents the model as a visible machine in its real-time action, and not as a static, textbook-like picture of language execution and machine behavior (see figure 1). Through this dynamic, visible machine, novices (1) can observe how program statements are executed in an animated way, (2) can see hidden and internal changes in some conceptual parts of the underlying computer, such as the memory space, and (3) can relate problem solving with the properties of the machine they are interacting with.

To be able to (1) analyze partial solution steps as they are provided by the user during the guided phase, and thus (2) explicitly guide novices in the process of putting together program statements to solve problems, DISCOVER, like the Lisp Tutor and GIL [5], supports model-tracing [1] based diagnosis. However, unlike these systems, our system utilizes goals and plans (not a production system) to represent the knowledge of its domain and expertise [4], and (2) supports an ability to give delayed feedback on error by increasing the grain size of automatic diagnosis to a complete program statement (not just a single symbol) and a limited capability

algorithm cannot detect conceptual difference which does not appear in decision trees explicitly. If decision trees with diverse structures for an input file are constructed, such conceptual differences can be detected and the above problem can be solved. Since the decision tree space in which an input file creates is very large, decision trees constructed at random tend to be inappropriate trees structurally. On the other hand, GA can carry out efficient search by keeping or improving the quality of decision trees. Therefore, we apply GA to construct the decision trees.

In our approach, three operators, crossover, mutation and selection, are used in GA. Crossover and mutation are carried out as in the case of standard GA. The decision tree set which should survive to the next generation needs to have the ability of classifying examples efficiently and to have diverse structures as described above. Therefore, selection is carried out under following two indexes; error rate and mutual distance. When examples are classified into a single class efficiently in each leaf, error rate gets larger. Smaller value in this index is better. When each structure in a decision trees set is more diverse, mutual distance among decision trees gets larger, and larger value in this index is better. Decision trees with smaller error rate survive in the first stage and mutual distance is calculated for each decision tree set which is constructed as the combination of the trees. Finally the decision tree set with the largest mutual distance becomes the initial population in the next generation. For further details of the algorithms, see reference [6].

4. EXPERIMENT FOR EVALUATION

A prototype system is developed on the UNIX workstation. Each component of the system in Figure 1 is written in C language. As an example, the motor diagnosis case is evaluated. In this case, two persons give their knowledge expressed by thirty examples, which are composed of six attributes, two or three values and five classes, respectively. The artificial conceptual

(1) Type(a) in attributes
1) A: Noise B: Stench (1)
(2) Type(b) in attributes
1) Temperature (4)
2) Amplitude (2)
3) Vibration (1)
4) Frequency (1)
5) Current (1)
(3) Type(a) in values
Nothing
(4) Type(b) in values
1) attr: Amplitude -> val: Normal (1)
2) attr: Current -> val: Stable (1)

Figure 2. The result by using ID3 algorithm

(1) Type(a) in attributes
1) A: Noise B: Stench (6)
(2) Type(b) in attributes
1) Vibration (49)
2) Frequency (9)
3) Temperature (3)
4) Current (2)
(3) Type(a) in values
1) attr: Amplitude -> val: inc Normal (1)
(4) Type(b) in values
1) attr: Amplitude -> val: Normal (2)
2) attr: Vibration -> val: No (2)
3) attr: Vibration -> val: Yes (1)
4) attr: Frequency-> val: High (1)

Figure 3. The result by using GA

difference, including what could not be detected in the system with ID3 algorithm alone, were given to the system to evaluate the ability to detect conceptual difference. GA carried out until the hundredth generation. Figure 2 shows the result with ID3 algorithm and figure 3 shows the one with GA. The detected artificial conceptual difference is expressed in italics.

In ID3 algorithm, type(a) conceptual difference in values was not detected. On the other hand, it detected in GA. The priority for the candidates was also improved on the whole in GA, type(b) in attributes especially. It followed from this result that the candidate which seems to be concern as conceptual difference was intensified and that noisy candidate was restrained by using diverse structures. It was confirmed by these results that the performance improves by adopting GA as decision tree producing algorithm.

5. CONCLUSIONS

In this paper, we proposed interactive interfaces to detect conceptual difference by constructing decision trees with diverse structures for knowledge acquisition and confirmed that the performance of the system with diverse structures gets better than the one with a single structure.

As a problem for the system, using diverse structures results in the increase of the number of candidates. It is necessary to improve the accuracy of the detection algorithm for conceptual difference. In future, we will tackle the problem to improve the system.

REFERENCES

1. H. Ueda and S. Kunifuji, "GRAPE : Knowledge Acquisition Support Groupware for the Classification-Choice Problem", *IEEE/SICE Int. Workshop on Emerging Technologies for Factory Automation*, pp.111-127, 1992.
2. K. Hori and S. Ohsuga, "Toward Computer Aided Creation", *Proc. Of PRICAI'90*, p.607-612, 1990.
3. K. Hori, "A system for aiding creative concept formation", *IEEE Transactions on Systems, Man, and Cybernetics*, Vol.24, No.6, pp.882-894, 1994.
4. J. R. Quinlan, "Induction of decision trees", *Machine Learning*, Vol.l, No.l, pp.8l-l06, 1986.
5. T. Kondo, N. Saiwaki, H. Tsujimoto and S. Nishida, "Design of Interactive Interfaces to Detect Conceptual Difference", *5th IEEE International Workshop on Robot and Human Communication*, Tsukuba, pp.525-530, 1996.
6. T. Kondo, N. Saiwaki, H. Tsujimoto and S. Nishida, "A Method of Detecting Conceptual Difference among Different People on the basis of Diverse Structures", *9th SICE Symposium on Decentralized Autonomous Systems*, pp.125-130, 1997.

Transformation of Human-to-Human Interaction into Asymmetric and Lightweight Computer-Mediated Interaction

Amane Nakajima, Fumio Ando, and Younosuke Furui

IBM Research, Tokyo Research Laboratory,
1623-14 Shimotsuruma, Yamato, Kanagawa 242, Japan

This paper explains an approach of designing collaboration systems. The approach consists of two phases. In the first phase, analysis of the real work is done, and in the second phase, the processes of the real work are transformed to those of a computerized system. The transformation is done using the H-C-C-H model, where 'H' represents a human, and 'C' a computer. The H-C-C-H model is useful for analyzing and describing each user's view of the system. This paper describes how the approach is used in the development of a remote loan contract system. By using the approach, a new type of collaboration, an asymmetric and lightweight collaboration, is found and realized.

1. INTRODUCTION

Business processes in the real world include several kinds of human activity, such as face-to-face conversation, telephone conversation, data input using a personal computer, and form filling. Hence, computerizing human business processes requires analysis of each step of the current business process, classification of the steps from a viewpoint of human interaction, and design of a computer system that effectively supports the business processes. However, many realtime human interaction systems are based on a technology-oriented approach and are thus unable to support the target work effectively. An example is a desktop conferencing system [1,2]. Desktop conferencing systems provide realtime motion video communication and shared workspace. The desktop conferencing functions are used in real systems [3,4], but the usage is simple; video phone and sharing of work space. This is because their usage is decided only from a technological viewpoint.

Our objective is to create a new human-to-human interaction system that represents the target business processes very well. For this purpose, we took a process-oriented approach, using new collaboration technologies for asymmetric and lightweight human-to-human interaction. This paper analyzes specific business processes of loan contracting between a customer and a clerk. It also describes how we transformed the processes into those of a computer-based system, and explains the implementation of the system.

2. ANALYSIS OF WORK

To achieve the goal, we analyzed the loan contracting procedure, classified the work into four types, and extracted seven steps.

The types of work are:

1. Identification of a customer by a clerk,
2. Completion of a form by the customer,
3. Correction of the form and interview by the clerk,
4. Review of customer data by the clerk.

The seven steps are:

1. Identification of the customer through ID cards, face check, and telephone calls by the clerk,
2. Completion of an application form by the customer,
3. Correction of the application form and interview by the clerk,
4. Review of the data by the clerk,
5. Completion of a contract form by the customer,
6. Correction of the contract form by the clerk,
7. Presentation of a magnetic card and a copy of the contract form to the customer by the clerk.

Some steps are done cooperatively by the customer and the clerk, but their roles and the degree of involvement are different.

Through analysis of the work and steps, we found that

1. The loan contracting processes include two-way and one-way human interaction.
2. Not only face-to-face interaction, but also lightweight interaction exists in the processes.
3. Nonverbal communication and observation play an important role.
4. A clerk does not necessarily take care of the customer all the time; the clerk and customer sometimes act separately.
5. Customers prefer non-face-to-face interaction if possible.

On the basis of this analysis, we built a computer system [5] that effectively supports the business processes, and implemented the seven steps. The system consists of a self-service terminal in an unmanned branch, and a clerk terminal connected to the self-service terminal through a network. A customer uses the self-service terminal for data input and interaction with a remote clerk. However, The interaction with the clerk is avoided as much as possible; the customer has psychological barrier to making a loan contract. Non-face-to-face interaction eliminates the barrier. The clerk reviews the data, controls the self-service terminal, and accesses the database in the host computer. To realize the system, we developed new collaboration functions for asymmetric and lightweight interaction.

3. H-C-C-H MODEL

To transform the real business processes into computer-based ones, we used an H-C-C-H model. Here 'H' represents a human, and 'C' a computer. Figure 1 shows the loan contract system and H-C-C-H model. In the collaboration system, two or more humans are interacting with each other using the computer system. Each human uses a computer, and the computers are connected by a network. Hence, the system configuration including the users is represented as H-C-C-H. The connection between H-C is human-computer interaction, and the connection between C-C is computer communication. When interactions between humans are very natural, they are not aware of computers: Users regard the interactions as

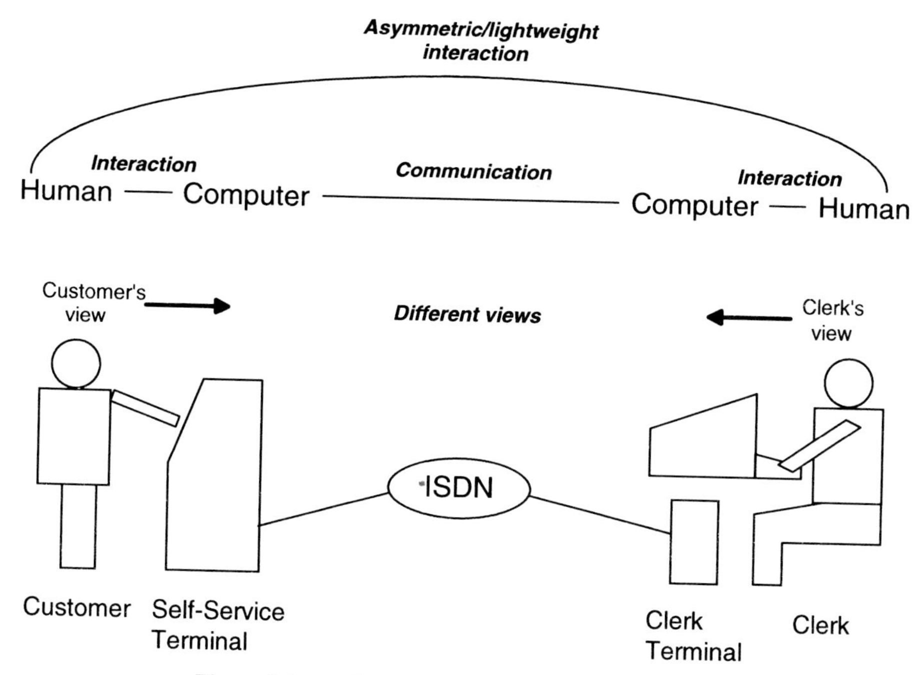

Figure 1. Loan Contract System and H-C-C-H Model.

H-H. When a customer is not aware of a remote clerk's support, the customer regards the interaction as H-C, not H-C-C-H. Thus, the user's view depends on his awareness of a remote human. In addition, a customer's view and a clerk's view are not always the same. We call the collaboration in which user's views are not same *asymmetric collaboration.* The view includes a system view such as H-C-C-H, user interface such as appearance of windows, and data view. We also call collaboration in which interactions between humans are not very tight *lightweight collaboration.*

We have used the H-C-C-H model as follows. We first analyzed the human-to-human interactions and human's work processes. From the analysis, we found that a customer and a clerk sometimes do their own work even when they are sitting face-to-face. We also learned that the strength of interaction changes according to the work process or the situation. We, then, designed the system so that (i) the work process in the system keeps the clerk's interaction patterns as much as possible and (ii) the work process reduces the customer's interaction with a clerk as much as possible. This is because we learned that customers do not like face-to-face interactions with a clerk.

Let us show two examples of asymmetric collaboration as the results of the approach. In the second of the above seven steps, a customer inputs personal data into the self-service terminal. As far as he is aware, he is only interacting with the terminal, but actually a remote clerk sometimes receives data from the self-service terminal, reviews the data, and uses one-way motion video and audio communication to monitor the customer and his behavior. In this step, the customer's view and the clerk's view are asymmetric. The customer's view is H-C, because he thinks that he is using the terminal without any interaction with a remote site. On the other hand, the clerk's view is H-C-C when she obtains data from the self-service terminal and reviews them. When she monitors the customer, her view is H-C-C-H, because she uses one-way, nonverbal communication with the customer.

Another asymmetric views exist in the form completion and correction steps. When a customer fills in a paper form, its image data are obtained with a scanner and sent to the clerk terminal. If the clerk finds an error, she selects one of several classified error patterns. The kind of pattern is sent to the self-service terminal, which plays back the corresponding error correction guidance with text, animation, and voice. Thus, the customer's view is H-C, but the clerk's view is H-C-C. In this case, a clerk not only receives data from the self-service terminal, but also sends a command to initiate the error correction process. Hence, the direction of data flow is two-way. If the correction process does not work well, the clerk and customer use a two-way video conference function. In this case, the views for both persons are H-C-C-H.

4. CONCLUSIONS

The interaction methods describe in Section 3 are designed in line with the analysis of the business processes and the requirements of customers and clerks. With the process-oriented analysis and H-C-C-H model, we have built a new human interaction system that supports asymmetric and lightweight collaboration between humans. Our uniqueness resides in the change of interaction patterns, and asymmetry of interaction patters. The system sometimes provides H-C-C view to a clerk, and it also provides H-C-C-H view to the clerk at another time. At the form correction time, a customer has the view of H-C, but the clerk has the view of H-C-C at that time.

One of measurements of the approach is the acceptance of the system in the real world. The system has been very well accepted by its end users and finance companies, and has won many contracts. We believe that our H-C-C-H approach, and asymmetric and lightweight collaboration are useful to many other systems.

REFERENCES

1. S. R. Ahuja, J. R. Ensor, and D. N. Horn, The Rapport Multimedia Conferencing System, Proc. ACM Conf. Office Info. Syst. (1988), pp.1-8.
2. K. Watabe, S. Sakata, K. Maeno, H. Fukuoka, and T. Ohmori, Multimedia Desktop Conferencing System: MERMAID, Trans. Info. Proc. Soc. Japan, Vol.32, No. 9 (1991), pp. 1200-1209.
3. L. Orozco-Barbosa, A. Karmouch, N. D. Georganas, and M. Goldberg, A Multimedia Interhospital Communications System for Medical Consultations, IEEE J. Selected Areas in Comm., Vol. 10, No. 7 (1992), pp.1145-1157.
4. A. P. Karduck, A. Geiser, and T. Gutekunst, Multimedia Technology in Banking, IEEE Multimedia Magazine, Vol. 4, No. 3 (1996), pp. 82-86.
5. A. Nakajima, F. Ando, and Y. Furui, Multimedia Communication and Collaboration for Remote Loan Contracting, Proc. IEEE GLOBEOM'96 (1996), pp. 882-887.

those which are connected via machine-machine communications. Objects that require human interactions become data elements for subsequent analysis. Individual object definitions contain clues for identifying classes of human-computer interactions envisioned by the object designer. In their simplest form, these interactions require human operators/users to edit values and parameters, enter values or parameters, and observe object values or parameters displayed by the computer.

Developers are next asked to define strings of related human-computer interactions that act on objects. These strings, or work flows, are depicted using graphical decision-action diagrams. Individual human-computer interactions are defined as discrete tasks and diagrammed within boxes or diamonds, to represent actions or decisions, respectively. Tasks are connected with directional vectors to determine the sequence or flow of human-computer interactions through the entire work flow. Work flows thus defined mimic the iterative and often recursive nature of human-computer interactions using today's GUI designs. The two steps so far identified in the HMI methodology define the human-computer interaction without committing to a specific GUI construct or paradigm. This means that the same analysis can support multiple GUI implementations, advantageous when considering a requirement to develop applications for the same functionality using Java programming (for web-based applications) and X/Motif programming (for traditional Unix environments).

In the third step, developers prepare paper prototypes of proposed GUIs. Paper prototyping permits design of GUIs without the personal investment of time and labor required to produce a full GUI prototype. This allows developers and users to establish usability while the GUI is still just a paper mockup, prior to the commitment of programming resources to its design.

Once basic usability is achieved with the paper prototype, developers proceed to GUI implementation. This may involve application of a commercial GUI builder tool, C++ coding (or other commercial software development tools), classes or libraries of GUI "widgets," or a combination.

Usability testing completes the process. In this step, end users interact with the prototypes in tests that provide data on errors made in attempting to use the GUIs, assistance needed, and subjectively rated acceptability of the GUIs. Operator/user input is sought as a basis for iterative refinement of the work flows and the GUIs themselves.

2. AN APPLICATION OF THE HMI METHODOLOGY

The HMI methodology was developed by human factors specialists as part of an interdisciplinary team developing a major information system. This article reports the results of a preliminary usability assessment on the GUIs from one of its custom software tools.

2.1 Method

The development of the custom software tool (a Motif editor of the ECS data dictionary) considered here included all of the steps of the HMI methodology. Human factors specialists assisted the software designers and programmers, analyzing the object models to identify and characterize the human-computer interactions involved in using the tool. The team used the data derived from the analysis to develop and refine work flows reflecting typical sequences of user interactions with the tool. From the work flow analysis, the team derived initial screen layouts, capturing them in paper prototypes. Three representatives of the operations community who will use the tool reviewed the paper prototypes, and their feedback guided refinements. Programmers then used a GUI builder to create an operating executable prototype. Finally, the usability of the executable prototype was assessed in preliminary tests by the three representatives of the operator/user community. For the assessment, the representatives were provided with a brief explanation of the functions of the tool. No demonstration of the prototypes was provided. Immediately following the explanation, the group of representatives was given access to the prototype and assigned three sample tasks selected to be approximately equivalent in complexity and difficulty. Operator interactions and comments for each task were recorded, as well as instances in which *assistance* was requested or in which unrequested *prompts* were given.

2.2 Results and Discussion

The data were examined to identify the number of human-computer interactions for each of the three tasks. Interactions were categorized as *correct* (actions necessary to achieve the objectives of the task), *critical errors* (actions or omissions requiring corrective action necessary to achieve the objectives of the tasks), and *discovery errors* (actions apparently taken to understand the tool). Table 2 summarizes the interactions. Because there was only one critical error during the entire test, in Task 2, error rates were combined for the analysis.

Table 2
Interactions with the prototype GUIs

Task	Total Interac-tions	Critical Errors	Discovery Errors	Assist-ance Requests	Prompts
1	38	0	10	0	0
2	24	1	5	0	3
3	18	0	1	0	0

Figure 2 shows the accuracy rates for the tasks and a combined rate for the entire test. Operator/user comments reflected general satisfaction with the interface, noting its ease of use and likelihood that it would be mastered with very little training. They suggested no change in interface layout, and only minor refinements (e.g., content of tab labels).

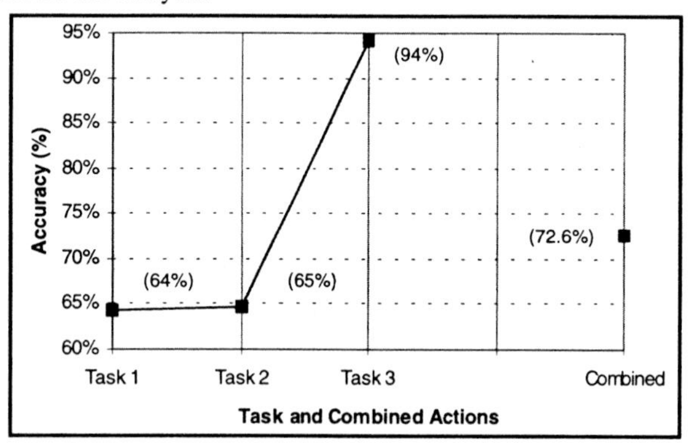

Figure 2. Task and combined accuracy rates

3. CONCLUSION

The results of the usability assessment indicate the effectiveness of the HMI methodology. If HMI concerns in system design are raised only after the design is mature, there is little hope for significant human factors influence on the design. The proposed HMI methodology integrates fundamental human factors heuristics throughout the development to assist the development team in maintaining a focus on user concerns that may otherwise be obscured by biases. It incorporates early and sustained human factors influence in the development cycle for maximum impact on usability, which can be a measure of its success.

A Highly Extensible Graphical User Interface in a Software Development Environment

Yasuhiro Sugiyama

Department of Computer Science, Nihon University
Koriyama, 963 Japan

1. Introduction

This paper presents an overview of the graphical user interface (GUI) of our software development environment (SDE). In our SDE, every software artifact is an object that is an instance of its class. The tools to manipulate these artifacts, the repository to store them, and the user interface to manipulate them are integrated around the artifacts by the classes of the artifacts. Classes uniformly determine the nature and behavior of the software artifacts in our repository as well as our GUI.

2. The Roles of Classes in GUI

A basic role of a class is to determine the nature of its instances, such as a binary application, a word-processor document, or a C source program. Classes also determine the *attributes* of software artifacts, such as their names, their owners, and their access privileges. Our SDE offers a pre-defined class library which includes classes of common software artifacts. However, users are allowed and encouraged to define and use their own classes. New classes can be defined either from scratch or by defining sub-classes of existing classes. Please note, in our SDE, typical *users* are *software engineers*. I will use these two words interchangeably.

The use of object-orientation in a repository to represent the structure and inter-relationship of objects is not a new idea in recent SDE's [5]. The revolution of our system, however, is that we have extended the role of classes to include the description of the characteristics and behavior of objects in the GUI. Figure 1 shows the roles of classes in our GUI.

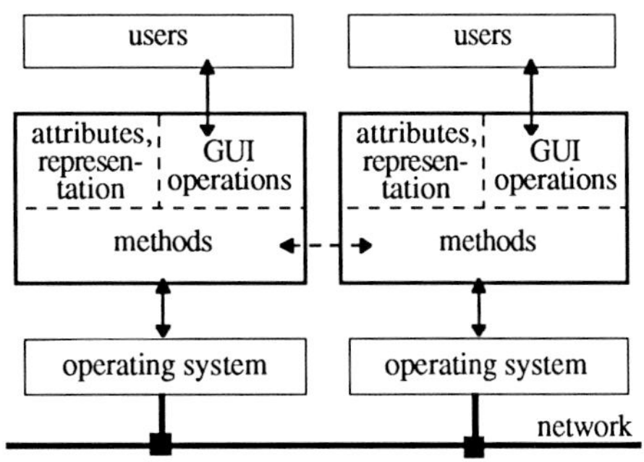

Figure 1. Roles of classes

The first role of a class in our GUI is to determine the *representation* of its instances. Software artifacts are shown as graphical objects in our GUI. The class of an object determines the style, like an icon or a window, and the design of the graphical representation of the object. Figure 2 shows a snapshot of our GUI. Five objects are shown in four icon styles. Each icon style denotes a different class. For example, *bank.c* belongs to *Source_File* class.

The second role of a class is to determine the collection of *methods* that instances of the class can execute. Users will send requests to the instances to invoke their methods. Examples include a method to compile a source program, and a method to invoke a text editor on the source program.

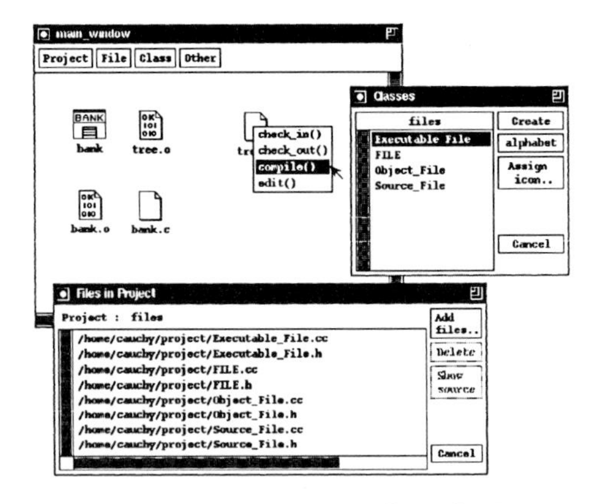

Figure 2. A snapshot of our GUI

Methods can be overloaded, that is, a single method may be implemented differently in different classes. A typical compile method on C source programs will invoke a C compiler, while the same method on Pascal source programs may invoke a Pascal compiler.

Objects may include methods to communicate to each other to synchronize their activities. For instance, in Figure 2, *check_in* and *check_out* operations are provided to avoid simultaneous update of a single object by two or more users. Objects may be distributed over a network. The communication among objects will take place through the underlying operating system and the network.

Methods can be associated with preconditions. Two or more methods with a same name, but with exclusive preconditions, can be defined. Methods with preconditions become active only when the preconditions are met. Preconditions of a method are stated in terms of the attributes of the class that includes the method. Consequently, the implementation of a method on an object can dynamically change as the values of the attributes of the object change.

The third role of a class is to determine how the instances of the class understand and process users' requests to them. Users rely on GUI operations, such as mouse clicks and drags, to send requests to the instances to invoke their methods. The way in which the instances respond to a GUI operation is defined in the method associated with the GUI operation. For example, the method associated with a mouse click on an application program object may start executing the program. The same method on an object of some other class may pop-up a menu which includes a list of available methods on the object, so that users can select one of the methods in the list to execute. In Figure 2, the method for a mouse click is implemented in the latter way, as you see the menu is popping up on the icon.

3. Implementation

Our SDE is being implemented on top of the UNIX environment. We use C++ and X-toolkit for the implementation. Figure 3 shows the system structure of our SDE. The three major components of our SDE are: GUI, GUI engine, and the object base. Through our GUI, users see the underlying UNIX files as if they are objects stored in the object base. The GUI engine instantiates objects based upon their classes. Although instantiated objects are primarily

UNIX files, their attributes and other information, which are not part of the UNIX files, are stored in the object base.

Users manipulate the objects by GUI operations through the GUI. The GUI engine accepts users' requests through the GUI, and will invoke appropriate methods to process the requests.

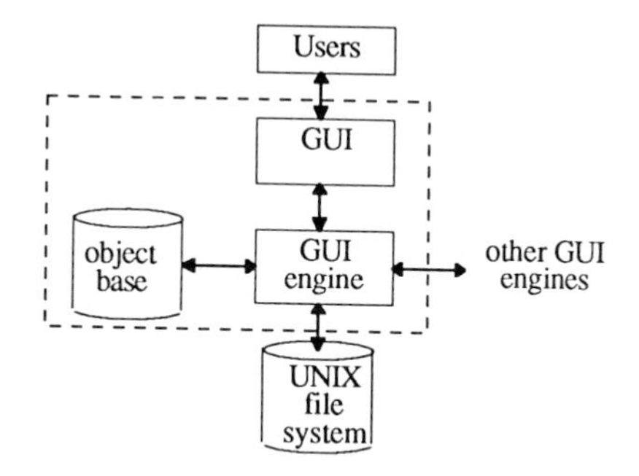

Figure 3. The system structure

Classes are defined in our notation which is an extension of C++. B-shell or C-shell scripts may be directly written in methods. Methods are executed interpretively by the GUI engine. Two or more GUI engines may run simultaneously on different computers connected by a network. GUI engines communicate to each other for task synchronization.

4. Related Works

Our efforts have been motivated by our research on process centered software engineering environments (PSEE's) and the investigation of various file and/or window managers. Our approach is to allow software engineers to participate in the execution of the software process through the user interface of the PSEE. Although the notion of the software process plays an important role in our system, I do not have space here to discuss it due to the space limitation. A description can be found in [4]. In this paper I will focus on user interface issues.

UNIX shells, like C-shell and B-shell, have been common user interface in the UNIX environment. The UNIX shells offer strong scripting capability that allows users to define new commands, but they lack ease of use. Recently, quite a few window managers that run on the X-window system have been introduced. Examples include *twm* and *fvwm*. These window managers try to achieve the scripting capability of the UNIX shells and the ease of use of GUI in a single place. Users define their own pop-up menus, and write shell scripts to process the commands in the menus. However, menus can be defined only on a limited set of GUI objects, such as the root window and window title bars. Moreover, the lack of the class concept in these window managers makes them hard for users to write and reuse the description of menus.

Apple Macintosh's operating system includes *finder* which works as both file and window managers. In *finder*, each document (file) is associated with a *signature* and a *file type* [1], which are four-byte text strings. The *signature* denotes the application that produced the document, while the *file type* indicates the kind of the document, like a binary file or a text file. Applications use the *signature* and *file type* to determine how to process the document. One of the critical drawbacks of Macintosh's approach is that users are not allowed to define their own *signatures* and *file types*. Users are not allowed to define methods associated with *signatures* and *file types*, because they are not classes. Each application, not a user, defines its *signature* and the semantics of the associated *file types*.

Finally, I would like to point out that an object-oriented graphical class library and our GUI with class libraries are not identical. Recently, quite a few object-oriented graphical libraries, like X-toolkit and tcl/tk [3], are available. Software developers may increase their productivity using object-orientation in these object-oriented graphical libraries, but the resulting products do not necessarily offer an object-oriented user interface to their users. The strength of our system is that it offers an object-oriented GUI that is highly and easily customizable by virtue of classes.

5. Conclusion

The use of classes in our SDE, which I outlined here, realized the strong scripting capability of the UNIX shells in our GUI which offers ease of use. It also realized a mechanism to overload methods, and a mechanism to dynamically change the behavior of artifacts in the GUI. Moreover, the use of classes allows novice programmers to reuse software development experts' knowledge.

The use of classes also allows users to easily augment the capability of the GUI by defining new classes. For example, we have defined a class *workspace* that supports group activities in software development. An instance of *workspace*, which is called a workspace, is a composite object, like a directory, that includes other objects. It is originally shown as an icon and the objects in the workspace are hidden from the outside. Only the registered users are allowed to open up the icon to a window, and then, to access the objects in the workspace. Workspaces isolate private works from the outside to improve the efficiency and accuracy of the works. Workspaces also support inter-workspace communication for task synchronization. We also defined a class *remoteObject* that allows transparent access to files on remote computers through the underlying network. Instances of *remoteObject* on a computer can be mounted on workspaces on some other remote workstations. Mounted instances can be accessed as if they are local objects in the workspaces.

One of the drawbacks of our system is that our GUI asks the users to write classes in compensation for high customizability. Writing classes is not a easy task for novice programmers. However, this can be easily overcome by providing common class libraries, in addition to *workspace* and *remoteObject*. Novice users are allowed to use our GUI without programming by using the predefined class libraries.

We are currently working on moving our system from C++ to Java [2] in all respects. We are rewriting our system in Java. We also plan to accept Java notation for the description of GUI objects. Particularly, Java is suitable for our GUI engine because Java is an interpretive language and it has strong network support.

References

[1] Apple Computer, *Inside Macintosh Volume III*, 1985, Addison-Wesley.

[2] http://www.javasoft.com/

[3] Ousterhout, J. O., *Tcl and Tk Toolkit*, Addison-Wesley, 1994

[4] Sugiyama, Y. and E. Horowitz, *Building Your Own Software Development Environments*, Software Engineering Journal, vol.5, no.6, pages 317-331, IEE, September 1991.

[5] Tichy, W. F., editor, *Configuration Management*, Wiley & Sons, 1994

The Effect of Internet Delay on the Design of Distributed Multimedia Documents

Andrew Sears and Michael S. Borella

School of Computer Science, DePaul University, 243 S. Wabash Avenue, Chicago, IL 60604, USA

The advent of the World-Wide Web (WWW) as a popular method for providing and disseminating information has made it imperative that designers understand the impact of Internet delays on users of distributed information systems. The effect of computer response time on users has been well-studied, but Internet delay is less predictable, more variable, and longer than that experienced on non-networked computers. We discuss the causes of Internet and WWW delay, and suggest some simple guidelines for minimizing the impact of these delays. A tool which helps document designers cope with these delays is also described.

1. INTRODUCTION

Distributed multimedia documents are becoming increasingly common. Each document typically consists of a collection of text, graphics, animation, video, or audio stored in one or more physical locations that can be viewed from a variety of physical locations. This includes documents located on a central server within one company, and viewed by employees located in one building, a collection of buildings in a single location, or locations scattered around the globe. It also includes documents placed on the Internet to be viewed by potential customers, clients, or anyone that happens to find them.

1.1. Designing distributed documents

Designing traditional documents can be a complex task. Ensuring that users can find the necessary information is the primary concern, which depends on numerous factors including typography, layout, the use of graphics, and navigation. Designing multimedia documents becomes even more complex. Additional issues include when and where to use animation, video and sound, as well as providing an appropriate level of interaction. Designing multimedia documents that will be viewed from remote locations involves all of these issues, plus at least one extra concern. Designers must consider the delivery delays introduced by the network as documents are retrieved. Traditional documents are located temporally near the user. The user requests a document and it appears quickly. Large documents may take longer, but delays are predictable and usually small. Since the user controls the hardware, changes can be made to ensure adequate delivery speed.

Distributed documents often rely on hardware that is not controlled by the user. In addition, many users do not have the resources necessary to significantly improve the hardware they do own. Whether the user connects to the Internet with modem, or via a high-speed corporate link, delays can be substantial due to network congestion. Although hardware is regularly upgraded, the number of Internet users has almost doubled each year since 1970 and these users are transferring more, larger files than in the past. Thus, long Internet delays will be a relevant issue for many years to come.

Currently, the World-Wide Web (WWW) is the most popular method for disseminating information over the Internet. Many corporations and organizations are investing large quantities of time and money developing a "web presence." The resulting web sites often include many graphical images. While graphics provide information effectively, many exist for aesthetic reasons only and a recent study [1] indicates user perceptions are influenced by both the type of media (text or graphics) being used and the delays users experience. While documents that included text and graphics were preferred when delays were short, plain text was preferred when delays were longer. The researchers concluded that users appear to want web sites to include graphics, unless the graphics cause delays users consider too long.

1.2. Internet Delay

The Internet is an interconnection of autonomous systems, each of which is operated by a different organization (e.g., a corporation or university). While an Internet backbone supports high speed data transfers, the connection between individual systems and the backbone is usually at a much slower speed. For example, in a recent survey over 70% of the respondents indicated that their primary connection to the Internet was at 28.8Kbps or slower [2].

Internet delay is caused by a number of factors, some can be controlled by the end user, but many cannot. In the context of the WWW, we consider the total delay to be the time between when the user requests a document and when that document is displayed for use. This delay consists of the following components:

- Transmission delay: the time necessary to actually transmit a file on a link. Messages experience a transmission delay at each intermediate step (hop) of their path from source to destination. The delay is proportional to the size of the message and inversely proportional to the bit rate of the link that it is being transmitted on.
- Propagation delay: the time a message spends traversing a link. Typically a few dozen milliseconds for terrestrial networks, this is less critical than other sources of delay.
- Queueing delay: the time a message spends waiting to be serviced by the server or an intermediate hop. This is usually the largest and least predictable component of the delay.
- Server processing delay is the time the server spends processing a client request. Although this is usually negligible in comparison to server queueing delay, it can be substantial if documents are created dynamically or rely on information stored on other servers.
- Client processing delay: the time the client spends presenting the document to the user.

The user has the ability to improve, few, if any, of these delay factors. A user can reduce transmission delay by upgrading their modem or network connection or reduce the client processing delay by upgrading their client host with more memory or a faster processor. Designers can reduce transmission delays by reducing the size of the documents they create. Queueing, the most dominant component of the delay, cannot be improved by an individual. Queueing delays vary based on the time of day, day of week, and week of the year. Since the majority of Internet delay is beyond the control of individual users, and Internet traffic continues to increase rapidly, document designers must learn to minimize the impact this delay has on users.

2. NETWORK DELAY AND DESIGN DECISIONS

One obvious suggestion is to limit the use of graphics, video, and other media unless they add substantially to the quality of the site. There are a number of other simple guidelines that can help designers reduce the impact network delays have on users. This section highlights several guidelines that are not regularly employed by document designers.

2.1. Always define the height and width of images and other objects

HTML is rendered starting at the top of the screen. If the browser encounters an object, but does not know its size, the object must be retrieved before lay out can proceed. Browsers employ a number of strategies to reduce this problem, including displaying the images as small icons and redrawing the entire screen after the image has been retrieved, but simply defining the height and width of every object eliminates this problem.

2.2. Reuse graphics if possible

Designers often include graphical banners throughout a site. One particularly noteworthy example is a site that lists numerous awards given by the organization annually. Each award is described in a paragraph preceded by a graphical banner. All banners are identical, except for the name of the award. Unfortunately, this forces users to wait for over a dozen unique graphics to load when they visit this page. Using the same graphic, and using HTML to indicate the name of the award, would reducing the loading time substantially.

2.3. Choose image formats carefully

The Web supports a variety of formats for graphics, but GIF and JPEG are the most common. While there are no concrete rules, the two formats tend to work better on different types of images. The GIF format works best with "cartoon-like" images containing large sections of exactly the same color. The JPEG format works better for "real-life" images like photographs. The JPEG format allows designers to trade image quality for document size. Compressing the image results in a smaller file, but lowers the quality of the resulting image.

2.4. Using IP addresses can decrease delays

Although using domain names (e.g., www.cs.depaul.edu) in URLs makes them easier to read, it can also add unnecessary delays. Every time a link is selected the domain name (www.cs.depaul.edu) must be converted to an IP address (140.192.33.5) before the document can be retrieved. If the conversion is not already known, the corresponding IP address must be determined by using the Internet's Domain Name Service which may add a substantial delay. If your document contains numerous links to additional domains, that may not already be known by the users' systems, using IP addresses instead of domain names may reduce delays.

2.6. Take advantage of parallelism

When users request a document, Web clients (browsers) retrieve the main HTML document, parse it, and then retrieve the supporting subdocuments (graphics, audio, video, etc.). Most browsers will load multiple subdocuments, often four at a time, in parallel. This, combined with the impact of document size on transmission and queueing delays, provides an opportunity for designers to reduce network delays by dividing large graphics into multiple smaller graphics. The following numbers are based on six extensive periods where documents were automatically retrieved from a variety of web sites with the assumption that up to four documents are loaded simultaneously. Under these conditions, using several smaller images would reduce the delay by over 200msec when the original image was approximately 4k in size and more than 750msec when the original image was approximately 128k. The specific benefits will depend on a number of factors including the speed of the users' network connection and current network conditions.

3. LETTING DESIGNERS EXPERIENCE NETWORK DELAYS

In addition to these guidelines, several tools have been developed that can assist designers as they deal with network delays. Designers often develop documents on their own computer or on a high-speed LAN. Graphics, animation, and other eye-catching effects are often employed to make the site more attractive. When the designers view the documents, or give demonstrations to management, everything looks fine. However, when the documents are placed on the Internet, where delays are longer and less predictable, attractive graphics become bottlenecks and audio, animation, and video that were supposed to grab the users' attention are slow and let users become distracted while they wait for documents to download.

The problem is that designers view their creations with artificially short delays. Allowing designers to view their documents, while experiencing realistic network delays, can reduce this problem. WANDS is a set of tools that does allow designers to experience realistic network delays while viewing the documents they create, using one computer that does not have to be connected to a network [3]. An instrumented server automatically delays documents based on a number of factors including document size and the network conditions being simulated. This allows designers to view their documents with any browser while experiencing realistic delays. Using these tools, designers can, for the first time, experience their documents as if they were across the street, the country, or an ocean.

4. CONCLUSIONS

Internet traffic continues to grow at a remarkable pace indicating that network delays will remain relevant. Therefore, designers need to understand the causes of these delays and how to create documents for this environment. In this paper, we provide several simple guidelines that can reduce the impact of network delays, many of which are currently ignored when documents are created for the WWW. In addition, we introduce a tool that helps designers deal more effectively with delays that cannot be eliminated.

REFERENCES

1. Sears, A., Jacko, J. A., & Borella, M. S. (1997) Internet delay effects: How users perceive quality, organization, and ease of use of information. *Proceedings of the ACM Conference on Human Factors In Computing Systems (CHI'97)*, 2, 353-354.
2. GVU's Sixth WWW User Survey. http://www.cc.gatech.edu/gvu/user_surveys/
3. Sears, A., & Borella, M. S. (1997) WANDS: Tools for designing and testing distributed documents. *Proceedings of the ACM Conference on Human Factors In Computing Systems (CHI'97)*, 2, 327-328.

The Effect of Internet Delay on the Perceived Quality of Information

Andrew Sears[a], Julie A. Jacko[b], and Michael S. Borella[a]

[a]School of Computer Science, DePaul University, 243 S. Wabash Avenue, Chicago, Illinois 60604, USA

[b]Department of Industrial and Systems Engineering, Florida International University, University Park, Miami, Florida 33199, USA

From the early days of personal computing, system response times have been shown to significantly affect user frustration, annoyance, and perception of the usability of computer systems. Despite reports that complaints about system response time on the Internet are prevalent, the effects of Internet delays on users have not been assessed in an empirical manner. This research is the first to empirically investigate how network latency affects the perceived usability of information provided on the Internet. Analyses of covariance were utilized to establish statistically significant relationships between delay length, the manner in which information is presented, and various facets of perceived usability of the Internet.

1. INTRODUCTION

People around the world have turned to the Internet to accomplish a vast array of tasks quickly and accurately. In turn, companies are rushing to establish an on-line presence by creating corporate World-Wide Web (WWW) sites describing their organizations and products. A 1996 estimate of the number of unique Uniform Resource Locators (URLs) in existence was in excess of 50 million [1].

This explosion in Internet use has prompted web site designers to make inquiries about fundamental usability issues. Some of the most common inquiries made by designers are: how long are Internet users willing to wait for distributed documents to load? and for which types of documents are users more willing to wait? The impact of system response time has been addressed in the context of computer terminal and personal computer use, but has not been addressed in the context of Internet use.

While the Internet is being used more frequently to provide or locate information, little is known about how perceived usability of this information is influenced by Internet delays. In a study conducted by Lightner, Bose, & Salvendy [2], a questionnaire was submitted to the on-line community requesting that respondents indicate what they like least about the WWW. Respondents indicated that speed of data access was the most disliked feature of the WWW. Thus, perceived latency is a real concern of the average user. Web servers, web clients, and network congestion are some of the causes of delays. Web servers can take a long time to process a request if they are bombarded with requests. Web clients can contribute to delay when they parse the retrieved data and display it for the user [3]. Network congestion can also add significantly to the total delay. However, the specific cause of the delay is generally considered irrelevant by the average user.

Perceived usability may not be influenced only by delay length. It may also be affected by the types of documents users are waiting for. While all information is transmitted in the same way, different media often result in much larger quantities of information being transferred. Therefore, perceived usability may be influenced not only by delay length but also by the types of information that users find themselves waiting for.

This research addresses how system response time affects users' perceptions of the usability of web sites. By addressing this issue, site designers may develop a better idea of how to best design their sites so that they are viewed more favorably.

2. OBJECTIVE

This research differs from previous research concerning human reactions to computer response times in several important ways. Previous research demonstrated that response times can have a dramatic impact on users, but focused on delays introduced by individual computers. As computers got faster, response time was viewed as less important. The current research focuses on response delays introduced by network latencies. These delays are longer and more variable than those previously studied. More importantly, unlike the delays studied previously, network latencies are often beyond the control of any individual or organization. As a result, the ability to understand how these delays influence user perceptions is important.

The researchers hypothesize that perceived usability of web sites is highly dependent upon delay length. Delay length will interact with the way that information is presented at specific sites to influence users' perceptions of Internet sites. Furthermore, it is hypothesized that specific user characteristics like exposure to the Internet, age, and command of the English language will influence perceived usability of web sites.

3. METHOD

3.1 Subjects and Experimental Design

127 subjects were recruited for participation in the study. The subjects were enrolled either at DePaul University or Florida International University. The subjects were not paid for their participation. Participants were randomly but evenly divided into six groups that resulted from the factorial combination of a 2x3 design in which the factors were type of document (text or text plus graphics) and document delay (short, medium, or long). The dependent variables were the subjects' responses to two questionnaires questionnaires.

3.2 Experimental Tasks and Procedure

The distributed documents used in the study were constructed based upon an established WWW site that contained both text and graphics. The authors created a second, text only, version of the site. To create the text only version, graphical banners were converted to HTML and photographs were deleted and related text was adjusted accordingly. Each subject was randomly assigned a Web site that included either text only, or text and graphics at one of three levels of delay: short (mean=385msec/subdocument), medium (mean=2210msec), or long (mean=3600msec). The subjects were not aware that six different conditions were being investigated.

and accelerate while looking ahead at a 17" monitor with a driving scene. Subjects were required to respond to the simulated roadway by steering and interacting with other cars on the highway. They were required to keep the vehicle on the road and between the driving lanes. A second PC was used to present variable message signs. A customized presentation of consecutive VMS displays were presented at random intervals. The viewing angle, text size and background color were designed to simulate actual viewing conditions. Sixty messages were shown to each subject with combinations of size and case.

2.3 Procedure

Subjects received a brief training session where they became familiar with the driving task and the VMS presentations. During data collection, random VMS messages were displayed with a fixed five second exposure time at random intervals. Messages included the location of a traffic incident and a suggested alternate route. Subjects were tested to determine legibility and recall. Subjects were asked to read the message aloud to measure legibility. After the message was removed, subjects were asked to report the incident location and the direction of recommended action to determine the level of recall of the salient information. The entire session lasted about 1.5 hours per subject. Subjects were given breaks every fifteen minutes to prevent visual fatigue and boredom.

3. RESULTS

A randomized block design was used to test the effects of letter size and number of phases on legibility and recall. Letter case (capital, lowercase and sentence) was fully randomized within blocks of letter size (5x7 single phase and 6x9 two phase). An Analysis of Variance was used to evaluate the effects of case and phase on both legibility and recall of the VMS text.

There was a significant effect ($p<0.05$) of letter case on legibility and recall. Capital letters were more likely to be read correctly and were recalled better than other cases. Sentence case was more likely to be read correctly than lower case, but was not recalled better. Figures 1 and 2 show the magnitudes of these differences.

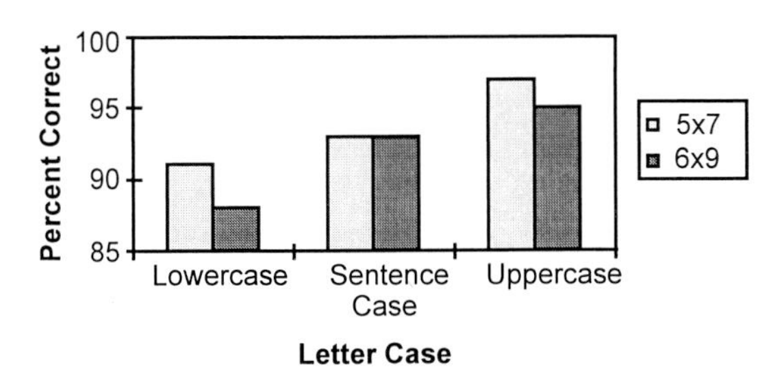

Figure 1. Percent of VMS messages *read* correctly

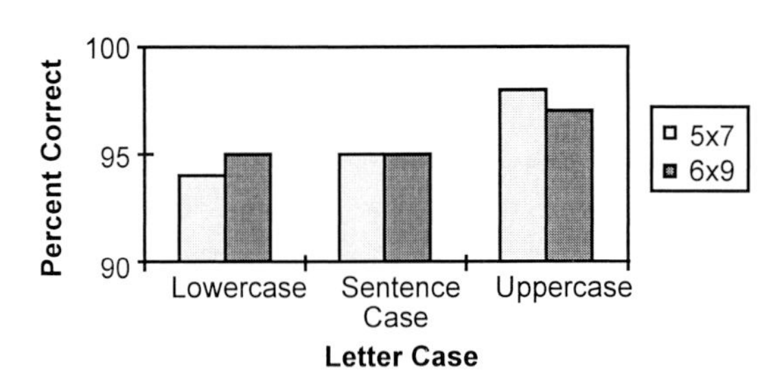

Figure 2. Percent of VMS messages *recalled* correctly

5. DISCUSSION

These results support a guideline recommending the use of capital letters for VMS signs. Contrary to expectations, however, there was no statistical proof that larger font size in a two phase message improves comprehension or recall. One explanation for this result is that for short messages made up of just a few key-words, capital letters catch the most attention and expedite reading, but the large font occupies most of the available display area, degrading the distinctive envelope necessary for the quick recognition of characters. Larger fonts are presumably visible from further away, but not necessarily more distinguishable. And furthermore, since the larger text cannot be presented on a single screen, the adoption of two-phase messages is required. The tradeoff between the decreased exposure time needed for two-phase messages and the benefit of the increased letter size has been partially shown by the results. Upper case, 6x9, two phase messages had a poor success rate for legibility and low outcome for recall.

Using the results of this study, VMS manufacturers can construct signs with a greater chance of being read and recalled correctly. This will enhance the effectiveness of VMS, allowing better traffic management within the framework of an ITS network.

REFERENCES

Dingus, T.A. and Hulse M.C. (1993). Human Factors Research Recommendations for the Development of design guidelines for Advanced Traveler Information Systems. *Proceedings of the 36th Annual Meeting of the Human Factors Society* (pp.1067-1071). Santa Monica, CA: Human Factors Society.

Poulton, E. (1967) Searching for newspaper headlines printed in capitals or lower-case letters. *Journal of Applied Psychology*. v.51, pp. 417-425.

Woodson, W.E. (1981). *Human Factors Design Handbook.* New York, NY: McGraw-Hill.

Safety Management: Some Issues and Limitations

Vincent G. Duffy

Department of Industrial Engineering and Engineering Management, Hong Kong University of Science and Technology, Clear Water Bay, Kowloon, Hong Kong, vduffy@uxmail.ust.hk

This research considers the issues for integrating voice recognition tools as a possible interface to improve safety in the workplace. Such tools enable improved access to the computerized workplace and security. However there are limitations as well. Suggestions are included for engineering managers who would want to assess the required human and organizational support to enable such technologies to succeed in advance of implementation.

1. INTRODUCTION

The focus of this research is on improving safety in the workplace with voice recognition as the interface to computer controlled processes. Subjects can be tested in real and simulated industrial environments to begin to see the capabilities. However, such demonstrations also highlight the current limitations. In light of the pending Occupational Safety and Health Bill in Hong Kong, employers will want to be confident that their response will improve safety in the workplace. They will not just want to spend money for new 'safety' technologies that are not effective. How should an employer decide such issues in advance?

The purpose of most work related to interfaces of this type would be to compare existing interfaces with improved ones in order to assess the improvements. However, typically such measures for real applications related to safety can not be obtained until after the implementation. Suggestions are included for trying to assess the required human and organizational support needed in cooperation with or in advance of implementation. Assessments and models of this type in the future are needed to improve the strategies for safety management as well as enable such technologies to succeed.

2. BACKGROUND

The following sections describe how voice recognition tools can improve access to the computerized workplace, provide security not available to normal key input computer systems and highlight the current limitations as well.

2.1 Improve Access and Affordability

Recently software tools of this type for PCs have become available at a reasonable cost. These inexpensive systems can recognize the voice input based on a template created by the user and then map the voice command with the equivalent keystroke to execute the command. The 'trained' system does not care in what language the commands are spoken. Typically computer control limits the access by the workforce. This type of tool in a workplace can improve access of the workers.

Voice recognition tools improve access for non-English speaking people as well as those with lack of familiarity with computers, lack of typing skills or with a handicap. Some previous systems required that the user remember all of the available commands. More recent systems, such as InCube, show a list of commands on the screen for user reference.

2.2 Security and Reduction of System Resource Requirements

The uniqueness of the voice of each person lends the possibility of guarding access to information. This has potential for use in computer data as well as that intended to be accessed by phone. It has been shown, as well, that natural language conventions, such as a pause, can be used to prompt a user to respond early and thus save system resources [1]. This is particularly important for systems intended to be accessed by many users at once, such as in telecommunications.

2.3 Limitations

Still, the most common problem is that of voice recognition errors. These errors can come in the form of unrecognized commands with no result except the need for the user to repeat the command. Voice commands can also be mistakenly recognized with unintended results. Both types of errors cause the user to recognize the error and respond in order to achieve the desired result. This usually causes frustration for the user.

3. APPLICATIONS AND PERCEIVED BENEFIT

For such consumer products as answering machines, it has been shown that there is a clear benefit when using voice control [2]. Remote control with voice reduces the need for recall of codes for remote checking of the device. Some success has been achieved in customer access of bank account information and some banking services by phone. Local communities have expectation that wearable computers will play an increasing role in public safety issues from police to emergency medical treatment [3].

Material handling in factories and in the lumber industry have shown improvement in data collection methods using such hands-free technologies [4]. One lumber company reported the voice recognition system reduced workers fear of accidents in addition to improving data collection and that the system contributed to the improved accident rate, counting 705 days without lost time due to accident [5].

4. SAFETY MANAGEMENT

The criteria managers should use when presented with new technologies that are supposed to improve safety in the workplace is not always clear. This is a critical component for determining effective safety management strategies. It is believed that the model proposed by Majchrzak [6] can be used as a basis for predicting the required human and organizational efforts that success of such technologies is believed to require. The success in using this model will depend on the ability to assess properly the value of improved perception in the work environment [7]. This would be needed to show the relationship between perceived improvement and actual improvement.

5. TESTING THE INTERFACE: IMPACT ON SAFETY

It is difficult to directly measure impact on safety. Such measures can only come after implementation. The best we may hope for in planning for safety in the workplace is that we can, in the future, predict the success of such technologies in the workplace. As was described in the previous section, it is likely that the human and organizational support required for such technologies can be determined so as to allow engineering managers responsible for safety management to enable success by planning for it. Even without being able to directly measure the safety improvement in the real world situations, it is expected that such systems can give an indication of their impact on safety through assessment of the following : user satisfaction, transfer of training and perceived safety improvement.

5.1 User Satisfaction

One previous study shows less favorable attitudes toward voice recognition tools for problem solving tasks that require one to work at a desk to complete the task [8]. However, it is anticipated that when the task demands physical work, the cognitive requirements will be different and the user may find it more satisfying to use this type of interface. It is expected that user satisfaction will give some indication of the willingness to use a system in the real environment. Such willingness is expected to give indication of the success of the system in improving safety.

5.2 Transfer of Training

A central objective for the use of virtual environments is that they allow the user to be immersed in an environment that is not normally or easily experienced. A proposed model for virtual reality (VR) based training includes VR technology, task analysis and considers issues in human-computer interaction to enable assessment of the transfer of training from simulated industrial environments to real environments [9]. This model could be used to enhance transfer of training to a real environment from a simulated system using voice input for a system.

5.3 Impact on Safety

Those who are familiar with their job can give a good perspective on the potential for system and safety improvement through the use of voice recognition tools. Useful results may require subjects who have experience with the task. Perceptions of impact on safety can be informative as one piece of information in determining the effectiveness of such systems. It will be necessary to continue efforts to show relationship between such perceptions and the performance in future.

6. CONCLUSIONS

Current research underway at the Hong Kong University of Science and Technology Department of Industrial Engineering and Engineering Management is involving student teams to study the improvements due to such voice recognition interfaces through assessments of user perceptions, the ability to show transfer of training for a person in a simulated environment, as well as gaining understanding from engineering managers. Those managers of most interest during the research will be those who have been responsible for safety management efforts who are either currently using such technology or who might consider the use of such technologies in the future. It is hoped that such assessments of voice recognition interfaces and the development of technology implementation models of this type for safety can improve strategies for safety managers as well as enable such technologies to succeed in improving safety in the workplace.

REFERENCES

1. Brems, D., Rabin, M., and Waggett, J.L. (1995) “Using Natural Language Conventions in the User Interface Design of Automatic Speech Recognition Systems”, *Human Factors*, 37 (2) 265-282.
2. Gamm, S. and Haeb-Umbach, R. (1995) “User Interface Design of Voice Controlled Consumer Electronics”, *Philips Journal of Research*, 49 (4) 439-454.
3. Guyette, J. (1996) “Speech recognition having a say in safety issues”, *Automatic I.D. News*, 12 (10) September, 57.
4. Moore, B. (1996) “Automatic data collection: now hear this” *Material Handling Engineering*, August, 28.
5. Davis, A.W. (1996) “Safety and productivity improvements with voice terminals”, *Professional Safety*, 41 (3) March, 37-39.
6. Majchrzak, A. (1992) “Management of technological and organizational change”, in Salvendy, G., The Handbook of Industrial Engineering, second edition, John Wiley & Sons, Inc.: New York, 767-797.
7. Duffy, V., Su, C.J., Hon, C.L., Finney, C. (1997) "Safety implementation in manufacturing: using virtual reality for computer aided manufacturing and material handling", *IEA'97-From Experience to Innovation, 13th Triennial Conference of the International Ergonomics Association*, Special Symposium on Safety Management at IEA’97, June 29-July 4, Tampere, Finland, in press.
8. Molnar, K.K. Kletke, M.G. (1996) “The impacts on user performance and satisfaction of a voice-based front-end interface for a standard software tool”, *International Journal of Human-Computer Systems*, 45, 287-303.
9. Su, C.J., Lin, F.H., Ye, L., Finney, C.M., and Duffy, V.G. (1997) “Industrial training using virtual reality”, *HCI International ‘97, 7th International Conference on Human-Computer Interaction jointly with 13th Symposium on Human Interface (Japan)*, August 24-29, San Francisco Hilton and Towers, San Francisco, California USA.

Modular Dialogue Units: A Software Architecture for Programming Human-Computer Dialogues

Manuel A. Pérez-Quiñones[a] and John L. Sibert[b]

[a]Center for Computing Research and Development, Universidad de Puerto Rico-Mayagüez, Mayagüez, PR 00680 (perezm@acm.org)

[b]Department of Electrical Engineering and Computer Science, The George Washington University, Washington, DC 20052 (sibert@seas.gwu.edu)

A software architecture is presented that uses structural properties of human dialogues to express human-computer dialogues. It allows control requests from the user to negotiate the flow of control of the dialogue, without being restricted to a stack-based control architecture. The architecture uses a notation with syntactical parts that separate the normal flow of the dialogue from the communication of control requests (e.g. interruptions). This results in a more modular design that has several advantages over previous dialogue models.

1. INTRODUCTION

Dialogue control request is the process by which a user regains control of the dialogue flow in a human-computer dialogue. The transfer of control is usually done via a negotiation between the user and the system (e.g. Are you sure you want to interrupt action X?). These requests and the ensuing negotiations are very important parts of a user interface dialogue, nevertheless their implementation has usually been done in an *ad hoc* way. Two possible reasons are that the implementation of this negotiation cannot be done cleanly on a single stack control architecture, and user interface notations require more expressiveness over the representation of control information.

In this paper we present a logical software architecture, called modular dialogue units (MDU), that allows programming dialogue level issues, such as user-initiated interruption and cancellation requests [2]. The architecture has been implemented using Scheme continuations to express the control abstractions. It has the following characteristics: it separates interaction technique specification from dialogue, thus making the representation of dialogue structure explicit without compromising dialogue independence; it uses structural properties based on human conversations; and uses a procedural notation that is suitable for either programming, or code generation.

2. PREVIOUS WORK

The software architecture presented in this paper has much in common with structural models of user interface design, such as [1; 3; 4]. In general, all structural models (including the one described in this paper) build the dialogue structure by hierarchically combining smaller pieces in a predetermined way. The restrictions on how the pieces are put together is what gives structural approaches some advantages over other design notations. These restrictions help to improve the consistency of the designs built with them.

Most of the other dialogue programming models are based in part on an input-process-output batch processing model. The MDU model is based on the structure of conversations. In our work, we have defined the parts to be similar (if not identical) to human conversational units (initiative-response pairs).

3. MODULAR DIALOGUE UNITS MODEL

The MDUs use a procedural representation, analogous to functions in several ways. They are defined as functions, called by other MDUs, and return a value. This representation has several syntactically delineated parts, the initiative and response clauses are briefly discussed here. Other parts are defined to fulfill other run-time goals; details on these can be found elsewhere [2].

The MDUs provide separate representation for data and control information exchange, yet the two are encapsulated in a dialogue unit. Request-handlers are used for implementing the control requests, with syntactically separate code from the normal dialogue flow.

In the example shown in Figure 1, the system prompts the user for his/her name. If the user enters the name and presses OK, a *response* event is sent with the name entered (the data in this example) as an argument to the waiting dialogue. This closes the dialogue and control returns to the calling dialogue. If, on the other hand, the user clicks Cancel, a *request* event is sent to the dialogue with 'Cancel' as an argument. If a request-handler was defined for a cancel request, it would be used to handle the request. If no handler was defined there, then the calling-dialogue will be checked for a handler that matches the argument ('Cancel' in this case). The process continues up the chain ending with a set of global handlers, where a default handler could be defined that terminates the dialogue.

In this example, it is the job of the dialogue designer to separate the two flows

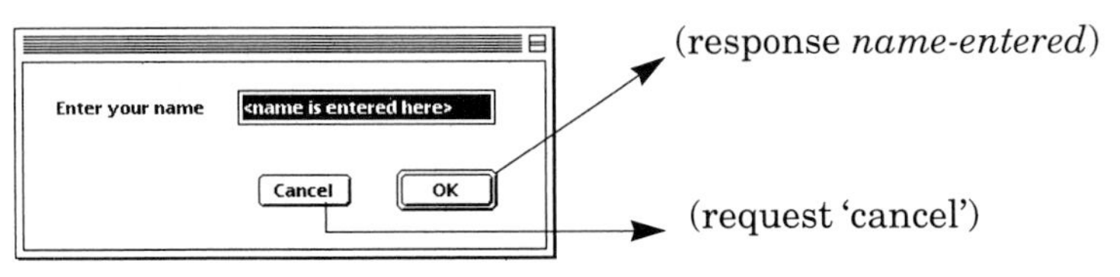

Figure 1. Different data and control channels

into separate channels of communication. The interaction technique of clicking the Cancel button generates a 'request cancel' event and a click on the Ok button generates a 'response' event with the data entered in the text field. Note that it is easy to add a new interaction technique, for example pressing the return key, to signal a response. This is done independently of the structure of the dialogue of the example. This shows how the dialogue notation is independent of the interaction techniques used.

3.1. Initiative and Response Clauses

The dialogue unit `prompt-for-name`, shown below in Figure 2, has three clauses: initiative, response and request-handlers clause. The system-initiative clause is called by the system when this dialogue unit is invoked. This clause is executed like a function call with its return value being discarded. After this function returns, the dialogue thread is suspended until an interaction technique generates a call to *response*, as described in the previous example. The call to *response* causes a call to the user-response clause. This clause also behaves like a function. The arguments passed to the response clause are values entered with the interaction technique. In the example above, clicking on the Ok button would cause a call to the user-response clause with the value entered in the type-in field. The value returned by the user-response clause is the value returned by the dialogue unit `prompt-for-name`.

3.2. Request-handlers

Control requests made by the user are programmed using request-handlers. These are somewhat analogous to exception handlers. They are lexically defined inside dialogue-units and are inherited via the call-chain to implement modular handling of control requests.

The MDU's use of request-handlers provides call-by-call expressiveness of user control requests. Handlers can be specified locally within each MDU (Figure 2), with each call to a subdialogue (Figure 3), or at a global level (Figure 4). This allows the programmer more control over the possible flow of the dialogue negotiations than would otherwise be possible with an exception handling system.

Request handlers allow growing the dialogue call tree without returning from lower levels. This allows repairs and negotiations to occur as subdialogues where

```
(define-dialogue-unit prompt-for-name
  (system-initiative ()
    (open-window window-id))
  (user-response (value-entered)
    value-entered)
  (request-handler
    (on-request cancel
      (close-window self)
      (terminate-dialogue))))
```

Figure 2. Dialogue Unit with a local request-handler

```
(subdialogue prompt-for-name
  "Enter your name"
  (on-request cancel
       (terminate-dialogue)))
```

Figure 3. Example of a call-specific handler

```
(on-request cancel
     (if (eq (subdialogue yes-or-no "Want to cancel?") 'yes)
          (terminate-dialogue)))
          ;; does nothing if answer is no
```

Figure 4. Example of a global handler

they are requested, as shown in the sample handlers of Figure 3 & 4. Also, these negotiations could be defined at higher levels, thus allowing some consistency on how the negotiation is handled without having to re-specify the handlers in every place where they are used. This enhances reuse of lower and middle levels of the dialogue structure definitions without compromising their functionality.

4. CONCLUSIONS

In this paper we have briefly described a software architecture for programming human-computer dialogues. It has the following characteristics: it separates interaction technique specification from dialogue, thus making the representation of dialogue structure explicit without compromising dialogue independence; it uses structural properties based on human conversations (initiative response pairs); and it uses a procedural notation suitable for programming. This notation has been implemented using Scheme continuations. The most important contribution of this software architecture, and what sets it apart from other dialogue models, is the separation of data and control channels. The data channel provides a communication medium for the normal flow of control of the dialogue. The control channel is used to program meta-dialogues and the negotiation of control requests, such as interruptions.

REFERENCES

1. Kasik, D. J., Lund, M. A., & Ramsey, H. W. (1989). Reflections on Using a UIMS for Complex Applications. IEEE Software, 6(1), 54-61.
2. Pérez-Quiñones, Manuel A. (1996). Conversational Collaboration in User-Initiated Interruption and Cancellation Requests. Doctoral Dissertation, The George Washington University, Washington, D.C.
3. van den Bos, J., Plasmeijer, M. J., & Hartel, P. H. (1983). Input-Output Tools: A Language Facility for Interactive and Real-Time Systems. IEEE Transactions on Software Engineering, SE-9(3), 247-259.
4. Wood, C. A., & Gray, P. D. (1992). User Interface-Application Communication in the Chimera User Interface Management System. Software-Practice and Experience, 22(1), 63-84.

Optimizing Cut-and-Paste Operations in Directed-Graph Editing

Bertrand Ibrahim, Ph.D.

Computer Science, University of Geneva, Général Dufour 24, 1211 Geneva 4, Switzerland.
bertrand.ibrahim@cui.unige.ch - http://cuiwww.unige.ch/eao/www/Bertrand.html

This paper describes the implementation strategy for the common editing primitives - select, copy, cut and paste - in a graphical editor for a visual language based on directed graphs. By focusing on minimizing the number of user actions to move parts of a graph structure around the overall structure, this strategy gives the user as much control as possible over which edges will be reconstructed automatically after a cut or paste operation. To minimize the cognitive load on the user, we clearly define the concepts of starting nodes, entry nodes, stopping nodes and exit nodes in the selection. Based on these concepts, we explain how the editing primitives work to try to reconstruct connections between the selection and the rest of the graph.

1. Introduction

Visual languages are fundamentally non-linear. Most graphical items do not have just one predecessor and one successor, and cut or paste operations are thus much less straightforward than the corresponding operations performed on linear text. With linear text there is **one** selection with **two** ends, with a clear flow from one end to the other. Cutting the selection is a rather simple operation, except in text processing systems, in which paragraphs usually have associated characteristics. One difficulty is to chose the characteristics of the new paragraph formed by the combination of the two initially separate paragraphs, as some characteristics might be mutually exclusive. The choice made by the editing program is often disconcerting for the user. Implementing select/cut/copy/paste primitives for non-linear representations requires even more careful consideration of how the visual structure should evolve after each operation to maintain its consistency while reducing as much as possible the cognitive load on the user.

2. Related work

Few articles have been published about cut-and-paste strategies for structural editors. In their AVI'94 paper, Citrin et al. [1] describe various ones, including a style strategy developed with the Escalante [2] system. Their approach is based on heuristics, called style rules, that allow the editor to reconnect edges and nodes automatically with a fixed strategy hardwired in the implementation. This innovative approach, however, leaves little room for user hints to the system as to which connections should be privileged and which should be discarded. The strategy that we describe below allows the user to control how the selection is to be cut from the overall structure and how the paste buffer is to be reconnected to the rest of the graph in a paste operation.

3. Important characteristics of the language

The visual formalism we use has some particularities that have to be taken into account when implementing a select/copy/cut/paste strategy:

- Nodes are typed and each type has a different visual representation. Nodes cannot overlap.

• For nodes of most types, spatial position is semantically irrelevant, but for nodes of a specific type, some semantics have been attached to their adjacency to other nodes of the same type

• To simplify, we will assume only two types of nodes, represented by a circle and a rectangle respectively. Circular nodes can be spatially positioned independently from other nodes and can have at most one successor, while rectangular nodes can be vertically adjacent to each other and potentially have multiple edges stemming from their sides, connecting them to multiple successors (Figure 1).

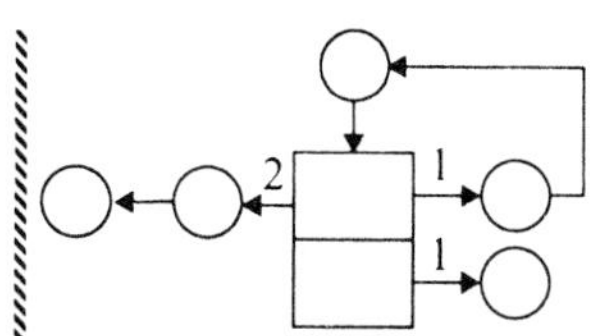

Figure 1. example of a directed graph combining spatially independent nodes, and nodes the adjacency of which is semantically significant.

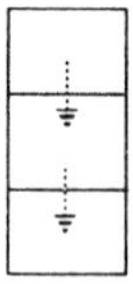

• Edges can be labeled. A default label is assumed when none is provided. It is assumed that an invisible edge connects two adjacent rectangular nodes, going from the one above to the one underneath, as in Figure 2.

Figure 2. invisible edges, represented here as dotted arrows, are assumed between adjacent rectangular nodes.

We do not use a regular graphical editor, but a structural editor that tries to save the user as many actions as possible; e.g. the user rarely has to draw edges or place nodes. Indeed, by placing the insertion point (represented by a ⊠ in the following figures) on one side of an existing node, the user defines the direction of a new node's insertion as well as the origin of the edge connecting the current node to the newly created one. Edges are thus automatically added (with paths automatically determined) by the editor unless the user places the insertion point in an empty area. Similarly, the insertion point is moved to the newly created node, on the same side as the previous node, keeping the same insertion direction. The user, in a single operation, can insert a new edge and a new node as well as move the insertion point to the newly created node (Figure 3). A simple click on the appropriate side of an existing node or empty area will change the direction and/or place of insertion. Manual addition of edges is possible but much less frequent, in practice, than automatic insertion.

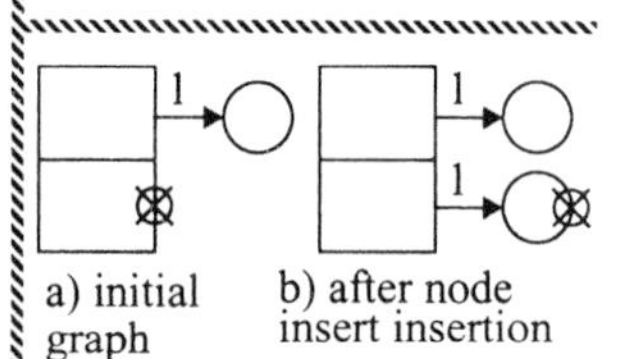

a) initial graph　b) after node insert insertion

Figure 3. In one operation, the user can add a new edge, a new node and move the insertion point.

4. Selecting, cutting, copying and pasting

Selection in a directed graph: Any number of edges and nodes may be simultaneously selected. In our drawings, selected edges are represented with thicker lines and selected nodes have black backgrounds. Selection can occur by clicking on individual edges and nodes, or by adjusting the size of a selection rectangle, as it is done in many WYSIWYG editors. Successive selections are cumulative and the user must explicitly cancel the current selection to start a new one. The selection rectangle is designed to ease pasting the selection elsewhere in the graph, based on which edges get selected, and where in the logical structure the selection will be pasted. The rectangular selection works as shown in Figures 4 to 7. In most cases, the edges originating from a rectangular node overlapping the

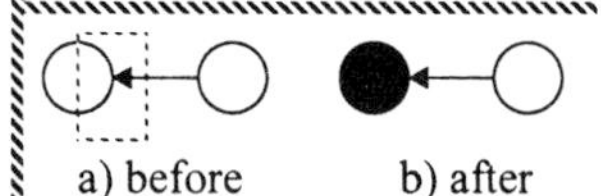

a) before　b) after

Figure 4. Any node overlapping the selection rectangle gets selected.

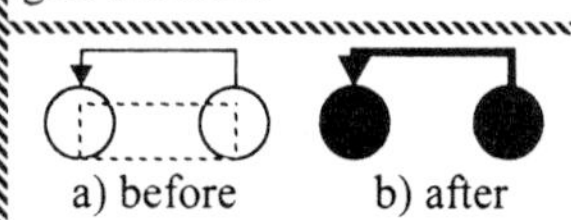

a) before　b) after

Figure 5. :Any edge connecting two nodes overlapping the selection rectangle gets selected, regardless of the path of the edge.

selection rectangle will get selected (Figure 6), except when a rectangular node has only one successor, and that successor does not overlap the selection rectangle, the edge between the rectangular node and its successor does not get selected, so as to preserve the connectedness of the graph, were the selection to be deleted (Figure 7).

We consider four categories of nodes in the selection: selection entry nodes, selection starting nodes, selection exit nodes and selection stopping nodes. A selection **entry node** has at least one unselected edge pointing to it. A selection **starting node** has no predecessor at all or has unselected predecessors but no unselected edge pointing to it. A selection **exit node** is one from which at least one unselected edge originates. A selection **stopping node** has either no successor or no unselected edge originating from it. These four categories are illustrated in Figure 8 (consider each group of connected nodes as a separate graph). The selection heuristics try to maximize the number of cases where the selection has a single entry node and a single exit node.

Cutting the selection: The cut operation removes all selected nodes and edges from the graph, potentially splitting it in multiple fragments. To save the user as much manual reconnection of the various fragments, the graphical editor tries to connect together nodes that were directly connected to the selection: in selections with single entry and exit nodes, all unselected edges pointing to the entry node are redirected to an exit node's successor reachable through the unselected edge that has the lowest label (the invisible edge between two adjacent rectangular nodes is always assumed to have a lower label than that of any visible edge, e.g. Figure 9). With more than one entry node or exit node, it is best not to guess at predecessor / successor connections. The editor will thus not try to reconnect them after the cut operation. All the edges connecting nodes outside the selection with nodes within it will be removed. Figure 10 illustrates how the user might have to complete the rectangular selection process with the selection or de-selection of individual edges to reach the stage where the cut operation will give the desired result.

Copying the selection: Both cut and copy operations result in putting the selection in a paste buffer. The buffer includes not only information about the nodes belonging to the selection and their connecting edges, but also information about which nodes are entry, starting, exit and stopping nodes. This additional information is used for the cut operation, as well as for the paste operation.

Pasting the selection: Pasting selections in empty spaces, using no other node than those in the paste buffer is trivial. Pasting a selection within an existing graph requires checking that the insertion point is in a place compatible with the paste buffer content. This "compatibility" takes

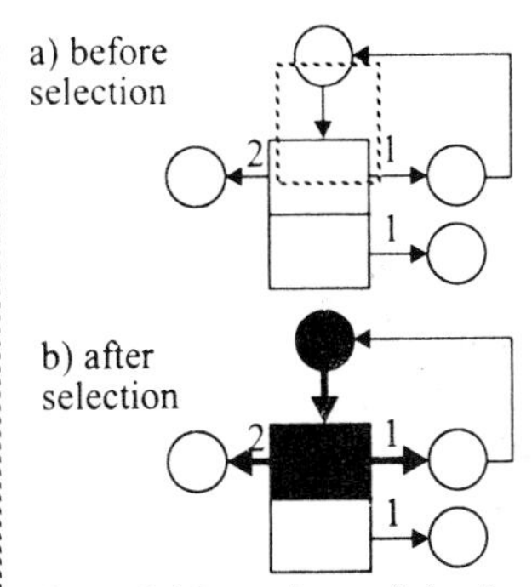

Figure 6. Most edges originating from the side of a rectangular node overlapping the selection rectangle get selected.

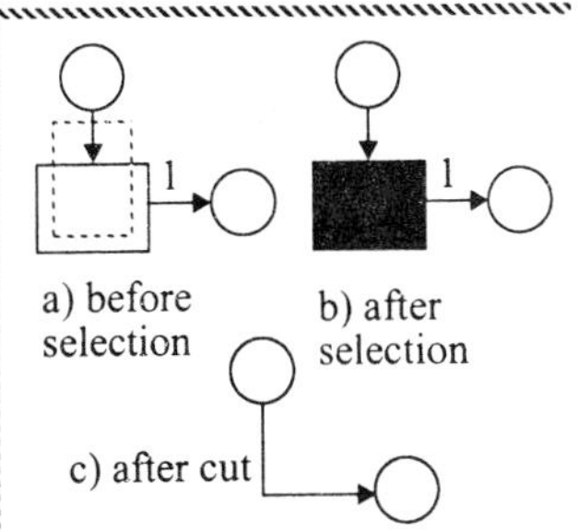

Figure 7. Selection of a rectangular node, but no edge selected. After cut, the predecessor gets connected to the successor.

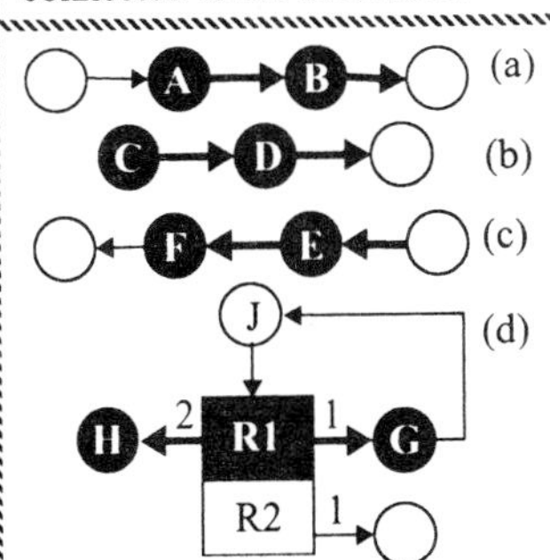

Figure 8. A & R1 are entry nodes, C & E starting node. B, F,G & R1 are exit nodes, D & H stopping nodes. (R1 is an exit node because the invisible edge connecting it to R2 is not selected)

into account entry and exit nodes of the paste buffer, as well as the insertion point position. The starting and stopping nodes of the paste buffer are not taken into account here, since it is precisely those nodes that the user didn't want to see connected. In our case, there are five main possibilities for pasting the buffer within the graph structure:

• The insertion point is on a node that has no successor (or on the side of a rectangular node with no side-successor). Here the paste buffer is inserted under the condition that it has only one entry node. The node holding the insertion point will automatically connect to that entry node.

• The insertion point is on a circular node, or on the side of a rectangular node, with a successor: the paste buffer is inserted there if it has only one entry node and one exit node. The node holding the insertion point will automatically connect to the entry node, and the exit node will automatically connect to the successor of the node holding the insertion point (Figure 11).

• The insertion point is between two adjacent rectangular nodes: In this case, the paste buffer is inserted there if it has only one entry node and one exit node, and both are rectangular nodes. The node above the insertion point will be above the entry node and the one below the insertion point will be below the exit node.

• The insertion point is at the top of a cascade of adjacent rectangular nodes: In this case, the paste buffer is inserted there if it has only one entry node and one exit node, and the latter is a rectangular node. The predecessor of the top rectangular node will have the entry node as its new successor, and the top rectangular node will be below the exit node

• Similarly, if the insertion point is at the bottom of a cascade of adjacent rectangular nodes, the paste buffer is inserted there if it has one rectangular entry node.

Otherwise, the insertion request will be rejected, forcing the user to move the insertion point to a valid place.

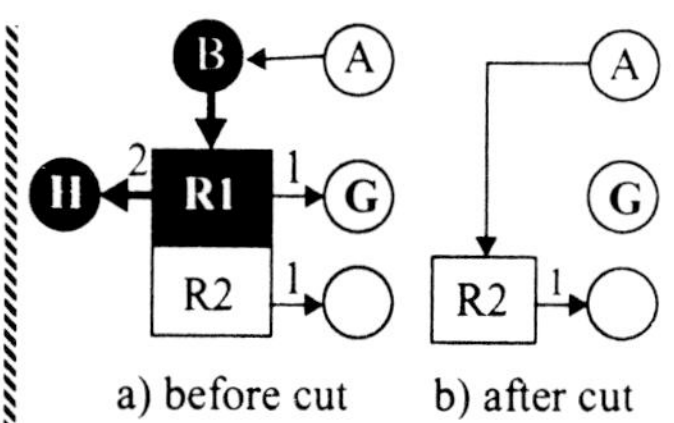

Figure 9. B is the entry node, R1 is the exit node with two unselected successors: G and R2. After the cut the edge from A to B is redirected to R2 and G is disconnected because the edge R1-R2 is considered to have a label lower than R1-G.

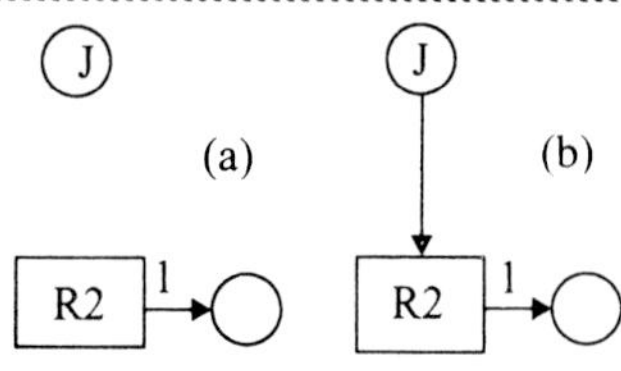

Figure 10. (a) Result of the cut on graph of Figure 8-d. (b) Result of the cut on the same graph, were the edge between G and J selected.

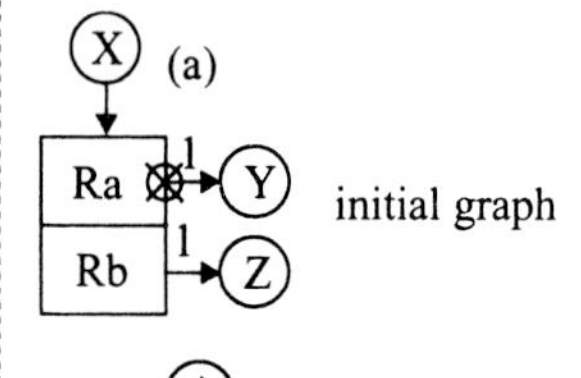

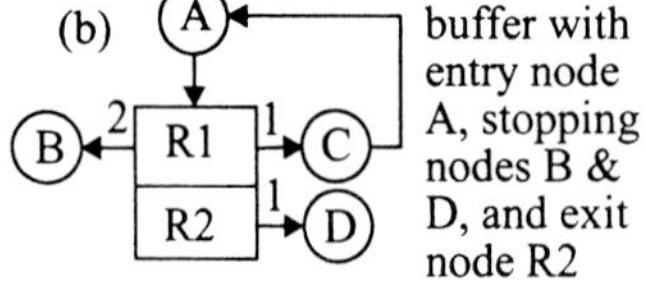

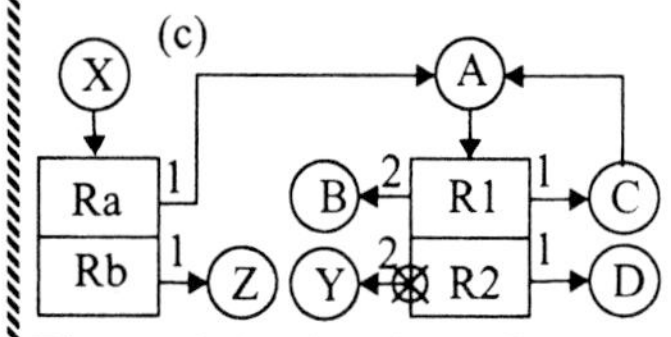

Figure 13. Pasting the buffer (b) in graph (a) results in graph (c).

REFERENCES

1. Wayne Citrin, Daniel Brodsky, Jeffrey McWhirter; "Style-Based Cut-and-Paste in Graphical Editors"; proceedings of the Workshop on Advanced Visual Interfaces, AVI '94, Bari, Italy, June 1-4, 1994, pp 105-112.

2. Jeffrey D. McWhirter, Zulah K. F. Eckert, Gary J. Nutt; "Building Visual Language Applications with Escalante"; Technical report, University of Colorado, Boulder, CS Department, CU-CS-655-93, Oct. 93, 78 pages.

New Modeling / Analyzing Architecture and Simulation mechanism for Complex Structure

Tsuneo Sawa[a] and Masafusa Yasuda[b]

[a]Department. of Management and Information Science, Meio University
1220-1 ,Bimata, Nago, Okinawa, 965 Japan

[b]Information Technologies - 21
2-4-10 Nogata, Nakano-ku , Tokyo, 165 Japan

As we approach the end of this century we face upheavals in society related to knowledge, intelligence and resources. In our opinion, Integration of science/engineering intelligence and social/cultural intelligence will be a central theme. We also think that Information Technology will play a major role in supporting and realizing this integration.

1. GMA - Global Model Architecture

GMA(Global Model Architecture) is a conceptual information architecture that aims at our thinking support and group decision making support. It's based on (1) integrated dictionary system which is composed of natural language, (2) multi-dimensional screen system. Figure-1 shows this model.

With GMA conceptual model, we can monitor social, economic and management changes and gather our response, behavior pattern. And also we can refine these response and pattern. Data mining is a part of our system.
GMA's another theme is a visualization/representation mechanism. The current window system on personal computers is a useful and helpful means of visualizing information. However this mechanism is not sufficient to display/represent intelligence or know-how. We need multi-modal and multi-window mechanism which displays separate information but harmonized with each other.
We called this mechanism "Mandala architecture". "Mandala" is a famous Buddhist word. It simultaneously expresses our universe and its structure using by visual methodology.

GMA model is our knowledge collection and representation architecture for "Generalized Logic Modeling and Simulation System".

Figure-1. **GMA(Global Model Architecture) Concept & GMAIS(GMA Information System)**

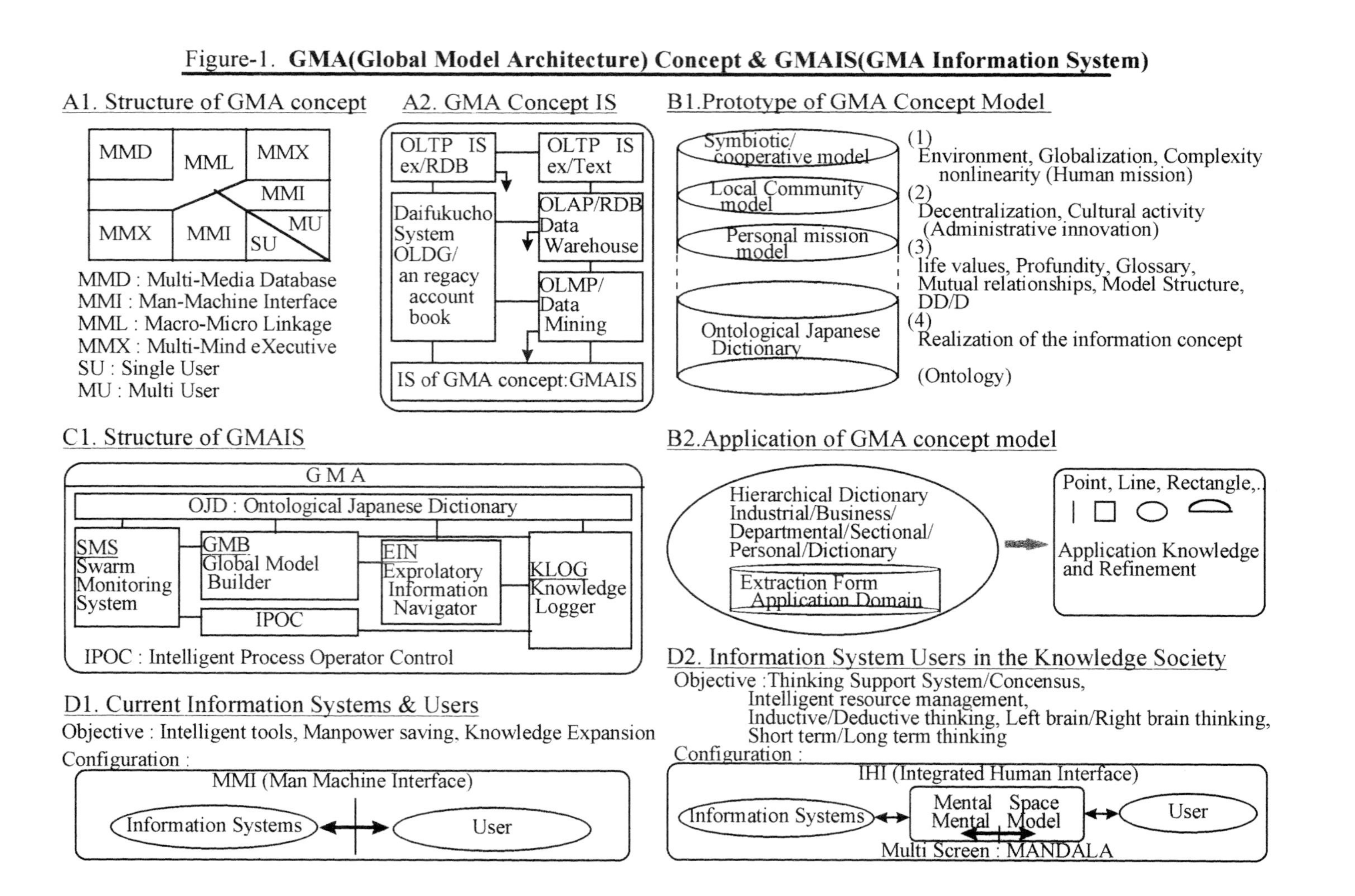

2.Generalized Logic Modeling and Simulation System

2.1.Background

There are many modeling and simulation systems but they can't treat complicatedly structured world model. In the reality every society or organization, even our own bodies and organs, is highly structured. For example, an enterprise is composed of organizations, an industry is composed of many individual enterprises and/or companies and a nation is composed of many industries. Finally, organizations such as the United Nations are composed of member nations. In this way, the structure of our real world consist of "Nested Structures".

This nested structure has multi-dimensional and many layered elements in it's inside. Today, Datawarehousing and Data Mining technologies have been introduced in the area of Decision Support Systems(DSS). Data visualization techniques have also improved. But the fundamental problems - lack of a modeling mechanism for complex/compound structure - still remain

Over 10 years, we have studied this architecture and methodology and have established a new architecture and interface mechanism to model, simulate, analyze and understand this "Complex Structure" - "Generalized Logic Modeling and Simulation System". This original idea was developed as "Structured Matrices" in Germany to appreciate the cost structure of chemical plant. We expand this idea to business application area, especially management information science area.

The business process is a good example of a highly nested and networked organization. We apply our system to business process modeling/analyzing and simulation tool for business planning and control. Many companies apply it to profit and cost control and gain profits. Profit or cost management is a very important area in the virtual corporation environment in particular, but currently there is no software or descriptive mechanism. In time dependent simulations such as well known System Dynamics, we introduce a nested structure.

2.2.Structured Matrices

Figure-2 shows principle of Structured Matrix. Structured Matrix is composed by 3 parts, such as (1) Up Side Part, (2) Left Side Part and (3) Center Part. Each part is divided by cell.
In figure-2 V_i shows vector and M_i shows matrix.

Figure-2. Principle of Structured Matrix

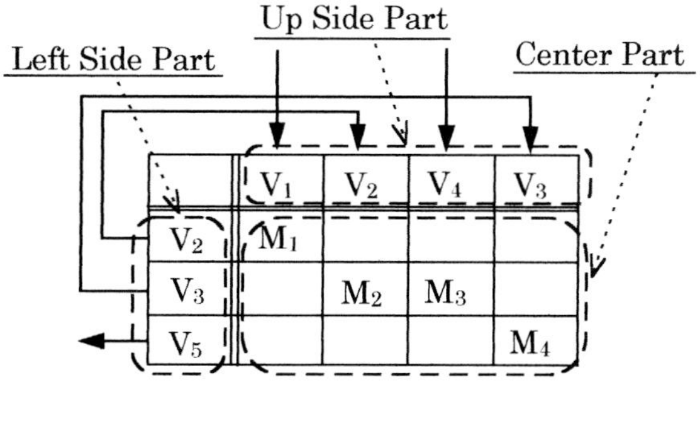

In figure-2 V_1 and V_3 are outside variables and the other vectors of upside part are computed from this matrix. Those vectors and matrices are composed relation of linear equations as follow.

$$V_2 = V_1 \times M_1$$
$$V_3 = V_2 \times M_2 + V_4 \times M_3$$
$$V_5 = V_3 \times M_4$$

2.3.Generalized Logic Base System

In Structured Matrix, we can replace vectors to generalized variables.
Also we can replace matrices in center part to generalized functions. Thus we can write these relations as follows.

$$X_2 = f_1(X_1)$$
$$X_3 = f_2(X_2) + f_3(X_4)$$
$$X_4 = f_4(X_3)$$

Figure-3. Principle of Generalized Logic Base

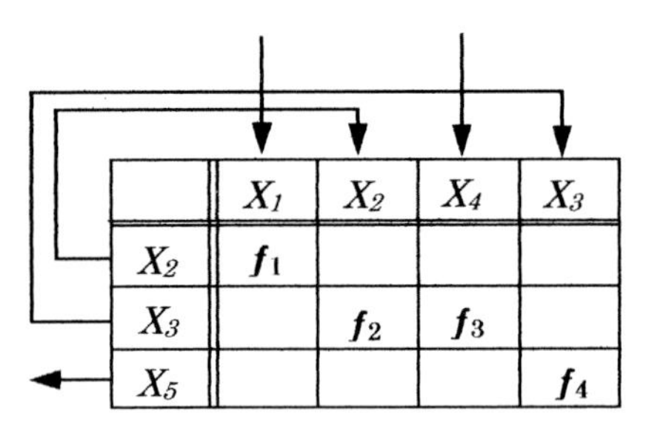

And more we expand "f_i" (function or process) can replace to another structure. Figure-4 shows this nested structures.

By using this "MANDALA " structure we can express complex structure and each relations. Also we can connect separated structure on the network, such as division's models, company's models, industrial models, and so on.

Figure-4. Principle of Multiple layer

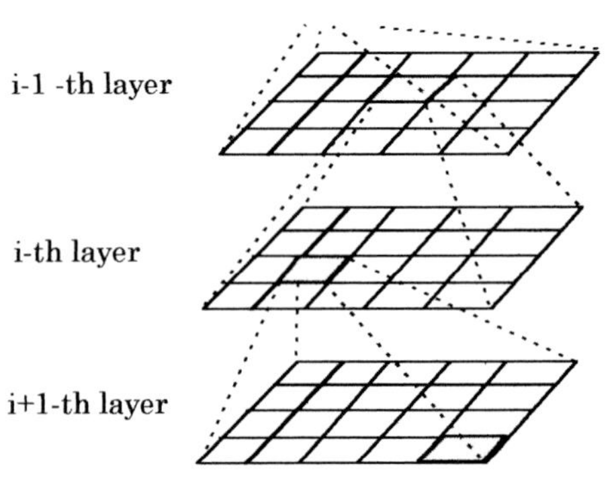

Our next study goal is (1) time series simulation capability,
(2) self reproduction and self transformation mechanism.
Almost modeling and simulation systems do not change their parameter and/or model itself. We introduce knowledge system, such as GMA, or rule based mechanism to describe and operate this "change" mechanism.

Designing the User Interface for a Natural Spoken Dialogue System

Susan J. Boyce

AT&T
101 Crawfords Corner Road, Rm. 2J-323, Holmdel, NJ 07733, USA

1. INTRODUCTION

Technology advances in Automatic Speech Recognition (ASR) and Natural Language Understanding (NLU) in recent years have brought us closer to achieving the goal of communicating with machines via unconstrained, natural speech. Until recently, most applications of speech recognition technology required that the user know a restricted set of command words. In contrast, the research system that I will describe can understand and act upon fluently spoken language. This markedly changes the nature of the dialogue between the human and the computer. The objective of this research is to evaluate user interface design alternatives for these new natural spoken dialogues between humans and machines.

The experimental spoken dialogue system upon which this research is based was designed to automatically route calls to a telephone operator by classifying callers' responses to an open-ended "How may I help you?" prompt. The system itself is described in detail in [Gorin, 1994]. The goal of this experimental system is to classify responses as one of 17 call types so that the call can then be routed to an appropriate destination. For example, if a person said "*Can I reverse the charges?*" then the appropriate action is to connect them to an automated subsystem which processes collect calls. If the request was "*I can't understand my bill*" then the call should be routed to the business office.

The first phase of research involved analyzing conversations between callers and human operators to gain a better understanding of how callers express their requests to operators and how operators elicit clarifications. This allowed us to incorporate elements of the human-human dialogue in our human-machine dialogue. However, not all aspects of human-computer dialogue can be modeled after human-human dialogue. Some elements of the human-computer dialogue are necessary simply because the automated system does not have all of the capabilities of a human listener. These elements required that we empirically test alternatives to determine the best design.

2. ISSUES IN DESIGNING A NATURAL CALL ROUTING DIALOGUE

In the call routing system, there were several aspects of the dialogue for which have little or no correspondence in human to human conversation.

2.1. Initial Greeting

A frequently-voiced concern about designing very natural human-computer dialogues is that early in the interaction, users are likely to assume that the system has greater capabilities than it actually has, and therefore attempt to speak in a manner that the system has little probability of understanding. Designing the right initial greeting is necessary to appropriately set user expectations.

2.2. Confirmations

Since any automated system will not perfectly interpret the user's speech, it is often necessary to confirm what the caller said. The system repeats what it thinks was said, giving the caller a chance to confirm or deny the system's interpretation. The confirmation strategy can be adjusted contingent on the confidence estimate provided by the spoken language understanding (SLU) subsystem. One design difficulty is that a human operator is not likely to say a phrase such as "*Do you need area code information?*" in such a context. The challenge is to ask the question in a way that more closely resembles the way a human might ask for confirmation and to elicit a meaningful user response.

2.3. Disambiguating an Utterance

In some cases, the automated system is going to come up with more than one interpretation of the user's speech. There are several ways that the system could ask for clarifying information. One way to do this would be to simply ask the user, "*Do you want A or B*?". Another way might be to ask a yes/no question "*Do you want A?*" The most appropriate strategy depends on the relative confidence levels of likely interpretations and which is most effective at completing the dialogue quickly and naturally.

2.4. Reprompts

In cases where there is very low confidence from the SLU subsystem it is better to ask for the user to repeat his or her request than to confirm. I refer to this step as a reprompt. Often in human-computer dialogues this step takes the form of "*Sorry, please repeat*". The system admits culpability but does not provide any information as to what went wrong. In contrast, human listeners have a wider repertoire of responses available to them to communicate to the speaker which elements they didn't understand in an utterance. A challenge for human-computer dialogue is to intelligently mimic the devices humans use in order to communicate to the user how the conversation has failed so that the user can provide useful inputs.

2.5. When to Bail out

Sometimes a human-computer dialogue experiences repeated breakdowns, as shown by either low system confidence or repeated responses of "no" to confirmation prompts. In such cases, a human should be brought in to complete the transaction. The design question is how to decide when a situation calls for human intervention. The correct criteria might be number of recognition errors in a row, or it might be the number of errors that have occurred overall in the dialogue. It seems likely that the right answer to this question may very well depend on characteristics of the user population and of the task being accomplished by the automated system. The goal should be to bailout before the user's frustration causes a negative perception of the system.

3. METHOD

Experiments were conducted using a "Wizard of Oz" methodology [Gould, 1983]. In such a study, the speech recognition and natural language understanding components of the system are simulated by the experimenter, a fact that is hidden from the user. The experimenter simulates an error or a correct response by the automated system by pushing the appropriate key on a computer that is controlling which system prompts get played back across the telephone to the caller. To conduct these experiments, the Wizard of Oz system was tested with volunteers.

4. RESULTS AND DISCUSSION

4.1. Confirmation Experiment

The necessity of a confirmation step makes the dialogue unlike human-human communication. For this reason we experimented with alternative ways to confirm the machine's understanding of the user's request. Two methods were evaluated. The first was the *explicit confirmation*, "Do you need X?". An alternative method of confirmation was suggested by some research I had previously done on confirming strings of digits, such as phone numbers, for automated systems. Rather than explicitly asking the user the question, the system posits an interpretation (*"You need X."*) to which the user can either say yes, no, or be silent where silence is interpreted as agreement. I call this the *implicit confirmation* method.

For 259 callers to the simulation, the explicit confirmation strategy was used; 145 were given correct feedback, 114 were given incorrect feedback (to simulate classification errors). 233 callers heard the implicit confirmation strategy; 119 received correct feedback and 114 were given incorrect feedback. The strategies were evaluated by comparing the probability of successfully repairing the dialogue from a simulated error.

Both strategies were very successful if the machine had correctly interpreted the user's speech. However, users were somewhat less successful at repairing errors in the dialogue when the implicit confirmation was used as compared to the explicit confirmation (Table 1).

Table 1
Callers responses to Explicit and Implicit Confirmations

	Explicit Confirmation (Do you want X?)	Implicit Confirmation (You want X.)
"Yes..." for Correct Feedback	91%	93%
"No..." for Incorrect Feedback	74%	63%

For trials in which a recognition error was simulated, only 63% of the users corrected the error by saying "No" then optionally following this with the correct information. Furthermore, 15% of the callers with Implicit Confirmation failed to correct the error at all, as compared with only 6% of calls that received Explicit Confirmation. It is possible that these users were unclear about how to interrupt the system to correct the error. These data suggest that the explicit confirmation method, albeit possibly less natural, is the more robust strategy to pursue given that more of the errors get correctly repaired. However, it is possible that with more experience callers would learn how to use the implicit confirmation method.

Reprompt Experiment. When the system is unable to interpret the caller's request it is necessary to reprompt. Several different wordings of reprompts were evaluated. They fell into two categories: ones in the general form of an apology followed by a restatement of the original prompt and others that included an explicit statement that the response was not understood. The goal of this experiment was to develop a reprompt that produced clearer, more concise speech from the user for the second utterance. This will give the system the highest probability of correctly classifying the call on the second try.

Each of these reprompt wordings was tested with 44 users. There were no significant differences in caller behavior to the different types of reprompts, so the data have been combined in Table 2.

Table 2

Classification of user responses to reprompts	Percent
Exact or almost exact repeat of initial utterance	37%
Shorter utterance on second try	31%
Longer utterance on second try	20%
Rephrased, same length	12%

Users as a whole did not adopt a single strategy and they did not all have the same mental model of the system's capabilities. Some of the strategies adopted by the users are likely to produce a more successful dialogue than others. The goal of the dialogue system is to guide users toward the successful strategies. Thus, a subject for future research is to explore reprompting strategies that are more explicit about the reason for the communication failure.

A great deal of empirical work remains to be done to further address the best strategy for reprompting, as well as to address the other issues outlined in this paper. I have proposed that the best designs for human-machine dialogues will be ones that very closely model human to human communication. However, I have found that there are necessary elements of a dialogue which are difficult to model after human-human communication. Technology improvements and empirical research in the future will result in easy to use, likable spoken natural dialogue systems.

REFERENCES

Gorin, A. L., Henck, H., Rose, R., Miller, L. (1994). Spoken Language Acquisition for Automated Call Routing, *Proc. ICSLP,* 1483-1485, Yokohama (Sept 1994).

Gould, J. D., Conti, J., & Hovanyecz, T. (1983) Composing letters with a simulated listening typewriter. *Communications of the ACM.* vol. 26, pp. 295-308.

Allowing Multiple Experts to Revise a Thesaurus Database

Norifumi Nishikawa and Hiroshi Tsuji

Systems Development Laboratory, Hitachi, Ltd.
8-3-45 Nankouhigashi, Suminoe, Osaka 559 Japan.
e-mail : {nisikawa, tsuji}@sdl.hitachi.co.jp

A thesaurus contains a variety of terms in many categories, so multiple experts should join as a team to systematically maintain it. In this paper, we describe a thesaurus management system that allows multiple experts to maintain a thesaurus database.

1. INTRODUCTION

In a modern computer network environment where users access a variety of databases from individual terminals, Information search opportunities have been increasing. For both a keyword-based index search and a full-text search [1], users need to supply the terms for the information search. A thesaurus indicates the relationship between terms and is often used to help remind us of suitable terms [2, 5]. Since the terms that appear in documents change from day to day, a thesaurus should be maintained regularly and/or continuously.

In a traditional thesaurus management system [3, 4], an authorized system manager checks the consistency of the thesaurus structure and updates the thesaurus. However, when designing the functions of such a system, the prestages of creating updated drafts by multiple experts and of updated draft checking by specialists are not considered. Therefore, it takes a lot of time to systematically maintain a thesaurus because the system does not support complex task such as a temporal saving of updated drafts or checking of updated drafts by specialists.

2. REQUIREMENTS FOR A THESAURUS MANAGEMENT SYSTEM

A thesaurus illustrates three types of relationships between terms [2]: equivalent (USE/UF), hierarchical (BT/NT), and associative (RT).

Generally, a thesaurus contains a variety of terms in many categories, so multiple experts should join as a team to systematically maintain a thesaurus. During thesaurus maintenance, for experts on the maintenance team create updated drafts. There are also specialists for each category who check these updated drafts. Note that the categories in a thesaurus are not mutually exclusive, so some experts need to maintain a thesaurus by obtaining updated drafts from the other specialists. We analyzed the tasks involved in thesaurus refinement and categorized them as follows:

(a) each expert creates drafts by referring to the drafts of the other experts or by forecasting the results of updating the thesaurus, allowing for some inconsistent drafts;
(b) specialists eliminate the inconsistent drafts, and refine the updated drafts;

(c) an authorized person accepts the refined draft and updates the thesaurus database.

According to this analysis, a thesaurus management system that supports the preparatory stage of thesaurus maintenance must meet the following requirements:

(1) displays the functions of maintenance status;

(2) includes temporary update functions that introduce updated drafts, thus allowing the database to be in an inconsistent state;

(3) data validation functions at three levels, i.e., a draft creation check, a lump check for consistency among the created drafts and an interactive check by an authority.

3. DESIGN OF THE THESAURUS MANAGEMENT SYSTEM

3.1 Overview

Let us describe the construction of the thesaurus management system, which is called Theater. Theater consists of an interactive environment for referencing and an interactive environment for updating. To manage drafts created by experts, Theater introduces blocks of data called *Updated Draft Data* (UDD). A UDD is created and stored into a database that is separate from the thesaurus data. Such data management allows users to store temporally inconsistent data in a database. Figure 1 shows a functional overview of Theater.

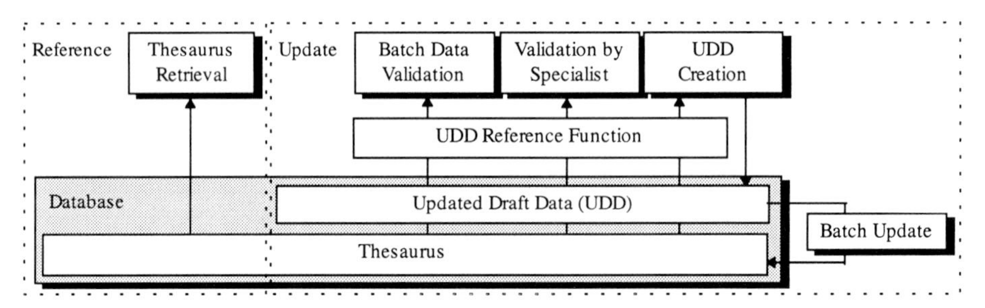

Figure 1. Functional Overview of Theater

3.1 Reference environment

The interactive environment will be described for reference. Theater supports keyword search of a thesaurus and navigation among terms. It has two types of displacements for the retrieved results: a detailed displacement and a listed displacement. In the detailed displacement, all the attributes of a term and its related terms are displayed. In the listed displacement, an alphabetical list of terms, a hierarchical tree of terms, and sibling terms having the same broader term(s) are displayed.

3.2 Structure of a UDD

The interactive environment for updating is constructed by the UDD creation function which is a validation function by a specialist and a batch data validation program, and a batch function that updates the original thesaurus using authorized drafts. A UDD is constructed from *Updated Record* (UR) and *Updated Draft* (UD). UD is a unit of thesaurus maintenance for each UDD. UR keeps track of operations by using a term identifier, and the results are saved as either a *Term Modification Record* or a *Relation Modification Record*. These records are created for each modification operation, linked from a UD, and maintain the identifier of a UD. Figure 2 shows the data structure of a UDD.

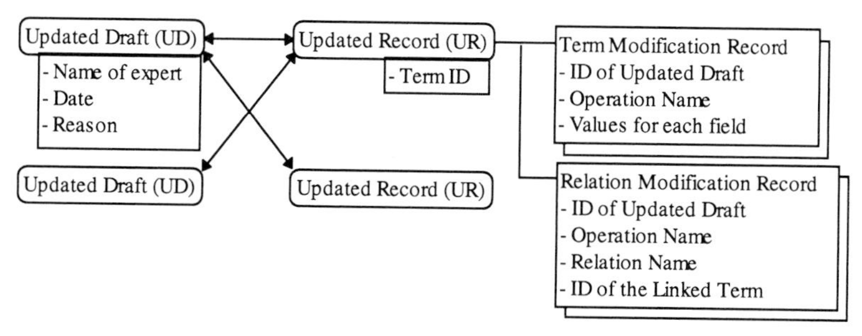

Figure 2. Data Structure of UDD

3.3 UDD reference function

Theater provides two types of UDD reference functions: status view and result view. For a status view, Theater presents the updated status of terms along with an operation name; and the results are shown using a detailed displacement and a listed displacement. If the UDD for the term is in a database, the status view function reads both the original thesaurus data and the UDD, then merges them and displays the modified values along with operation name.

For a result view, Theater presents the updated results for the specified UD without actually updating the original thesaurus.

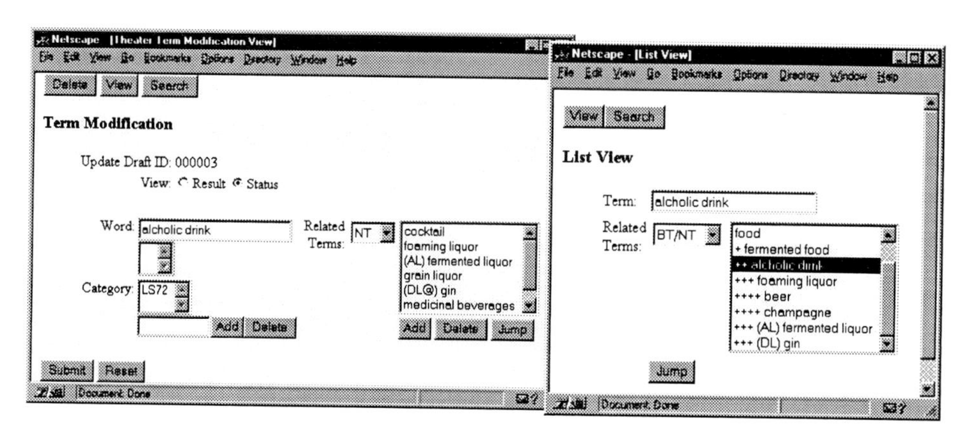

Figure 3. Status view

3.4 UDD Creation

An expert creates a UDD by modifing the term in the editing window that shows the original term. If the term has another UDD, the editing window shows its status by using the status view function. Even if there are UDDs that conflict with each other, Theater stores all of them.

If an expert creates a UDD for the purpose of modifing a relation between terms or for deleting a term, the modification will affect multiple (linked) terms. In such a case, Theater will automatically create or update the UR of every other term. For example, an expert adds a new relation between terms T1 and T2 from the editing window of term T1. Theater first creates the UR that maintains the relation adding operation for T1, then creates the UR that maintains the relation adding operation for T2. Experts can refine or delete these UDD later. Figure 3 shows the status view for a UDD that will remove the link between alcholic drink and gin. The UDD for gin is automatically created.

3.5 Validation of a UDD

Theater provides data validation functions at three levels, i.e., a draft creation check, a lump check for consistency among created drafts and an interactive check by specialists. A draft creation check is done when a UDD is created; it only checks the data format of a UDD. If there is an error in a UDD, users will be promptly notified and the UDD will not be stored.

A lump check for consistency is done as a batch data validation after all UDD are created. A result view function is used for this check. If an inconsistent status exists among any of the UDD, error records against those UDD will be created.

After a lump check, an interactive check by specialists is done. Each specialist checks the UDD of each discipline. For this function, Theater shows a UDD as both a status view and a result view. To point out an error shown in these views, a specialist selects that term. Theater will show a list of UDDs for that term and the UDD of terms with the same hierarchy.

4. CONCLUSION

Theater provides users with the following functions for thesaurus maintenance:

(1) management of inconsistent UDDs that need to be judged for consistency by specialists;

(2) confirmation of results and detection of drafts that conflict when a draft is created;

(3) checking of data validation for each maintenance stage.

As a result, Theater supports all stages of thesaurus maintenance including the interactive creation of updated drafts by many experts and checking by specialists, and it improves the efficiency of systematic thesaurus maintenance.

REFERENCES

1. Dawana T. Dewire, Text Management, McGraw-Hill, Inc., 1944.
2. Jean Aitchison and Alan Gilchrist, Thesaurus Construction Method, Maruzen, 1989.
3. Roy Rada, Maintaining Thesauri and Metathesauri, Int. Classif. 17, no. 3/4, pp.158-164, 1990.
4. STRIDE: Questans thesaurus management system, http://www.questans.co.uk/p10012.html, 1995.
5. JICST, Thesaurus of the Japan Information Center of Science and Technology (In Japanese), 1993.

CyberForum on InterSpace

Toru Sadakata[a], Takashi Yamana[a], Tetsuo Tajiri[a],
Hisashi Hadeishi[b], Yoshinori Okuvo[b] and Yoshiaki Tamamura[b]

[a]NTT Human Interface Laboratories
1-1 Hikari no Oka Yokosuka-shi Kanagawa 239 JAPAN

[b]NTT Software Corporation
3-43-16 Shiba Minato-ku Tokyo 105 JAPAN

Abstract

The multi-user networked virtual world system "InterSpace" is described. This system's main purpose is to enhance the user's communication activities. In InterSpace, real world information is embedded in the shared virtual world as a combination of video and CG images. Users can observe and access this information by simply looking at and approaching embedded objects.

We produce CyberForum, a trial service based on the InterSpace platform. Also, we provide new functions for a virtual forum. The concepts of InterSpace and CyberForum is discussed.

1. INTRODUCTION

Today, many systems use virtual reality (VR) technologies in their user interfaces. Most systems are stand-alone "virtual world" simulators. Some systems (Habitat[1], SIMNET[2], and DIVE[3]) were developed to construct multi-user shared virtual worlds for entertainment, training, conferencing and collaborative work. These systems use the sensation of total immersion as the basic user interface. The virtual worlds of these systems have no relation to the real world and are system-specific ideal models of the real world. Users communicate within and feel totally immersed in the computer generated virtual world.

This paper describes the many-user networked virtual-world system "InterSpace." InterSpace uses VR technology as the user-interface medium, but the system's main purpose is to enhance the user's real world communication activities. In InterSpace, many users and services can coexist in the same shared virtual world. InterSpace connects the real worlds of the users to the real worlds of the services and other users. InterSpace mediates between real worlds or real spaces. The name InterSpace reflects this aspect.

For this study, a virtual exhibition was achieved, using the InterSpace system. We named this exhibition "CyberForum". In CyberForum, participants can freely browse around the exhibits. The expositors can give an explanation in virtual space.

E-mail: sadakata@nttvdt.hil.ntt.co.jp

2. BASIC CONCEPT BEHIND INTERSPACE

In the InterSpace system, users inhabit the "real world" and drive their images in the "virtual world" with their "avatars." To intentionally meet or casually encounter in the virtual world is equivalent to meeting in the real world through visual communication. The real world is embedded in the virtual world as a combination of video and CG images. Captured video images of users are used to generate their avatars. The avatar uses a real video-based representation. The visual image of the avatar represents the user's real appearance. We propose a combination of a CG-based virtual world structure and video-based objects as a new concept for a shared and interactive virtual world system for multiple users. These avatars, called video avatars are shown in Figure 1. Users can communicate with each other and with each service by simply looking and approaching each object. This architecture is realized through the linkage of a distributed and synchronized world model (CG) and video, audio, and data communication networks.

3. INTERSPACE SYSTEM

The InterSpace system consists of an InterSpace server, facial video server, voice server and user terminals. Terminals and servers are connected via networks, typically N-ISDN lines. A diagram of this system is shown in Figure 2. The InterSpace server gathers all positional information from user terminals, keeps current user locations and directions, and redistributes this information to the terminals. User terminals use this information to generate their local view of the virtual world. The facial video server gathers users facial images and redistributes them to user terminals. User terminals use these video images to generate video avatars. The voice server controls voice communication, establishing a voice link between users and mixing users voices for group/conference talk. When avatars are separated by small distances, the corresponding users can speak to each other. The InterSpace server and facial video server are run on workstations. The voice server uses our special hardware. A user terminal consists of a high-end multi-media PC, network adapter (e.g., ISDN router), video camera, navigating device (e.g.,. mouse, joystick, or etc.) and video capture card. For voice communication, there are many options. To maintain quality, normal telephones are used for the CyberForum experiment.

Figure 1. Avatars

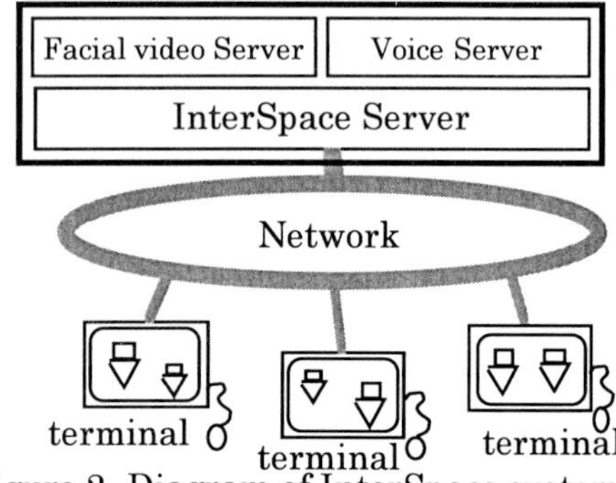

Figure 2. Diagram of InterSpace system

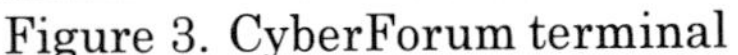
Figure 3. CyberForum terminal

Figure 4. Exhibition room

4. CYBERFORUM

We have produced CyberForum, an exhibition using a 3D virtual world. The Cyber Forum system combines both the InterSpace technology[4] and video-on-demand technology.[5] One of the user terminals is shown in Figure 3. In CyberForum, many exhibits are displayed on walls in the virtual world [Figure 4]. Participants at CyberForum are able to get information while walking through. Expositors are waiting for them, and explain their exhibits from remote locations. Participants can converse with them like in the real world. Moreover, participants may encounter each other unexpectedly and begin to talk. Individual actions are visible to other users, so others can look at their behavior.

Furthermore, the video-on-demand system delivers video data via the network. This system can offer a high-quality videos because it uses the MPEG video coding scheme. Moreover, lots of users can access the video server at the same time due to the use of high multiple readout technology.

By combining the InterSpace and Video-on-demand systems, people can not only converse using voice and facial images, but can also get detailed information about exhibits through video.

For this study, we designed new functions for the virtual forum

4.1 Display panel

Figure 5 shows a typical display panel. In a virtual world, texture resolution is limited for various reasons. Therefore, we designed display panels in which participants can turn over pages by clicking an arrow button.

Also, participants using this system for the first time usually can't move their avatar to the exact locations they wish, because moving in the virtual world requires a little training. Thus, we designed "foot marks." If users click these marks, they can move to the best viewing position for that display panel.

4.2 Table

When participants meet each other in the virtual world , it is difficult for them to move "face to face." When the number of people increases, it is especially difficult to line up in the best position. This is because the virtual world has sensitive movement capabilities which users can't control easily. We tried to add limitations to user movement for convenience. The table shown in Figure 6 is prepared in virtual space. When participants approach the foot marks drawn on the floor surrounding this table, they are drawn to the table. It designed so that people are located at the best "face to face" position for the conversation.

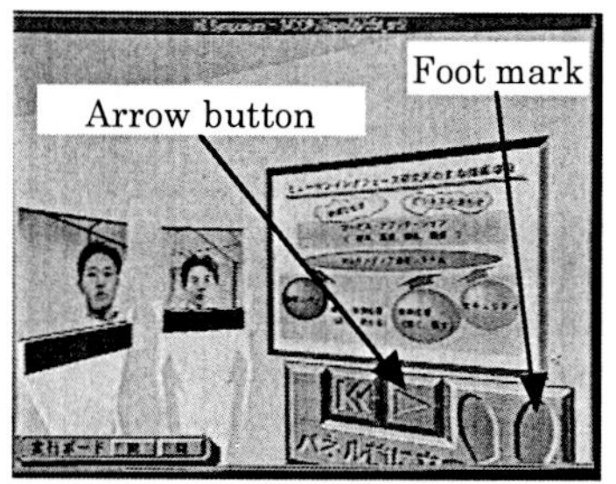

Figure 5. Display panel

4.3 Escort mode

In a real exhibition, the exhibits are explained well by a host guiding the guests. Escort mode realizes this situation in the virtual world. Using Escort mode, two or more participants are able to share an identical screen and walk around with host participants. To do this, the host sends his position information to others directly and controls others' positions to be the same as own position. Moreover, the back mirror appears in the upper part of the screen and displays facial images. So, it is possible to talk while seeing each others' faces.

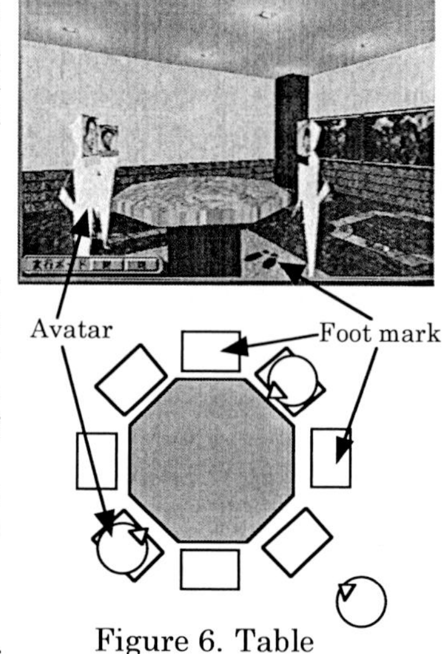

Figure 6. Table

5. CONCLUSION

We have introduced InterSpace and Cyber Forum. Inter-participants communication in the virtual forum was made smooth by the new functions achieved. The annoyances caused in a virtual world have been decreased for users doing free information stroll.

Functions to support inter-paerson communication in addition will be examined in future.

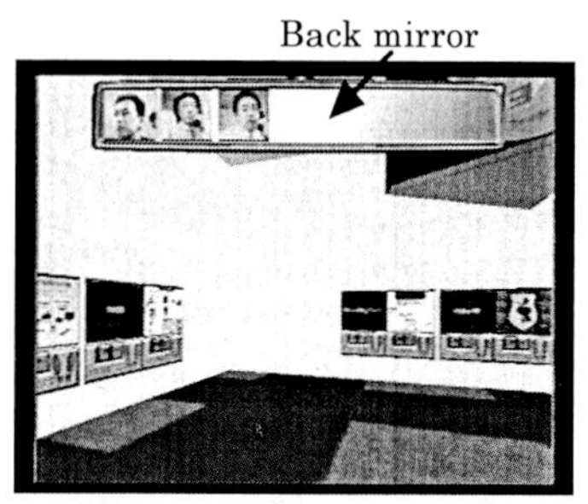

Figure 7. Escort mode

REFERENCES

1. Morningstar, C. and Farmer, F. R., "The Lessons of Lucasfilm's Habitat," in Cyberspace: First Steps, MIT Press, Cambridge, 1992.
2. Alluisi, E. A., "The Development of Technology for Collective Training: SIMNET a Case History," Human Factors, 33, 3, pp. 343-362, Jun. 1991.
3. Fahlen, L. E., Brown, C. G., Stahl, O. and Carlsson, C., "A Space Based Model for User Interaction in Shared Synthetic Environments", Proc. INTERCHI'93, pp. 43-48, Apr. 1993.
4. Sugawara,S. et al., "InterSpace: Networked Virtual World for Visual Communication", IEICE Trans. Inf. & Syst., Vol. E77-D, No. 12, pp. 1344-1349, Dec. 1994.
5. Kimiyama,H. et al., "System Architecture for Middle-Scale Video-on-Demando Service," Proc. IS&T/SPIE Electronic Imaging '96, Vol.2670, pp. 226-273, Feb. 1996.

User interface modeling for the design of administrative information systems

Magnus Lif

Uppsala University, Center for Human-Computer Studies, Lägerhyddvägen 18,
SE-752 37 Uppsala, Sweden.

This paper describes a method whose aim is to determine what features of the user's work that could later be directly applicable when designing the human-computer-interface. The method is really a complement to existing methodologies for information analysis. It presumes an existing, object-oriented data model. The modeling is performed in a user-centered manner with users working in collaborative sessions with designers and software engineers. The method is to be employed for in-house development projects in the administrative domain, where all users can be characterized as skilled and where the information system is used several hours per day.

1. INTRODUCTION

There are several methods for the analysis of the users' work that supplies *necessary* information for the design of the user interface. We have, however, observed that designers have particular difficulties utilizing the results of these analyses during the design process. Today's methods simply *do not give enough support* for the design decisions that have to be realized. We believe that some important aspects are missing in traditional information analysis. Making a design decision is undoubtedly a complex process. This process can, however, be enhanced if the designer has knowledge regarding the specific information the users require and *how* they use this information. We have previously presented a method for the analysis of information utilization where important aspects are captured through observation interviews with users while they perform their actual work [1]. In this paper, we present a method that identifies similar aspects during user-centered modeling sessions.

2. USER INTERFACE MODELING

With this method for user interface modeling, the users' work is analyzed in terms of what the designer needs to know to be able to make "optimal" design decisions. These aspects are identified during sessions led by a modeling leader, where users, designers and software engineers work in cooperation. It is especially useful when a user-centered modeling technique has been used to specify the data model.

User interface modeling is performed in four consecutive phases specifying *work tasks*, *workspaces*, needed *classes* and needed *attributes* and *operations*. In each phase the different aspects are captured during modeling sessions. The modeling leader must be able to lead the discussion without losing focus and allow all members of the group to contribute with their knowledge. A white board, slides or "post-it notes" can be used to guide and hold the discussion on track. The results of the modeling is documented in pre-defined forms.

An important concept in this method is the *work task*, defined as *a continuous moment of work, performed by one individual to reach a specific goal*. Such a work task usually involves information search, reading, writing and a judgment process terminated by a decision. For the user, it is important that all information and tools needed to perform a work task are simultaneously present on the screen [2]. It is also essential that the user can easily switch between tasks. To overcome some of these potential problems this method is based on a workspace metaphor [3]. Each workspace in the information system contains all information and tools needed to perform one or a few work tasks. It is important that a work task can be fully completed without having to switch between workspaces. With a design based on this metaphor, the user is given an overview of his/her work.

2.1. Specification of the work tasks

In the first phase the work tasks and the *actors* are identified. An actor represents a category of users that interacts with the system [4]. A user can play the role of one or several different actors and an actor can perform one or several work tasks.

An important cue when identifying the work tasks is the decision terminating the judgment process. Examples of such decisions are sending or not sending a form to a customer, or asking a colleague for consultation in a specific case.

The results from this phase is documented as a list of all work tasks performed by the potential end-users, and for each work task the corresponding actors.

2.2. Specification of the work spaces

Each actor performs tasks in one or a few workspaces. In each workspace, all information and tools needed to perform a work task should be available. It is usually possible to perform one or a few work tasks in a workspace.

When identifying the workspaces to be included in the user interface, the results from the previous phase is essential. The following "rules" may be useful:

- Work tasks that may have to be performed in parallel *should* be grouped in the same workspace
- Work tasks that naturally adhere *could* be grouped in the same workspace
- Work tasks that share the same informational needs *may* be grouped in the same workspace
- The same work task can be performed in several workspaces, if needed.

The results from this phase is documented as a list of all workspaces, and for each workspace the corresponding actors, and for each actor the work tasks performed in the workspace.

2.3. Specification of the classes

In each workspace, a set of objects has to be available. The classes in the data model corresponding to these objects are specified, and the approximate number of instances of each class that has to be available is also defined.

In the documentation, a class is represented with a box and each instance of a class is represented with a vertical line. See figure 1. One class may contain other classes. This documentation is made for every work task in each work space. Notice that no actual design of the interface has been performed yet.

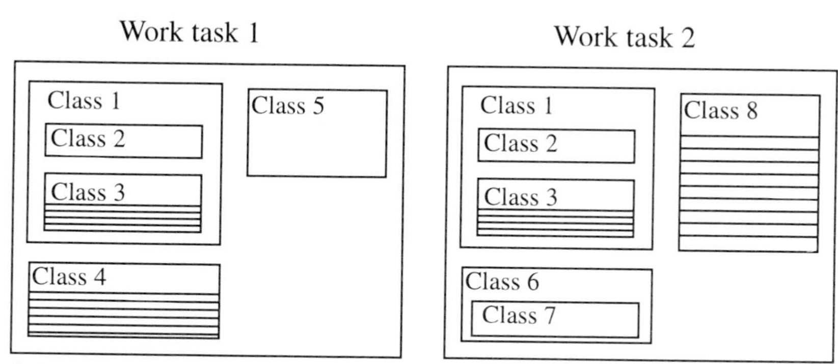

Figure 1. Documenting the classes needed for each work task in a work space.

2.4. Specification of attributes and operations

In this final phase, important features of the *attributes* (information) and *operations* (behavior) in each class are documented. The main purpose is to establish the attributes and operations that are needed when performing the different work tasks. The importance of an attribute is defined by giving each attribute a *priority*. Attributes with priority 1 contain important information that has to be visible on the screen at all times. Attributes with priority 2 are less important but should be visible at all times if possible. They may, however, be hidden if there is insufficient space on the screen. Priority X denotes "extra" important. These attributes can be emphasized using a visual cue (e.g., font, style, or shape). The *field type* defines whether the attribute can be edited or not. *Description* includes other relevant information that is not documented in the data model, such as possible status, default values etc.

The results from this phase is documented as two distinct lists: (a) for each class all attributes with their corresponding priority, field type and description are specified and (b) the operations needed in each class together with a brief description.

2.5. Design

The result of the user interface modeling is used as a basis for design decisions on how to display and arrange objects on the screen, how to choose information to display on the screen all the time and the less important information that can be hidden. It can also be used when deciding which information to emphasize and how to use visual cues.

3. RESULTS

The method has been used in development projects within the Swedish National Tax Board. In one of these projects, all members of the modeling group were given questionnaires to answer. In this group there were five end-users, one method leader, two software engineers and designers and one supervisor. The questions were divided into two parts. The first measured the ease-of-use and the second part measured how useful the documentation was when making the design decisions. A four-point scale was used for each question, where four indicated the highest rating. The mean values for ease-of-use and for practical utility were 2.94 ($V = 0.35$) and 4.00 ($V = 0$), respectively.

4. DISCUSSION

According the results the method was useful in the evaluated project. Even though the questionnaire measured only the subjects satisfaction with the method, there are reasons to believe that the method may have practical utility in the design decision process.

In our work within different development projects, we have observed that the identification of work tasks is indeed a complicated process. One reason for this is that the users are not always able to fully describe their work because they perform parts of it automatically. Another reason is the difficulties in predicting the future work situation.

In the future we will continue to evaluate this method within different projects and to pursue its further development.

REFERENCES

1. Gulliksen, J., Lif, M., Lind, M., Nygren, E., & Sandblad, B. (In press). Analysis of Information Utilization. International Journal of Human-Computer Interaction, Ablex Publishing Corporation, Norwood, New Jersey.

2. Lind, M. (1991). Effects of Sequential and Simultaneous Presentations of Information. CMD Report 19/91, Center for Human-Computer Studies, Uppsala University, Sweden. http://www.cmd.uu.se/papers/19

3. Henderson, A., & Card, S.K. (1986). Rooms: The Use of Multiple Virtual Workspaces to Reduce Space Contention in a Window-Based Graphical User Interface. ACM Transactions on Graphics, Vol. 5, No. 3, pp. 211-243.

4. Jacobsson, I., Cristerson, M., Jonsson, P., & Övergaard, G. (1993) Object-Oriented Software Engineering. A Use Case Driven Approach., Addison-Wesley Publishing Company, Wokingham, England

Direct interaction with flexible material models

Monica Bordegoni, Giancarlo Frugoli and Caterina Rizzi

KAEMaRT Group
Department of Industrial Engineering, University of Parma
Viale delle Scienze, 43100 Parma, Italy

This paper presents novel interaction techniques and tools to support direct manipulation of flexible material models. We discuss interaction techniques and tools at two different levels: definition of the object/environment model, and manipulation of the digital environment as it was real to modify and validate the model.

1. INTRODUCTION

Since 80's our group gained experience with CAD/CAM systems usability. From the beginning our attention has been focused on defining and implementing user interfaces suitable for CAD/CAM systems, addressing the problem from the end-user's point of view. Our objective was to provide the end-users with tools simulating their usual working environment. In the last decade, CAD/CAM vendors, having realized the key role of the user interface, put a lot of efforts in remaking their packages, while the end-users, being more used with software systems, improved their technical background and increased their level of competence with regard to software technologies. In such a context, we are currently facing the issue of human-computer interaction within systems for flexible materials modeling and behavior simulation. In this paper, we propose novel interaction techniques and tools based on advanced technologies, in order to enhance and simplify the interaction with flexible material models.

Some related works [1] can be found in literature, but most of them are oriented to animation. The objective of our research in the field of modeling and simulation of flexible materials is to define an interactive and graphical simulation environment for supporting the automation of industrial processes dealing with the manipulation of deformable products. Within this area, we have identified HCI issues at two different levels:

- interaction techniques and tools to create/define the object/environment model;
- interaction techniques to manipulate the digital environment as it was real, i.e. to modify and validate the model.

In the following, these issues will be discussed together with some examples.

2. INTERACTION TECHNIQUES FOR DEFINING THE MODEL

In this chapter we present interaction techniques for the representation of nonrigid materials making use of a discrete model, named *particle-based* model [2].

Using this model, a nonrigid object is represented as a set of mechanical elements (particles) and forces acting among them, representing the macro-behavior of the material (e.g. elasticity, plasticity). For example a piece of fabric can be modeled as a grid of particles with a distribution of forces, like bending, stretching and repelling (Figure 1b). Different materials (e.g., fabrics, dough, wires) are modeled using different:

- particle meshes;
- particle masses;
- sets of inter-particle forces.

In such a context, we have used a software prototype, named *SoftWorld* [3], that has been developed at our labs. It provides a graphical environment to simulate the handling machinery integrated with the dynamic and static behavior of the material.

The definition of a correct model, that is particle mesh and physical parameters, is the crucial point. In fact, the particle-based model is mathematically suitable for modeling flexible materials, but does not match the user's 'mental model' of the object. It is therefore necessary to develop interaction techniques and tools that allow the definition of flexible products, abstracting as much as possible from the underlying mathematical model.

We have identified three levels of interaction, that correspond to different classes of users and levels of abstraction. At the highest level, the end-user should be able to create a flexible object and simulate its behavior, without having to deal with complex mathematical details. To this end, we have planned to develop a *Material Database*, which will provide a set of predefined materials subdivided by classes of objects (e.g., fabrics, bulks). Each material type encapsulates the information about discretization rules and distribution of internal forces and constraints for the specific material, with values for related parameters.

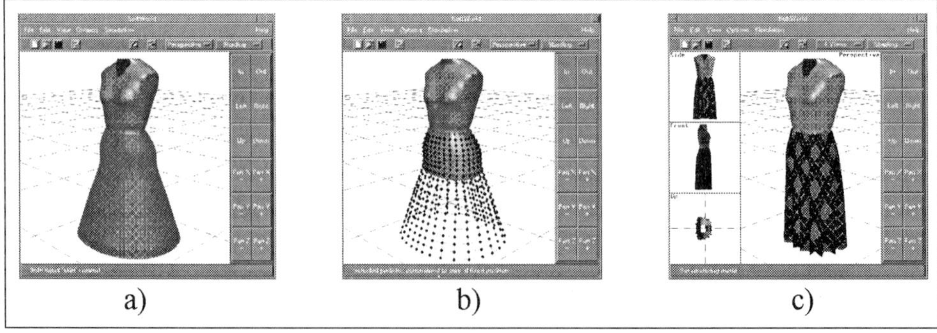

Figure 1. a) geometrical model of a skirt; b) distribution of particles and forces (based on the specified material); c) final result of the simulation.

Starting from a pure geometrical model of the object, the end-user has only to select a particular material from the Material Database, and the system automatically generates the mathematical model of the object. Figure 1 portrays the sequence of operations.

At an intermediate level, the user is more involved with the details of the model. A tool, named *Material Editor*, will be included in the system. Its aim is to allow the definition of new flexible materials to extend the Material Database. A new material can be defined from scratch or by modifying an existing one. The Material Editor makes available a set of rules for discretizing and creating a distribution of different types of forces.

The lowest level of interaction is provided by the library APIs, which allow defining new rules for the discretization and the distribution of internal forces. However, it requires high programming skills and a general view of the problem.

3. INTERACTIONS TECHNIQUES FOR MANIPULATING THE MODEL

The second HCI issue we have addressed concerns the interaction with the digital model of a flexible object. In fact, once the model is created, we want the user to be able to manipulate it, in order to validate and modify it. The user might want to test the material assigned to the model, to verify how close it is with respect to a real object made of that material.

A pure visual interaction with the model is not appropriate for achieving our goal. In fact, visual interaction only provides information about the graphical appearance of the object, such as color, pattern, shape, position and orientation in space, but does not provide any clue about its physical characteristics, such as its hardness. Moreover, the manipulation of a 3D object based only on visual perception of the object and on the use of common input devices is not natural for the user and does not convey enough information about the object.

What we propose then is the use of haptic interaction, besides the visual one, to overcome the mentioned issues. This choice is also supported by the fact that users are multi-sensory, so that it is sensible to provide interaction facilities that allow them to perceive digital objects through multiple senses [4]. We considered the possibility to interact with the digital model through two haptic devices (two Phantom devices [5]) the user wears on two fingers.

Several examples of interaction with the flexible materials simulator through the use of a haptic device can be thought of. Suppose we want to define a new class of materials for wires and cables. Once modeled the new material through the Material Editor, the user might check the bending property of the cable, or bundle of cables, by simply holding one of its ends with the fingers wearing the haptic devices. Figure 2 shows the simulation of wires behavior when grasped.

Another example concerns the possibility of modifying a pence of the model of a skirt by directly pulling up the upper part of the skirt. Last example refers to the capability of modifying a simulated object, such as a piece of clay, by molding its graphical and tactile appearance, instead of acting on its geometrical representation. All these examples refer to real problems met in several industrial processes, where validation through the use of digital mock-ups of the products is crucial in order to save time and money and adopt a time-to-market philosophy.

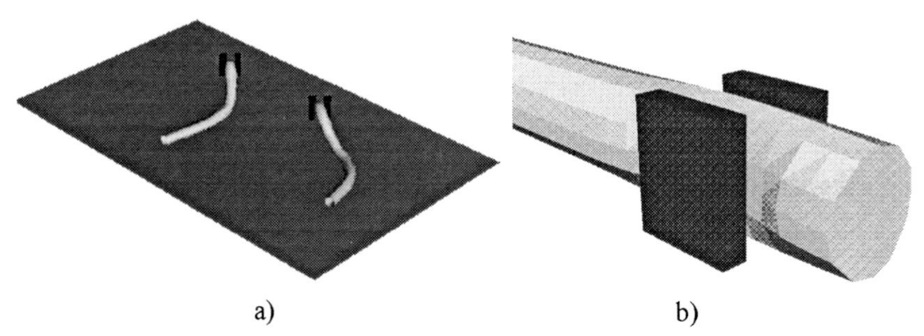

a) b)

Figure 2. a) simulation of wire grasping; b) a detail.

4. CONCLUSIONS

In this paper we have discussed some HCI issues regarding the definition and manipulation of nonrigid material models. At present we are experimenting them within the framework of two industrial projects funded by the European Union. The former is the Brite/Euram Project n. BE1564 SKILL-MART: it considers the skilled manipulation of wires and infusion bags by means of robotics systems. The latter is the Brite/Euram project n. BE3542 MASCOT: it deals with garment design process and aims at developing a 3D graphical environment to design new base garment.

First results have demonstrated that the interaction techniques and tools proposed enhance and simplify the interaction with flexible material models. This encouraged us to go on with the experimentation even if some technical problems are still open, such as computation time.

REFERENCES

1. Thalmann D., Magnenat Thalmann N., Werner H.M., User Interface for fashion design, in Proceedings of the 1993 International Conference on Computer Graphics (1993).
2. Witkin A., Particle system dynamics, in Proceedings of the 19th International Conference on Computer Graphics and Interactive Techniques (ACM Siggraph 92) Vol.II, (1992) 26–31.
3. Denti P., Dragoni P., Frugoli G., Rizzi C., SoftWorld: a system to simulate flexible products behaviour in industrial processes, in Proceedings of the 8th European Simulation Symposium (ESS 96), Vol.II, (1996) 235–239.
4. Bordegoni M., Enhancing human computer communication with multimodal interaction techniques, in Proceedings of the 1st International Conference on Applied Ergonomics (ICAE 96), (1996) 986–991.
5. Massie T.H., Initial haptic explorations with the Phantom: virtual touch through point interaction, Master thesis, Massachusetts Institute of Technology, (February 1996).

Reducing Operator Mental Load through Dynamic Icon Interfaces and Process Notice

Badi Boussoffara, Peter F. Elzer

Institute of Process- and Production Control Technology, Technical University of Clausthal, Julius - Albert -Strasse 6, D-38678 Clausthal-Zellerfeld, Germany.

This paper introduces a new form of process visualization, called Dynamic Icon Interface (DII). It supports direct perception and reduces operators´ mental load during information processing. To design DII neither the exact mathematical relationships between the process variables nor a complex preprocessing of process values is required. Furthermore a mnemotechnical aid called "Process Notice" will be discussed. This "Process Notice" allows operators to filter, to "pick" and to "pin up" relevant information while navigating in a complex information space.

1. INTRODUCTION

To support the exchange of information between the operator and the process and to facilitate the navigation in the complex process information space there exist many techniques for process visualization. Some of them are:

- Intelligent graphical interfaces based on P&I-Diagrams and rule-based support systems. One example is the GRADIENT [1] (GRAphical DIalogue ENvironmenT). Some important features of this interface are "rolling map" and "zooming" facilities.

- "Mutilevel Flow Models" (MFM) [2] that allow the visualization of mass and energy information flows at different levels of abstraction.

- "Ecological Interfaces" [3] that enable visualization of normally unvisible relationships between process variables.

In this paper the DII [4, 5] will be compared to other techniques of visualization. Furthermore the "Process Notice" will be discussed.

2. PROCESS VISUALIZATION THROUGH DYNAMIC ICONS INTERFACES

DII are based on dynamic icons (cf. figure 1) that allow direct perception of information of the behaviour of process variables. Since the icons directly reflect the behaviour of process variables, the operators can also perceive directly, whether a level in a tank, a temperature, a pressure, a gain of a valve etc. increases or decreases.

In order to build an interface, the icons will be connected to each other considering the functionality of the system. Figure 2 shows DII of a high pressure preheater which is part of a coal fired power plant.

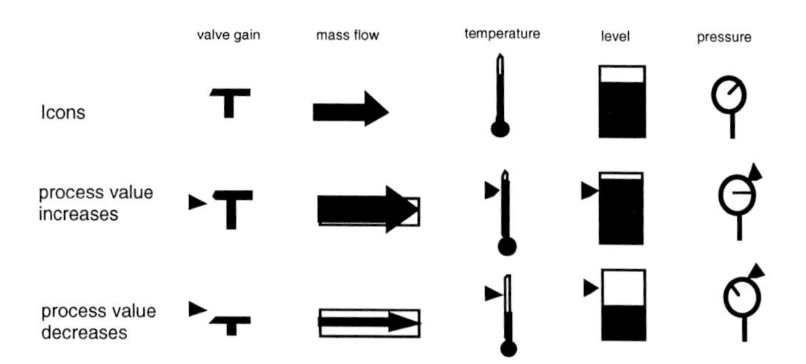

Figure 1. Some examples of Dynamic Icons

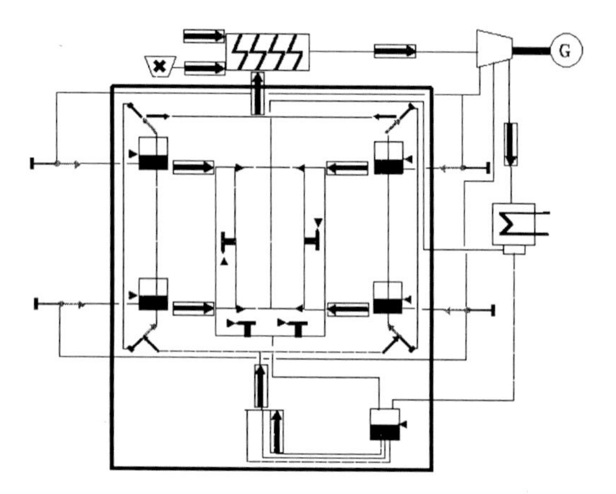

Figure 2. DII of the high pressure preheater in the normal state.

In order to reduce the amount of information presented to operators and to locate a disturbance an abstract overview describing the feedwater-steam-circulation provides operators with initial information about a disturbed subsystem. To every main subsystem a dot is assigned. The size of a dot indicates the location of a disturbance of the respective subsystem (cf. figure 3).

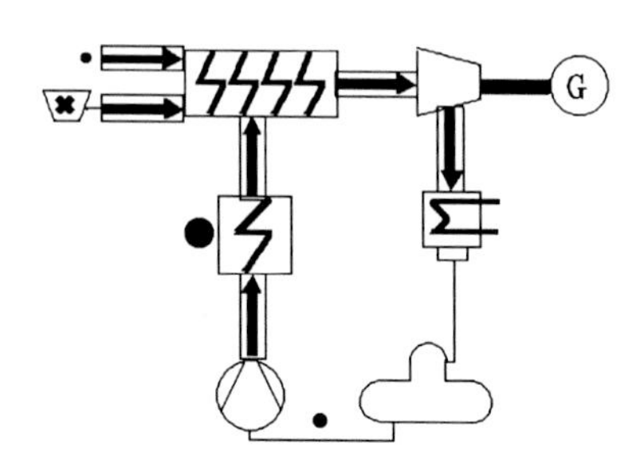

Figure 3. Abstract overview of the power plant showing a disturbed subsystem.

In order to get more detailled information about occuring disturbances operators can zoom into the respective subsystem. Figure 4 shows a detailed view of the disturbed high pressure preheater. It contains all the information allowing the current disturbance to be identified and evaluated (cf. figure 4).

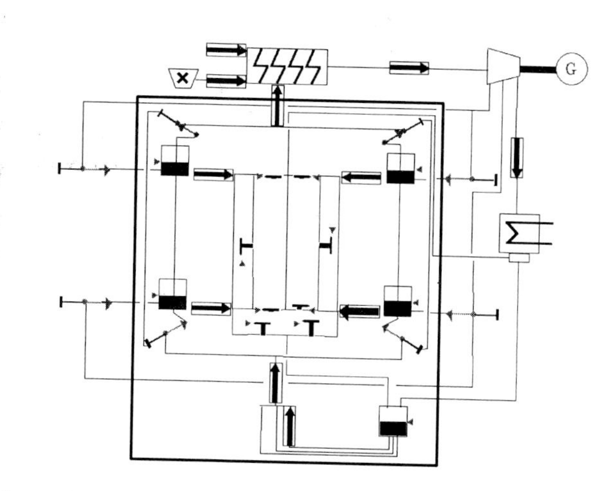

Figure 4. The DII of the high pressure preheater during a leakage

In comparison to "Ecological Interfaces" the exact mathematical relationships between process variables are not required for the construction of a DII. In comparison to the P&I-Diagrams and the MFM a "mental detour" can be avoided. A "mental detour" is given when operators firstly have to identify the deviating process values. Secondly, process values have to be assigned to their appropriate process components, and thirdly they have to be compared mentally with their corresponding setpoints. There are at least three mental operations that have to be done before operators can describe the state of a process. Using DII, mental operations will be reduced to a minimum (e.g. it is sufficient to detect deviating process values).

3 THE "PROCESS NOTICE"

In order to identify and to analyse a disturbance, the system offers operators an aid called "Process Notice" [4, 5]. It allows them to "pick" and to "pin up" relevant information while navigating through a complex information space. Such an aid is very helpful, because it supports operators during the search for relevant information and its processing. Furthermore the "Process Notice" relieves the operators' short term memory.

The "Process Notice" is based on polar diagrams [6]. In contrast to polar diagrams the advantage of the "Process Notice" is that it is not predefined by the designer. The "Process Notice" can be generated on-line by the operator [4, 5]. As operators have different views and levels of knowledge, the "Process Notice" is an important aid for any operator. Furthermore, the "Process Notice" facilitates the search for information in a complex information space, because during the analysis of disturbances the search path is not bounded, but very large (e.g. changes of displays). The "Process Notice" allows operators to memorize different information from different displays without redisplaying information of disturbed subsystems (e.g. graphical displays). The "Process Notice" avoids unnecessary switching between displays to get past and perhaps "forgotten" information. The "Process Notice" represents a mnemotechnical aid and has the features of a filter [4, 5]. Furthermore the "Process Notice" allows operators to define their own working and decision context.

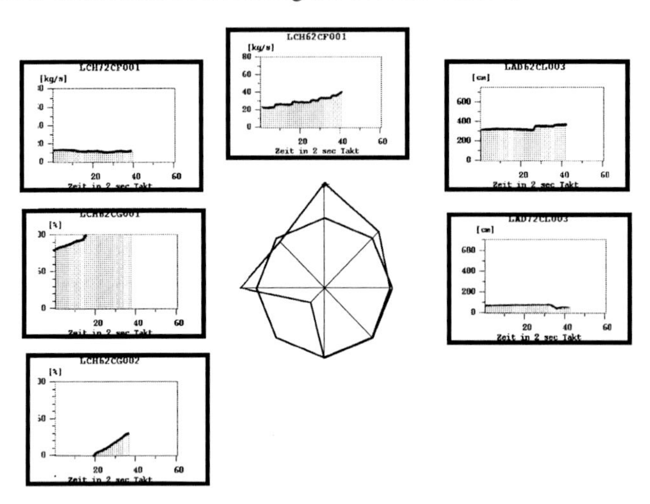

Figure 5. A possible set-up of the Process Notice.

4. CONCLUSION

A new form of process visualization has been introduced and compared to other techniques of visualization as well as a mnemotechnical aid. The combination of those supports operators to get the right information at the right time and facilitates the information processing during identification and evaluation of a process state.

REFERENCES

1. Elzer, P.; Weisang, C.; Zinser, K. (1989): Knowledge-Based System Support for Operator Tasks in S&C Environments. In: IEEE International Conference on Systems, Man and Cybernetics. Cambridge, Mass., USA.
2. Lind, M. (1991): Representations and Abstractions for Interface Design Using Multilevel Flow Modelling. In: Human Computer Interaction and Complex Systems, Eds. George R. S. Weir, J. L. Alty. Academic Press: Harcourt Brace Jovanovich, Publishers, pp. 223-243.
3. Vicente, K.J.; Christoffersen, K.; Pereklita, A. (1995): Supporting Operator Problem Solving Through Ecological Interface Design. In: IEEE Trans. On Systems, Man, And Cybernetics, Vol. 25, No. 4 , pp. 529-545.
4. Boussoffara, B. (1996): Ein Verfahren zur Bereitstellung von Betriebserfahrungen zur Unterstützung der Bediener technischer Anlagen. Dissertation. Technical University of Clausthal. Clausthal: Papierflieger Verlag.
5. Boussoffara, B.; P. Elzer (1997): A New Approach Supporting Operators to Evaluate Situations in Supervisory Control. In: 4th IFAC International Workshop on Algorithms and Architectures for Real-Time Control, Vilamoura, Portugal. (Preprints) pp. 223-228.
6. Woods, D. D.; Wise, J. A; Hanes, L. F (1981): An Evaluation of Nuclear Power Plant Safety Parameter Display Systems. In: 5th Annual Meeting of the Human Factors Society, pp. 1001-1014.

Information Retrieval Supported by Rich and Redundant Indices

Kenji SATOH, Susumu AKAMINE, Kazunori MURAKI, Akitoshi OKUMURA

Information Technology Research Labs, NEC Corporation, 1-1, Miyazaki 4-Chome, Miyamae-Ku, Kawasaki, Kanagawa 216 Japan

1. INTRODUCTION

The need for text sharing is becoming increasingly important as the popularity of the Internet increases. While some groupware supports local-area text sharing[1] and research has progressed in digital-library projects to put ordinary books online[2], pamphlets and reports are still commonly delivered to office in paper form, to be filed and kept on the shelves.

The OCR technology for scanning such pamphlets and reports seems to work very well for language based on a simple alphabet, but not for languages having many characters, such as Japanese, and current technologies for correcting OCR texts by dictionary-matching are less than entirely effective[3,4]. Correcting OCR errors by hand is, of course, both time-consuming and costly.

In fact, however, the bulk of such pamphlets and reports do not really need to be put online in such a form as can be edited or otherwise word-processed. They simply need to be readable for reference purposes. In such cases they can just be scanned and input as "text-image" objects, to be retrieved as needed and read as displayed. The only difficulty in such an approach is the difficulty of storing them in such a manner as permits them to be searched out effectively later. In this paper, I propose indexing methods to be used on such image data, and searching methods using the resulting indices.

2. INDEX METHODS FOR IMAGE DATA

We propose three different types of indexing methods on image data.

a) Indexing characters in the text as produced by OCR (and including possible errors).
b) Indexing layout information regarding figures, tables, etc.
c) Indexing strings found in text that has been produced while referring to the image data.

In type a), because OCR texts contain errors, some query expansion methods for dealing with errors will be helpful in searches. One example of this is a wildcard expansion for each character of a query term; “コンピュータ” (computer) → “ンピュータ+コ？ピュータ+コン？ュータ+コンピ？ータ+コンピュ？タ+コンピュー” (‘？’ will match any single character.) Another query expansion method is the replacement of the respective halves of two-character-combination queries with characters they closely resemble; “内蔵”(built-in) → “内威+内減+…+丙蔵+…” Query term expansion with a thesaurus is also effective, as well as query expansion methods used for information retrieval in non-OCR produced texts.

In type b), layout information includes the number of figures and/or tables on a page, the location on the page of figures and/or tables (left-side or right-side, top or bottom), the type of column format, etc. These indices are effective when a user recalls a rough image of a page of a pamphlet or report.

In type c), this index grows dynamically whenever the stored image data is referred to in the production of new texts.

These indices are complementary; i.e. their use in combination results in more successful retrievals than does the use of any one of them alone.

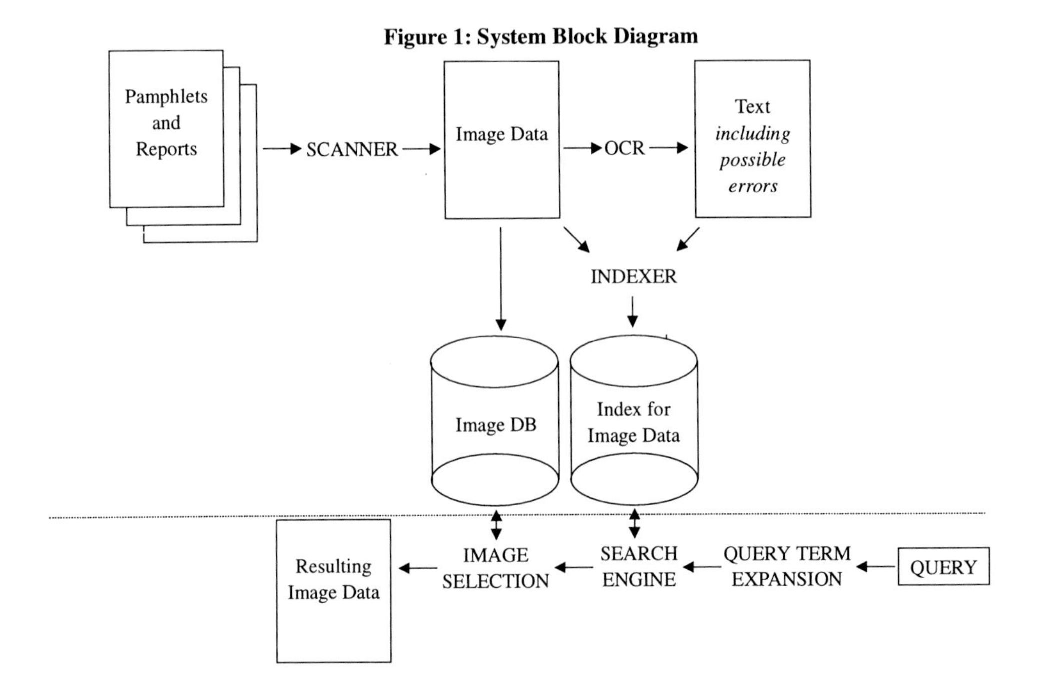

Figure 1: System Block Diagram

3. EVALUATION

We evaluated the effects of query term expansions with type a) indices. We used two types of image data: 1) 188 pages of uncorrected OCR-produced text, and 2) 25 pages each containing both OCR-produced text and newly produced error-free text. The probability of correct choice was calculated for the first type; precision and recall were evaluated for the second.

We chose two words for keywords from each page of a text-image to be used in separate queries, one a 'two-character word' and the other a 'three-or-more-character word.' Character replacement was applied only to 'two-character words', while wildcard expansion was applied to each character in the 'three-or-more-character words.'

Figure 1 shows block diagram of the system used for this evaluation.

4. DISCUSSION

The results of the evaluation are shown in Tables 1 and 2. The probability of 'two-character word' retrieval is higher than the probability of that for a 'three-or-more-character word,' i.e. the probability that one of the characters in a 'three-or-more-character word' has been misread by the OCR is higher than that one in a 'two-character word' has been misread.

Character replacement increased retrieval accuracy by only 4%, wildcard expansion increased it by 20%. Character replacement might be more effective if conducted with more than the top 10 characters that we used, since characters in OCR texts have a greater number of error patterns than 10. Neither of these two expansion methods deals with the common 'insert character error' and 'delete character error' problems of OCRs.

Table1: Probability of correct choice for uncorrected OCR-produced text

	Two-character word *Character replacement*	**Three-or-more character word** *Wildcard expansion*
No query term expansion	0.91	0.77
With query term expansion	0.93	0.86

Table2: Precision / recall for containing both OCR-produced and error-free text

	Two-character word *Character replacement*	**Three-or-more-character word** *Wildcard expansion*
No query term expansion	1.00/0.77	1.00/0.68
With query term expansion	0.96/0.81	1.00/0.88

5. CONCLUSION

We have proposed indexing methods for use with OCR images and methods for searching out those indexed images. With these methods, such texts as pamphlets and reports can easily be input into computers and searched for when they are needed, without any requirement for OCR text correction. We have also evaluated the influence of query term expansion methods on the probability for erroneous OCR text searches, and confirmed that wildcard expansion is effective.

We plan to implement all proposed indexing methods in a system, and to evaluate them in actual use.

REFERENCES

[1] Peter J. Nurnberg, Richard Furuta, John Leggett, Catherine C. Marshall, Frank M Shipmann III: Digital Libraries: Issues and Architectures, Digital Libraries '95, June 1995.

[2] Kenji SATOH and Kazunori MURAKI: Penstation for Idea Processing, Natural Language Processing Pacific Rim Symposium (NLPRS '93), Dec 1993.

[3] MARUKAWA, FUJISAWA, SHIMA: A Study for Information Retrieval Methods Which Permit Ambiguity of Recognition Outputs, Journal of IEICE, Vol.J79-D-2, No.5, 1996.

[4] USA, TANAKA: Realization of String Search Module for Text Image Data contains Japanese Characters, IPSJ Information Media 19-1, 1996.

Retrieving and Transmitting Real World Oriented Information by Simple Interaction through Augmented Electrical Stick

Soichiro Iga and Michiaki Yasumura [a]

[a]Keio University, Graduate School of Media and Governance
5322 Endo, Fujisawa, Kanagawa, 252 JAPAN
e-mail: igaiga@sfc.keio.ac.jp

We propose a novel system called "Stick World" which can transmit and retrieve information through the electrical stick device. By tapping two pads attached on the stick shaped device, the users can easily retrieve and transmit information in the real world situations such as certain location and time.

1. INTRODUCTION

We usually communicate with others in various situations. For instance, you may ask the way to the station, you could talk about the movie you saw last night, and so on. And it is useful to exchange views with others in many cases.

On the other hand, as a personal computer getting smaller in its size, it can be utilized not only in the office, but also in the real space. Recent researches on real world oriented interface aim at the use of a personal computer in the real world [1][5]. Although, the users can retrieve useful information related to the real world such as location and time, most researches focus on information retrieval only and information transmission is not taken into consideration.

In this paper, we propose a novel system called "Stick World" which can both transmit and retrieve real world information only by simple interaction with an electrical stick device. By tapping the stick, the user can exchange information with others in a certain real world situation. Information archives which are attached to the real space will then be accessed by this device.

2. STICK WORLD PROTOTYPE SYSTEM

We have designed a prototype system called "Stick World" which enables the retrieving and transmitting of real world oriented computer information by simple interaction which is to tap the stick device in the real space.

2.1. Merits of Stick World

The merit of the stick shaped interface is that it can handle information by simple interaction of tapping on the real world objects. By this interaction metaphor, it is easy for the users to handle interface and to understand the relation between real world

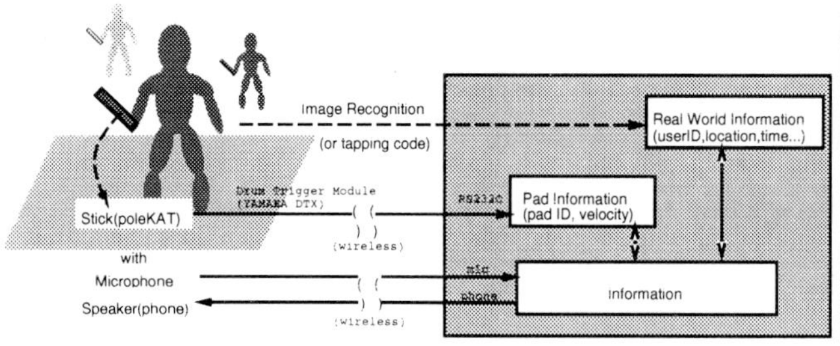

(a) This figure shows the system architecture of the system component. The user taps the stick device on the real space for receiving real world oriented information. The user then can transmit original real world oriented information also by tapping the stick pad. Here, the real world oriented information archives has been produced.

(b) Two pads are attached on the stick device Upper pad is for retrieving and lower is for transmitting. The user can easily handle information.

Figure 1. Stick World System Architecture

situation and the retrieved and transmitted information. And the user can feel the actual feedback by applying stick to the real world object.

2.2. System Architecture

The system architecture is appeared in Fig.1(a). Stick part is constructed by musical drum trigger pad (*KAT poleKAT*), a microphone, and a headphone or a small speaker (see Fig.1(b)). The trigger signal of the drum pad is sent to the drum trigger module (*Yamaha DTX*). There are two pads on a stick. One is for retrieving information and the other is for transmitting information. A signal from the stick trigger and the user's speech is sent to the host computer through wireless device, and corresponding information which is related to the real world situation is sent back to the user by sound.

2.3. Recognizing the real world situation

The user's location is recognized by two ways; by tapping the stick device and by image recognition technique. By assigning unique numbers for each recognition point in the real space, the user can input the number by hitting the device. This technique is used when the device is used in wide area. Image recognition technique can also be used for detecting user's location. User's location is captured by two movable cameras(*Canon VC-C1*×2), and the position in the limited space can be recognized. The recognition configuration is adjustable according to the size of area.

2.4. Handling multiple layered information

As the stick is structured by the musical drum trigger, it can recognize the strength of the user's tap. We mapped this hit strength function to handle layered information. The user can get rough information by a soft tap, and obtain detailed information by a strong

(a) Sharing information in real space: the user can transmit information in the real space, and the other user can retrieve that information while in the same real world situation.

(b) Exchanging information between multiple users: the users can exchange their information database by simply tapping the stick each other.

(c) Seamless interaction with WWW: information handled by Stick World system can also be obtained through the Web browser.

Figure 2. "Stick World" system in use

tap. The system can recognize about 64 steps in current device.

3. APPLICATIONS

- **Sharing information in real space:** The user can attach information on the real world space by tapping the system in the real world. Then the transmitted information can be shared by other users (see Fig.2(a)).
- **Interaction between multiple users:** When multiple users are in the same real world situation, the users can exchange information by manipulating system each other (see Fig.2(b)).
- **Seamless interaction with World Wide Web:** Information which is transmitted through the system can be obtained in WWW environment(see Fig.2(c)). Real world information can also be handled through the network.

4. EVALUATION AND DISCUSSIONS

We have conducted a preliminary evaluation by a simple information retrieval and transmitting application. The system has been tested by 10 users.

Recognition rate of detecting the location of the user's position by tapping the stick device was about 90 % in wireless configuration and it can be used practically. For image recognition, the user's position is quantized in about 6×6 and recognition rate is about 70 %

For retrieving and transmitting multi-layered information by tapping, although the system can handle about 64 steps, the users were able to handle about 3 steps. Practicing process may be needed in handling information by the strength of the tap before using the system.

The users' opinion indicates that our system provides the simple and easy way of handling real world oriented information. Furthur, it was easy for the users to understand the retrieving and transmitting metaphor which uses two pads on the stick device. Current stick device is rather heavy (about 2kg in weight), so that much lighter device should be to take place. On going research to apply GPS for transmitting part will enable our system to broaden the possible size of area the system can handle.

5. RELATED WORKS

NaviCam provides a visual information retrieval by the see-through handy monitor [5]. With our system, the user can transmit and retrieve information by very simple interaction and information can be exchanged by multiple users.

Various kinds of small computer and wearable interfaces are developed [1][6]. Our system can be used as an interface for these mobile computers. Visual information is able to be handled by combining these wearable interfaces in our future works.

6. CONCLUSION

We have proposed the system called "Stick World" which enables the users to retrieve and transmit real world oriented information only by the simple interaction. We have conducted a preliminary evaluation and obtained a result which indicates that our system can be used in various application areas such as sharing information with people, entertainment, and aid for disabled users.

REFERENCES

1. Feiner, S., Macintyre, B., Seligmann, D., Resnick, M., Weiser, M. and Wellner, P., Computer Augmented Environments: Back to the Real World, In *Comm. ACM*, Vol.36, No.7, pp.52-97, 1993.
2. Iga, S., Yasumura, M., Interacting with Real Objects: Real Object Interface and Transferred Object Interface, In Proceedings of the 6th International Conference on Human-Computer Interaction, (20B)pp.231-236, 1995.
3. Iga, S. and Yasumura, M., Stick World: Retrieving and Transmitting Real World Oriented Information through Electrical Stick, In *Proceedings of Human Interface Symposium'96 (HIS96)*, 1996. (In Japanese)
4. Iga, S., Yasumura, M., Real Object Remote Controller: Applying Real World Affordance to Computer Interfaces, In Proceedings of the First International Conference on Multimodal Interface(ICMI'96), pp.18-22, 1996.
5. Rekimoto, J. and Nagao, K., The World through the Computer: Computer Augmented Interaction with Real World Environments, In *Proceedings of Eighth Annual Symposium on User Interface Software and Technology (UIST'95)*, pp.29-36, ACM, 1995.
6. Starner, T., Rhodes, B., Healey, J., Russel, K.B., Levine, J., Pentland, A., Wearable Computing and Augmented Reality, In *MIT Media Lab Technical Report* No.355, Nov.1995.

Evaluation of a Theory-Based Display Guide

Anker Helms Jørgensen[a] and Jon May[b]

[a]Psychological Laboratory, Copenhagen University, Njalsgade 88, DK-2300 Copenhagen S, Denmark. E-mail: anker@axp.psl.ku.dk.

[b]Jon May, Department of Psychology, University of Sheffield, Western Bank, Sheffield S10 2TP, United Kingdom. E-mail: jon.may@sheffield.ac.uk.

1. INTRODUCTION

As graphical user interfaces are becoming increasingly widespread and complex, the accompanying need for guidance for designers has increased. A large number of user interface design guides do exist. Some of these are general and cover all aspects of user interface design (Mayhew, 1992; Minasi, 1994), some address the interface from a graphics point of view (Wagner, 1988; Mullet and Sano, 1995), some are specific platform style guides (Anon, 1995), while others cover a range of platforms (Marcus et al., 1995). These guides are indispensable and meet user interface designers' needs in various ways. However, none of them provide an underlying cognitive theory that offers a unifying set of concepts and principles that span varities of interface features, such as representation (e.g. iconic, textual) and modality (e.g. acoustic, visual) and cognitive features such as attention and recognition.

An interface design guide "Structuring Interfaces: A Psychological Guide" (May et al., 1996) - in the following the *Guide* - has been developed in order fill in this gap. The guide is intended to support user interface designers to use psychological principles to construct the appearance of computer interface objects, their arrangements on the display, their behaviour, and their relationship to the users' tasks. The Guide is based on the cognitive architecture ICS: Interactive Cognitive Subsystems developed by Barnard (1987). ICS encompasses nine interacting subsystems spanning from perception over various internal representations to bodily expressions. This paper reports an evaluation study of the Guide, where a class of HCI students participated. The purpose of the study was firstly to explore the feasibility of theory-based design guidance for user interface design and secondly to enhance the Guide and the way it is taught.

2. THE DISPLAY GUIDE

The Guide comes as a 100+ page document organized in eight chapters. First the structure of visual scenes are introduced as a basis for two principal techniques: Structure Diagrams (that help decompose visual scenes) and Transition Path Diagrams (that help analyze shifts of attention). Two key concepts in attention are then introduced (pragmatic and psychological subject). Next the Guide describes the cognitive architecture ICS, basic concepts in visual perception (e.g. groupings, proximity, collocation) and the interactions between visual search, visual structure, and task knowledge. Finally, the Guide presents dynamic changes in visual structure and how sound and vision blend in multimodal interfaces. The Guide is abundantly illustrated and presents numerous graphical and visual examples. In addition, every chapter concludes with a set of exercises.

3. EVALUATION STUDY

The evaluation study took place in an educational setting at the Psychological Laboratory at Copenhagen University. Fifteen graduate computer science and psychology students doing a three-term HCI-course run by Anker Helms Jørgensen participated. Eleven of the fifteen participants had professional experience in software design (average 3.5 years) and user interface design (average 2.1 years). Three of the participants worked individually and the remaining twelve worked in five groups.

The study took place over a 4-week period in the autumn 1996. Jon May gave the students four 2-hour multimedia-supported lectures in English over two weeks. Ideally the study should have compared teaching outcomes of the Guide to those of other teaching materials or textbooks, e.g. (Mayhew, 1992) or (Mullet and Sano, 1995). This was, however, not possible due to the teaching setting constraints.

The Guide had been tested in earlier pilot tests, focussing on the comprehensibility of the exposition and how the text supported the exercises (Aboulafia et al, 1995). In the present study we intended to asses the Guide based on real design problems and not only on the design problems in the Guide. Therefore, we asked the participants before the 4-week period to collect real design problems. This enabled us to relate the scope and contents of the Guide to the participants' initial perceptions of problems in user interface. All students handed in design problems pertinent to the scope and contents of the Guide, consisting typically of 2-3 pages of text accompanied by one or two screen shots. These problems were distributed among the participants and used as the main assignment.

We collected five types af data: time records, answers to exercises, answers to the main assignment, ratings and questionnaire answers. These types are described briefly in the following. We asked the particpants to keep a record of the time spent; on average the individual participant spent 31 hours in total during the four weeks. The participants did a number of exercises from the Guide, drilling basic techniques and concepts. These were discussed in class and also handed in and later marked by Jon May. In the principal assignment runnning over two weeks, the participants resolved the seven user interface design problems. Their answers were marked by Jon May. We also collected ratings of the usability and utility of the techniques and concepts of the Guide in connection with the main assignment. Finally, we asked the participants to fill in a questionnaire on five themes: utility and usability of the Guide; the setting in the HCI course; the use of English in the teaching and in writing; the course of the teaching; and the exercises and the assignment.

4. RESULTS, DISCUSSION AND CONCLUSIONS

In this section we will give a brief presentation of the results of the study. As space doesn't allow lengthy presentation of the qualitative data gathered in the main assignment and as these analyses are not completed, we will only summarize these results and focus on the results from the other data sources.

4.1 Exercises and main assignment

Assimilating the basic concepts and techniques of the Guide was a prerequistite for a proper evaluation of the Guide based on the real design problems. We found that almost all the participants did the exercises with few and minor errors. Thus the basic concepts and techniques were assimilated satisfactorily.

As the design problems handed in by the participants was the main evaluative basis in the study, Jon May analyzed them, thereby establishing the "correct" way to handle the design problems. This firstly encompassed an identification of the design problems at hand and secondly a specification of appropriate use of the two types of diagrams, appropriate use of concepts from the underlying cognitive architecture ICS, etc. This analysis was later used as a key to the subsequent marking of the participants' answers to the assignments. We should note in passing that ideally the marking should have been performed by independant raters. However, as the Guide is novel it was not possible to find collegues sufficiently knowledgeable of the Guide.

The marking resulted in two answers being excellent, four being good, and two being satisfactory. The excellent answers were featured by in-depth analyses, backed by comprehensive argumentation using the concepts from the Guide and competent application of the techniques. The good answers produced good points and analyses, but in some cases stopped short of using a relevant technique or failed to apply concepts to aspects of the design problem where they could have been helpful. The satisfactory answers provided relevant analyses but failed in many cases to introduce and apply relevant concepts and techniques.

4.2 Ratings and questionnaire

The ratings of the utility of the Guide centered on four aspects of the main assignment. The answers were given on a Likert scale from 1 *(obvious, easy, or good)* to 5 *(unclear, hard or bad)*. The first aspect was how obvious the causes of the usability problems were. This is important as it is a precondition for letting the Guide "go to work" on the design problem. This was assessed fairly easy: the mean of the ratings was 1.6 (s.d. 0.95). The second aspect was how easy it would be to find a resolution of the problem without using any method or technique. The mean was 2.4 (s.d. 1.09); thus it wasn't too difficult without the Guide. The third and fourth aspect addressed the utility and usability of the concepts and techniques. The means were here 2.2 (s.d. 1.04) and 3.0 (s.d. 1.19), respectively. This witnesses some experienced difficulties with the Guide upon the part of the participants.

The final questionnaire addressed the evaluation study of the Guide as embedded in the HCI tecahing setting. The answers were given on a Likert scale from 1 *(disagree completely)* to 5 *(agree completely)*. The questionnaire focussed on five areas as follows.

First, teaching and writing in English was fine: average 4.5 (s.d. 0.87). This is an important point for Danish participants whose first language is not English.

Second, the setting in the HCI course was fine (workload, change, relevance of Guide): average 4.2 (s.d. 0.85);. Thus the Guide and teaching fitted well in the HCI course.

Third, the utility and usability of the Guide was appropriate: average 3.9 (s.d. 0.87). Thus the participants found the outcome of the teaching and their work with the exercises and main assignment worthwhile and relevant to their work.

Fourth, the teaching was appropriate: average 3.7 (s.d. 0.83) covering dialogue with Jon May, sufficient and timely feedback, use of illustrative examples, appropriate use of illustrations, and appropriate use of equipment.

Fifth, the exercises and assignments were acceptable: average 3.2 (s.d. 1.25). This measure has a fairly large standard variation compared to the other areas. It covers the two lowest scores: How easy it was to find a design problem for the main assignment: average 2.9 (s.d. 1.44) and how clear it was to apply the Guide to the design problems: average 2.7 (s.d. 1.28). The former probably mirrors the uncertainty experienced by the participants in trying to find appropriate design problems based on our written instructions. These were deliberately phrased in general terms in order to avoid positive bias towards the concepts and techniques of the Guide. Nevertheless all the design problems handed in were pertinent to the scope and contents of the Guide. As to the latter (how clear it was to apply the Guide to the design problems), this probably reflects the participants' uncertainty in applying the concepts and techniques to the design problems. It may also well cover the participants' serious difficulties in deciding on the scope and applicability of the various concepts and techniques to be applied. Nevertheless, as mentioned above, the marking of their answers showed that they all had applied the concepts and techniques satisfactorily or better.

4.3 Conclusion

In conclusion, the Guide seems a viable approach to theory-based design guidance for user interface design as most of the participants were able to draw on the underlying cognitive architecture in enriching the analyses of the design problems and their resolution.

REFERENCES

Aboulafia, A., Jørgensen, A. H., May, J., Scott, S. and Barnard P. Transfer and Assay of the guide 'Designing Displays': A preliminary Report. ESPRIT Basic Research Action 7040: AMODEUS II Report TA/WP 35, 1995.

Anon: The Windows Interface Guidelines for Software Design, Microsoft Press, 1995.

Barnard, P.: Cognitive Resources and the Learning of Human-Computer Dialogs. In Carroll, J.M.: Interfacing Thoughts - Cognitive Aspects of Human-Computer Interaction. MIT Press, 1987, pp. 112-157.

Marcus, A., Smilonich, N., and Thompson, L.: The Cross-GUI Handbook. Addison-Wesley, 1995.

May, J., Scott, S., and Barnard, P. Structuring Interfaces: A Psychological Guide, 3rd ed., Dept. of Psychology, University of Sheffield, 1996.

Mayhew, D.J.: Principles and Guidelines in Software User Interface Design, Prentice-Hall, 1992.

Minasi, M.: Secrets of Effective GUI Design. SYBEX, 1994.

Mullet, K. og Sano, D.: Designing Visual Interfaces, SunSoft Press, 1995.

Wagner, E.: The Computer Display Designer's Handbook. Chartwell-Bratt, 1988.

Potential key stress factors among academic library assistants

Anne Morris and Sarah Holmes

Department of Information and Library Studies, Loughborough University, Loughborough, Leics, LE11 3TU, England.

Although the cost of stress can be high for both individuals and organizations, surprisingly little research has been undertaken to determine the possible effects of stress on library staff. The research described here aimed to fill this gap by identifying potential sources of stress among library assistants in three academic libraries of varying sizes. The potential stressors at work found were primarily work overload, lack of adequate emergency back-up procedures, insufficient pay levels and lack of promotion opportunity. The results disproved the common belief that stress only effects people in important, highly-paid jobs.

1. INTRODUCTION

Stress is the bodies' biochemical response to a threatening stimulus. Different stimuli will cause people to react differently; people may blush, jump or hide the reaction under a confident mask, but the biochemical response is the same - an increase in heartbeat and breathing rates, increased secretion of stomach acid and the release of certain hormones. An increased flow of adrenalin and a rise in cholesterol and blood sugar cause a surge of energy to the muscles, whilst the stomach becomes inactive. Emotionally, frustration, anger, excitement or anxiety may be experienced. The experience of stress through work is 'associated with exposure to particular conditions of work, both physical and psychosocial, and workers' realization that they are having difficulty in coping with important aspects of their work situation', (Cox, 1993).

The cost of stress at work can be high for both the individual and the organization. A certain degree of stress is needed so that apathy and boredom do not set in and in order that stimulation and challenge are maximized. However, if the negative aspects of stress becomes prevalent, then detrimental effects on physical and psychological health will come into play, with symptoms such as high blood pressure, stomach ulcers, disinterest and irritability, to name but a few. Such effects could mean a decline in productivity, increased absenteeism or a high turnover of labour for the organization.

Over the last ten years the library profession has changed drastically. Academic librarians have had to develop strategies to cope with an increase in the demand and range of services they offer together with major technological changes and budget reductions. Student numbers have increased rapidly as have expectations of service provision and accountability. Access to automated circulation systems, CD-ROM technology and a variety of electronic media, including the Internet, and e-mail are now expected by most students and staff. Many academic libraries have undergone restructuring to adapt to the change in size, the type of service provision and to technological advances. All this change has resulted in a number of potential sources of stress among library staff.

Surprisingly little research has been carried out on stress and its possible effect on library staff. Only one major study has been carried out in Great Britain (Hodges, 1990) and this examined stress levels among middle managers in public libraries. This study concluded that the main sources of stress were: feeling undervalued in the library organization, feeling the library service is undervalued, increasing competitiveness among staff, bureaucratic inertia, poor management, low pay, lack of communication and having to deal with changes both inside the library and inside local government. Some studies have been conducted in the USA but these have been predominately on the subject of reference librarians and have investigated burnout rather than stress. No research to-date has investigated the potential key stress factors among low paid academic library assistants. The research described in this paper aimed to fill this gap.

2. METHODOLOGY OF RESEARCH

Questionnaires were distributed to all library assistants (121) at three academic libraries representing different sized establishments. The questionnaire focused on stressors which could be affecting the individual at work. It was divided into three sections: personality factors, working environment and conditions, and the home environment.

The main purpose of the personality section, which largely comprised attitude statements, was to establish whether library assistants display Type 'A' or Type 'B' stress responses - Type 'A' responses being a potential endogenic source of stress for the individual concerned. Previous studies have very rarely dealt with this aspect despite their constant portrayal of library assistants in the media as quiet, 'unstressed' individuals. As Neville (1981) states, "every individual brings to his job a package of stressors, and the distribution or degree of intensity of each determines the ability to cope with stressful conditions". Stress is not an automatic outcome of an organizational set-up and Neville emphasized the need to look at each individual personality to determine whether that person will perceive a particular environmental stressor as a cause of stimulation (Eustress) or of stress (Distress).

The section on the working environment and conditions concentrated on factors intrinsic to the job such as work overload/underload and contact with information technology, interpersonal relationships at work with colleagues and supervisors, their role at work in terms of ambiguity, conflict and responsibilities, job prospects particularly promotion appointments, status, and levels of pay, and factors regarding management such as communication and feedback. Questions were also asked about environmental conditions in the workplace.

The section dealing with the home environment was short but aimed to identify any pressures outside work that could contribute to their current stress level. Personal details about each respondent were also collected such as age, time in service and current position within the library. Anonymity was guaranteed.

3. RESULTS AND DISCUSSION

A response rate of approximately 60% was obtained across the three sites. The library assistants who responded varied in age between 17-60 years. Site A tended to employ younger assistants than the other two sites but, overall, there was a fair distribution of staff throughout the age ranges. Most of the staff had worked at Sites A and B longer than 6 years, 53% and 70% respectively, but Site C had a high turnover of staff; only 34% had worked there longer than 6 years.

3.1. Personality

People with Type 'A' personality are much more susceptible to stress (Friedman & Rosenman, 1974). Type 'A' traits include the inability to relax, propensity for getting headaches and colds, feeling like there is not enough time, perfectionism, the need to have work recognized, impatience, importance of work, hard driving disposition, achievement, tendency to hide feelings, competitiveness, desire to work through breaks, career-mindedness, intolerance, ambition and so on. The converse is true for people with a Type 'B' personality who tend to be much more relaxed individuals. In general, most people will fall somewhere between the two extremes (HMSO, 1987), and this is what was found to be true in the sample taken.

The results showed that absolute Type 'A' and Type 'B' personalities were not prevalent among the library assistants. Most fell in between the two types and displayed traits from both. There was a small bias towards Type 'A' characteristics at Site A but it was insufficient to suggest that any one library had a set of workers who would be more prone to stress than the others. Consequently, there are no grounds to suggest that library assistants can be stereotyped or that their personality is a key factor in attracting stress.

3.2. Stressors in the work environment

The responses from Site A, the smallest library serving the largest student community with less library assistants than its fellow sites, indicated a more stressful working environment. Volume of work, having to do more than one job at a time and too much physical work were regarded as the key potential stressors. These were closely followed by inadequate supervisor communication and feedback. Library assistants at this site appeared to be struggling to cope with the work expected of them with little chance for communication.

Fifty percent of the medium sized library respondents at Site B were not happy to remain library assistants and there were significant levels who found automation and learning new skills stressful. This could imply that library assistants here found the constant need to evolve and adapt a major cause of stress.

Site C, the largest library surveyed, had home factors and work factors as potential stressors. With 62.1% suffering from outside pressures, it could be the case that these exacerbated work pressures. A key potential stressor at this site was that of insufficient responsibilities, however, it also scored the highest stress score for 'having to do more than one job at a time'. Perhaps the implication here is that a key stressor for library assistants at this site is having to do lots of possibly repetitive, unstimulating jobs which they do not perceive as commensurate with their experience, qualifications or skills.

Key potential stressors at work which proved common throughout the sample sites were that of work overload both quantitative and qualitative, lack of adequate emergency back-up procedures, insufficient pay levels and the lack of promotion opportunity. The physical conditions in which they all had to work were also pinpointed as potential stressors, for example, inadequate control of air conditioning and temperature, and desks which were ergonomically incorrect resulting in many complaints of physical distress.

3.3. Stressors in the home environment

Key home pressures were of course, common to all libraries and fell into these main areas: family problems, relationship problems, financial and health worries. The majority purported to family pressures.

4. CONCLUSIONS

Technological advances are now an inevitability of the library scene as it attempts to keep up with the information boom with great expectations coming from both funding bodies and the clients they serve. Libraries are having to operate at an optimum level regardless of cuts in resources and finances. Many of the potential stressors identified in the study can be attributed to change and development. Though each library assistant is an individual in their stress response as each organization, change is the common catalyst which turns potential positive stressors into potential key 'distess'ors. As Bayman noted, stress is not "symptomatic of problems in the workplace. It is the end result of the individual and organization coming under pressure..." (Bayman, 1992).

The study disproved the common belief that stress only affects people in important, highly-paid jobs. It showed that library assistants, who are at the bottom in terms of their career ladder, are also exposed to stress. The stress findings have important implications for library managers who have a duty under UK law to ensure the safety and health of their employees in every aspect related to work, this includes stress which "should be treated as any other health hazard", (HSE, 1995).

REFERENCES

Bayman, A. (1992) *Stress management in libraries*, unpublished Master's thesis, Loughborough University, England.

Cox, T. (1993) *Stress research and stress management: putting theory to work, HSE Contract Research Report No.61/1993*, Sheffield, UK: HSE books.

Friedman, M. & Rosenman, R. (1974) *Type A: your behaviour and your heart*, NY: Knoff.

HMSO (1987) *Understanding stress: part one*, London, UK: HMSO.

Hodges, J.E. (1990) Stress in the library, *Library Association Record*, 92(10), 751, 753-754.

HSE (1995) *Stress at work - a guide for employers*, HS (G) 116, Sheffield, UK: HSE books.

Neville, S. (1981) Job stress and burnout: occupational hazards for services staff. *College and Research Libraries*, 42(3), 242-247.

Physiological Analysis of Entrainment in Face-to-Face Communication

Tomio Watanabe and Masashi Okubo

Faculty of Computer Science and System Engineering, Okayama Prefectural University
Soja, Okayama 719-11, JAPAN

The physiological entrainment of respiration between talkers is researched on the basis of burst-pause of speech, respiration and heart beat-to-beat interval as indices from the viewpoint that respiratory entrainment would be a physiological main factor to make communication smooth in face-to-face interaction. The existence of respiratory entrainment is demonstrated by the cross-correlation analysis of the indices. The evaluation of respiratory entrainment could be applied to the evaluation of face-to-face interaction support systems such as a teleconference system.

1. INTRODUCTION

Entrainment in communication, which is the coherently related synchrony of independent biorhythm, plays an important role in the smooth exchange of information in face-to-face communication. We have proposed an entrainment model on the basis of the analysis of entrainment between a speaker's voice and eye-blink, and a listener's eye-blink and nodding in face-to-face interactions using image and sound analysis([1]-[3]). The model would be applicable to a variety of human-computer interactions where visual and auditory feedback is possible and would lead to a more user-friendly interface. However, the model paid no attention to the entrainment of emotional state in communication, not considering that the entrainment which forms the biological relation between talkers would exist in the emotional state reflected in physiological measurements[4].

In this paper, to clarify the existence of physiological entrainment in communication and to apply the mechanism to human-computer interaction, the physiological entrainment of respiration between talkers is evaluated on the basis of burst-pause of speech, respiration and heart beat-to-beat interval as indices from the viewpoint that respiratory entrainment would be a physiological main factor to make communication smooth in face-to-face communication. From the relationships which are formed among the indices, the existence of respiratory entrainment in face-to-face communication is demonstrated.

2. METHOD

Figure 1 shows the set up of the experiment. Two subjects faced across a table and then one subject talked to the other, each subject assuming the role of a speaker and a listener, under two different conditions: face-to-face communication (FFC) and non-face-to-face communication (NFFC) without face-to-face across a partitioning screen. The experiment was performed for a total of 12 minutes by repeating two times alternatively 3 minutes in FFC and 3 minutes in NFFC. The speaker's voice and both electrocardiograms (ECGs) and respirations of the speaker and the listener during the experiment were recorded simultaneously by a DAT data recorder

(TEAC RD-130TE) through a multitelemeter system (NIHON KODENWEB-5000) in order to measure the speaker's burst-pause of speech, respiration and heartbeat, and the listener's respiration and heartbeat. Talkers' behavior was also recorded on the same frames with a video editing system (SONY FXE-100) by two CCD cameras, one for the speaker and the other for the listener. The subjects consist of six male student speakers and one male student listener. The contents of speech concerned job hunting in which the listener was assumed to be interested.

Speech, respiration and ECG signals were simultaneously digitized by a 12 bit analog-to-digital converter (SDS Dasbox) at a sampling rate of 1 kHz, and then stored in a computer (SUN SPARC station). Talkspurt or silence was discriminated in each 10 ms, by whether the speech power level was over or under a threshold value; this threshold value was 12 dB plus the background noise level presented in the silence parts. The short silence duration before an unvoiced consonant is, however, caused by inertia of the vocal organ. This duration is therefore included in the talkspurt duration from the standpoint of respiration. Hence, the burst-pause of speech was generated with a fill-in value of 120 ms where the fill-in operation converted silence durations less than or equal to 120 ms into talkspurt. The heart beat-to-beat interval of cardiac cycle, or R-R interval (RRI), was measured by detecting the peaks of ECG. The time sequence of RRI was calculated in each 10 ms using the cubic spline interpolation. Respiration data were also calculated on the average in each 10 ms by the thoracic picking-up corresponding to its expansion and contraction.

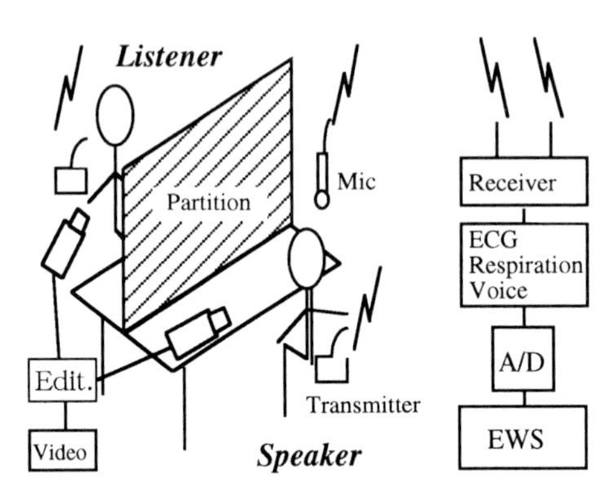

Figure 1. Set up of the experiment.

3. RESULT

A representative example of time sequences of burst-pause of speech, respiration and RRI of a speaker himself is shown in figure 2. The relations between these parameters were evaluated using cross-correlation function. The analyzing time is 120 s in the region where the time lag is within 5 s on the basis of the listener's respiration. Because there exists negative relations in both cross-correlations with respiration in figure 2, burst-pause of speech and RRI are closely related with respiration in the speaker. Figure 3 shows a representative example of the cross-correlations between the same speaker's burst-pause of speech in figure 2 and a listener's respiration in FFC and NFFC. The significant negative peak is noted at time lag 1.0 s in FFC compared with that in NFFC. Figure 4 shows the evaluation of respiratory entrainment between the same speaker and listener in figure 3. In the cross-correlations between the speaker's respiration and eyeblow inside, and makes wrinkles between left and right eyeblow. But those muscles can't pull down eyeblow and make eye thin. So new muscle "Orbicularis oculi" model is appended. Normal muslce is located between born node and fascia node. But Orbicularis oculi has irregular style which is attached between fascia nodes in a ring feature. Orbicularis oculi has 8 linear muscles which approximate ring muscle. Contracion of ring muscle makes eye thin.

3. Recognition for Motion Capturing

Marker is attached on a feature point in the face to measure and model facial expression. This feature point is chosen for each muscle from the grid point in face model which gives the biggest movement when contracting the muscle. 16 feature points in forehead area, and 26 feature points in mouth area are chosen.

3.1. Neural Net Structure

Layered neural network realize a mapping from feature points' movement to muscle parameters. 4 layer structure is chosen to get the non-linear perfomance well.

Feature point movement has 2 dimension, so the number of input layer unit is double number of feature points. Number of output layer unit is number of muscle in each sub-area. The number of unit in hidden layer is decided heuristically. For a forehead sub-area, neural network consists of 16 units in input layer, 20 units in hidden (Second, Third) layer and 8 units in output layer. For mouth aub-area, neural network consists of 52 units in input layer, 60 units in hidden (Second, Third) layer and 28 units in output layer.

3.2. Learning Patterns

Learning patterns are composed of the individual contraction of each muscle and its combination. In case of individual motion, contraction of each muscle between maximum strength and neutral is quantized into 11. In combination case, we create 6 basic facial expression consisting of Anger, Disgust, Fear, Happiness, Sadness and Surprise, and quantize the each difference from neutral also into 11. So total number of learning patterns is 143 in each forehead sub-area.

In mouth area, each muscle contraction does not happen individually. So all learning patterns are composed of combinations. Learning pattern has basic mouth shape for vowel "a", "i", "u", "e" and "o", and nasal consonant "n" as closed mouth shape. Also 6 basic expression is appended as same as the eye area and jaw rotation is specially introduced. So totally, 13 actions is selected for training and they are also quantized into 11 steps. The number of learning patterns is 143 for mouth sub-area. Each pattern is composed of the data pair of muscle parameters vector and feature points movement vector.

3.3. Normalization

In order to absorb an individual variation of human face, each feature point movement is normalized by the standard length to decide the facial geometrical feature. Both forhead area and mouth area have local axis normalized by the local standard length.

4. Evaluation

Actually, we attach marker on real human's face and get movement of marker from 2D image captured by camera when any aribitral expession is appearing. After normalization, the movement value of marker is given to neural network and facial image is re-generated using facial muscle model on the parameter from neural network output. Example images are shown in Plate 2. Mouth shape speaking "u" and pushing out lips have big variation in depth as shown in figure 10. Muscle parameter is decided only from 2D image, but 3D facial image is well regenerated. By comparing a regenerated image and an original one, it appears that regenerated image gives

weaker impression than original one in
well reproducted. And some exceptions h
by any muscle combination in our mode

5. Conclusion

An automatic estimation method of 3D
is presented in this paper. Parameter conv
is fitted to the target person's face precis
within the combinations of basic expressi
fitting tool to the face image by manual o
all grid points precisely in the real face in
on its initial location or target person and is
So the correspondence between real face a

Reference

[1] Hajime Sera, Shigeo Morishima, Der
Mouth Shape Control", Proceedings of Ro
pp. 207-212(1996)
[2] Yuencheng Lee, Demetri Terzopoulos
Animation", Proceedings of SIGGRAPH '9
[3] Shigeo Morishima, etc. "Life-Like, E
#25, Siggraph (1996)

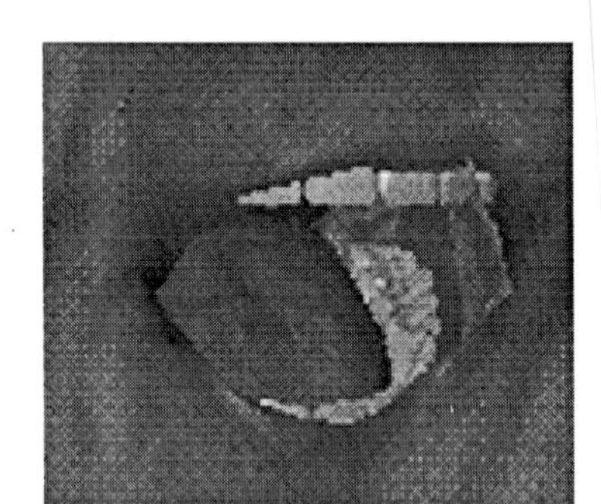

Plate 2 Mouth Model

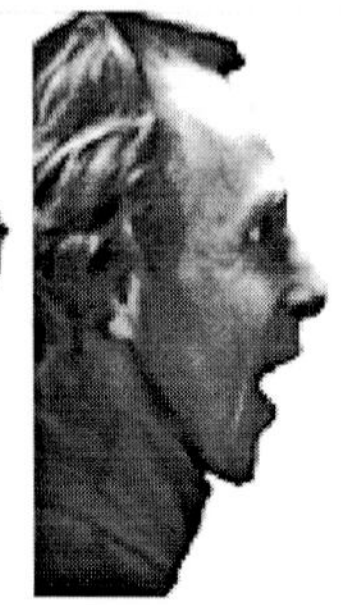

Plate 3 Synthesized Expressin
a) Frontal View b) Side View

lowered.

In TV conferencing, the camera can be preprogrammed to the seat locations of individual participants, to the bulletin board, to the document desk, etc., and camera angle could be automatically adjusted to the location of talker by pressing the number key dedicated to the location. Conflict in selection might occur sometimes, it could be managed on first come first out base. Some system uses voice key to select camera location. It is not always convenient that the camera is always fixed to the talker however.

In case of process monitoring, preprogrammed fixed location image and remote camera control with video feedback will be used alternatively.

2.4. *Image switching*.

When several cameras are installed and operating in parallel, it is necessary to select one or a few images to be sent to other sites. Very quick response is promised when image switching is used. But the talker of conferences frequently forget about the need of switching. An dedicated attendant would be employed to avoid this trouble, briefings between talkers and the attendant are require in advance to the meeting.

The more casual switching is expected, if the switching is operable at the presentation site. For example, the image can be switched to document camera, when a participant shows a document to the document camera, or to the one, when some-body stood by the bulletin board.

2.5. *Manual change of display materials*.

When many presentation materials are to be shown, manual change of material is most satisfactory for the talker. The materials can be changed quickly at the appropriate timing. In many case, a participant has to move to the document camera, or some camera to show an object. The requirement to move is certainly a barrier, but once the talker have taken the place, it is easy to continue presentation.

Once in an experiment, the participants had equal opportunity of selecting the local camera control and manual presentations. Many participants took liberty of using manual presentation.

An alternative way of manual change is presenting materials from archive of computer systems. But presenting documents from the archive might have the impression indirect. Manual presentation of document to the document camera might give the impression of direct operation.

3. MENTAL MODEL OF CAMERA CONTROL

When people start conferencing, they are ready at least to talk and hear. We might say they have the talk and the hear mind. However, the strength of the talk and the hear mind might different from one to another. Some might have stronger and others might have weaker talk mind. But the strength of the talk and hear mind itself cannot be measured directly. One possible measure to evaluate strength is to measure actions. The strength of the talk mind will be reflected to the number of the talk. There is no proper measure reflecting the hear mind, but for example the number neglecting the response to a question or a request might be a measure of weak hear mind.

The concept of watch and show mind was introduced in explaining behaviors of TV conference participants (Zhang et al., 1995).

When it is compared with the talk and the hear mind, the mind to show and to watch is not so strong in the normal TV conferencing. In the situations (Takada, Tamura & Shibuya, 1996) shown in Fig. 2, participants A and B are required to solve some tasks scanning the information board in the other's room.

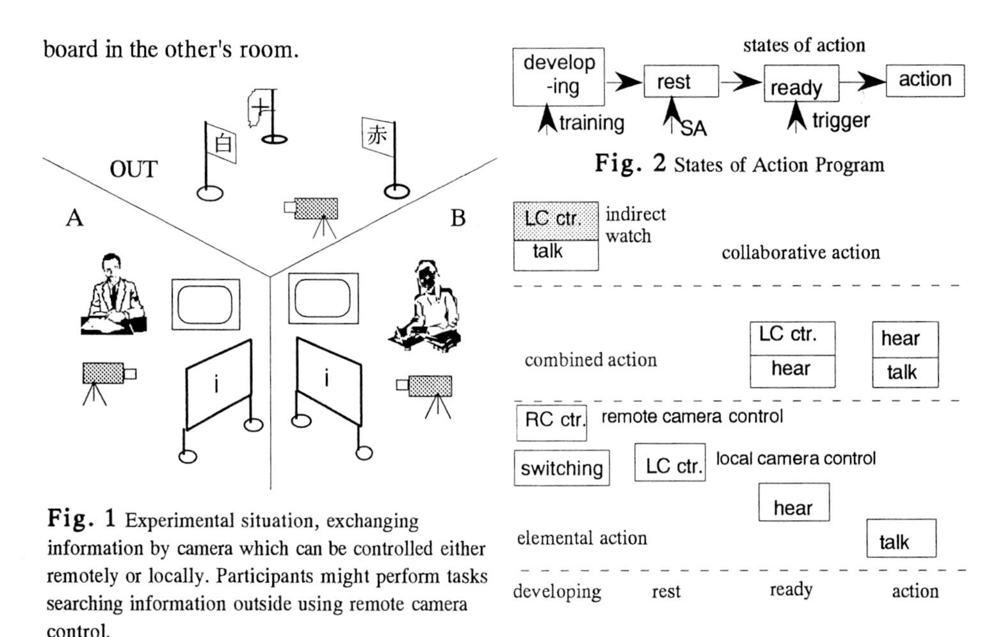

Fig. 1 Experimental situation, exchanging information by camera which can be controlled either remotely or locally. Participants might perform tasks searching information outside using remote camera control.

Fig. 2 States of Action Program

Fig. 3 Structure of Action Control

But the task could be solved operating the camera in his own room locally. The operation will be much faster and exact, since he can use visual guidance and the visual feedback is much faster. Most participants nevertheless took liberty of operating the remote camera. Frequent use of remote camera is reflecting the strong watch mind.

The watch mind could be fulfilled by an indirect method. It is done as follows. One participant asks another in the other site to move the camera locally to a certain angle, so that certain necessary data might be found in the scene. The indirect method is the only way to solve the problem when remote camera control is not available in the systems. In this situation, if a participant in one room had very week show mind, the task become very difficult to accomplish.

When participants A and B are talking and searching some data in the outside, they certainly might show strong watch actions by controlling the remote camera. Takada et al(1996) reported that in case of cooperative work of solving the problems in search of data, natural division of roles is observed. For example one in two will be operating camera control continuously and others are talking more frequently.

Although some evidences are showing weak show mind and strong watch mind, in case of message transfer conference the talker shows apparent show action.

Although some evidences are showing weak show mind and strong watch mind, in case of message transfer conference the talker shows apparent show action.

Fig. 3 is showing the states of action control. Four states of action control are assumed. On the right most is the action level. Only one action can be performed at one time. An action may terminate by the end of fixed motor program or by the removal of in coming stimuli, which will return the action program to the ready state.

Several action programs are waiting at ready state, and will be brought into operation as soon as certain appropriate trigger is given. Some priority rules might be established to select one trigger in case more than two triggers are given at the same time or one after another. Some actions have to wait some time until the foregoing action is finished.

Still larger storage of action programs are stacked in the rest state. Actions in the rest state will be brought into ready state by the situation awareness. For example, the task or the situation requires one to operate a camera, camera control program of action will be set into ready state.

Some actions are not well established. They need training before they could be used. For example, camera control and image switching are foreign to many participants at the beginning. Thus, even though they are aware of the need, they dare not to do such actions, unless they have made some exercise in advance.

Some actions could be combined and organized so that they could be performed at the same time. For example, in general the hear and the talk actions are assumed to be different program. So people talks continuously when sending a message, and they listen to message continuously. They are unidirectional. Daily conversations may be the combination of hear and talk actions, but they are already independent from the unidirectional action program. Thus when one encounters message recorder after dialing, he has to change quickly the action program from bidirectional to unidirectional program. The first row from the bottom in Fig.8 is showing the elemental actions, while the second row from the bottom is showing the combined actions.

In our experiment, some participants are self-satisfied in controlling camera while others are debating. Some skill are necessary to control camera or switch images timely from one talker to another. The operator as well as other participants are in favor of timely control of image. This type of action is the combination of the hear and the local camera control actions.

At the third row from the bottom in Fig. 8 is showing the action program of indirect camera control. It is the combination of talk action and the local camera control. But the latter is the one in the other site. This type of action programs may be called collaborative, and two or a group of people in the different site have to know their intent and skill each other before the program is brought into action.

4. CONCLUSION

Some evidences related to camera control in image communi-cation are described. Mental model to explain the behaviors related to camera control is proposed.

References

K. Takada, H. Tamura, Y. Shibuya(1996) On the Camera Control in the TV Conference, Human Interface N&R, vol. 11, pp. 477-482

H. Tamura, S. Choi, K. Kamada, Y. Shibuya (1994) Representation of Mental Model of Media Users and the Application to TV Conference, Progress in Human Interface, vol. 3, pp. 31-38.

H. Tamura, S. Choi(1996) Verbal and non-verbal behaviors in face to face and TV conferences, in Gorayska & Mey, Cognitive Technology, pp. 361-374, Elsevier

R. Zhang, H. Tamura, Y. Shibuya(1995) Communication Behaviors Related to Camera Control in TV Conference, Human Interface, vol. 11, pp. 941-644.

Quick Address Search System with Handwriting Using Character Transition Information

Keiko Gunji Soushirou Kuzunuki Koyo Katsura
Hitachi Ltd. Hitachi Research Laboratory

1. ABSTRACT

We have developed an address search system for pen-based computers which can guess the correct address immediately, even if an unclear or insufficient address is given. The system overcomes problems with handwritten entrees regarding miss recognizing and miss writing. In order to speed up reference to an address dictionary, our system searches for the correct address using not only the address dictionary, but also a bi-gram (transition information) index. When an insufficient address is given, the system can guess the correct address immediately, by referring to the address dictionary using the bi-gram index. Compared with the existing address input system, our system can halve the time to input addresses.

2. INTRODUCTION

Pen-based computers are suitable for insurance businesses and transportation businesses because of their portability. In these businesses, workers often input a customer's address. Because the address has many characters, simplified inputting of the address is very important. The following two address input methods are currently available. (a) Straightforward Handwriting Method: the user inputs the whole address by handwriting, (b) Zip Code Method: the user inputs a zip code by handwriting, then the system shows a list of candidate addresses, and one is selected from the list. With the Straightforward Handwriting Method, users have to write many characters. With the Zip Code Method, only three or five characters are necessary, but the zip code has to be looked up in advance. Therefore both systems are inconvenient. We propose the Quick Address Search System which can guess the correct address immediately when an insufficient address is given.

3. ISSUES OF ADDRESS SEARCH SYSTEM

When an address is input, incorrect characters might be entered, or the system might not be able to recognize the input characters correctly, also the address input system might be given insufficient addresses. An example is shown here for insufficient characters for the Japanese address "東京都 新宿区 神楽坂" ("東京都" is a prefecture name, and "新宿区" is a city name, and "神楽坂" is a street name. Let us suppose the prefecture name "東京都" is "p11 p12 p13", and the city name "新宿区" is "c11 c12 c13", the street name "神楽坂" is "s11 s12 s13", and "p11","p12", ... ,"s13" are Japanese characters.)

(a) only street name is given ： "s11 s12 s13"
(b) some input characters are missing ： "s11 s12 □" or "□ s12 s13"
(c) some input characters are incorrect ： "s11 s12 sei"
(d) some characters are miss recognized ： "s11 s12 ser"
(e) some characters are disordered ： "s11 s12 c11 c12" (In Japanese, the prefecture name is written first, the city name second, and the street name last.)

We aimed at developing the address search system which can guess the candidate address, even when an insufficient address like above is given. To develop the system, we had to develop the following two methods.

(1) a method to retrieve candidate addresses with insufficient address words

(2) a method to assign a correctness priority to the candidate addresses

4. QUICK ADDRESS SEARCH SYSTEM

Figure 1 shows the structure of Quick Address Search System. In order to guess the correct address, our system uses the address dictionary and the character transition index. The address dictionary is a tree of the address words. The character transition index is a table of character transition information and a pointer to address words in the address dictionary. The character transition index is in alphabetical order to speed up referencing to the address dictionary. We explain the process of the system next. If a word is written which is part of an address (e.g.,"神楽坂": "s11 s12 s13"), the system recognizes the word. The system divides the word "s11 s12 s13" into a set of bi-grams (e.g.,"s11 s12" and "s12 s13"), and uses the character transition index to refer to the address words "s11 s12 s13" and "s11 s12 sk3" immediately. "s11 s12 s13" and "s11 s12 s13" are street names. If the street name "s11 s12 s13" is chosen, address words in higher hierarchy levels "c11 c12 c13" (city name) and "p11 p12 p13" (prefecture name) are fixed, and the system gets the candidate address "p11 p12 p13 c11 c12 c13 s11 s12 s13": "東京都 新宿区 神楽坂" immediately.

We compared the speed and the size of our system with the Straightforward Handwriting System. The latter system has only an address dictionary, and the dictionary size is 1.5MB. The system has to compare the input characters with every character in the address dictionary. In worst case, the system compares the input characters up to 750,000 times. On the other hand, our system has the address dictionary and the character transition index, and their total size is 3.5MB. The system refers to the address dictionary using the character transition index, which is in alphabetic order. Even in the worst case, the system compares the input characters with the dictionary characters only 36 times. With our system, the correct address is gotten within a practical time.

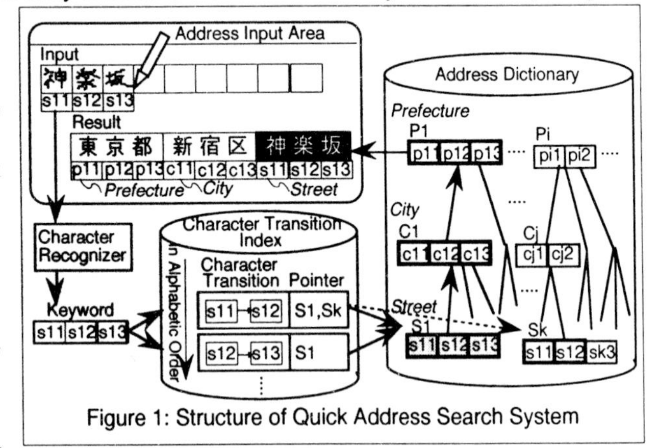

Figure 1: Structure of Quick Address Search System

4.1 Method to retrieve candidate address with insufficient address words

Figure 2 shows insufficient address words (input) and the character transition information and the candidate addresses. Let us suppose that the user wants to input the address "p11 p12 p13 c11 c12 c13 s11 s12 s13", but instead input an insufficient word "s11 s12 se". Existing systems cannot guess the correct address, because an address including "s11 s12 se" does not exist. But our system can guess candidate addresses including the correct address using the character transition index. Our system divides the input word "s11 s12 se" into a set of bi-grams "s11 s12" and "s12 se". Since "s11 s12" is a portion of the correct address "p11 p12 p13 c11 c12 c13 s11 s12 s13", the system can get the correct address. Our system is very useful because it can get the correct address even if insufficient words are given. As many candidates are supplied because our system searches for the address using character transition information, it is important to give a correctness priority to the candidate addresses.

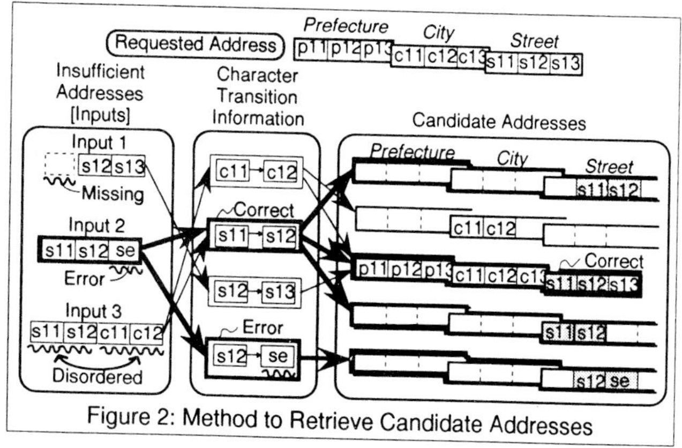

Figure 2: Method to Retrieve Candidate Addresses

4.2 Method to give a correctness priority to the candidate addresses

Giving correctness priority to candidate addresses is described in Figure 3. Three factors are used. The first factor is the number of address characters which match input characters, because generally the candidate address which has more of these characters matching tends to be the correct address. The second factor is the hierarchy level of the word which matches input characters, because generally a high level address is used more often and has a bigger population. The candidate address which has a high level hierarchy word which matches the input character is taken to be the correct address. The last factor is the position of the address characters which match input characters, because generally the head of address words is more often entered than the tail of them.

We explain an example for deciding the order for the candidates (A1:"p11 p12 p13 c11 c12 c13 s11 s12 s13", A2:"p11 p12 p13... c11 c12 c13... s21 s22 s23...", A3:"p21 p22 p23... c21 c22 c23... s11 s12 s23...", A4:"p31 p32 p33... c31 c32 c33... s31 s11 s12...") which are obtained with the input word "s11 s12 c11 c12". Using the above factors, we decide A1:"p11 p12 p13 c11 c12 c13 s11 s12 s13" to be the first candidate, because it has 4 characters matching input characters "s11 s12 c11 c12", and the others have only 2 characters. Next, we decide A2: "p11 p12 p13... c11 c12 c13... s21 s22 s23..." to be the second, because "c11 c12 c13..." whose characters are equal to the input characters "s11 s12 c11 c12" is a city class, and the others are a street class. Next, we decide A3: "p21 p22 p23... c21 c22 c23... s11 s12 s23..." to be the third and A4: "p31 p32 p33... c31c32 c33... s31 s11 s12..." to be the fourth, because the input "s11 s12" is the head of "s11 s12 s23...", but it is not the head of "s31 s11 s12...".

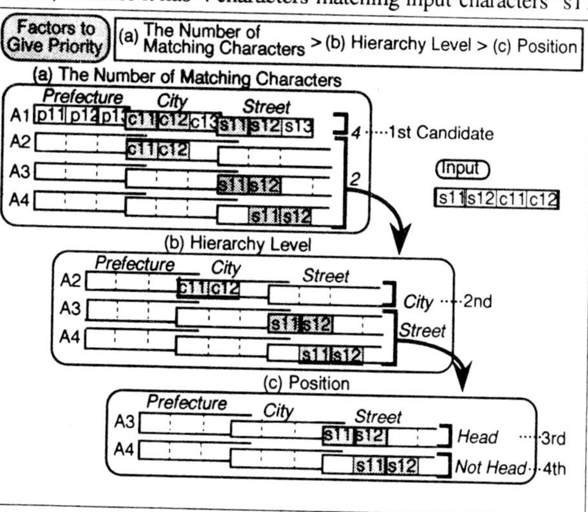

With our system, the correct address can be selected easily, because our system shows the candidate addresses in the correctness priority order.

5. EXPERIMENT RESULTS

We tested the Quick Address Search System. And we evaluated it with the two existing systems, Straightforward Handwriting System and Zip Code System. We prepared 10 sample addresses. We asked 10 persons to input these addresses with all three systems. If there were any mistakes, users were asked to correct them. We measured the operation time and then obtained the users' impressions.

The operation times to input an address were 59s with Straightforward Handwriting System, 99s with Zip Code System, and only 29s with Quick Address Search System. Breakdown of the time use is shown in Figure 4. With Straightforward Handwriting System, it took 32 seconds to input an address by handwriting. With Quick Address Search System, it took only 9 seconds to input keywords by handwriting, because the system accepts insufficient input. With Zip Code System, it took 30 seconds to select the correct address. With Quick Address Search System, it took only 6 seconds to select the correct address, because candidate addresses are shown according to priority.

We gathered users' impressions through questionnaires. Users scored the system from -3 to +3 and gave reasons why. The scores were as follows. (a) Straightforward Handwriting System: -1.4 points (bad). (b) Zip Code System: -2 points (bad). (c) Quick Address Search System: +1.6 points (good).

The reasons were as follow

[bad(-)]

(a) Straightforward Hand writing System

- have to be input many characters by hand writing
- wrongly recognized results have to be corrected

(b) Zip Code System

- the zip code has to be looked up in advance

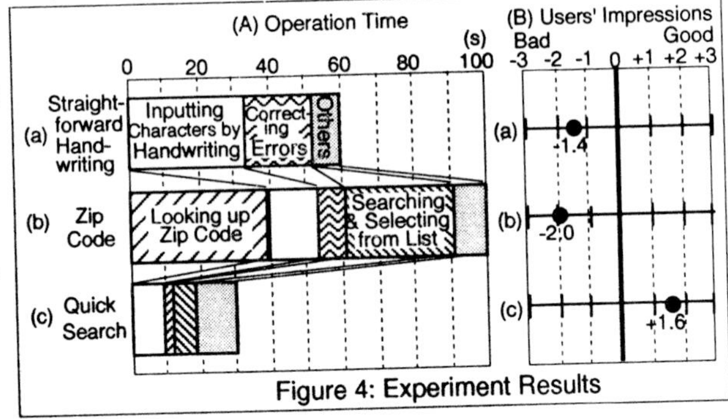

Figure 4: Experiment Results

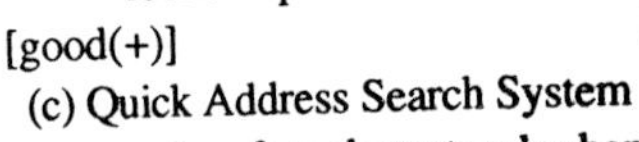

[good(+)]

(c) Quick Address Search System

- only a few characters by handwriting are needed

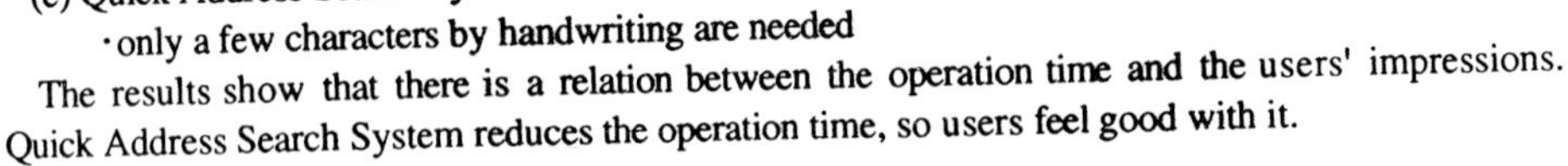

The results show that there is a relation between the operation time and the users' impressions. Quick Address Search System reduces the operation time, so users feel good with it.

6. CONCLUSION

We have proposed the Quick Address Search System which can guess the correct address immediately, even when an unclear or insufficient address is given. In the future, we will develop Quick Search System for general databases, and provided a useful user interface.

REFERENCES

[1] M. Nakagawa, K. Machii, N. Kato, T. Souya: Lazy Recognition as a Principle of Pen Interfaces, INTERCHI'93 Adjunct Proc. 89-90 (1993.4).

[2] M. Nakagawa, N. Kato, K. Machii, T. Souya: Principles of Pen Interface Design for Creative Work, Proc. ICDAR'93, 718-721 (1993.10).

A System for Eliciting and Helping to Score Test Answers Input through a Handwriting Interface

Toshio SOUYA

Tokyo Research Laboratory, IBM Research
1623-14, Shimotsuruma, Yamato-shi, Kanagawa-ken 242, Japan.

Printed tests are still efficient media for education. Reproduction of such tests on computer systems would combine the efficiency of paper tests and the functionality of computers. This paper presents a system that provides an environment for answering and helping to score tests input through a pen interface, which is the most suitable for CAI systems.

1. Introduction

The chief merit of pen input, we believe, is that it requires no practice and consequently does not distract the user's attention to the input task itself. To take advantage of this merit, we designed a system for eliciting and helping to score test answers input through a handwriting interface. This combines a "lazy recognition scheme" [1,2] with a "tacitly expected input area" to completely eliminate restrictons on the input area from the user's point of view. The basic design of the system and a prototype are described below.

2. Basic design

First, we decided to support the "test paper method" as its learning system. The main object of this method is let candidates solve questions on the question sheet itself.

Among tasks included in a CAI system, we focus on solving questions (eliciting answers to a test from a candidate) and scoring answers, which can be facilitated by converting the test papers into digital data. Since digital data can be exchanged through networks, digital test papers can be exchanged more easily than sheets of printed papers.

2.1 Lazy recognition scheme

The lazy recognition scheme is simple. Basically, it means "show nothing but ink marks when writing." Recognition results are not shown at the time of writing, but only when required by the user. In this respect, the scheme is the same as what is called "differed recognition," except that its main aim is not to disturb the writer's flow of thought. Recognition results should be shown after a natural break in concentration on the written content. In this way, showing the results will not disturb the user's flow of thought and will allow them to concentrate on the results.

This lazy scheme can be applied naturally to a system for eliciting and helping to score test answers. Solving questions and scoring correspond to writing and recognizing, respectively.

Only inking is done at the time of answering. After the user has finished answering, the scoring helper, which includes a recognition process, is invoked. Thus, the test system take advantage of the merits of the lazy recognition scheme, namely,

(1) Students can concentrate on the questions while answering.

(2) They can write their answers anywhere on the displayed test papers.

(3) The scoring helper function can use global features of the input ink marks.

2.2 Tacitly expected input area

As mentioned above, students can choose where to write their answers freely, as they can when they are answering on a piece of paper. However, when they write an answer on a test paper, students are implicitly required to write a certain (correct) answer in a certain place. In most test papers, answers are accepted only if they meet this condition. We call the area in which an answer should be written the "tacitly expected input area".

When tacitly expected input areas are used, each answer consisting of sampled ink track data can be assigned to a particular question, because the inks are segmented as strokes (traces made by a pen from the time it is placed on a surface to the time it is lifted off the surface) when they are written. This segmentation cannot be expected in off-line systems (e.g., OCR systems).

3. A prototype system

To test the basic design and the judgment function, we created a prototype system.

For this prototype, we chose mathematics as the domain for the tests, because:

1. It requires answer formats(two-dimensional formal arrangements of written objects) that are difficult to input by conventional methods.

2. It requires only numeric characters and a few other symbols. Such a small set of characters is relatively easy for an engine to recognize.

The prototype was designed to have both a test-answering environment and a scoring-helper function.

3.1 UI system for test answering

The system provides a "pencil mode" and an "eraser mode" for writing. In the pencil mode, the user can write in the same way as with an ordinary pencil. The movements of the input pen are recorded as strokes, and its tracks are shown as ink marks on the display. The eraser mode is used to erase ink marks on the display. The user can erase such marks in the same way as with an ordinary eraser.

3.2 Scoring helper for mathematics

A correct answer to a question is expressed as a tree structure of various answer patterns (we call a set of tree data a "right-answer tree") using these answer forms. An example of a right-answer tree is shown in Figure 1.

A tree is constructed by the top-down method. The pattern at the top shows the structure of the whole answer. Details are shown in sub-trees. In the case of mathematical tests, the leaves of the trees must be strings of numerals. Other patterns (e.g., fractions, mixed numbers, etc.) have sub-trees under them.

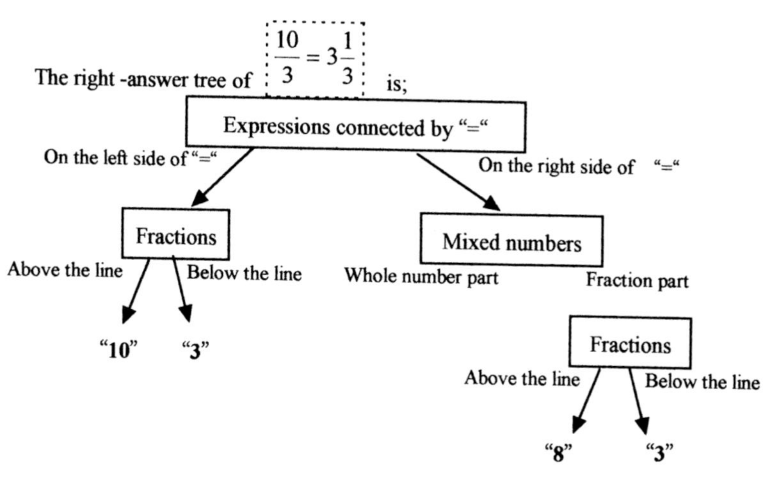

Figure 1. Answer tree of $\frac{2}{3}+\frac{8}{3}=?$

A judgment processor analyzes the structure of pen strokes assigned to a question and traces a right-answer tree corresponding to it from the top down. All pen strokes are analyzed as consisting of the pattern at the top of the tree. As a result, groups of strokes are analyzed as substructures. In this way, strokes are assigned to substructures one by one. If all of the strokes assigned to a question have been analyzed and found to be as in the right-answer tree, the answer is judged to be right. Otherwise - for instance, in the event of a failure to analyze the answer as having the required format, or of a numerical string's not being recognized as in the tree - the answer is judged to be wrong.

Example views of the judgment results are shown in Figures 2.

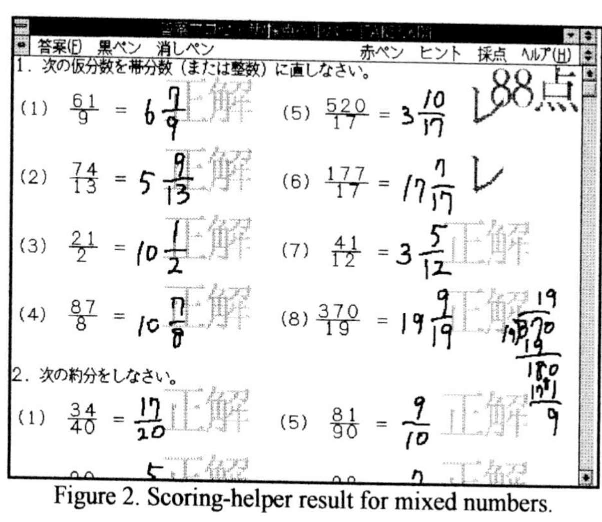

Figure 2. Scoring-helper result for mixed numbers.

4. A trial

To evaluate this prototype, we observed how it was used by 10 subjects. Subjects (all of them are employees of our laboratory, although the questions are for elementary-school students) solved 118 questions including calculations on paper (addition, multiplication, and division), additions of fractions, reductions, and equations of the first degree.

The following results were observed:

1. All the subjects started answering questions after simple instructions.
2. All the subjects finished in about 30 minutes.

The first result suggests that the UI was accepted naturally. We may say that the UI helped all subjects to maintain concentration, since they solved all the questions at one sitting.

We have not attempted to evaluate the judgment accuracy quantitatively. This would be preferable, but the complexity of answer arrangement affects the accuracy. There is scope for improvement.

5. Conclusion

This paper has presented the basic design of a system for eliciting and helping to score test answers input through a handwriting interface. The system uses a “lazy recognition scheme” and “tacitly expected input area.” A prototype of the system for mathematical tests and concrete processes has also been described.

Several merits were observed in a trial of its UI system. Improving in the scoring helper is a task for the future.

Although the prototype employs a top-down method to analyze written answers, it will be worth considering using a layout analyzer that has more flexibility. We are also interested in applying the system to other subjects besides mathematics.

References

[1] T. Souya et al., “Handwriting Interface for a Large Character Set,” Proc. of 5th Handwriting Conf. of IGS (1991).

[2] M. Nakagawa et al., “Principles of Pen Interface Design for Creatve Task,” IEEE, Proc. of 2nd ICDAR (1993).

[3] H. Sasaguri et al., “An ITS Guidiance in Fraction Calculation -- A Development of a Handwriting Interface” (in Japanese), technical report of the IEICE, ET95-31(1994).

[4] K. Tsushima et al., “Computer Algebra System for Mathematical Education A Handwriting Interface Using Tablet,” Asia Pacific Information Technology in Training & Education Conference & Exhibition 94 (APPTITE ’94) (1994).

[5] M. Matsumoto et al., “Handwritten Formula Inputing System and Handwritten Formula Editing”(in Japanese), Technical report of the IEICE, ET93-26 (1993).

[6] A. Murase et al., “The Implementation of METAH, a Recognition System for On-line Handwritten Mathematical Expressions” (in Japanese), Proc. of IPSJ 46th Annual Conf. 4H-5 (1993).

[7] K. Toyokawa et al., “An On-line Character Recognition System for Effective Japanese Input,” IEEE, Proc. of 2nd ICDAR (1993).

Form input system by pen interface

Yukiko Nishimura and Masaki Nakagawa

Nakagawa Lab., Department of Computer Science,
Tokyo University of Agriculture and Technology
2-24-16 Naka-cho, Koganei, Tokyo, 184, Japan.
e-mail: nishimura@hands.ei.tuat.ac.jp

This paper describes the prototype of a form-input system by pen interface. The system is designed to provide easy preparation of forms by reading in an empty form with a scanner, recognizing frames in the form and displaying them on a display integrated tablet, and then allowing people to input data with a pen.

1. INTRODUCTION

Up until now if we wanted to draw up a neat and tidy form, we had to create entries using word processors or software tools, print them out and then cut and paste the entries to the form. Otherwise, we had to reproduce the entire form using word processors and input data in it. These methods have merits such as being able to produce neat documents or carry out computer processing of their contents. However, these have been very labor-intensive tasks.

On the other hand, the method that allows a user or an operator to draw up a form by writing entries neatly by hand, reads it in with a scanner, and then applies pattern recognition to the entries has been researched and developed on a commercial basis. However, this also has problems such as errors related to the pattern recognition and the tedious task of correcting misrecognized entries.

We propose a system which makes it easy to carry out the computerized preparation of forms by reading in an empty form with a scanner, recognizing frames in the form and displaying them on a display integrated tablet, and then allowing people to input data with a pen, a keyboard or a mouse. Some forms require handwriting such as signatures which can be easily inputted by a pen.

With pattern recognition, or even without it for field labels or entries, several degrees of automation and high document quality are realized. Recognition errors may occur if pattern recognition is applied, but they are verified by the user and alternative choice selection or rewriting can be done easily with a pen. This is much easier than the pure OCR based form reader where unreadable forms must be rewritten and inputted again or errors must be corrected by the operator using a keyboard.

This system can be used on a network, that is, it enables a person to obtain a form template via a network, and then return only the entries. Considering the current development of networks, this facility may have a great potential.

2. HARDWARE SYSTEM CONFIGURATION

The hardware configuration of our system is shown in Figure 1.

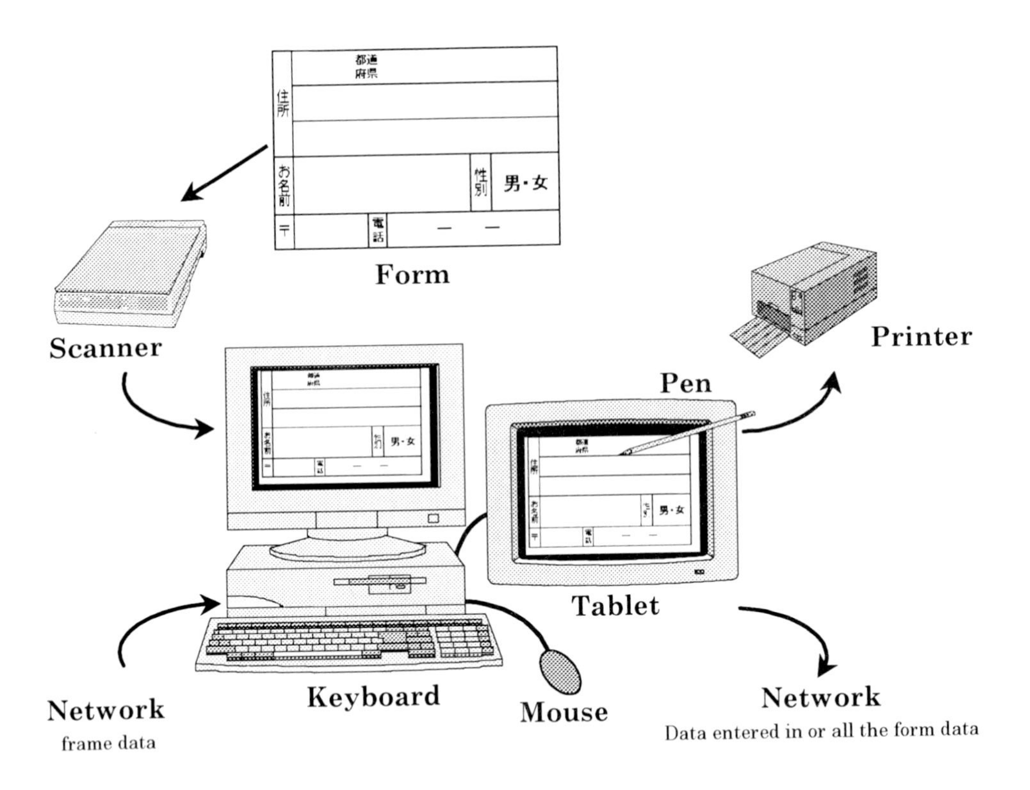

Figure 1. Work flow and system configuration.

3. BASIC DESIGN OF THE FORM INPUT SYSTEM

This system has been designed according to the following requirements:

(1) two ways to input data

Some forms require handwriting such as signatures. Therefore, we think the system should make it possible to input both handwritten patterns and character codes obtained by invoking pattern recognition.

(2) two ways to process labels

If the system does not try to recognize labels of the fields in a form, the problem of errors in the pattern recognition does not occur. The system can produce the document without reducing output quality very badly when the system reads in a form with a scanner, revises only its position and inclination, merges its image with entered data and prints it out. This is particularly true if it is a printed matter. On the other hand, if the system recognizes field labels in a form, the system can insert some entries into each appropriate frame automatically. These are fixed entries such as name, address, telephone number and so on. Therefore, we consider that it would be very effective if a user can choose whether the label recognition process is applied or not.

(3) two modes of printing out

If a form has some special paper quality or it includes objects not reproducible such as seals, printing must be made on the form. Consequently, the system must provide the mode to print entries on the original form so that the contents fits in the fields as well as the mode to print out the whole form on a blank paper.

(4) usage on environment of network

On a network environment, it is not necessary to read in a form with a scanner. If a form data available through a network is image data, the system converts it into code data by recognizing the form. If it is coded already, the system can skip the form recognition. On the network environment, it is possible to return the completed form through the network. By returning only the filled-in data, the cost for communication and processing can be saved.

Figure 2 shows the flow of process in the system to satisfy the above requirements.

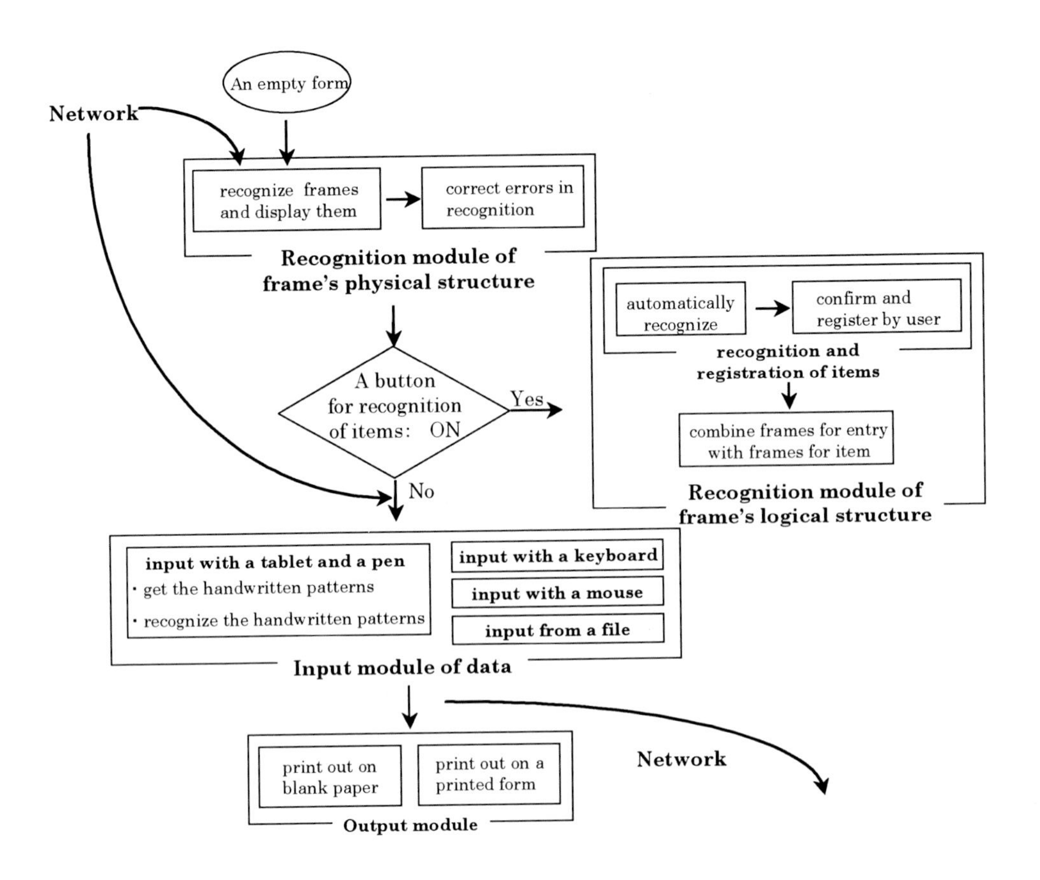

Figure 2. Process flow in the system.

4. CONCLUSION

This paper has described a form-input system. The system makes it easy to carry out the computerized preparation of forms. We are going to evaluate this initial prototype to improve its functions and ease of use.

This research is being supported by the Advanced Software Enrichment Project of IPA under MITI of Japan.

Prototyping of Digital Ink E-Mail System Based on a Common Ink Format

Hiroshi Tanaka, Naoki Kato*, Masaki Nakagawa*

Fujitsu Laboratories Ltd., 64 Nishiwaki, Ohkubo-cho, Akashi, Hyogo 674 Japan
Email: tanaka@flab.fujitsu.co.jp

*Tokyo University of Agriculture and Technology, 2-24-16 Naka-cho, Koganei, Tokyo 184, Japan

This paper describes the prototyping of a digital ink e-mail system based on a common ink format. This e-mail system can deal with handwritten figures and text strings on a same electronic paper.

Digital ink is useful to express one's message on electronic paper (digital paper). Because digital ink is similar to real ink used on real paper, it is easy to use even for the end user. But on existing e-mail systems, most messages are written in text only. We think that the e-mail system must be made more user-friendly and useful if digital-ink is to be easily used on it. To deal with digital ink on an e-mail system, we have designed a new document format "HandsDraw" which can be used for text, simple figures and digital ink. This format is so simple that it is easy to implement software which can deal with it on any computer system. We also have made a prototype digital ink e-mail system. This system uses "HandsDraw" to exchange handwritten messages by e-mail.

We think the digital ink e-mail system allows e-mail users to express their messages more quickly, directly and intuitively.

1. INTRODUCTION

1.1. Using Digital Ink on E-mail

On real paper, handwritten figures are often used by people to express their idea. However, most of the messages are just written as text on digital paper (Figure 1). To use figures on digital paper like on real paper, digital ink is necessary. Digital ink makes digital paper as easy and intuitive to use as real paper.

Now, digital paper is used for remote communications via e-mail. E-mail allows faster communications between people over long distances than letters. This is one of the most valuable aspects of e-mail communication. Digital ink could make an e-mail more powerful and useful tool for remote communications.

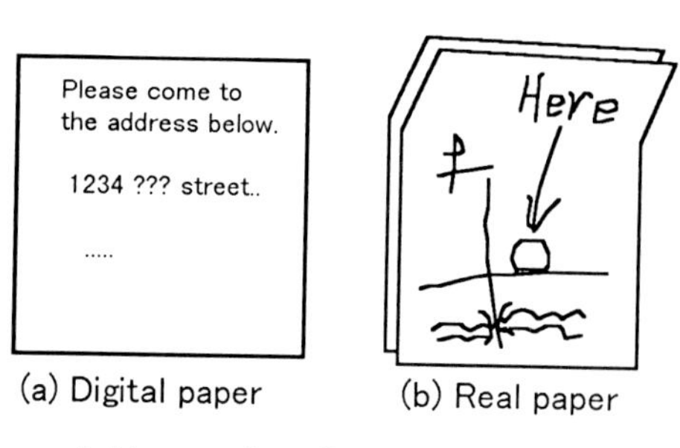

Figure 1. Messages on former digital paper and real paper.

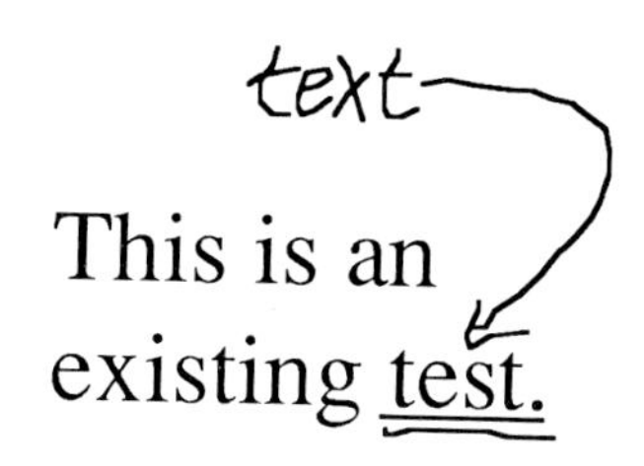

Figure 2. Handwritten comment on an existing text.

1.2. Ink Format

Digital ink is not very popular in network communications because there are no effective common data formats which deal with digital ink. In the following description, we refer to the document data format which can deal with digital ink as the "ink format."

An ink format is considered to be "effective" if it is useful and suitable for use in digital ink e-mail messages. On an e-mail system, a message sent by a user may often be edited by another user. So the ink format used in e-mail systems must be easily editable. For example, Adobe PostScript is a very powerful document format (language) used to express images on digital paper, but a document described in PostScript may be very difficult to edit after it has been created. As another example, image data formats such as GIF or JPEG can be used to express handwritten image data. But such image data formats can become very large when expressing digital papers. Also the image format is not very useful for handwritten comments on an existing text message (Figure 2).

The "common" ink format is free and easy to use for anyone. The specification of the ink format must be completely open and must be simple enough to implement easily by any programmer who has any normal programming skills.

While there are some existing document formats that can deal with digital ink, there are no "effective" and "common" ink formats. Therefore, we have designed a new ink format called "HandsDraw."

2. HANDSDRAW

2.1. Overview of HandsDraw

HandsDraw is a kind of "draw" data format. A HandsDraw document has several pages, and there are some primitive objects placed on each page. Each primitive object represents a specific figure (Figure 3). Digital ink is represented as a group of handwritten stroke objects.

The HandsDraw document is described by text strings (Figure 4). Each description of primitive figures is written in sequence according to the order of the layer in the document page. In other words, an object on the lower layer is described earlier in the HandsDraw description. In Figure 4, the "Oline" is place don a lower layer than "Oval."

2.2. Preservation of Time Information

At the time when a document is created, the order of object layers is the same as the order in which the objects were created. But after modification of the object layer, the order may be different. Because the handwritten stroke order is important (as used in character recognition), time information must be preserved itself independent of object layer information.

HandsDraw has a structural field and a data field in each page area. In the structural field, object creation order is preserved away from the object data.

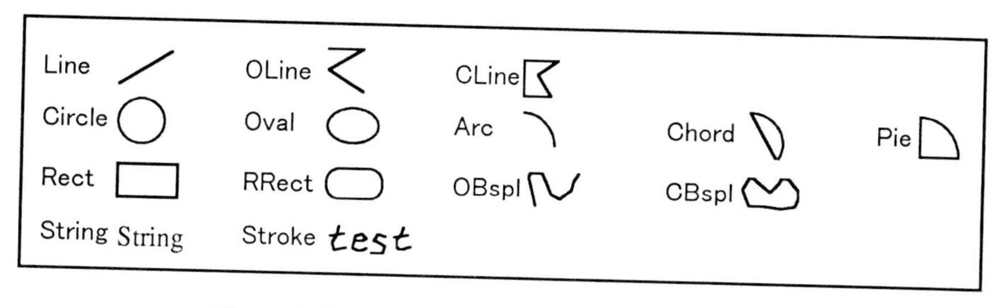

Figure 3. Examples of primitive objects.

```
Page
OLine [1 10 20 black white][1 20]
  [5 0 0 30 5 40 27 12 22 -10 70]
End OLine
Oval [2 70 30 black orange][1 20]
  [10 15 60 25 150]
End Oval

End Page
```

Figure 4. Example of HandsDraw description.

2.3. Preservation of String Width

Because HandsDraw may be used in many kinds of situations, some character fonts used in a HandsDraw document may not exist on a computer system. If another font is used, the character position may be changed. As shown in Figure 5, while a circle is drawn over the character "A" on the original (Courier font) string, the circle is at "st" on the next (Times Roman font) string. If a user wants to point out "a" by circling it, the Times Roman example does not agree.

For instance, Adobe PDF files can include font information in the document itself, so it can avoid this difficulty. But this font information is too heavy to send by e-mail.

HandsDraw solves the problem by keeping the whole string width. In the third example (revised), the string width is equal to the width of the first string (Courier). The string width is adjusted by the gaps between each character. Of course, each character position is not completely fixed, but is in almost the same as the original position.

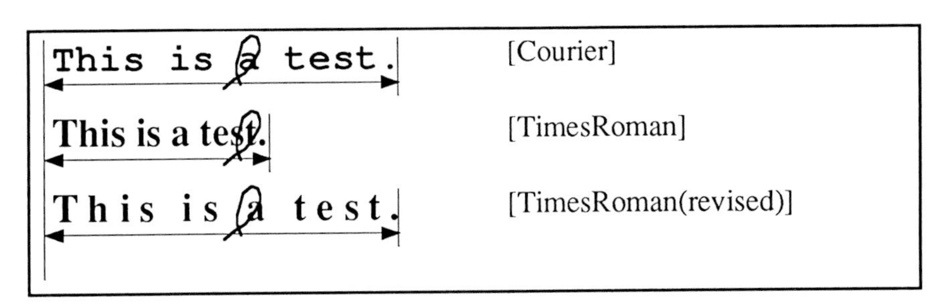

Figure 5. Font Width Revision

3. PROTOTYPING OF DIGITAL INK E-MAIL SYSTEM

We have made a prototype program for digital ink e-mail systems. This program can send a HandsDraw document to the mail server by SMTP (simple mail transfer protocol; rfc821) and receive it by POP (post office protocol; rfc1081). Usually, an ink e-mail message is encoded by MIME to tell the recipient that the message is written by HandsDraw.

On the Internet, there are some mail environments which cannot keep MIME headers. In such cases, binary files cannot be sent via the Internet. But HandsDraw mail can be sent in such environments because they are written only in text.

CONCLUSIONS

We have designed a new document format "HandsDraw" which can include digital ink with text and simple figures on the same digital paper. The main feature of the format is that it is completely open and is simple to implement easily by any programmer.

We have made a digital ink e-mail system using this format. We think that this format will make e-mail systems more friendly and useful for end users.

ACKNOWLEDGMENTS

This research is being partially supported by the Advanced Software Enrichment Project of IPA under MITI of Japan.

REFERENCES

1. "HandsDraw" specifications (http://hands.ei.tuat.ac.jp/).
2. Tanaka, H., Kato, N. and Nakagawa, M., "A Trial of Hands-Writing Environment on WWW Page," *Proceedings of the 12th Symposium on Human Interface (HIS 96)*, 22-25 October 1996.
3. Adobe Systems Incorporated, "Portable Document Format Reference Manual," Addison Wesley; 1993 (ISBN 0-201-62628-4).

Strategies for Integrating and Separating Pen-Based Operational States

S. Navaneetha Krishnan and Shinji Moriya

Department of Information and Communication Engineering, Tokyo Denki University, Tokyo, JAPAN. E-mail: krishnan@cck.dendai.ac.jp, moriya@c.dendai.ac.jp

Abstract

This paper focusses on pen-based operational states -- operand specification, menu opening, menu selection and execution -- and shows that there exist eight strategies for integrating and separating these operational states. An experiment was conducted to determine which of these eight strategies and menu shape are better, and statistically significant differences were found among these strategies, and among the two menu shapes.

1. INTRODUCTION

This paper focusses on integration and separation of pen-based operational states. Here the "operational states" are operand specification, menu opening, menu selection and execution. "Integration" means performing two or more operational states without lifting the pen-tip (tip of stylus pen) from the tablet surface, and "separation" is lifting the pen-tip after operational state(s).

In [1] the authors talked about performing all operational states without lifting the pen-tip, such as when editing (delete, move, etc.) data during discussions on big-sized pen-computers. While this would speed up operations, users might wish to lift the pen-tip at some point. We aimed to determine at what locations users perform" pen-up" and subsequent "pen-down", so that this knowledge can be used to design interfaces that reduce pointing errors during subsequent pen-downs, speed up overall operations, while allowing users to freely integrate and separate.

While research such as [2]-[4] target various aspects of integration and separation, this research is the first to -- (i) focus on strategies for integrating and separating operational states, (ii) show that there exist eight such strategies, (iii) perform experiment regarding these strategies.

2. INTEGRATION AND SEPARATION OF OPERATIONAL STATES

Figure 1 illustrates the integration and separation of operational states, by simulating the situation of jotting down the main points of this paper. Here "Compress" operation is used as an example. The labels "State 1" to "State 4" denote the four operational states and "Transition 1" to "Transition 3" denote the three state-transition locations at which action (such as keeping the pen-tip still for a brief period) enables transition from one operational state to the next. The handwritten text in Figure 1 is actually input "ink-data" (sequence of coordinate points input by writing on the tablet surface with stylus pen). The mock-up of pen was pasted later on.

In State 1 of Figure 1, the user specifies the operand by constructing a "rubber-band" between the points labeled ① and ②. Immediately after operand specification, in State 2 the "pie menu" opens at location ② with its centre exactly coinciding with the bottom-right corner of rubber-band. Without lifting the pen-tip from tablet surface, in State 3 he/she drags the pen-tip

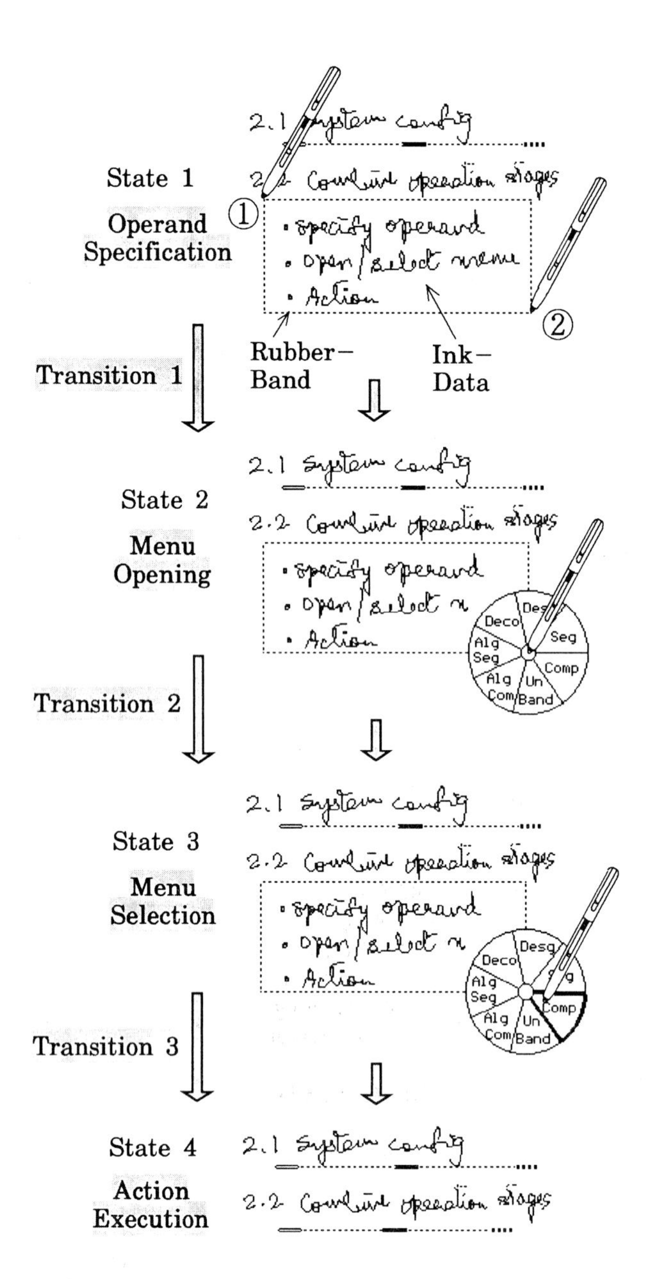

Figure 1. Integration and separation of pen-based operational states "State 1" to State 4". "Integration" is done by keeping the pen-tip down at the three state-transition locations, and "separation" is done by lifting the pen-tip at these locations.

to "Comp" menu (abbreviation of "Compress"), and lifts the pen-tip from the tablet surface. The moment pen-tip is lifted, execution ("compression" of ink-data) takes place in Stage 4.

In Figure 1, transition from State 1 to State 2 was achieved by keeping the pen-tip down at "Transition 1" location. That is, State 1 and State 2 are "integrated". In fact, all the operational states in Figure 1 are integrated since the pen-tip was lifted only after execution. If the pen-tip is lifted at location ②, then State 1 and State 2 are "separated". In this paper, when the pen-tip is lifted on reaching ②, the rubber-band remains visible with its bottom-right corner shown as thickened lines. The moment the pen-tip is put down on this thickened portion, the pie menu opens, and the operation continues.

In this paper, integration and separation are defined based on whether or not the pen-tip is lifted at the three state-transition locations. Though operational states could be integrated and separated in various ways, this paper focusses on "pen-up", because lifting the pen-tip and keeping it down can be said to be the most fundamental among pen-based operations, and which users can easily understand.

Figure 1 showed an example of "one-operand operations" (having one operand) where users specify the operand and select the menu. In this paper, we designed and implemented integration and separation for "two-operand" operations also,

Such an analysis and evaluation of the global process is made possible by the storage and representation mechanisms of our CSCW system. The information is stored in different files depending on the user, phase or session date. The systematic storage method permits the professor and/or the student to analyze the whole process, interactions among students, prosodic information associated to handwriting etc. Besides that, one can "replay" a whole session, or continue from a given phase of a session.

Another subject studied in this paper is the use of gestures, annotations and drawings during individual and collaborative writing. This form of interaction, typically used in daily life, conveys a lot of meta-textual information in an informal and efficient way. In terms of protocol, not only the action but also its modifiers and arguments are included in the gesture. Due to the variety of gestures reported in the literature, we performed a detailed study of such actions especially during the edition and revision phases, among University students. This study leads to a codification and therefore recognition of gestures and annotations, that are really employed by the users.

The proposed system was implemented in a TCP/IP network, that included a variety of operating systems (Unix and Microsoft for Pen Computing), or hardware (fixed PC's and workstations with a transparent tablet or notebooks with Touchpen capabilities), in order to test its applicability to different environments. The test was initially performed in a University course, that deals with Techniques of Composition, in the Faculty of Education, although more tests are planned for students of primary and secondary education. The obtained results confirmed our expectations about the importance of improved human-to-human and human-computer interaction. It is important to state that, during the experimental work, we faced several problems typical of a cooperative environment and of the use of small size tablets. In fact, a major issue in all multi-user interfaces is the limited screen space, of increased importance in pen-based computers [6]. Several windows have to be present in the workspace, besides the fact that especially cooperative phases require a large space.

On the other hand, *user awareness* is especially important in cooperative work, since for example the following questions are raised:

- *In what window region are the others working?,*
- *How can a user knows what non-visible parts of the workspace have already been used?,*
- *How we can distinguish the action of each user?*

The solutions adopted in PENCACOLAS are:

- Using the paradigm of a notebook with various pages, four graphic markers are used. The first three correspond to the page in which a certain user is currently located and the fourth one indicates how many pages have currently written on them. Scrollbars help in the displacement within the window, but only in the sense of changing a page.
- Cursors of different type identify the user and the action (write, erase, ...).
- In any phase, any student or the teacher can observe the work of the others, either for orientation or help purposes. Since these windows are read-only, these visualization windows were reduced to the 2/3 of the original size, without disrupting normal work.

In Figure 1 we can observe the cooperative planning window of a real session.

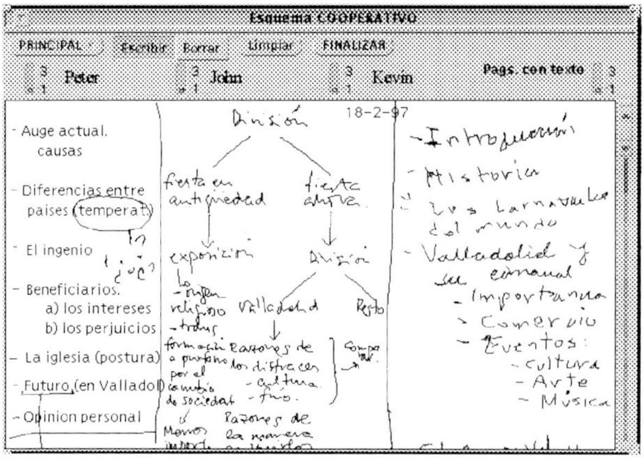

Figure 1. The Cooperative Planning Phase.

3. FUTURE WORK

Current research tasks deal with management of the reduced drawing space of the pen-computers, comparison with a Java-based system, and improvement of the intelligent agents and gesture recognition modules, according to the experience obtained till now. A classification of the events generated during cooperative work can help us in establishing dynamics links among them, relating actions or concepts. Then, interaction among users and the posterior analysis are greatly facilitated. A multiagent platform, called MAST, is being currently employed to analyze, design and implement intelligent agents in a Web-Java environment.

REFERENCES

1. Streitz, N. A., DOLPHIN: Integrated Meeting Support across Local and Remote Desktop Environments and LiveBoards, in Proceedings of CSCW'94, pp. 345-358, USA, 1994.
2. Ellis, C. A., Gibbs, S. J., and Rein, G. L., Groupware: Some issues and experiences. Communications of the ACM, Vol. 34, No. 1, pp. 38-58, January 1991.
3. Baecker, R.M., et al., The user-centered iterative design of collaborative writing software. In Proceedings of the InterCHI'93, pp. 399-405, 1993.
4. Mitchell, A., Posner, I., and Baecker, R., Learning to write together Using Groupware. In Proceedings of CHI'95 ACM, 1995, available in http://www.acm.org/sigchi/chi95/.
5. Rodden, T., and Smith, G., An access model for shared interfaces. Collaborative Computing. Vol. 1, No. 2, pp. 109-126, June 1994.
6. Kimura, T. D., A Pen-Based Prosodic User Interface for Schoolchildren, *IEEE Multimedia*, Vol. 4, No. 3, pp. 48-55, 1996.

The conceptual framework of Preference-based Design

Masato Ujigawa

Chief Researcher, Research and Development Institute, Takenaka Corporation, 1-5-1 Ohtsuka, Inzai, Chiba, Japan

The importance of preference and attractiveness has been increasing, not only in product design but also in the design of all computer-related systems. Although they are the contact points between computer systems and the human spirit, little attention has been given to them in academic fields. Since 1991, the author has been holding interdisciplinary meetings to discuss theories and methods for creating attractive things.

Effective theories and methods were classified and a table of technical systems for preference-based design was made. Since the start-up of the Cyber Laboratory for Preference-based Design Forum in September 1996 to continue discussions through the Internet, some researchers of computer-human interface have joined the laboratory, and two modeling techniques have been developed there. On a side note, prototyping was added in the table of technical systems for preference-based design as a method of designing.

1. INTRODUCTION

Following years of competition in price and quality, we are entering a new age when attractiveness of products has become very important. To develop attractive products and systems, the author proposed to formulate rational approaches for "preference-based design." Preference is an attitude of humans, and attractiveness is an attribute of products or systems. They are different aspects of the same phenomenon.

Preference is based on the user's value system, and is related to many fields, including psychology, sociology, and art. There are many kind of problems and issues related to attractiveness that arise when creating new products. The author conducted some workshops in Tokyo entitled 'Miryoku Engineering Forum' for the purpose of discussing these problems and gathering effective theories and methods. ('Miryoku' is a Japanese word meaning 'attractiveness.' Here, we use 'preference-based design' in place of 'miryoku engineering'.) This paper introduces basic concepts and the "Cyber Laboratory for Preference-based Design Forum."

2. A FRAMEWORK OF PREFERENCE-BASED DESIGN

At the 'Miryoku Engineering Forums,' some phenomena related to attractiveness or preferences, methods to grasp major attributes of consumers' product selection, and actual cases of successful creation of attractive products have been introduced and discussed.

For example, a researcher of land planning introduced an attraction coefficient of gravity model, in which with the decline of traffic friction caused by the development of transportation and communication networks, attractiveness will have greater influence on people's activities and land use. An environmental psychologist showed a new interview method based on clinical psychology to grasp attractive attributes. Conjoint analysis was introduced by a marketing researcher, using a method which can measure contributions of attributes to the overall attractiveness of a product.

Lectures in concepts and methods of modeling of social systems were given by a professor of systems engineering. In addition, case studies of skiing, the structure of interpersonal relations, falling in love at first sight, and the allure of mountain climbing and skiing have often been presented, as have successful developments of resorts and new automobile models. Through these discussions, a system of approaches for creating attractive products was created (see Table 1). This system is composed of three areas: basic theories, modeling (techniques for research and analysis), and designing.

Table 1
A system of aproachs for creating attractive products

AREA		DISTINGUISHING FEATURE	THEORY AND METHOD
1. Basic theories	Definition		
	Recognition	Insclusive Recognition	
		Pattern Recognition	
	Learning		
	Value system	Fashion, Character goods	Reference groups
2. Modeling	Structure identification	In-depth Interview	Evaluation Grid Method
			Paired Comparison
	Parameter	Composite Effect	Regression Analysis, Conjoint Analysis
	identification	Segmentation	Cluster Analysis
3.Designing	Planning	On-site Thinking	Scenario making
	Design strategy	Positioning	Cognitive Map
	Materializing	Prototiping	User participation

2.1 Basic theories

The basic theories' field includes the definition of attractiveness, the mechanism of recognition, learning, value systems etc., the relative study fields of which are psychology, cognitive psychology and sociology.

2.2 Modeling

In systems engineering, there are two processes for modeling: *structure identification* and *parameter identification*, which are also used to classify methods and techniques for research and analysis. The former determines the elements of a model and the relationships between these elements. The latter process examines relationships quantitatively.

(1) Techniques for structure identification

In-depth interviews, the evaluation grid method, paired comparison, and interpretive structural modeling are used for structure identification. Consumers' actual behavior is often much different from what their answers to survey questions would indicate. Therefore, methods must be devised to provide real evaluation items for attractiveness. Kelly (1955), a clinical psychologist, developed the *repertory grid method* originally to grasp the mechanism of people's understanding and recognition of their surroundings, especially human relationships, in the 1960's. This method is conducted through a personal interview where a person is asked what the similarities and dissimilarities are between the objects A and B. From their response, how and with what kind of unit the objects are recognized becomes apparent.

Sanui, a Japanese researcher who had learnt the *repertory grid method*, improved it in two ways. First, in the comparison of objects to be assessed, people are asked to answer what is good or bad, and what they like or dislike about them. Second, the meaning or conditions of their answer to the questions are clarified through additional questions. This makes it possible to grasp the mechanism of their construct system hierarchically. This method is called the *evaluation grid method*.

(2) Techniques for parameter identification

For parameter identification, regression analysis, conjoint analysis, and analytic hierarchy process are applicable. The method frequently adopted in conventional studies on consumer behavior was mainly the multi attribute attitude approach. In this method, first, the reasons for consumers' selection of products were obtained by means of an opinion poll. Then, some attributes of the product were selected based on these reasons, and the consumers were asked to rate the level of importance of these attributes. Finally, the importance of each attribute was obtained by weighing and accumulating each evaluation of the attributes. However, the accuracy of prediction made by this method is usually very low since consumers tend not to explicate their real feelings in opinion polls.

The idea of applying conjoint analysis to the field of marketing research first appeared in the 1970's. In the *conjoint approach*, consumers are not directly asked their opinions on the weight of attributes. Instead, the consumers' shopping behavior or preferences are observed and then the reasons for such behavior or preferences are determined. In this method, an overall evaluation is made first, then the effect of each attribute is deduced. This is a method to analyze the causal structure of consumers' preferences.

2.3 Designing

Cognitive map, scenario making, and on-site thinking are examples of methods for design.

Cognitive map is adopted widely as an effective method to compare existing products and to consider characteristics of new products.

3. CYBER LABORATORY FOR PREFERENCE-BASED DESIGN

The system of approaches for creating attractive products is not completed, and will be revised as more knowledge and information becomes available on preferences.

To integrate this knowledge and information, a network called "Cyber Laboratory for Preference-based Design Forum" was started in September 1996 by the participants of the 'Miryoku Engineering Forum'. A mailing list system and home page are made available on the Internet and off-line workshops have been held on a mostly monthly basis. Usually, the evaluation grid method needs personal interviews. To reduce the time and trouble of personal interviews, E-mail interviews were tested as an activity of the Cyber Laboratory. Conjoint analysis also requires interviews, which were attempted using WWW.

4. TO REVISE THE SYSTEM OF APPROACHES FOR CREATING ATTRACTIVE PRODUCTS

Participants in the earlier 'Miryoku Engineering forum' in 1991 consisted of researchers from land planning, environmental psychology and marketing research. Members' specialities later increased to cover policy planning, visual environment, car design, architecture and statistics. After the Cyber Laboratory for Preference-based Design Forum started, some computer-human interface researchers joined. Prototyping was also recognized as an important method to realize user's needs. Even in urban planning prototyping will be adopted by using virtual reality techniques.

The creation of attractive products or systems is an internationally-universal theme. The integration of knowledge and understanding should therefore certainly be significant.

REFERENCES

1. Kelly,G.A.:The Psychology of Personnal Constructs, Vol. 1& 2, W.W.Norton (1955)
2. Junichiro Sanui:Visualization of users' requirements : Introduction of the Evaluation Grid Method. in Proceedings of the 3rd Design & Decision Support Systems in Architecture & Urban Planning Conference. Vol.1, pp.365-374. (1996)
3. Green, P. E. and Rao, V. R., Conjoint measurement for quantifying judgmental data. Journal of Marketing Research, 8 (1971)
4. M.Sugeno and T.Yasukawa, A Fuzzy-Logic Based Approach to Qualitative Modeling, IEEE Transactuins on Fuzzy Syatems, 1-1, 7/31(1993)
5. Miryoku Engineering Forum: Miryoku Engineering. Kaibundo (1992)(Japanese)

Revealing of preference structure by the Evaluation Grid Method.

Junichiro Sanui[a] and Gen Maruyama[b]

[a]Senior Researcher, Nissan Research Center, Nissan Motor Co., LTD.
1 Natushima-cho, Yokosuka 237, Japan.

[b]Technology Development Dept. 1, Taisei Corporation
Sanken Bldg., 25-1, Hyakunin-cho 3-chome, Shinjuku-ku, Tokyo 169, Japan

The Evaluation Grid Method (EGM), a semi-structured interview method to reveal people's preference structure proposed in 1986 [1], has stimulated a number of academic and business studies in Japan [2]. The first part of this paper explains the procedure of EGM in detail, while the second part examines the authors' attempts to raise the efficiency of EGM by using e-mail instead of face-to-face interviews.

1. INTRODUCTION

With the maturation of the consumer market, users' concepts of good design have shifted in many areas from design that provides comfort or convenience to that which is attractive. However, these concepts are rarely conveyed to designers, whose own ideas are often incompatible. The diversity of ideas among designers themselves adds to the problem. Several qualitative methods such as open-ended interviews, group interviews, and in-depth interviews have been proposed to enable direct communication between designers and users. Nonetheless, they have turned out to be inefficient mainly because of their heavy reliance on the ability and interpersonal skills of the interviewer. Moreover, there is no methodological protection against a biased outcome caused by the subjectivity of the interviewer.

2. THE EVALUATION GRID METHOD

The Evaluation Grid Method (EGM) has been developed as an extension of Kelly's Repertory Grid Method [3] to overcome the above mentioned shortcomings of the conventional interview methods. A general outline of EGM is shown in Figure 1. EGM proceeds in two stages. First, given a pair of design solutions, or elements, the participants are asked to state why they consider one element to be preferable to the other. Second, these reasons (i.e., requirements) are organized into a hierarchical diagram with the aid of guided questions to elicit antecedents and consequents. The procedure can be likened to laddering. By accumulating individual diagrams, we can obtain a group diagram on which the strength of links is indicated by the frequency of cases.

3. AN EGM EXPERIMENT BY E-MAIL

3.1. Research on gas station preferences

As stated above, the largest merit of EGM is that users' preference structure can be elicited as they are. However, as long as EGM is carried out in face-to-face mode, the number of participants is inevitably limited within a given research budget. Thus, the interviewees are restricted by time, and interviewers have to conduct personal interviews on an individual basis.

This section introduces EGM research on gas station preferences conducted by e-mail among voluntary participants. We adopted e-mail in order to improve the efficiency of EGM without losing its merits. The procedure was as follows. Participants were asked to rank 5 gas stations they frequently use according to their overall preferences and to state the reasons why these stations were so ranked. This was the elicitation of the original components of the preference structure. Upon receiving a participant's completed questionnaire, the interviewer sent another e-mail questionnaire to confirm the meaning of the stated reasons and to obtain the antecedents and consequents of the preference components obtained in the primary questionnaire. Since there are no restrictions of time and space in e-mail, we could ask these further questions even during the process of

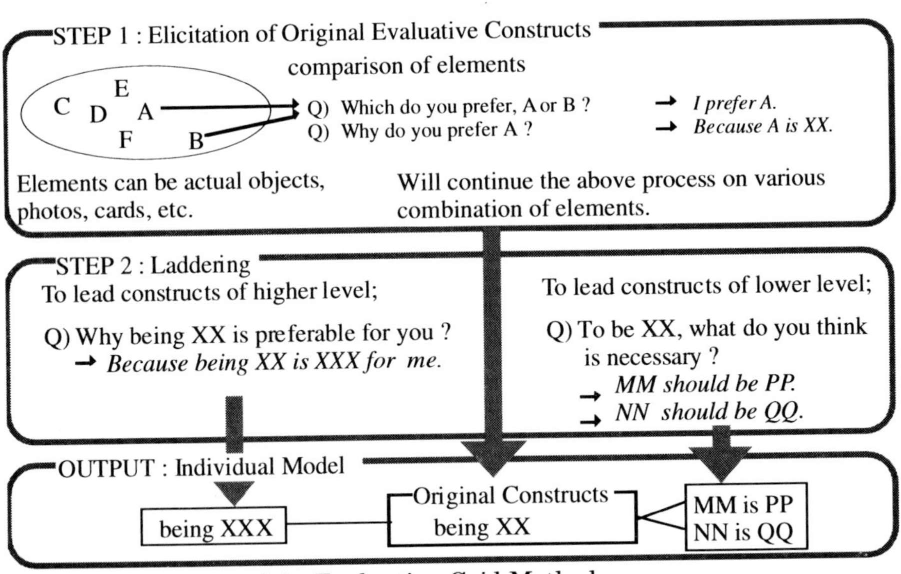

Figure 1. An outline of the Evaluation Grid Method

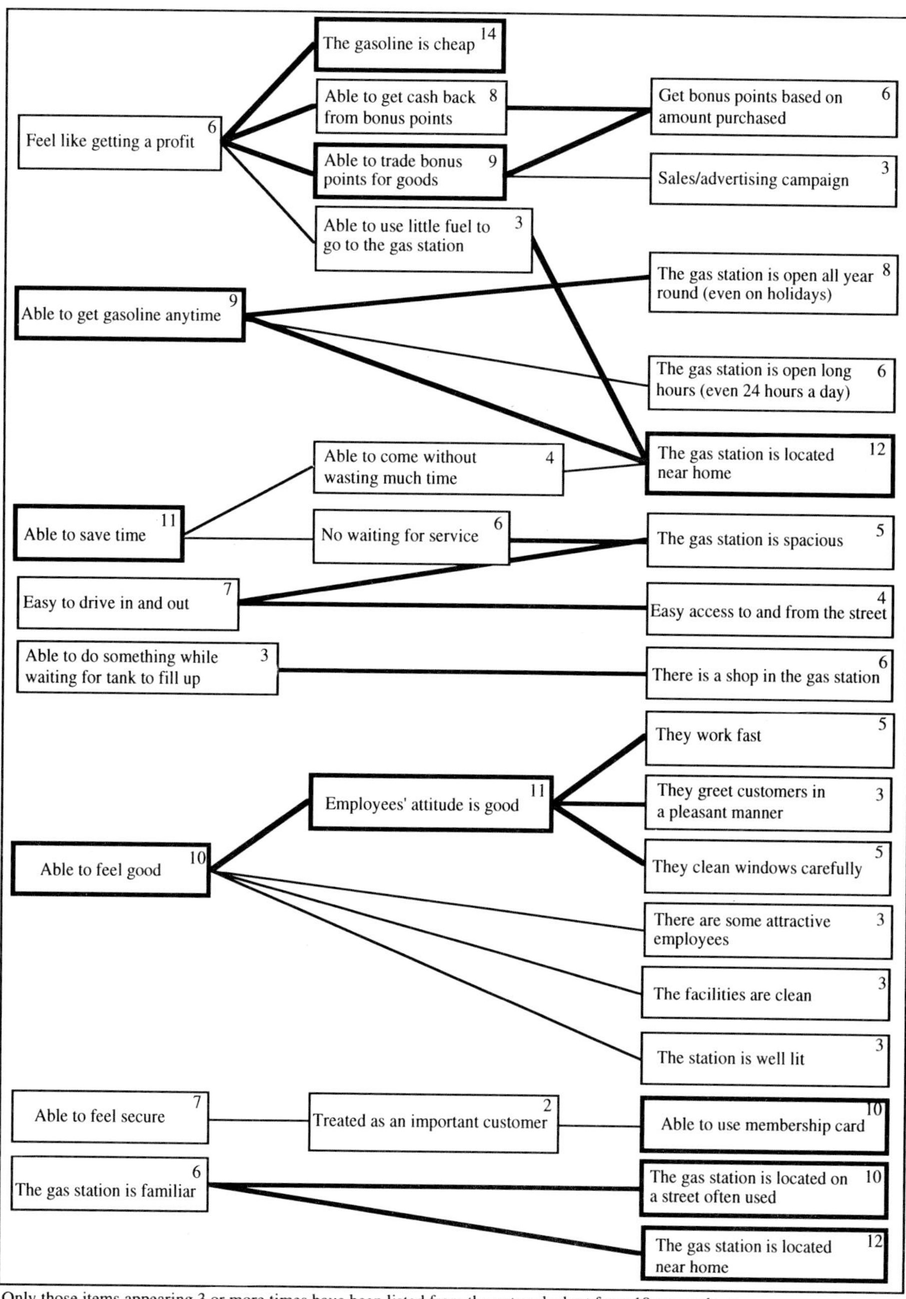

Only those items appearing 3 or more times have been listed from the network chart from 18 respondents.
The numbers in the boxes represent the number of times the respective items appears in the responses.
The items in the boxes with bold outlines appeared at least 9 times, while the bold connecting lines represent at least 3 responses.

Figure 2. A hierarical diagram of gas-station preferences

analysis. This was an unexpected merit of adopting e-mail.

There were 18 participants responding to the questionnaire. By accumulating each participant's diagram, a gas station preference structure of the group was produced, as shown in Figure 2. This integrated model presents the extent and structure of users' viewpoints in evaluating gas stations. In other words, this model can present an overview in which one can check the strong and weak points of an existing gas station to find the needs which are still unsatisfied.

3.2. Problems with using EGM in an e-mail network

The quality of the results was comparable to that of ordinary EGM in face-to-face interview mode even though the interviewer's labor has been drastically reduced. However, a few drawbacks were reported. For instance, typing responses was troublesome to some participants (ordinarily, the interviewer would have written them down). A more serious problem, however, was the difficulty in expressing or understanding subtle nuances by e-mail. In such a method with no personal contact, we cannot read subtle nuances from the respondent's tone of voice and way of answering as we can in a face-to-face interview. Therefore, since interviewers are required to "read between the lines," responses in this study were confirmed by the additional questionnaire. At the same time, interviewers must pay special attention to using concrete, explicit expressions, particularly in the questionnaire for the confirmation.

4. CONCLUSION

E-mail is becoming increasingly popular in Japan. Sooner or later, it will be a very common communications medium. Despite its shortcomings, this experiment confirmed the high potential of the combination of EGM and e-mail as an efficient tool to bridge the gap of ideas between designers and users. We will further develop this method as a practical tool for conducting qualitative surveys on a large number of people in various design fields.

REFERENCES

1. Sanui, J. and Inui, M. (1986) Phenomenological approach to the evaluation of places: A study on the construct system associated with the place evaluation (1), Journal of Architecture, Planning and Environmental Engineering, No. 367, pp. 15-22. (in Japanese)
2. Junichiro SANUI (1996) Visualization of users' requirements : Introduction of the Evaluation Grid Method.
   In Proceedings of the 3rd Design & Decision Support Systems in Architecture & Urban Planning Conference. Vol.1, pp.365-374.
3. Kelly, G. A. (1955) The Psychology of Personal Constructs, Vols. 1 and 2. W. W. Norton, New York.

Conjoint Analysis of Consumer Preferences in Cyber Space

Makoto Mizuno

R&D Division, Hakuhodo Inc., 3-4-1 Shibaura, Minato-ku, Tokyo 108, Japan

The present work demonstrates the feasibility of conjoint analysis on the World Wide Web (WWW) which could provide more accurate information if the task is sufficiently small. However, a more advanced method is needed to allow heterogeneity in the consumer preferences.

1. INTRODUCTION

Conjoint analysis is a method for decomposing preference for a product to part-worth for each of its attributes under the premise that consumers can reveal their preference for a whole product rather than each attribute. Usually, respondents are presented a set of product profiles as bundles of attribute-levels and are asked to choose, rank or rate the profiles according to their preference. By means of a regression-like method, the part-worth for each attribute is estimated so as to fit the responses observed. (See [1] for details.)

Recent progress in information technology offers ample opportunities for conjoint analysis. The use of WWW will help to dramatically reduce time and cost. Full-motion pictures or virtual reality will enrich the reality of conjoint tasks like some stand-alone non-conjoint systems. (See [2] [3] for example.) Those systems can be extend to be linked to cyber space if necessary. Also, their analytical capability can be enhanced by incorporated conjoint analysis into them.

2. HYPOTHESES

When conducting conjoint analysis in cyber space, one of the most important problems is how to motivate respondents to answer all the questions. If respondents feel bored, they will skip to another site immediately since they are prone to mentally overloaded by conjoint analysis. Moreover, even if respondents do answer all the questions, their responses could end up being random or fictitious, deteriorating the validity of the results.

Thus, on the one hand, practical consideration leads to a hypothesis:

H1. As the number of choices in a respondent's conjoint task increases, the validity of the model will decrease.

On the other hand, the statistical consideration leads to the contradicting hypothesis:

H2. As the number of choices in a respondent's conjoint task increases, the validity of the model will increase.

3. EXPERIMENT

We conducted an experimental study on WWW to test the above hypotheses. Visitors to the site of the "Preference-Based Design Forum" were invited to engage in a conjoint task on gas stations. If they accepted, they were presented a set of pairs of profiles describing hypothetical gas stations and were asked to choose the preferable one from each pair. About 200 people participated in the task from November 1996 to January 1997.

3.1 Attributes, levels and profiles

The following attributes were selected through a discussion in a mailing list of the Forum: 1) price discount rate of gasoline, 2) frequent user program, 3) service of employees. The levels of each attribute are shown in Table 1.

An orthogonal array provided 9 profiles, resulting in 36 possible pairs. Since this number was too large for a respondent, the pairs were split into two sets of 18 pairs each. One of these sets was presented to a respondent according to whether the last figure in his/her license number is odd or even.

Table 1.
The attributes and levels used in the experiment

Attribute	*Level 1*	*Level 2*	*Level 3*
Price Discount Rate of Gasoline	0%	5%	10%
Frequent User Program	Not Available	Available	
Service of Employees	Self Service	Average	Friendly and Careful

3.2 Estimation

After excluding incomplete responses, we analyzed the pooled responses by the logit choice model (See [4] for details). As shown in the last column of Table 2, the parameters estimated for all the responses suggest that when choosing a gas station, the average driver tends to consider the price of gasoline most important, followed by 'friendly and careful' service.

Table 2.
The results of conjoint analysis

Depth of Pairwise Comparison	*6*	*9*	*12*	*15*	*18*
Parameters					
Price Discount Rate	.6345	.5987	.5459	.4764	.4302
of Gasoline	(15.297)	(18.232)	(23.732)	(26.963)	(29.502)
Frequency Program	1.0514	.5012	.5981	.0458	.1699
- Available	(5.180)	(3.880)	(5.401)	(.564)	(2.358)
Service of Employees	-1.0034	-.6213	-.3186	-.3312	-.5692
- Self Service	(-4.076)	(-5.187)	(-3.273)	(-3.682)	(-7.079)
Service of Employees	1.1155	1.5893	.9688	.9983	.6899
- Friendly and Careful	(5.624)	(10.224)	(8.700)	(10.193)	(8.259)
Log Likelihood	-224.2	-436.5	-681.6	-978.1	-1204.1
Internal Validity					
Rooted Mean Square Error	.2129	.2673	.3017	.3294	.3358
Hit Rate	.9534	.9088	.8678	.8248	.8369
External Validity					
Rooted Mean Square Error	.3645	.3929	.4463	.3722	-
Hit Rate	.7505	.8142	.6740	.8076	-

Note: The values were set to zero for the 1st level of each attribute, "Frequency Program-Not Available" and "Service of Employees-Average." "Price discount rate of gasoline" is treated as the continuous variable in the estimation. Figures in parentheses are asymptotic t values.

3.3 Validation

Validity is measured by two indices: the rooted mean square error (RMSE) and the hit rate (HIT). The former is an index of discrepancy between actual choices and predicted choice probabilities, while the latter is the proportion of successfully predicted choices to total choices. For our purpose, we repeated analysis varying in the depth of pairwise comparisons used for parameter estimation: first 6 pairs, 9 pairs, 12 pairs and 15 pairs.

There are two notions of validity. In the first notion, internal validity indicates how the data used for estimation can be replicated by the model estimated. Table 2 shows that as the depth of pairwise comparisons increases, both indices of the internal validity deteriorate. In other words, the responses will become internally less consistent with increasing depth of pair comparison.

In the second notion, external validity indicates how the model estimated can predict the data *not* used for estimation, that is hold-out data. Except for one case where all pairs are used for estimation, the last three pairs were hold out.

The results of external validation, unlike internal validation, are not straightforward. RMSE seems to be an inverted U-shape function of the depth of pairwise comparison; HIT seems relatively to decrease with the depth. Yet, at least a shorter pairwise comparison shows rather high external validity.

4. CONCLUSION

Our experiment illustrates that conjoint analysis basically can be implemented on WWW and provides as accurate prediction of responses as traditional conjoint analysis does. *H1* was supported more clearly than *H2* in internal validity and more weakly in external validity. Reducing the number of comparisons per respondent, therefore, is desirable both for keeping respondents motivated and for obtaining accurate information on consumer choice.

We might see different results if we adopted an individual-level or a small segment-level analysis, which is recommended when we should take into account the heterogeneity of consumer preference. (See [6].) When we use the data from a few respondents for estimation, the estimation will possibly yield unreliable estimation if the number of comparisons per respondent is small. On the other hand, the motivation of respondents would decrease if the number of comparisons per respondent is large. To solve this dilemma, Katahira and Mizuno [5] have proposed a method of pairwise comparison that gives more accurate information under a limited number of comparisons. It can be incorporated into the conjoint system on WWW.

In conclusion, there is much room for improvement in our system. The current interface is tremendously far from the ideal level. No wonder introducing multimedia technology into the system is fruitful. In the near future, conjoint analysis will be incorporated as a means for measuring the customer's preference in a broader range of information systems in cyber space where commerce and research will converged.

REFERENCES

1. P. E. Green, D. S. Tull and G. Albaum, Research for Marketing Decisions, 5th edition, Prentice-Hall, 1988.
2. G. L. Urban, B. D. Weinberg, and J. R. Hauser, Journal of Marketing, 60, January (1996) 47-60.
3. R. R. Burke, Harvard Business Review, March-April (1996) 120-131.
4. G. S. Maddala, Limited-dependent and qualitative variables in econometrics, Cambridge University Press, 1983.
5. H. Katahira and M. Mizuno, An individualized/interactive conjoint analysis, mimeo, 1997.
6. P. E. Green and A. M. Krieger, Management Science, 42 (1996) 850-867.

Interactive Support for Decision Making

Noriyuki Matsuda[a] and Ken Nakamura[b]

[a]Institute of Policy and Planning Sciences, University of Tsukuba, 1-1-1 Tennou-dai, Tsukuba, Ibaraki 305, Japan

[b]Nissan Motor Co., Ltd., Nissan Technical Center, 560-2 Okatsukoku, Atsugi, Kanagawa 243-01, Japan

1. INTRODUCTION

People often rely on their "kansei" in making judgments and decisions in practical situations where not only the available information is limited, but the heuristics tend to be highly intuitive. Although this may provide satisfying results, the quality of these results can be improved with an appropriate supporting tool in an interactive environment.

The purpose of the present paper is to demonstrate the potential use of AHP (Analytic Hierarchy Process) proposed by Saaty (1977, 1980) in the selection of a new automobile design. It is our hope that our method, when fully developed, can assist the decision making of both experts and lay people.

1.1. AHP

Saaty's AHP proceeds in three stages: (a) hierarchical structuring of evaluation items, (b) pair-wise weighting of items of the same class and the evaluation of objects under study, and, (c) computation of priorities among the objects which is used to make a final decision.

Although AHP is appealing to practitioners, one of its most severe impediments is its need for repetitive comparisons, particularly when the number of items and classes is large. Our visual-programming approach should lessen the mental strain in this respect without violating the original premise underlying the pairwise weighting.

1.2. Visual Authoring Tool

Of the other visual authoring tools available, we chose Oracle Media Objects for its interactive versatility in stimulus presentation and data collection. Moreover, its programming environment is suitable for implementing AHP and feeding back its results.

2. METHOD

2.1. Subjects and Stimuli

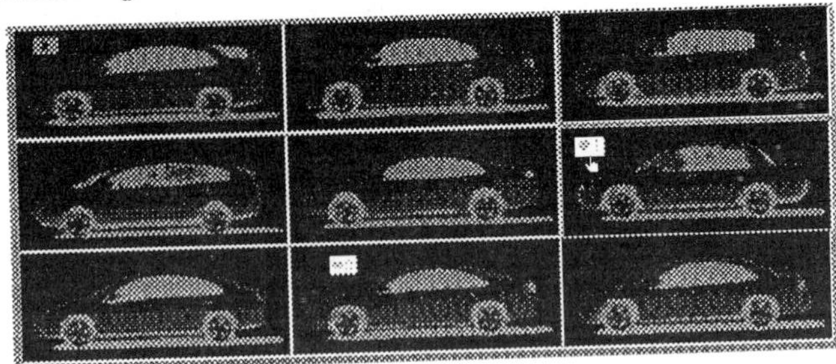

Figure 1. An example of the stimuli set
(The graphic quality is intentionally lowered.)

Subjects. The subjects who participated in the experiment were 33 recruited consumers (Group P), 33 employees of an auto-maker including designers (Group N), and 25 design majors (Group SD) and 28 other majors (Group SS) from the University of Tsukuba. The experi ment was individually conducted on PowerMacs with human assistance.

Stimuli. A side and front view of a prototype was designed for a forthcoming model of a passenger car of particular interest. Then, alternative designs were digitally generated by varying the degree of wedgedness and other critical elements. There were two such sets for the side views (Sets a and c) and one for the front view (Set b) presented to the subjects as stimuli in the order of Sets a, c and b.

2.2. Procedure

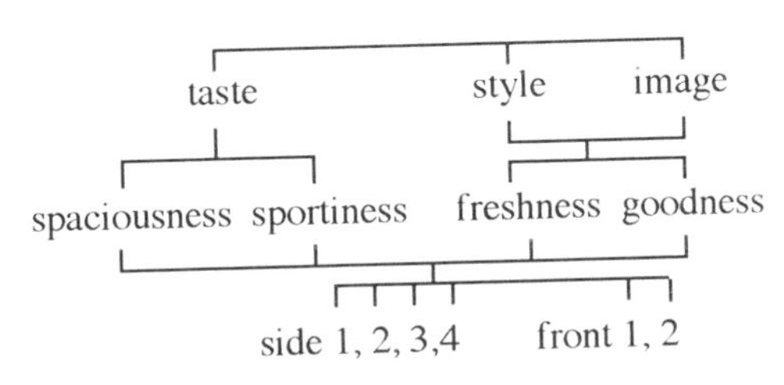

Figure 2. Evaluation items in AHP

After practice sessions intended to activate awareness of model changes, the subjects selected in two stages the best side- and front-view designs from among the alternative sets that would succeed the early and present models also shown on the display.

The selection in the first stage was direct in that the subjects identified their choices by marking the picture with special objects for the best, second best and worst candidates (Figure 1).

In the second stage, the best designs were determined by AHP from the candidates chosen in the first stage—there were four and two such candidates for the side and front views, respectively. The subjects evaluated them according to the item hierarchy in Figure 2 upwardly along 1-dimensional scales (Figure 3). A 2-dimensional version was employed for the style and image subclasses (Figure 4). The subjects were not informed in advance of the priority computation by AHP.

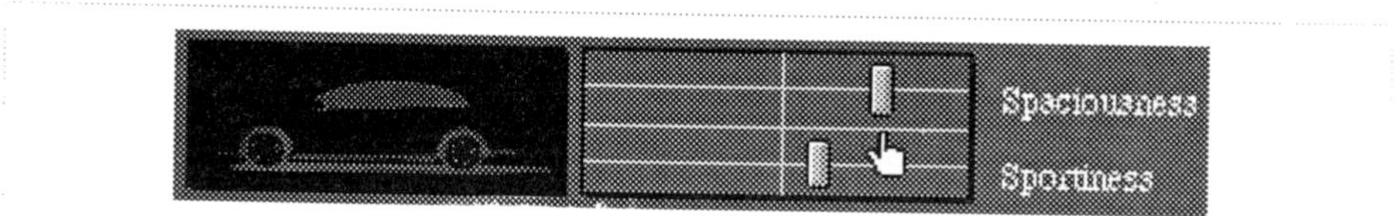

Figure 3. 1-Dimensional scaling

The candidateş with the highest AHP priority in each view were subsequently presented to the subject subsequently as their own best choices, about which they answered multiple-choice questions regarding their impressions of five aspects of the best choices. The questions were broken to a quasi-conditional form with three precedent and two consequent aspects.

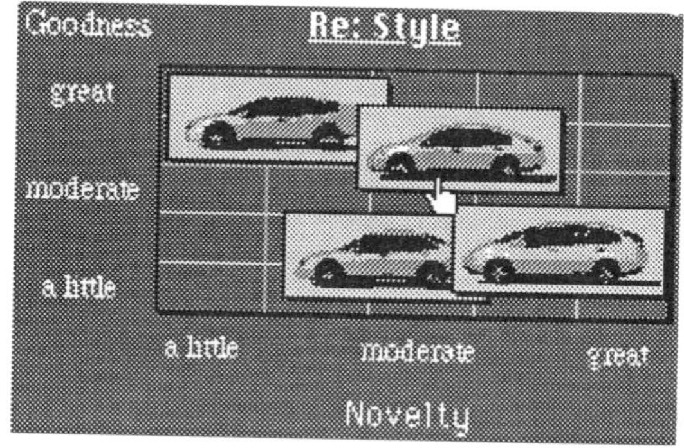

Figure 4. 2-Dimensional scaling

3. RESULTS

Some preliminary results are reported below.

3.1. The primacy of the prototypes

(1) *Stage 1.* The side-view prototype was most frequently chosen by all subject groups as best among the alternative in both sets with two exceptions. Its modality in the relative frequency ranged from 21.1 to 42.4 percent. Even when it was second modal (Set a of Groups SG and SS), it received no less than 20% of the votes. Whether it was the top or the second primary choice, the percentages were higher in Set c in all groups. Hence, the primacy was relative to the alternatives to be compared.

The primacy of the front-view prototype was limited only to Group N. It was the second primary choice in Groups P (21.2%) and SS (17.9%). In the latter group, it was tied with two other alternatives. In Group SG, it received only 8.0% support.

Neither the side nor the front prototype was the worst choice in any group.

(2) *Stage 2.* In contrast to the aforementioned tendency, primacy was observed only for the side view as the modal and the second modal choice in Groups P (33.4%) and SS (21.8%), respectively. Although the front prototypes were the second modal choice, the percentage was generally low (<17.9%).

3.2. Reversal of the best candidates

Since the subjects had been uninformed of the selection through AHP, a reversal of the best designs was well anticipated between the two stages of selection. As shown in Table 1, in 35.7 to 52.0% of the cases, the second best candidates in the first stage became the best in the second stage. Though not statistically tested, there was some interaction between *view* and *group*.

Table 1. Proportion of reversals by group.

View	Group P (33)	Group N (33)	Group SG (25)	Group SS (28)
Side	48.5	42.4	52.0	35.7
Front	51.5	36.4	40.0	35.7

The number of subjects is shown in parentheses.

4. DISCUSSION

The appeal of the prototypes turned out to be weaker than what the designers had expected in generality and in extent. Prototypes' primacy was limited to the side view and they never became the majority's choice. Moreover, the explicit verbal/numerical ratings in AHP weakened their appeal.

The presentation of the best designs selected by AHP was a pleasant surprise to many subjects, especially when preference reversal occurred. They became more interested in their own decision making processes. This observation encourages us to further develop the current method toward a fully interactive one in which the subjects can repeat the process until they are satisfied. One promising way to achieve this goal is the implementation of the internal and external anchoring of Matsuda and Namatame (1995).

REFERENCES

Matsuda, N. and Namatame, M. (1995). Interactive measurement of hierarchically related consumers' images. *Behaviormetrica*, 22:129-143.

Saaty, T.L. (1977). A scaling method for priorities on hierarchical structures. *Journal of Mathematical Psychology*, 15:234-281.

Saaty, T.L. (1980). *The Analytic Hierarchical Process*. New York: McGraw-Hill.

A Design and Implementation of Cyber Laboratory

Tomoyuki Tsunoda

Chief Researcher, the Institute of Social Environment Systems, Inc.
2-5-22 Suido, Bunkyo, Tokyo, 112, Japan

1. INTRODUCTION

In these past few years, in Japan, in the same trend as the rest of the world, the user environment for the Internet has rapidly been fully fitted out. As of December, 1996 there were 11,831 domains and the number of host computers connected to the Internet surpassed 710,000 [1]. The commercial service of the Internet, which began in 1994, has developed to more than 700 service providers in March of 1997 [2]. The author has started with the making of a new word "Cyber Laboratory" to grope with the new styles of research organization imagined to be necessary for the highly connected network society [3]. This paper introduces the concept of "Cyber Laboratory" and an example of a pilot model of it, and reports current progress of implementation of a cyber laboratory for the *Preference-based Design Forum.*

2. DEFINITION OF CYBER LABORATORY

Cyber laboratory is defined as "a research organization using an electronic network to its optimum capability for the research activities."

Not all the research activities are necessarily performed on the Net. Activities on the Net and off-line activities are not a multiple-choice problem, but rather in a mutually beneficial relationship (Figure 1).

On-line works > Off-line works

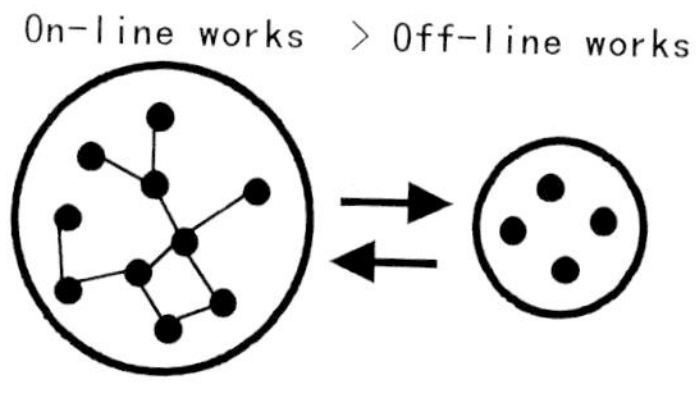

Figure 1. Concept of Cyber Laboratory.

3. CYBER LABORATORY PILOT MODEL

3.1 Survey of Network Usage

A survey was made of network (Internet, BBS, Intranet) using the methods as outlined in Table 1, to determine the present use and the future anticipated use of technology and services.

Table 1
Methods Used to Survey Network Usage

Survey Points	Survey Method
What is being done	Net surfing, Web & E-mail interviews, News & document surveys
What is possible	News & document surveys
What is hoped for	Web & E-mail interviews, Metaphors [4], Analogies [4]

The web interviews were carried out by putting questionnaires on the web and getting replies by e-mail. When further questions were arisen from replies received, they were asked by e-mail additionally. Thus, the web interviews opened out the continual e-mail interviews. As is also intended by the Why Method [4] and the Evaluation Grid Method [5], just by asking the user a question once, you are unable to grasp the potential requirements of users, but by asking again and again these can be realized. The effects of these continual questions were high, and even though it was realized that certain requirements were not possible at the present moment, it could be seen that users held real hopes, such as "wanting to use it for better decision making" or "co-authoring of research papers." Also, by these continual questions, the hierarchical structure of services and effects of cyber laboratory was clarified easier, such as, expression using the special characteristics of multi-media (video, animation, etc.) → many different types of expression being possible → becoming easier to understand → obtaining mutual understanding → activating discussions.

3.2 Pilot Model

Figure 2 shows a concept model as designed by the author as the pilot model of a cyber laboratory. The author worked on a model of hypothetical political science research activity, but it is designed to be have the services necessary across the board in any research organization. When setting up the facilities, it would presumably be necessary to add various facilities and services according to its organizational characteristics, for example a testing room in a chemical research facility.

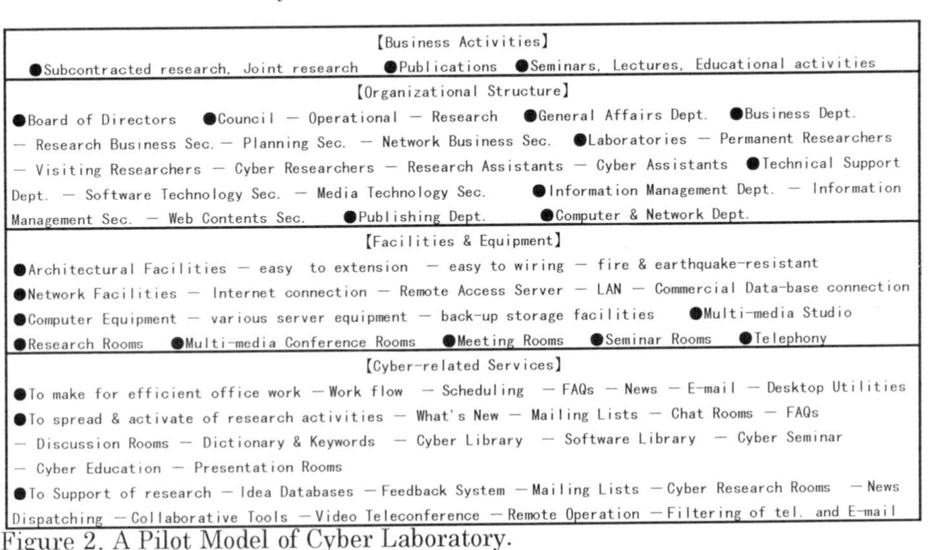

Figure 2. A Pilot Model of Cyber Laboratory.

In this model, there are cyber researchers and cyber assistants on the network, and the researchers in the building and people co-operating from the outside harmoniously develop their research activities. From this point, it is possible to set-up a wider team than before. Also, there are many staff such as the network engineers, the software engineers, the multi-media engineers and the contents creators all supporting the activities of the researchers. The service with the object of making for "efficient office work" is made up of various group-ware functions, and is actually implemented in the organization's closed network, Intranet, but can also be accessed from remote. The services with objects of both the "spread & activation of research activities" and "support of research" appear on the network, Internet. The object of "support of research" is to support research activities at all levels from digging up research themes to developing research systems, various research activities, the public disclosure of research results and feedback of response from someone.

In the pilot model, various services are included focusing on the size of the scale and the magnitude of effects the cyber laboratory is supposed to have. In order to realize this, many things can be applied depending on the scale and activities of the research organization. For example, one could choose to place the Web server and researchers outside, and the set-up could be a "garage laboratory" with just 1 or 2 people.

4. CYBER LABORATORY APPLIED MODEL

In order to set a cyber laboratory up most cost-efficiently, it should be designed taking into account the necessity of the facilities and equipment shown in the pilot model depending on actual needs. The author has up to now applied from the pilot model shown in Figure 2 and designed, (a) a cyber laboratory as a catalyst-function for a research zone [3], and (b) a cyber laboratory for a small-scale academic or research group such as the *Preference-based Design Forum* (Figure 3).

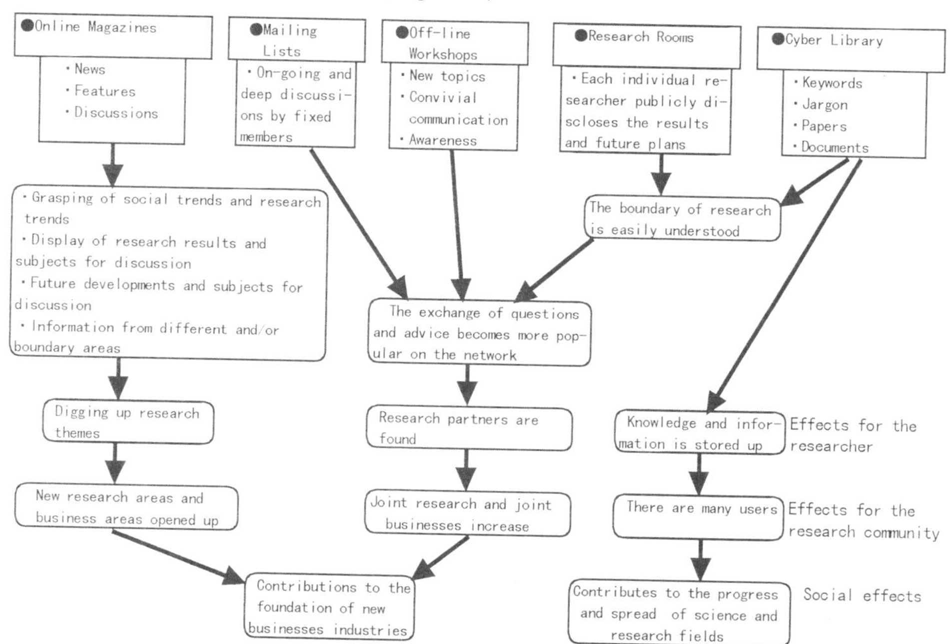

Figure 3. Design Example for a Small-Scale Research Group.

5. IMPLEMENTATION OF A CYBER LABORATORY

Activity of the cyber laboratory for the *Preference-based Design Forum*(PDF) was begun making reference to Figure 3. The mailing list was begun in January, 1996 with 5 members. This increased to 56 people by March, 1997. The total number of mail up until the 31st of March, 1997 was 791. Table 2 shows the breakdown of the mailing list. Out of 402 discussion groups, 21 continued with more than 10 comment chains. "Keywords" and "Research Rooms" are implemented on the Web (http://www. ises-web. com/miryoku/). Since the 5th of July, 1996, 1,086 people visited the Web site (as at the 6th of April, 1997). Up to now 8 people have joined the mailing list from the Web, and these people are active in their discussions.

Table 2
Breakdown of PDF Mailing List

Area	No. of Mailings
Discussion	402
Off-line Workshop planning, communications (for 7 times)	132
Office communications and operating consultations	111
Research Works (e.g. E-mail interview)	98
Self-introductions, news, forwarded mail	48

6. FEASIBLE EFFECTS OF A CYBER LABORATORY

From the author's experience up to now (from making the cyber laboratory pilot model then putting it into practice), he believe that if a research organization builds its activities into a cyber laboratory, generally the following effects will be found:

(a) Effects for the researcher:

・Productivity will increase. ・The results will be more substantial. ・New research opportunities will come up. ・Themes more demanded will be found.

(b) Effects for the research institution:

・Activity can begin with a small amount of capital. ・The main facility functions can be kept to a minimum. ・Operations can be expanded. ・It will be easier to keep researchers.

(c) Social effects:

・An increase in social knowledge assets. ・Academic and research activities will become more active. ・It will lead to new businesses and industries. ・It will make working from home possible. ・It will assist employment. ・It will make a contribution to local development.

REFERENCES

[1] ftp://ftp.nic.ad.jp/jpnic/statistics/

[2] "Internet" magazine, May 1997, Impress Corporation, Japan, 1997

[3] Tomoyuki Tsunoda, Masato Ujigawa, et al., "Cyber Laboratory", Engineering Advancement Association of Japan, Japan, 1997(Japanese)

[4] B. van Gundy Jr., "Techniques of Structured Problem Solving, Second Edition" , Van Nostrand Reinhold, New York, 1988

[5] Junichiro Sanui, "Visualization of Users' Requirements: Introduction of the Evaluation Grid Method", the proceedings of the 3rd Design & Decision Support Systems in Architecture & Urban Planning Conference, Vol. 1, pp 365 - 374, 1996

example of a continuous tracking system. Hutchinson, et al. (1989) reported eye-gaze operation recreational software including blackjack and music composition. Calhoun, et al. (1986) made simple switch selections by looking at intended switches in a simulator cockpit. Eye movements can also control weapons sighting systems (e.g., Nicholson, 1966; Dick, 1980). Frey, et al. (1990) reported a fully eye-movement controlled text entry task for disability applications, using continuous letter prediction based upon probabilistic models.

Cognitive experimental-task design

In our labs, we investigate moderately complex tasks using several, simultaneous physiological measurement approaches. By holding the task constant, and investigating EEG and eye movement data in parallel, it is possible to map physiological data to reasonably well understood cognitive mechanisms. Consistent with Sternman and Mann (1995), we will utilize trend analysis, keeping our physiological recordings tightly coupled to cognitive events. By defining cognitive terms relative to the task domain, we can compare the abilities of different recording methodologies and data handling methods to capture and predict the same cognitive phenomenon. By investigating physiological data on a task-relative timescale (as opposed to averaging data and running the risk of losing reactions to task events) we will be better able to map data signatures to cognitive events.

Complex performance in environments such as process control, aircraft piloting, airspace monitoring, and even driving, involves a mix of cognitive processes and complex coordination of learned and adaptive actions and behaviors. Cognitive psychologists are just beginning to understand and model performance in tasks of even moderate complexity. The promise of adaptively controlled task environments such as aircraft cockpits must await better understanding of cognitive resources used to accomplish such behavior and the kinds of information we can obtain about deployment of these cognitive resources through physiological recording.

Ideal tasks for investigations of relationships between physiological measures and cognitive mechanisms must involve several controllable performance aspects including vigilance/visual sampling, motor output, and a mental coordination of resources and contents. Further, to capture the difficulty of sampling from realistically complex environments, subjects must be presented with a reasonably complex displays. Our initial experiments will focus on individuals abilities to monitor keep track of changing information as one would in an air traffic control. Tasks of this sort will allow us to investigate the visual events such as scan paths and dwell duration, and provide data concerning working-memory performance under easily controlled conditions of load. We will also have control over response types. We extend the keeping-track methodology (Detweiler, Hess, and Ellis, 1996) to include investigation of participants' abilities to integrate information in memory and the environment. Integration and alternation between perception and memory have been discussed as examples of controlled attention switching (Dark, 1990; Weber, Burt & Noll, 1986). These extensions will afford investigation of physiological indicators of controlled processing and attention, issues central to discussions of workload and task difficulty.

CONCLUSIONS

Advancement in the design and implementation of adaptive human-computer interfaces will require the cooperative efforts of researchers in computer science, engineering, physiology, psychology and design. To realize the vision of computers and performance environments which actively monitor and respond to human states, we must first understand what the available data sources (e.g., EEG, ocular and performance measures) can tell us about those states. We must also consider the targets of adaptation by asking questions about what makes performance hard in certain environments and what can be done to improve performance unobtrusively. Our research goals are to investigate cognitively rich yet experimentally tractable tasks in order to map physiological measures onto cognitive phenomena discovered to make up task performance.

Acknowledgment: This supported in part by grants from NASA Langley.

REFERENCES

Calhoun, G.L., Janson, W.P., and Arbak, C.J. (1986), Use of eye control to select switches, Proceedings of the 30th Annual Meeting of the Human Factors Society, Santa Monica, CA: HFS, pp. 154-158.

Dark, V. (1990). Switching between memory and perception: Moving attention or memory retrieval. Memory & Cognition, 18, 119-127.

Detweiler, M.C., Hess, S.M., & Ellis, R.D. (in press). The effects of display layout on keeping track of visual-spatial information. In W. Rogers, A.D. Fisk, & N. Walker (Eds.), Aging and skilled performance. Hillsdale, NJ: Erlbaum.

Dick, A.O. (1980), Instrument scanning and controlling: Using eye movement data to understand pilot behavior and strategies, NASA/Langley Research Center Technical Report NASA-CR-3306, NTIS-N*)-31039.

Frey, L.A., White, K.P., and Hutchinson, T.E. (1990), Eye-Gaze Word Processing, IEEE Transactions on Systems, Man, and Cybernetics, 20(4): 944-950.

Hutchinson, T.E., White, K.P., Martin, W.N., Reichert, K.C., and Frey, L.A. (1989), Human-Computer Interaction using Eye-Gaze Input, IEEE Transactions on Systems, Man, and Cybernetics, 19(6): 1527-1534.

Just, M.A., and Carpenter, P.A. (1976), Eye fixations and cognitive processes, Cognitive Psychology, 8: 441-480.

Kotchoubey, B., Schleichert, H., Lutzenberg, W., & Birbaumer, N. (1996) A new method for two-dimensional self-regulation of EEG: Toward non-motor communication. *Psychophysiology,*

Makeig, S., Jung, T., & Sejnowski, T. (1996) Using feedforward neural networks to monitor alertness from changes in EEG correlation and coherence. In D. Touretzky, M. Mozer, & M. Hasselmo (Eds.) *Advances in neural information processing systems 8.* Cambridge MA: MIT Press.

Myer, G.A. (1992), The EyeMouse, Technological Horizons in Education Journal: Zenith Data Systems Supplement, January 1992, pp. 13-15.

Nicholson, R.M. (1966), The feasibility of a helmet-mounted sight as a control device, Human Factors, 8: 417-425.

Pope, A.T. Bogart, E.H., and Bartolome, D.S. (1995). Biocybernetic system evaluates indices of operator engagement in automated task. Biological Psychology, 40, 187-195.

Rouse, W.B. (1989). Adaptive aiding for human/computer control. Human Factors, 30, 431-443.

Schryver, J.C., and Goldberg, J.H. (1993), Eye-gaze and intent: Application in 3D interface control, Proceedings of the HCI Conference, Orlando, Fl.

Sternman, A.T. Bogart, E.H., and Bartolome, D.S. (1995). Biocybernetic system evaluates indices of operator engagement in automated task. Biological Psychology, 40, 187-195.

Weber, R.J., Burt, D.B. and Noll, N.C. (1986). Attention Switching between perception and memory. Memory & Cognition, 14, 238-245.

Wolpaw, J., & McFarland, D. (1994) Multichannel EEG-based brain-computer communication. *EEG and Clinical Neurophysiology, 90,* 444-449.

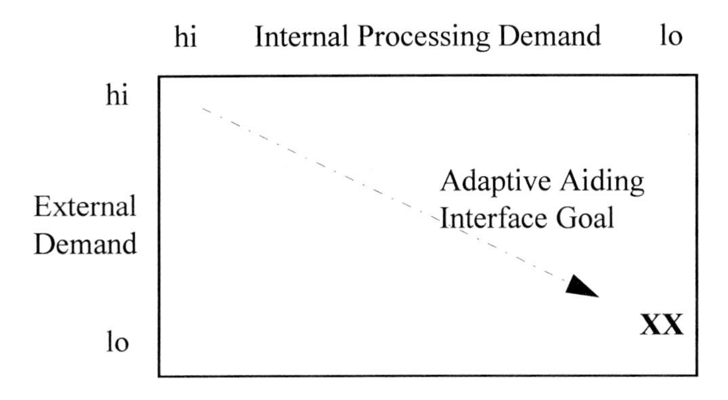

Figure 2. The goal of adaptive aiding interface design is to reduce both internal and external processing demand on the actor relative to task activities.

Efforts to adaptively aid knowledge based processing should focus on assisting the actor in generating new semantics for framing the situation, problem, or solution plan. Aiding aimed at shifting internal K-based work to Rule-based processing should concentrate on how to assist recognition of immediately relevant states using current semantics for the task domain. The aiding of S-based development requires increasing cue saliency specified in actions taken over a wide range of task and state conditions such that the actor can pickup affordances and regions of "safe" activity.

The Word Process Framework applies to teams of actors or actor-agent combinations, as well as individual an individual actor-machine agent dyad. Further, the framework is neutral to different method for producing adaptive interfaces, such as computational based methods (e.g. model-based, statistical, neural net, associate technology) or representational design methods (e.g. ecological interface design concepts).

3. CONCLUSIONS

In summary, the Work Process Framework emphasizes the need to development a thorough, principled analysis for domain work from a process orientation in order to determine useful areas of adaptive aiding and to understand general structural requirements for aiding to meet the demands of procedural, embedded problem solving, and off-line problem solving activity forms. An analysis of what methods to use to provide aiding and to fashion the surface structure of the interface should not be initiated until after this task process analysis is completed. The framework serves as a guide and analysis tool for the adaptive interface designer, one that should be used before and interactively with other methods to identify, select, and shape both he surface and deep structure of the adaptive interface.

REFERENCES

Boyd, J.R. (1987). *A Discourse on Winning and Losing*, Air University Library, Maxwell AFB Report no. MU 43947.

Gibson, J.J. (1979). *The Ecological Approach to Visual Perception.* Houghton-Mifflin: Boston.

Neisser, U. (1976). *Cognition and Reality*, San Franisco: W.H. Freeman.

Hollnagel, E. and Woods, D.D. (1983). "Cognitive Systems Engineering: New Wine in New Bottles," *International Journal of Man-Machine Studies*, 18, 583-600.

Hollnagel, E., Mancini, G., and Woods, D.D., eds, (1988). *Cognitive Engineering in Complex Dynamic Worlds*, New York: Academic Press.

Norman, D. (1986). Cognitive engineering, in Norman, D. and S. Draper (eds.), *User Centered System Design*, Hillsdale NJ: Lawrence Erlbaum Associates, pp. 31-61.

Norman, D. (1984). Cognitive engineering principles in the design of human-computer interfaces, in G. Salvendy (ed.), *Human-Computer Interaction*, Amsterdam: Elsevier Science Publishers, pp. 11-16.

Rasmussen, J. (1986), *Information Processing and Human-Machine Interaction: An Approach to Cognitive Engineering*, New York: North-Holland.

Rasmussen, J., Pejtersen, A.M., and Goodstein, L.M. (1994). *Cognitive systems Engineering,* New York: John Wiley & Sons.

Woods, D.D., and Roth, E.M. (1988). "Cognitive Systems Engineering," in M Helander (ed), *Handbook of Human-Computer Interaction*, New York: Elsevier Science Publishers, pp. 3-43.

Aid methodology for designing adaptive human computer interfaces for supervision systems

Charles Santoni - Elisabeth Furtado - Philippe François
DIAM-IUSPIM, Domaine universitaire de St Jérôme, Avenue Escadrille Normandie-Niemen, 13397 - Marseille Cedex 20 (FRANCE) - Tel : (33) 04 91 05 60 14 - Fax : (33) 04 91 05 60 33
E-mail : charles.santoni@iuspim.u-3mrs.fr

1. INTRODUCTION

Industrial processes and exploitation conditions (control, supervision, maintenance...) are getting increasingly complex every day. Faced with this problem, the work of a supervision interactive system designer is becoming more and more difficult, because he has to take into account both the application and interaction parts of the system. The operator's work becomes harder, as well, since he has to process a very large quantity of information coming from the process : to understand, to sort, to select, to analyze them, in order to implement the control and supervision actions. It is clear that communication between the operator and the system must be achieved in an efficient, non ambiguous and safe way.

2. OBJECTIVES

The current HCI designing tools for supervision systems require an important effort from the designer who has to apprehend and handle many models (process model, task model, operator model...) with different formalisms and to perform himself the translation from one to the other, taking into account ergonomic recommendations. Also, the resulting interfaces are not often evaluated from an ergonomic point of view and do not take into account the human cognitive system. Furthermore, these tools involve structure and behavior implementation of the interfaces, because most of them do not use architecture model. In that case, those tools take the risk of resulting into a bad modular decomposition, which could impede the iterative development of the interactive system.

The adaptive HCI designing methodology which we present here, tends to reduce the designing effort while decreasing the using effort of interactive system. This method is based on a conceptual modelization of the interactive system architecture. It uses a similar formalism at every designing phase, which makes it possible to take into account both the ergonomic recommendations and the human operator's cognitive system.

3. AN AID METHODOLOGY FOR DESIGNING ADAPTIVE INTERFACES

The methodology we have developed is a method for assisting the designer, which allows an automatic translation between one model and an another, using a similar formalism, without the designer's intervention (cf. figure #1).

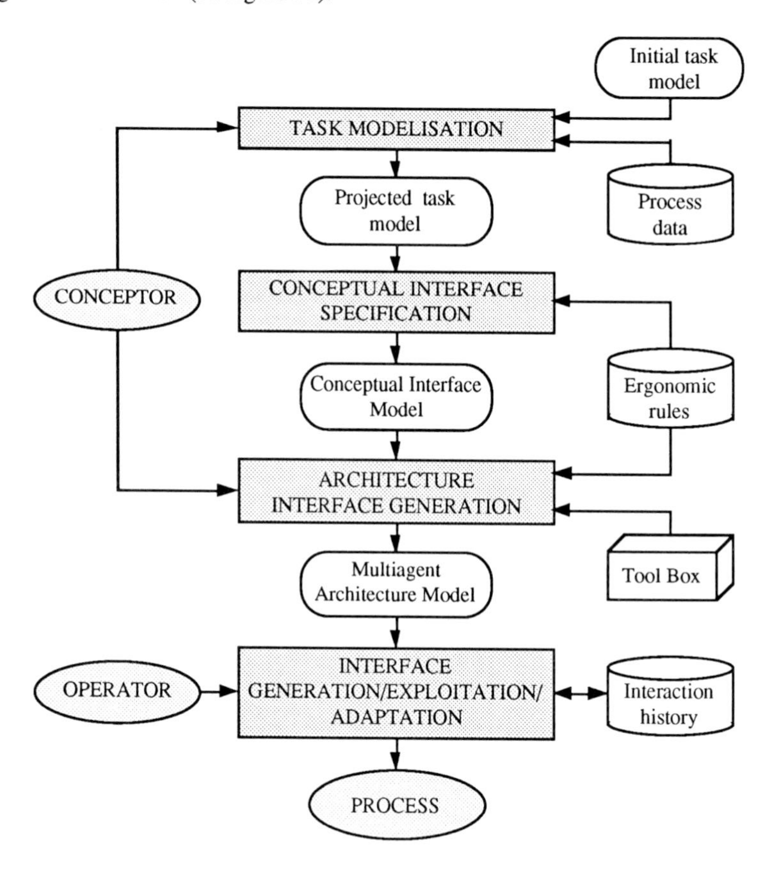

Figure #1. Principle of the aid methodology designing

3.1. The task modelisation phase

The assistance consists in providing the designer with a basic generic model, named *Initial Task Model*, which represents the application skeleton. This model takes into account the operator's cognitive model in the problem-oriented phase. It is based on the supervision task model defined by Rasmussen [1] and transcribed into the MAD model formalism [2] [3]. The initial task model has to be completed by the designer according to the application in order to obtain the *Projected Task Model*. This model describes the operator's interaction tasks within the supervision system.

3.2. The conceptual interface generation phase

Our method makes it possible to automatically translate the projected task model into an interaction model which constitutes the *Conceptual Interface Model* with a formalism which is close to the MAD formalism. This model is made up from the projected task model by the application of ergonomic rules. Then, some basic interaction spaces (welcoming space, edition space, saving space...) are added, as well as their relations and links [4].

3.3. The architecture generation phase

The interactive system architecture model could then be automatically generated from the conceptual interface model [5]. We have chosen the PAC multiagent formalism [6] to produce the architecture interface model, because it gives the advantage of making it possible to manage both the interaction objects and the interfacing system's dynamics. To obtain this real system, the designer must choose the real interaction objects, by linking the PAC model presentation agents and the tool boxes he has at his disposal.

The interface system could then be automatically generated by the implementation of each PAC agents : the control agents are responsible for the dialog control, the abstraction agents are responsible for communication between the interface and the application tasks and the presentation agents are responsible for the presentation of the interactive objects.

4. THE ADAPTATION PROCESS

In order to reduce the using effort, it is essential to take into account the operator's cognitive model from the designing phases, as well as the ergonomic recommendations [7]. Nowadays, there is no operator model [8] which is non ambiguous, reliable and defined by a formal formalism. The purpose of the work presented here is not to define an operator model, but consists in implementing, from the designing phase, the mechanisms which make it possible to analyze the supervision operator's behavior, when faced with current problems during the operator's task running.

We have limited this work to the analysis of the operator's behavior, when process failures are detected. The method consists in characterizing, for each detected failure, the operator's experience level and behavior when he is trying to correct it. This characterization is made up of stereotypes [9] which are deduced from an interaction history. It makes it possible to compute the operator's experience level in order to configure and set the decision support system, which determines the assistance tasks which are necessary to correct a failure (action plans, alarms...) according to the operator's level.

The ergonomic recommendations [10] for developing an interactive system, are taken into account under the form of rules, during the translation between the projected task model and the conceptual interface model, as well as during the choice of the interaction objects. These rules make it possible to respect ergonomic principles, such as the limitation of the number of objects involved during the system's identification, the limitation of the information required for the implementation of the application tasks...

5. CONCLUSION

The aid methodology for designing adaptive human computer interfaces presented here makes it possible to reduce both the designing efforts end the using efforts. To reduce the designing efforts, the designer has at his disposal a description support for the operator's interaction tasks through the initial task model, as well as a prototype of the whole interactive system through the conceptual interface model. Furthermore, the interface design process is mainly automatic. To reduce the using efforts, this process takes into account the operator's cognitive model and ergonomic recommendations.

REFERENCES

1. J. Rasmussen, *Skills, rules, knowledge, signal, signs, symbols and others distinctions in human performance models*, IEEE SMC, 13(3), pp 257-266. 1983
2. C. Pierret-Golbreich, I. Delouis, D.L. Scapin, *Un outil d'acquisition et de représentation des tâches orienté objet*, Rapport de recherche INRIA 1063, France. 1989
3. S. Sebillotte, *From users' task knowledge to high-level interface specification*, Journal of human-computer interaction 6(1), pp 1-15. 1994
4. E. Furtado, *Etude et réalisation d'une méthode de conception d'interfaces adaptatives à partir de spécifications conceptuelle*, PhD Thesis, Aix-Marseille III University. Mars 1997
5. Ch. Santoni, E. Furtado, Ph. François, *Towards adaptive UIMS for supervision systems*, IEEE SMC, pp 2598-2603, Vancouver. 1995
6. J. Coutaz, *Architectural design for user interfaces*, ESEC'91, pp 7-22. 1991
7. M. Schneider-Hufschmidt, T. Kühme, U. Malinowski, *Adaptive user interfaces : Principles and practice*, Elsevier science publishers, North Holland. 1993
8. A. Montoy, *Building a user model for self-adaptive menu-based interfaces*, User Modeling'94, pp 15-19. 1994
9. Ch. Santoni, E. Furtado, Ph. François, *Adaptive human computer interfaces for supervision systems*, HCI International'95, Tokyo, pp 1077-1083. 1995
10. F. Moussa, Contribution à la conception ergonomique des interfaces de supervision dans les procédés industriels : application au système Ergo-Conceptor, PhD Thesis, Valenciennes University. 1992

Toward An Adaptive Command Line Interface

Brian D. Davison and Haym Hirsh

Department of Computer Science, Rutgers, The State University of New Jersey
New Brunswick, New Jersey 08903, USA

This paper explores different mechanisms for predicting the next command to be used for the UNIX command-line shell. We have collected command histories from 77 people, and have calculated the predictive accuracy for each of five methods over this dataset. The algorithm with the best performance has an average online predictive accuracy of up to 45%.

1. Introduction

This paper describes research concerned with improving an interface by making it adaptive — changing over time as it learns more about the user. We hypothesize that general machine-learning methods can recognize regularities in human-computer activities without requiring specialized knowledge about users or applications. To test this hypothesis, we have collected and analyzed a sizable set of UNIX command histories. This paper[1] discusses how the data was collected and how a sampling of algorithms performs on it.

There has been a range of work developing systems that recognize regularities in the usage of a computer. Yoshida and Motoda's [2] investigation of the use of machine learning to predict a user's next command is the most similar to the work reported here, but uses a different estimation of performance, different algorithms, and has a very small dataset. Also similar is work in programming by demonstration [3], such as Cypher's Eager [4], which recognizes and automates simple repetitions in user actions in a graphical interface. Finally, Lesh and Etzioni [5] also consider UNIX commands in plan recognition.

2. Approach and Methodology

We captured command histories from 77 people at Rutgers University. Two of these were faculty (including the second author), five were graduate students (including the first author), and the rest were undergraduates in an Internet programming course. Collection periods ranged from two months to over six months. These histories represent the primary work performed online by the authors and the participating graduate students. The undergraduate data was collected over two months on computer systems dedicated to their programming projects, and not their primary systems, and so the patterns of commands selected may reflect the orientation of their use (i.e., editing and compilation rather than email).

[1] An expanded version of this paper is available [1].

We collected data unobtrusively, by causing the UNIX shell at its closing to record the command history in a time-stamped file. This method was used to minimize potential interference with the user's activities.

This paper describes the application of a total of five methods to this data:

- C4.5 [6] is a well-used decision-tree learner developed in the machine-learning community, with demonstrated excellent performance over a wide variety of problems. Each command formed the basis for a single training example containing as its two attributes the two previous commands, and whose label was the command itself.
- The hypothetical "Omniscient" predictor correctly predicts all commands, providing that the desired command was used previously (under the assumption that, with no prior knowledge, no learning method would be able to predict the command). We use this as upper-bound on the potential predictive accuracy of any learner.
- In contrast,"Most Recent Command" (MRC) predicts the same command that was just executed. This method provides a useful baseline method to calibrate results.
- Slightly more sophisticated is "Most Frequent Command" (MFC), which predicts the command that previously occurred most frequently in the user's interactions.
- The last algorithm, Prefix, looks at the sequence of immediately preceding commands in the current shell session, and attempts to find the longest such sequence in the user's history, and predicts the command that followed that sequence.

With the exception of C4.5 (for which we used the distributed "off the shelf" software), each of the algorithms was implemented in Perl. Each algorithm was constrained to use a maximum-sized window of past commands in predicting the next command, where our experiments used window sizes n=100, 500, and 1000. Thus, for example, to predict the next command MFC predicts the command that occurred most frequently in the last 100 (or 500 or 1000) commands.

The data for this domain is inherently sequential, and thus it is inappropriate to use cross-validation and similar evaluation methods that effectively take a random sample of commands from a user's history and use them to predict the remaining commands. Yoshida and Motoda [2] evaluate their ability to predict the last third of commands using the first two-thirds,[2] but this could worsen predictive accuracy on the last third if users' behaviors are a "moving target", changing over time. Further, this eliminates two-thirds of the data from use in computing predictive accuracy. We therefore compute predictive accuracies for each method in an "online" fashion, by predicting each command in sequence using the preceding n commands in the user's history. We computed individual user accuracies and then averaged these results (the "macroaverage"), as well as the average overall prediction accuracy — number of correct predictions divided by total commands for all users (the "microaverage").

[2]For comparison purposes, we note that on average C4.5 achieved 32±12% (macroaverage) accuracy on the last third of each user's data using this evaluation method.

3. Experimental Results

The 77 subjects had history sizes ranging from 15 commands to a maximum of 34,940 commands, and had an average of 2,184 commands. The average number of distinct commands per user was 77.1, and the average command length was 3.77 characters.

Figure 1 demonstrates the macroaverage performance of each algorithm across the range of training-window sizes. Table 1 shows the performance for each algorithm with a training window size of 1000.

The results given so far have all been in terms of predictive accuracy. While this is a common and useful metric, it has some limitations. It does not, for example, measure the extent to which user productivity is improved (or harmed) by a shell with a prediction component. For example, if a user has defined many single-character aliases, then even high accuracy in prediction need not result in a significant improvement in a user's efficiency if it is measured in number of keystrokes, since both the user's command and the command-inserting may require only a single keystroke. Additionally, the overhead of managing a history and computing predictions from it may require too much time for an online system.

In the real world, user productivity is the primary concern. Since the data collected was from unmodified, non-predictive command shells, we can only speculate as to the improvements in productivity afforded by the methods presented here. However, in the best case C4.5's predictions were correct up to 45% of the time. Recall that the lengths of the histories on average were 2184 commands, and the average command was 3.77 characters long. Assuming that a correct prediction could be inserted with a single character, and that commands recorded in the study were all typed explicitly (not using any historical shortcuts or command completion), this method would have saved just over 31% of the keystrokes typed, which is very close to the 33% expected if predictability were independent of command length.

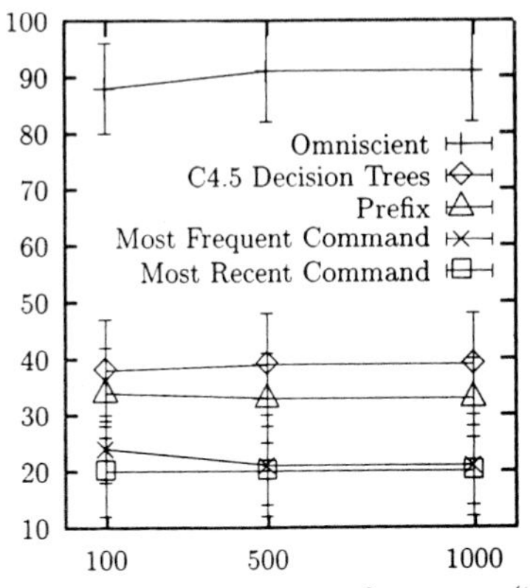

Figure 1. Macroaverage performance (in percent) vs. training-window size.

Table 1
Online predictive accuracy (for training-window size 1000).

algorithm	macroaverage	microaverage
Omniscient	91±9%	96±21%
C4.5	39±9%	45±50%
Prefix	33±7%	36±48%
MFC	21±7%	18±39%
MRC	20±8%	30±46%

4. A Prototype Shell: ilash

We have implemented a simple prototype that incorporates some of the ideas described in this paper. Called `ilash` (Inductive Learning Apprentice SHell), it is an extension to the UNIX shell `tcsh` that allows the user to present the predicted command in the prompt string, and to insert the predicted command with a single keystroke.[3] It uses only two attributes — the two previous commands issued. Learning is performed by the decision-tree learner C4.5. This prototype is described further elsewhere [7].

While this initial version of `ilash` incorporates a strong learner, C4.5, it does make other tradeoffs. Decision trees are not created or updated online — a new decision tree must be built explicitly by the user, commonly just once a day. This choice allows the shell to keep its natural responsiveness, but reduces prediction accuracy.

Our user studies did not include use of this prototype, for a number of reasons. First, doing so would have required a commitment to a single prediction method (at the very least, for each user), and we wanted to explore a range of methods. Second, we wanted to be able to separate as much as possible effects of the interface from the quality of competing prediction methods. On the other hand, the system's predictions, when provided to the user through the interface, could influence the user's next action, potentially changing the sequence of commands executed.

5. Summary

This paper has described a number of methods for predicting a user's next command in a UNIX command-line shell. We found that relatively straightforward, knowledge-free methods were able to correctly predict the next command that the user would execute up to 45% of the time, potentially reducing the keystrokes needed by close to one third. The best performance was achieved using C4.5 to learn prediction rules based on two previous commands, and a training window of either 500 or 1000 past commands.

REFERENCES

1. Brian D. Davison and Haym Hirsh. Experiments in UNIX command prediction. Technical Report ML-TR-40, Department of Computer Science, Rutgers University, 1997.
2. Kenichi Yoshida and Hiroshi Motoda. Automated user modeling for intelligent interface. *International Journal of Human-Computer Interaction*, 8(3):237–258, 1996.
3. Allen Cypher, editor. *Watch What I Do: Programming by Demonstration.* MIT Press, Cambridge, MA, 1993.
4. Allen Cypher. Eager: Programming repetitive tasks by demonstration. In Cypher [3], pages 204–217.
5. Neal Lesh and Oren Etzioni. A sound and fast goal recognizer. In *Proceedings of the Fourteenth International Joint Conference on Artificial Intelligence.* Morgan Kaufmann, 1995.
6. J. Ross Quinlan. *C4.5: Programs for Machine Learning.* Morgan Kaufmann, San Mateo, CA, 1993.
7. Haym Hirsh and Brian D. Davison. An adaptive UNIX command-line assistant. In *Proceedings of the First International Conference on Autonomous Agents.* ACM Press, 1997.

[3]The modifications to `tcsh` that formed our prototype were implemented by Mark Limotte.

Integrating Human Factors with Software Engineering for Human-Computer Interaction

John Long

Ergonomics & HCI Unit, University College London,
26 Bedford Way, London WC1H 0AP, UK

This paper suggests how the integration of Human Factors with Software Engineering might be conceived in terms of the emergent discipline of Human-Computer Interaction. It also illustrates some attempts to integrate Human Factors with Software Engineering. Finally, the paper considers the need to construct, in the longer term, a unitary discipline of Human-Computer Interaction.

1. INTRODUCTION

We are in the midst of an information revolution driven by computing technology. If once the main issue was the computer and 'getting it to work at all', now the main issue is human-computer interaction and 'getting computers to work for us effectively'. To this end, there is a need to integrate Human Factors (HF) and Software Engineering (SE), the two current sub-disciplines of Human-Computer Interaction (HCI). Integration is required to make present HCI design knowledge and practices more effective. There are two requirements for such integration as concerns: (i) the discipline of HCI; and (ii) the general design problem, which that discipline is intended to solve.

2. HCI DISCIPLINE INTEGRATION OF HF WITH SE

Following Long and Dowell (1989), a framework for the discipline of HCI, which integrates HF and SE can be expressed as: the use of HCI knowledge to support practices, seeking solutions to the general problem of HCI, as humans and computers interacting to perform effective work. HCI knowledge is constituted of HF and SE knowledge respectively, supporting HF and SE practices. Those practices address respectively the HF general problem of humans interacting with computers and the SE general problem of computers interacting with humans, in both cases to perform effective work. This framework supports HCI discipline integration of HF and SE, inasmuch as HF and SE design knowledge and practices are related to the same expression of the HCI general design problem. Consider next the technical requirement of integration in terms of that general design problem of HCI.

3. HCI GENERAL DESIGN PROBLEM INTEGRATION OF HF WITH SE

Dowell and Long (1989) have proposed a conception for HCI, appropriate for expressing technically its general design problem, which supports the integration of HF and SE. In the conception, an application domain (of an HCI worksystem) is where work originates, is performed and has its consequences. It comprises one or more objects constituted of attributes (which have values). Task goals express a requirement for change in the value of these attributes, and goals are allocated to worksystems by organisations. A domain is distinct from, and delimits, a worksystem. A worksystem comprises at least two separate, but interacting, sub-systems – of human behaviours interacting with computer behaviours. These behaviours are supported by mutually exclusive human structures and computer structures, and are executed to perform work effectively. Effectiveness is expressed by the concept of performance, that is, how well a worksystem performs its goals – that is, task quality, and the system costs that are incurred in so doing. Costs are incurred by both the human and the computer and are structural and behavioural.

The general problem of HCI would thus be:

design {H} and {C}
such that {H} interacting with {C} (that is, HF)
and {C} interacting with {H} (that is, SE)
= $\{S\}P_A=P_D$
where {S} is a worksystem
and $P_D = fn\{Q_D,K_D\}$
Q_D expresses the desired quality of work
K_D expresses the acceptable costs incurred by the worksystem (both human, that is, HF and computer, that is, SE).

4. ASSUMPTIONS CONCERNING HF INTEGRATION WITH SE

In this paper so far, it is assumed that: (i) HCI is composed of HF and SE (and is not at present a unitary discipline); (ii) HF integration is explicit (and not implicit); (iii) HF is integrated with SE (and not SE with HF); (iv) HF is integrated with SE as concerns HCI (and not simply conjoined); and HF is integrated with SE technically (and not informally).

Once integrated, it is assumed HF will provide better design support to SE by: (i) making good HF deficits of SE methods, tools and techniques; (ii) becoming associated with established SE methods, etc; (iii) relating to more advanced SE techniques; (iv) bringing HF and SE closer together in the design process.

5. ILLUSTRATIONS OF HF INTEGRATION

Denley et al (1993) have attempted to integrate HF substantive knowledge in the form of guidelines. They formalised the expression of their guidelines: in a prescriptive manner; in line with the HCI conception expressed earlier, that is with reference to both human and computer, and computer and human interactions; and with some indication of an associated guarantee. So, the guidelines as well as having a triggering and task attribute set, also include a user and computer (service) attribute set. The attribute sets together are accompanied by: a desired performance; the prescription; and a rationale. Their guidelines concern service integration for integrated broadband communication.

Lim and Long (1994) have attempted to integrate HF methodological knowledge in the form of a method for usability engineering (MUSE). MUSE supports an HF design process, complementary to an SE process. For example, MUSE has been configured for, and applied in an integrated fashion with, Jackson System Development (JSD) and Structured System Analysis and Design Method (SSADM). Hence, the design of humans interacting with computers, and computers interacting with humans, is addressed, both to perform effective work.

Stork, Middlemass and Long (1995) and Middlemass, Stork and Long (1995) have attempted to integrate HF practice by integrating substantive knowledge, in the form of guidelines, and HF methodological knowledge, in the form of MUSE, to solve specific HCI design problems. Respectively, the design problems concerned energy management in the home and networked recreational facilities booking. In both cases, the design solutions addressed humans interacting with computers, and computers interacting with humans, to perform effective work.

6. HCI AS A UNITARY DISCIPLINE

Rather than decomposing HCI knowledge, practices and general technical problems into HF and SE, it would be possible to conceive of a unitary discipline without such decompositions. There is not much evidence, at present, of attempts to form such a unitary discipline. However, Colbert et al (1995) have conducted a unitary HCI evaluation of a laboratory amphibious off-load planning system which assessed task quality, user (human) costs, and computer costs. There was no separate HF and SE evaluation, only a unitary HCI evaluation.

7. CONCLUSION

There is a need for more effective human-computer interactions, practices to support their design, and knowledge to support those practices. Integrating HF with SE for HCI is essential for meeting that need. Although some progress has been made, much remains to be accomplished.

REFERENCES

1. Colbert, M., Long, J. and Dowell, J. Integrating Human Factors with Software Engineering Evaluations: An Illustration with Reference to a Military Planning System. In *Proc. PACIS '95*, Singapore.
2. Denley, I., Hedman, L., Papadopoulos, K., Hill, B., Clarke, A., Hine, N. and Whitefield, A. Usability Principles for Service Design. In *Computers, Communication and Usability: Design Issues, Research and Methods* (Byerley, P., Barnard, P. and May, J. eds), Elsevier, Amsterdam, Netherlands, 1993.
3. Dowell, J. and Long, J. Towards a Conception for an Engineering Discipline of Human Factors. *Ergonomics* (32:11), 1989.
4. Lim, K.Y. and Long, J.B. *The MUSE Method for Usability Engineering*. Cambridge University Press, UK, 1994.
5. Long, J. and Dowell, J. Conceptions for the Discipline of HCI: Craft, Applied Science and Engineering. In *Proc. Fifth Conference of the BCS HCI SIG*, A. Sutcliffe and L. Macaulay (eds), Cambridge University Press, Cambridge, UK, 1989.
6. Middlemass, J., Stork, A. and Long, J. Applying a Structured Method for Usability Engineering to Recreational Facilities Booking User Requirements: A Successful Case Study. In Proc. *Eurographics*, Bonas, France, 1995.
7. Stork, A., Middlemass, J. and Long, J. Applying a Structured Method to Domestic Energy Management User Requirements: A Successful Case Study. In *Proc. HCI '95* , Huddersfield, UK.

Microsoft Interactive Media Products: Worldwide Usability and Design Practices

John Geoffrey Corso

Microsoft Usability Group, Microsoft Corporation, One Microsoft Way, Redmond, WA 98052, USA

1. INTRODUCTION

Many product teams at Microsoft Corporation are internationalizing their product line. Typically, native-speaking international program managers sit with the core product team while the product is being designed and developed to provide feedback on the core product specification and documentation for culturally inappropriate issues. In addition, many international program managers are responsible for writing the product specification for the product that they are responsible for releasing on schedule.

There are many challenges in designing software for an international market. One of these challenges is designing a product that is easy to use. Over the past several years, Microsoft usability engineers have been conducting usability studies in Europe and Japan. This article describes our experience usability testing two international multimedia products.

2. INTERNATIONAL USABILITY TESTING

The Microsoft Usability Group has responded to the challenge of usability testing international products differently in Eastern and Western countries. For example, in Europe, a contract-based HCI infrastructure exists that supports international usability testing making it convenient to test with local people. In Japan, the HCI infrastructure is still growing, and it is expensive to contract usability work. Consequently, the Microsoft Word team, with the help of the Microsoft Usability Group, has built a usability lab in Tokyo to support the design and evaluation of Microsoft products for the Japanese market.

2.1. Multimedia Product 1

In evaluating the usability of Multimedia Product 1 (M1) internationally, we chose to replicate usability tests that we had already conducted on the US version of M1 in Redmond, Washington. A contractor was chosen to evaluate the usability of the German (G), French, (F) and International English (IE) versions. To replicate the usability tests as closely as possible, the contractor used the same test materials that had been used to evaluate the US version. The test participants were registered Microsoft Windows users. Participants were asked to accomplish tasks with a low fidelity, localized prototype. Each prototype included country-appropriate content and translated text in the user interface (UI). The participants were asked to complete the tasks a second time with a higher fidelity build of the US product. The higher fidelity build was used to help the participants visualize the final product.

In analyzing the results from the international tests, it became clear that many of the same issues had been identified with the US product in the Microsoft Usability Labs. Specifically, as illustrated in Table 1, 47 - 75% of the issues identified with the German, French and International English tests were the same issues identified with the US product that had not been resolved.

Table 1
Usability Issue Analysis of M1 Test Results

Product UI	Usability Issues	German	French	Inter. English
Old UI	Known issues	75%	53%	47%
Old UI	New issues	25%	33%	52%
New UI	New issues	0%	13%	0%
		100%	99%	99%

In addition, many newly uncovered usability issues were identified with the German, French, and International English products that had not been identified in previous usability tests of the US product. There are likely to be numerous variables influencing these data. Clearly, usability testing a product with a sample from a new population is likely to uncover previously unobserved issues. Likewise, simply evaluating the products with additional participants is likely to uncover some new issues (Virzi, 1992).

Finally, all in all, the German, French, and International English prototypes were similar to the US product. Given that there was very little new design work beyond translating the UI and inserting appropriate content, it's not surprising that few new usability issues were observed with the new design work.

2.2. Multimedia Product 2

A Japanese-speaking intern working in the Microsoft Usability Group helped evaluate the usability of the Japanese (J) version of Multimedia Product 2 (M2). No attempt was made to replicate the usability tests of the US product because of technological constraints. The test participants were recruited through a temporary work agency in Tokyo and paid for their time. The usability test took place in the Microsoft Word Usability Labs in Tokyo. Participants were asked to accomplish tasks with a fairly high fidelity build of the product. The build included translated content and text labels in the UI

In analyzing the results of the usability issues, it became clear once again that several of the issues had been previously identified with the US product in the Microsoft Usability Labs in Redmond, Washington. Specifically, as illustrated in Table 2, 36% of the issues identified with the Japanese build were the same issues identified with the US product that had not been resolved.

Table 2
Usability Issue Analysis of M2 Test Results

Product UI	Usability Issues	Japanese
Old UI	Known issues	36%
Old UI	New issues	11%
New UI	New issues	53%
		100%

In addition, a few new usability issues were identified with JM2 that had not been identified in previous usability testing of the US product. The US product had been usability tested 12 times before the usability test of JM2 took place, so it is possible that a high percentage of the usability issues with the design of the US product had been identified prior to the JM2 study.

Although the Japanese prototype somewhat resembled the US product, there were significant design changes to support searching with 4 character sets, no spacing between characters, etc. Given this new functionality and design work, it's not surprising that 53% of the usability issues are new issues.

3. CONCLUSION

The usability data from usability testing German, French, and International English M1 have had a large impact on the core code of the product line including the US product. It is still too early to state how much influence the data from the usability test of the JM2 will have on the core product code. However, once a usability issue becomes an issue in more than one market, it becomes a higher priority issue that should be addressed in the core code. This is one way usability engineers can drive the internationalization of products. In some cases, it appears that the usability engineer may be able to significantly improve the usability of international products by ensuring that usability issues identified with the US version of the product are resolved successfully. Finally, although we have not had the opportunity yet, it is important to test the design solutions in appropriate target markets to ensure the solutions actually resolve (or at least reduce the severity of) usability issues and have not created any new usability issues.

REFERENCE

Virzi, Robert A. (1992). Refining the test phase of usability evaluation: How many subjects is enough. Human Factors, 34(4), 457-468.

A closed-loop approach for integrating human factors into system development: a case study involving a distributed database system

R. H.-Y. So, C. M. Finney, M. Tseng, C.J. Su, and B.P. Yen

Department of Industrial Engineering and Engineering Management, Hong Kong University of Science and Technology, Clear Water Bay, NT, Hong Kong

The benefit of integrating human factors into system development is clear: it improves the usability of the system. However, what is less clear is the methodology used to achieve such a goal. This paper focuses on the combinations of different methods in integrating human factors into the development of a distributed database system. The benefits of a combined program of clients' preference anticipation, ergonomics checklists, and usability testings are reported and discussed.

1. INTRODUCTION

1.1. Review of traditional methods to enhance the usability of a software system

Different methods and implementation plans have been proposed to incorporate human factors into the design stage of a software. (e.g. ergonomics checklist: Brown, 1988; usability experiment: Coleman *et al.*, 1985). While some authors expressed a preference for a particular method (e.g. Schell (1986) put an emphasis on usability testing), the trend is towards a combined approach of checklists, testing and evaluating procedures (Mayhew, 1992). Also, an early human factors involvement in a software development cycle was reported to yield better usability than a late involvement (Miller and Stimart, 1994).

1.2. Review of the MUSE concept for Usability Engineering

Lim and Long (1994) argued that early involvement of human factors considerations alone could not solve the usability problem of a software product. In order to ensure good usability, any early human factors involvement had to be continued throughout the development cycle. To facilitate this, a Method for Usability Engineering (MUSE) framework for integrating human factors into a software development

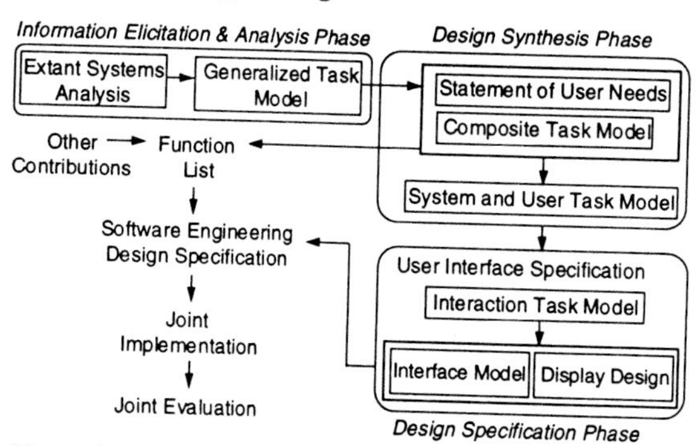

Figure 1 A structured framework for Human Factors Design (adapted from Lim and Long, 1994).

cycle was published (Figure 1). This method specifies three phases of integration to ensure the considerations of human factors issues through out the whole development cycle.

2. A CLOSED-LOOP APPROACH FOR INTEGRATING HUMAN FACTORS INTO SYSTEM DEVELOPMENT

A review of the literature suggests that the traditional way of testing a software late in its development cycle is not effective. Instead, human factors considerations should be integrated into the design process at an early stage. From a control system point-of-view, the traditional approach is an *open-loop* approach. In industries where software development time is constantly under pressure to be reduced, a usability testing conducted at the late stage of a development cycle can easily be forced into a mere 'rubber-stamping' exercise. Therefore, there is usually no room for the development team to improve the software system. In this case, no information is being fed back, hence, it is an *open-loop* structure. The integrated approach, however, is a *closed-loop* approach because testing is conducted continuously through out the development cycles and the information is constantly used to improve the software. The concept of the closed-loop approach is illustrated in Figure 2. This concept is similar to the argument by Lim and Long (1994) that human factors considerations have to be integrated throughout the development cycle.

3. THE CASE STUDY

3.1. Background: clients' requirement

A proto-type of a graphic database system existed for the communications between the clothing accessory producers and the clothing manufacturers in Hong Kong. Since the textile industry is highly seasonal, such a database will need to be constantly updated. The Department of Industrial Engineering and Engineering Management, at the Hong Kong University of Science and Technology, proposed to implement an industry-wide communication infrastructure to link and update this proto-type database system. The infrastructure utilized the Internet to maintain effective communication among buyers, merchandisers, suppliers, and manufacturers. The aim was to support the global business applications of Hong Kong's textile and apparel industry.

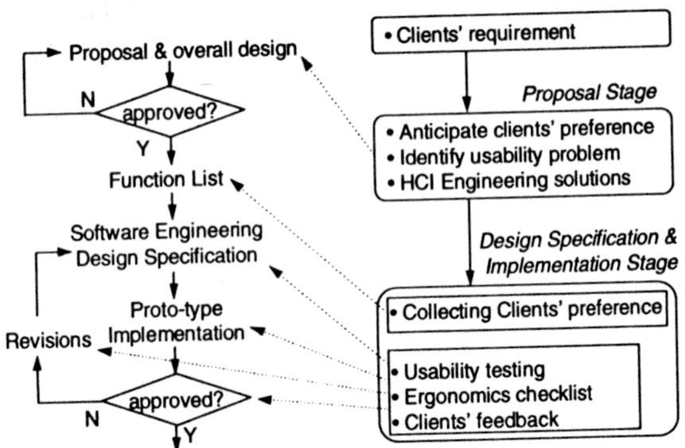

Figure 2 Illustration of the closed-loop approach to integrate human factors into system development.

3.2. The proposal stage

Clients' preference anticipation & identification of usability problems. The intended users of this system would include executive managers, merchandisers, and designers in the textile and apparel industry. Their expertise would be in textile merchandising rather than in computing, therefore the graphic database program must be 'easy to use'. In addition, common to any data retrieval system, 'response time' of the database would also be critical.

HCI engineering solutions to the usability problems. With the closed-loop approach, the human factors consideration was integrated at the earliest stage of the development cycle: during the project proposal. The proposal addressed both the 'easy to use' and the 'response time' issues identified as critical to the success of the project:

(i) for the 'easy to use' issue, the proposal stated that HCI technology would be applied in the design of user interface of the database system;

(ii) for the 'response time' issue, a distributed database structure was adopted. Instead of using the Internet for on-line access, the Internet was used only for off-line updating of the database. In so doing, the database would be stored locally in a computer hard-disk and the data retrieval time would be small.

3.3. The design and implementation stage

Continuous clients' preference monitoring and formulation of a function list. In order to facilitate the continuous communications between the clients and the software designers, a steering committee was set up. This committee had included directors from the textile industry and representatives from textile-related government authorities. In addition, a user forum was formed which consisted of about twenty textile companies. These companies were visited and feedback on the functional requirements were collected.

Ergonomics checklist. A literature survey was conducted to formulate an ergonomics checklist. Normally, this checklist would have been used by the design team to produce the software design specifications. In this case, since a proto-type was already available, a usability testing on the prototype database system was conducted first.

Usability testing. A usability testing concerning the response time was conducted on the proto-type database system. The tests used four typical search tasks, 10 subjects, and a Pentium (75 MHz) computer. Four database response lag conditions were used: baseline lag plus 0, 2, 4, and 6 seconds. Laboratory testings indicated that the use of a 486 PC (66 MHz) instead of a Pentium would introduce an additional lag of about 4 seconds. After completing the four tasks as quickly as possible, the subjects were asked to give a subjective rating on the acceptability of the system response time using a 11-point scale (0: too slow; 5: acceptable; 10: too fast). Inspection of Figure 3 shows that the median subjective rating of the baseline response time is '6' which is acceptable. A Wilcoxon matched-pair sign ranked test showed that the subjective acceptance ratings reduce significantly with additional time lags of 4 seconds or more ($p < 0.05$). In view of future expansion of the database and the possible use of a slower computer, it was decided to compress all the graphics image files from 'bmp' to 'jpg' format. This would reduce the database loading time. The compression ratio was carefully chosen after further testings on the image quality. A

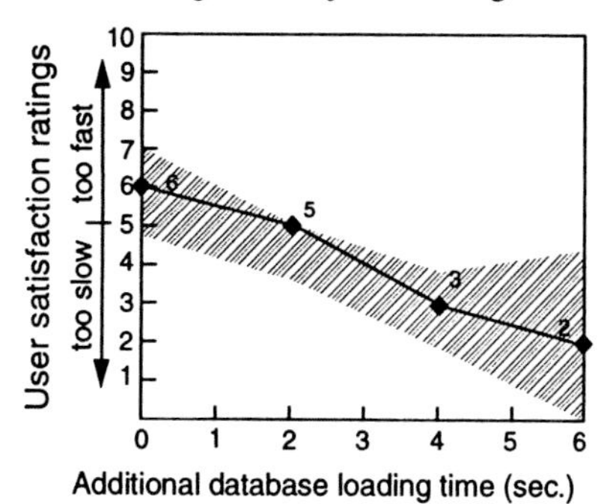

Figure 3 Acceptance ratings of different system response time (median and inter-quartile ranges of 10 subjects).

reduction of about 75% in graphics image file size was achieved.

User-interface modification. Using the proto-type system, the time required to complete a typical task was about 25 seconds and about 20% of this time was computation time (including the database search time). That is, 80% of the total task time was related to the users. This 80% consisted of reading time, thinking time, user response time, time to make mistakes, and time to recover from mistakes (Mayhew, 1992). This suggests that in order to optimize the task completion time, a better user-interface would be a logical starting point. The ergonomics checklist was used to revise the screen layout. At the time of writing, testings and clients' feedback sessions are being conducted to evaluate the benefits of the revision.

3.4. Benefits and problems

The benefits of integrating human factors into the system development started in the proposal stage. Potential human factors problems (e.g. the response time) were identified and appropriate solutions were included in the proposal. Through this integration, the awareness of human factors issues within the whole development team had been raised. During the development process, there had been some priority conflicts between the human factors engineers and the software engineers. In every case, the conflicts were resolved by clear customer-centered priority from the top management team. This suggests that one of the key components of a successful integration of human factors into system development is the commitment from the top management.

4. FINAL REMARKS

This paper presents the implementation of a closed-loop approach to integrate human factors into the development of a distributed database system. The essence of the integration was an early and continuous involvement of human factors engineering throughout the development cycle. The importance of commitment to such an integration from the top management was also highlighted.

REFERENCES

Brown, C. (1988) Human-Computer Interface Design guidelines. Norwood, NJ: Ablex.

Coleman, W.D., Williges, R.C., and Wixon, D.R. (1985) Collecting Detailed User Evaluations of Software Interfaces. Proceedings of Human Factors and Ergonomics Society 29th Annual Meeting, Santa Monica, CA. pp. 240-244.

Lim, K.Y. and Long, J. (1994) The MUSE Method for Usability Engineering. Cambridge University Press. ISBN 0-521-47494-9.

Mayhew, D.J. (1992) Principles and Guidelines in Software User Interface Design. New Jersey: Prentice Hall.

Miller, M. A. and Stimart, R.P. (1994) The User Interface Design Process: The Good, the Bad, and We Did What We Could in Two Weeks. Proceedings of Human Factors and Ergonomics Society 38th Annual Meeting, pp. 305-309.

Schell, D.A. (1986) Usability Testing of Screen Design: Beyond Standards, Principles, and Guidelines. Proceedings of Human Factors and Ergonomics Society 30th Annual Meeting pp.1212-1215.

Usability engineering and software engineering: How do they relate?

Simon Hakiel

IBM UK Laboratories
Hursley Park, Winchester, Hants SO21 2JN, United Kingdom
hakiel@hursley.ibm.com

Despite continuing developments in the disciplines of Human-Computer Interaction and Usability Engineering, software products continue to fall short of their usage requirements in the field. One approach to resolving this problem is to recognise that design for use, or product design, invokes very different skills from those represented by software engineering. This paper examines implications for the relationship between usability engineering and software engineering deliverables that follow from this distinction.

1 INTRODUCTION

The HCI community has considerable experience and literature relating to methods and techniques for user-centred design. We also have specifications of how to apply user-centred design processes to software development. The problem to be addressed to improve ease of use of software, therefore, is not that we do not know what do. The problem is that we do not routinely do what we know. Our experience and knowledge are not yet routinely applied to design specifically for quality of use. This paper explores the relationship between usability engineering and software engineering processes in practice, with the aim of facilitating the effective deployment of both during the development of software products.

2 USABILITY ENGINEERING APPROACHES TO EASE-OF-USE

Approaches to improving ease of use of software products address two major issues: the development and industrialisation of methods, techniques and tools to support usability engineering, and the application of usability engineering methods and practice to software product development. In view of the abundance and relative sophistication of available methods for usability engineering, we do not consider the major inhibitor of design for use to be the lack of methods for usability engineering. Our concern here is with approaches to the integration of usability engineering into the design and development of software products.

Application of usability design methods to software development has been described in terms of several software development process models. Among the most comprehensive are STUDIO (Browne, 1994), in which a usability engineering process is described with reference to software engineering activities in the context of SSADM (Ashworth and Goodland, 1990; Brown, 1994 p24), and MUSE (Lim & Long, 1994) in which the software engineering context, and the notation used, is derived from Jackson Structured Design (Jackson, 1983).

In these, and in other comparable approaches, key software development process steps are identified: typically, requirements analysis, high level design, detailed design, implementation, test, and delivery. For each of these, corresponding usability activities or deliverables, are identified and aligned with the corresponding deliverables in a software development process. Accordingly, the requirements for a product include usage requirements, high level designs include designs for user tasks, and detailed designs include specifications of user interfaces (which can include packaging, information, service, packaging, information, service support, etc.). It is intended that explicit representation of usability engineering activities and their deliverables in the process model and, therefore, in a development plan, will ensure that ease of use issues will be systematically addressed. In fact, the contribution made by more recent developments in usability engineering (e.g. Nielsen, 1993) is the detailed specification of the activities and deliverables that are required to achieve ease of use. The requirement that usage related activities must be represented in software development processes was recognised much earlier (Bennet, 1984; Shackel 1986b).

However, even in software development organisations that include usage related activities in their development processes, and that employ usability engineers, delivering ease of use is still a problem. What, then, inhibits the deployment of usability engineering resources even when they are available?

3 THE VIEW FROM SOFTWARE DEVELOPMENT

Consider the software engineer's view of the development process. Unsurprisingly the primary focus of software development organisations is on the development of program code. Software engineering activities and deliverables are concerned with the efficiency of software production and the quality of the code that is produced. In a software engineering context, software design is concerned with methods and techniques through which this can be achieved, with reference to software architectures, algorithms, data structures, functional organisation of processes, and to the principles and practice of project management.

It is here that we find the source of our problem. The processes typically defined for software design and development: requirements, high level design, detail design, implementation, verification and validation, invoke the same principles and processes as those described for usability engineering, but they apply to very different domains.

Software engineering design deliverables are statements about code, at various levels of abstraction. Even when the deliverables in a particular implementation of a software design process are described in terms of functional abstractions, the abstractions typically refer to architectural relations between software subsystems.

With rigorous application of software engineering methods we can, and do, produce defect free code. However, in its context of use, this perfect code can still fail our customers. The software engineering methods that lead to the code produced are not immediately concerned with the problems that software is intended to solve for our customers. Rationales for function implemented by the software to be produced are not addressed as a software design issue but are supposedly given in the requirements specification which drives a software design activity.

In other words, from a strictly software engineering perspective, a customer's problem is supposed to have been solved before a software development process is started. The role of the software engineering is to translate a solution specification into a programming specification and to implement and verify program deliverables against that specification.

In summary, the design and development steps of a usability engineering process should not be fitted alongside the apparently corresponding steps of a software engineering process. Design for use must precede design for implementation, and its final deliverable contributes not to the software programming specification, but to the initial software requirement specification.

To the extent that this is not recognised in the organisational implementation of a software development organisation, it is unsurprising that there is a requirements "crisis", seen in findings (Vintner and Poulsen, 1996) that as many as 36% of all reported defects are attributable to problems with requirements specifications, while the remaining 64% are distributed over the eight other problem categories described by Beizer (1990). Of these, 60% can be identified as usability errors, 20% relate to working with third party software and only 15% are identified as domain and functionality related errors.

History Lesson: The Web Discovers User Interface Design

Aaron Marcus, President, Aaron Marcus and Associates, Inc.
West Coast Office: 1144 65th Street, Suite F, Emeryville, CA 94608-11053 USA
T: 510-601-0994, F: 510-547-6125, E: Aaron@AMandA.com, W: www.AMandA.com

ABSTRACT

By transferring established principles of user interface design and providing tools or templates that support good practice ,industry professionals may improve Web design more rapidly and avoid the problems of past generations of technology evolution.

1. INTRODUCTION

The Web has grown rapidly in the number and diversity of its users in the past few years. In that same time, the technology has evolved quickly to provide more elaborate graphics, sound, and interaction. A current shift is from a "browse and search" orientation to an "interview and push" strategy for bringing information, advertising, and entertainment, to large audiences. Web products and services targeting users in industry, government, education, and consumer markets will compete with each other for the users' attention, time, and decision-making. To be successful in delivering useful information that benefits most people, the Web will need to communicate ever larger amounts of data and functions through well-designed user interfaces.

As with many technology innovations over the past thirty years, developers have focused too much on the engineering issues, and not enough on the communication challenges. The situation now is similar to the past advances of computer graphics technology in acquiring color, fonts, chart/diagram graphics, animation, three-dimensional display, graphical user interfaces (GUIs) for productivity tools, multimedia/CD-ROM support, etc. In almost every previous introductory period, usually lasting about three to five years, promoters made extravagant claims and companies produced much poorly-designed material. The poor quality derived from several factors: (1) professional designers were not involved in the design of the displays and (2) principles of good communication from the past were ignored in the search for paradigms favored by the new technology.

One seeming oversight in Web discussions is the general lack of attention to user-interface design of the Web. User interfaces for the Web both benefit and suffer from the freedom of developers to design novel forms without consistency across the Web. Much of the industry seems oblivious to the relevance of traditional or "classical" GUIs of the 1980s and 90s. The traditional components of GUIs viewed from a communication perspective still provide a useful basis for analysis [5,6,7,8]:

Metaphors are fundamental concepts communicated through terms, images, sounds, kinesthetic actions, etc. that are easily recognized, understood, and remembered. The *mental model* is the appropriate organization and representation of data, functions, work tasks or play activities, and roles. Navigation refers to efficient movement among data, functions, tasks/activities, and roles depicted in the model. *Appearance* refers to the appealing perceptual characteristics that efficiently convey all information to the user. Interaction consists of the input and output techniques that operate efficiently and provide an appealing perceptual experience.

2. TWO HOMEPAGE EXAMPLES OF WEB DESIGN

The Web provides a significant challenge to those concerned about traditional concepts of communication quality [6,8]. Figures 1 and 2 illustrate typical design issues in Web homepages for a data resource and a “Web ‘zine”.

Figure 1: Nynex Interactive Yellow Pages Homepage

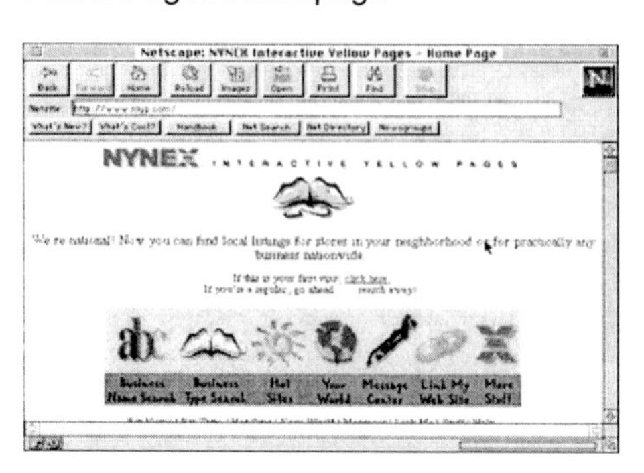

Figure 2: Hotwired Homepage

Figure 1 shows a home page from the Nynex Interactive Yellow Pages Website, which exhibits typical design problems.

The metaphorical references include redundant, ambiguous, and/or obscure terminology. For example, what Nynex calls “hot”, Netscape calls “cool;” both buttons refer the user to sites the corporate sponsors deem noteworthy. These terms at opposite ends of the temperature range are likely to cause confusion among some users. Another example is The Nynex label More Stuff which is ambiguous.

An example of mental model confusion is to be found in the button labeled Business Name Search, which seems to be different from the Business Type Search. In fact both buttons lead to a form-like screen filled with many data fields in a nearly identical array but differing by one data field; the Business Type Search contains the additional data field for the kind of business. A typical navigation limitation is that no immediate overview diagram or cascading-menus technique is available to reveal more of the content structure thereby enabling a user, especially a frequent visitor, to navigate more efficiently.

Appearance inadequacies (in a valid attempt at product differentiation for marketing purposes) include the decision to use black, cursive labels on a green background for the primary Nynex buttons. The letter forms and figure-field color combination badly degrade the legibility of the labels. Three typical problems of Web-oriented interaction design are (1) The layout is one in which controls for the windowing environment, Netscape, and Nynex all compete, in part redundantly, for screen area and the user's attention, (2) the visual signals of what portions of the screen are, or are not, selectable controls are ambiguous and (3) the nature of the controls (e.g., inclusive vs. exclusive selection) is ambiguous.

Figure 2 shows a home page from the Hotwired Website intended to attract potential readers. The GUI poses additional challenges to good design. This single initial screen powerfully intermixes the metaphorical references of information (e.g., "Signal" and "?"), with entertainment (e.g., "Piazza" and "Renaissance") and transaction (e.g., "Coin" and the acknowledgment of a corporate sponsor at the bottom of the window). This merging of messages is a radical departure from classical computer-based tools, yet is as familiar an experience as the advertising imprinted on a writing pen or the mixture of advertisements and score information at a sports event.

An additional issue in Figure 2 is the ambiguity, obscurity, and exotic allusiveness of the label references. What exactly is denoted by the "patrician" items labeled Piazza and Renaissance in the context of such pedestrian functional terms like Signal and Coin? Should a GUI be hard to interpret even by native-language speakers? In this Website, the allure of joining an elite group seems to turn traditional human factors concerns upside down.

While these examples are admittedly versions long since revised and of limited scope, they reveal typical questionable design decisions, or at least challenges to the assumptions that GUIs should be easy to learn and use.

DISCOVERING THE PAST TO INVENT THE FUTURE

Upon examining several recent publications about Web design [2,3,4,10], one finds that terms like "user interface", "interface," "metaphors," or "mental model" may not even appear in the index of the book; or, if they do, the section devoted to user-interface design may be merely one or two pages in length. Citations to established theoreticians, practitioners, and commentators about user-interface design (e.g., Baecker, Buxton, Grudin, Shneiderman, etc. [1, 8, 11]) are generally absent from their bibliographies. Topics of long-standing interest to the ACM/SIGCHI community seem to be missing from tables of contents and indices (e.g., internationalization [1], usability, productivity, and collaborative use [8]). These documents exhibit a collective amnesia about the research and practice of the last thirty years. Web developers testing the new technology seem reluctant to acknowledge the potential value of previous communication media. What might be done to improve this situation? For one thing, Web designers, developers, and critics would benefit from acknowledging the typical GUI components of Web artifacts. The use of common terminology may make it easier to learn from the theory, principles, and evaluation of well-designed classical GUIs. Presumably future guidebooks would relate to and reference some of the established sources.

However, even if this were the case, it is unlikely that most Web developers will have an opportunity to consult these professional guidebooks. The Web is, potentially, a dramatically democratic means of communication (as if every literate person in the fifteenth century had the means to publish books); at the same time, large corporations will expand intranets and sources of pushed information. It does not seem reasonable that most of the people involved will have studied user interface design. A more effective means to achieve better design seems to be building good practice into the very building tools and templates made available to every developer. Apple Macintosh developers employed this strategy in the mid 1980s with considerable success to encourage consistency and adherence to standards among third-party software developers. By making it easier for novice and occasional designers to use well-designed default conditions and components, or "templates", it becomes feasible to nurture superior practice in Web design that derives benefit from the past experience of several decades of GUI design.

REFERENCES

1. Del Galdo, Elise, and Jakob Nielsen, *International User Interfaces*, Jonathan Wiley and Sons, Inc., New York, 1996.

2. DiNucci, Darcy, with Maria Giudice and Lynne Stiles, *Elements of Web Design*, Peachpit Press, Berkeley, 1997.

3. Horton, William, Lee Taylor, Arthur Ignacio, and Nancy L. Loft, *Web Page Design Cookbook*, John Wiley and Sons, Inc., New York, 1996.

4. Marcus, Aaron, and Andries van Dam, "User Interface Developments for the Nineties," *IEEE Computer*, Vol. 24, No. 9, September 1991, pp. 49-57.

5. Marcus, Aaron, *Graphic Design for Electronic Documents and User Interfaces,* Addison-Wesley, Reading, 1992.

6. Marcus, Aaron, Nicholas Smilonich, and Lynne Thompson, *The Cross GUI-Handbook for Multiplatform User Interface Design*, Addison-Wesley, Reading, 1994.

7. Marcus, Aaron, "Principles of Effective Visual Communication for Graphical User Interface Design," in Baecker *et al, Readings in Human-Computer Interaction*, Morgan-Kaufman, San Francisco, 1995, pp. 425-468.

8. Marcus, Aaron, "Graphical User Interface Design for the Web," in McCoy, John, ed., *Mastering Web Design*, Sybex, Alameda, CA, 1996, pp. 315-330.

9. Sano, Darrell, *Designing Large-Scale Web Sites: A Visual Design Methodology*, John Wiley and Sons, Inc., New York, 1996.

10. Shneiderman, Ben, *Designing the User-Interface*, Third Edition, Addison Wesley Longman, Reading, 1997.

11. Weinman, Lynda, *Deconstructing Web Graphics*, New Riders Publishing, Indianapolis, IN, 1996.

User-Centered Design at AT&T Labs

D. James Dooling

AT&T Labs, Room 1L-309, 101 Crawfords Corner Road, Holmdel NJ 07733, USA

The theme of this session is "theory and practice". As manager of a very applied organization, I'm here to emphasize the "practice" part of the equation and to engage others in discussions about the relevance of "theory" to the work that we do.

I manage the User-Centered Design (UCD) Department at AT&T Labs. This is a team of approximately 35 professionals who do user interface work across a wide range of projects at AT&T. We represent about a third of AT&T's behavioral science resources. While a few do research, most behavioral scientists at AT&T Labs work on the development of services for AT&T's customers or management systems to run AT&T's network.

User-Centered Design has a long history at AT&T and can be dated to 1945 when John Karlin was hired as the first experimental psychologist to do human factors work in industry (Hanson, 1983). Karlin's human factors department had significant impact on the old Bell System and played a leading role in defining human factors engineering in American industry. The company has changed a lot since then, and so has the practice of user-centered design. But there is still a very strong legacy of UCD at AT&T.

To those of you who think of AT&T Labs as focused on voice services (and wonder why AT&T Labs is here talking about HCI) I would point out that AT&T was also one of the first companies in the world to computerize its operations and there is a long tradition of doing HCI work for the legions of people who manage and maintain our network and provide service to our customers. And, of course, AT&T is now a major player in offering internet services both as an access provider and as a platform for electronic commerce.

1. USER-CENTERED DESIGN ISSUES

This paper is about managing user-centered design. I've organized it around a few key issues, the sorts of questions that any large company has to answer one way or another (perhaps by default) in trying to figure out how to provide excellent user interfaces for its customers. The views I express are my own; needless to say, not all of my colleagues back at AT&T will agree with all of the views that I express here.

1.1 Central vs. Distributed Organization

When human factors began to grow rapidly at AT&T in the 1970's an important decision was made to decentralize by putting groups of user-centered design experts inside various business units instead of in one large, central organization. This allowed UCD people to be better focused on business needs and yielded a number of outstanding successes on specific projects. It made us less bureaucratic, more responsive.

The current trend, however, is to consolidate UCD resources into a large organization. This solves several problems that have arisen over the years: We are able to maintain a critical mass for support services, such as a usability lab; we have more flexibility in assigning staff with particular skills and interests to projects; we have a professional community where people help each other and learn from each other; the professional community helps keep good people in the UCD profession, instead of becoming engineers or programmers; a large organization is in a better position to promote UCD with senior management; and we provide career paths for people who want to go into management and keep an affiliation with UCD work. In my view, centralizing UCD is the way to go.

But how do we keep a centralized organization from becoming an unresponsive bureaucracy? That is the topic of the next section.

1.2. Top-Down vs. Project Funding.

There was a day at AT&T when all UCD work was funded from a central, corporate fund. At the beginning of the year, a manager knew the budget and didn't have to worry about funding. This made the manager's job easier and it also had the advantage of supporting longer term work.

But today, most of our work is funded on a project by project basis. If the business leader of a project wants our work, they pay us for it. This funding model makes us more responsive to business priorities—and it also represents a tangible definition of our value to the corporation. If we have projects waiting in line to fund our work, we are obviously doing something of value to the business.

While I am convinced that funding on a project basis is the right model, there does need to be some funding to do cross-business standards work or to do some longer term work that takes us into new areas. The work of Blanchard (1997) on standards is a good example of work that requires corporate funding. The paper by Lehder et al.(1997) on international user interfaces is an example of a project where we invested corporate funds two years ago and are now reaping the benefits for specifically funded projects. Central funding of 20-30% of the total budget seems to be the right amount to me.

1.3 Staffing User-Centered Design

The tradition that I represent at AT&T is to staff human factors positions with PhD psychologists, some with specific human factors training, but many others with general academic research skills. Just about every behavioral scientist on our staff also has software skills and is adept at rapid prototyping. For the most part, we have found better results in hiring psychologists who know software, rather than hiring computer scientists who know user interfaces. Whether from the nature of their training or from self-selection, behavioral scientists tend to do a better job of representing the user.

1.4 How We Work

On a typical project, there is a UCD specialist working with a team of engineers and computer scientists who are developing a service or support system. If we are brought in early enough, we do standard usability methodology: task analysis, user needs analysis, usability objectives, iterative prototyping, usability testing, etc., (normally adapting the processes to schedule and budget constraints). On some projects we are responsible for writing the user requirements and this, in turn, drives the service definition. There are also, however, other projects where we are brought in late. There, we typically use "expert judgment" to identify and suggest fixes for the most important problems. When we do this well we are likely to become members of the team for the next release of the application—from the beginning—and can do our job better.

Within this model it is necessary to have expert staff who are flexible and who have a large bag of skills that can be applied as conditions warrant. Because very few projects have the same dynamics, we do not have specialists for the various stages of user-centered design. People do tend to specialize, however, in various content domains such as voice messaging, network management systems, etc.

1.5 User-Centered Design Process

At AT&T Labs we have a well-defined process for ensuring excellence in user-centered design. It's called the *AT&T Best Practice on Usability Engineering*. (It is not available for outside distribution because it contains proprietary case studies from actual projects.) The *Best Practice* contains information that will be familiar to anyone who has made a professional study of user-centered design processes and it is accompanied by a *Metrics Handbook* that defines usability metrics most relevant to our business.

The *Best Practice* is a guideline, not a corporate policy. And it is a rare project that attempts to follow all of its precepts. The reality, therefore, is figuring out how to boil down the methodology to its essence and ensure that it is followed in an appropriate way. To accomplish this, we need very talented people who can customize UCD processes on a project by project basis.

1.6 Selling User-Centered Design

What "works" for us in selling user-centered design? First of all, personal experience—a track record of success for the particular manager or organization. Secondly, data. There is nothing better than to have good data comparing a user interface that we designed with others from competitors; or to have data on significant cost savings due to a well-designed operations support system. We have plenty of testimonials and success stories from prior projects and these are necessary conditions. But they're not sufficient.

Another factor is executive perception. Executives do not need to be sold on "ease of use". They believe that it is very important—but they greatly underestimate what it takes to get excellent user interfaces into our services and management systems. Much of the "selling" that we do, therefore, is in the form of educating managers on what it takes to do excellent UCD. Managing the expectations of executives is, therefore, an important part of what I do as a manager.

2.0 CONCLUSION

The theme of this session is bridging the gap between theory and practice. Looking at this from my vantage point in industry, the solution I see is to get more academic effort focused on specific applied problems. When we design a new voice mail system, for example, we'd like to be able to draw on a large empirical body of data on user interactions with voice mail. In the absence of such data, we rely on person-to-person technology transfer. What works well for us is summer internships or part-time jobs with nearby graduate schools to get people with the latest academic knowledge focused on our problems. What also works well is to form teams within AT&T Labs to work on specific forward-looking projects. In fact, some of the very best research ever done was motivated by practical problems. For example, it was a team of AT&T researchers who were trying to replace vacuum tubes for our switching equipment who invented the transistor.

3.0 REFERENCES

Blanchard, H. E . International Standards on Human-Computer Interaction: What Is Out There and How Will It Be Implemented? Paper presented at *HCI International '97*, San Francisco, 1997.

Boyce, S. J. Designing the User Interface for a Natural Spoken Dialog System. Paper presented at *HCI International '97*, San Francisco, 1997.

Hanson, B. L. A Brief History of Applied Behavioral Science at Bell Laboratories. *The Bell System Technical Journal*, Vol. 62, No. 6, July-August 1983.

Lehder, D. Z., Alvarez, M. G., Aykin, N., & Friedman-Muller, K. Globalization: Meeting the Challenge. Paper presented at *HCI International '97*, San Francisco, 1997.

A New Design Concept and Method Based on Ergonomics and Kansei Engineering and So On

Toshiki Yamaoka

Design Center, TOSHIBA CORPORATION
Shibaura 1-1-1, Minato-ku, Tokyo, Japan

1. Learn the structure in a daily life

As user interface designers always design and construct user interface under the condition in the computer world, they can not produce a lot of ideas about new user interface. However, as human is a part of nature and we live in the real world, we can learn the strucure or the relation between human and artificial objects in a daily life or between human and nature. They show us good examples on how to convey information to user efficiently.

Although user interface designers have their own design concepts and methods, these are not common, efficient and systematic. These are based on their designer's experience, not on ergonomics, cognitive psychology or other science. Most of user interface designers who were trained as industrial designers in Japan know well user behavior by their personal experience.

However, if user interface designers know the theory based on ergonomics, cognitive psychology, etc., they can create a new user interface design or paradigm using the theory in addition to their personal experience. Namely we can produce a new human-machine interface by means of ergonomics, cognitive psychology, etc. from the viewpoint of the relation between human and artificial objects in nature. And analytic and holistic design approach are needed to do user interface design. User interface designers can not make a mistake by the analytic design method when constructing an user interface. The holistic design approach means user interface designers should give consideration to the five aspects of the Human Machine Interface (mentioned in the next chapter) like viewing distance, illumination, operating time,etc. related to constructing user interface.

2. A new design concept and method

2.1. The real world

It is a very useful method to make use of the structure of one's surroundings and one's behavior in a daily life as mentioned in chapter one when we design an user interface. What kind of behavior is needed when we go to a department store to buy a man's suit for example? At first, We plan a strategy on how to reach the department store. After deciding to use a train, we can go to a station using visual cues like a pedestrian crossing, a bridge and a highest building in a town and so on. Visual cues also help us decide which train is available at the station near the department store. We can know the floor where a man' suit is sold by reading a directory board on the first floor, and look at the sign which directs to the man's suit corner on the floor. In this way, a visual cue is a very important guide in our daily behavior. The structure of GUI or CUI in the computer world is based on this daily behavior. Hence, the real world behavior influences the concept and design of user interface in the computer world. In the computer world, using metaphor is a very efficient method to convey information from the real world. Metaphor provide users with information on how to operate the computer properly.

Thus, we should refer to the behavior and concept in the real world when we design user interface of products.

2.2. User interface design items

It is logical to examine daily life examples to derive design principles for GUI design. By collecting bad user interface design samples encountered in daily life, it is possible to arrive at solutions that ultimately form the foundation of new user interface guidelines. Approximately 500 bad user interface samples were collected by the author and university students. These samples cover a wide range of items ranging from stationary, furniture, home appliances to public facilities.

The collected samples were classfied according to their function. In addition, they are futher categorized under the five aspects of Human Machine Interface (HMI). The five aspects of the HMI are : (1) the physical level, (2) the brain level, (3) the time level,
(4) the environmental level, (5) the organisational level.

A hierarchy of items classfied under the physical level was established by using the ISM method.

1) danger, 2) operational direction of the control, 3) amount of space for movement,
4) spacing between controls, 5) location of controls, 6) ease of maintenance,
7) size of control, 8) anthropometric fit, 9) location of the center of gravity,
10) ease of access and exit, 11) protrusion

A hierarchy of items classified under the brain level is established by using the ISM method.

1) Mapping between indicator and the operating area,
2) presentation of information, 3) ease of information retrieval,
4) classification of information, 5) frequency of operation, 6) procedure of operation,
7) wording, 8) easy to see, 9) feedback, 10) relationship between functions,
11) affordance, 12) time left

The physical level and brain level are very important aspect of user interface design. Six main practical design items for constructing user interface are selected from user interface design items mentioned above.

They are :

1.wording, 2.visual cue, 3.mapping, 4.consistency, 5.feed back,
6.informing users of the general idea of the machine structure

Particularly a visual cue is a very important concept in guiding the operation of a computer.

2.3.Mimic the structure of paragraph to design GUI

In the paragraph theory, a paragraph is divided into a topic sentence,subtopic sentence, conclusion sentence and two elements that indicate the structure and logic among sentences. This theory is also used to design an operational screen like ATM, a ticket machine, and other appratus. As we use any conjunctions and adverbs like "therefore","because" and so on in a paragraph to clear the relation between sentences, a visual cue in GUI has the same function in the screens of the computer world.

2.4. Three design principles in the visualization of information

The user interface designer can visualize any information easily by three design principles[3]. The design principles are "emphasis","simplicity" and "consistency".

"Emphasis" means easy reception of information or quick information detection. A design should be done to attract attention at first from the viewpoint of the detection. The primary external factors for attracting attention are as the following [2] :
(1) the nature of the stimulus, (2) the size of the stimulus, (3) the strength of the stimulus,
(4) the duration of the stimulus, (5) the repetition, (6) the variation of the stimulus,
(7) the position of the stimulus, (8) the originality of the stimulus
This emphasis means the prioritization of the displayed information. The important information should attract attention by occupying a large area and caused by the size and the strength of the stimulus.

"Simplicity" means an arrangement of information or achievement of visibility.
The information can be conveyed easily by the "simplicity" principle. After obtaining the important prioritized information emphasized, the simplicity aspect of the information and its layout become important.

"Consistency" means consistency of information or understanding of contents. The above two design principles are directly involved with the construction of the design of screen format. Context is a method which makes conceptually driven processing easier and can be thought of affecting not just the verbal aspect but also the visual aspect of information as well. The consistency means context for the relationship between screens, consistency of constructing displayed information , consistency of word usage, consistency of important visual elements and the like.

3. Analytic design method and holistic design method

The old design procedure of user interface designer were vague and not systematic in general. A new method of user interface design should be done step by step according to an analytical procedure which involves systematic steps to design. A new design procedure to recommend is the following.

(1)The clarification of system goal
(2)Constructing or understanding the system
(3)Constructing a concept regarding the user interface design
(4)The functional allocation of human and machine
(5)Target user is predicted. The disabled are also considered.
Identifying a target user's behavior and mental model.
(6)Task analysis: contents of task is decided.
(7)The clarification of human requirement
human requirement : What is the important item for user?
(8)Taking the environmental level into consideration.
The environmental factors: illumination, air conditioning, noise, vibration and so on.
For example, user interface designers take the contrast in consideration if a CRT or LCD

on machine is used under brilliant illumination

(9)Taking the time level into consideration.

(10)The posture of user is predicted by a task analysis.

(11)The viewing distance between user and display (machine) is decided from location of computer on a desk.

(12)The apprppriate size and font on a display like CRT,LCD and so on is decided from the viewing distance.

(13)The style of screens is decided by the method of Kansei engineering.

Kansei engineering provides general seven design items for all objects like products,screen and so on.

The seven design items[4] are as follows.

1) a new conbination of objects 2) an unexpected design 3) a color 4) a form
5) a function and convenience 6) a feeling of a material 7) a fitness

The style of screens is designed by a multiple regression equation in which the criterion variable and predictor variables are the style and the seven design items.

(14)Constructuring information (Designing user interface)

1)Arranging and determining information to be displayed
2)Constructing grouping and hierarchy of information
3)Determining priority of imformation to be displayed
4)Determining order of displaying information

(15)Visualization (or making audible) of information

Three design principles for visualization : emphasis, simplicity, consistency

1) "Emphasis" : easy reception of information or quick information detection.
2) "Simplicity" : an arrangement of information or achievement of visibility
3) "Consistency" : consistency of information or understanding of contents

(16)The user interface designs are evaluated by AHP(Analytic Hierarchy Process) and Fuzzy measure/Fuzzy integration.

(17) Taking the organisational level into consideration.

User interface design should be examined how to be operated under working conditions and organizations.

After these analytic design approach, the holistic design approach should be done by unifying all design factors mentioned above using six main design items and the five aspects of the HMI under the concept. The holistic design approach will be promoted by general design training, not engineering. The ability to be done the holistic design approach is enhanced by observing structural relation between human and artificial objects in nature.

REFERENCES

1. Toshiki Yamaoka: Structuring elements of the operational screen by paragraph theory and so on, Human Interface, N&R,1996,vol.11
2. Masaichi Turuta: Jiko no Shinri (Psychology of Accident),pp20-21,Chuko Shinsho,1968
3. Toshiki Yamaoka,Hiroshi Tamura:Information display method and process considerations in the Tron/GUI,proceedings of the tenth tron project international symposium,1993
4. Toshiki Yamaoka: Examining kansei design method,proceedings of 2nd Kansei Engineering Symposium, The Japan Science Council,1996

Toward Rehabilitation Cognitive Engineering
-- Gap between theory and practice in the human interface of information processing devices for people with disabilities --

Akira Okamoto

R&D Group, Ricoh Co., Ltd.
3-6, 1-Chome Nakamagome, Ohta-ku Tokyo 143, Japan

1. INTRODUCTION

In recent years, the progress of information processing technology has greatly improved the performance of equipment such as personal computers(PC's), fax machines, and so on. Moreover, much research has been done on the ease of use of these machines in the field of cognitive engineering. In these circumstances, it is widely expected that good theories in this field should contribute to the improvement in human/machine interfaces.

Is it always the case that a good cognitive theory has wide applicability? Unfortunately not. We have to admit that there is an extensive gap between a theory and its practice. One reason for the gap is that no theory can cover a multiplicity of human behaviors, either physical or perceptual.

This problem becomes more serious when designing interfaces for people with disabilities. The person who is not disabled can often compensate for a poor interface; people with disabilities may not be able to. It is critically important that user interfaces meet these special needs. Today, major theories of human cognition are based on the behavior of subjects without disabilities, which means that little theoretical research has been done on the cognitive behavior of disabled people. It might be possible to design the same type of human interfaces for people with and without disabilities, but what we need here is to custom-design interfaces adapted to their differing capabilities and living conditions.

As a result, much empirical but little theoretical work has been done on human interfaces for them. Therefore, we must say that there is a big gap between theory and practice in this field.

2. PRACTICE COUNTS MORE THAN THEORY IN THE AREA OF ENGINEERING FOR PEOPLE WITH DISABILITIES

Because disabled people vary in the extent and type of their disabilities more practice than theory is needed in human interface design for these people. In general, the range of differences observed between two disabled people is much greater than between two people without disabilities. General theories, therefore, are not of much help in meeting an individual's differing needs. The only way to provide satisfying human interfaces is to attend to each person's needs one by one.

This section describes two case studies which we have recently undertaken. These cases provide the examples which suggest the importance of (1) direct observation about problems in usability encountered by people with disabilities and (2) meeting the needs of individuals.

Based on our observations and analysis of the needs of two people with severe physical disability, we have made simple modifications of existing devices. As shown below, these modifications are not the outcome of any cognitive theory, but they have turned out to be very helpful for the two people.

2.1 A Reading Machine Employing an Electronic Filing System

There are many cases where people with severe physical disability don't read books or newspapers, not because they don't want to, but because they can not turn the pages. We have, therefore, developed a prototype of a reading machine which employs Ricoh's optical disk electronic filing system (here-after called "EF") originally designed for clerical use[1].

To design a human interface for this machine, we have explored the needs of our "research partner", i.e., the patient with muscular dystrophy who is confined to bed with a respirator. He is able to move his right thumb two or three millimeters, and indeed he has been operating PC's using a one-switch input device. Fortunately, in the EF, all the operations can be handled using a mouse. Thus we have modified an existing one-switch mouse emulator for PC's to be used for EF. Employing this mouse emulator, he can freely operate the EF to read books or newspapers stored as optical files.

Additional modification has been made to the EF to enable the person lying in bed to see the whole screen. The LCD (Liquid Crystal Display) has been separated from the main body so as to be fixed right over his face. To accomplish this, we lengthened the cables while shielding electromagnetism and made an iron frame for mounting the LCD.

The availability of this reading machine has changed his approach to reading. He could read only one book per year before, but now he reads four or five booksper month, e.g., novels, magazines, operation manuals of PC's and so on, thereby finding a new way of life.

2.2 An Auto-Scan Type Input Device with a "REVERSE" Function

The auto-scan type input device is used by people with severe cerebral palsy to operate their PC's. In the auto-scan input device, the alphabet, numerals and symbols are shown on the computer display and a cursor scans each "key" automatically in sequence. The user is only required to push the switch when a cursor falls on the "key" they want to input.

The disadvantage to this input method is that they have to patiently wait for the scanning by the cursor. The user's frustration will be increased if they miss the character because then they have to wait until the next scanning. This is because scanning is always uni-directional. To solve this problem, we have attached "REV(REVERSE)" keys on the top line of the key display[2]. When the "REV" key is selected by the user, the cursor begins scanning in the opposite direction. For example, when the cursor passes the target character, the user can back the cursor up using this "REV" key. Moreover, the user can judge which direction is more suitable for a quick access to the next target character and use "REV" key if needed.

Our experimental results show that this kind of flexibility can not only give the user a sense of being in control but also reduce his operation time and typographical errors.

usability engineering in Japan may be related to the historical background of the computer industry. The details are shown in Table 1.

Table 1. Situational differences of the usability engineering in Japan and the U.S.

	Japan	U.S.
1. Consciousness about the usability	Most of the stakeholders, including the engineers, are concerned with only the positive aspects of the usability.	At least, the engineers are realizing the importance of the non-negative aspects as well as the positive aspects.
2. Background	Until recently, Japanese computer industry has been just following what was developed in the U.S. and has been focused on much more performance and many more functions.	Computer industry in the U.S. has always been creating new concepts based on the analysis of the user and their evaluation.
3. Consumer	Because of the homogeneity of the people, they have not been so much aware of the differences among them and could not establish their own independence as the user.	Because of the diversity among the people and the consciousness of the difference, they have established their own criteria for everything including the computer.
4. Usability engineering activity	Only small groups can be found in some of the computer-related company.	Usability engineering activity is an everyday thing and there are many more staffs compared to the Japanese company.
5. How to obtain the usability staff	It is not easy because the psychology course at university is mainly focused on the theoretical issues and the students have less chance to realize the relation to HCI.	Multi-disciplinary education is common and the students have more chances to realize the relationship between the psychology and the HCI than in Japan.

4. POSSIBLE SOLUTION

4.1. Consciousness on the part of the corporate

Because of the fact that the variety of functions the computer can provide to the user has somewhat been saturated and the performance is also reaching the sufficient level for the ordinary usage, the computer industry seems to be going to the unseen future. In such a situation, what will make the product more attractive may be the completedness of the usability of the product including the non-negative aspects as well as the positive aspects. Besides that, the computers with high cost-performance is coming from various countries now and Japan will have to change its approach from that of 10 years ago. Hence the stakeholders in the computer industry will have to realize the importance of the completedness of the usability of the product now.

4.2. Consciousness on the part of the user

The users will have to realize the importance of the completedness of the usability. Most of them must have had experiences that they have realized the difficulty of use, inefficiency of use or unpleasantness of use of the product, after they started its actual use. They should remember what they have experienced and a kind of social propaganda will better be done. If the consumers starts to have their own criteria for the product, the quality of the product will then be improved, thus they could receive the benefit.

4.3. International standards

The international standard such as ISO 13407 will have some positive influences over the computer industry and the computer market especially in Japan. Because Japan is a vertically-structured society, what was determined as the standard at the government level will have a strong effect over the whole industry.

Because there are differences among different countries, the solution for the future usability activity will not be single. Because the usability activity is strongly connected to the organizational structure of the society, we will have to make efforts to solve the dilemma of our own without any preceding example.

REFERENCES

1. Nielsen, J. (1993). Usability Engineering, AP Professional.
2. Kurosu, M. (1996). The structure of the usability concept, N&R of SIGHI, SICE Japan, Vol. 11.

Understanding HCI Requirements: Expertise and Assistance to Imagery Analysis

I.S.MacLeod [a], A.J.McClumpha [b] and E.Koritsas [b]

[a]Aerosystems International, West Hendford, Yeovil, Somerset, BA20 2AL, UK

[b]DERA CHS, Farnborough, Hampshire, GU14 6TD, UK

1. Introduction

Reconnaissance, Surveillance and Target Acquisition (RSTA) systems are increasingly based on advanced technology and are required to handle a diversity of sensors to exploit the associated forms of collected imagery. The future real time collection and dissemination of imagery will require co-operative use of multi-user workstations that have the capacity to display a variety of different forms of imagery (for example, Synthetic Aperture Radar (SAR), Moving Target Indicator (MTI) Radar, Electro Optic (EO) Sensors, and Infra Red (IR)). To illustrate the variety of imagery that the Image Analyst (IA) has to analyse, Figures 1, 2 and 3 are provided in illustration. Each figure has some explanatory text included with the title.

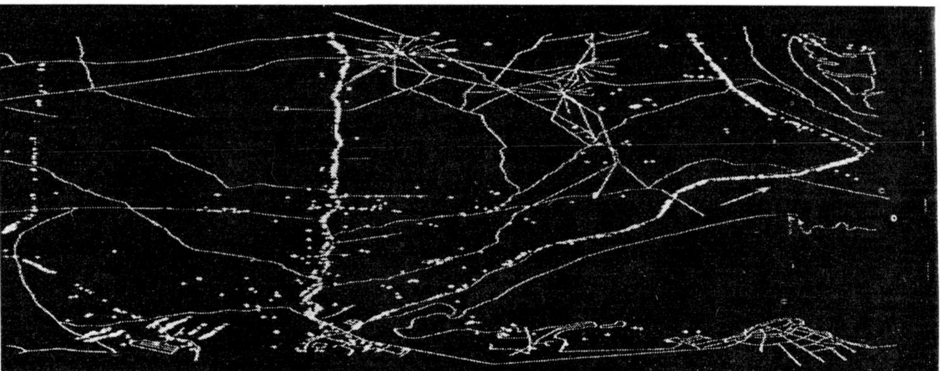

Figure One - An MTI Scene. (Thick lines and dots are MTI returns, thin lines depict roads and coastlines. It can be readily appreciated that improved map cultural detail, contrasted against MTI returns, would assist analysis of this image).

Currently, technology offers important assistance for the collection, display, storage, display and dissemination of imagery. However, only minimal technical assistance is given to the IA who is required to analyse and exploit the presented imagery. It is clear that in current 'technologically sophisticated systems' there is only limited support to the skilled performance of the cognitive and perceptual analyses that are required to interpret imagery.

Furthermore, the introduction of this new technology has changed the nature and methods of imagery analysis and the associated work practices that the IA needs to undertake. These new work practices are also being relocated into new work domains where the maintenance of situational awareness is becoming more important (Reference 1). However, few studies have addressed the problem of taking a user-centred perspective to both qualitatively and quantitatively explore the cognitive and perceptual aspects of imagery analysis.

The work reported in this paper is part of an ongoing Human Factors (HF) programme to better understand the cognitive and perceptual aspects of the imagery exploitation tasks so that guidance can be offered to make efficient use of technology and to enhance the productivity of the analyst. In recent years the UK Centre of Human Sciences (DERA CHS) has conducted several research concerning the use of VDU and computer technology to assist the process of image analysis (References 2 & 3). These studies attempt to identify and scope the expertise involved in the human cognitive and perceptual tasks that are crucial to the efficient and effective analysis of images.

The objective of this paper is to present and discuss a few of the findings from studies and to indicate their to the better understanding of aspects of human performance in the analysis of imagery. This improved understanding will help provide a more user-centred basis for the future determination of HCI standards and guidelines to facilitate performance in image analysis.

2. Guideline Requirements

Essentially, guidelines should give advice on what information is needed, its presentation, form, timeliness, and the rate of its display. Many HCI related guidelines have been produced in recent years, their content depending on the subject area and the agency commissioning the guideline work. However, no HCI guidelines have been produced that are focused on the display and analysis of imagery.

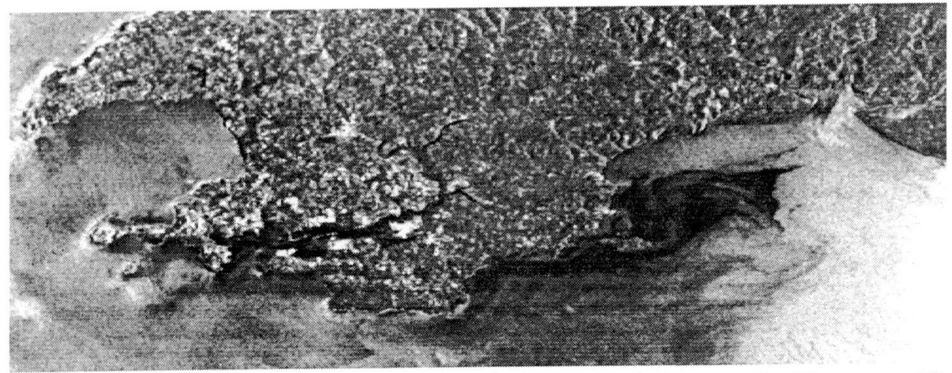

Figure 2 - SAR Example - Sea Empress Oil Spill near Milford Haven, UK. (The course 30 metre resolution of this satellite image still allows meaningful depiction of the spill size).

One influential book produced on the problems of image analysis and target detection, and still pertinent to this day, covered the problems of image analysis from the perspective of human perception and the many diverse influences on that perception (Reference 4). However, this book failed to attempt any delineation of the cognitive contribution to analysis.

It follows that prior to the production of any HCI guidelines related to imagery analysis, the fundamentals of the associated imagery analyst's cognitive based expertise must be discovered, and pertinent analysis tasks understood. This work is essential to produce the source material necessary for the production of guidelines or for the design of HCI based tools to assist the IA's analysis tasks.

Part of the problem is the determination of the form of skill requirements and strategies most useful to the performance of IA tasks. These skill requirements should be considered with regard to personnel selection, training, and the appropriateness or level of assistance given to the IA through system and HCI design. In the UK, the selection process for the IA is primarily based on vision capabilities (stereoscopic vision is deemed to be essential) and aspects of applied intelligence. UK based ab initio training of IAs emphasises a procedural approach to image analysis; but in contrast, recent HF studies have shown that an important manifestation of IA expertise is an almost automatic approach to analysis tasks. Another manifestation of expertise is the variety of approaches used to the analysis of imaged targets. These findings are all of importance to post ab initio continuation training.

Therefore, as well as studying the nature of the expertise required of the future IA (for some discussion on the nature of skill and expertise see References 5 & 6), it is necessary to study the work factors that influence the application of that expertise. These are the factors that should be the primary influence on HCI design. Examination of this topic requires the use of a variety of approaches and techniques. Domain experts have participated in a number of studies to explore the performance based aspects of the cognitive and perceptual interpretation of imagery. These studies have been conducted using equipments currently operational within the UK Royal Air Force.

3. Methods and Implications

The recent approaches to the problem have covered the following areas: classification of IA skill; classification of targets; understanding of the implications to image analysis of imagery types; classification of IA tasks.

These approaches have been supported through a combined use of many methods, mainly qualitative in nature. They have included the use of: concurrent and retrospective verbal protocol analyses; repertory grid techniques; subjective workload assessments; method of limits; concept mapping; cognitive probes; individual and group debriefs.

The studies have indicated that there are important differences in the way that domain experts with differing levels of expertise carry out the task of target interpretation. For example, experts do not apply similar analysis procedures to each imaged target but individually vary the procedure depending upon the particular characteristics of the target, its familiarity, and the form of the imagery. This finding is important because it demonstrates that there is not likely to be a simple explanation of how to provide technological support to the domain expert for imagery exploitation. The finding also emphasises the need to establish a cognitive based assessment of the nature of image analysis tasks. Furthermore, it is also clear that the form of required system support to the IA may need to be qualitatively and quantitatively different between IAs with different levels of experience.

In addition, there is an important difference between the usage of an image based HCI and an alpha-numeric HCI. With image based HCI the input is the image, the work is primarily conducted on the displayed image, but the output is normally alpha numeric. With alpha numeric based HCI the input and outputs are normally based on alpha-numerics, and possibly associated graphics, and the work is normally only associated with the displayed information.

It has been found that the greater the experience of the IA the more they require all image manipulation controls to be through mouse/rollball and cursor so as to allow continuity of attention and the analysis through an image display occupying the whole of the display area. Novices and the less skilled IAs initially prefer a set piece screen control panel to occupy part of the display in support of their more procedural approach to image analysis. This preference rapidly diminishes with experience. All IAs prefer the use of icons as a method over menus to select ancillary work applications.

Figure 3 - IR Image of an Oil Refinery (Various levels of liquid in the tanks is discernible as is the increasing distortion of the image towards the horizon).

4. Conclusion

The skills and developed expertise of the IA are largely cognitive and have yet to be fully understood. Modern technology assists in the display and development of the imagery but gives little assistance to the actual analyses of images by the IA. Implications for the development of suitable HCI for the IA are that the displayed image is considered to be sacrosanct and that the image manifestation must make optimum use of the available display area. Therefore, the system/image control facilities should accommodate as little of the display space as possible and that any additional analysis facilities should be accessed as appropriate through simple Icon selection. However, the less expert IA still needs dedicated screen and image manipulation controls as a stepping stone towards their expertise with the use of the analysis system and the exploitation of various forms of imagery through that system. Future guidelines for imagery HCI should approach these points.

References

1. MacLeod, I.S., Taylor, R. M & Davies, C.D. (1995) Perspectives on the Appreciation of Team Situational Awareness, in *Proceedings of the International Conference on Experimental Analysis and Measurement of Situation Awareness*, Embry-Riddle Aeronautical University Press, Florida.

2. McClumpha, A.J. & MacLeod, I.S., (1996), *Skills and Understanding in Imagery Exploitation*, 1996 SAR/SLAR Steering Committee Meeting, Berlin, Germany, 17-19th October 1996.

3. MacLeod, I.S. (1996), *Final Report: Analysis of IAs Cognitive and Perceptual Activities*, undertaken under UK DERA CHS Contract No CHS3/5038. (Aerosystems International 0543S/1/TR.1-1 dated June 1996).

4. Jones, D.B., Freitag, M., & Coolyer, S.C., (1974), Air to Ground Target Acquisition (TA) Source Book: Review of Literature, Office of Naval Research, NR 196/121.

5. Boy, G (1995) Knowledge Elicitation for the Design of Software Agents, in *Handbook of Human Computer Interaction*, 2nd edition, Helander, M. & Landauer, T. (Eds), Elsevier Science Pub., North Holland.

6. Ericsson, K.A. & Staszewski, J.J. (1989), Skilled memory and Expertise: Mechanisms of Exceptional Performance, in *Complex Information Processing: The Impact of Herbert A. Simon*, Klahr, D. & Kotovsky, K. (Eds), Lawrence Erlbaum, NJ.

User-centred design: the application of the LUCID interface design method

Andy Smith & Lynne Dunckley

Department of Computing, University of Luton. Park Square,
Luton, Bedfordshire. LU1 3JU, UK.
Tel : +44 (0)1582 34111, Fax: +44 (0)1582 489212
Email : andy.smith@luton.ac.uk

The Logical User Centred Interface Design (LUCID) method aims to provide a user-centred environment through which not only is usability enhanced but, with respect to identified criteria is actually maximised. This paper describes the method, discusses the key design and implementation issues and is illustrated by descriptions of a number of implementations.

1. USER CENTRED INTERFACE DESIGN

1.1 Introduction

Whilst the underpinning elements of user-centred design have been well explored, the levels of user involvement evidenced to date indicates only limited success in modifying current commercial practice to enhance user-centredness. In response to this we have been undertaking the development of user-centred tools and techniques with the objective of making these as accessible as practicable to commercial designers, particularly in the arena of interface design. In addition we were influenced by Hartson & Boehm-Davis's (1993) review of user interface development processes and methodologies and their conclusion that 'while considerable progress has been made there still remains the need for real breakthroughs'; in particular to focus on iterative evaluation techniques that lead to a convergence on an improved design. In this paper we present an account of our progress in the development of the Logical User Centred Interface Design (LUCID) method. LUCID is a method derived from Taguchi techniques (Taguchi, 1986) and incorporates the philosophy of Total Quality Management within the context of interface design.

1.2 Current methods

Most developers agree that effective user engagement is generally a vital ingredient in the successful take up of Information Technology projects. However there seems to be a lack of clarity about how, in practice, user engagement can be enhanced and developed into a fully user-centred approach. Prior to the development of LUCID we undertook a major survey of UK commercial organisations (Smith and Dunckley, 1995) in order to elicit approaches to systems design. The survey showed that whilst the underpinning elements of UCD have been well explored and the levels of user involvement identified there has been only limited success in modifying current commercial practice to enhance user-centredness.

2. THE LUCID METHOD

The rationale behind the LUCID method is to develop a more systematic and rigorous user-centred procedure for developing and optimising the user interface. LUCID is founded on the following underlying principles:

- *adoption of a user-centred development approach*, whereby users and developers can work together on the identification of factors which effect usability,
- *effective use of iterative and parallel prototyping*, so as to facilitate participation, maximise effectiveness in interface evaluation whilst minimising development time,
- *integration of Taguchi techniques,* thereby providing the rigour necessary for the identification of the optimum interface,
- *systematic and logical integration of techniques,* so that the method can be applied by commercial interface designers.

The LUCID method, the basic phases of which are outlined in Figure 1, is an evolutionary technique but, through a crucial parallel phase, places an emphasis on the design, rather than the iterative testing of the interface. LUCID provides the following information:
What is the optimum interface design?
What factors and associated levels give the optimum design?
What is the expected quality characteristic (usability metric) for the optimum interface?

2.1. Phase 1

In Phase 1, *Analysis*, the first step is to perform an initial socio-technical analysis of the application domain. From this it will be possible to establish appropriate participative structures in which users and developers can work together in a user-centred environment. Pulling on a range of participative analysis techniques the full range of user requirements are identified. Within a user-centred development process these requirements are of three types:

- *functional*, to perform the tasks in the operational situation
- *physical*, to perform the tasks in a manner well suited to the characteristics of the user
- *aspirational*, to support the personal goals of the user.

2.2. Phase 2

In Phase 2, *Exploratory interface investigation*, users and developers jointly identify the issues which constitute usability and will enhance acceptability. Based upon the user requirements identified in Phase 1 these issues lead to the specification of a range of factors, each with a number of different potential values, or levels. Examples of design factors would be menu nesting and method of help provision. The key factors of design are crucial to the LUCID method. This is where the HCI specialist's experience plays an important role. It is not intended to investigate factors which do not influence the quality characteristic or to set levels which are not realistic design options. The objective is to set the specifications of the contributing design factors to improve quality and reduce software maintenance costs. The desire is to find the design which is as immune as possible to environmental effects. When it appears that there are too many controllable factors whose levels can be set freely then it is necessary:

- firstly to exclude all factors that cannot be tested or measured
- secondly to concentrate on the factors which seem important to the design team.

It is also necessary at this stage to derive a number of quality characteristics (usability metrics) which will later be used in interface testing. For any given interface an overarching combined quality characteristic can be established. The participative environment in which this work is undertaken is enhanced through the development of exploratory interfaces.

Phase		Activities
1	Analysis	• initial analysis • establishment of participative structures • user requirements capture
2	Exploratory interface investigation	• developers and *users as developers* identify key factors, and levels which determine usability • identification of performance criteria (metrics) • exploratory prototypes developed
3	Taguchi interface design	• investigate interaction between factors • selection of orthogonal array • design of experiments
4	Taguchi interface testing	• user testing of parallel prototypes • analysis of results • selection of optimum design
5	Interface refinement	• confirmatory testing • further iterative refinement

Figure 1 Phases in the LUCID Method

2.3. Phase 3

The technique for systematic investigation of factors is based on the factorial design method first introduced by Fisher in the 1920's and extensively applied in agrarian and social sciences. In many situations full factorial analysis will involve many experiments. For example, three factors each with two possible values, or levels, would involve only 2^3, or eight, experiments but fifteen two level factors would involve 32,768 experiments. Fractional factorial experiments can be used to simplify the investigation by looking at only a fraction of all the possible experiments. Taguchi has simplified and standardised the procedure by providing a framework for design, based on orthogonal arrays and also a simplified analysis of results which makes the fractional factorial design accessible to non-statisticians. The orthogonal arrays are independent: every column covers one factor and has the same number of occurrences of each level as every other column. This ensures that any differences in the results are only due to the change of factors. Orthogonal arrays are balanced because there are always an equal number of occurrences of each level in every column.

Full details of the Taguchi process can be found elsewhere (Taguchi, 1986). However, the basic concept is that once the design team has agreed on the quality characteristic, interface factors and levels an orthogonal array can be selected. The array then provides the detailed specification of the variant interfaces (the parallel prototypes) assigning the levels to each factor. Orthogonal arrays are a set of tables devised by Taguchi and are used to determine the minimum number of experiments (in our case prototype interfaces) and their input factors. The orthogonal arrays are systematically named as $L_A(B^C)$ where A is the maximum number of experiments (rows), B is the number of levels and C is the number of factors (columns). The smallest array, an L_4, can investigate up to three factors and requires four experiments.

								Outer user array Condition $L_4(2^3)$			
							Factor	1	2	3	4
							X	1	2	2	1
							Y	1	2	1	2
							Z	1	1	2	2
	Inner interface array Factor $L_8(2^7)$							Repetitions			
Cond.	A	B	C	D	E	F	G	R_1	R_2	R_3	R_4
1	1	1	1	1	1	1	1	111	121	212	221
2	1	1	1	2	2	2	2	111	121	212	221
3	1	2	2	1	1	2	2	111	121	212	221
4	1	2	2	2	2	1	1	111	121	212	221
5	2	1	2	1	2	1	2	111	121	212	221
6	2	1	2	2	1	2	1	111	121	212	221
7	2	2	1	1	2	2	1	111	121	212	221
8	2	2	1	2	1	1	2	111	121	212	221

Figure 2 Example Inner and Outer Orthogonal Array

In some situations factors in the design may influence each other and may not be independent. The Taguchi method will allow the design study to investigate both the input factors and the suspected interactions between the factors. A particular feature of LUCID is that it can not only take account of factors internal to the interface itself but can also account for diversity in the external environment in general and the user population in particular. Principally we focus on those external factors which will arise from variations in the users' skills, experience, motivation and attitudes. It should be possible therefore to develop a range of interfaces to meet differing user groups particularly for shared interfaces. Taguchi provides a mechanism to investigate both internal and external factors through both an inner and outer orthogonal array. Figure 2 illustrates the first two orthogonal arrays, the outer L_4 array which would deal with up to three factors and the inner L_8 array which can cope with up to 7 factors. The outer array is set up at right angles to the inner array of control factors. In this example three external factors (X, Y and Z) existing at two levels were considered important and therefore an L_4 outer array was used in conjunction with an L_8 array for the seven internal factors (A to G). An L_4 array requires four experiments. Therefore each trial of the L_8 must be repeated four times for each of the possible four external combinations. The outer array, with four combinations of the three external factors, tests each of the 8 trial interfaces four times.

2.4. Phase 4

Having identified and developed the necessary interfaces indicated from the orthogonal arrays the quality characteristic for each is measured through usability testing. In order to determine the optimum interface (probably not one of those produced) two approaches are available:

- non statisticians can easily identify a *first-shot* optimum interface by calculating the average metric for each factor at each level,
- within the *detailed analysis* stage the results can be subjected to a detailed statistical analysis using Taguchi's adaptation of ANOVA.

2.5. Phase 5

Following the identification of the optimum interface a confirmatory test is undertaken to ensure that the performance characteristic is indeed maximum and agrees with the value indicated within the detailed analysis. Further iterative refinement is possible to take account of any factors which were not previously considered.

3. APPLICATIONS OF THE LUCID METHOD

The LUCID method was originated by Smith & Dunckley in 1993. During 1994 a large pilot study was carried out using the proposed method which provided a controlled environment in which to evaluate the feasibility and validity of the method. The application selected for the first study was an interface for a large scale database developed with ORACLE tools. The application was based upon requirements and data from a commercial organisation. Three factors were identified (screen design, screen colour, method of help provision) each having two levels. As a result an L_4 array was required. A single quality characteristic (percentage of useful time) formed the basis for analysis. Results of this pilot study (Smith and Dunckley, 1996) showed that the method was realistic, produced encouragingly reliable results and was sufficiently robust to be used in less controlled commercial environments. The method predicted an optimum interface with percentage of useful time of 75.17 and the confirmatory test gave 75.20.

A further study, funded as part of a UK Department of Trade and Industry Teaching Company Scheme, has also been completed with a UK commercial SME using an interface based on a Windows NT application This was a particularly interesting trial as the application involved different types of interface to a help system with remote users. In this application an L_8 array was required as seven factors were identified including one interaction. A combined quality characteristic based upon *percentage of useful time*, *number of button presses* and *satisfaction rating* was the basis for analysis. The most significant factors were found to be *nesting level of menu* and *window size*. The results of these further studies (Smith et al, 1997) confirmed the earlier promising results. The predicted performance of the optimum interface was within 3% of the confirmatory test.

The implementation of LUCID to shared interfaces is best described as *work in progress*. In an illustrative case study in which remote users in differing cultures are assumed to differ in three external factors: *cognitive decision making*, *field dependency* and *skill level*, each existing at two levels and in which seven internal (interface) factors, each with two levels, have been identified. An L_4 outer array is required in conjunction with an L_8 inner array as shown in Figure 2. In the situation of shared interfaces with diversity within the user group the method can be applied at each level of standardisation as identified by Day (1996):

- for global interfaces LUCID should provide the optimum *'cultureless standard'*,
- for international interfaces LUCID should provide a mechanism whereby customisation for different cultures can be systematised and optimised,
- for local interfaces the repeated application of the method (using internal factors only) within each culture should determine the optimum situation.

4. CONCLUSION

The discipline of human-computer interaction is maturing rapidly and a wide number of interface design methods have been developed within the last decade. The different methods can be classified in a number of ways such as by their *user focus, formality in design* and *iterative nature* (Smith, 1997). The LUCID method is proving to offer an effective way of maintaining a high level of all these issues. In situations in which the rigour of the approach can be substantiated by the payback received it offers the potential not only to identify the single optimum interface but to specify the best one for a range of user groups and social / cultural environments.

REFERENCES

Day, D. (1996), Cultural Bases of Interface Acceptance, in Sasse, M. A., Cunningham, R. J. and Winder, R. L. (eds.), People and Computers XI, Proceedings of HCI-96, Springer.

Hartson, H. R. & Boehm-Davis, D. (1993), 'User Interface Development Processes and Methodologies', *Behaviour and IT* **12** (2), 98-114.

Smith, A. (1997), Human Computer Factors, McGraw-Hill

Smith, A. and Dunckley, L. (1995), Human Factors in Software Development - current practice in the UK, in Nordby, K. et al, Human-computer Interaction, proceedings of INTERACT-95, Chapman and Hall

Smith, A. and Dunckley, L. (1996), Towards the Total Quality Interface, in Sasse, M. A., Cunningham, R. J. and Winder, R. L. (eds.), People and Computers XI, Proceedings of HCI-96, Springer.

Smith, A., Dunckley, L., Burkhardt, D., Murkett, A., Eason, K. and Church, J. (1997), Developing the optimum help system using the LUCID method, in Proceedings of INTERACT-97, Chapman and Hall.

Taguchi, G. (1986), *Introduction to Quality Engineering*, Asian Productivity Organisation.

Developing Usability in the Software Life-Cycle: Experiences from a Large European Government Software Contract

Anna Giannetti, Rossana Mignosi, and Giuseppe Taglialatela *

SOGEI S.p.A.- Quality Management - Software Usability Support Group - Via M. Carucci 99, 00143 Rome - ITALY

This paper illustrates the experience gained from the introduction of user-oriented software production and quality control processes at SOGEI SpA to improve *Usability and Quality in Use* of interactive software systems and products.

The validity of this experience has also been widely confirmed both by the almost full compliance of our *software production processes* to ISO DIS 13407 *"Human Centered Design for Interactive Systems*" which represents the international reference for suppliers of interactive software products and, furthermore, by the introduction of our *Evaluation module for Usability and Quality in Use* in the new norm ISO CD 14598_6 *"Software Product Evaluation: Evaluation Modules*" which already represents the international reference for software evaluation in medium and large IT supply contracts in Europe.

1.CORPORATE SCENARIO

SOGEI is a large IT company (staffed around 2,000 people) which manages the Inland Revenue Service on the behalf of the Italian Ministry of Finance and it represents far the largest outsourcing contract within the Italian Government. The company has always been very keen on improving and consolidating its software production processes as well as quality control and assurance methods and tools and it has long been ISO 9000 certified.

The company has also invested in a proprietary software development methodology (namely DAFNE - DAta and Functions NEtworking), supported by automated tools, whose software life cycle phases, activities and products are considered the main corporate cultural and organizational assets.

The continuous contact with more than 200,000 Ministry employees and over millions of taxpayers and the ever-growing demand of efficiency and effectiveness in Government processes sustained by the newly established AIPA- *Authority for Information Technology in Italian Public Administration*, has made clearer and clearer how much usability and quality in use of a software product was crucial for the core business of the company.

---

* This project was supported by European Commission and SOGEI SpA., STET-FINSIEL holding. The authors thank Donata Ghiglia, Cinzia Stortone, Sebastiano Bagnara, Laura Binucci and Alessandro Bianchi for their high level support. Correspondances to: Anna Giannetti, Head of the Software Usability Support Group, tel. +39-6-5025-2659, fax. +39-6-50957201, email: anna@sogei.it.

2. CORPORATE PROGRAM GOALS AND INTENTIONS

A corporate funded program called "*Usability for the new Graphical User Interfaces*" was launched at the beginning of 1995. To fully understand the program, some major characteristics of the corporate scenario should be taken into account, namely the very high interactive products/services development rate (more than 1,000 system analysts and programmers involved) and a very significant cultural shift from mainframe CUI-based to client/server GUI-based applications, involving a growing variety of heterogeneous development and target platforms.

The program therefore pointed to four major directions:

1. improving *gradually* products and services quality profile through a newly defined process for sustaining, enforcing, tracking, and controlling Usability and Quality in Use in every single phase of the software life-cycle, both in production and quality control processes.

2. sustaining the shift from a *transaction-oriented* system design approach to an effective and efficient deployment of Graphical User Interfaces environments and Object-Oriented Modeling approaches, with relevant end-user involvement throughout the whole cycle, by means of different prototyping and user requirement validation processes, ranging from scenario-based to GUI-oriented or more functional oriented prototyping.

3. experimenting an European approach to *Usability and Quality in Use evaluation and measurement,* developed in the context of the ESPRIT Research Programme, called MUSiC -*Measuring Usability Standard in Computing*, which includes methods and tools for usability context analysis, laboratory testing (Performance Measurement Method) and user satisfaction questionnaires (Software Usability Measurement Inventory), with the help of a European funding of project such as MAPI- *MUSiC Assisted Process Improvement* [1].

4. highlighting new organisational solutions within the Public Administration and AIPA to ensure end-user involvement, commitment and participation in software development and quality control/testing.

A *bottom-up* approach has been adopted so as to, first, provide production methods and tools which were more directly related to user interface development and coding, such as Corporate Styleguide for GUI, GUI inspection checklists, User Interface Code Inspection automatic tools, and, later, methods and tools for higher level specification and design, such as reusable use cases, scenario-based prototyping methods and support tools and User Requirements validation procedures. Eventually, a more comprehensive usability testing, including laboratory equipment, has been introduced, inspired by the MUSiC approach.

An Analysing and Modelling Tool Kit for Human-Computer Interaction

M. Rauterberg & M. Fjeld

Institute for Hygiene and Applied Physiology (IHA)
Swiss Federal Institute of Technology (ETH)
Clausiusstrasse 25, CH-8092 Zurich, SWITZERLAND

A tool kit has been developed to analyze the empirical data of the interactive task solving behaviour described in a finite discrete state space (e.g., human-computer interaction), helping the human factors engineer to design a good interactive system. The observable sequences of decisions and actions produced by users contain much information about (1) the mental model of the user, (2) the individual problem solving strategies for a given task, and (3) the underlying decision structure. AMME (Automatic Mental Model Evaluator), the presented analysing tool kit, handles the recorded decision and action sequences and automatically provides (1) an extracted net description of the task dependent "device" model, (2) a complete state transition matrix, and (3) different quantitative measures of the decision behaviour.

1. THE BASIC IDEA

The basic idea is to use Petri nets to model task solving behaviour. Major operations (or mappings) between Petri nets are abstraction, embedding and folding. The *folding operation* is the corner stone of our approach. A *process* (or: sequence) serves as input for this operation. An *elementary process* is the shortest meaningful part of a sequence: (s') -> [t'] -> (s") where s' is the pre-state and s" the post-state of the transition t'. Folding a process means mapping S-elements onto S-elements, T-elements onto T-elements while keeping the flow (F-) structure constant. This gives the structure of the performance net (see Figure 1), where each state corresponds to a system context, each transition to a system operation. The aim of the folding operation is to reduce the number of S- and T-elements of an observed task solving process to a minimum, giving the logical *task structure* (or: net). Hence, folding a process means extracting the embedded net structure while neglecting the amount of repetitions as well as the sequential order of actions.

2. THE ARCHITECTURE OF THE ANALYSING TOOL KIT

The whole system of our analysing tool kit consists of seven different programs: (1) An interactive *dialog system* with a logging feature, generating the task solving process description. This description should be automatically transformed to a logfile with an appropriate syntactical structure (see Table 3). However, a logfile can also be hand written by the investigator (e.g., based on protocols of observations).
(2) The *net generation program* AMME, extracting the interactive process sequence and calculating different quantitative measures of the generated net. AMME needs three input files: (i) a complete system description on an appropriate level of granularity (e.g., the state list and the pre-/post-state matrix in Table 1), (ii) the interactive process description (e.g., the logfile in Table 3), and (iii) a support file for the graphic output ("defaultp.ps" is part of the tool kit). AMME produces five different output files: (i) a *protocol file* (*.pro") with different quantitative measures of the process and of the extracted net, (ii) a *Petri net description file* ("*.net") in a readable form for the Petri net simulator PACE, (iii) a plain text file ("*.ptf") with the *connec-*

tivity matrix for KNOT, (iv) a plain text file ("*.mkv") with the *probability matrix* for the Marcov chain analysing software SEQUENZ, and (v) a PostScript file ("*.ps") to print the *net graphic* for pattern matching 'by hand'.
(3) The *Petri net simulator* PACE, being a commercial product. It is implemented in Smalltalk 80 and consists of a graphical editor and an interactive simulator with graphical animation. PACE can deal with hierarchical nets, refinement of T- and S-elements, timed Petri nets as well as stochastic Petri nets. Smalltalk 80 standard classes are available for token attributes.
(4) The *net analysing program* KNOT, computing the similarity between pairs of nets. With the multidimensional scaling (MDS) module of KNOT (Kruskal non-metric MDS algorithm) we can compute a MDS solution for any set of nets.
(5) The Marcov analysing software SEQUENZ, offering a method to compare user triggered sequences. These sequences are transformed into first-order Marcov-chains. Similarity between such lattices can be directly obtained by summation of the differences between lattice-cells. Also the resulting distances provide an input to the MDS models.
(6) Any Postscript interpreter (e.g., Ghostscript) that can read and print the output file *.ps.
(7) Any text processing software supporting pure ASCII files, *.pro.

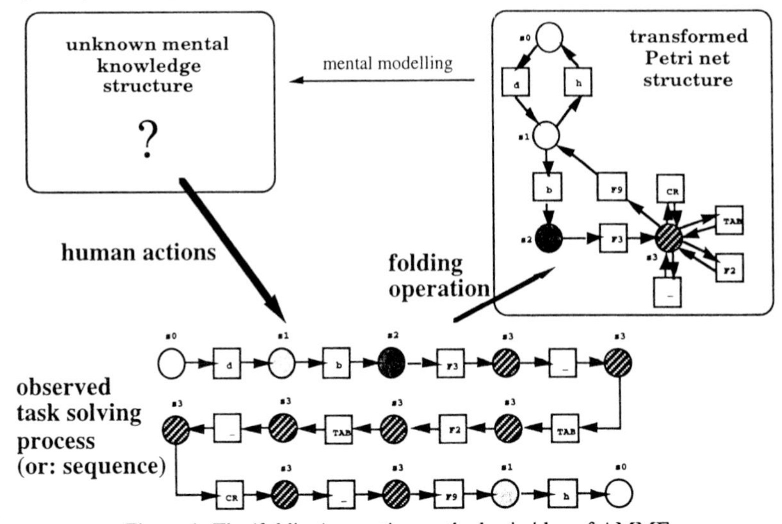

Figure 1: The 'folding' operation as the basic idea of AMME.

The current version of AMME is restricted to process descriptions that can be traced in a *finite, discrete state space* with an upper limit of different states. A further restriction is the constrained syntax of the logfiles that serve as input for AMME. To transform a given logfile to the appropriate form, several tools can be applied; like Coco/R, YACC, or any other tool that can convert text strings into other formats. AMME is freeware and available for IBM or compatible PCs (with MsWindows ≥3.0) via Internet: http://www.ifap.bepr.ethz.ch/~rauter/amme.html

3. HOW TO PROCEED

First, the usability engineer has to describe the action space of the user in a quite simple syntax; the example in Table 1 shows one possible description. Table 2 describes what has happened on the screen. Table 3 shows the content of the logfile produced by the task solving process in Table 2.

variety of questions. The literature indicated that we were likely to get varying results from different techniques.

Altogether two groups of participants used four different techniques. The first group used the *individual walkthrough* in which the two prototypes and hard copies of the screen were provided to each subject. Participants were asked to respond to direct questions about the interface.

The second group used a combination of three techniques. The first was the *empirical evaluation* in which each subject performed ten goal-oriented tasks using the prototypes in a controlled experiment and data were collected using a video camera and time and accuracy collection tools embedded in the prototypes.

The second technique was the *heuristic evaluation* in which subjects used a set of usability guidelines to identify problems with each prototype.

The third technique was the *group walkthrough* in which subjects viewed the prototypes on a large monitor and were guided through the tasks by a moderator with data collection via video camera and notes.

4. *Data Analysis*—From the data analysis an assessment of each prototype's strengths and weaknesses was reached. From this, the development team met and reached consensus on the interface design for the final product which encompassed portions of both prototypes.
5. *Interface Specification Development*—The results of the usability tests were used to generate a specification to document the common characteristics of the controls and interfaces of the final product. This interface specification guided t he development of the third dynamic prototype.
6. *New Prototype*—A final prototype was developed t hat implemented the results of the previous usability tests and conformed to the specification.
7. *Final Testing*—The final prototype was subjected to a final usability test in order to ensure that new problems had not been introduced. This was a *pluralistic walkthrough* in which a group of 18, consisting of end users, human factors experts, and developers viewed the prototype. Group responses were recorded and the moderator strove for consensus on user interface comments, which were then implemented in the actual software.

The design team for Product 2 did not perform a formal usability test. Prototyping was static and minimal. The product's functional requirements were not formally documented, and were developed by soliciting input from a selected group of users, many of whom were in-house experts. A user interface specification was not developed; however, a single programmer developed the majority of the code, providing some standardization of interface across the product.

3. RESULTS

To assess whether the extensive usability testing effort increased software durability, and was, therefore, cost effective, we gathered two measures.

The first measure represents the proportion of user comments that included requests for changes to the user interface. To estimate this measure, we evaluated the databases of user comments to determine the proportion of those comments that would affect the "look" of the software. For Product 1, user comments were actively solicited through personal contact, email requests, and frequent interaction with a steering committee of representative users. For Product 2, solicitation of user comments was primarily passive; using limited surveys, email and user contact. User comments were categorized as user interface improvement, bug identification, or process improvement comments. Process improvement comments included suggestions for additional functionality.

Table 2. Results of User Comments

User Comment	Product 1	Product 2
User Interface Improvement	8%	50%
Bug Identification	38%	45%
Process Improvement	54%	5%

Our second measure of interest concerned the amount of relative effort expended in developing new versions of each product. We were interested in measuring the proportion of programmer hours spent on interface changes since the initial release on the Windows platform. We excluded time spent fixing bugs. We also excluded time spent making changes needed to incorporate new functionality since the extent of product upgrades were not comparable; however, we would expect the effort required to implement upgrades on a durable product to be less.

Product 1 used 5% of their time reprogramming the user interface, whereas Product 2 expended 47% of all programmer hours reprogramming the interface. This percentage of time is only a small part of the total cost of changing an interface. Documentation, marketing materials, and training materials are also affected by significant interface changes.

4. SUMMARY

We acknowledge that there are many good reasons to update an interface (e.g., user requests). However, we believe that the majority of user interface changes are made for bad reasons. One of these is a desire to rush to market. In our limited study, the usability testing caused a nine month delay in getting to market. In some industries, this time is unacceptable; however, we would argue that it is better to be nine months late with a good product than on time with a product that leads to customer dissatisfaction. A second bad reason is to extend an existing interface to incorporate new functionality that could have been predicted.

We believe that the results of our study form a convincing case for justifying usability testing based on the long term cost benefits.

REFERENCES

1. Dahl (now Archer), S.G., Allender, L., Kelley, T. and Adkins, R. (1995). A Process for Transitioning Complex Software to the Windows Environment: The Transition of MANPRINT Tools from DOS to Windows. Proceedings of the Human Factors Society 39th Annual Meeting.

Evaluating REAL Users, using REAL Software, performing REAL Tasks, in REAL Contexts

Ilona Posner, Ronald Baecker, and Alex Mitchell*

Collaborative Multimedia Research Group — DGP, University of Toronto, 10 Kings College Rd. #4306, Toronto ON M5S 1A1 CANADA
* ISS - Institute of Systems Science, National University of Singapore, Heng Mui Keng Terrace, Kent Ridge, Singapore 119597.

Our group carries out research on collaborative multimedia. We design, build, and test prototype software to aid people in working together on tasks such as writing, making movies, using the Internet, and managing information. We continually face the question: How do we study *real users* working with *real software* to perform *real tasks* in *real work contexts* over *real time frames?*

Evaluation methodologies from human-computer interaction and human factors (reviewed in Chapter 2 of Baecker et al., 1995) provide only modest assistance with this question. *Usability testing* (Nielsen, 1994) tends to be carried on in a laboratory on relatively prescribed tasks of limited duration. *Usability inspection* (Nielsen and Mack, 1994) makes use of the judgments of experts who typically examine the interface for relatively brief periods of time out of a real work context. *Contextual inquiry* (Holtzblatt and Jones, 1993) stresses real users in a real work context, but tends to focus on employing insights about work process in the design process.

None of these methodologies address our needs. Ideally, in order to gather the most information about real system usage, we would be:

- omnipresent
- able to remember and reconstruct everything we see and hear including precise details about user actions and system responses
- so unobtrusive that we had no effect on the phenomenon we are trying to observe.

This is impossible; the question is how best to approximate this at a reasonable cost and with minimal interference to the work going on.

This paper first reviews some case studies in which we have tackled this problem over the past few years. We provide brief descriptions of the study details, data collected, analyses performed, and the problems encountered. The paper concludes with a summary of recommendations derived from our studies.

1. CASE STUDIES

Collaborative Writing Study — Prejudice Project: We organized an after school program for grade six students who worked in groups to produce a magazine about prejudice (Mitchell et al., 1995). Two groups of 4 students worked together at four networked computers during twelve one-hour weekly sessions. Students worked on-line using collaborative writing software; they also worked on-paper in small groups. We collected large volumes of data including: 100 hours of video (4 video cameras - 2 on people 2 on screens, audio recording enhanced using 4 microphones and an audio mixer), electronic records of documents, marked up paper documents, questionnaires, individual interviews, two teachers' blind evaluations of the final documents, and teachers' evaluations of the students performance in class. We performed qualitative data analysis by having 2 judges annotate 30 hours of video. Some of the pragmatic problems encountered during this project included hardware-software incompatibility, unexpected network conflicts and delays, organization of digital records across different machines, and unsynchronized computer clocks complicating digital record keeping.

Multimedia Authoring Study — MAD Camp: We ran a multimedia summer camp for grade 7 students working together to create motion pictures (Posner et al, 1997). The camp was run for two one-week sessions, lasting 5.5 hours per day over five days. Twelve campers attended each session working in groups of 3 campers and one counselor. Data collected included: four questionnaires administered throughout the camp sessions, paper diaries of group activities, audio journals recording group activities, video records of the moviemaking process (32 hours of videotape), paper artifacts, digital records (1.2 Gigabytes per group for 8 groups), group discussions, expert ratings of movie quality (technical, creative, and overall categories). Quantitative analysis of the data focused on software preference, movie structures, process information, counselor effects, and movie quality ratings. Qualitative analysis involved viewing of 12 hours of specially selected videotape for examination of the movie process, counselors' instruction and feedback effects, and determinants of success. Problems encountered on this project included hardware and software reliability, variability of counseling techniques, and counselor biases towards technology.

Information Visualization Study — TimeStore 2.0: A study of the usage of a time-based email management and visualization system (Silver, 1996). TimeStore 2.0 was used periodically by Eudora users during a four week evaluation period. Data collected included: program instrumentation or logging allowing recording and playback of user interactions, "thinking aloud" audio recordings documenting the context of each usage, and follow up interviews with users. Qualitative data analysis was conducted by playing back user interactions captured by the logging data and the "think aloud" recordings. One problem encountered was user selection — all users worked in the HCI group of one computer company, had prior knowledge about this system, and had preconceived ideas about interactive prototypes in general; these expert interaction designers were unable to focus on the usefulness of the system concept and instead primarily focused on the system's usability problems. Another problem concerned synchronization of audio recordings of user interactions and their corresponding digital records; without common time stamps on these, the synchronization becomes significantly more complex as the amount of data increases.

Information Management Study — TimeStore 3.0: A study of the usage of a redesigned time-based email management and visualization system (Yiu, 1997). Users used TimeStore 3.0 daily in their regular environments during a three-week evaluation period. Data collected include think-aloud sessions with screen and audio capture by the users' computers using Microsoft Camcorder (Microsoft, 1997), and weekly interviews with users. Qualitative data analysis was performed by viewing real-time play back of movie files containing the users' on-screen interaction and think-aloud audio. One problem encountered in this methodology was the lack of consistency in the think-aloud protocols; this was most evident from minimal commentary during routine operations. The limited functionality of the software, which for example did not provide a spell-check for email messages, lead to users adopting alternative methods for composing email messages, and consequently reducing their use of the prototype. Finally, since users were in charge of running Camcorder and storing space intensive movies of their interactions, storage space limitations reduced the total data recorded for analysis.

The following table (Table 1) summarizes and compares the evaluation methodologies that were used in the above studies. User experiences were captured using interviews, questionnaires, and journals, while interactions details were recorded using artifact capture, video, logging, and think-aloud recordings. Synchronization of recordings and logs was done mostly manually.

the study, where they performed a series of seven different randomized tasks with the medical device simulation prototype. Performance data and subjective opinions regarding usability were collected. In particular, NASA-TLX was used to assess mental effort, an in-house 16 item usability questionnaire was used for perceived usability and excessive keystrokes was employed as a measure of error. Time durations on each screen and total time to completion, were used as an assessment of speed. Five key conclusions were derived from the data during trials.

- Speed improved while *mental effort* and *errors* did not.
- *Mental effort* and *errors* were related to perceived Usability while no Time effect was found.
- The *"HELP"* key was used for **abbreviations** or wording issues.
- *Errors* were related to **editing** tasks.
- *Mental effort* was influenced by **decision-making** and **keypad** input issues.

Based on the data analysis, a model of the medical device usability was proposed to understand the impact of the system features on perceived usability.

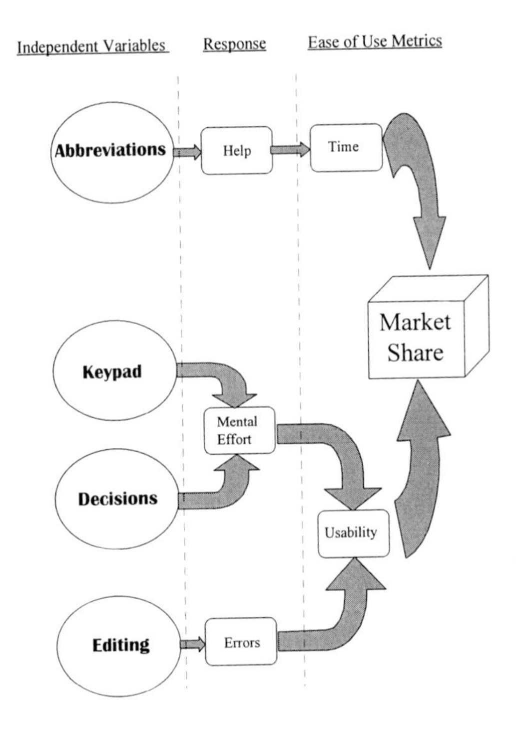

From this data, the team examined each of the usability problems identified in empirical testing and heuristics, weighed the costs of addressing the issue versus the potential usability benefit and decided upon a set of design changes. In particular, usability issues related to keypad size could not be changed due to customer requirements. Given the limited screen size, abbreviations were also left intact. The method for editing and the decision flow process were modified based on the usability issues identified in the user testing, round one.

3.5. Empirical Testing (Round Two)

The agreed-upon design changes were implemented in the computer simulation and a second set of users were tested to validate the impact of modifications. In this group, nine naive subjects and five experienced subjects participated in the experiment. As expected, for the screens selected for modification by the team, usability improvements were achieved. For those screens in which modification of usability issues were deemed not feasible, usability improvements were not achieved. The experienced group performed the tasks in the same manner as the naive group, but were faster and incurred less mental effort. The key findings are summarized below.

- Design changes to address **decision-making** issues reduced *mental effort*, as predicted.
- No changes to "speed" factors were made, hence no overall Time improvements realized.
- Specific screens modified by the team showed speed improvements. Those left unchanged did not.
- For the design changes made, data from empirical testing improved the system.
- Questionnaire responses/subjective opinion for assessing Usability was not sensitive to differences from design changes.

4.0 CONCLUSIONS

At present, the design has been finalized and will soon move to prototype and alpha-site testing phase. The human factors effort made significant contributions to usability of the product while not increasing design cycle time. It is anticipated that this effort will save design time by reducing the re-work traditionally required at the alpha and beta testing phases due to usability problems and allows the design team to optimally shape training manuals and customer service programs. In addition, the process of user centered design served as a tool to help the design team make explicit the product intent, expected customers, intended uses and trade-offs required.

Networking and Urban Forms in the Electronic Age – Towards Flexible Location Decisions in Stockholm

Reza Kazemian

Department of Architecture and Town Planning, Division of Design Methodologies, Royal Institute of Technology, S-100 44 Stockholm, Sweden*

This is a preliminary outcome of an ongoing international project dealing with the impacts of information technologies on urban forms. It focuses on the development of IT and the (re)localization of dwellings and workplaces in the Stockholm city-regions.

1. INTRODUCTION

Electronic networking while is increasing time, place and work flexibility, becoming important factor for the transformation of cityscapes. This new mode of development, though at an early stage of diffusion, shows signs of being able to alter many vital social, cultural and spatial relationships. It is embedding new conditions for many companies and individuals to become "footloose" (Sassen, 1991). Out-sourcing routine office information-handling operations to remote areas and teleworking in flextime are becoming the accepted concepts in "Post-Fordist" work and production management philosophies (Forester, 1989).

The emerging development is causing new tensions between "space of flows" and "space of places" (Castells, 1996). It is creating a new compartmentalized spatial structure of technologically "favored" and "less-favored" urban structure, engendering crucial problems for planners to choose locations with certainty and to reconcile the two diverging spatial logics: localities and nets. "The dominant tendency is toward a horizon of networked, ahistorical space of flows, aiming at imposing its logic over scattered, segmented places, increasingly unrelated to each other, less and less able to share cultural codes. Unless cultural *and physical* bridges are deliberately built between these two forms of space, we may be heading toward life in parallel universes whose times cannot meet because they are warped into different dimensions of a social hyperspace" (Castells, 1996).

In the context of Stockholm, the main research problems are to know how this city and its regional hinterland are adapting to the flows of the new information and communication technologies? What kinds of initiatives can be made for sustainable-flexible location transformations? How can the potential of the information technologies be diffused at this early stage?

---

* The author is an associate professor at the Royal Institute of Technology in Stockholm and the recipient of the 1997 Postdoctoral Research Fellowship from the Japan Society for the Promotion of Science (JSPS). He is working with a research team at the Waseda University in Tokyo to develop a computer-aided methodology to study the impacts of IT on city forms.

2. STOCKHOLM IN INFORMATIONAL WRAPS

Over the past two decades, information technologies, though continually on the verge of turmoil, have become important changing forces in urban-regional settings in advanced capitalist systems. It is expected that by the year 2000 telecommunications industry is to reach 7 per cent of the Gross Domestic Product of Western Europe compared to 2 per cent in 1984; and more than 60 per cent of all employees are to be supported by IT (Mulgan, 1991; Graham, 1996).

Since the early 1990s, Sweden have become a leading country in Europe that both in development and application of the new means of communication and information exchange took serious steps forwards. The use of information technologies is growing very fast among different social categories. Today, Sweden has the most rapid growth per capita of Personal Computer in the world.

Stockholm has a long tradition in welding informational infrastructure in urban-regional settings. At the end of the nineteenth century, the city had more telephone per capita than large European cities like London, Paris and Berlin. Today, inhabitants of the Stockholm city region ranks at the top in the world considering the inhabitants access to Internet, fax and mobile telephone. About 60 percent of the Stockholm population are using computer in their daily activities. A great part of the European Internet networks go via Stockholm. More than 10 percent of the population in Stockholm in 1996 were working with information Technologies. Information-based sectors are considered to be the most important source of economic development for the Stockholm city-region in the near future. There are already signs indicating the impacts of the IT on the existing work and housing locations; transforming them into a new type of socio-spatial order, following both recentralization and decentralization patterns.

After the building boom in the 1980s a period of recession in building activities occurred in Stockholm. Ideas that Stockholm is "ready-built" have gained momentum and created a doubtful attitude among the construction companies. At the same time, the number of urban renewal schemes increased and became a dominant part of construction activities in Stockholm.

Despite the lower rate of new building production substantial transformations are taking place in Stockholm. One such trend is the gradual replacement of outmoded ports and industrial activities with residential and new types of work-places and houses in central parts of Stockholm. "Left-over" plots of land are being reconsidered for urban development schemes, for upgrading and densification of the urban landscape. The city core is becoming more attractive for a growing number of people as a cultural foci, as a place for shopping or meeting in restaurants, cafeterias and pubs. More and more, higher income citizens are moving into the Stockholm's inner areas as a more secure and favorable for living. At the same time a demographic transformation is taking place in Stockholm, especially in semi-peripheral suburbs from the 1950s and '60s, and in related regional hinterland. For some neighborhoods and communities the new transformation mean adapting to new technological and economic niches with renewed prospects for development and prosperity; for others they mean economic decline, social adversity and a struggle to find a basic means of survival. Several peripheral suburbs around Stockholm are losing the functional necessities that brought them into being. The public services and employment opportunities are deteriorating and the areas are becoming less favorable for living. Concentration of low income people, unemployed and immigrants to certain suburbs are contributing to socio-spatial tensions and to an increase in ethnic, economic, cultural and spatial segregations.

Some suburban communities built upon the spatial logic of Modernism are undergoing functional changes. They are evolving from once dormitory towns to become sites for secondary headquarters, the back-offices of non-production services and information-intensive manufacturing enterprises. These suburbs, such as Kista, Husby, Akalla, in the northern part of Stockholm now provide homes and workplaces for both well-paid professionals and re-skilled low-waged, blue collar service- and informational workers. While these areas are "favoring" a large portion of the low-income, unskilled and unemployed segments of these suburban inhabitants are facing the threat of a new wave of poverty, segregation, displacement and the lose of place identity.

Since the early 1990s, the local and national governments together with many manufacturing and service companies have tried to direct and promote the infrastructure potentialities of the Stockholm city region to combat the negative aspects of transformation and to speed up the tendencies towards information society. Currently, several pilot teleworking projects are being experienced, testing the application of telematic devices in the daily activities, such as for elderly and disables in private- and municipality-owned dwelling areas, in modern suburbs, small towns and villages around Stockholm in order to control the demographical transformation. Applying teleworking schedule for the employees of municipality in the harbor town of Nynäshamn and the Telematic Scheme in Vällingby suburb are among these projects being carried through a cooperation of several partners including academia, private and public housing agencies and local governments.

Particularly important are the infrastructural IT schemes at the county and regional levels where the impacts are more profound. The Stockholm County Council's 1991 Regional Plan, a regional strategy scheme mainly for land use and infrastructure in the county, is incorporated to development strategic plans of other counties and municipalities around the Mälar Lake. The Regional Plan of 1991 shows a 15 year development program. It is coordinated by the Mälar Regional Council comprising the representatives from municipalities around the Mälar Lake, established in 1989. It is estimated that over $7,6 billion will be invested in developing transportation and communication systems in the Stockholm and Mälar Region. The ambition is to limit the travel time required to reach any part of the region to less than two hours. Great emphasis is placed on the expansion of telecommunication networks and the rapid development of a high-speed electronic infrastructure. This scheme is important because it is affecting a large region consisted of 51 municipalities with nearly 2.5 million population (30% of Swede's population). Already, over 50% of Sweden's research and university training activities are located in Stockholm-Mälar region. The Scheme is designed to mainly favor several multinationals such as ABB, Ericsson, Astra, Pharmacia, Atlas Copco and Electrolux. These enterprises have their origins in the region and it is thought that this gigantic infrastructural investment would eliminate the relocation wave of these companies and that these multinationals might continue to run a large proportion of their business operations in the region. However, the scheme can have substantial impacts on the relocation of homes and workplaces in the region as well as on the societal and spatial fabrics of many less-favored rural communities, towns and cities at the national level. There are already the traceable signs of social and spatial transformations that either directly or indirectly are connected to this regional strategic plan. If nothing special happens in other regions, this expansive and critical scheme would have a great negative effect on localities at national level.

3. NEW CONTINUUMS

The choice of appropriate locations is an important part of urban planning and it is becoming a vital issue in emerging informational societies. Well thought-out location decisions might bring about new opportunities for synergetic urban planning with symbiotic coexistence of electronic nets and localities but, in contrast, inappropriate location decisions can result in irreversible social, cultural, and spatial damages.

Some visions for the future of location strategies in Stockholm city-region will continue to be guided by utopian, dystopian, futuristic, simplistic, linear, and confusing "technology-will-fix" attitudes. Nevertheless, compared to the rigid and constraining urban design methodologies of the 1960s, '70s and '80s, the new mode of development promises greater potential for creating places in sync with current concepts and realities, places of integrated cultural, social and spatial diversities. Planners now employ information and communication technologies in placemaking, help them to replace, rebalance and redensify urban spaces, reducing urban traffic congestion and allowing a more for culturally and socially effective use of places. But this step should be taken with a great deal of patient and carefulness. Planners are also experimenting with socio-spatial multifunctionality that combine employment opportunities, cultural activity, social diversity, recreation, and housing in order to preserve the decaying community life and to answer to the problems compartmentalized cityscape.

It is widely accepted that the diffusion rate of new information and communication technologies should be in pace with the democratic principle of giving citizens access to and control over the flow of information to increase their awareness and making the planning process more participatory and its solutions more contextual. The *democratization* of information is needed, and this should be as the primary source of planning inspiration. It is a challenging task for planners to reconcile different spatial logics, to discover how new opportunities can best be diffused, success be achieved, and pitfalls be avoided at this early stage. Our success will surely depend on our ability to form a well-thought out and far-sighted strategy for location decisions and placemaking.

REFERENCES

Castells, Manuel (1996) *The Rise of the Network Society*, Cambridge and Oxford: Blackwell Publishers.

Forester, David (1989) "The Myth of the Electronic Cottage," Forester (ed.) *Computers in the Human Context,* Oxford: Blackwell Publishers.

Graham, Stephen and Marvin, Simon (1996) *Telecommunication and the City: Electronic Spaces, Urban Places*, London and New York: Routledge.

Mulgan, G. J. (1991) *Communication and Control: Networks and the New Economies of Communication*, Oxford: Polity Press.

Sassen, Saskia (1991) *The Global city: New York, London, Tokyo*, Princeton and New Jersey: Princeton University Press.

Asking Users About What They Mean: Two Experiments & Results

Hervé Blanchon[a] and Laurel Fais[b]

[a]GETA-CLIPS, BP 53, 38041 Grenoble Cedex, France. herve.blanchon@imag.fr

[b]ATR-ITL, 2-2 Hikaridai, Seika cho, Soraku gun, 619-02 Kyoto, Japan. fais@itl.atr.co.jp

1. INTRODUCTION

1.1 Situation

Natural language (spoken and/or written) is an attractive modality for human-machine interaction. As stated in [8]: speech requires no training; is fast; and requires little attention. Text may be attractive when the utterances are short, when speech is not mandatory, or when speech may be annoying to those surrounding the user. Currently foreseeable applications using a natural language interface include multi-modal drawing tools [3, 7, 11], on-line travel information [5] (and more generally, on-line information retrieval [6]), oral control systems, and finally interpreting communication systems [8, 10].

1.2 Interest

We think that there is a real need to fill the gap between a "toy" and a real-scale application with a component that can overcome difficulties arising while analyzing natural language. Interactive disambiguation of the input is proposed as a reliable solution, which aims to produce more robust, fault-tolerant, and user-friendly software integrated with natural language processing components [1].

The interactive disambiguation process is then a crucial part of the system. How natural and easy it is to use are the preconditions to the success of this idea. There is, thus, a real need to experiment with the design of such a process to be able to propose a wording of the questions used in the disambiguation interaction which will be understandable to users.

1.3 Presentation

What we investigate here is the understandability of the disambiguation dialogues that can be produced by the method described in [2].

In the two experiments for which we give results here, the subjects read a text. They were interrupted to answer questions when an ambiguous sentence was read.

In the next part of this paper we give the results of a pilot experiment and discuss its implications. In the last part, we describe a second experiment and results. In conclusion we draw implications of these results for automatic interactive disambiguation.

2. PILOT EXPERIMENT

2.1 Experimental materials

Two classes of dialogues were used. This required two groups of 12 subjects each. The text and the ambiguities to be solved were the same for each group.

We designed a text that contained a set of 35 ambiguous sentences. The ambiguities in the sentences were selected from naturally occurring ambiguities in a corpus of spontaneous conversation collected at ATR, Japan [4]. The text itself was made up of two different stories. The 35 ambiguities were distributed evenly over seven categories of ambiguities, i.e., five examples of each category.

Of the five examples in each category, there were two sentences with easy interpretations and three sentences with hard ones. "Easy" interpretations corresponded to the most frequent or most salient interpretation of a given sequence of words and "hard" interpretations corresponded to an unusual, though possible, interpretation of a given sequence of words. The "easy" interpretation tended to be the sense that would "pop up" first in someone's mind for that sequence of words; the hard one was a much less likely interpretation. For example, the easy interpretation of "I want to check in to the hotel" is "I want to register at the hotel;" the hard one is "I want to investigate the hotel."

2.2 Lessons learned

In the course of conducting the pilot experiment and discussing their impressions with subjects afterwards, we learned a number of things that affected the design of the subsequent experiment.

Subjects frequently commented on how unnatural the text seemed to be. There were two reasons why the text sounded unnatural. First, it included actual spoken English examples in written form, surrounded by (made-up) written context. Transcriptions of spoken English often sound unnatural, especially embedded in written text. Second, some of the "hard" interpretations were ones that, in real life, only a computer would have trouble understanding. In trying to motivate these difficult interpretations, unnatural text was produced.

2.3 Recommendations

We made a number of changes to the format for the second experiment based on our experience in the pilot. We realized that using hard interpretations in the pilot experiment was a mistake. It made the text sound unnatural and made the task more difficult for the subjects, clouding the real issue: how well they could respond to the different wordings of the dialogues.

We also changed the arrangement of the screen so that subjects could check back to the text to confirm their understanding of the ambiguity involved. Subjects complained that they could not do this in the pilot; this was also an unnecessary obstacle to the accomplishment of the task.

In the disambiguation interaction, subjects could chose as the correct interpretation, one of two possibilities given in the dialogue box, or they could chose "no answer." While the "no answer" option gave some interesting results, it also made it difficult to see clearly the trends in how subjects answered the questions. For that reason, we designed the next experiment as a forced-choice task.

3. SECOND EXPERIMENT

For this second experiment we chose to ask questions using both a textual mode and spoken one. The spoken mode was added since it seems to be a promising modality [9, 12] in the context of either textual or spoken input (interpreting communication, for example). [Is this OK? I wasn't quite sure I understood this sentence.]

3.1 Setting

For this experiment, the experimenter and the subjects were separated, sitting on either side of a partition. They communicated through head sets (microphone, headphones).

The subject was asked to read aloud, slowly and carefully, a text displayed inside a text window , and pause between each sentence. The scrolling of the text window was controlled by the experimenter (i.e., the text windows of the subject and the experimenter were synchronized).

3.2 Experimental materials

Two classes of dialogues and two modalities were used. This required four groups of subjects; there were fifteen subjects in each group. The text and the ambiguities to be solved were the same for each group.

The text to be read was entirely made up, that is, it didn't contain examples from the "real" corpus as in the pilot experiment. It consisted of three different stories and contained 35 ambiguous sentences.

Two sets of questions were prepared: one human-like, i.e., as if a human were explaining the ambiguities, and one machine-like, i.e., as if the system were generating the explanations, similar to those in the pilot experiment. The contents of the textual and spoken dialogues were the same. The analysis of the results is divided into three parts: statistical analysis, behavioral analysis, and the post-experiment questionnaire analysis.

3.3 Statistical analysis

3.3.1 *Analysis of data actually collected*

If we use the actual answers collected, the basic result is that the difference between the responses to the human-like dialogues and those to the machine-like dialogues is significant ($p<.05$) ; the human-like dialogues were easier to answer. This is affected by two questions in the machine setting which were problematic. We did not present the interpretation choices for these questions consistently with others for the same ambiguities. We will see in another analysis below that the difference is not significant.

It appears that in both machine-like and human-like phrasings, the performance of the subjects tends to be better with text questions, but we can't draw any definitive conclusion since the differences between spoken and textual dialogues are not significant.

3.3.2 *Filtered analysis: problematic questions disregarded*

In this case there is no significant difference between the subjects' performance in the machine-like dialogue settings and in the human-like dialogue settings.

Subjects seem to show better performance for textual dialogues than for spoken dialogues for the human-like phrasings; however, there is no difference at all between text and speech for the machine-like phrasings. Again, the differences are not significant so no definitive conclusion can be drawn.

3.3.3 *Projected analysis: problematic questions corrected*

In this third way of looking at the data, we excluded results for one question from the results for the machine-like dialogues and adjusted the answers to second one according to what we conjecture the answers would have been if the question had been labeled correctly. In this third case, there is again no significant difference between the results in the machine-like and human-like dialogue settings.

In this analysis, the results for spoken and textual dialogues were different for the machine-like and human-like phrasings. The difference is again not significant; thus no definitive conclusion can be drawn.

4. CONCLUSION AND PERSPECTIVES

If we allow for the problematic questions we see that there were no significant differences according to the style (machine, or human) of the presentation of the disambiguation dialogues, and no significant differences according to the modality (spoken or textual). The former result is essential to the success of an automatic interactive disambiguation program. We have seen that subjects are able to interpret the dialogues when presented in human-like, i.e., natural, phrasing, but it is not likely that automatically generated dialogues can be so natural. Therefore, it is critical that users be able to interpret the type of dialogues that machines are likely to be able to generate. The results reported here show that this is indeed the case.

We also investigated whether spoken or textual dialogues would be easier to understand. This is a design question; it affects how an automatic system will be designed, but is not crucial to the system. The results found here, as well as comments made by some of the participants about wanting to have text instead of speech, suggest that one design feature for an interactive disambiguation system should be the option for users to choose in which modality they would like to have the dialogues presented. According to our results, both modalities are understandable.

Although the "repeat" option was not extensively used in the spoken setting, it is still necessary to include it for cases where users cannot understand the dialogue after the first hearing. Other suggestions made by the subjects can be easily implemented. For example, more of the context of the ambiguity can be included in the dialogue; this would also support the most frequent strategy used by the subjects in determining their responses, i.e., the use of context. In addition, spoken utterances can be made shorter and faster. How best to use intonation in the spoken presentation of disambiguation dialogues is an open and interesting question.

It will be also necessary to run an experiment using as the text, one provided by the subjects themselves. This may be the only way to have a better analysis of the interactive disambiguation methodology we have proposed.

REFERENCES

[1] Blanchon, H. Clarification: Towards more User-Friendly Natural Language Human-Computer Interaction. in Proceedings of Poster session of HCI'95, 6th International Conference on Human-Computer Interaction. (Yokohama, Japan, July 9-14, 1995), vol. 1/1 : 42-42.

[2] Blanchon, H. Interactive Disambiguation of Natural Language Input: a Methodology and two Implementations for French & English. in Proceedings of IJCAI-97. 23-29 August, 1997) : to be published.

[3] Caelen, J. Multimodal Human-Computer Interaction. in Fundamentals of Speech Synthesis and Speech Recognition. John Wiley & Sons. New York. 1994.

[4] Fais, L. and Blanchon, H. Ambiguities in Task-oriented Dialogues. in Proceedings of MIDDIM'96. (Le col de porte, Isère, France, 12-14 Août 1996), vol. 1/1 : 263-275.

[5] Goddeau, D., Brill, E., Glass, J., Pao, C., Philips, M., Polifroni, J., Seneff, S. and Zue, V. GALAXY: a Human-Language Interface to On-Line Travel Information. in Proceedings of ICSLP 94. (Yokohama, Japan, September 18-22, 1994), vol. 2/4 : 707-710.

[6] Haddock, N. J. Multimodal Database Query. in Proceedings of Coling-92. (Nantes, France, 23-28 juillet 1992), vol. 4/4 : 1274-1278.

[7] Hiyoshi, M. and Shimazu, H. Drawing Pictures with Natural Language and Direct Manipulation. in Proceedings of Coling-94. (Kyoto, Japan, August 5-9, 1994), vol. 2/2 : 722-726.

[8] Kay, M., Gawron, J. M. and Norvig, P. Verbmobil: A Translation System for Face-to-Face Dialog. CSLI lecture note no 33. Center for the Study of Language and Information, Stanford, CA. 1994.

[9] Lehiste, I., Olive, J. P. and Streeter, L. A. Role of duration in disambiguating syntactically ambiguous sentences. Journal of the Acoustical Society of America. 60, 5, 1199-1202.

[10] Morimoto, T., Suzuki, M., Takezawa, T., Kikui, G. i., Nagata, M. and Tomokiyo, M. A Spoken Language Translation System: SL-TRANS2. in Proceedings of Coling-92. (Nantes, France, 23-28 juillet 1992), vol. 3/4 : 1048-1052.

[11] Nishimoto, T., Shida, N., Kobayashi, T. and Shirai, K. Multimodal Drawing Tool Using Speech, Mouse and Key-Board. in Proceedings of ICSLP 94. (Yokohama, Japan, September 18-22, 1994), vol. 3/4 : 1287-1290.

[12] O'Shaughnessy, D. Specifying accent marks in French text for teletext and speech synthesis. International Journal of Man-Machine Studies. 31, 4, 405-414.

Ergonomic evaluation of some Brazilian hypertext systems

Stephania Padovani[a/b] and Anamaria de Moraes, Ph.D.[a]

[a] Pontifical Catholic University, Master Program in Design
Rua Marquês de São Vicente, 225 GÁVEA Rio de Janeiro RJ BRASIL
Tel.: +55 21 529 9418 / Fax.: +55 21 527 1907 / E.mail: moraergo@rdc.puc.rio.br

[b] Faculdade Carioca, UNIVIR - Virtual University
Av. Paulo de Frontin, 568 RIO COMPRIDO Rio de Janeiro RJ BRASIL
Tel +55 21 502 1001 / Fax.: +55 21 527 1907 / E.mail: steph@carioca.br

1. INTRODUCTION

As long as information technology develops itself, the number of people who use or depend on this technology also increases. The multimedia systems which present information using a hypertext structure are an useful example of the insertion of this technology on our daily life.

The fact that multimedia is now available to a great number of users, with extremely different profiles, reaffirms the necessity of a clear interface to facilitate the handling, reading and comprehension of the information displayed.

2. ADVANTAGES AND DEFICIENCIES OF A NEW MEDIA

Hypertexts appear as an alternative to printed text since they provide advantages such as large quantity of information stored, easiness of handling, speed and non-linear access to information.

On the other hand, there are aspects inherent to hypermedia philosophy which are responsible for the user's disorientation. The great number of paths, and the easiness to choose new paths inside the document cause the "getting lost" syndrome where the user does not have a clear idea of location with respect to the material and is unable to predict a path to the information required.

Another problem is the direct transposition of contents from the printed media to the electronic one. The main difficulties are related to visibility, legibility, comprehensibility and navigation, making difficult not only the apprehension but also the comprehension of the information.

3. OBJECTIVES

- Characterize the hypertext system and its features;
- Investigate the conflicts experienced by people navigating hypertext systems;
- Identify the difficulties and constraints related to the problems identified;
- Propose ergonomic recommendations for their solution.

4. PROCEDURES

First of all, 11 hypertexts were selected to be analyzed (software helps, instruction manuals, enterprise annual reports, scientific magazines, design and architecture magazines and encyclopedias). These systems were observed and characterized using a characterization protocol which included information about the content, audio visual features, navigation and information presentation.

Some topics of the characterization protocol were the following:
- title and content;
- audio-visual features (text, images, video, etc);
- navigation by mouse or keyboard;
- links description;
- help to identify the links
- feedback indicating the activation of a link;
- navigation aids;
- location inside the document;
- page layout (scroll bar or separated screens);
- windows (multi-windows, cascade, tile, pop-up windows);
- typography used for text, titles, links (serif, contrast, size, style);
- text and background colors.

After the characterization, the conflicts experienced by people using these systems were investigated. All hypertexts were observed in order to identify these problems. After the identification, problems were classified according to usability, utility and readability criteria.

Utility involves the software ability which allows the user to reach his main objectives and achieve his tasks. The utility problems are related to the *functional characteristics of the computerized system. Usability* involves the software ability which allows the user reach easily his interaction goals. Usability problems are related to *man-computer dialogue* (Moraes apud Scapin, 1993). *Information* problems refer to page layout and information presentation, arranged into similarity or proximity (screen legibility considering the design of characters, leading, text justification, the use of colors for text and background.

Analyzing the problems, we could verify the difficulties and constraints the user faced and propose ergonomic recommendations for their solution.

Table 1 shows the problems related to one of the hypertexts while Figure 1 and Figure 2 illustrate some of these problems.

Table 1

Name: *Ciência Hoje hipertexto nº. 11*
Usability problems
* The cursor assumes the same appearance in different situations: a "little hand" appears both to indicate a link and a scroll bar. The problem is that this "little hand" remains appearing over the scrolling text area making highlighted words look like links. * The magazine logo is used as a navigator but since there is no indication or instruction, users have great difficulty in identifying the navigator. * Non-familiar expressions are widely used: zoom, move, size, close (all written in English, while the whole hypertext is written in Portuguese) * Instructions to close windows appear at the end of the text, so the user is not allowed to close them until he reaches the end of the document. * Confusing instructions: the software says that one should just click to view a pop-up window. What actually happens is that the user has to keep pressing the mouse button to view the window.
Utility problems
* Navigation incompatibility: the "back" button goes back only within a certain chapter, it does not allow the user to return to the main menu unless the chapter has got only a single screen. Thus, the user keeps going back and forward as if he were "locked" into that chapter.
Information presentation problems
* The "zoom" option changes only the window size, instead of changing text size. * The colors applied to text and background making reading the texts a very difficult task. * "Sans serif" typography is applied to long texts, making reading uncomfortable. * Very narrow columns and small scrolling texts cause an extremely fragmented reading.

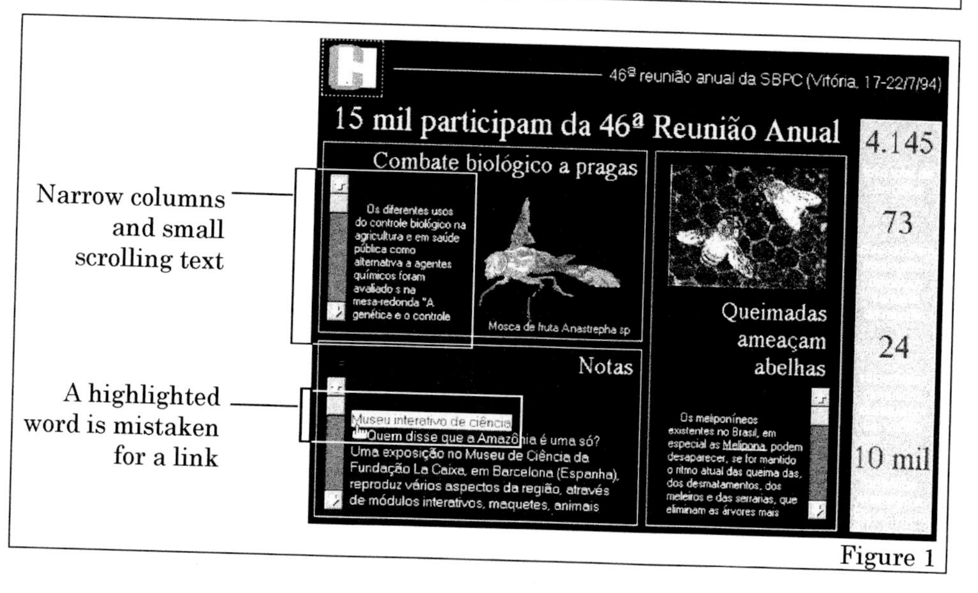

Figure 1

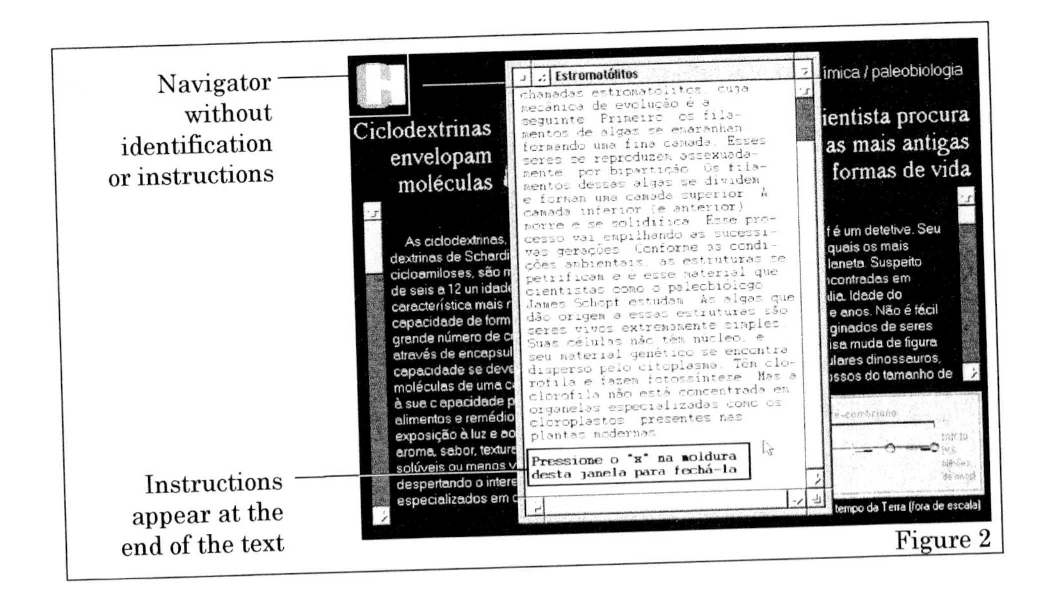

Figure 2

5. DISCUSSION OF RESULTS

We could notice that although the hypertexts selected had different functions and purposes, they presented quite similar features. This indicates that hypermedia is already developing its own identity. Some quite interesting new features could be found on those magazines which were already created using the electronic media resources, differing from those whose content and visual identity were transferred from printed issues.

In relation to the problems identified, the great majority had to deal with utility and usability while informational problems were not very critical. What we can conclude is that the information presentation is being well developed, but the fundamental flaws that are inherent in the user interface to such systems were not considered yet.

Finally, the great number of difficulties and constraints verified, show that utility and usability requirements have been neglected by those who are responsible for the hypertexts interface project.

REFERENCES

A. Dillon, Designing Usable Electronic Text: Ergonomic Aspects of Human Information Usage, Taylor & Francis, London, 1994.

A. de Moraes et al, Navigating in Shopping Information Systems: an Ergonomic Approach, Proceedings of the IEA World Conference 1995, Rio de Janeiro, 1995.

P.A. Smith and J.R, Wilson, Navigating in hypertext through virtual environments, Applied Ergonomics (24), Butterworth-Heinemann, Oxford, 1993.

Design and Usability Evaluation of a Novice User-Oriented Control Panel for Lighting and Air Conditioning

Kazunari MORIMOTO[a], Takao KUROKAWA[a], Masahito HATA[a], Noriyuki KUSHIRO[b] and Masahiro INOUE[b]

[a] Faculty of Engineering and Design, Kyoto Institute of Technology, Matsugasaki, Sakyo-ku, Kyoto 606, JAPAN

[b] Living Environment Systems Laboratory, Mitsubishi Electric Corporation, Ofuna, Kamakura, Kanagawa 247, JAPAN

Abstract

The objective of this paper is to design a control panel for novice users based upon new concepts for lighting and air conditioning. First we investigated the contents of tasks and the terms used in commercial sheets and control panels of lighting and air conditioning. Second we made up design concepts of new panels for novice users, and designed four panels separately by iterative design method and one panel by the parallel design method. Usability tests carried out those five control panels. Thirteen subjects were instructed to do four tasks in the experiment. Parameters for assessing usability were operating time, error rates, the number of times referring to the users' manual and subjective ratings. The results showed that the panel designed by parallel design method was superior to the other panels in operating time, error rate and the number of referring to operating manuals.

1. INTRODUCTION

Lighting and air conditioning of a large business building have been controlled by experts with separate control panels installed in a central control room. They can not change local states of lighting and air conditioning in a building and can not accept various requests from persons dispersed widely. In order to control environmental states in relatively small sections like rooms and floors and to eliminate the amount of energy consumption it is necessary to provide control panels which supervise a small number of units and can easily be operated by any employee of each tenant company or firm. Designing such panels needs novel concepts in both hardware and software.

Design activity is a complex task. In order to improve the productivity of designers and the quality of design results, a considerable number of design methods have been developed. An iterative design method that is the most conventional is time-consuming, because designers make some design versions and carry out usability tests on them before getting the best one for release. In order to reduce time for development, a parallel design method has been proposed which works out design alternatives independently and combines them into a single design [1].

Our aim is to design a control panel for novice users based upon new concepts for lighting and air conditioning through iterative and parallel design methods.

2. METHOD

Through the examination of technical terms we found so many cases where a single term was used with different meanings and where two different terms had the same meaning. We, therefore, avoided such complicated terminology and adopted as more non-technical words as possible in designing the panels for novice users. We specified the functions necessary to a unified panel for control of lighting and air conditioning in a local section of a building based on the analysis of control tasks. Four control panels P1, P2, P3 and P4 were designed independently and separately. The panel P1 was composed of the panel P1(L) for lighting and the panel P1(A) for air conditioning. The initial screen images of the P2 and P3 are shown in Figure 1 and Figure 2, respectively. The main design concepts of P1, P2 and P3 were menu, tabular and graphical presentation, respectively. On the other hand the main design concept of P4 was the coordination of menu and graphical objects as a result of an evaluation by a cognitive walkthrough method[2]. These panels were tested on their usability. Thereafter another panel PM was designed by merging the superior designs from the four proceedings' panels. The main design concepts of the panel PM are mainly to keep consistency of screen layouts, to be able to take a view of much information in a screen, and to use metaphor of rooms and folders. The typical screen of the panel PM is shown in Figure 3. Finally usability test was also carried out with the panel PM.

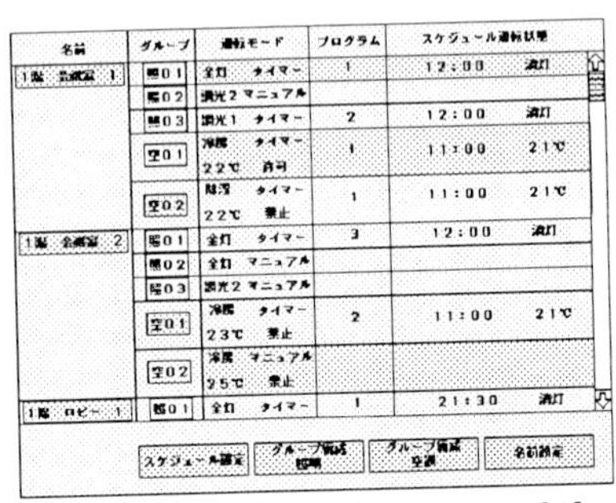

Figure 1. Typical screen of the panel P2

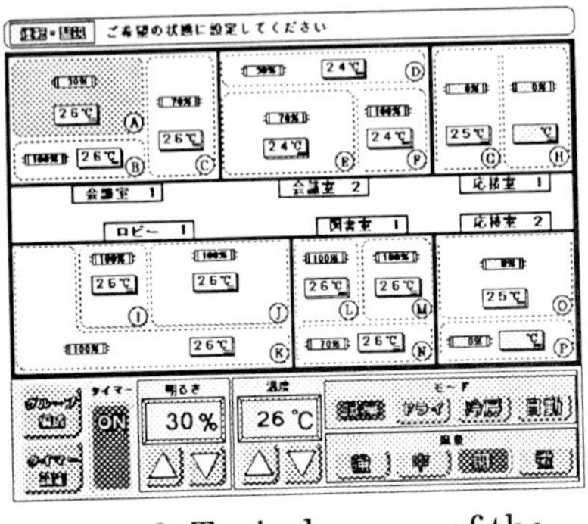

Figure 2. Typical screen of the panel P3

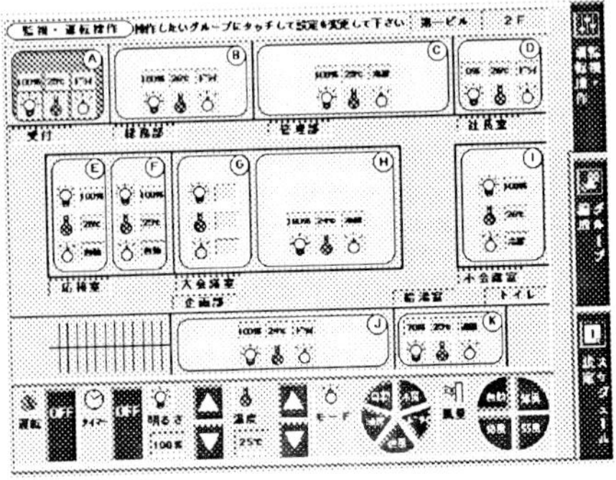

Figure 3. Typical screen of the panel PM

Thirteen subjects participated in the usability experiment with each panel. Operating manuals of the panels was previously given to the subjects. Therefor they could learn how to use the panels by the manuals. In the experiment they were instructed to do four tasks watching and controlling states of lighting and air conditioning in several rooms, grouping the units, and scheduling control for a day or a week.

Verbal protocols and behavior of the subjects were recorded using a video camera. Whenever one of the tasks was over, subjects were asked to evaluate comprehensibility of operation. Parameters used for assessing usability were operating time, error rates, the number of times referring to the users' manual and subjective ratings.

3. RESULTS AND DISCUSSIONS

Firstly usability was examined among P1, P2, P3 and P4. Average operating times with P3 and P4 were shorter than those with P1 and P2 as shown in Figure 4. Significant difference of the operating time between P1 and P4 was obtained. Error rates (Figure 5) and the number of times to refer to the manual (Figure 6) in P3 and P4 were significantly lower than those in P1 and P2.

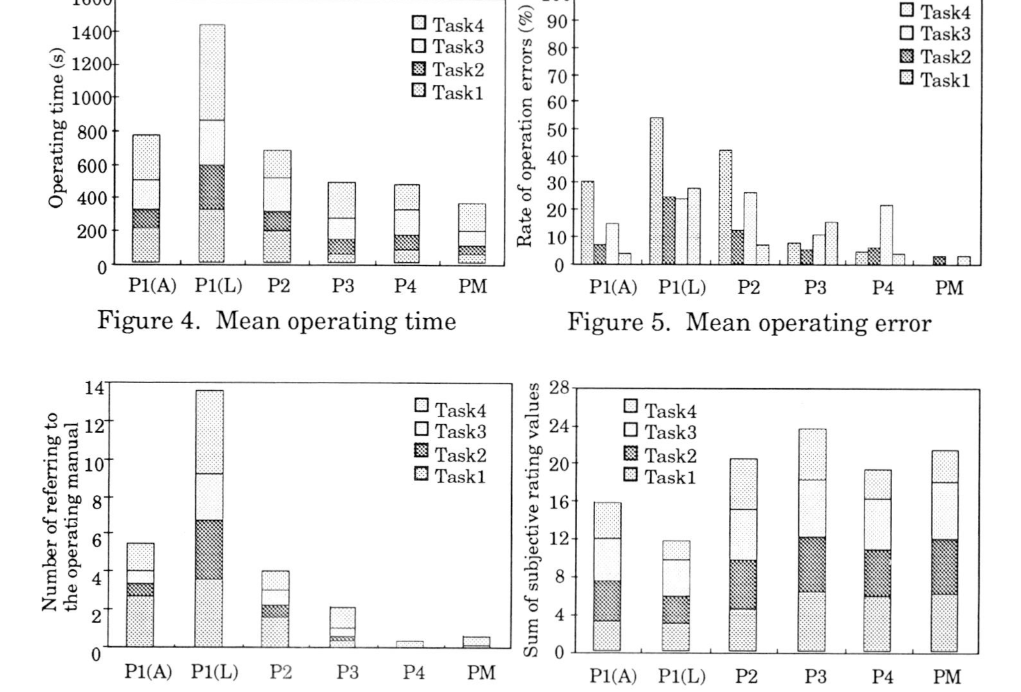

Figure 4. Mean operating time

Figure 5. Mean operating error

Figure 6. Number of referring to the operating manual

Figure 7. Subjective rating values

Furthermore subjective ratings of P3 and P4 were higher than those of P1 and P2 as shown in Figure 7. It was cleared that usability of P3 and P4 was superior to that of P1 and P2.

Secondly, therefore, comparison of usability among P3, P4 and PM was carried out. Average operating time with PM was significantly shorter than that with P3 ($t=2.018$, $p<0.05$) and P4 ($t=2.537$, $p<0.05$), however difference of the operating time between P4 and PM was not significant. Subjective ratings of P3 were higher than those of PM, however the difference was not significant. On the other hand error rates in PM was significantly fewer than that in P4 ($t=3.799$, $p<0.05$).

Finally the number of the times of referring to the users' manual in PM was fewer than that in P3. However the number of the times of referring to the manual in P4 and PM was little as shown in Figure 6. These evidences mean that novice users did not need the users' manuals to operate the PM.

Techniques and tools are needed to provide novice users with awareness of design scenarios and to enable their active participation in the selection of an appropriate interface. We adopted a room metaphor and represented icons of lighting and air conditioning in PM. Metaphors constrain and mislead users [3], however literal metaphors for rooms provide novice and experienced users fruitful usability [4]. The parallel design method was useful to find suitable interfaces and to make design faster. Besides we have to assert the subjective rating values on ease-to-use depend on the contents of a task.

4. CONCLUSIONS

We designed and tested four panels with different design concepts and one panel with merging a best design of those panels in considering terminology for novice users and in analyzing the characteristics of operating tasks. We could obtain the following conclusions. The merged panel was slightly inferior to the one panel designed independently in subjective ratings. However the merged panel was apparently superior to the others in reducing the operation times and the number of errors and of times of refer to manual. For novice users it is especially important that they can operate a panel without assistance and with avoiding an error.

REFERENCES

1. J. Nielsen and M.J. Faber, Improving system usability through parallel design, Computer, **29**, 2 (1996) 29-35.
2. R.G. Bias, The pluralistic usability walkthrough: Coordinated empathies, J. Nielsen and R.L. Mack (eds), Usability Inspection Methods, John Wiely & Sons, (1994) 63-76.
3. D. Gentner and J. Nielsen, The anti-mac interface, Communications of the ACM, **39**, 8 (1996) 70-82.
4. K. Morimoto, T. Kurokawa, M. Inoue and N. Kushiro, Design and usability test on control panels for lighting and air conditioning: Comparison between novice and experienced users, Proceedings of The 11th Symposium on Human Interface (1995) 769-774, in Japanese.

Usability in the Perspective of Work Environment

K. D. Keller
Institute of Communication, Aalborg University
Langagervej 8, DK-9220 Aalborg, Denmark

Notions of interactive computer systems' usability have predominantly been common sense ideas and conjectures. A phenomenological understanding of psychosocial work environment points to some basic principles of usability in work settings: The computer system should appear to the users like a media, a tool and a model. This approach is exemplified by a case of R&D, an action research project with a particular attention to the users' experience and practice.

1. Getting serious about usability in HCI

Questions concerning the usability of interactive computer systems have been emphasised in HCI research and development since the mid-eighties. It has become quite clear that computer based information and communication systems have to be viewed (i.e. defined, analysed, designed and discussed) as social systems as well as computer systems. However, the attention to usability has been marked by pragmatic attitudes due to the lack of scientific theories and concepts of usability. Consequently, usability has often been confused with utility, if not reduced to technological facilities or mistaken for simple user statements of 'satisfaction'. In short, the focus of research and development on usability has much too often been 'easy solutions' and 'guidelines' rather than solid grounding in theory and practice.

Undoubted, the main reason for this situation is that (traditional) confidence in cognitive science as the potential theoretical foundation for HCI research has decreased while the need for practical solutions to essential and urgent problems of HCI development has grown. Phenomenology offers an alternative to cognitive science as well as pragmatic 'trial and error' (Keller 1995). Indeed, the focus on usability in HCI would be more detailed in the practice and experience oriented perspectives of a phenomenological approach. While these perspectives allow us to thematize the artifact more thoroughly in the ways which it actually appears to the users, the design of the computer system which operates below this appearance can be discussed in such perspectives as well.

It is necessary to distinguish between various fields of applied information and communication systems in accord with the very different perspectives of artifacts' usability in social practices: work, learning, service and amusement. For now, we shall only discuss usability in work settings. On theoretical grounds, it has been suggested that this topic basically concerns the soundness of the psychosocial work environment, i.e. the work conditions' latitude for the unfolding of social identity, qualifications and control throughout the work performance (Keller 1994). A phenomenological comprehension of psychosocial work environment points to 'social identity' as the background, 'qualifications' as the foreground, and 'control' as the figure of our structured practice and experience of work performance.

2. Conceiving a Computer System in a Psychosocial Work Environment

The disposition work at the traffic control centre of Scandinavian Airlines (SAS) in Copenhagen is about surveying the ongoing execution of all the company's planned air traffic, international and local, and correct in the best possible way any problem or deviation that might occur. The following discussion is based upon an investigation of the work organization which was leading to the conceptual design of a training system for flight disposition work (Jensen, Keller & Thorne 1990). The research methods employed were participating observations, semi-structured interviews and group discussions.

The traffic control centre had suggested to us to design an 'expert system' in order to artificially 'augment' the abilities of new disposition workers. However, our investigation of the work performance did elucidate that there are good reasons why attaining the skills of a competent disposition worker usually requires several years of training. Therefore, we proposed a training system to help building up human problem solving experience.

2.1. Control and the Computer as a Model

The topic of *control* makes up a thematic figure in any work performance. Fulfilling the disposition work requires the control of a work process under traffic conditions that can be very complicated and unpredictable. Even for the treatment of fairly simple problems, it is required to vacillate between dynamic and static views of the traffic in flexible and creative ways. Very often, the solving of a specific problem, e.g. the delay of a single flight, has to be integrated with other necessary and possible changes of the flight plan. So, it is quite essential that the sociotechnical work system is arranged to easily maintain a general overview throughout the work processes and to smoothly apply it in any situation.

Therefore, (out-plotted) schedules on paper are preferred as the most important tools for the actual treatment of the disposition problems. The schedules on paper (an overview plan and a disposition plan) with the corrections made by pencil and rubber give the best overview and the best access to experimental dispositions. This accords closely with observations made by Hughes, Randall & Shapiro (1992).

Another example of how control is maintained under changing work conditions is that the thresholds of problem handling are raised in cases where the problems are heaping up. For example, one of the monitors is an 'alarm-monitor' showing all aircraft delays of more than 9 minutes. In critical situations the worker may choose to raise alarm limit, e.g. to 30 minutes.

When the possibility of constructing a computer system for training in flight disposition work was discussed with the users, the advance of control was pursued at a general level by deciding that the solutions and strategies of the training system for new flight disposition workers should *model* the established praxis of the skilled workers so far as possible. In any foreseeable future, a training system could hardly be enhanced into an expert system component which would make the disposition work into an interactive use of a computer system solely. The out-plotted overview plan and disposition plan would still be substantial tools because of the larger scope and flexibility for different kinds of handing the plans in the manual versions.

2.2. Qualifications and the Computer as a Tool

The *qualifications* of the disposition workers indicate that the labour demands a considerable expertise. It must be noticed that the core of the disposition work, i.e. the very treatment of unforeseen difficulties by the execution of the traffic plan, is not computer supported. The existing computer based information and communication system does not generate any suggestions of dispositions to remedy the occurring problems in the traffic. In other words, the flight disposition work is marked by the unfolding of knowledge-intensive and creative competencies.

It became evident that the workers do not predominantly think in aircrafts and stations (airports), like we do as newcomers, but in the slings (i.e. the chains of flights) of various types. In the treatment of the traffic problems they employ some specific abstractions which characterize their professional stance.

As regards the qualification aspect of the work environment the general requirement to a computer system is that it should function like *a tool* for the users. A typical deficiency by the work in control rooms is that the demands do not at all correspond to workers' qualifications, but make up a dynamic combination of very low demands most of the time and very high demands in critical situations, both sides of which are straining. To some extent, a training system may be a remedy to that, because it can be used as a test system for the exercise on difficult traffic problems. Thus, low-demand time may be used more constructively by investigating various hypothetical situations and strategies.

2.3. Social Identity and the Computer as a Media

As it was mentioned previously, *social identity* forms the background of the structuring of work processes and situations. This aspect has a lot to do with social relations and collective procedures of working. In our study social identity was clearly manifested in a number of ways, including the distinctive vocabulary and language usage in the communication between the disposition workers as well as in their communication with colleagues in other parts of the company. In accord with the findings of Heath and Luff (1991) as well as Suchman and Trigg (1991), we noticed that collective attention to an emerging problem which a colleague takes up is a significant phenomenon in the work processes.

The character of a flexible *media*, which an information and communication system should have in order to correspond to the regards of social identity, is not so obvious in our case. The disposition workers are a small group and work in the same room most of the time, so the internal communication is predominantly verbal. Directed towards problem solving, the training system was neither intended to substitute nor predominantly destined to mediate human interaction. Still, the usage of the training system would be embedded in the performance of disposition work as regards the educational basis and the experimentation with test cases. So, the system's conceptual design had to reflect the collective competencies and practices in disposition work, and in so far it would be available as a media for communicating these competencies and practices.

We did identify many patterns of heuristics for solving the typical traffic problems which turned out to be collectively shared. When the disposition workers are going to correct disturbances in the traffic they choose among and test different solutions, utilising their experience in aircraft disposition together with their sociocultural background (including common sense and creative imagination). We abstracted general heuristics for each of the two main categories of disposition problems, delay and diversion of aircrafts. Our explicit descriptions of these heuristics were recognized when presented to experienced disposition workers.

Furthermore, we found that the collectively established procedures and heuristics of the disposition work are integrated with various individual preferences of tasks accomplishment. Just as little as the collective procedures prevent individual initiative and creativity, should the training system prescribe particular routines which might reduce the latitude or limit the self-confidence of the workers. To keep composed and maintain an overview when the worst known traffic problems occur is acknowledged as elegant working. Efficient work in these situations, i.e. making fast decisions that does not show blemish when scrutinised afterwards, is known to require the competence of 10-15 years' work experience. The idea of the training system as a media reflects that the actual problem solving in disposition work has to be anchored in the workers' experience and practice, and that the computer system should *serve* the human communication and reflection of this experience and practice, rather than enforce any structure on it.

3. Conclusion

The study presented above indicates the importance of taking psychosocial work environment into consideration by the reorganization of work conditions with new computer systems. From a phenomenological point of view, the usability of an interactive computer system is more than socio-economic and sociotechnical matters of utility: Usability means that the computer system functions in smooth prolongation of human work, rather than being restricting, demanding, or straining to the work performance.

REFERENCES

C. Heath & P. Luff, 1991: Collaborative Activity and Technological Design. *Proceedings of the 2nd European Conference on Computer Supported Cooperative Work*, pp. 65-80. Kluwer Academic Publishers.

J.A. Hughes, D. Randall & D. Shapiro, 1992: Faltering from Ethnography to Design. *Proceedings of the Conference on Computer Supported Cooperative Work*, pp. 115-122. ACM Press.

H.S. Jensen, K. Keller & M. Thorne, 1990: *Knowledge Engineering with Personal Computers*. Institute of Computer and Systems Sciences, Copenhagen Business School.

K. Keller, 1994: Conditions for Computer-Supported Cooperative Work. *Technology Studies* 1, 2 pp. 242-269.

K. Keller, 1995: *Datamatstøttet samarbejde på fænomenologisk grundlag*. [Computer Supported Cooperative Work on a Phenomenological Foundation]. Copenhagen: Samfundslitteratur.

L.A. Suchman & R. Trigg, 1991: Understanding Practice. In: J. Greenbaum & M. Kyng: *Design at Work*, pp. 65-89. Lawrence Erlbaum Associates.

Experimental Method for Usability Test of Industrial Plant Operation System

Hirokazu Nishitani, Taketoshi Kurooka, Teiji Kitajima and Chie Satoh

Graduate School of Information Science, Nara Institute of Science and Technology,
8916-5 Takayama, Ikoma, 630-01, Japan

We proposed an experimental method for the usability-test of the plant operation control system called DCS. The method includes an experimental procedure to acquire data and a comprehensive operation analysis to determine obstacles related to human interface. A check list was also made to support identifying problems generated from complicated interaction between human and system.

1. INTRODUCTION

The CRT operation based on the DCS is commonly used in the process industry. The DCS saved space and labor in the control room. However, human operators must skillfully interact with the plant operation control system to be able to execute their advanced tasks such as monitoring for fault detection, diagnosis and emergency operation. This situation has generated new problems with relation to human interface in the supervisory control scheme [1,2].

A systematic approach, consisting of analyzing the operational performances and considering countermeasures based on measured data is desired to solve problems relating to human-machine interaction in the CRT operation. We can obtain much useful data on human-machine interaction through the usability test. We applied both a simple observation method and a method with protocol analysis to discover issues to be improved in the graphic panel system [3]. The experimental results show that the protocol analysis is a promising method to identify problems, although it is time-consuming.

In this paper, an experimental method for the usability test using the protocol analysis is proposed to acquire data based on the subjects' behavior in a virtual plant operation environment and to discover where the obstacles lie. The data can be put to good use for the improvement of human interfaces and development of operator support systems including training systems.

2. EXPERIMENTAL ENVIRONMENT AND PROCEDURE

2.1. Experimental environment [3,4]

A newer DCS was chosen as the subject of study. The system is composed of an operator console with an upper and a lower CRTs and a field control station. A training simulator of a boiler plant installed in the control station was used for experimental study. Four video cameras record four synchronized scenes of the upper and lower CRTs, the keyboard and the subject's face. This video image with its audio recording is used for protocol analysis of CRT operation. Figure 1 illustrates equipment items for protocol analysis. An experimenter presses the subject for his thoughts on how to solve the task while another observes and records the subject's behavior. At the same time, invocation of operational panels, the occurrence of alarms and annunciators, and actions for intervention such as the touching of the screen by the subject are recorded in the system.

The subject who operates the boiler simulator must be skilled in its operation to a certain extent. We ask all the subjects to master the operational sequences under eight kinds of malfunction scenarios, which are used for fundamental training of DCS. In the experiment, some new malfunctions are implemented to examine the usability of the installed panel system which are essential for monitoring and intervention. The members of the technical subcommittee for Human Interface in Plant Operation of the 143 Process Systems Engineering Committee in the JSPS, co-operated with us to make the experiments.

2.2. Experimental procedure

The experimental procedure is summarized in Fig.2. Malfunction scenarios are planned to achieve the objectives of the experiment, e.g., to test the usability of subsystems such as the panel system. The behavior of the plant under a malfunction is described with an explanation of the occurring phenomena in the plant including control systems. We proposed a new chart to represent a sequence of cognitive processes of an operating person. Figure 3 illustrates a step composed of the scope of observation, data acquisition, and proof of a proposition for judgement. This chart can be made both for the standard operational sequence expected and the actual operational sequence practiced by a subject. An example of the standard operational sequence under a burner-header-pressure-sensor failure is shown in Fig.4.

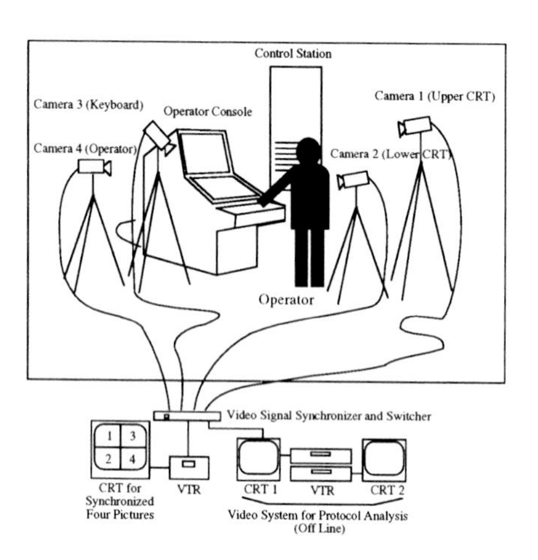

Figure 1. Experimental equipment

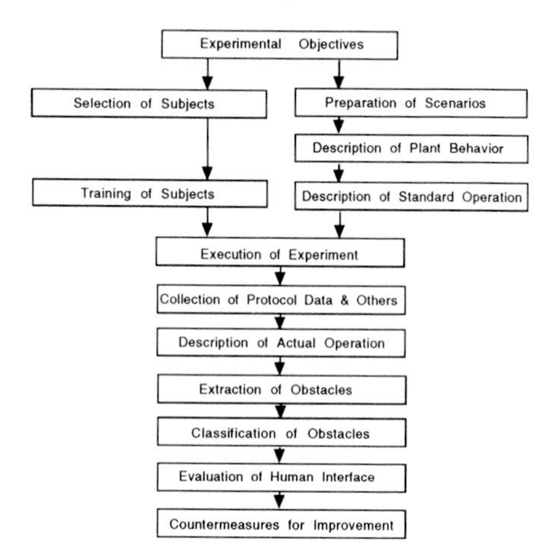

Figure 2. Experimental procedure

3. OPERATION ANALYSIS

The observed data is summarized in a table called operation history table. It is composed of the regional time, the time elapsed from the occurrence of a malfunction, the arc name in the actual operational sequence chart, verbal protocol data written down from the audiovisual records, historical data such as alarms and annunciators, operational panel number opened on both screens, the panel which the subject focuses on, control actions recorded in the system such as touches to the screen, experimenters' observations and remarks and a reference number to look up problems.

Comparison between two operational sequence charts benefits to identify deviation by the subject from the standard operational sequence and to extract obstacles related to human interface. Discrepancies give hints of controversial points to be examined in detail. Other remarkable points in the chart are places of alarm occurrence, places taking a long time, and places taking the same path.

Each point at issue is characterized by a stage of the sequence of cognitive processes and the place where the responsibility lies, i.e., the system including the human interface or the human. The results of operation analysis are put into tabular form called an operation analysis table. In the table each problem is characterized by the following items: reference number of problems; elapsed time from the occurrence of the malfunction; arc number in the actual operational path; contents of the problem; analyst's view; key word of the problem; problem classification.

The operation analysis with protocol data is an effective method to analyze the subjects' behavior but it takes time. We made a check list to support identifying problems generated from complicated interaction between human and system. Problems are generated from both sides of human-machine interaction; one is on the part of the system and the other is on the part of the human. The former is subdivided into the following three classes:

I-1: Deficit of data; I-2: Ease of access to data; I-3: Quality of display.

The latter is subdivided as follows:

H-1: Essential knowledge; H-2: Operational skill;

H-3: Ability to use the already acquired knowledge in practice.

Each class includes many common factors, which are represented by key words. These factors were summarized as a check list to provide clues to find problems.

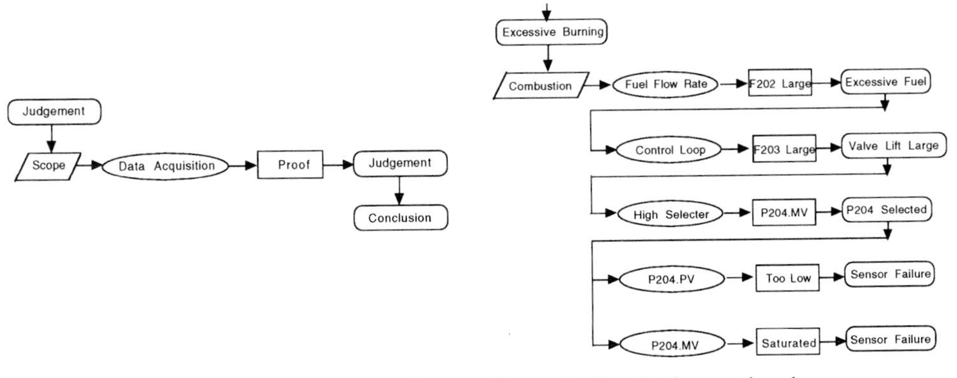

Figure 3. Representation of plant operation

Figure 4. Standard operational sequence

Table 1 Result of operation analysis

Ref. No.	Elapsed time	Arc No.	Contents of problem	Analyst's view	Problem	Classification
1	01:04	P3	Subject isn't familiar with trend analysis for oxygen concentration	Subject lacks complete understanding of cause and effect in plant behavior	Familiarity of control loops	H-2
					Ease of understanding of control loops	I-3
2	02:48 ~03:15	P7	Subject doesn't recognize importance of alarms of L202.PV.LO and L202.PV.LL	L202.PV.LL implies possibility of destruction of boiler drum	Interpretation of alarms	H-3
					Display of alarm importance	I-1
3	03:55	P8	Subject doesn't notice cause of BFP trip	Subject is unfamiliar with BFP trip under new cause	Ability to use already acquired knowledge in practice	H-3
				System should show possibility of BFP trip based on drum pressure	Display of cause of equipment failure	I-1
4	04:44	P12	Subject forgets control loop switching	System should show changes in control loops	Familiarity of control loop switching	H-2
					Display of loop switching	I-1
.....						

4. EVALUATION OF HUMAN MACHINE INTERACTION

An example of the operation analysis table for a burner-header-pressure-sensor failure is shown in Table 1. In many experiments system's problems and human's problems exist together for the point at issue, e.g., familiarity of control loops (H-2) and ease of understanding of control loops (I-3), skill of alarm handling (H-2) and display of alarm importance (I-1), etc. This characterizes the close interaction between human and system in the CRT operation. The results can be used to improve both the human interface and operator training and education.

5. CONCLUSION AND FUTURE DEVELOPMENT

We proposed a systematic approach to grasp the obstacles in a large and complex system of plant operation. Now we are accumulating all the experimental results in the form of an operation history table and operation analysis table in a data base. Using these data we are going to investigate various computational models proposed formally and to develop a human behavior model, which can be used as a virtual subject for interface evaluation of the plant control systems.

REFERENCES

1. H. Nishitani, Jr. of Process Control, Vol.6 (1996) 111.
2. H. Nishitani, AIChE Symposium Series No.312 "Intelligent Systems in Process Engineering, " (1996) 209.
3. H. Nishitani, T. Fujiwara, T. Kurooka, T. Kitajima and M. Fukuda, Jr. of Plant Human Factors Society of Japan, Vol.1 (1996) 26.
4. H. Nishitani, T. Kurooka, T. Kitajima, ibid., Vol.2 (1997) In print.

An Informal Usability Study of a Videoconferencing System

B. Auernheimer, C. Chongcharoen, and S. Arritt

Department of Computer Science, California State University
Fresno CA 93740 USA

This paper describes informal usability studies of using low cost collaborative technologies to support activities typical of a "distance learning" software engineering course.

1. INTRODUCTION

Distance learning is a point of contact for software practitioners and academics [1]. Instructional collaborative activities are a challenge for physically distributed students. A goal of this project is to informally evaluate the utility of some low-cost collaborative technologies in typical software engineering distance education situations.

A secondary goal is to explore the relationship between software inspections using collaborative tools and software "testability".

This work was done in three parts. The first two parts were informal studies of two collaborative systems, Collage and CU-SeeMe, on "neutral content" (i.e., not software engineering) tasks. The third part of this research is using video collaboration in code inspections, and relating the results to the notion of software testability.

2. PREVIOUS WORK

Universities are increasingly providing education and training to students at different sites and possibly at different times. A significant part of many software engineering courses is software development by teams of students. Even courses that do not require team projects may require student collaboration [2]. Although some low cost collaboration tools do not support asynchronous collaboration, they have the promise of allowing students to work together synchronously. A typical software engineering group activity is a code inspection.

Distributed software inspections have been explored. The simplest systems are free, text-based MUDs tuned to software engineering group tasks [3]. More sophisticated systems like Collaborative Software Inspection (CSI) supported "geographically distributed" inspections [4]. CSI allowed both synchronous and asynchronous inspection activities, as well as collection of usage statistics. The CSI-supported meetings were "as effective as face-to-face meetings" and "fault correlation" was aided by the "electronic support" [4]. It is particularly interesting that the self-reported time spent becoming familiar with the CSI tools averaged only 34 minutes.

CSRS (Collaborative Software Review System) supported an incremental model of software development [5]. Unlike CSI, CSRS allowed only asynchronous reviews. Like CSI, CSRS also provided the collection of statistics on the amount of time and issues noted per reviewer and module. It is particularly interesting that CSRS's data collection capabilities can be used to reduce "free-riding" since the inspection moderator has available each participant's data. Detecting weak performance is a goal of both managers and course instructors.

To minimize the logistics of equipment set-up and the recruitment of users, this research uses two-person inspections. A two person inspection is a compromise that might come about in a distance learning environment since it minimizes the "same time" problem. A study of two-person inspection teams showed an increase in programming speed, and appears to be "more effective at improving the performance of the slower programmers" [6].

In addition to informally exploring the usability of common collaborative tools to software inspections, a second goal is to relate the users' experiences to the "testability" [7, 8] of the inspected code. By analyzing each line of code's *sensitivity* (a prediction of the probability that a fault will cause a failure in the software at a particular location under a specific input distribution), a program's overall testability can be determined. Sensitivity is calculated for each line of code by considering the execution probability of the line of code, the probability that an infection occurs at that line of code, and the probability of infected data propagating from that line of code. We are curious whether there is a relationship between sensitivity and our inspectors' experience.

3. PART 0

The first pilot study used NCSA's Collage to prototype the test environment and to reveal weaknesses in test procedures. Collage allows real-time collaboration across hardware platforms. Users can share simulations, text, graphics, and animation sequences with colleagues. Ten computer science students were asked to use Collage to play a simple game. A game was chosen that was not familiar to the students, yet is easy to play. This was a neutral content task. For a more sophisticated use of a game task in the development of CSCW applications see [9].

Each user had a Sun Sparctation for this task. Since the computers were in the same room, participants could talk during the task. Also, the test proctor was one of the participants. A session began with the proctor explaining that four windows would be used: the main Collage window with the participants list, a chat box, a window with the rules of the game, and the task board window (a shared whiteboard). Since the users could hear each other, the chat box was rarely used.

Results of a post-task questionnaire showed that users were satisfied with the ease of error correction and response time. Users were very pleased with the overall system, with nine of 10 marking four or five on a five point scale. Open-ended questions revealed that users desired a feature to highlight the latest update made to the shared whiteboard – getting lost in a rapidly changing shared workspace was a problem. Line drawing with a mouse also frustrated the users. The proctor's general observations were also recorded. The most notable was that although users chose their pen colors, some color combinations proved to be annoying to the users by the end of the game. Finally, the proctor also observed that many users did not read the rules!

Now, to conduct the analysis it is necessary to have a model which allows us to divide the components of a multimedia application as exactly as possible, which means atomising and studying its relationships, behaviour, etc. as if it were semiotic [9]. The model for our work is the HDM, and the main notions that we apply from the point of view of dynamism or behaviour and structure are: links, structural links, index links, guided tours, collections, etc. (for more information about such concept consult [10]). For example, within access design, the notion of the collection is employed. The kernel of any multimedia system is an access structure. This can be thought of as a network in which reference links associate locations (nodes) in texts, files of structured data, maps, 3-D models, and so on. A link in an access structure has a direction (one-way or two-way) and a label. Obviously, two-way links consume more storage but permit bi-directional navigation. The principal types of access structures are: linear, circular, tree, free-form network, and graphic [11]. One of the main advantages of the HDM is that its structures are valid for high-level design.

The steps taken in the elaboration of our method can be summarised as follows: once the universe of study has been generated and the chart of evaluation has been drafted, we go on to test the chart with the universe of the study and determine the main components. Then an initial group of heuristic evaluation criteria is established and refined, as is a group of metrics. Subsequently, we generate a map or a three-dimensional diagram of "The Space of Heuristic Evaluation" [12], which consists of an Entity of Interest Plane (EIP) and an Evaluation Space (ES):

$$EIP = \{Content, Structure, Dynamics, Presentation\} \times \{in\text{-}the\text{-}large, in\text{-}the\text{-}small\} \quad (1)$$
$$ES = EIP \times \{criteria_1\ criteria_2\ criteria_3 \ldots criteria_n\} \quad (2)$$

The next step is the validation of the universe of the study (also tested this method with the WWW). The final step is the interpretation of the results reached.

Knowing the types of the resources used in the universe each CD-ROM analysed helps in the elaboration of our method, because it clarifies the specific entities to be studied and the type of metric to be applied (descriptive statistics) [6–8–12]: binary of presence (yes = 1, no = 0), mode, range, midrange, minimum values, maximum and average values, standard deviation, median deviation, standard error, and frequency. For example, if we want to examine the richness of the access to information in the hyperbase (in-the-small) of the whole application, one of the specific entities will be the available browsing methods.

CONCLUSION

The method and techniques presented here provide a way to break away from the tendency, widespread among certain manufacturers of multimedia applications, of not taking into account the evaluation of the design of the system. Multimedia product design today is prone to a sort of centralism that respects neither the cultural factor nor the idiosyncrasies of different users, as is investigated in the are of international user interfaces. Furthermore, we should bear in mind that the previous experience of users of interactive systems necessitates different combinations of modes, concerning navigation, content, access to the structure of the hyperbase, and presentation of information, which together constitute the richness of the multimedia application. These basic components must be regarded as interrelated parts of a whole in order to achieve a successful user/computer communication process. As well as to CD-ROM's, the method presented can also be applied to the Internet (we checked the method to the WWW, with very positive results). Furthermore, the method has been tested on the creation of a distance university education system (the first virtual campus of an open university based in Barcelona), the production period was barely one a month and with excellent results to date. Our method and techniques evaluate design by combining social sciences with computer science, and since they do not have extensive equipment or staffing needs for their implementation, can result in savings in both time and money.

ACKNOWLEDGMENTS

The author would like to thank Franca Garzotto (Hypermedia Open Center. Politecnico di Milano) for her valuable comments, and Eladio Domínguez Murillo (Department of Computer Sciences and Engineering Systems. Universidad de Zaragoza) for his support and advice.

REFERENCES

1. N. Fenton, Software Metrics: A Rigorous Approach, Chapman & Hall, London, 1991.
2. F. Garzotto, L., Mainetti, P., Paolini, Designing Modal Hypermedia Applications, Proc. Hypertext 97, Southampton, ACM Press (1997) 38-47.
3. J. Nielsen, Usability Engineering, Academic Press, Massachusetts, 1993.
4. F. Garzotto, L., Mainetti, P., Paolini, Multimedia Design, Analysis, and Evaluation Issues, Communications of the ACM, Vol. 38, No. 8 (1995) 74-86.
5. F. Garzotto and P. Paolini, Design and Usability Criteria for Multimedia Application Evaluation, Proc. First International Workshop on Evaluation Methods and Quality Criteria for Multimedia Applications, ACM Multimedia, San Francisco (1995) 1-6.
6. F. Cipolla-Ficarra, et al., Towards a Set of Rules for Evaluating Interactive Multimedia Products from the Viewpoint of their Content, Structure, Dynamics, and Layout, Proc. First International Workshop on Evaluation Methods and Quality Criteria for Multimedia Applications, ACM Multimedia, San Francisco (1995) 7-15.
7. F. Cipolla-Ficarra, Evaluation and communication techniques in multimedia product design for on the net university education, Multimedia on the Net, Springer-Verlag, Vienna (1996) 151-165.
8. F. Cipolla-Ficarra, A User Evaluation of Hypermedia Iconography, Proc. Compugraphics, GRASP, Paris (1996) 182-191.
9. F. Cipolla-Ficarra, A Method that Improves the Design of Hypermedia: Semiotics, Proc. IWHD, Springer-Verlag, Montpellier (1995) 249-250.
10. F. Garzotto, P. Paolini, D., Schwabe, HDM: A Model-Based Approach to Hypertext Application Design, ACM Trans. on Information Systems, Vol. 11, No. 1 (1993) 1-2.
11. W. Mitchell and M. McCullough, Digital Design Media, ITP, New York, 1995.
12. F. Cipolla-Ficarra, Evaluation of Multimedia Components, Proc. IEEE Multimedia Systems '97, Ottawa (1997) in print.

Multimedia Communication for family

Asako KIMURA, Hirokazu KATO, Seiji INOKUCHI

Osaka University, Department of Systems Engineering,
1-3 Machikaneyama-cho, Toyonaka 560, JAPAN

ABSTRACT

In this paper, we describe two functions that support close communications between users working outside and their families at home. These functions are human face tracking by camera and remote control of home electrical appliances from outside. The face tracking function enables the person of the other end of the line to move freely in their house and always puts his/her face in the center of a display while they are communicating. The remote control of home electrical appliances helps the communication with user's children, old parents and handicapped family who can not operate home electrical appliances by themselves.

1. INTRODUCTION

Conventional video-conference systems were expensive and need massive amounts of bandwidth network. Therefore they have been used only for formal purposes such as tele-conferences in campanies and in TV programs. So most of researches about video-conference have been made to support communications not for informal but for formal purposes like CSCW[1]. Today multimedia techniques have, however, been advanced. After few years fiber cable networks will be spread, so that multimedia communications will become ubiquitous, and all of us will be able to use multimedia communication system for informal purpose. The aim of our study is to support such informal purpose communications.

We aim at family communications in daily life. Today, both husband and wife are customed to work outside. They do not have enough time to talk and play with their children and to take care of their old parents. Therefore the system by which people can communicate closely with their family is needed.

There are two ways to support family communications. One is to make the work-at-home system popular, that allows users to stay at home and to communicate with their family. Another is to develop a system that allows users to communicate from their office with their family at home. Few companies have established work-at-home policies, and some jobs are difficult to be made at home. We are, therefore, going to construct a system to support family communications on the basis of the latter way.

2. OUR APPROACH

What kind of elements will be required in the system? Because of network mediation, the multimedia communication carries less information than face-to-face communication. The multimedia communication lacks;

a) dead angle information of camera and sound position information,
b) information of smell, touch and taste,
c) physical effects caused by user's action in the communication space.

So the functions to send these lacking information are important for the multimedia communication to supply close communication. We are planning to realize the multimedia communication system for family by adding these elements one by one.

In the beginning, we picked up two functions. One is a tracking function to track the face of the person at the end of the line by camera. Second element is a remote control function of home electrical appliances in user's house from outside. The face tracking function keeps the face of the person of the other end of the line at the center of a display while they are communicating. The remote control of home electrical appliances helps their children, old parents, handicapped family who cannot operate the appliances by themselves. We will introduce ideas to realize these functions below.

3. FACE TRACKING

Face tracking processing consists of following two steps; First is the detection of face position from input image. Second is a process of panning and tilting camera to the center of face position. There have been many studies of human face detection using such as template matching[2], neural network[3] and color information[4][5]. We have tried to detect face area by skin color information[6]. Since we are going to use this function for multimedia communication, high accuracy for detection is not always necessary, but real-time operation and

(a) Background image

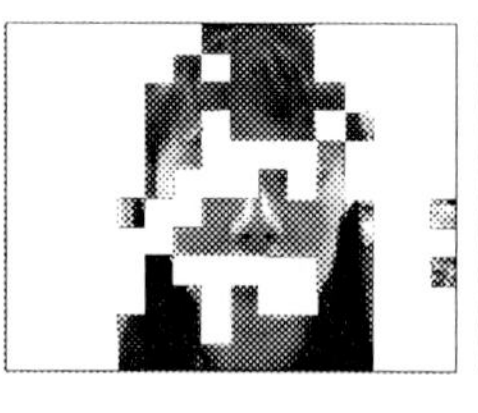
(b) User's face image

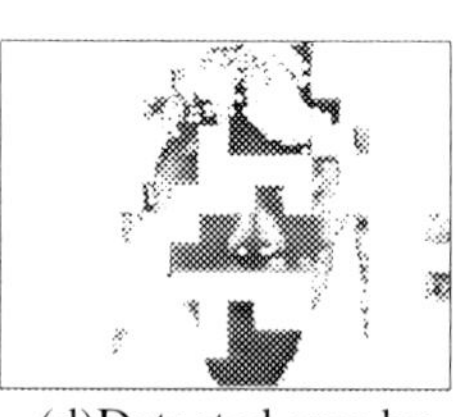
(c) Difference area

(d)Detected area by skin color dictionary

Figure 1. Learning of user's face color information.

(a) Input image

(b) Skin color area

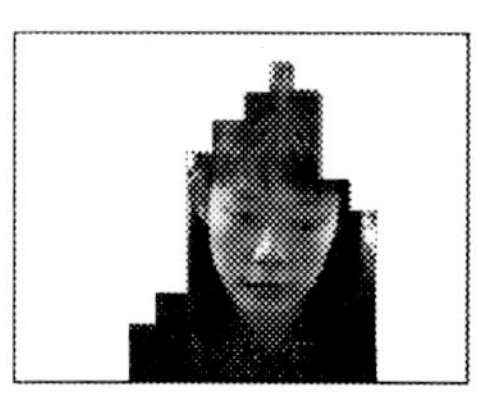
(c) Difference area

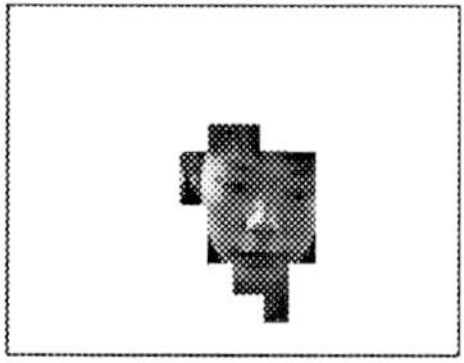
(d)Result of face detection

Figure 2. Flow of face detection.

robustness to light condition, skin color and background are rather important. The face detection methods we propose are shown below.

1. Beforehand we prepare skin color dictionary using skin data of many people. We also store background images according to camera angles (Figure 1a).
2. Before starting face detection, we take one image of user's face to learn his/her skin color and the lighting condition (Figure 1b). First we detect user's body area using difference of this image from the background (Figure 1c), and get skin color information from the body area using skin color dictionary (Figure 1d). We set up a suitable "skin-detect-function" from the skin color information to detect face area robustly to various users and environment.
3. The skin color area is detected using the skin-detect-function (Figure 2b) from input image (Figure 2a). In our system the largest skin color area is judged as face. However, if background has same color objects with face, the judgment sometimes will be wrong. So we also use difference area from background (Figure 2c). Figure 2d shows the result of AND operation of skin color areas and the difference areas, and the area is judged as a face finally.

To track human face, a camera pans and tilts to keep the face area to the center of image. This system has achieved a rate of about 30 frames/second using a SGI-indy workstation and a Canon VC-C1 camera.

4. REMOTE CONTROL OF HOME ELECTRICAL APPLIANCES

Home-automation system is one example of remote control system of home electrical appliances (Figure 3). Conventional home-automation system uses telephone as its interface, which has less affordance for operation.

The affordance, especially visual affordance, is very important because it is able to provide users with much information about operation[7][8]. We built a prototype system[9] in which an interface can be designed using multimedia such as image, sound and character by use of WWW. But affordance is different for each individual. Since the home page is constructed by software, the interface can be selected or built adaptively to each user.

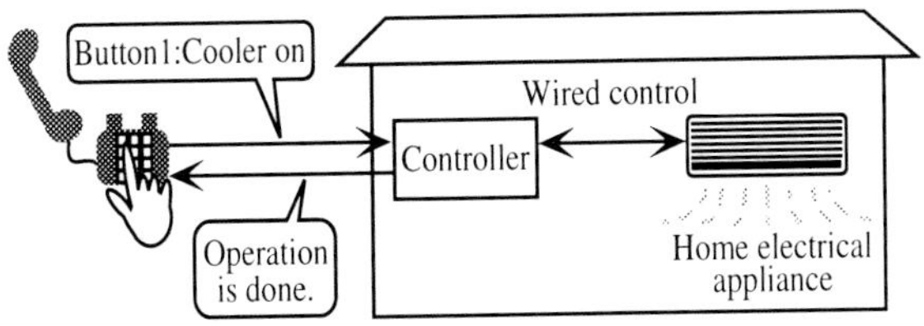

Figure 3. Conventional home-automation system.

In the conventional home-automation system, all home electrical appliances are controlled through wire. In our newly proposed system, there is no wire between controller and each appliance by use of infrared remote controller.

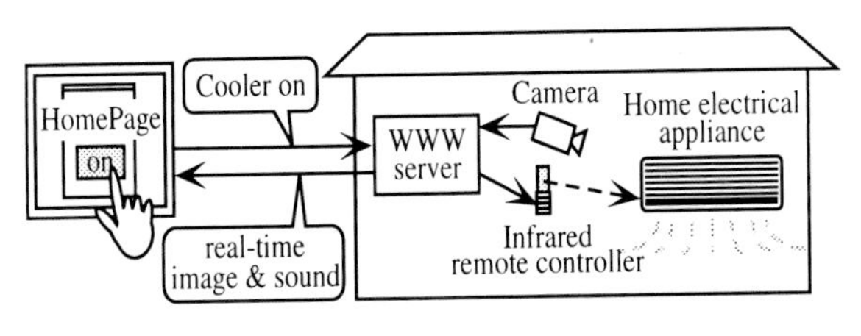

Figure 4. Overview of our prototype system.

Figure 4 shows an overview of our system. First, the user browses a remote control home page in a server computer in his/her house, enters passwords and selects operation button from his office. The selected operation is changed to infrared signals at the server, and the signal is sent to appliances from infrared remote controller. All the details and the results are taken with camera and shown in the conference window.

Figure 5 is an example of interface of this system. It shows "light on/off operation". User selects a home electrical appliance at (a) and operation button at (b). The result of operation can be cofirmed at (c). By visualizing all operations and confirmations, the user can get more affordance than telephone.

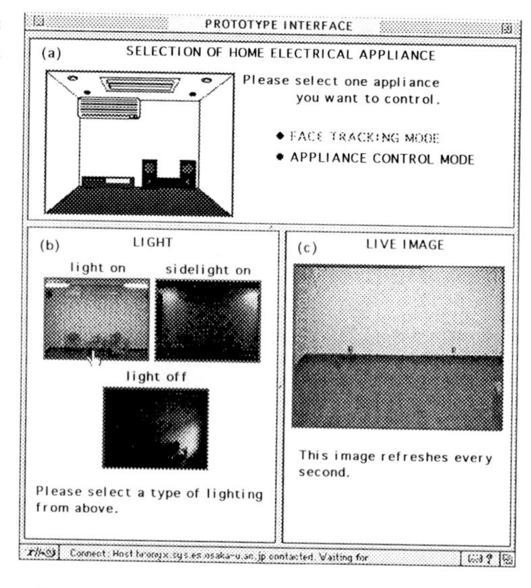

Figure 5. Interface using WWW home page.

5. CONCLUSION

We realized two technologies that support close communications between users working outside and their family at home. One is human face tracking function that provides us with robust and real-time operation using skin color information. Another is remote control function of home electrical appliances from outside using WWW that makes the interface more visual.

For our future work we have to evaluate the usability of this system totally. And it is necessary to realize more functions to support multimedia communication for family, for example, a function that users can get more informations of their family. We are also planning to realize this system on PIM(Portable Information Media), to make a portable system to be carried everywhere.

REFERENCES

1. David Brittan: "Being There -The Promise of Multimedia Communications," CSCW'96, p57-65,1996.
2. R. Brunelli and T. Poggio, "Face recognition: features versus templates," IEEE Trans. Pattern Analysis and Machine Intelligence, Vol.15, No.10, Oct.1993, pp.1042-1052.
3. Henry A.Rowley, Shumeet Baluja, Takeo Kanade: "Human Face Detection in Visual Scenes," November, 1995, CMU-CS-95-158R.
4. Jie Yang, Alex Waibel: "Tracking Human Faces in Real-Time," November, 1995, CMU-CS-95-210.
5. Qian CHEN, Haiyuan WU and Kunihiro CHIHARA: "Real-time Face Detection," ACCV'95, VolII pp.479-483,1995.
6. Asako Kimura, Hirokazu Kato, Seiji Inokuchi:" Tracking of Face Image using Skin Color Information," Technical Report of IEICE, HIP96-12 (1996-06), pp65-70.(In Japanese)
7. P. Johnson: "Human Computer Interaction," McGraw-Hill International,1992.
8. Donald A. Norman: "The psychology of everyday things," , New York,1988.
9. Asako Kimura, Atsushi Nakazawa, Hirokazu Kato, Seiji Inokuchi: "Control of Home Electrical Appliances using WWW and Infrared Remote Control," Proceedinfgs of Twelfth Symposium on Human Interface,1996, pp199-204. (In Japanese)

Computer-Generated Presentation Scripts for Life-Like Interface Agents

Elisabeth André, Jochen Müller, Thomas Rist
DFKI GmbH, Stuhlsatzenhausweg 3, D-66123 Saarbrücken, Germany
email: {andre,mueller,rist}@dfki.uni-sb.de*

In this paper, we describe a life-like interface agent which presents material to the user following a script. Since the manual creation of such scripts is tedious and error-prone, we have developed a plan-based approach to automate this process. We describe two sample applications in which our approach has been used.

1. OBJECTIVE

In this paper, we describe a new system that uses a life-like character, the so-called PPP Persona, to present multimedia material to the user. This material has been either automatically generated or fetched from the web and modified if necessary. Our work was motivated by the following two key observations:

(a) There is a rapidly growing number of applications in which the manual creation of multimedia/hypermedia presentations is no longer feasible. In order to meet the specific needs of the individual presentation consumer, information has to be communicated fast and flexibly. Since it would involve immense effort to manually design documents for a large number of potential users and to hold them on stock, it is more reasonable to automatically generate *personalized* documents on the fly.

(b) The next major step in the evolution of interfaces is very likely to be a shift from the window-based desktop towards more *personalized* user interfaces in which life-like characters play the role of a communication assistant.

To meet (a) and (b), we have developed a new method for automated presentation design which enables not only the synthesis of multimedia documents (e.g. text passages and graphics), but also the generation of a presentation script which defines the behavior of the presentation agent PPP Persona.

2. APPROACH

In order to build up multimedia presentations automatically, we have developed principles for describing the structure of coherent text-picture combinations (cf. [1]). Essentially, these principles are based on

*This work has been supported by the BMBF under the grants ITW 9400 7 and 9701 0.

- the generalization of speech act theory to the broader context of communication with multiple media, and
- the extension of RST (Rhetorical Structure Theory) to capture relations that occur not only between presentation parts realized within a particular medium but also those between parts conveyed by different media.

Since we consider the generation of multimedia presentation as a goal-directed activity, it seemed appropriate to implement a goal-driven, top-down planning approach. The planning component receives as input a communicative goal (for instance, *the user should be able to localize the internal parts of a modem*) and a set of generation parameters, such as target group, presentation objective, resource limitations, and target language. The task of the component is to select parts of a knowledge base and to transform them into a multimedia presentation structure. Whereas the root node of such a presentation structure corresponds to a more or less complex communicative goal, such as *describing a technical device*, the leaf nodes are elementary retrieval or generation acts, currently for text, graphics, animations and gestures. Design knowledge is represented by so-called presentation strategies which encode knowledge on: (1) how to select relevant content, (2) how to structure selected content, and finally (3) which medium to use for conveying a content.

In order to cope with the dynamic nature of presentations made by an animated agent, several extensions became necessary:

- *the specification of qualitative and quantitative temporal constraints in the presentation strategies*
  Qualitative constraints are represented in an "Allen-style" fashion which allows for the specification of thirteen temporal relationships between two named intervals, e.g. *(Speak1 (During) Point2)*. Quantitative constraints appear as metric (in)equalities, e.g. *(5 $\leq$ Duration Point2)*. For more details, see [2].
- *the development of a mechanism for building up presentation schedules*
  To temporally coordinate presentation acts, the presentation planner has been combined with a temporal reasoner which is based on MATS (Metric/Allen Time System, cf. [3]). During the presentation planning process, PPP determines the transitive closure over all qualitative constraints and computes numeric ranges over interval endpoints and their difference. After that, a schedule is built up by resolving all disjunctions and computing a total temporal order. Since the temporal behavior of presentation acts may be unpredictable at design time, the schedule will be refined at runtime by adding new metric constraints to the constraint network.

The realization of the Persona Server follows the client/server paradigm; i.e, client applications can send requests for the execution of presentation tasks to the server (cf. [4]). However, to ensure that the Persona exhibits life-like qualities, the Persona Server enables not only the execution of presentation tasks, but also implements a basic behavior independent of the applications it serves. This basic behavior comprises: reactive behaviors on sensed events, idle-time acts and low-level navigation acts. The Persona's behavior is coordinated by a so-called behavior monitor which determines the next action to be executed and decomposes it into elementary postures. These postures are forwarded to a character composer which selects the corresponding frames (video frames or drawn images) from an indexed data-base, and forwards the display commands to the window system.

3. APPLICATIONS

X11-Implementations of our systems are available for Sun Sparc and Silicon Graphics workstations. The WWW version requires a browser which runs Java applets. Our system has been tested in several applications two of which are shown in Figure 1.

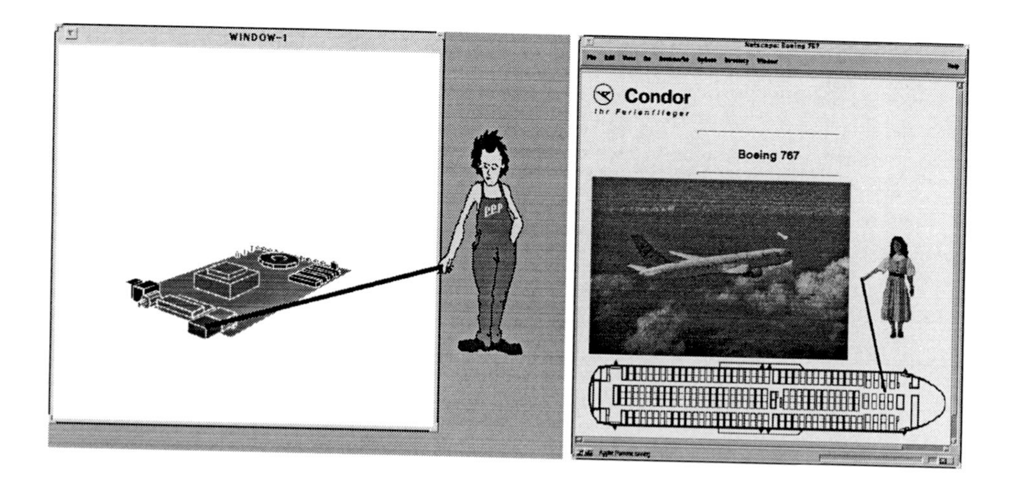

Figure 1. PPP Examples

The first application scenario deals with instructions for the maintenance, and repair of technical devices, such as modems. In the ideal case, instructions are customized to the individual needs of a particular user. Doing this manually, however, is not feasible for many applications. It would require to prepare an exponential number of presentations in advance and to hold them on stock. In contrast to the labor-intensive and thus costly manual authoring of instructions, the PPP system generates on the fly both all the material to be presented (such as textual explanations, illustrations, and animation sequences) and a plan for presenting the material in a temporally coordinated manner.

For illustration, let's consider the task of explaining how to operate a modem. That is, we start the PPP system with the presentation task: *Explain modem*. Starting from this high-level task, the system produces an audio visual presentation given by the interface agent. The screen shot on the left-hand side was taken from this presentation. At the beginning, a window was created that contains a depiction of the modem's circuit board. After the window has appeared on the screen, the PPP Persona verbally informs the user that there is a socket for the phoneline cable. Since several modem parts are visible from the chosen viewing angle, the system decided to resolve the ambiguity by means of a labeling act. While a conventional graphics display would rely on text labels, the PPP Persona enables the realization of dynamic annotation forms as well. In the example, it points to the socket and utters "This is the phoneline socket" (using a speech synthesizer). One advantage of this method over static annotations is that the system can influence the temporal order in which the user processes a graphical depiction. It is even possible to combine both methods since the PPP Persona can also place textual labels on the illustrations before the user's eyes. After that, the Persona describes the remaining actions to be carried out (not shown in the illustration).

The next scenario is taken from a study that we conducted for an airline. This time, the PPP Persona acts as a virtual travel agent that presents travelling packages, configured on the fly from a continuously changing stream of online data (available flights, seats, accommodations and so on). The screen shot on the right-hand side shows the PPP Persona pointing to a picture of an aircraft and suggesting that the user should book a seat in the business class. The screen shot also shows that the appearance of the Persona is not restricted to cartoon characters only. Similar to the option of choosing a favourite text font, a system may provide the user with alternative characters in order to meet his/her individual preferences. This time, the presentation system personifies itself as a "real" person (one of the authors) composed of grabbed video material.

Our current implementations provide a good starting point for future work on animated interface agents. Possible directions include: *augmentations* with respect to the Persona's behavior, the *evaluation* of possible appearances and behaviors of the Persona through different users/user groups, and *further applications* (e.g. Persona can adopt the role of a synthesized shop assistant or investment consultant on the web, or may participate in an audio visual teleconference on behalf of a human user).

REFERENCES

1. E. André and T. Rist. The Design of Illustrated Documents as a Planning Task. In M. Maybury, editor, *Intelligent Multimedia Interfaces*, pages 94–116. AAAI Press, 1993.
2. E. André and T. Rist. Coping with temporal constraints in multimedia presentation planning. In *Proc. of AAAI-96*, volume 1, pages 142–147, Portland, Oregon, 1996.
3. H. A. Kautz and P. B. Ladkin. Integrating metric and qualitative temporal reasoning. In *Proc. of AAAI-91*, pages 241–246, 1991.
4. T. Rist, E. André, and J. Müller. Adding Animated Presentation Agents to the Interface. In *Proceedings of the 1997 International Conference on Intelligent User Interfaces*, pages 79–86, Orlando, Florida, 1997.

Supporting Task Performance: Is Text or Video Better?

Brian E. Norris and William B.L. Wong

Multimedia Systems Research Laboratory, Department of Information Science, University of Otago, P.O. Box 56, Dunedin, New Zealand

1. INTRODUCTION

Multimedia technology allows a variety of the presentation formats to portray instructions for performing a task. These formats include the use of text, graphics, video, aural, used singly or in combination (Kawin, 1992; Hills, 1984; Newton, 1990; Bailey, 1996). As part of research at the Multimedia Systems Research Laboratory to identify a syntax for the use of multimedia elements, an experiment was conducted to determine whether the use text or video representations of task instructions was more effective at communicating task instructions (Norris, 1996). This paper reports on the outcome of that study.

The repair and assembly environment of a local whiteware manufacturer provided the study domain. The task chosen for the study was the replacement of a heating element in a cooktop oven. As there were no task instructions available from the manufacturer, the study was conducted in two phases: Phase I was a cognitive task analysis of service technicians to determine the steps as well as the cues and considerations of the assembly task; and in Phase II we evaluated the text and video representation of the task instructions. The next sections briefly describe the methodology and the results from the experiment.

2. METHODOLOGY

In the first phase of the study, the Cognitive Task Analysis (CTA) resulted in the identification of important cues and considerations that service technicians used to replace the elements of the cooktop (Klein, 1993; Vicente, 1995). Two techniques were used in the CTA: The first was controlled observation in which the service persons were observed performing the task to identify the overt steps in the process. 12 steps were identified in the process. The second technique was a retrospective interview using cognitive probes to identify important cues and considerations of the service persons during the assembly of the heating element. These cues, e.g. little tricks to pop open the back of the oven, were incorporated into the instructions. In the second phase, the experiment consisted of two sets of trials to evaluate the text and video formats (Dumas & Reddish, 1993). A deliberately simplified user interface that was common to both trials was used to present the instructions for the 12 task steps. The instructions were presented in the same locations on the user interface as either text or as video.

As this was a pilot study ten participants were involved in the two trials. There were a total of six males and four females, divided equally between the two test conditions. They were all undergraduate students with little or no computing background. None of the participants had any prior experience in replacing cooktop elements.

In each trial, the participants were asked to observe the text or video-based instructions for one step and then to perform that step before proceeding tot he next step. Participants were told they could refer or re-play the instructions for each step as many times as they considered necessary in order for them to correctly perform the task. The participants' performance was observed by the experimenter sitting in the same room, and was measured on the following dimensions (Nielsen, 1993): (i) Efficiency, (ii) Number of errors, and (iii) Learnability.

Efficiency was measured in terms of time taken to perform each step of the task. For each step the time that was recorded was in two parts. The first part was the time the participant took to look at the information on the screen. This time is termed "time looking." The second part was the time taken to actually perform the step. This component of time is referred to as "time doing." An error is defined as any action that deviated from that specified in the task instruction. The participant was observed performing the individual steps of the task, and any time the participant did not perform the task exactly as the computer system had instructed, this was counted as an error. Learnability was measured in terms of repeats, i.e. the number of times a participant referred to the instructions for the step in order to carry out that step. If the participant turned to look at the instructions on the screen again, this was counted as a repeat.

3. RESULTS

The results from the experiment indicate that the video format was more effective than the text format for supporting task performance. This is briefly reported below.

3.1. Efficiency

The average total time (time looking plus time doing) for completing the entire task was less for the text format than for the video format. But on closer examination, the results indicate that actual task execution is better with video instructions than with text. Figure 1 shows that while it took longer to view the video instructions than the text instructions, participants using video were able to complete their tasks in 12% less time than those presented with text. While not a significantly larger advantage, the other measures to be presented next suggest there is a quality difference in task performance through the use of video.

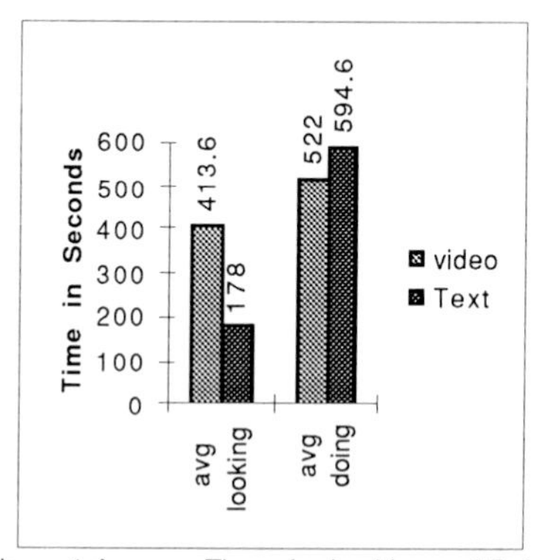

Figure 1 Average Times for Looking and Doing

3.2. Number of errors

Errors give an indication of how well the instructions for the element replacement procedure is understood and replicated. Participants presented with the video instructions made an average of 2.6 errors during the entire trial, compared with an average of 5.6 errors committed by participants who were presented with text instructions. See Figure 2.

3.3. Learnability

Measuring how quickly a participant learned the replacement procedure is the purpose of this dimension. Participants who were presented with video instructions needed to refer to the video only on the average 0.8 times during the entire 12-step task. Whereas participants presented with text needed to refer to the instructions on an average of 3.8 times, or about four times more often than the participants who used the video instructions. Figure 3 illustrates this.

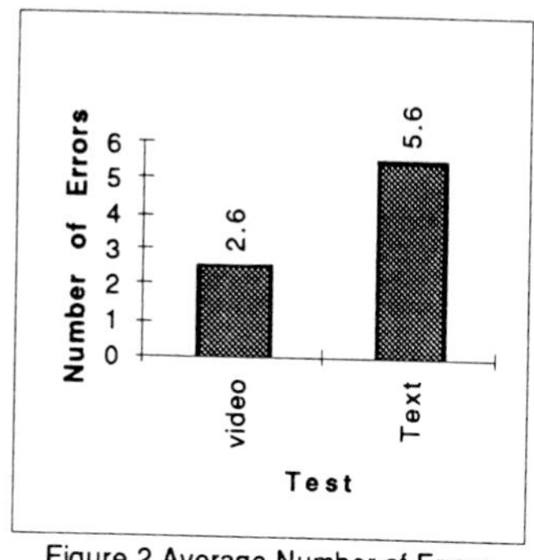

Figure 2 Average Number of Errors

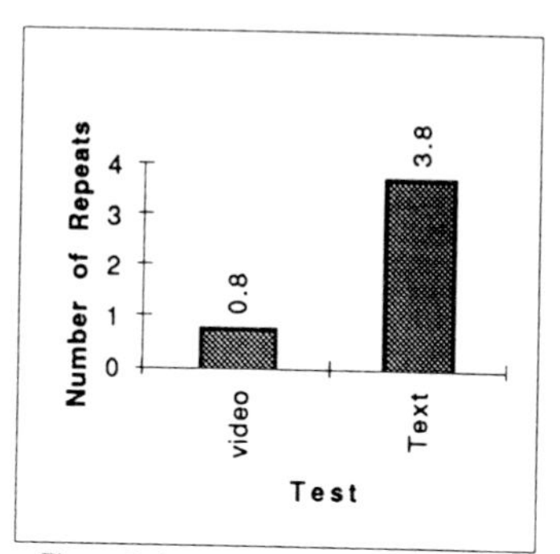

Figure 3 Average Number of Repeats

4. DISCUSSION

Although the results from the experiment seem to suggest that higher efficiency can be obtained through the presentation of text-based instructions, the effectiveness of the two presentation formats must also be viewed in terms of process quality. Text-based instructions resulted in more than twice the error rate experienced than when using video instructions. While we have not investigated the cause of the difference in error rates, it is plausible that the errors are a result of different cognitive processing: Text requires more decoding and interpretation, while video relies on the more powerful perceptual-cognitive systems to understand how a procedure is carried out. The number of repeats represent how quickly one is able to learn the procedure for replacing the heating element. Learning procedures appears to be about four times faster with video instruction than text. Again, the cause of this learning advantage was not investigated in the study, but the findings suggest that procedural type information is better assimilated by a viewer by showing (video) than by describing (text).

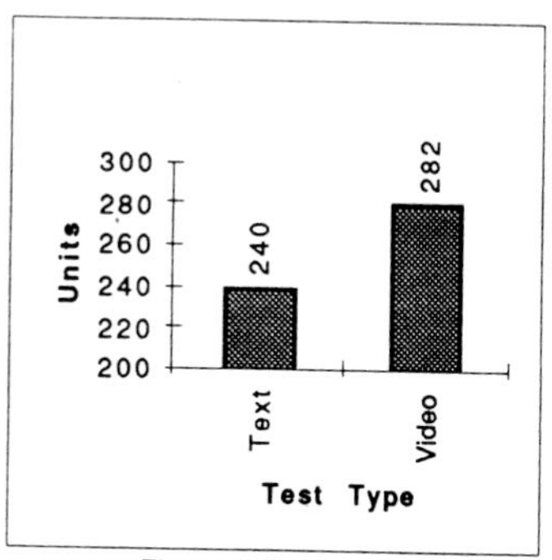

Figure 5 Productivity

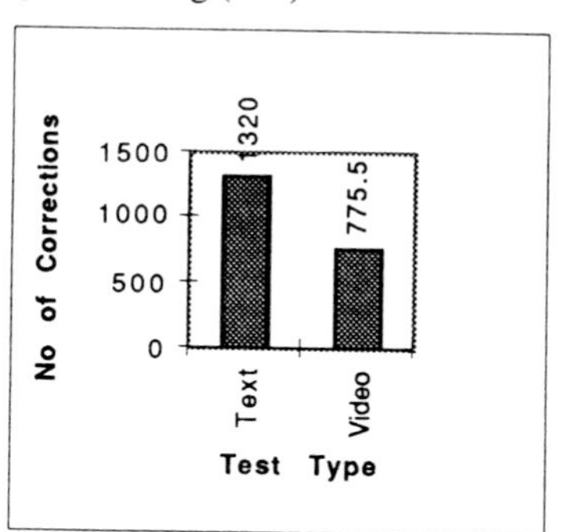

Figure 6 Corrections

To facilitate discussions with our clients at the whiteware manufacturing plant, these results were translated into more meaningful figures to them: Production per week, Corrections needed per week, and Time between interruptions. These measures suggested that a worker working with text-based instructions would produce about 240 units while his or her counterpart working with video-based instruction would produce about 280 units (Figure 5). This is a general improvement in production capacity. This representation was relevant to our clients as their factory floor workers are rotated through different jobs during the week, and hence some re-learning will be needed. The real gains to our client are, however, not in increased production but in the dramatically fewer corrections as Figure 6 suggests. Based on the same number of units produced, the number of corrections needed per week per worker using text was estimated at 1300 corrections, while the expected number of corrections for the worker using video is estimated at 800 corrections. This translates into reduced warranty and product recall costs. Another factor that affects productivity is the time between interruptions. It appears that workers using text-based systems can expect to interrupt their task to refer to the instructions by as much as once every 160 seconds. However, workers using a video-based system are expected to be interrupted only once every 650 seconds to review their instructions. While these figures do not account for learning, they have been useful in demonstrating the differences between text- and video-based task support.

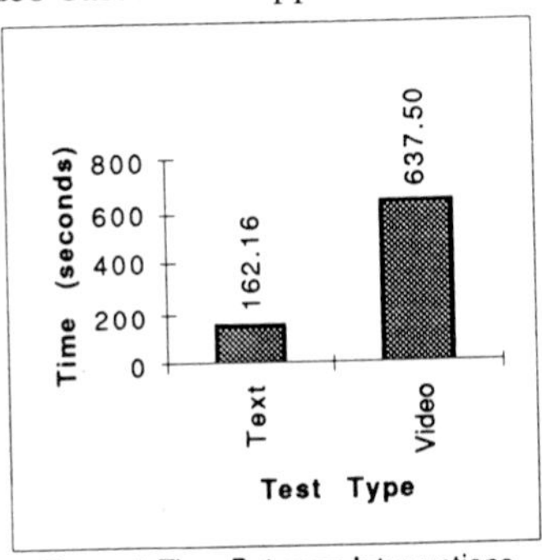

Figure 7 Time Between Interruptions

5. CONCLUSION

While the samples used were small, the results are indicative of better task support performance using video than text-based instruction. More work with larger samples is being planned to improve the generalisability of these results. These planned studies would also investigate why such differences in performance exists, and how these results could be explained within the context of a multimedia syntax.

REFERENCES

Bailey, R.W. (1996). *Human Performance Engineering: Designing High Quality Professional User Interfaces for Computer Products, Applications and Systems. Third Edition.* NJ: Prentice-Hall, Inc.

Dumas, J.S. & Redish, J.C. (1993). *A Practical Guide to Usability Testing.* NJ: Ablex Publishing Corporation.

Hills, P.J. (1984). *Video Production in Education and Training.* Kent: Croom Helm Ltd.

Kawin, B.F. (1992). *How Movies Work.* CA: University of California Press.

Klein, G. (1993). *Naturalistic Decision Making: Implications for Design.* OH: Klein Associates, Inc.

Newton, D.P. (1990). *Teaching with Text: Choosing, Preparing and Using Textual Materials for Instruction.* London: Kogan Page Limited.

Nielsen, J. (1993). *Usability Engineering.* Academic Press Inc.

Norris, B. (1996). *A comparative analysis of a Task Support Environment: Text vs. Video. A Pilot Study.* Unpublished dissertation, Department of Information Science, Otago University.

Vicente, K.J. (1995). Task Analysis, Cognitive Task Analysis, Cognitive Work Analysis: What's the Difference? *Proceedings of the Human Factors and Ergonomics Society 39th Annual Meeting*, pp. 534-537.

Mobile multimedia communication: a task- and user-centered approach to future systems development

F.W.G. van den Anker and A.G. Arnold
Department of Psychology of Work and Organization, Delft University of Technology, WTM/ AOP, De Vries van Heystplantsoen 2, 2628 RZ, Delft, the Netherlands

ABSTRACT

In the future, multimedia will also serve the interests of mobile users, thus extending the electronic superhighway to work situations that lack a wired infrastructure. Companies with mobile workers (e.g. in traffic and transport, field service engineering, emergency services) could substantially improve safety and efficiency with multimedia communications. However, communication media differ widely in terms of interaction and modalities. Therefore, choices have to be made between media, that suit the particular 'context of use'. It is hypothesized that the use of 'full' multi-media will not be suitable for every work situation. This article identifies the work situations in which full multimedia applications can be useful. From different sources the conclusion is drawn that interactive, multi-modal ('rich') media can best be applied to perform non-routine tasks. However, the empirical results have not shown improved task performance so far. A more task- and user-centered approach to communication is proposed in this article. This approach is primarily based on action theory.

1. THEORETICAL BACKGROUND

Too often communication systems are developed without full consideration of the work context in which they will be used. The results of such an approach may turn out to be catastrophic, as was shown by a project that involved the computerization of control room operations in two ambulance services, those of London and Manchester. The technology-driven approach failed, while the user-centered design approach was successful (Wastell & Cooper, 1996). In this paper the work and interaction model by Andriessen (1996) is used, which takes into account technological as well as user, task and organizational aspects (see Figure 1).

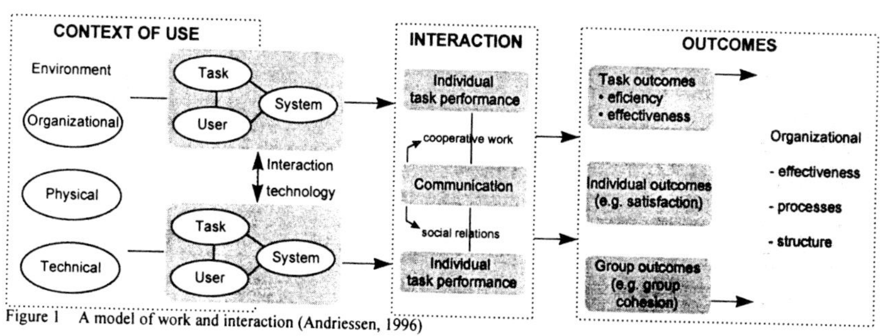

Figure 1 A model of work and interaction (Andriessen, 1996)

In this model, task performance and social interaction are determined by the degree of fit of the system in the 'context of use' (the organizational, physical and technical environment and the task and user characteristics). The specific interaction between the system and the users performing a task results in three types of outcomes: task (e.g. effectiveness, efficiency), individual (e.g. satisfaction) and group outcomes (e.g. cohesion). Furthermore, the interaction has impact on the organization as a whole, in terms of effectiveness, processes and structure.

2. PROBLEM DEFINITION AND METHOD

The choice for a particular media configuration must fit the particular 'context of use', supporting task performance and interaction in such a way that task, personal, group and organizational outcomes will be enhanced. Communication media differ widely in aspects like interaction and representational modalities. The use of interactive, multi-media (including video) communication will not be suitable for every work situation. On the basis of literature study and expert interviews we have tried to identify potentially useful applications of full mobile multimedia communication.

3. RESULTS

3.1 Media richness theory

In literature the concept of multimedia is unclear. In the first place, a medium is defined as a *way of communication*, e.g. E-mail, video-conferencing, telephone or face-to-face contact. Secondly, a medium is defined as *a way of presenting information*: text/ data, graphics, animation, images, still and moving video, sound/ speech. Instead of focussing on one definition of multimedia, the concept of 'media/ information richness' (Daft & Lengel,1984) can be used to integrate the different definitions. Interactive, multi-modal ('expressive') media are called 'rich'.
According to Information Richness Theory, there is a match between the richness of media and the complexity of phenomena that are the subjects of communication through the medium. Simple phenomena can be handled via lean media, whereas complex or equivocal phenomena require rich media for effective communication (Rice,1992). Rich media can be useful in unstructured work environments, in which information processing is less formalized and requires more improvization. This conclusion is also supported by the practical experiences of experts from the scientific as well as practical fields of mobile communication, multimedia and telematics and transport.

3.2 Practice

As a result of the interviews with experts several potential areas of application for mobile multimedia communication were identified: fleet management, logistics, traffic control, surveillance and video-monitoring, emergency services (police, fire-brigade, ambulance services) and field service engineering. Incident- and accident management in the areas of emergency response - especially remote assistance in ambulance services - as well as remote support in service engineering were rated as most suitable for the interactive use of a combination of text/ data, audio and video communication. Text- and data-messages, as in Electronic Data Interchange, can be used to streamline the standard work processes. However, multimedia was seen as most suitable for non-standard work situations, *i.e.* when deviations from the normal procedures occur, which are difficult to put into words or into structured data or text-messages.

3.3. Limited effects

However, there is little evidence that rich media improve the *performance* of complex, non-routine tasks. Rich media rather support social activities, and may only be effective when the personal relationship is important for the task at hand (Gale,1990). This could be due to the limited functionality that has been given to video. Audio-visual communication is mainly used to substitute face-to-face contact by video-conferencing. The function of video in this case is to create context-awareness: to show the non-verbal context - gestures and facial expressions - of the spoken word (Van der Velden,1995). There are few examples of research into other possiblities of interactive video-communication. Much emphasis has been laid on rich media as a means to support interpersonal interaction and communication, as a goal in itself. However, within work domains like service engineering and incident management, criteria like the effectiveness, efficiency and safety of task performance are crucial. To identify useful applications of mobile multimedia communication, we need to model and design work and multimedia in a way that is more oriented towards these criteria.

3.4 An alternative approach

A starting point for a more task- and user-oriented approach to the development of interactive systems can be found in 'action theory' (Hacker,1986; Rasmussen,1983), a cognitive psychological framework of human work behavior and task performance, which considers people's behavior as goal-oriented. The theory has the advantage that it relates task complexity to user experiences and skills. Three levels of behavior regulation are distinguished: skill-, rule- and knowledge-based. At the skill- and rule-based levels no extensive information and explanation is needed, because skills and action procedures are available. But at the knowledge-based level it is not immediately clear how to solve a problem. We think multimedia communication will be most useful in supporting task performance at this level, at which knowledge falls short. This supports the idea that rich media are most suitable to solve unstructured problems (Rice,1992). At this higher intellectual or cognitive level sytem and/ or human support is important. Here, mobile multimedia communication can be used for collaborative task performance, offering remote assistance to local task performance. However, to support the effectiveness and efficiency of task performance in this situation, multimedia/ video communication should be considered as a means to show task-related information or video-as-data (Nardi et al., 1996). Rather than showing 'talking heads' through video-conferencing, the object of task performance (e.g. a machine or a patient) and the actions of the worker towards the object can be made visible via multimedia communication, e.g. to a remote expert.

4. DISCUSSION

To be of use for mobile workers, the future development of systems for multimedia communication should be more oriented towards supporting task performance, instead of communication as a goal in itself. At the application level a move from 'talking heads' video towards 'video-as-data' (Nardi et al.,1996) could be a step forward.

However, the question is how to proceed in developing systems for specific work settings. Here, the Action Facilitation Design Method (AFDM) (Arnold, in press), based on 'action theory', may have a guiding function in the whole development process. However, the user criteria the AFDM produces in terms of effectiveness and efficiency, are based on individual task performance and computer interaction. The AFDM needs to be elaborated towards the specifics of multimedia communication and co-operative work (as shown in Figure 1).

5. CONCLUSION

Enhancing work performance requires a more task- and user-oriented approach to the development of multimedia systems. To suit the specific 'context of use', choices must be made between media, defined in terms of ways of interaction and representational modalities. In this paper we tried to make clear that the so-called 'rich' media (highly interactive, multimodal media) can be useful for performing non-routine tasks. However, the focus on substituting interpersonal communication by electronic media, e.g. video-conferencing, has obstructed the effective use of multimedia applications. As an alternative we combined 'action theory' and the concept of 'video-as-data' (Nardi et al., 1996), as a first step in bridging the gap between multimedia communication and efficient and effective task performance in mobile work settings.

REFERENCES

Andriessen, J.H.E. (1996), The why, how and what to evaluate of interaction technology: a review and proposed integration. In: P.J. Thomas (Ed.), *CSCW Requirements and evaluation.* London: Springer-Verlag.

Arnold, A.G. (in press), *Action facilitation and interface evaluation.* Delft University.

Daft, R.L. & R.H. Lengel (1984), Information richness: a new approach to managerial behavior and organization design. *Research in organizational behavior*, 6, 191-233.

Gale, S. (1990), Human aspects of interactive multimedia communication. *Interacting with computers, 2 (2)*, 175-189.

Hacker, W. (1986), *Arbeitspsychologie - Psychische Regulation von Arbeitstätigkeiten.* Berlin: VEB Deutscher Verlag der Wissenschaften.

Nardi, B.A., A. Kuchinsky, S. Whittaker, R. Leichner & H. Schwarz (1996), Video-as-data: technical and social aspects of a collaborative multimedia application. *Computer Supported Cooperative Work, 4*, 73-100.

Rasmussen, J. (1983), Skills, rules, and knowledge; signals, signs and symbols, and other distinctions in human performance models. *IEEE Transactions on Systems, Man and Cybernetics, SMC-13*, 257-266.

Rice, R.E. (1992), Task analyzability, use of new media, and effectiveness: a multi-site exploration of media richness. *Organization Science, 3 (4)*, 475-500.

Van der Velden, J.M. (1995), *Samenwerken op afstand: twee longitudinale laboratoriumstudies naar effecten van mediagebruik op groepsfunctioneren.* Delft: Universitaire Pers.

Wastell, D. & C.L. Cooper (1996), Stress and technological innovation: a comparative study of design practices and implementation strategies. *European Journal of Work and Organizational Psychology*, 5(3), 377-397.

Individual User Differences in Data Model Comprehension

J.C.Nordbotten[a] and M.E.Crosby[b]

[a] Dept. of Information Science, University of Bergen, N-5020 Bergen, Norway

[b] Dept. of Information and Computer Science, University of Hawaii, Honolulu HI 96822

A variety of graphic styles are used for presentation of both dynamic and structural models in the design, implementation, use, and maintenance of information systems. To date, there has been little focus on the legibility of these models. Our study of the effect of graphic style on data model legibility shows significant variation in model comprehension between experiment participants and between the graphic styles used to present the models.

1. BACKGROUND AND MOTIVATION

Graphic models of information system components and constraints are a recommended communication tool for: confirmation of system requirements, implementation design specifications, documentation for system maintenance, and support of user-system interactions. It has been shown that graphic models can be easier to comprehend than their textual or tabular equivalents.

However, the lack of standards for graphic representations has lead to syntactically very different model styles, as illustrated by the 3 equivalent data models shown in Figure 1. One experiment suggests no differences in comprehension of 2 syntactically different, entity-relationship data models [Kim'95]. Other experiments have questioned the assumption of the intuitiveness or ease of interpretation of graphic models [Gill'93, Petr'95, Pree'83].

We have rerun an experiment designed to study the effect of graphic model style on model legibility [Nord'97]. We have observed significant variations in data model comprehension, defined as the ability to give a complete description of the components of a graphic data model. Our studies indicate that, even with training, graphic data models are difficult to comprehend.

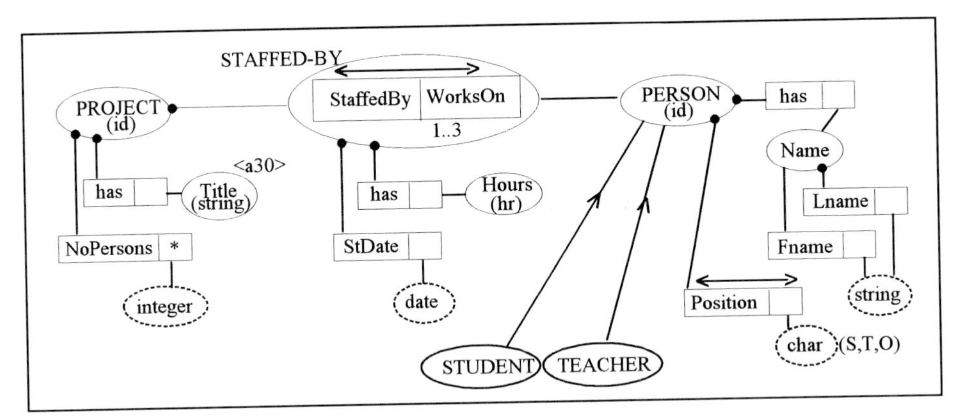

Figure 1a.. A highly graphic model using a simplified NIAM syntax [Nijs89]

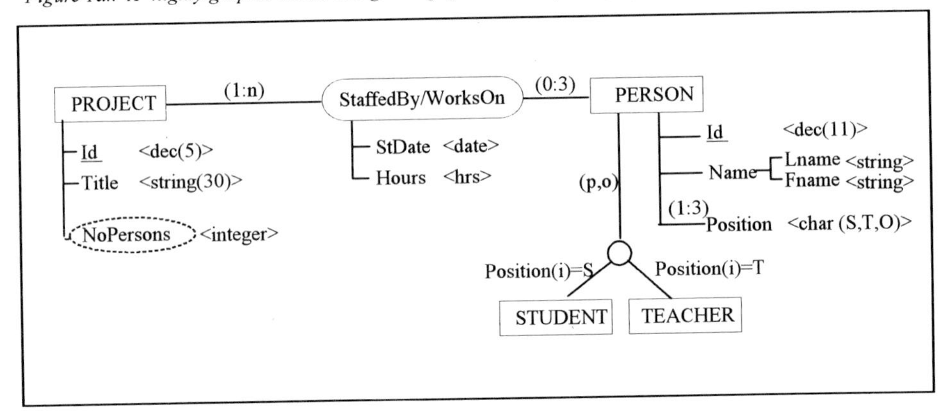

Figure 1b. A minimally graphic model using the SSM syntax [Nord93]

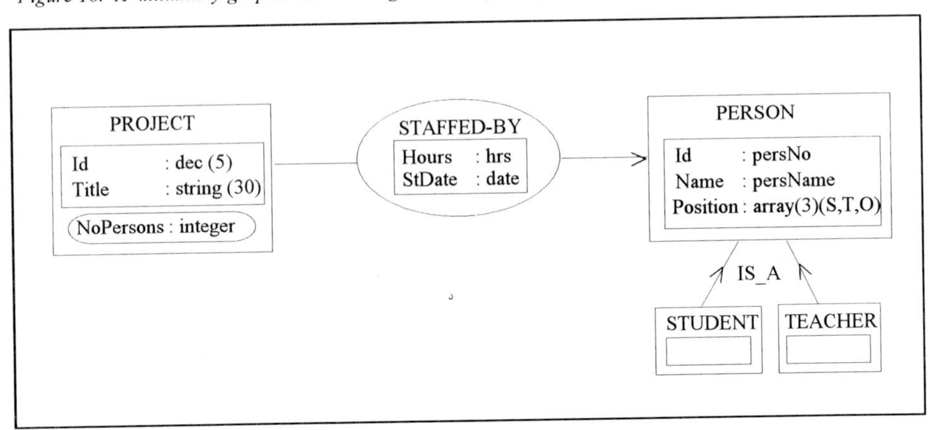

Figure 1c. An embedded graphic model using a simplified OODM syntax [Catt94]

Figure 1. Equivalent data models showing graphic syntax variation

Evaluating Multi-user Interfaces (EMI)

Martha E. Crosby and David N. Chin

Department of Information and Computer Sciences, University of Hawaii
2565 The Mall, Keller 319 Honolulu, Hawaii 96822, USA

Abstract

The present research utilized eye-tracking methodology to investigate whether or not embedded information is more difficult to find than unembedded information and whether or not embedded information is more likely to be misinterpreted than unembedded information. Results from this study suggest that adjacent rather than embedded information may be preferable in the design of complex multi-user interfaces.

1. Introduction

Computers are capable of instantly providing large amounts of information to users. Unfortunately, in the time it takes time to sift through volumes of data to find what is needed, relevant data may escape the users' notice. In time-critical applications such as crisis management, delays from search and overlooked information can have serious consequences. EMI is software that supports multiple team members collaborating in a crisis management situation. Figure 1 shows an example of a map generated for a fire captain dispatching fire trucks. EMI provides team members with customized user interfaces based on a user model of the user's current role and task.

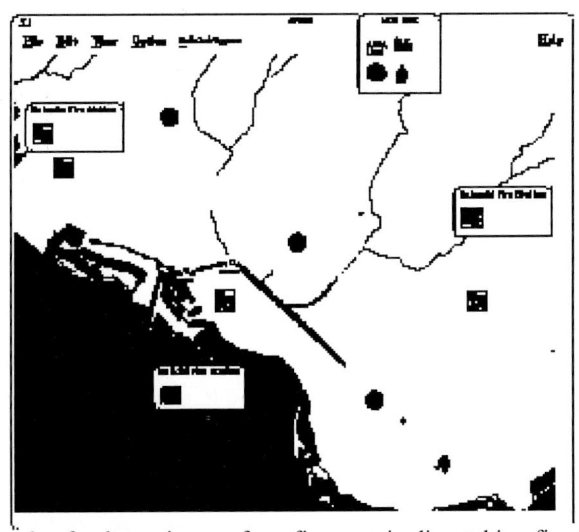

Figure 1. An example of a thematic map for a fire captain dispatching fire trucks.

For example, a fire captain (the user's role) dispatching fire trucks (the user's task) sees a map showing the locations of fires, roads, fire stations, and fire trucks. If the fire captain switches to the task of ensuring adequate water supplies for fighting a particular fire, then fire stations and fire trucks are replaced with water hydrants (with pressure readings) and pump trucks (with capacities). In contrast, an ambulance dispatcher selecting an ambulance to send to an accident site sees an entirely different map, one which shows the region's roads, the accident's location, hospitals, and all ambulance locations and availability. These different maps are created automatically by EMI based on the user's role and task.

In order to reduce the users' information load by presenting only data that is relevant to the user's current task, more information is needed about how users' locate salient information in a multi-user interface. Although crisis management involves many other complex cognitive tasks such as situation awareness and decision making, the focus of this study is visual search or the process of finding particular data within the visual field.

Several studies have reported that search times are related to the density or complexity of the background [1, 3, 4, 5]. Training improves performance when distractors and targets are switched between training sessions as reported by Rogers, Lee, and Fisk, [6]. Williams [7] studied the probability of fixating on objects of different size, shape and color. He concluded that when two or more target characteristics were specified, fixations were generally based on a single characteristic. He proposed that the specification of a target creates a perceptual structure that the searcher explores and the study of visual fixations is the study of the perceptual structure. Eye scanning patterns provide rich collections of data about how people view computer displays. Crosby and Peterson [2] used an eye-movement monitor to investigate how students searched lists of sorted and unsorted information.

2. The Experiment

The present research utilized eye-tracking methodology to investigate the relative effects of embedding information typically used in crisis management maps. In order to conserve space on crisis management maps, critical information is often embedded. Specific questions addressed by this study is whether or not embedded information is more difficult to find than unembedded information and whether or not embedded information is more likely to be misinterpreted than unembedded information.

We chose to experimentally manipulate increasing numbers of both adjacent and concentric squares . Specifically, we recorded the fixation patterns of volunteers who viewed fourteen scenes of 1, 2, 3, 4, 5, 6, or 7 unembedded and embedded squares. The participants were 31 undergraduates (27 males and 4 females) from the Department of Information and Computer Sciences at the University of Hawaii. Each participant viewed the 2 conditions of 7 squares or 14 scenes. The two conditions of squares were intermixed Each set of scenes was preceded by an instructions which stated that, as quickly as possible without making errors, the participant identify the number of squares in the scenes that followed. The experimenter verified that the participant understood the instructions before proceeding, and then instructed the computer to display the next scene after each response. The participants' answers were recorded by the experimenter and on audio tapes.

Eye movement data was collected by an Applied Science Laboratories Model 1996 Eye Movement Monitor (ASL) controlled by a host microcomputer. The ASL eye movement monitor is equipped with an infrared camera that captures reflections from the subject's retina. The low intensity (1.5 ft. lambert) infrared eye illuminator produces no hazard or discomfort to the subjects. The eye-movement monitor provides up to sixty samples per second of eye-fixation positions and pupil-dilation measurements, which is collected by software on the host microcomputer. Fixations were defined by the occurrence of at least 3

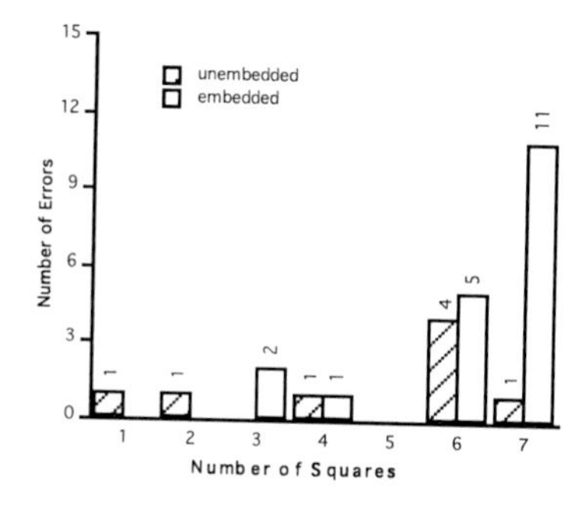

Figure 2. Total Errors for Viewing Squares

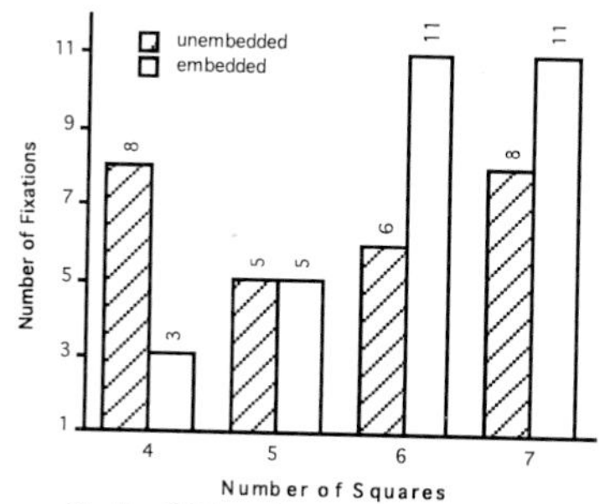

Figure 3. # of Fixations for Subject 8

Figure 4. Unembedded Squares, Subject 8

Figure 5. Embedded Squares, Subject 8

consecutive point locations within a 10 by 18 pixel area. The participant's eyes were calibrated in relation to the screen at the beginning, middle, and end of experiment.

A significant difference in the viewing time between the unembedded and the embedded figures was found by an ANOVA F=6.5, $p< 0.002$. For unembedded squares, the mean increment per additional item was 12.6 msec in the 1-3 range vs. 505.5 msec in the 5-7 range, t (28)= 3.19, $p< 0.002$. For embedded squares, the corresponding mean increments were 191.7 msec in the 1-3 range and 651.9 msec in the 5-7 range, t (12)=4.38, $p < 0.001$. Figure 2 shows the total number of errors for the embedded and unembedded squares. Notice the dramatic increase in the number of participants making errors (22%) for the case of 7 embedded squares. The number of errors, however, was not significant for any squares in the unembedded condition. This was also true when there were 6 or fewer embedded squares. Not only was the tendency toward errors different for the two conditions, participants viewed the embedded squares differently than the unembedded squares. Figure 3 is an example of typical fixation counts. Most participants, did not increase the number of fixations for unembedded squares as the number of squares increased. Fixations increased significantly, however, for 6 and 7 embedded squares. Figures 4 and 5 show typical viewing patterns for unembedded squares and embedded squares. Notice, the similar patterns for 4 and 7 unembedded squares. The number of fixations as well as the number of errors increase significantly for 6 and 7 embedded squares. The nature of the fixation patterns that contribute to increases in processing time when attentional demands are high appears to vary with the spatial characteristics of the scenes. In the case of the squares, the slower processing times seem to stem primarily from the difficulty of differentiating items from their neighbors; thus when squares are embedded, people don't need to scan very widely to find them but they do need to look carefully to tell apart one square from the next. Several embedded square appear not only take more processing time but are more likely to cause errors. These results suggest that adjacent rather than embedded information may be preferable in the design of complex multi-user interfaces

Acknowledgements

This study was supported in part by NSF grant no. IRI93-09711 awarded to M. E. Crosby.

References

1. Bloomfield, J.. (1972). “Visual Search in Complex Fields : Size Differences Between Target Disc and Surrounding Discs.” in *Human Factors.* **14** (2), pp. 139-148.
2. Crosby, M. and W. Peterson. (1991). “Using eye movements to classify search strategies.” In Proceedings *of the Human Factors Society* **2**, pp. 1476-1480.
3. Drury, C., M. Clement and R. Clement. (1978). “The Effect of Area, Density, and Number of Background Characters on Visual Search.” In *Human Factors.* **20**(5), pp. 597-602.
4. Eriksen, C.. (1952). “Object location in a complex perceptual field.” In *Journal of Experimental Psychology.* **45** (3) pp. 124-132.
5. Monk, T. and B. Brown. (1975). “The Effect Of Target Surround Density on Visual Search Performance.” In *Human Factors.* **17**(4) pp. 356-360.
6. Rogers, W. A., M. D. Lee, and A. D. Fisk. (1995). “Contextual Effects on General Learning, Feature Learning, and Attention Strengthening in Visual Search.” In *Human Factors* **37**(1), pp. 158-172.
7. Williams, L.. (1966). “The Effect of Target Specification on Objects Fixated During Visual Search.” In *Perception and Psychophysics* **1**.

A Multicast Algorithm in a Distributed Multimedia Conference System

Anna Hać and Dongmei Wang

Department of Electrical Engineering, University of Hawaii at Manoa, Honolulu, Hawaii 96822, U.S.A.

ABSTRACT

A multicast algorithm in a distributed multimedia conference system uses efficient routing by transmitting multicast packets across fewer links in the network. The algorithm is used to set up multipoint connections for broadband high-speed network.

1. INTRODUCTION

A multicast algorithm provides selective delivery to sparse group via sender (source) and group (destinations) specific multicast tree, to avoid the overhead of transmitting a multicast packet across unnecessary links. The algorithm considers factors which affect routing decision, and keeps the traffic on fewer links, leaving the bandwidth for other transmissions.

In the multicast routing method a packet is transmitted through the tree shaped path, which is rooted in the source and all branches terminate at destinations. The multicast packets are copied in the branching nodes [1].

A number of algorithms used in local area networks with multicast destinations are presented in [2]. Reference [3] presents a multicast source routing scheme in packet-switched networks. A kernel implementation for high-performance multicast communication is described in [4]. A distributed algorithm for resource management in a hierarchical network is introduced in [5]. A distributed multicast algorithm with congestion control and message routing is presented in [6].

2. MULTICAST ALGORITHM

We present a multicast algorithm based on the dynamic priority search technique to build a multicast spanning tree, called Dynamic-Priority-Spanning-Tree (*DPST*) algorithm. The *DPST* algorithm can reduce congestion and increase the path finding rate by using fewer links in the spanning tree than the other algorithms. The principle of *DPST* algorithm is that instead of finding the shortest path between the source and the multiple destinations, it always chooses paths that can make the total number of links (or hops) in multicast tree as few as possible, therefore leave the bandwidth for other transmissions. Intuitively, fewer links implies less bandwidth used and reduces congestion. The *DPST* algorithm also considers the balanced use of the network by limiting the duplication of multicast packets in the nodes to minimize congestion.

The *DPST* multicast algorithm uses the source routing strategy to establish the tree of paths from the source to the multicasting destination. The algorithm allows for transfer of messages from the source to the multicasting destination by using routing and message duplication in network switches.

The multicast tree is rooted in the source and all branches terminate at destinations. The packets are duplicated only when two branches of the path diverge. Postponing the duplication of packets in this way saves network bandwidth due to transmission of fewer packets. If a single link leads to multiple destinations, only a single copy of the packet traverses the link; if necessary the packet will be duplicated later.

The *DPST* algorithm uses source to find the first closest destination, and then it uses the established path to find the next destination close to this path. Thus, the algorithm forms an uncompleted multicast tree, which is used to find another close destination that is not included in the tree. The uncompleted multicast tree is then updated and used to find the other destinations until all destinations have been included in the multicast tree.

In the networks with both multicast and non-multicast vertices, we modify the *DPST* routing algorithm. In this version, we avoid the multicast tree branching in the nodes that do not have the multicast capability.

The *DPST* algorithm can be applied to a packet switching network. The multicasting architecture considered here consists of a number of switches connected by trunks, or routers with links like in the TCP/IP internet. The source and destinations are connected to switches or routers by using lines.

In the internet the service is provided at the network level and the transport-level protocol similar to the Transmission Control Protocol (TCP) or User Datagram Protocol (UDP). Since point to point link-state routing scheme in TCP/IP provides a link state database, which is essentially a dynamic map of the internetwork, *DPST* algorithm can be implemented in this environment.

Reference [6] suggests an architecture for virtual circuit network which considers the balance of switches, trunks and links occupancy. Since this architecture can control congestion, we adapt this design when applying *DPST* algorithm to a virtual circuit network.

The *DPST* algorithm calculation is performed in real-time (on-demand). When a multicast packet with a given source and destination is created, the source switch calculates the multicast tree, using the network map. The *DPST* algorithm guarantees that the multicast packet is not forwarded any farther than it is necessary. In a datagram network, the results of these on-demand tree calculations are then cached in memory for later use by the subsequent matching packets. In a virtual circuit network, the path in multicast tree is reserved immediately for transmission.

3. PERFORMANCE AND COST ANALYSIS

There are basically three costs incurred in large network routing. These costs include the storage required in each network node, the computation required to calculate optimal paths, and the bandwidth consumed by the path update message.

The bandwidth consumed by the path update message is an important technique and has been discussed extensively in the literature. The cost of storage of *DPST* algorithm is approximately the same as of the other algorithms, for instance, the link state multicast algorithm [2]. We focus only on the computation cost of the *DPST* algorithm.

In a sparse network such as the most existing internetworks, the computation time of *DPST* algorithm is $O(M(E+V)\log V)$, where E is the number of links in the network, M is the number of destinations, and V is the number of vertices.

We prove that the cost of generating a spanning tree in *DPST* algorithm is not higher than in the Shortest-Path-Spanning-Tree (*SPST*) algorithm.

Lemma 1: If the path $P(source,...,i,..,d)$ from the source to destination d in *DPST* multicast tree is not the same as the shortest-path $SP(source,...,d)$; i is a node in multicast tree $T(D_i)$ $(D_i \subset (D-d))$, in which d is included in *DPST* multicast tree $T(D_{i+1})$ $(D_i = D_i + d)$ by path $SP(i,...,d)$; then the cost of path $P(source,...,i,..,d)$ is

$$cost(P(source,...,i,..,d)) = cost(P(source,...,i)) + cost(SP(i,...,d)).$$

where $cost(SP(i,...,d)) < cost(SP(source,...,d))$.

Proof :

Let us assume that Lemma 1 is not true, and $cost(SP(i,...,d)) > cost(SP(source,...,d))$.

We know that before d is appended to *DPST* multicast tree, both the source and i are in *DPST* multicast tree $T(D_i)$ $(D_i \subset (D-d))$. When using $T(D_i)$ as the input to find destination d, d should be appended to the shortest path from $T(D_i)$ to d. If $cost(SP(source,...,d)) < cost(SP(i,...,d))$, d should follow the path $SP(source,...,d)$ instead of the path $SP(i,...,d)$. This contradicts the condition that d is not in $SP(source,...,d)$. Therefore, the assumption $cost(SP(i,...,d)) > cost(SP(source,...,d))$ is not true, and by contradiction, Lemma 1 holds.

□

Lemma 2 : If there is a path $P(source,...,i,...,d)$ from the source to destination d in *DPST* multicast tree that is not the same as the shortest path $SP(source,...,d)$, then *DPST* algorithm guarantees lower cost than the cost in *SPST* algorithm.

Proof :

Let $T_{DPST}(D_i)$ $(D_i = D-d)$ denote the *DPST* multicast tree before d is found. Because $P(source,...,i)$ is in $T_{DPST}(D_i)$, then $cost(P(source,...,i))$ is already included in $cost(T_{DPST}(D_i))$. Thus, the cost for *DPST* multicast tree is $cost(T_{DPST}(D)) = cost(T_{DPST}(D_i)) + cost(SP(i,...,d))$. On the other hand, the cost in *SPST* multicast tree is $cost(T_{SPST}(D)) = cost(T_{SPST}(D_i)) + cost(SP(source,...,i,...,d))$. Since $T_{DPST}(D_i))$ is the same as $T_{SPST}(D_i)$ from the condition that only the path from the source to d is not the same as the shortest path, and in Lemma 1 we proved that $cost(SP(i,...,d)) < cost(SP(source,...,i,...,d))$, thus, $cost(T_{DPST}(D)) < cost(T_{SPST}(D))$. Lemma 2 holds.

□

Theorem 1: The cost of generating a spanning tree in *DPST* algorithm is not higher than in *SPST* algorithm.

Proof :

From Lemmas 1 and 2, *DPST* algorithms generate either the spanning tree different from that in *SPST* algorithm, which makes the total cost of *DPST* multicast tree lower than the total cost of *SPST* multicast tree; or the spanning tree is the same as the *SPST* multicast tree, which is the worst case for *DPST* algorithm. This guarantees that *DPST* algorithm does not cost more than *SPST* algorithm. Theorem 1 holds.

□

Finding the minimum cost multicast path is similar to solving the Steiner tree problem, and the heuristic Minimum Spanning Tree (*MST*) algorithm [1, 2] can obtain close to the optimal worst case bound. Our *DPST* algorithm has the same worst case bound as *MST* algorithm.

The multimedia multicast connections are dynamic and complex. The multicast routing algorithm should solve the multicast connection in the networks where some nodes do not have the multicast capability. The algorithm should also restrict the concentration of packet duplication in some nodes to avoid the congestion caused by too many copy operations in a node. The *MST* algorithm can not solve these problems in a simple and inexpensive way as does our *DPST* algorithm.

4. SIMULATION EXPERIMENTS AND RESULTS

Network topology is constructed by a random graph model [7], which has some of the characteristics of the real network. *N* nodes are randomly distributed over the rectangular coordinate grid. Each node is placed in the location with integer coordinates. Links are introduced between pairs of nodes with a probability that depends on the distance between the nodes. Table 1 shows the average results for a 100 node network with multicast vertices. The link costs for networks in Table 1 are generated by randomly selecting a number from 1 to 9. This Table indicates that *DPST* algorithm has closer to optimum results than *MST* algorithm for an average case.

Table 1
Performance for a 100 node network with the link cost between 1.0 to 9.0

Destination	DPST cost	MST cost	SPST cost
4	34.87	35.01	45.96
5	35.96	36.02	46.33
8	47.86	48.13	61.74
10	53.38	53.64	79.03
15	66.23	66.73	84.56
20	77.07	77.87	99.00
25	88.96	89.29	109.36

The results show that *DPST* algorithm uses fewer links, thus, reduces the cost of the multicast path in comparison with *SPST* algorithm, especially in large size of destination groups. The *DPST* algorithm also performs better than *MST* algorithm in finding the minimum cost multicast path in an average case. The *DPST* algorithm is applicable to the networks where some nodes do not have the multicast capability.

5. CONCLUSION

A multicast algorithm considers both the efficient routing by transmitting the multicast packets across few links, and the balanced use of the network. The computation time for the algorithm is not expensive, and the performance of the algorithm is analyzed, compared with the other algorithms, and proven better.

REFERENCES

1. J. Moy, Multicast Routing Extensions for OSPF, Communications of the ACM, Vol. 37, No. 8, pp. 61-66, August 1994.
2. S. Deering, and D. Cheriton, Multicast Routing in Datagram Internetwork and Extended LANs, ACM Trans. Comput. Syst., Vol. 8, No. 2, pp. 85-110, May 1990.
3. T. Yum, and M. Chen, Multicast Source Routing in Packet-Switched Networks, Proc. IEEE INFOCOM, pp. 1284-1288, 1990.
4. J. Gait, A Kernel for High-Performance Multicast Communication, IEEE Transactions on Computers, Vol. 38, No. 2, pp. 218-226, Feb. 1989.
5. A. Hać, Resource Management in a Hierarchical Network, Proc. SBT/IEEE International Telecommunications Symposium, pp. 563-567, Sept. 3-6, 1990.
6. A. Hać, Distributed Multicasting Algorithm with Congestion Control and Message Routing, Journal of Systems and Software, Vol. 24, No. 1, pp. 49-65, 1994.
7. C. Chow, On Multicast Path Finding Algorithms, Proc. IEEE INFOCOM, pp. 1274-1283, 1991.

An Object-Oriented Uncertainty Retrieval Approach for Graphics Databases

Stephen Y. Itoga, Xiangdong Ke, and Ya Liu
Information & Computer Sciences Department, University of Hawaii at Manoa
Email: itoga@hawaii.edu

Abstract

This paper describes an object-oriented approach with uncertainty features for graphics database retrieval. It provides control on the retrieval range for the objects to be searched, and the object behaviors to be queried. It also provides confidence information on the query result, and supports precise query. The proposed approach has been implemented in the systems GOURD and GOURD Server.

Introduction

Much work has been done on management approaches for numerical and text oriented information, however, these traditional database methods are not appropriate for graphics oriented data. Recently, manufacturers and developers of commercial databases try to provide support for spatial data. For example, Microsoft has provided OLE (Object Linking and Embedding) in Access and Foxpro, which can store and access images as objects. Also Paradox, Ingres, Oracle, etc. provide support for storing images in their database products. Although these measures offer some support for spatial data, they haven't completely solved the spatial data handling problem. The reason is that these measures only provide image display and simple operations. Besides software manufacturers, researchers are also involved in this area [1, 2, 4]. For example, Chang et. al. have discussed their method based on 2D string theories. Guttman proposes another approach based on R-tree. Nievergelt and Widmayer have proposed the guard file approach on restricted set of spatial objects. Aref and Samet have also extended the relational model to handle queries based on spatial information. There is much interest in improving the performance of these environments for such applications like Geographical Information Systems and Interactive Multimedia Educational Systems.

For computer graphics, there are the related issues of data sharing and efficient reuse of these resources [1, 3]. These issues include two activities:

• Given a graphics database of images and animation, an end-user would like to specify the images and animation he wants.

• When a user needs to create new images and animation, he may want to modify an existing project to minimize production costs. Thus, there is a need to identify previous projects.

2. Spatial and Spatial-Temporal Relationships

In computer graphics, the focus of attention is restricted to the problems and issues of representing three dimensional scenes in two dimensional space. In order to keep things in perspective, we concentrate on organizing and accessing standard graphical terms and features as the graphic object behaviors. These object behaviors are best defined as spatial and spatial-temporal relationships[1, 2].

For better understanding of the relationships describing the spatial and temporal attributes in the graphics, let us first define a few basic concepts. A *scene* is defined to be a finite set of objects with specific spatial relationships among them. A scene may have a view described by a camera object or by default. An *animation project* is a set of scenes meant to represent key frames in the animation position and orientation. A single image is treated as an animation with only one scene.

Based on the above discussion, we define two types of relationships. ***Spatial relationships*** are unary and binary relationships that deal with the absolute and relative spatial scene positions of objects. For example, a scene may be described as having a palm in front of a rainbow. ***Spatial-temporal relationships*** are unary and binary relationships that are assertions about the temporal properties of spatial relationships. For instance, an animation project may have a ball moving from left to right. Examples of these relationships are giving in Table 1.

Table 1. Spatial relationships and spatial-temporal relationships.

Spatial relationships
  Unary relationships: at the top (bottom, left, right, center) (of the scene).
  Binary relationships: (object A) above (object B), below, to the left, to the right,
  in front of, behind of, and coincident.
Spatial-temporal relationships
  Unary relationships: moving toward right (of the scene), moving to center.
  Binary relationships: (object A) moving from left toward (object B).

For distance-related spatial and spatial-temporal relationships, we build their uncertainty computation models according to the factors related to both distance and object sizes. By the uncertainty retrieval model for spatial and spatial-temporal relationships, users can control their retrieval range on spatial and spatial-temporal relationships, and perform precise query on them.

3. Indexing

A graphics database consists of two database: one is a physical database, and the other is an index database [1, 5]. The physical database consists of the raw image data which are the original raster files and the related text files. This database can not provide useful information during query resolution. The indexing scheme is compatible with current object-oriented graphic design, and supports simple semantic interpretations of 3D graphics querying concepts.

An animation index contains global features of an animation and the information on the changes along the time axis that are inherent in the component scenes. An image index is composed of its objects and global features. An object index contains the graphical object information for querying resolution. The contents of the index database are extracted from project files during preprocessing. In query resolution, only the index database provides the system the related information for query resolution.

4. Fuzzy Object Retrieval

Our graphics database retrieval system supports fuzzy object retrieval. Conventional systems require that objects match exactly the objects in the spatial data. This causes several problems for users. The first problem is that users must know the pre-defined objects in the spatial data, which is a burden for users. The second problem is that if a user's object definition is not consistent with the system's object definition, the retrieval will fail. The third problem is that users can not compose a query with similar objects. For example, if a user wants to find an animation with a "ball" or objects similar to a "ball", the certainty object approach can not work in this situation

To address these problems, we use a fuzzy object retrieval model [5]. This model uses fuzzy object retrieval instead of certainty object retrieval. It provides similarity object retrieval, and enables users to control the search range of the objects to be retrieved. The model is based on the concept of an object hierarchy. The hierarchy is described as an object hierarchy tree. When a user gives a fuzzy object, the tree will be partitioned into some subtrees according to the fuzzy factor of the fuzzy object. The retrieval on the fuzzy object is processed among the partitioned subtrees.

5. The Systems

The architecture of GOURD is shown in Figure 5:

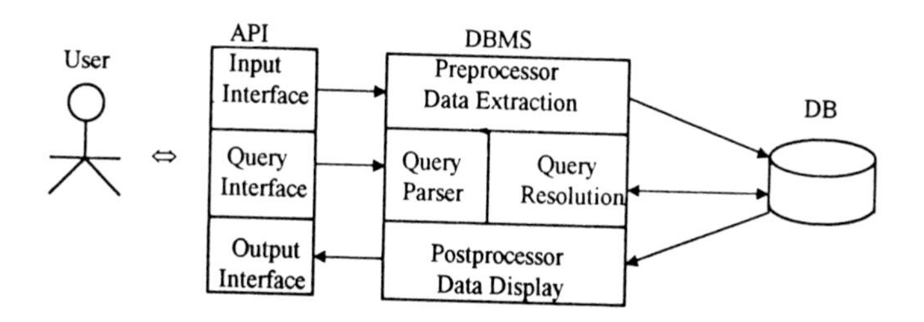

Figure 5. The Architecture of GOURD

In Figure 5, GOURD consists of the Application Interface (API), DBMS, and Database. The Application Interface(API) includes Input Interface for helping users to input data into the graphics database, Query Interface for helping users to compose queries, and Output Interface for displaying query results. The DBMS consists of the Preprocessor, Query Parser, Query Resolution, and Postprocessor. The Preprocessor puts graphic binary files and their related text files into the Database. The Query Parser parses queries, and Query Resolution decomposes the internal representation tuples into basic operation functions, and evaluates these functions. The Postprocessor collects query results for displaying. The Database includes a physical database and an index database. The physical database stores the graphic binary files, and the related text files. The index database stores the extracted graphic objects' information for query resolution.

GOURD Server has the same functions as GOURD. Its client part is responsible for composing a query, sending the query to the server part, receiving query results from the server, and displaying query result. Its server part is responsible for receiving queries from clients, then resolving queries, and sending outputs to clients. GOURD Server can increase flexibility and data sharing for an enterprise, and supports graphics databases retrieval through the Internet.

6. Conclusion

We have discussed our approach for graphics database retrieval, that is an object-oriented approach with uncertainty features. The main characteristics of this research can be listed as follows:

- An object-oriented approach for graphics databases
- Uncertainty retrieval on graphic objects and their behaviors, which includes:
    - A fuzzy object model for object retrieval
    - An uncertainty retrieval model on graphic object behaviors.

By our approach, users can control the retrival range on graphic objects, and graphical object behaviors. It also helps users to query graphics databases in "natural" terminology and provides a confidence value for the query results.

Reference

[1] Ya Liu, *GDB: A Graphic Database Indexing and Retrieving By Visual Relations*. A PhD Dissertation, University of Hawaii at Manoa, U.S.A., 1995.

[2] Liu, Y. and S. Itoga, *A Graphic Database Query Language*. Proceedings of the 1995 Pacific Workshop on Distributed Multimedia Systems, Knowledge Systems Institute, pp. 68-75.

[3] Earl Cox, *The Fuzzy System Handbook: A Practitioner's Guide to Building, Using and Maintaining Fuzzy Systems*. A Division of Harcourt Brace & Company, 1994.

[4] Venkat N. Gudivada and Vijay V. Raghavan, *Content-Based Image Retrieval Systems*. Computer, Vol. 28, No. 9, September 1995, pp.18-22.

[5] Xiangdong Ke, *GOURD: An Object-Oriented Uncertainty Retrieval Approach for Graphics Databases*. A PhD Dissertation, University of Hawaii at Manoa, U.S.A., 1996.

[6] Edward Yourdon, *Object-Oriented Systems Design, An Integrated Approach*. Yourdon Press, Prentice Hall Building, 1994.

[7] A. Pentland, R.W. Picard, and S.Sclaroff, Photobook: *Tools for Contented-Based Manipulation of Image Databases*. Proceedings of Storage and Retrieval for Image and Video Databases II, Vol. 2, 185, SPIE, Bellingham, Wash., 1994, pp.34-47.

Multimedia communication and technology: a semiotic perspective

H.C. Purchase

School of Information Technology, University of Queensland, Australia

A new perspective on the definition of multimedia is proposed: one which considers the classification of the text with respect to semiotic categories and additional dimensions suggested by complex communication devices. Communication in both the visual and aural modalities is classified in a taxonomy based on the nature of the message, rather than on the nature of the technology used to communicate it.

1. INTRODUCTION

The initial classifications within semiology (the science of representational systems) were made in an era of simple communication devices. As technology advances, it is appropriate to reconsider these classifications and extend them to include additional dimensions that are suggested by increasingly complex methods of communication. By separating the classification of the nature of the message to be communicated from the nature of the device, the focus is on the importance of appropriate matching of text with technology.

The aim of this paper is to provide a definition of multimedia that is based on an extension to a common classification of semiotic representational systems, in both the visual and aural modalities. The semiotic terms used are defined and a syntactic classification of representational systems in both the visual and aural modalities proposed. Multimedia is ultimately defined independently of communication devices.

2. SEMIOTIC TERMINOLOGY

For the purposes of this document, the following semiotic definitions are used. A *sign* is an intimate relation between an object and a concept; while the nature of the bond between the two components of a sign is irrelevant, a *symbol* is a particular category of sign, where the relationship between its object and interpretant is arbitrary. A *representational system* is an organisation of patterns of particular signs which comprises a system of meaning. A *message* is a semantically coherent sequence of statements, where a *statement* is a syntactically correct combination of signs in a representational system.

3. A TAXONOMY OF REPRESENTATIONAL SYSTEMS

Traditional classifications of representational systems are based on the use of simple communication devices like paper or static displays. A further dimension is proposed: that of the organisation (or syntax) of the representational system. Five syntactic methods of arranging icons and symbols are proposed: individual, linguistic, schematic, temporal, and network. They are defined below, and examples are given for both the visual and aural modalities.

3.2 The Visual Modality

Representational systems in the visual modality can be identified for each of the five syntactic categories:

- *Individual:* This is the simplest arrangement: there is only a single object to communicate a single concept. Examples: a photograph, a road sign, a word.
- *Linguistic:* Linguistic systems place the objects in a purely sequential manner, and interpretation of the message depends on the objects being considered one-by-one in this linear arrangement. Examples: a cartoon strip, a paragraph.
- *Schematic:* Schematic representational systems use spatial indicators to show the structure of the information. They represent relationships between the concepts associated with the individual objects in a two-dimensional or three-dimensional manner, according to a conventional code. Examples: taxonomic diagrams, iconic charts.
- *Temporal:* As its name suggests, time is important to this arrangement: the message cannot be interpreted if only a snapshot is taken. This arrangement is obviously related to the linear representational systems defined above, as a requirement of linear texts is that they be interpreted temporally. Temporal texts, however, do not exist outside of the temporal dimension, while linguistic texts have a physical quality. Examples : video, animation.

The linguistic, spatial and temporal categories, while important for describing the arrangement of the objects in distinct representational systems, may also be used to describe the organisation of a text that is made up of a number of smaller *mini-texts*. The mini-texts which comprise a text need themselves to be arranged according to a syntax (linguistic, spatial or temporal), and may be in the same or different representational systems. For example, an instructional video (which uses a temporal syntax) may include the following mini-texts:

* film of a lecturer explaining a problem (temporal syntax),
* some photographs (individual syntax),
* and some written paragraphs (linguistic syntax).

The final syntactic category (*network*) does not describe the syntax for individual objects, but is only used for the arrangement of mini-texts, and, unlike the first four categories, there is no restriction on the order in which the user receives the mini-texts.

- *Network:* The most obvious feature of this syntax is the implicit lack of linearity: mini-texts are connected together in a network structure, with related mini-texts linked with each other. Most importantly, there is no predefined sequence of receiving the entire text. Thus, in the interpretation of the text, a mini-text may be followed by any one of the other mini-texts that it is associated with. Examples: interactive video, hypertext.

3.3 The Aural Modality

In any single modality, the messages received must be considered independent of each other, as the receiver can only focus the relevant sense on one message at a time. With the introduction of other modalities, it becomes possible (and indeed, sometimes essential), for more than one message to be transmitted at once, as the receiver can now receive messages through each of the differing modalities.

- *Individual:* It is difficult to define the individual objects of aural communication without taking the temporal dimension into account. If the duration and the possibility of decomposition of the communication is considered, individual aural objects may be defined as very brief, atomic, aural texts. Examples: a synthesised machine "whirr" sound which indicates that an inference engine is working on a problem, the "beep" sound that occurs when mail arrives in a electronic mailbox.
- *Linguistic:* The distinction between *linguistic* and *temporal* made above is that temporal texts only exist in the temporal dimension, and do not have a physical form. As this is the case for *any* text in the aural modality, the linguistic syntactic category is not appropriate here.
- *Schematic:* In order to consider a schematic arrangement in the aural modality, a physical space dimension needs to be introduced. An example is the differing locations of warning sounds in an aeroplane cockpit.
- *Temporal:* This is obviously the most common syntax in the aural modality. Examples: audio recordings, fire alarms, spoken conversation.
- *Network:* Like in the visual modality, the network arrangement means that the receiver can choose which part of the message is to be transmitted at any time. Examples: audio cassettes used in language laboratories, touch-tone telephone menus of recorded spoken information.

  In addition, this network syntax in the aural modality is closely related to audio support for networked visual communication, for example, as the sound tracks for interactive video or interactive animation.

4. A DEFINITION OF MULTIMEDIA

Within this framework, it is proposed that multimedia communication can be defined as *the production, transmission, and interpretation of a text which comprises more than one mini-text, with the condition that more than one representational system in either modality is used in the entire text.* These mini-texts need to have an organisation themselves: any of the syntactic categories of linguistic, schematic, temporal or network may be used. The organisation of these mini-texts does not extend to an individual syntax, however, but with the introduction of additional modalities, the individual syntax can be extended to a syntax of *juxtoposition*, where mini-texts of differing representational systems *and in different modalities* are presented at the same time.

This definition of multimedia is much broader than that usually employed: more traditional definitions are firmly tied to the nature of the device, rather than to the nature of the message to be communicated. Examples of multimedia communication under this definition include:

* A wall poster that includes a photograph, some written paragraphs, and a map.
* A guided tour of an art gallery.
* A hypertext system which emits a "beep" whenever a link is traversed.
* A picture of a waterfall, accompanied by the sound of running water.

5. CONCLUSION

This paper provides a sound terminology and theory with which we may communicate about multimedia systems. A novel framework for the definition of multimedia is proposed, based on traditional semiotic classifications of communication, and different syntactic methods: both the visual and aural modalities are analysed and classified. Multimedia is thus defined not in terms of the technology that transmits it, but in terms of the representational systems and syntax that it employs. By separating text from technology, we can more easily consider the potential of other devices for multimedia communication. Ultimately, we would like to ensure that the message is matched appropriately with the technology, a matching that needs to be done with respect to the semiotic nature of both message and device. In addition, the ease of both creation and interpretation of a message which is to be transmitted using a particular device needs to be considered, and this initial perspective on representational systems provides a secure basis for this future investigation.

REFERENCES

[1] BRUNER J.S. On cognitive growth. In BRUNER J.S., OLVER R.R., AND GREENFIELD P.M., editors, *Studies in Cognitive Growth.* Wiley, 1966.

[2] DE SAUSSURE F. *A Course in General Linguistics.* Fontana, 1959.

[3] MERCER N. AND EDWARDS D. Ground-rules for mutual understanding: a social psychological approach to classroom knowledge. In MAYOR B.M. AND PUGH A.K., editors, *Language, Communication and Education.* Croom Helm, 1987.

[4] PEIRCE C.S. *Semiotic and significs: the correspondence between Charles S. Peirce and Victoria Lady Welby.* Indiana University Press, 1977. Edited by HARDWICK C.S.

[5] SHNEIDERMAN B. *Designing the user interface.* Addison Wesley, 2nd edition, 1992.

Acknowledgement:
I am grateful to Bob Colomb for commenting on drafts of this paper.

ISO 14915: A Standard on Multimedia User Interface Design

Dr. Klaus-Peter Fähnrich and Franz Koller

Fraunhofer Institut für Arbeitswirtschaft und Organisation (FhG-IAO),
Nobelstr. 12, D-70569 Stuttgart, Germany
email: Klaus-Peter.Faehnrich@iao.fhg.de, Franz.Koller@iao.fhg.de

Multimedia user interfaces have been gaining a rapidly growing importance on the market. In its early days multimedia was mainly technology-driven without paying special attention to the ergonomic design aspects of the respective user interface. On the other hand current ergonomic standards for graphical user interfaces fail to support the dynamics of multimedia nor do they cover the special presentation and interaction techniques offered by multimedia. For these reasons ISO is at present elaborating a standard that gives recommendations on design and usage of different media. The standard will consider the effects from combining various types of media and the extended interaction possibilities provided by the individual media. The recommendations will guide designers and developers of multimedia user interfaces to choose appropriate media for certain environments and different purposes. Evaluators may use the standard to assess the usability of multimedia applications.

ISO 14915 will comprise four parts: Multimedia Design Principles, Multimedia Control and Navigation, Selection of Media and Media Combination, and Domain-Specific Multimedia Aspects. The structure and intended content of Parts 1-3 will be presented in the following chapters.

1. DESIGN PRINCIPLES

For the design of multimedia applications and systems the general ergonomic principles as stated in ISO 9241 Part 10 apply. These refer to the design and evaluation of visual display terminals (VDTs) and comprise:

- suitability for the task
- self-descriptiveness
- controllability
- conformity with user expectations
- error tolerance
- suitability for individualisation and
- suitability for learning.

On top of these rather basic principles ISO 14915 addresses the particular characteristics of multimedia and its application domains. The following principles are discussed:

1.1. Suitability for Perceptibility

In the context of ISO 14915 the term perceptibility means that the information which is pertinent or to be conveyed to the user can be easily perceived. This property is particularly important for multimedia systems because the complexity of the presentation may be high, the information conveyed by dynamic media is transient, and several media may be presented simultaneously. In order to achieve perceptibility it is necessary that

- the pertinent information can be easily discriminated from other information presented simultaneously by the same or a different medium
- the quality and presentation characteristics (like resolution or audio volume) are suitable for the purpose of the presentation
- the perception channels of the user are not overloaded by too much information being presented simultaneously, either through a single medium or a combination of media (multiple dynamic visual presentations are particularly critical in this respect)
- the work load resulting from orientation, navigation or manipulation activities does not hinder the uptake of the information pertinent to the user's goals.

1.2. Suitability for Exploration

Explorability means that the user can use the system and find relevant or interesting information with little or no prior knowledge of the particular system or the type and extent of the information stored in the system. Explorability has three main aspects:

- Navigation in the system should be achieved in an easy, consistent and transparent way. The user should always be able to return to significant points in the navigation structure previously visited in order to visit a different part of that structure.
- The information to be conveyed by the system should be structured in such a way that the user can identify the different parts of the contents and their relation with each other.
- The user should be provided with appropriate search and navigation aids in order to determine quickly whether the system contains the information wanted and how it can be accessed.

1.3. Aesthetic Quality

Multimedia systems should possess an appropriate aesthetic quality. This design principle is of particular importance in contexts which are not task-related. The information should be offered in a way that supports the respective communication goal, which in addition to conveying information can comprise the communication of emotion, identity or appreciation.

In some cases this principle may conflict with other more task-related principles. Then a suitable trade-off should be achieved based on knowledge of the user, the context, the task and the purpose for conveying the information.

1.4. Suitability for Engagement

Multimedia systems should stimulate and engage the user by being motivating, entertaining, and by allowing the user to interact directly with concrete representations, thus exploring different possibilities. In order to stimulate user engagement the user interface may be hidden behind the information to be conveyed or the task to be performed. The design of the content to be conveyed is another factor that exerts influence on user engagement: highly interesting contents are likely to achieve strong user engagement.

2. CONTROL AND NAVIGATION

Part 2 of ISO 14915 addresses two issues: the control of media and navigation in multimedia applications. With reference to control possible elementary tasks in multimedia systems are classified and shortly described. Elementary tasks are logical activities associated with multimedia applications.

The tasks are grouped into five areas:

- media control
- accessibility of controls
- attribute control
- controllability of media sources
- impact on general operations.

The main navigation-related topics comprise:

- navigation controls
- navigation aids
- searching and retrieving in multimedia.

The navigation-related topics are described more in detail in the contribution „Navigation Design for Interactive Multimedia“ of the present volume.

3. MULTIMEDIA IN HUMAN COMPUTER INTERFACES

The use of different media should enhance the quality of communication by widening the communication channel. However the increasing bandwidth of the information flow between humans and computers requires a better understanding of information characteristics, how these relate to media characteristics, and to models of tasks, users, and environment. Therefore part 3 of ISO 14915 discusses issues related to the selection and combination of media.

Recommendation for the use of media is given under the following headings:

- suitability for the task
- consider the context
- use redundancy for critical information
- avoid semantic conflicts
- avoid perceptual channel conflicts
- support different points of view.

4. FUTURE WORK

ISO 14915 is still in an early stage and has to pass the ISO voting procedures which help improve the quality of the individual parts. Parts 1 - 3 describe general ergonomic requirements and give recommendations which should hold valid for all multimedia application domains. To cope with the requirements of specific application domains - such as computer based training or interactive TV - a fourth part is intended which will provide the necessary recommendations.

5. REFERENCES

ISO-Norm 9241: Ergonomic Requirements for Office Work with Visual Display Terminals, Berlin: Beuth Verlag.

Sutcliffe, A.G. & Faraday, P.M.: Designing Presentation in Multimedia Interfaces. Proceedings CHI '94 ACM, 1994, pp. 92-98.

Working Draft ISO 14915 Part 1 - 3, ISO TC159/SC4/WG5.

Multimedia HCI Addressed by ETSI

Martin Böcker

Siemens AG, Private Communication Systems, Communication Terminals, User Interfaces, Siemens PN KE TI3, Hofmannstr. 51, D-81359 Munich, Germany. martin.boecker@pn.siemens.de

The design of interactive multimedia applications and services with communications functionality places certain Human Factors (HF) requirements on the design. Given the growing trend towards integrating traditionally computer-based and telecommunications applications and services, international standards bodies have started to issue Human Factors standards and recommendations for the design of this still relatively novel type of user interface. This paper provides an overview of current activities related to multimedia design of the Technical Committee Human Factors of ETSI, the European Telecommunications Standards Institute. Following a brief introduction to ETSI's history and aims, multimedia-relevant work items and documents are described.

1. ETSI'S MISSION AND ORGANISATION

ETSI, the European Telecommunications Standards Institute, was set up in 1988 by the European Council of Ministers by Council Directive. ETSI's aim is to develop telecommunications standards for Europe. The institute was born recognising that „a panEuropean telecommunication infrastructure with full interoperability was the only basis on which a European market for telecommunications equipment and services could thrive“ [1]. ETSI's main focus, therefore, is on accelerating the process of technical harmonisation. Since ETSI was founded, most countries of the European Union moved from national monopoly network operation and service provision to liberalisation and competition. ETSI is adapting to this changing environment and tries to anticipate new developments.

Even though ETSI was founded on an EU initiative, membership to ETSI is not restricted to Union members. Today, most European countries are represented in the organisation. ETSI is recognised as maybe the only forum in the telecommunications arena which brings together all key players including administrations, manufacturers, public network operators, private service providers, users and the research community. ETSI is a so-called regional standards body, it does, however, co-operate closely with international (global) organisations such as ISO, ISO-IEC, ITU-T.

ETSI has already achieved much in trying to create harmonised telecommunications systems, e.g. in ISDN, broadband, mobile communication and satellite communication. ETSI's main products are European Telecommunications Standards (ETSs) which are recognised across Europe and which may be used as a Technical Basis for Regulation within the European Union. In 1995 alone, ETSI published 503 ETSs together with 101 Interim Telecommunications Standards (I-ETSs) and 196 European Telecommunication Reports (ETRs, which are not binding but provide guidance and background comment on matters related to but outside the scope of ETSs).

One recent development reflected in ETSI's work programme is the merging of the computer, telecommunications, and consumer electronics industries. Therefore, ETSI is dedicating two Technical Committees (ETSI Project Multimedia Terminals and Applications and the Global Multimedia Mobility Co-ordination Group) to the issue of multimedia communications and services.

2. ETSI'S TECHNICAL COMMITTEE HUMAN FACTORS (TC HF)

ETSI's Technical Committee Human Factors (TC HF) is one of some 24 Technical Committees and ETSI projects. According to TC HF's Mission Statement, the committee is responsible for user interface standards for telecommunications equipment and services including terminal and service procedures, user guidance and presentation of instructions, symbols, experimental evaluation and methods of usability testing and requirements for People with Special Needs (PSN). The three former subcommittees have been dissolved and all work is currently being conducted in the Technical Committee.

Past areas of interest include telephony and videotelephony user procedures and interfaces (for mobile and corded as well as analogue and ISDN), UPT, PSN requirements and interface design, and usability testing. The results of these work items have been published in the form of ETSs and ETRs. Current work items include the issue of multimedia communications, symbols, telephone service tones, and numbering.

3. CURRENT WORK ON MULTIMEDIA ISSUES WITHIN TC HF

Within TC HF, work on multimedia issues is closely co-ordinated with the ETSI Multimedia Project (Multimedia Terminals and Applications). Two work items are exclusively dealing with the design of multimedia applications and services. In addition, due to merging technologies, the issue of videocommunications is regarded as closely related to multimedia design.

3.1. Human Factors Aspects of Multimedia Telecommunications (ETR 160)

ETR 160 specifies HF recommendations for telecommunications services that support multimedia applications. The report takes an HF approach to multimedia service design focusing on media-specific issues such as media types and quality levels and the suitability of different media for specific tasks, and on HF issues arising from media combinations.

Multimedia service design issues are addressed including the use of metaphors (advantages and disadvantages, choosing an appropriate metaphor) and handling addressing functionality (call set-up and termination, response types, navigation, and charging). Special attention is given to the impact of bandwidth restriction on media types and quality levels, and on the synchronisation of media objects. Hypermedia issues covered by ETR 160 include the design of suitable links and navigation issues in hypermedia space.

Finally, the implications of multimedia services on the required network capabilities is discussed in terms of quality levels, response times, required bit rates for specific media types and charging.

3.2. Human Factors Issues in Multimedia Information Retrieval Services MIRS (DTR/HF-01037)

The report "Human Factors Issues in Multimedia Information Retrieval Services MIRS" is currently available in a stable draft format and is expected to be published within 1997. It is aimed at manufacturers, network operators and other developers and providers of MIRS as well as at service and user interface designers. It gives advice on the development and qualifications of user control procedures for MIRS and provides guidelines for the development process of multimedia user interfaces for retrieval services. The report is based on the consideration that both novice and expert users of MIRS are expected to be able to use those services. Therefore, adequate navigation tools are required to allow users to cope with the complexity of manipulating and retrieving the desired information.

The central concern of the report is on navigation in MIRS and, in particular, on how the user interface design can ensure the effective, efficient and satisfactory retrieval of the desired information. Guidance is given on the support of user tasks including the connection to a remote service, information search, navigation and browsing, and the presentation of information. The pros and cons of different information organisation structures (linear, tree structures, databases) are discussed and navigation aids and procedures are introduced (e.g. addressing the problems of avoiding "getting lost in hyperspace", gaining an overview of what is there, and re-finding a particular piece of information). A section supporting alternative views of the information space suggests that the MIRS interface ideally supports different user models for exploring the information.

The last section provides guidelines on interface design issues pertaining to MIRS including authentication, security and access control to remote MIRS, system response time, user guidance and support, and terminal equipment.

3.3. Videocommunications in Multimedia

Many multimedia environments contain videocommunications components (e.g. CSCW applications). For this reason, the issue of videocommunications is part of ETSI's multimedia-relevant work items. The following ETSI reports on videotelephony and videoconferencing address issues that are relevant for multimedia user interface design.

ETR 297 "Human Factors in videotelephony". This report deals with HF interface design issues of videotelephony. The recommendations given pertain to desktop videotelephony as well as to videotelephony integrated into multimedia services and devices. Call control functions such as dialling and control of service mode are covered with a special emphasis on ISDN videocommunications. Video-related functions discussed include incoming video indication, control of camera and selfview, and video hardware considerations (camera positioning, field of view, lens, iris and focus control, lighting, colour correction and screen adjustability). In addition, issues related to the display refresh and frame rates are addressed. Audio-related functions considered include audio modes and functions (e.g. audio transmission on/off), and audio channel bandwidths. Finally, recommendations are given on the design of (onscreen and hardware) controls and indications including visual and acoustic indications, labelling, functions keys and control dialogues.

ETR 175 "Human Factors of multipoint videotelephony". This report focuses on the problem of camera selection in "switched video" conferences (i.e. only one person can be seen at any one time, as opposed to split-screen systems). The report discusses various technical options including a chairman and an autonomous mode (user-selected camera mode). A second issue dealt with in the report are setting-up procedures for multipoint calls.

TC-TR 005 "Study of ISDN videotelephony for conference interpreters". The empirical study presented in this report (see also [2]) addresses the issue of a minimum acceptable video quality (as expressed in bit rates) for professional use. The findings obtained from conference interpreters are to some degree representative for requirements in professional environments.

ETS 300 375 "Symbols for point-to-point videotelephony functions". This ETS lists eight symbols for use on videotelephony interfaces. The symbol set has been internationally evaluated and consists of symbols for service mode (videophone / telephone call), microphone on/off, camera on/off, selfview, still picture, document camera, loudspeaking and handsfree modes.

4. PLANNED WORK ON MULTIMEDIA

Future multimedia work items of ETSI TC HF include symbols for multimedia functions and services, and multimedia network access and control.

REFERENCES

1. Quotations from Documents accessible from ETSI's WWW-Server (www.etsi.fr).
2. Böcker, M. & Anderson, D. (1993) Remote conference interpreting using ISDN videotelephony: A requirements analysis and feasibility study. *Proceedings of the 37th Annual Meeting of the Human Factors and Ergonomics Society,* Seattle, Oct. 1993.

Navigation Design for Interactive Multimedia

Franz Koller and Andrea Wöhr

Fraunhofer Institut für Arbeitswirtschaft und Organisation (FhG-IAO),
Nobelstr. 12, D-70569 Stuttgart, Germany
e-mail: Franz.Koller@iao.fhg.de, Andrea.Woehr@iao.fhg.de

Some of the most crucial design factors for multimedia systems are navigation and orientation. Without the system's offering an intuitively understandable and easy to use navigation design the users will neither be able to efficiently operate the system, nor will they feel motivated to keep on using it.

Efficiency in the use of a system presupposes the users' understanding of and feeling of competence to handle the underlying navigation concept regardless of whether the user in question is a computer novice or an expert. The possibility to form a coherent mental model on the system's underlying structure is essential for the user in order to maintain orientation. Not knowing the answers to the four well-known questions of where he is located, where he came from, what possibilities for navigation are open to him as well as reminding his original task implies that the structures are not transparent enough to offer guidance and assure orientation. Such an experience will frustrate the user and thus lead to a diminishing acceptance of the system. In the worst case, the feeling of „being lost in hyperspace" might arise.

Good navigation concepts avail the user of the possibility to choose the individually suitable navigation and information selection structure while at the same time offering comprehensive and detailed help. For these reasons, ISO 14915 Part 2 gives recommendations on the control of media and navigation in multimedia applications.

1. NAVIGATION STRUCTURES

ISO 14915 differentiates between three basic structures: linear structure, tree/hierarchical structure and hypermedia structure. This classification forms the basis for describing and recommending certain navigation controls and organisation forms for multimedia systems.

When designing a multimedia system, all three structural approaches should be taken into account, and may be employed in parallel if appropriate for the task. Multiple ways to access information should be provided in order to accommodate as much as possible to the entire range of user needs and expectations.

1.1 Linear Structure

Linear structures impose a pre-defined sequence of usage and presentation. Such a strict order may be caused by specific requirements of the information itself (e.g. with indexes or an alphabetical order), or by assuming a specific order (e.g. chronologically). This form of organisation is adequate for information that is to be found in this order in real life and thus corresponds to user expectation.

A linear structure may also be employed to offer a certain type of users (e.g. beginners) a predefined navigation path through a system, for example in the form of a guided tour. Within the structure the user should be able to move backwards and forwards, advance in larger steps (e.g. page to page or other groups of chunks), and rush to the beginning or end of the entire structure.

1.2 Tree Structure

Tree structures usually comprise a more complex structure or organisation than linear structures, and are adequate when the information is complex, and can be clustered into logical units with regard to a certain subject which can be further refined. The most characteristic feature is the fact that the information is organised hierarchically, i.e. from the most general aspect to the most particular.

In addition to the navigation issues provided for linear structures the user should be able to go up and down in the structure, and to return to the first / main layer. The system should provide clear feedback on the current position in the tree at any moment.

1.3 Hypermedia Structure

This form of information organisation is adequate when there is no clear way to organise the information, or when there are many ways to adapt to different users' search needs. The user is then able to move forward and backward in the structure from any point. This form of organising information can be combined with any other structure. Users not accustomed to hypermedia structures may get confused by the apparent lack of structure. A training advice, explicit help procedures, or guidance information may be required to achieve adequate performance with the system. A typical hypermedia system consists of screens containing so-called "hot spots" that may consist of any type of media. On selection of such an "hot spot" the associated material will be displayed.

Hypermedia structures are essentially non-linear structures. Simple metaphors such as those used in other structures do not suffice for data in non-linear forms, and the designer should impose a recognisable structure or methodology to link the data items because hypermedia structures require relatively high cognitive and memory capacities.

2. NAVIGATION AIDS

Another important issue with respect to usability in hypermedia are navigation aids. The purpose of a navigation aid is to show what piece of information can be accessed with a certain navigation step and how this information will be presented, e.g. in the form of a video, music, a text or a graph. Guided tours are a good example for navigation aids on the structural characteristics of a system. In addition, overviews in the form of a map, fish eye view, or an index provide valuable information on the overall structure. Bookmarks help return to previously visited places the user thinks of as important. Information on type, size and kind of information is given by so-called visual cues. These may also indicate all previously visited links.

3. PRACTICAL EXAMPLES

In the following, two practical examples from the ACTS project MUSIST (Multimedia User Interfaces for interactive Services and TV; AC010) are briefly presented.

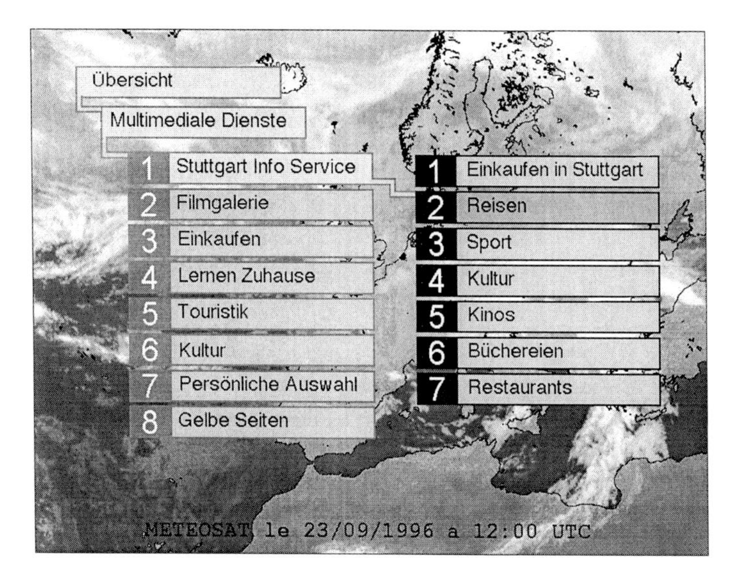

The first example illustrates a model for selection among several multimedia services. The headline displays the activated service to the user. The „pipelines“ connecting the different elements on the screens indicate the path the user has recurred before arriving at a certain position, thus supporting orientation. This orientation line from higher level menu to the lower entry changes according to the present focus highlight.

The actual entry is highlighted and easy to recognise by means of colour and „pipeline". The representation of the previously visited menu differs from the momentarily activated one. The previous (and higher) level is still available but coloured grey. The entry previously visited is in a pale yellow whereas the active entry appears in a richer tone.

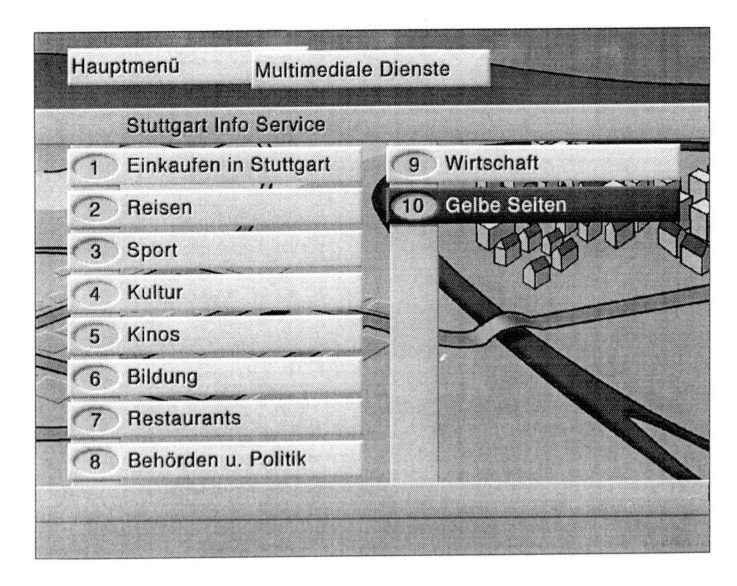

The second example shows the same model realised under a different concept. The previously visited levels appear as light grey-coloured headlines at the top of the screen. The activated menu („Stuttgart Info Service") and the respective menu entries occupy almost the entire screen; the entries from the higher level are no longer visible. The selected menu entry (no. 10: „Gelbe Seiten") is characterised by its dark grey colour.

REFERENCES

1. Berk, E. and Devlin, J. (1991). Hypertext/Hypermedia Handbook. McGraw-Hill, New York.
2. ETSI DTR/HF-01037 Human Factors (HF): Human Factors Issues in Multimedia Information Retrieval Services.
3. ISO 14915 Part 2. Working Draft ISO TC 159/SC4/WG5.
4. Nielsen, J. (1990). Hypertext and Hypermedia. Academic Press, London.
5. Ziegler, J. (1997). Interactive Techniques. In: Tucker Jr., Allen B. (ed.),. The Computer Science and Engineering Handbook. CRC Press, Florida, pp. 1531-1550.

Generating Adaptable Multimedia Software from Dynamic Object-Oriented Models: The *OBJECTWAND* Design Environment

Christian Märtin and Michael Humpl

Fachhochschule Augsburg, Fachbereich Informatik, Baumgartnerstraße 16, D-86161 Augsburg, Germany, E-mail: maertin@informatik.fh-augsburg.de

Abstract

OBJECTWAND is a design environment for modeling and generating high-quality multimedia applications. It enables close cooperation between human developers and automatic construction tools. The system exploits abstract object-oriented models, which specify the internal features and the static and dynamic interrelationships of application domain classes. An automated design process gradually refines models into object-oriented source code. At each development stage models may be adapted to user-specific requirements.

1. INTRODUCTION

Since 1990 there has been tremendous progress in the area of user interface design environments. Today several model-based systems offer tools for specification, simulation and automatic generation of the interactive parts of application software [1-5, 8, 9, 11, 13, 14]. Most of these systems either produce flexible UI software from detailed design models or software with restricted UI structure and behavior from abstract domain models. The first approach is time-consuming and requires human developers with advanced UI design and programming skills. The second approach is limited to application domains with only modest UI requirements.

As a next step, therefore, we must provide *both* improved tool usability *and* mechanisms for designer-system collaboration in those situations where generators alone cannot achieve the required UI quality and functionality. *OBJECTWAND*, which is discussed in this paper, was developed with that kind of cooperative design capabilities in mind. *OBJECTWAND*, the successor of *AME* [6,7], is a model-based CASE environment for interactive applications. It provides advanced generation facilities and interactive tools for designing real world multimedia applications. The system supports:

- definition of OMT-based domain models [10], including message-based communication structure and design-specific requirements (e.g. for video, sound and speech objects);
- *automatic translation* into OOD models, which specify UI structure, action hierarchy, abstract interaction objects, dynamic behavior, domain functionality links;
- interactive model refinement;
- automatic adaptation to requirements of target users and environments;
- simulation of layout prototypes;
- generation of production quality target code for Delphi; and
- model reuse and synthesis based on analysis and design pattern recognition.

2. LIFE CYCLE SUPPORT FOR DESIGNER-SYSTEM COOPERATION

ObjectWand implements a cooperative design automation life cycle (figure 1). The result of each design activity is represented as an object-oriented internal model using *A*pplication *S*ystem *O*bjects *(ASOs)* [7]. In any development state automatic generators and interactive tools can be used alternatively or in combination to refine the model.

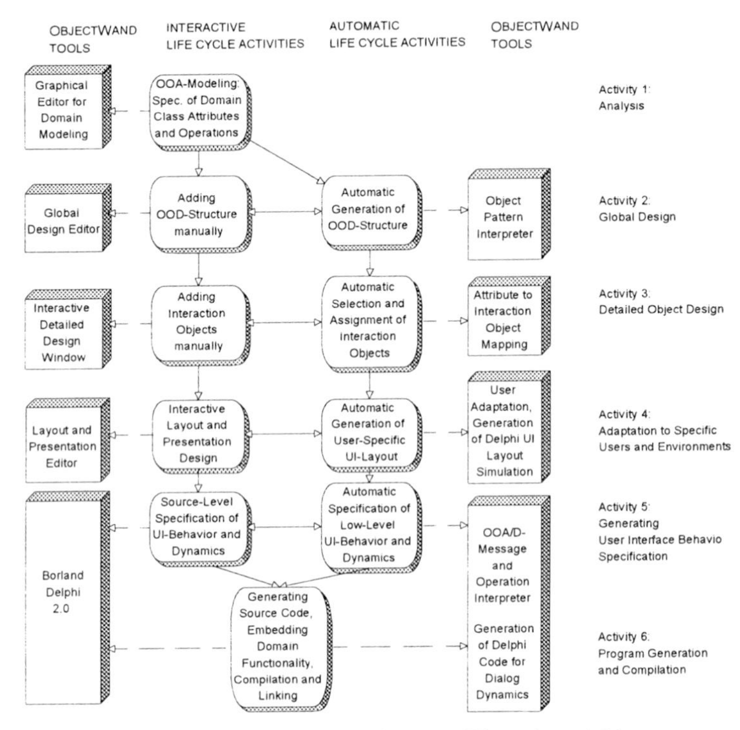

Figure 1. *ObjectWand* software development life cycle activities

An *ObjectWand*-session starts with the definition of a domain object model for the target application. Attributes and operations of the application domain classes and their relationships (inheritance, aggregation, association) may be defined. For representing interactions between domain classes inter-class message channels can be created. They are used for sending and receiving data and activation records between the operations of different classes and for sequence control. Structural and dynamic aspects are capsulated within a single model. Thus mapping problems during implementation can be avoided. At this stage the designer may also introduce detailed design objects or patterns [12] for modeling complex requirements, e.g. for multimedia resources. Figure 2 shows some available modeling primitives.

Many design patterns, however, can automatically be derived from analysis patterns in the domain model. An *object pattern interpreter* parses the aggregation, association and inheritance structure of the domain model and expands each known domain pattern to a pattern of related OOD classes. Such design patterns represent categories of complex UI structures like windows or dialog boxes.

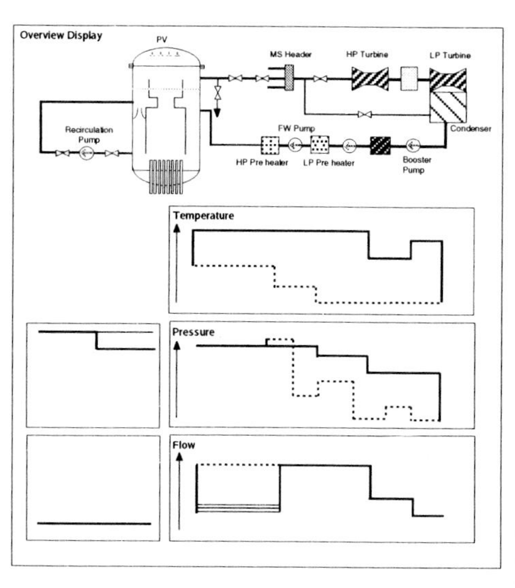

Figure.1 Example image of display mode; Higher level of information granularity

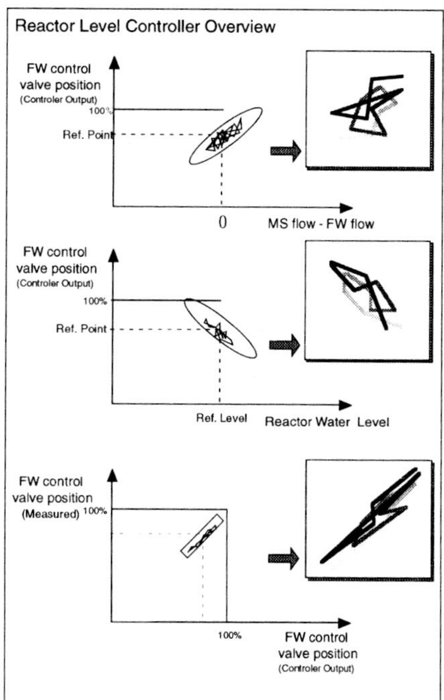

Figure.2 Example image of display mode; More detailed level

variation of the displayed information is simultaneously provided through intensity modification of the traces on the screen. Although the example, called structured functional display [SFD], is admittedly simple, it covers most of the perspectives of cognitive diversity addressed in 2. A precedent of the SFD display is a display mode called overview display [7] proposed by J. Paulsen. We certainly appreciate the design concept of the overview display. Our emphasis is , however, more on structural and systematic integration of diverse measurements rather than intuitive and graphical presentation of the dynamic behavior.

Only a limited number of subsystems of the plant were implemented in the prototype system at this stage. Also, further considerations are necessary to situation-adaptation of the diversified display modes. Nevertheless, the prototype system is informative enough to demonstrate the concept and potentiality of the multimodal display we proposed on the basis of the cognitive diversity.

4. RESULTS AND DISCUSSION

The prototype display is under a process of evaluation by a group of researchers and experienced plant operators. Even at this stage, we confirmed that the result of evaluation by operators is highly positive. This preliminary result is consistent with the experimental observation reported in [7] where user acceptance of overview display is studied. It is not surprising to know that the operators are inclined to support the systematically organized yet intuitive (i.e. graphical) information provision system over a detailed text-based or numerical messages. In our prototype system, we can reasonably expect, from a comparative viewpoint, that a user's capability of detection and comprehension of abnormal situations can be significantly enhanced by proper utilization of the diverse display modes. The freedom of navigating through functional display modes with different information granularity seems to be particularly helpful for users to formulate an interpretation of situation consistent with diverse observations. Although further efforts toward more extended and systematic evaluation are definitely needed, the positive acceptance of the prototype system is a strong support in pursuing such efforts.

5. CONCLUDING REMARKS

A prototype system of information provision system called structured functional display is proposed based on a design principle called cognitive diversity. The main aim of the design is to support problem-solving of operators of a large-scale, complex artifact facing to unfamiliar and unanticipated events. In order to meet the need, provision of plant data in diverse yet systematically organized (i.e. multimodal) manner is considered to be of crucial importance. A prototype system of the multimodal display, called structured functional display, was implemented by taking a simulator of a nuclear power plant as a target system. Through empirical evaluation by plant operators, the concept of structural functional display was judged to be promising. An extension of the project to cover wider range of nuclear plant systems is currently underway.

ACKNOWLEDGMENTS

The authors express their sincere thanks to J.Paulsen for valuable discussions. This project was supported by Grant-in-Aid for Scientific Research (B)(2)-07458095.

REFERENCES

1. K.J.Vicente and J.Rasmussen, Ecol. Psycol., , (1990), 207-250.
2. K.J.Vicente and J.Rasmussen, IEEE Trans. on SMC, 22, (1992), 589-606.
3. M. Kitamura, et al.; Proc. of the IMACS/SICE Int'nl Symp. on Robotics, Mechatronics and Manufacturing Systems, '92 Kobe, (1992) 831-836.
4. M.Kitamura, et al.; Proc. Topical Meeting on Nucl. Plant Instr., Contr. and Man-Machine Interface Tech., Oak Ridge, (1993), 427-434
5. L. Beltracchi and R Lindsay; Proc.8th Power Plant Dynamics, Contr. and Testing Symp., Knoxville (1992), 19.01-19.07
6. L.Chittaro, et al.; IEEE Trans. on SMC, 23, (1993), 1718-1751
7. J.Paulsen; Proc. CSEPC96, Cognitive Sys. Eng. in Proc. Contr., Kyoto, (1996), 79-84

Modeling and Evaluating Multimodal Interaction Styles

Boris E.R. de Ruyter **John H.M. de Vet**
IPO: Center for Research on User - System Interaction
P.O. Box 513, 5600 MB Eindhoven, The Netherlands - {ruyterb, devet}@ipo.tue.nl

1. INTRODUCTION

In the development of consumer electronics, the industry is faced with new challenges. Single domain systems such as audio equipment will be replaced in the near future by multi domain systems in which different media are integrated. Furthermore, multimodal interaction styles will be developed to operate these systems as smoothly as possible. We address these challenges by means of reusing artifacts and developing new artifacts that can be reused. The notion of artifact should be interpreted in the most broadest sense and includes software specifications and components, task models, and guidelines. By embracing the concept of reuse, it is clear that a monolithic approach to system development will not do. Instead, we must apply the principle of *separation of concerns*. Only by taking this approach we are able to produce artifacts that possess the required functionality to be reused in different application domains. Although the separation principle has been around for a long time, it is difficult to put it in practice. In particular, it is not easy to develop artifacts that abstract away from a particular application domain and yet can be tailored to solve a problem in a specific domain. Therefore, the quest for optimal reuse is still on.

2. DEFINING INTERACTION STYLES

In the Human-Computer Interaction (HCI) literature, definitions of interaction style are usually very informal. An interaction style can be loosely defined as the general way that users interact with a system. Typical examples include direct manipulation, menu-based, and form-filling user interfaces. It can be said that an interaction style characterizes the UI of the system. It "offers a cohesive way of organizing the system's functionality, of managing the user's input and of presenting information" ([8]). The term interaction style is used to include "the look (appearance) and feel (behavior) of interaction objects and associated interaction techniques, from a behavioral (user's) view" ([6]). Interaction styles offer different options for executing a user task, or likewise, offer different kinds of access to the system's functionality. An interaction style is described with three components: *interaction techniques, interaction structure*, and *conceptual operations*. These three components correspond roughly with the logical components recognized in the Seeheim model ([3]): presentation, dialogue control, and application interface. The main difference is that the Seeheim model is primarily used to address the specification of concrete, domain-dependent UIs, while the intended use of the interaction style decomposition is to capture the features of a UI in a domain-independent way. The three components of an interaction style are described as:

A *conceptual operation* is defined as a primitive action that either the user or the system can execute within a given interaction. This can be viewed as the pragmatic/semantic part of interaction: the domain-independent classification of user tasks. User tasks describe *what* the user can do with the system without actually prescribing *how*. Based on the flow of information, the following main types of conceptual operations can be distinguished: manipulation, perception, and communication. These main conceptual operations are defined and devised into subclasses adapted from [1] and [5].

An *interaction structure* is defined as the sequence of user actions and system reactions. It reflects the organization of options as they are presented to the user. The structure of this sequence can be classified as the operational procedure.

An *interaction technique* is defined as the way that physical input devices are used to facilitate user actions and output devices are used to present system reactions. It offers the physical control mechanisms offered to the user and the presentation techniques used by the system (partly based on [3] and [4]).

3. EVALUATING INTERACTION STYLES

The research question to be answered is: "*can we attribute manipulations on the level of the conceptual operation, interaction structure or interaction technique to the usability of an interface?*". Here we will use task effectiveness and task efficiency as quantitative usability indicators while user ratings are used as qualitative usability measures. The quantitative indicators are operationalized in the experimental design while the qualitative indicators are gathered in post-experimental questionnaires. Two types of interaction styles were implemented: a function key style using a remote control (RC-style) and a direct manipulation style (DM-style). Conceptual operations are kept constant for both types of interaction style while the interaction structure and interaction technique are varied across both interaction styles. A mode error is an action that is inappropriate (i.e. has no effect) given a specific context of the system. An invalid drop is the action where the user drops a dragged object on an invalid target. Mode errors cannot be made with the DM-style (see also [7]) while invalid drop actions cannot be made using the RC-style.

3.1. Method

Subjects. Twelve subjects, undergraduate students at the Eindhoven University of Technology, participated voluntarily in this evaluation.

Design. The experimental design was a 2 x 2 x 2 counterbalanced mixed design. The type of interaction style (Direct Manipulation or Remote Control), experimental session (first or second session) and interaction style order (first the DM-style and then the RC-style or vice versa) served as independent variables. The interaction style order was used as a between subject variable, while the type of interaction style and experimental session served as within subject variable. The experimental design is depicted in figure 1. The number of user actions (registered at keystroke level), the total task time needed, the number of task effectively completed (i.e. the number of predefined task goals that the user achieved during the experiment) and the number of errors (mode errors or invalid drop actions) were used as dependent variables.

	S1		S2	
	IS1	IS2	IS1	IS2
C1	block 1.1	block 2.1	block 3.1	block 4.1
C2	block 1.2	block 2.2	block 3.2	block 4.2

S1 = first session
S2 = second session
IS1 = RC-style
IS2 = DM-style
C1 = first IS1, then IS2
C2 = first IS2, then IS1

Figure 1. The experimental design.

Materials and equipment. Two interaction styles (RC-style and DM-style) were used during the experiment. For each interaction style the subjects were asked to complete five tasks. Both interaction styles were implemented as part of a simulated compact disc player's interface. The RC-style interaction style was controlled using an infrared remote control, while the DM-style was controlled using a mouse pointing device. The conceptual model implemented was the same for both simulations.

Procedure. Subjects were randomly assigned to the experimental design. The experimental procedure consisted of two sessions: an immediate and a delayed session. The second session was an exact replication of the first session but with a two day delay. The between subject variable was formed by the order in which the subjects used both interaction styles. For both styles the total task time, number of actions and task effectiveness were registered. For the RC-style the number of mode errors and for the DM-style the number of invalid drops were registered. Tasks are considered effective if the user achieves a predefined task goal. After each experimental block (see figure 1) the subjects are asked to fill in a post-experimental questionnaire.

3.2. Results and discussion

Due to the small sample size, non-parametric tests are used to evaluate the statistical significance of differences between groups.

Effectiveness. The number of tasks completed successfully was used as an effectiveness score. A statistical significant effect of type of interaction style on the number of effective tasks was found for the first experimental session. The DM-style resulted in more effectively completed tasks during the first experimental session.

Mode errors and invalid drop actions. When looking at the number of mode errors and invalid drop actions as equivalent error types for both interaction styles, we see that significantly less mode errors and invalid drop actions are made during the second experimental session when compared to the first session. A significant effect of the experimental condition on the number of mode errors in the first experimental session was found (see figure 2) It can be seen that the number of mode errors decreases for both experimental conditions in the second session, while the number of mode errors for experimental condition two is almost significantly lower in session one than for the experimental condition one in the same session. Looking at the effect of working with the DM-style over time in terms of the number of mode errors, we compare the number of mode errors for the RC-style in session one and session two. Again we find a significant effect.

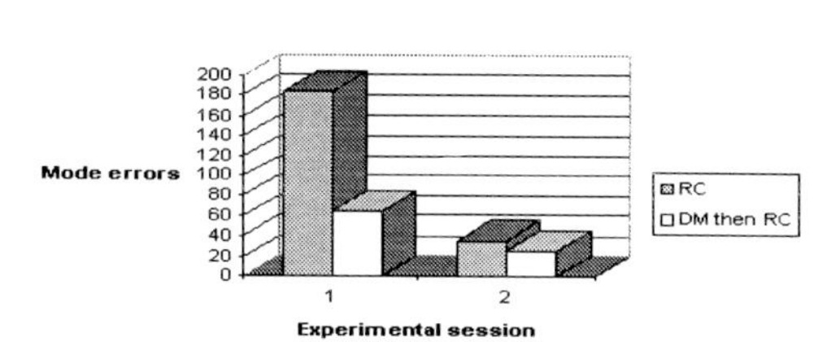

Figure 2. Mode errors in function of the experimental condition for two sessions.

Total task time. The type of interaction style has a main effect on the total task time. Based on the significant effect of the type of interaction style on the number of user actions, it is obvious that significantly less task time is needed for the DM-style when compared with the RC-style.

Presentation of the system. The effects of the presentation on the interaction with the system were rated using four statements that subjects could rate. The RC-style interface was judged as superior when compared to the DM-style interface. Again this difference is statistically not significant.

3.3. General discussion

The statistical results suggest the presence of a learning effect for both interaction styles: the number of user actions, the total task time needed to complete a task and the number of errors (mode errors or invalid drop actions) decrease in the second session. Although this effect is found for both types of interaction style, the results indicate the superiority of the DM-style in terms of the number of user actions, the total task time and the number of effectively completed tasks. More interesting is the effect demonstrated in figure 2. Since both the interaction technique and interaction structure are varied over the two types of interaction styles, the suggested learning effect can be attributed to the fixed conceptual operations for both interaction styles. In the context of this design it is suggested that users construct a conceptual model, when working with the DM-style, that can be used with the RC-style. With such a model, users can make significantly less mode errors when working with the RC-style than users who did not use the DM-style. Since the effect from working with the RC-style before using the DM-style on the amount of invalid drop actions is not significant we suggest that the DM-style is more powerful in building a conceptual model. Another interesting finding is the fact that users do not judge the presentation of both interfaces as a significant difference. The significant differences between both interaction styles can thus be attributed to the quantified characteristics of their implementation.

4. STATUS AND FUTURE WORK

Much of today's research has been focussed on the effect of very specific interaction styles ([2]). Instead of attributing usability effects to an interaction style we aim at attributing manipulations of very specific components of an interaction style to usability effects. Replication of the current research is needed in order to generalize these findings. Future work will concentrate on the translation from empirical findings into cognitive ergonomic guidelines. The work described in this paper has been applied in the realm of consumer electronics. The classification scheme and formalisms developed here have been successfully implemented for the specification of new multimodal applications. It is believed that our interpretation and application of interaction styles is viable for applications in other domains as well.

5. REFERENCES

1. Ahn, R.M.C., Beun, R.J., Borghuis, T., Bunt, H.C., & van Overveld, C.W.A.M. (1995). The DenK architecture: a fundamental approach to UIs. *Artificial Intelligence Review* 8, 431-445.
2. Benbasat, I. (1993). An experimental investigation of interface design alternatives: icon vs. text and direct manipulation vs. menus, *International Journal of Human Computer Studies*, 38, 369 - 402.
3. Buxton, W.A.S. (1983). Lexical and pragmatic considerations of input structures. *Computer Graphics* 17(1). ACM, 31-37.
4. Card, S.K., Mackinlay, J.D., & Robertson, G.G. (1990). The Design Space of Input Devices, *Proceedings. CHI'90*, 117-124.
5. Edmondson, W.H. (1993). A taxonomy of human behaviour and human-computer interaction. *Proceedings 5th International Conference on HCI*, 885-890.
6. Hix, D. and Hartson, H.R. *Developing User Interfaces: Ensuring usability through product & process*. John Wiley & Sons, 1993.
7. Monk, A. (1986). Mode errors: a user - centered analysis and some preventive measures using keying - contingent sound, *International Journal of Man - Machine studies*, 24, 313 - 327.
8. Newman, W.M. and Lamming, M.G. *Interactive System Design*. Addison-Wesley, 1995

Seven Rules in a Theory of Consistency

A. M. Olson

Department of Computer & Information Science, Indiana University-Purdue University, Indianapolis, IN 46202-5132, United States of America

Consistency in human-machine interfaces has been somewhat of a Holy Grail. Everyone believes it is worth pursuing, but there is no agreement on what it is or when it is desirable, cf. [1-3]. This is because, probably like the common cold, it is not one, integral concept, but many. People debate it, but because they shift inadvertently among the subtly interrelated ideas that compose it, conclusive definitions evade them.

The objective of the work here is to elucidate the concept of consistency and lay a logical foundation from which the user interface community can analyze it in whatever contexts it might arise. Better understanding will lead to more appropriate standards for applying it, and eventually, to its more effective use.

People recognize generally that the concept comprises several levels. The disparities observed among consistency's various interpretations [1], however, indicate that this classification is inadequate to account for all of its many facets. It is more accurate to describe it as a family of concepts that are interrelated. A proper understanding of these subconcepts and their relationships is essential to resolving the disagreements.

Although popularly viewed as a collection of useful, logical tools, the discipline that is mathematics actually involves characterizing relationships that define the structures of general systems. Thus, the concept of consistency has an inherently mathematical structure. But, what is it? The investigational approach that this work takes in analyzing this concept is to seek mathematical structures and relationships that match those found among its subconcepts. Executing this process is leading to a more exact understanding of what constitutes consistency. The mathematical characterizations found so far provide a precise way of enunciating the interrelationships for some of its subconcepts. The formal, logical structure that these reveal gives the cognitive engineer tools with which to establish standards on how to apply these subconcepts during interface design.

The discussion begins by examining closely some of the concepts underlying consistency. It introduces six rules based on principles of good design practice that are necessary for consistency. There follows a proof that these are equivalent to a mathematical condition on the syntactic relation between the user's goals and the machine's computations. A seventh rule imposes a semantic consistency condition. Finally, the discussion summarizes these in a concise mathematical principle and explores some ramifications.

Let us consider the situation in which a person needs to solve each of various problems; these goals form the person's goal space. The set of solutions to these problems forms the solution space, which the person connects to the goal space through some mental relation, M. On the other hand, achieving the goals involves finding a practicable mechanism, P, that associates the goal space with the solution space so that $P = M$. A machine has a predetermined combination of actions (i.e., computations) it can perform to solve a problem. The collection of such combinations forms its computation space. For simplicity here, assume that the machine is designed so that its computation space and the solution space are essentially identical. This rules out only the cases in which the machine cannot compute a solution of the user or can compute a worthless result.

The role of the human-machine interface is to provide a concrete mapping of the goal space into the computation space, i.e., provide an instance of P. The interface makes available a number of elementary actions on the machine's components, such as "press key" and "select menu item". These actions and components form the "symbols" in an alphabet of a language for communicating between human and machine. A sentence in this language is a combination of these symbols composed according to certain syntactic rules. It must also be meaningful to the user and the machine. The user "writes" a sentence by manipulating the interface components. Next, the machine translates this sentence into computations, which the machine's internal components perform. In effect, the interface implements the mapping between the user's goal space and the machine's computation space in two stages. The first maps from goal space to language, and the second maps from language to computation space.

Either of these two mappings may be the source of interface usability difficulties; for example, either may be one-many or many-one. To avoid the confusion that may arise by placing the user in ill-defined or ambiguous situations, it is necessary to follow six practical rules when designing an interface. The goal-to-sentence mapping requires three:

1) Each goal has a unique sentence that expresses it.
2) One sentence is not used to express more than one goal.
3) Each sentence expresses a goal.

The first insures that the language suffices to express the user's intentions. The "uniqueness" in it is necessary to avoid possibly confusing the user. Otherwise, by having more than a single means to accomplish a single goal, the very essence of consistency is lacking. If the interface does not adhere to the second, the machine has no way to know from the sentence which goal is intended; this is excessive use of consistency. The third avoids having sentences in the expressing language that the user must recognize as useless, which is a lack of consistency. The sentence-to-computation mapping requires the remaining three rules:

4) Not more than one sentence invokes a particular computation.
5) Each sentence invokes a unique computation.
6) For each computation, there is a sentence that invokes it via rule 5.

Rule 4 prevents the machine from assigning the same meaning to different sentences; otherwise, there is again, from the user's view, a lack of consistency. A given sentence must not cause the machine to do more than one computation, or else there is no way to specify just one of them. This is the reason for the "uniqueness" in rule 5. Violating this condition represents to the user an excessive use of consistency. The last rule insures that the invoking language is adequate for the machine.

All of these rationales derive from what the user views as being consistent in the interface. Three qualitatively different consistency "values" have emerged here: lacking in consistency, consistent, and excessively consistent. A binary-valued (is/is_not) concept is inadequate. Thus, it may be more fruitful generally to characterize consistency as belonging to a range of values rather than just to the binary values in common use.

What do these six rules, which are grounded in good design practices, imply about the relationship between the goal and sentence spaces in formal, mathematical terms? Rule 1 insures that the mapping of goals into sentences is formally a "function", say S. In this notation, $S(g)$ represents the sentence, s, that expresses the goal, g, i.e., $s = S(g)$. Rule 3 states that this equation must hold for all s in the language. Together, the second and third rules insure that there is a formal function, say G, from the sentences back to the goals. Rule 1 implies that $S(g) = S(G(S(g)))$ because both sides of this equality represent sentences for expressing g. This and the equation, $s = S(g)$, imply that

$s = S(G(s))$, which holds for all sentences, s, in the language.

Because rule 1 and the pair of rules (2, 3) are symmetric with respect to the terms "sentence" and "goal", a similar argument shows that

$g = G(S(g))$ for all goals, g, in the goal space.

These two equalities state that S and G are mutually inverse functions; that is, one "undoes" the other. A pair (S,G) of functions of this nature is called a "bijection" [4]. This property is a mathematical characterization of the "identicalness" of two structures as viewed from a general level.

Analogous reasoning demonstrates that the rules, 4-6, imply that the relationship between the sentence space and the computation space is a bijection. The relation between the goal space and the computation space obtained by composing these two bijections is yet another bijection. Therefore, the practical rules, 1-6, insure that the implementation, P, that the interface supplies is a bijection. This means that P associates these two spaces in such a way that, from the point of view of the user, they are "identical" in a precise, but general sense. There is no ambiguity or vagueness in the communication from the user to the machine. The relation, P, from the goal to computation is well defined syntactically.

Nevertheless, P could be wrong! These six rules say nothing that prevents the machine from computing an incorrect solution from a given sentence. An example occurs if the machine invokes a language sensitive editor, LSE, when the user enters the symbol, *ls*, intending to list the current directory. The rule below mandates that the semantics of *P* be consistent with the user's.

7) The interface associates with each sentence the computation that produces in fact the solution to the goal that the user expresses with the sentence.

It is not difficult to express this condition in terms of functions like *(S, G)*, but let us discuss here one of its consequences instead. It directly confronts a mistake that designers may make. They assume frequently that it is they, by means of the machine, who define the semantics of a user's situation. It is the user, nonetheless, who does this by selecting the solution space. Because this space varies and is generally unknown at interface design time, rule 7 effectively obligates the designer to make explicit efforts to insure that the user's and machine's interpretations of a given sentence are always the same.

The conclusion to which the reasoning above leads is that the following mathematical principle is a necessary condition for the interface to be consistent:

The machine's implementation of the relation from the user's goal space to the machine's computation space is a bijection that achieves the solution for each goal of the user.

One benefit of this formulation, besides its conciseness, is that the analyst may use it for formally analyzing a design. A practicable benefit is that the cognitive engineer can use it to fashion variations of the seven rules suitable for the particular interface to be designed. For example, to help meet the obligation that rule 7 imposes, the designer could create several, prefabricated designs, each of which satisfies the mathematical principle for a particular type of user or set of problems. Integrating these into a single, multimodal interface, accommodating the disjunctures at the overlapping design boundaries, seems an easier task, and less prone to error, for the cognitive engineer than building one from nothing.

An advantage of the investigational approach used here is that the search for formal mathematical characterizations of various consistency concepts leads the HCI researcher to examine and codify them carefully. This deepens and broadens the HCI community's understanding of them. Elucidation of additional and more subtle concepts of consistency is the goal of the continuing study here.

REFERENCES

1. Grudin, J. The case against user interface consistency. *Commun. ACM*, 32 (1989) 10, 1164-1173.
2. ________. Consistency standards and formal approaches to interface development and evaluation: a note on Wiecha, Bennett, Boies, Gould and Greene. *ACM Trans. Info. Sys.*, 10 (1992) 1, 103-111.
3. Wiecha, C. ITS and user interface consistency: a response to Grudin. *ACM Trans. Info. Sys.*, 10 (1992) 1, 112-114.
4. Gersting, J.L. *Mathematical Structures for Computer Science*, third ed., W.H. Freeman, New York, 1993.

Teaching Information Literacy in Technical Courses using WWW

Yu SHIBUYA, Gen KUWANA and Hiroshi TAMURA

Faculty of Engineering and Design, Kyoto Institute of Technology
Matsugasaki, Sakyo-ku, Kyoto city, Kyoto 606 JAPAN
[shibuya, gen, tamura]@hisol.dj.kit.ac.jp, http://www-hit.dj.kit.ac.jp/~[shibuya, gen, tamura]/

Experimental uses of WWW on teaching information literacy in technical courses were introduced and evaluated. Students' accesses during lecture were observed with proxy's log. Circulating and server-client brain writing system were also introduced.

1. INTRODUCTION

Information literacy is skill or knowledge to understand what is information or to handle information. Handling information is classified into following four categories or steps. They are (1) searching or selecting desired information; (2) processing and storing information; (3) creating new information; and (4) sending out information in proper expression. World Wide Web (WWW), which has rapidly spread in the world recently, is expected to facilitate teaching or learning the information literacy. A web page[1] contains links to 531 sites that are using the WWW for teaching and learning. In this paper, two experimental uses of WWW are introduced and several evaluations are made.

2. LOG ANALYSIS

Since the WWW might be regarded as the network linked hypermedia, it is a straightforward idea to use WWW for presentation instead of black board or transparency in technical courses. However, there are few evaluations for the effect of introducing WWW. In order to evaluate that effect, the access pattern of students are analyzed in this paper. An access log of web pages is used for this analysis.

In our lecture room, there are 43 workstations which construct a closed LAN. In order to access outer world or web pages, an proxy server is prepared. All accesses to web pages have to go through the proxy server. Their accesses were logged and used for analysis. Because of the caching function of the web browser, it is difficult to record all accesses perfectly. For example, if a user access to the same web page twice and the duration between two accesses is short, the browser uses a cached data and doesn't access to the web server. Of course, all first access is logged, so the tendency of the student access could be observed.

In the proxy's log, terminal name, access time, accessed URL are recorded. Usually, statistical analysis is made with this log file, e.g. total number of accesses on each page was counted. However, in this paper, the log was used to follow the accesses of each student during the WWW aided lecture.

From our log analyses, three characteristic access patterns were found. They were *serious*, *not-serious*, and *smart* access pattern. In a serious pattern, a student has only accessed to the web pages related to the lecture. While, in a not-serious pattern, a student has never accessed to the lecture pages but accessed to other web pages throughout the lecture time. Furthermore, in a smart pattern, a student has not only accessed to the lecture pages but also other pages suitably. The access patterns might be depend on each student's personality.

In the case of teaching some knowledge using web pages, accesses to non-related web pages are not desirable. However, such non-desirable access is not always bad because the frequent access might mean active behavior to get more information. Actually, the number of accesses in the smart or not-serious pattern is much greater than that in the serious pattern. From another point of view, these log analyses might be used to evaluate not only the student's access but also the lectures or lecturers themselves.

3. BRAIN WRITING

As an another use of WWW in our technical courses, a Brain Writing (BW) is introduced. BW is a collaborative idea writing to create a new idea by several persons. It is an improvement method of brainstorming. Problem finding skill is expected rather than problem solving skill in university courses. WWW might be effective to facilitate such finding skill of student. Another purpose of introducing BW is to establish a discussion space using the computer networks. In usual Japanese courses, students have little chance for discussion and they are only listening to teacher's talk. With reading others' idea and making comment to them, students might begin to communicate each other.

3.1. System Configuration and Procedure

Two types of electric brain writing system have been constructed on computer network and used in our technical courses for a few years. They are (1) circulating and (2) server-client BW. In both cases, before beginning BW, a teacher present some themes to students. In this presentation, students decide their theme about which they will write some proposal. Of course, they can propose new idea independently.

In circulating BW, participants write their idea and then transfer their idea to another through computer networks. As shown in Figure 1, an electric sheet is circulated among participants in fixed

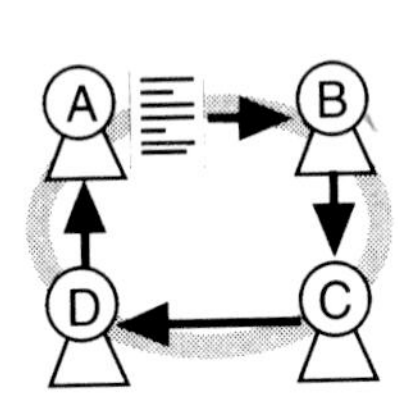

Figure 1.
Circulating BW

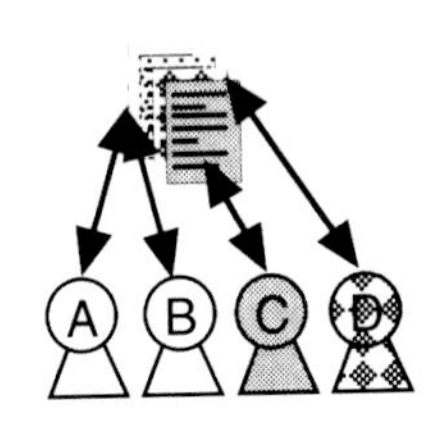

Figure 2.
Server-Client BW

order. Each participant can't select the target idea to which they are going to append their idea. While, in the server-client BW, written ideas are stored in the WWW or database server, and each participant can browse any ideas as shown in Figure 2. If a participant is interested in a certain idea, he or she can append idea to it. In this case, participants can select the target idea freely.

3.2. Circulating BW

Previously, circulating BW was compared with hand writing BW. In both cases, participants wrote their idea simultaneously, and then transferred their idea to another through computer networks or by hand. Our results showed that, in circulating BW, participants wrote more number of letters and took longer time than in hand writing BW. As to the contents of idea [2-4], originality and practical applicability of ideas in circulating BW were evaluated to be lower than that in hand writing BW. Input load with keyboard might be a disturbance to concentrate their attention to the idea production.

In order to reduce the input barrier, especially for the computer beginner, a typing practice is introduced. Furthermore, students are indicated to make their home pages on the WWW. By doing so, many students become to prefer to use computer. After above practice, many students don't dislike to use keyboard of computer.

3.3. Server-Client BW

Recently, server-client system of BW has been introduced and evaluated. Procedure of server-client BW is different from circulating BW. It is explained by following three steps.

1. Registration or Reentry

   In the beginning of the BW, participants are asked to make registration. In the registration, each participant enter student ID number, real name, nickname, affiliation, and e-mail address respectively. After this registration, participants use only their nickname to specify oneself. Once the registration has been made, participants just select their nickname from the list at the reentry.

2. Browsing

   After registration or reentry, participants can browse the articles which have been written until then. Firstly, they browse title only. When they select their interesting title, they can see the contents of that article.

3. Comment or Proposal

   If participants want to add their idea or to make comment, they can do so with web browser. Furthermore, if they want, they can propose new idea. Anyway, after writing, participants go back to the browsing step.

In our experiment, the number of articles written in server-client BW per student is less than that in circulating BW. Originality and practical applicability of server-client BW were evaluated to be lower than we expected. In order to explain this low productivity, flow of written idea was also recorded, that is the relation or structure of each idea was analyzed. In the server-client BW, the average number of added ideas was less than that in circulating BW. Furthermore, the level of idea structure of server-client BW was shallower than circulating BW. Most of ideas written in server-client BW were first idea or proposal, and some of their ideas were similar to each other. As a result, the productivity might decrease.

From this structure analysis, we also find the second idea is important to keep the BW active. If the second idea is good at expanding the first idea, there are many added ideas. In order to make good second idea and to keep the collaborative writing active, there must be a well experienced facilitator in the BW.

One of the most important thing to facilitate writing is selecting theme presented to students. If the theme was popular one, e.g. about famous talent on TV show, the number of browsing was increased. Furthermore, when the theme was familiar to them, e.g. when asked to rename their university's name, there were also many ideas.

In this system, students are given liberty of reporting their ideas, in principle, whenever and from wherever they like. Report arriving time was shown in Figure 3. In this case, the lecture was held from 19:30 to 21:00. Figure 3 shows that most of reporting was occurred around the lecture time and it is distributed to a half day. However, students wrote their idea mainly from the computer room of the university. The total number of reports is 99. 73% of them are reported from the computer room. 25% of them are from other rooms in our university. Only 2% articles are reported from outside of our university.

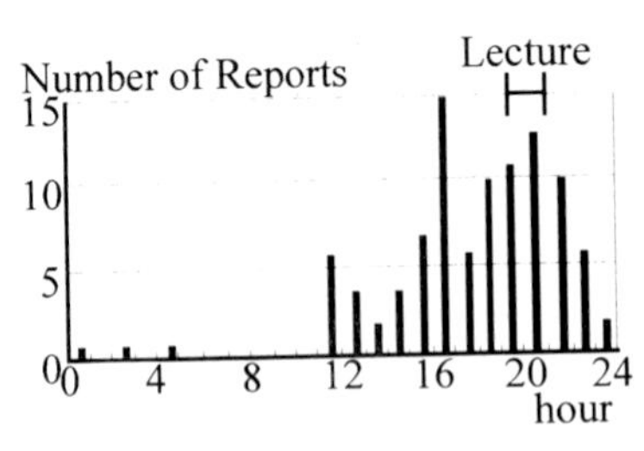

Figure 3. Report arriving time

4. CONCLUSION

Some experimental uses of WWW for teaching information literacy in university courses are introduced and evaluated. Two experimental system of BW have been used in our lecture and improved step by step. Not only the system but also the procedure of lecturing is also important. Further improvement and evaluation of our system will be made. WWW might have potential to cause a revolution in education. We want to propose effective use of WWW in education.

REFERENCES

1. http://www.mcli.dist.maricopa.edu/tl/
2. Diehl, M., Stroebe W., "Productivity loss in brainstorming groups: toward the solution of a riddle", J. Personality and Social Psychology, 53, pp.497-509, (1987).
3. Gallupe, R. B., Bastianutti, L., M., Cooper, W. H., "Unblocking brainstorms", J. Applied Psychology, 76, pp.137-142, (1991).
4. Hymes, C. M., Olson, G. M., "Unblocking brainstorming through the use of a simple group editor", Proc. CSCW'92, pp.99-106, (1992).

A Dynamic Linkage Method for Hypermedia and Its Design Support Tool

Yoshinori Hijikata, Naoki Saiwaki, Tetsuya Yoshida and Shogo Nishida

Department of Systems Engineering, Faculty of Engineering Science, Osaka University

1-3, Machikaneyama-cho, Toyonaka, Osaka 560 JAPAN

Abstract

Hypermedia systems have become increasingly popular as tools for user-driven learning and information retrieval[1][2]. This paper describes a dynamic linkage method which enables the adaptive modification of links in hypermedia dependent on users. It also explains a design support tool to construct network data with the method.

1. Introduction

With the development of broadband communication networks in recent years, hypermedia systems have been come into practical use. Their use have expanded to include the field of information retrieval represented by WWW, CAI(Computer Assisted Instruction), and so on.

The hypermedia system is basically a system by which users can freely scan the information network to retrieve the pieces of information which are structured by nodes and links. With the increasing volume of data, however, some problems are surfacing such as the accuracy of information retrieval, the static characteristic of hypermedia and the design support tool.

Some of these problems are being researched in the form of adaptive hypermedia or user adaptation[3][4]. But many problems still remain. For example, how to acquire user model information and realize its user adaptation, and how to support the network construction that the author and user desire.

This paper describes a dynamic linkage method that is one of the adaptive hypermedia and its design support tool.

2. Dynamic Linkage Method for Hypermedia[5]

2.1 System Outline

Our method realizes dynamic linkage by hiding some links based on the user model information. Both path history information and user parameter information are used in our system. Path history information shows how the user has selected the links in hypermedia network. User parameter information shows the user' s absolute level from a particular viewpoint and it is expressed in the numerical value. For hiding mechanism, we define a class on each node and apply rules expressed as logical equations. A rule that uses path history information is called path rule and a rule that uses user parameter information is called user rule. Furthermore, there exist two types of rules; one is a node rule which can be applied to the special node, and the other is a general rule which can be applied to the whole nodes. In using node rules, the author executes a complex link control on the node. On the other hand in using general rules, the author can describe the character of whole network and reduce his or her labor to construct the hypermedia data.

2.2 Logical Equation of a Rule

Path rule and user rule are represented as the following logical equations.

(1) Path rule

The node path rule is shown as:

$$C_{11}C_{12}\cdots C_{1h}+\cdots+C_{m1}C_{m2}\cdots C_{mh}=D_1,D_2,\cdots,D_n$$

The general path rule is shown as:

$$C_{11}C_{12}\cdots C_{1h}+\cdots+C_{m1}C_{m2}\cdots C_{mh}=C_1^J,C_2^J,\cdots,C_n^J$$

(2) User rule

The node user rule is shown as:

$$P_i**e:D_1,D_2,\cdots,D_n$$

The general user rule is shown as:

$$P_i**e:C_1^J,C_2^J,\cdots,C_n^J$$

C : Class
D : Id of node to be shown
C^J : Class of node to be shown
h : Number of history which will be refered
m : Number of path pattern
n : Number of id or class of node to be shown
P_i : User parameter id
e : Boundary number of user parameter
"**" represents one of the operations that are == (corresponds to =), << (<) and <= (≤).

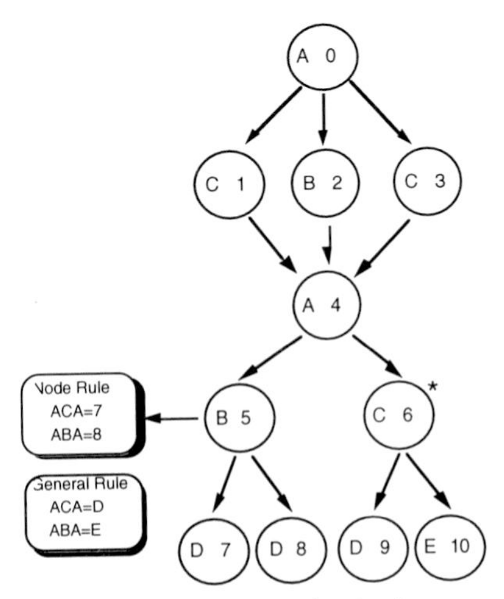

Fig. 1 An example of path rule

2.3 Example of Rule

Fig. 1 shows an example where links are hidden by path rules. There are two general rules and two node rules as shown in the figure. We consider the case when the user have come to node No.5. Node rules are defined on this node. If the user's path history is "ACA", the system shows the link to node No.7 and hides the link to node No.8. Next, we consider the case when the user comes to the node No.6, where a general rule is defined. If the user's path history is "ACA", the system shows the links to the node whose class is "D" and hides the links to other nodes.

The example of user rule and its behavior are also like the one of path rule.

3. Design Support Tool for the Author

3.1 Deadlock and Loop

By introducing the above method, the system itself seems to get complicated and increase the burden of the author who constructs the network data. Especially the behavior of links is difficult to expect. We consider that the following matters would be important.

(1) Deadlock

A Deadlock is the case when the user comes to one node and cannot proceed to any other node since all links are hidden due to the user model information.

(2) Loop

Loops in network may be created by the dynamic linkage mechanism without the user's notice.

3.2 Deadlock Detector

Deadlock detector is a tool to detect deadlocks in the network. We designed a deadlock detector. It pays attention to one node and analyzes the rules and network.

We will explain the case of path rule first. After the author appoints one node, the system starts to analyze the rules and the network and creates a deadlock detection table shown in Table.1. Then it applies the following equation to count the number of paths that may include deadlocks.

$$D = \sum_{i=0}^{C^H - 1} \overline{L_i} \cdot E_i$$

The algorithm for deadlock detector using path rule is as follows.

(1) Scan all path rules to acquire the maximum number of path history H.

(2) Initialize the deadlock detection table.

(3) Scan each node to check if the path in the rule exists in network and if the destination node in the rule exists as the next node. Then it gets the variable L_i and E_i.

(4) Acquire the number of deadlocks by the deadlock detection equation.

In the case of using user rule, the system scans all rules and checks if the destination nodes in the rule exist as the next node.

Table.1 Deadlock detection table

	No.	history	L:link	E:exist
All path history pattern	0			
	.	.	.	.
	.	.	.	.
	.	.	.	.
	H			

No : Line number
history : Path history pattern
L : 1 (The case if the links correspond to the path history exist.)
0 (The case if the links correspond to the path history don't exist.)
E : 1 (The case if the history corresponds to path history pattern exist.)
0 (The case if the history corresponds to path history pattern doesn't exist.)

3.3 Loop Detector

We designed a loop detector to deal with the issue of loops. When the system gets to a new node, it compares the class of the new node with the that of each node in the path history to check the existence of loops. When the length of loop gets longer, the time to detect loops gets longer. To cope with this problem, we also proposes a speedup method with the application of rules. There are some cases that some links are hidden by application of rules. We call this phenomenon "path cut". The example of it is shown as Fig. 2.

It is not necessary to search the network furthermore when path cut occurs, and this contributes to speeding up the loop detection.

The algorithm is as follows.

(1) Follow one link to reach another node.

(2) Scan the node ID in the path history to check if the current path makes a loop.

(3) Renew the path history.

(4) Apply the rules and control the links.

(5) Go to (1).

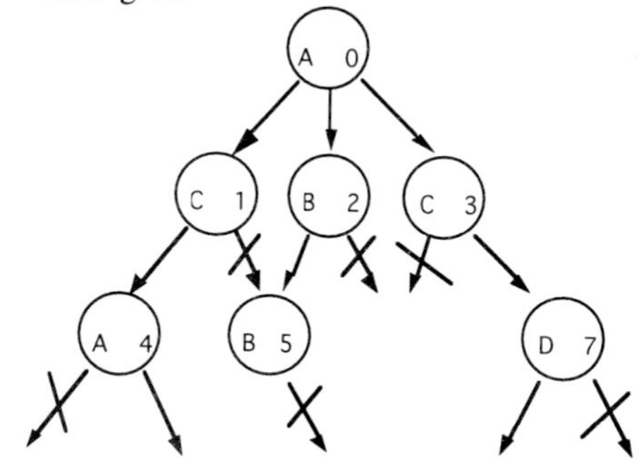

Fig. 2 An example of path rule

4. Prototype System and Evaluation

A prototype system of dynamic linkage for hypermedia was developed based on the above mechanism. The software is written in C language. Though the prototype system is a general-purpose framework, some experiments were conducted on CAI system for evaluation. An example of these experiments is shown in Fig.3. The following results were obtained from the experiments.

(1) It was confirmed that the proposed mechanism worked well, and user adaptability was realized by the method.

(2) The logical expression has a flexibility, and it is easy to change the adaptability mechanism by modifying the logical equations in the rules.

(3) As a problem for the system, we recognized that the design support tool for the author is necessary, since the whole behavior cannot be understood when the scale of the system gets large. Deadlocks and loops were observed in some network data.

A design support tool for the author is now under constructing.

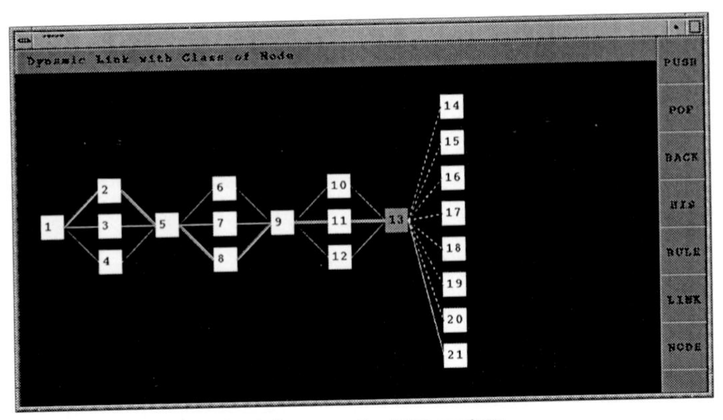

Fig. 3 An example of the system

5. Conclusions

This paper described a dynamic linkage method for hypermedia based on two kinds of user model information and its design support tool. We implemented the prototype system based on this method. According to the evaluation, the proposed mechanism worked well and the phenomenon such as deadlocks and loops were found.

References

[1] Jakob Nielsen, "HYPER Text & HYPER Media", ACADEMIC PRESS, 1989.

[2] Jeff Conklin, "Hypertext:An Introduction and Survey", in "Computer-Supported Cooperative Work", MORGAN KAUFMANN PUBLISHERS, pp.423-475, 1988.

[3] Polle T.Zellweger, "A Hypermedia Path Mechanism", Proceedings of Hypertext'89, pp.1-14, 1989.

[4] Ian Beaumont, "EDUCATIONAL APPLICATIONS OF ADAPTIVE HYPERMEDIA", Proceedings of Interact'95, pp.410-414, 1995.

[5] Yoshinori Hijikata, Naoki Saiwaki, Hiroaki Tsujimoto and Shogo Nishida, "A Dynamic Linkage Method for Hypermedia", Proceedings of Roman'96, pp.519-524, 1996.

Table Expander: hypertabular interaction with query results *

Giuseppe Santucci[a] and Laura Tarantino[b]

[a]Dipartimento di Informatica e Sistemistica, Universitá di Roma "La Sapienza", Via Salaria 113, I-00185 Roma, Italy

[b]Dipartimento di Ingegneria Elettrica, Universita degli Studi dell'Aquila, Poggio di Roio, I-67040 L'Aquila, Italy

We present the Table Expander, a visual tool for the hypertextual exploration of database query results. The tool is based on a data set organizer able to overcome typical problems in the interaction with tabular visualization of query results related to the mapping from the resulting relatiosn to graphical windows.

1. INTRODUCTION

Relational databases are widely diffused and accessed by diverse classes of users. Notwithstanding that the ultimate goal of the information consumer is the result of the seeking process and not the query formulation per se, most Data Base Management Systems front-ends provide effective (visual) support for the information *selection* while lacking adequate support for information *digestion* and *assimilation* (with some exceptions, as [1]). *Data set organizers* are necessary to help users make sense of retrieved information, by making patterns visible, capturing regularities, and allowing the construction of new information patterns from old. A static and a dynamic aspect can be singled out in such tools: the *static* aspect refers to the visualization techniques, while the *dynamic* aspect refers to the modalities offered to interact with the visual structures.

In particular, *tables* are widely used for the visualization of relational query results for their effectiveness in the analysis of structured alphanumeric data sets: they are a familiar data organization, and require low cost graphical representations. Tables may suffer from a number of problems that, if not adequately addressed, sensibly decrease their efficacy. A first kind of problems is *application independent* and pertains to the mapping from logical tables to graphical windows, while a second type is *application dependent* and relates to the content of tables as query results.

In this paper we discuss these two classes of problems along with our proposed solutions for overcoming them by acting mainly on dynamic aspects, in contrast with other approaches more focussed on static ones (e.g., [2]).

*Work partially supported by the Italian Ministry of University and Technological Research under the project MURST 60% "Advanced Interaction with Database Systems"

2. A HYPERTABULAR QUERY RESULT VISUALIZER

In our system, query results are organized as a set of interconnected displays explorable by the *Table Expander*, a hypertextual visual tool, to solve typical problems encountered during the interaction with tables.

Application independent problems When the table dimensions exceed the window dimension, the window is often regarded as a view panning over the table. This is equivalent to having a continuum of adjacent sub-tables, successively disclosed by means of scrolling steps. This may give raise to not meaningful sub-tables whenever a view clips out relevant portion of the structure (e.g., key attributes). The cause for this loss-of-contex effect can be found in the nature of the interaction provided by the scroll bars, purely syntactic because originated in the windowing system. The application must hence have as much control as possible over the interaction, without relying on system-specific tools. Furthermore, cognitive studies on display density show that it is not effective to force too much information into one (overloaded) display. It is preferable to map the total volume of information onto a set of smaller displays, each containing a closely related subset of information, explorable by means of navigation techniques.

Application dependent problems User requests often require the join of two or more relational tables, which, in general, may result in a table that is not in third normal form, exhibiting a tedious repetitions of values. The readability of the table is also decreased by the contiguity of unrelated data, such as attribute values gathered from different tables whose coupling may be meaningless.

To handle the mapping from logical tables to graphical windows, we introduce an interaction structure, called *hypertable*, based on the assumption that information should be presented on demand as a set of interconnected displays, dynamically generated on the basis of: window dimensions, metadata, and suitable exploration paradigms somehow captured by interdisplay links. In our framework this leads to the fragmentation of one (large) relation into a set of linked smaller tables. Our system adopts a visual query language based on a semantic model. The user can formulate queries by specifying either a path or a view on the entity-relationship schema (e.g., the view in Figure 1). The information stored into the semantic schema is exploited for (1) fragmenting the resulting table, and (2) interconnecting such fragments in a hypertabular manner, thus obtaining output presentations richer than flat sets of tuples.

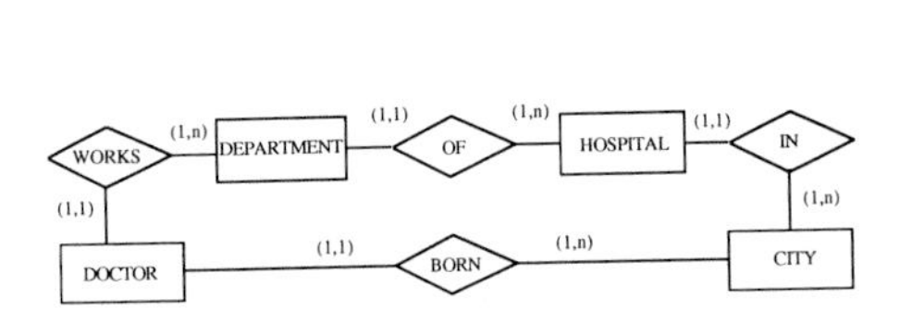

Figure 1. A sample initial hypertable.

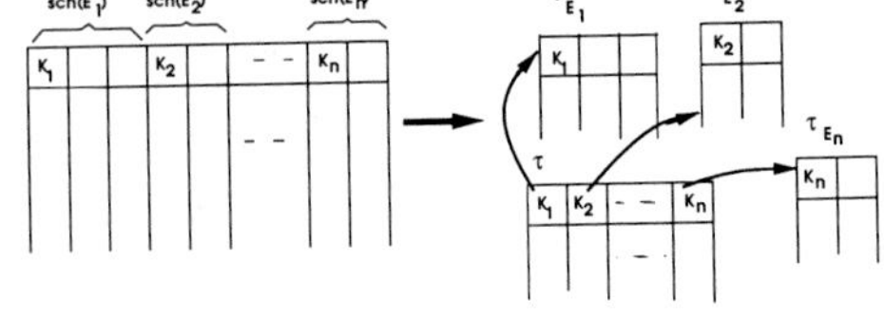

Figure 2. A sample initial hypertable.

A first fragmentation step is devoted to the elimination of the contiguity of unrelated data by clustering attributes belonging to the same entity. This process produces as many fragments τ_{E_i} as the number of entities involved in the query, in addition to a fragment τ representing the join and containing one key attribute for each entity. The result of such fragmentation is illustrated in Figure 2: given a query Q, if $E_1, E_2, \ldots, E_n$ are the entities belonging to a query path/view, in each τ_{E_i} we isolate all the attributes of the entity E_i, while in τ we retain one key attribute for each entity. The admissible interaction is defined by imposing an adjacency structure: each attribute in τ is source of a link pointing to an associated entity fragment τ_{E_i}.

By applying such procedure to our example, the fragment τ is visualized as the table $\mathcal{T}$ in Figure 3, showing many repeated values. The second fragmentation step is hence devoted to the elimination of repetitions, by acting on this table on the basis of the cardinalities of the relationships to devise a partial order of attributes. Note that $\mathcal{T}$ can be iteratively partitioned from left to right, on the basis of repeated values. This is due to the orderings chosen for placing the attributes: those in a one-to-many relationships with all the others (showing the greatest number of repetitions) appear first on the left, and the more we go to the right, the greater is the number of distinct values in a column. It makes sense to provide the distinct values of the first column as starting hints for the exploration of the result: the user may then select one of them to see the associated information. To enforce regularity, and let the user feel free to browse through the entire table, the remaining columns are maintained in the starting fragment $\mathcal{T}_{start}$ (see Figure 4), filled with the one of the tuples associated with the first column values.

CITY	HOSPITAL	DEPARTMENT	DOCTOR
ROME	GEMELLI	SURGERY	SPOCK
ROME	GEMELLI	SURGERY	ABBOT
ROME	GEMELLI	RADIOLOGY	ACTON
ROME	UMBERTO I	ONCOLOGY	TAFT
ROME	UMBERTO I	MATERNITY	UDALL
ROME	SAN GIOVANNI	SURGERY	LIDDEL
ROME	SAN GIOVANNI	SURGERY	LEE
VENICE	SAN MARCO	SURGERY	GALLUP
VENICE	SAN MARCO	SURGERY	GRAVES
VENICE	SAN MARCO	RADIOLOGY	NASH
VENICE	SAN MARCO	RADIOLOGY	BARTON
VENICE	POLICLINICO	MATERNITY	BELL
VENICE	POLICLINICO	ONCOLOGY	BENSON

Figure 3. A table displaying $\tau(T)$.

CITY	HOSPITAL	DEPARTMENT	DOCTOR
ROME	GEMELLI	SURGERY	SPOCK
VENICE	SAN MARCO	SURGERY	GALLUP

Figure 4. The starting fragment.

A method is now needed to allow the selective presentation of new data: for each distinct instance of the first column three additional fragments are defined and linked to it, with schemata $\langle Hospital, Department, Doctor\rangle$, $\langle Department, Doctor\rangle$, and $\langle Doctor\rangle$. Figure 5 shows the fragments associated with the value *Rome*. From the user point of view, these new tables can be viewed as *expansions* of $\mathcal{T}_{start}$. In practice, to expand $\mathcal{T}_{start}$, the user must select one instance of *City* and one attribute on its right (to identify one of the three links). In Figure 6 we see that each column is equipped with a set of buttons: *Expand* selects the starting column for the expansion, *Collapse* closes the performed expansion, and the right (left) arrow opens (closes) the entity fragment associated with the key attribute. Figure 6 shows the effect of the selection of *Rome* and its expansion starting from *Hospital*.

It should be intuitive that such an approach can be applied to any attribute and value of $\mathcal{T}_{start}$ as well as to any attribute and value of any new fragment. It must be observed that, in practice, new fragments are dynamically generated by the interface on-demand, only if the user explicitly asks to access such information.

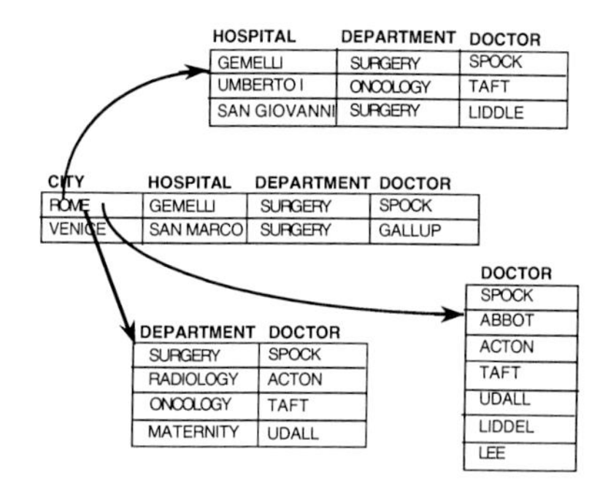

Figure 5. Subtables associated with Rome.

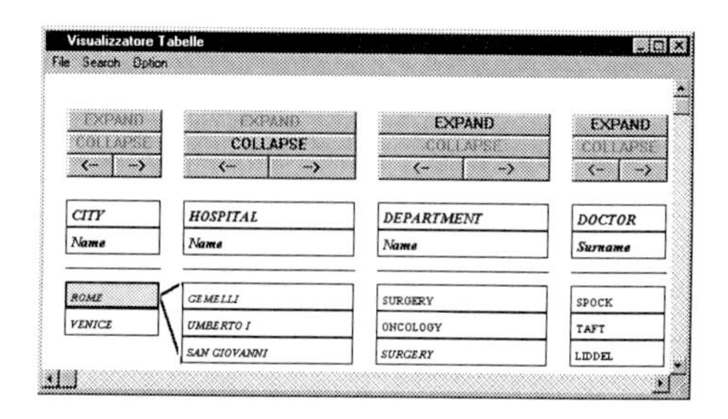

Figure 6. Expansion of the value Rome.

3. CONCLUSION

We presented a hypertextual framework for the interaction with tabular query results. The table content is rearranged as a set of smaller interconnected displays that avoids repetitions of values and contiguity of unrelated data (typically produced by the join of two or more relational tables). Result exploration is performed by following interdisplay links, which activate table expansions or contractions.

A prototype of the system has been implemented under the Unix operative system, using the C++ language and the XVT graphical toolkit, a set of libraries available for different platforms (Dos, Mac-OS, and Unix) that ensures a high degree of portability.

REFERENCES

1. C. Ahlberg and B. Shneiderman. Visual information seeking: Tight coupling of dynamic query filters with starfield displays. In *CHI '94*, pp. 151–164, ACM, New York, 1994.
2. R. Rao and S.K. Card. The table lens: Merging graphical and symbolic representation in an interactive focus + context visualization for tabular information. In *CHI 94*, pp. 318–322, 1994.
3. G. Santucci and L. Tarantino. To table or not to table: a hypertabular answer. *Special Issue on Information Visualization, Sigmod Record*, 25(4), 1996.

HyperBBS : Towards Smooth Navigation

Eikazu Niwano

NTT Information and Communication Systems Laboratories
Urbannet Mita 405, 3-10-1, Mita, Minato-ku, Tokyo, 108, Japan
E-mail niwano@mita.isp.ntt.co.jp

HyperBBS, real-world-oriented, agent-based and distributed hypermedia asynchronous group communication/navigation system which provide the integrated/fuzzy organization and retrieval functions is introduced. This paper describes the key concept " *Smoothness* ", and the AGCS/MGS model/architecture which is based on it. Tests conducted since 1992 are also reported.

1. INTRODUCTION

Recently, the need for messaging systems such as the Internet has grown. This has brought about a need for navigation systems such as the WWW.

Some navigation systems incorporate a cross between the hypermedia aspect of the WWW and Usenet News, such as HyperNews[1] and Netscape Corporation, which developed the most popular Web browser. Agent functions, called software robots, have been studied to support WWW information space searches and to make the network more intelligent.

These approaches are good, but are limited in terms of organizing and retrieving objects from points of view of integrity, flexibility, as follows

1) These systems are not structured and mapped based on the real world, so it is not easy for most people to understand the structure and to access or handle real world objects through the virtual world
2) Objects in the network are not organized enough to be easily retrieved
3) Target objects for organizing/retrieving are restricted and not expanded to service places such as BB(Bulletin Board) category or BB enough
4) Retrieval systems such as browsing, searching and filtering have not been integrated or synchronized enough
5) These systems' support of fuzzy matching is insufficient, so it is difficult to get the informa tion that has actually been requested or to know which results from the retrieval are important.
6) There are not enough agents to support the retrieval and organization system

Consequently, users have been unable to browse the network resources as easily as they desire. It is important to overcome these problems and be able to move around *co-existed* real and virtual world and use service/information objects in them *smoothly*, without being aware of boundaries or differences. The objective of this work is to construct such world in which real world can be augmented and extended by virtual world and vice versa, so the integrated worlds can be called "*Blended World*". Our approach, constructing the virtual world that mirrors the real-world and integrating organization and retrieval functions supported by agents is based on the philosophy " *co-existence and smoothness* ". The navigation system we developed, called *HyperBBS*, is a real-world-oriented, agent-based and distributed hypermedia system on the basis of *AGCS/MSG (asynchronous group communication system/messaging system) model/architecture*. These were designed and implemented in 1990-1991and slight modification has bee done until 1993. It has been tested many times since 1992. This paper describes its key concept "*smoothness*" as well as its design model and test results.

2. WHAT IS SMOOTHNESS ?

The key design concept and philosophy " *smoothness* " (differentiableness in mathematics) means not only continuity (seamlessness) but also non-singularity (*naturality*). It is very important for users to be able to be aware of and perform the following actions and operations to get to the desired services/information object, even if they don't know how to about the manner in advance.

The principle of design was based on *behavior pattern of the " body "* to get information in real world such as " move from a object to desired object based on those relation (link) and see, operate ", because it is one of the most primitive and so natural manner of human behavior patterns. And also " *smooth structure* " must be incorporated into the world to support it.

3. AGCS/MGS MODEL AND ARCHITECTURE

The following overview of the AGCS/MGS model/architecture shows how this concept is incorporated into the system and you can see how this system overcome the above problems respectively. First, the organization of the service/information objects is described towards *smooth structure*, followed by a discussion on how to retrieve the organized services/information objects towards *smooth navigation*. Roughly speaking, organization is important for service providers and retrieval is important for service users. The following design approach is based on the simple idea that all objects can be organized and retrieved by links between them. Finally, intelligent agents, which support the organization and retrieval, are discussed. More details on the functional and information models (object and structure models) , an overview of a simple method to incorporate the concept " *co-existence* " into the system, and more details on the intelligent agents are described in the previous papers[2,3].

3.1 Organizing Objects

To be retrieved easily objects in the virtual world must be organized and self-organized well. The most important and basic method for doing this is to implement them mirroring real-world objects. The domain, service object (*Category* : management unit, such as BB category, for the following linked activity objects , *Activity* : management unit, such as BB, for the following linked information objects), information object (*Item* : unit of information, such as BB message) and service providers/users object (*Entity* : user entity) are introduced as AGCS objects. The international standard of directory system X.500 DS, was applied to implement and manage those objects except for Item. That is, AGCS service objects are managed as X.500 hierarchical objects and have unique identifier that are recognized anywhere in the world as well as real-world. It allows users to move around the network with the same amount of flexibility as in the real world. Moreover, this directory system manages not only static attributes such as names, identifiers and keywords which characterize the objects, but also dynamic attributes such as keywords which appeared in a domain frequently. Information objects are managed at every service object Activity, in AGCS. As key techniques for organizing all these objects, dynamic (direct) /static (indirect) link, bidirectional link, multiple link, various types of link (upper[out] / lower[in], reference, relation) was used. The user can create static links easily by clicking the object of the peer (this was described later) as the destination (anchor). Moreover AGCS is based on the user agent and service agent model, and integrated and synchronized with those of X.400 MHS, so it can be synchronized with objects in interpersonal space[2,3]. This integrated messaging system model is called *MGS*. All protocol was described using ASN.1. The self-organization method is described in section 3.3 as part of intelligent agent technology.

3.2 Retrieving Objects

From the point of view of operating retrieval techniques such as browsing, searching and filtering, these techniques are basically the same, so they can be considered together and operated in unified manner. All browsing action is performed as " *list* " or " *search through a filter* " for linked objects, that is, any service user can move from one object to any searched linked object through a link. If a user wants to know referenced, related or pre-selected (as destination list) object, he/she can know some listed/searched objects as candidates for destinations using the "*navigator*". At that time, he/she can " *peer* " a desired object using the navigator only for

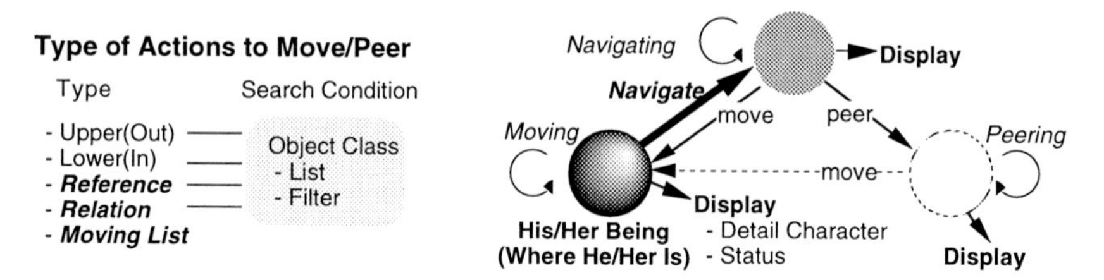

Figure1. Type of Actions and Transition Model to Move/Peer

looking at the object (investigating) and can " *move* " to it if he/she likes. The actions, type of links and search conditions, and the transition model to move and peer are shown in Fig. 1. As you see, it is important for user not to look at the objects, but to be able to act and handle objects as if he/she exists in virtual world actually. That is he exists in and move/peer around the world! Also search and filtering has only difference between its search domains concerning " creation time ", past or future respectively.

Any class of object and attribute value can be stating point for navigation and any class of object can be target or anchor. For example from name attribute value displayed by " detail " (shows the attributes of a object), " status (shows the status of a object) " menu, user can get appropriate " Entity " object as the creator's information or any service/information object created by the same creator. And "text" cut from a message, window can be starting node, because it can be regarded as keyword for search. In those two cases user can pre-define its search path as one of user's parameter (home, destination list, search path list, search condition for search by related link,). All navigation can be done as " select node and select action type (navigation/ link type). And targets can be filtered by pointing search conditions such as type of action (link), class of object (ex. Category,..) [target/anchor] and filter (base object [starting point], depth of link, keywords, creation time , etc).

For fuzzy retrieval, a weighted matching algorithm using similarity, based on fuzzy theory, vector space model, between sets of attribute was studied. With thesaurus it was implemented to calculate similarity between keywords and between sets of keyword. Users can attach free keywords or restricted keywords, by using fuzzy retrieval of thesaurus, to any objects easily . At initial stage in 1991, the algorithm is simple one, but it was improved as an extension of method proposed by E.Niwano[4] until 1993. Navigation from a object/text to objects through "relation" link or from filter using keywords to objects are performed based on this algorithm.

3.3 Intelligent Agents

Many type of intelligent agent such as *collection agent, extraction(weighted indexing[keyword]) agent, classifying agent, linking agent, filtering agent, media conversion agent, alerting agent* which are synchronized were implemented to self-organize and augment the objects. For example, a specified Category can collect information from other sites, extract indexing keywords, classify the information into lower Activities, filter it and forward it to other activities, or forward/alert it by mail/facsimile/pager/telephone after converting if accordingly .

4. USER INTERFACE

This section focuses on the guidelines for building UI for HyperBBS. The UI is based on the transition model described in section 3.2. How to represent it is one of the most important concerns for incorporating the concept of "*smoothness*" into the system. The guideline are

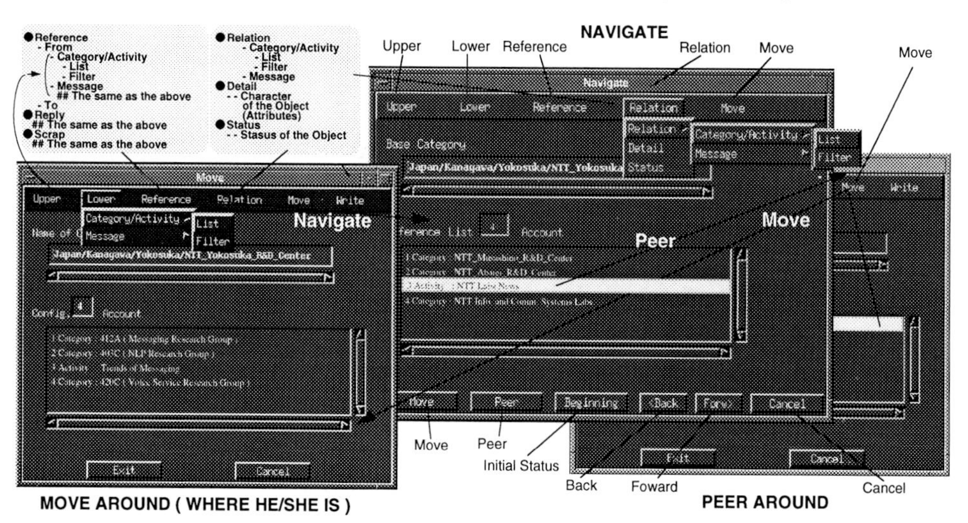

Figure 2. An Example of User Interface of HyperBBS

1) the objects based on hierarchical structure/physical metric (upper/lower links) should be always displayed on a "move (where he/she is located)" window
2) the objects based on other logical metric/distance (reference/related link) should be displayed on a "navigate" window (called "navigator") for the user's assistances
3) the objects which are only to be looked at should be displayed on a "peer" window
4) these windows should be presented as different windows
5) all actions for moving must always be visible and executable for any displayed functional/ information objects

Figure 2 shows a single example of a UI designed and implemented in 1992-1993 , as second version. It represents the concepts simply and guideline 1)- 5) were satisfied on the UI. . The windows for Category, Activity and Item are the same in first version of UI and these were separated in second version.

5. EVALUATIONS

Since 1992, the HyperBBS navigation system has been evaluated at a number of exhibitions and small-scale field trials, as well as a commercial service trial. These exhibitions and trials include the NTT AI&HI Forum 1992, a six-month trial of the VI&P Experimental Network 1993 and the NTT International Symposium 1994. The usability of the real-world-oriented structure and UI was tested as J-Mail, a small-scale Telescript™ agent system for three months in 1995, and at a commercial service trial as Paseo, a large-scale Telescript agent system for 11 months in 1996-1997. Evaluations for the NTT AI&HI Forum and the NTT International Symposium were conducted by interviewing visitors, while evaluations for the other trials were conducted through questionnaire.

The results of the interviews and questionnaires indicate that people think highly of the system, particularly of its real-world-oriented structure and linking functions. J-Mail and Paseo provide a more user friendly, realistic user interface[5,6] using MagicCap™ GUI platform, and its efficiency was confirmed by many service users. Among the various functions of this system, its static/dynamic linking and filtering facility is very highly praised.

The tests, however, also indicated that the following issues must be addressed. The complete system, including its precise operation, administration, and maintenance system, has not been tested on a large-scale, except for Paseo. Thus, a large-scale field study should be done for the rest of the functions, especially the linking functions. In the past motif-based interfaces were popular, but technology has improved recently, and more user friendly user interfaces such as Virtual Reality/VRML can be incorporated into the system. Some people have suggested using realistic multi modal interface with representation that can be command-based or menu-based and can provide for the various skill's of users. These are object-oriented systems, but except for J-Mail and Paseo, they have not been implemented with agent/object-oriented language yet. Thus, the rest functions need to be implemented by agent/object-oriented language and tested.

6. CONCLUSION

This paper describes the HyperBBS real-world-oriented, agent-based, distributed hypermedia asynchronous group communication/navigation system. We found that the key concept "smoothness" can be achieved by using manner of "move" to get services/information in real world and that HyperBBS overcome the problems pointed out in section 1 well.

The framework of this system can be expanded to general agent-based computer services, so currently this system is being redesigned based on these HyperBBS, Paseo works and current object-oriented technologies.

Acknowledgements

Special thanks are due to Kazuhiro Sora and Yoshikazu Onozato for their contribution to building and testing the X.500 directory system. I would also like to thank Kiyomi Maeda, for his comments on the draft version of this paper and for his encouragement, and Minoru Ohyama, Minoru Nakamura and Yasuo Sakama for their encouragement, all previous or current research director of NTT Information and Communication Systems Laboratories.

REFERENCES

1. HyperNews : http://union.ncsa.uiuc.edu/HyperNews/get/hypernews.html
2. E.Niwano et al. "Design of HyperBBS: Integrated Computer-Based Bulletin Board Systems that Realize Distributed Hypermedia Information Network", IPSJ SIG Notes, 92-GW 2-5 (1992) (in Japanese)
3. E.Niwano, "Intelligent Mail: Towards Seamless Access and Communication", Proc. of 15th International Symposium on Human Factors inTelecommunications (HFT '95) (1995)
4. E.Niwano," Quantification of Compound-Words Similality Using Fuzzy Matching Functions and Morphological Analysis Technique ", Proc. of the 45the Annual convention IPSJ (1992) (in Japanese)
5. E.Niwano et al. "Towards Construction of CyberCity with Agents in Symbiosis - Through Experience of Paseo Services - ", The 3rd symposium on Social Information Systems, The Society of Social Information Systems(1997)(in Japanese)
6. E.Niwano et al, "Evaluation of Large-Scale Mobile Multimedia Communication System Built on Telescript Agent Platform - The Paseo Service -", Proc. of IEEE Multimedia '97 (1997)

The 'history' as a cognitive tool for navigation in a hypertext system

Sissel Guttormsen Schär

Institute for Hygiene and Applied Physiology, Swiss Federal Institute of Technology, CH-8092 Zürich, Switzerland
email: guttormsen@iha.bepr.ethz.ch

1. Introduction

A current issue is how to optimise the use of hypertext programs. Hypertext programs are recognised by a network of information units, which open the possibility for free navigation by links between the different units. Free navigation allows users to apply different search methods according to their individual cognitive structure. This freedom is at the same time the biggest disadvantage of hypertext systems, because it causes disorientation problems and implies much cognitive load for the users. To deal with this problem, the idea has been to add structure to the hypertext system. The solution is normally to structure the hypertext document itself (e.g. by offering a good overview and indexes), and to offer different navigation tools which give the users more structured ways of navigating in the system (e.g. guided tours, bookmarks, keyword search). We suggest a supplementary approach: a balance should be found between structuring the hypertext system itself and the possibility of inducing cognitive structure in the users of the system. The users should be provided with cognitive tools which enhance the utilisation of a search mode suitable for the task they are solving.

Unstructured navigation in hypertext systems has been described as browsing; characterised by an explorative style. When browsing, users follow an idea in a rather unstructured way. Browsing may be a part of the disorientation problem as long as it is used as a search method for tasks demanding a more structured approach [1, 2]. Experiments in the field of human computer interaction show that the users can be influenced to learn in a certain mode by characteristics of the user interface [3-7]. Applied to hypertext navigation, the disorientation problem could be reduced by inducing the users to search in a more explicit and structured mode. Given that a minimum of external structure is offered, an explicit search mode can help the users focus their attention on the relevant aspects of the information. The mental model of the information is formed by the search mode. To put it simpler: if the users search in an unstructured and in mode it will be difficult to build a consistent mental model from the information. If their search mode is structured and in their control, the mental model will be accordingly structured.

An experiment was design to investigate the significance of an interface feature likely to induce a more explicit search mode.

2. Method

A hypertext system offering information about ecological use and production of paper was extended with a history of the users actions. Normally, when such histories are offered only a list of the pages previously visited is displayed. As there were several methods of arriving at a particular page, the history implemented in this hyper-text program also displayed which route the subjects had chosen. For every action the history displayed two columns: (1) *Your selection....* and (2) .... *has taken you to page.* An example of the information in the history can be seen in Figure 1.

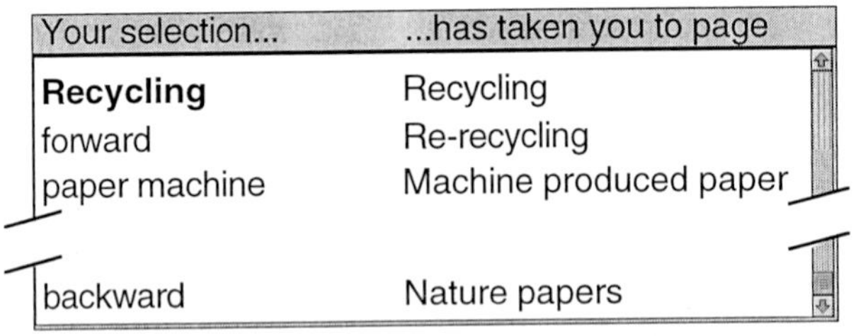

Your selection...	...has taken you to page
Recycling	Recycling
forward	Re-recycling
paper machine	Machine produced paper
backward	Nature papers

Figure 1: the information displayed in the history window

The hypertext system consisted of 148 pages, organised into 32 chapters. It offered both structured navigation (forwards/backwards within each chapter and index) and unstructured ways of navigation (free navigation by hyperlinks). All the subjects actions were registered in a logfile.

Two groups of subjects (15 in each group) were given 18 different search tasks (12 tasks were purely to search for single information, 6 tasks were inference tasks which required the subjects to collect information from several pages to be able to answer the questions). After each task the subjects were given a multiple choice question with four alternatives related to the search task.

3. Results

The users presented with the extended history searched in a more structured mode than the users who did not see the history. The differences in search mode were significant both as frequency of actions (the history groups visited less pages before they found the information), in thinking time between each action (the history group spent more time thinking between each action), and in the use of the external structure offered by the program (the history group used the structured search methods, like forward/backwards buttons and the index more than the non-history group, which navigated more by hyperlinks). Further, the results show a tendency for the history group to have acquired better knowledge about the domain; the subjects in this group had less wrong answers on the multiple choice questions. The main effects of interface can be seen in Table 1.

The complaints about the system also differed between the interface groups. This effect can be seen in Table 2.

Table 1
Main effects of interface: the effects on search mode and knowledge

	Interface			
Variables	History	Non-history	F	p
Mean number of pages per task	11.62	13.62	4.26	0.039
Mean thinking time (s) between each action	19.74	17.78	24.57	0.000
Hyperlink searches per task, (% of total actions per task)	19.93	26.82	23.12	0.000
Forwards/backwards per task, (% of total actions per task)	30.47	25.57	8.28	0.004
Wrong answers (% of total, Fischer's exact test)	17.41	22.22		0.097

Table 2
Main effects of interface: complaints about the system

	% Complaints	
Complaint	History	Non-history
None	24	10
Loss of orientation	29	70
Text (content/meaning)	41	10
Design	6	10

4. Discussion

The rational was that an extended history would induce a more explicit and structured search mode, and hence act as a cognitive tool because it would help the subjects focus on the relevant information. The expected consequence was less disorientation problems in the group with the extended history. The results support this.

The history as a cognitive tool resulted in a more structured searching mode with the result of reduced browsing to find the information. The extended history has reinforced the subjects use of the structured search methods they were offered and made them avoid the unstructured one (searching by hyper-links).

The cognitive structure induced by the extended history resulted in consequent use of structure in all parts of the system.

Several measures of performance indicate that the extended history provided a benefit for the users. This is seen both as a more consistent way of navigating and by less complaints about disorientation problems. The history group needed to navigate less in the system to find what they searched. The analysis of the complaints about the system between the groups indicate more cognitive control for the group with the extended history. The subjects in this group experienced less disorientation problems than the non-history group. Table 2 shows that the major complaint from the non-history group concerned loss of orientation, while the subjects in the history group complained more about system features, such as the content of the text and design aspects. The more structured mode did not result in higher costs on total time. Secondly, the knowledge measure, although not significant, encourages the interpretation that there was deeper knowledge acquired when the extended history was presented.

It is plausible to conclude that the history of the subjects actions in a hypertext system could act as a cognitive tool. Many other interface and software features can hypothetically also act as cognitive tools and induce different interaction modes. Which parts of the interface can induce an interaction mode must be evaluated according to each different program. In general, it is quite easy to induce the users of software to different interaction modes, and we have shown how easy the interaction mode affects the performance. Consequently, the cognitive effect of the interface on the interaction style should be paid more attention when developing software.

REFERENCES

1 R. McAleese, "Navigation and Browsing in Hypertext," in *Hypertext theory into practice*, R. McAleese, Ed. Oxford: BSP Intellect books, 1989, pp. 7-43.

2 D. Canter, R. Rivers, and G. Storrs, "Characterising user navigation through complex data structures," *Behaviour & Information Technology*, vol. 4, pp. 93-102, 1985.

3 S. Guttormsen Schär, "The influence of the user-interface on solving well- and ill-defined problems," *International Journal of Human Computer Studies*, vol. 44, pp. 1-18, 1996.

4 S. Guttormsen Schär, C. Schierz, F. Stoll, and H. Krueger, "The effect of the interface on the learning style in a simulation based learning situation," *International Journal of Human-Computer interaction*, in print.

5 K. Kunkel, M. Bannert, and P. W. Fach, "The influence of design decisions on the usability of direct manipulation user interfaces," *Behaviour & Information Technology*, vol. 14, pp. 93-106, 1995.

6 C.-I. Trudel and S. J. Payne, "Reflection and goal management in exploratory learning," *International Journal of Human-Computer Studies*, vol. 42, pp. 307-339, 1995.

7 G. B. Svendsen, "The influence of interface style on problem solving," *International Journal of Man-Machine Studies*, vol. 35, pp. 379-397, 1991.

parameters belonging to the same plant component are gathered around the mass center of the nodes, and then enframed with a box representing the component.

It may be hard to recognize a raw configuration generated as above due to crosses of links, too small node-node or node-link distances and too small link angles. Such visual interference is reduced by iterative move of nodes by the following algorithm.

1. Crosses of links are eliminated by searching for a node position that makes less crosses around the current node position.
2. Node positions are normalized to fit into the display screen.
3. If any node-link distance is less than a particular threshold δ, the distance is expanded by moving the node apart from the link.
4. If any node-node distance is less than the threshold δ, the distance is also expanded by moving one node apart from the other.
5. If any link angle is less than a particular threshold ϕ, the angle is expanded by moving one end node.

Move of nodes described above is carried out under the condition that no new crosses are generated. Since it may be impossible to eliminate all interference, the above procedure is iterated till the total degree of interference defined by

$$f = \frac{\min d_{ij}}{\delta} + \frac{\min D_{kl}}{\delta} + \frac{\min \theta_m}{\phi}$$

becomes minimum, where d_{ij}, D_{kl}, θ_m are respectively node-node distance, node-link distance and link angle. The threshold distance δ is now set to the square root of screen area per node, and the threshold angle ϕ to 90 degrees.

The whole process of knowledge model configuration is performed as follows.

1. Arrange nodes by PCA.
2. Normalize node positions to fit into the display screen.
3. Enframe nodes by component.
4. Reduce visual interference in terms of component.
5. Free nodes and reduce visual interference in terms of node.
6. Enframe nodes by component.
7. Reduce visual interference in terms of component again.

3. EXPERIMENT

An experiment was conducted to show the appropriateness of the second principle of the visualization that a graph configuration with less visual interference is superior for understanding the substance of display. Eight subjects were recruited and they operated a model plant simulation in two scenarios of unfamiliar component failures, but experimental runs for one subject were unsuccessful. The model plant is a simulation including five tanks, 21 valves and pipes connecting them. The subjects were first instructed with a text explaining the mechanism of plant behavior. Then they were instructed by oral explanation on the relation between the physical causality path and goals-means relation of standard operation procedures. During the latter instruction, stepwise animation on a graphical display of the relevant knowledge was shown to the subjects. The subjects were divided into two groups and each group was shown a display generated with or without reduction of visual interference. No specific names of plant components and parameters were used during the instruction in order to prevent the subjects from using a priori knowledge.

The operation board of the plant simulator includes indicators of tank pressures, tank levels, and alarm windows that inform abnormal component conditions in tank pressure or tank level. The subjects were asked to recover the plant from abnormal component conditions and to turn off alarms as soon as possible. Each operation run for a failure scenario took 20 minutes.

Since the subjects were asked to suppress alarms, total time intervals that any alarm was active were compared as a performance measure. The result is shown in Table 1, which shows the subjects instructed using a display with interference reduction showed superior operation performance ($F(1,5)=10.3$, $p<0.05$). The effect of failure scenario and the interaction between the failure scenario and display type were insignificant.

Table 1
Total time intervals (minutes) when alarms were active.

Subject	Without interference reduction		With interference reduction	
	Failure 1	Failure 2	Failure 1	Failure 2
1	45.6	50.2	7.7	17.8
2	44.6	32.3	19.9	30.5
3	34.8	22.5	22.6	19.4
4	22.0	49.4		
Average	36.8	38.6	16.7	22.6

4. CONCLUSION

A method was proposed to visualize a knowledge model used for plant operation in unfamiliar situations, in expectation that such a method will contribute to development of a training support system. The knowledge is displayed by arranging nodes of a graph according to the strength of relation between nodes and then rearranging them to reduce visual interference. It was demonstrated experimentally that the assumption of the visualization method proposed was appropriate. It is, however, open to a future research to prove that showing this type of display is effective for plant operators to acquire the relevant knowledge.

REFERENCES

1. J. Rasmussen, IEEE Trans. Systems, Man, and Cybernetics. MSC-13 (1983) 257.
2. Y. Furuhama, K. Furuta and S. Kondo, European Symp. Cognitive Science Approaches to Process Control (1995) 137.
3. D.H. Jonassen (ed.), Semantic Networking as Cognitive Tools, in Cognitive Tools for Learning, pp. 19, Springer, Berlin (1992).

A WWW Environment for Visualizing User Interactions with Java Applets

Jan Stelovsky and Martha E. Crosby

Department of ICS, University of Hawaii, 2565 The Mall, Honolulu, Hawaii, USA

Abstract

Experimenter's Workbench is an environment that supports the design, administration and analysis of experiments in the area of user-computer interaction. This environment is currently being rewritten in Java to support distributed experiments with World-Wide-Web-based applications. This paper describes Java interface and focuses on one of the visualization tools – the event-time diagrams.

1. Introduction

The Experimenter's Workbench is an open set of tools that is designed to support the experimenter in the tasks of conducting a controlled experiment. The workbench consists of three parts:

a) the "Administrator" simplifies the design and administration of the experiment,
b) the "Observer "collects the subject's interaction data, and
c) the "Analyzer" is a set of tools that visualize the collected data.

The Experimenter's Workbench has been originally implemented in HyperCard. The initial version was able to automatically enhance an arbitrary HyperCard application by the Observer code. In order to enhance the usefulness and applicability of the Experimenter's Workbench, we are currently rewriting all its components in Java. This approach ensures that the Observer can be inserted into an arbitrary Java program, whether a stand-alone application or an applet embedded in World-Wide-Web (WWW) site.

The Experimenter's Workbench is conceived as an open environment that can be extended in a several ways: new data collection forms can be added to the Administrator's inventory, the Observer can be enhanced by new data reduction procedures and the Analyzer can be augmented by new visualization tools.

After we briefly describe Administrator's functionality, we present the Observer's interface and aspects of its implementation. Then we describe in detail one of the visualization tools and discuss future enhancements and their application.

2. Administrator

The Administrator consists of a set of forms that are presented to the subject at the onset of the experiment. Typically, they elicit information about the subject, such as the demographic data or tests to determine the subject's cognitive style. The forms are implemented as an inventory of "cards" from which the experimenter can choose the most appropriate ones. Moreover, the experimenter can create entirely new forms or redesign forms by adding, deleting and rearranging items on the form. The new and modified forms can be added to the inventory. The Administrator's facilities are described in more detail in [1].

3. Observer

The Observer works behind the scenes during the experimental session and records the sequence of the user's interactions with the target application. The recorded data forms a log file of interaction events. Consider a simple application where the user can click the buttons "Play", "Pause", and "Stop", as well as quit the application by closing its window. Figure 1 shows a sample log file produced by the Observer.

```
subject:Subject 1
date:20-Mar-97
0,start
2757,java.awt.event.ActionEvent[ACTION_PERFORMED,cmd=Play] on button0
11869,java.awt.event.ActionEvent[ACTION_PERFORMED,cmd=Pause] on button2
12067,java.awt.event.ActionEvent[ACTION_PERFORMED,cmd=Play] on button0
12257,java.awt.event.ActionEvent[ACTION_PERFORMED,cmd=Pause] on button2
15098,java.awt.event.ActionEvent[ACTION_PERFORMED,cmd=Play] on button0
21234,java.awt.event.ActionEvent[ACTION_PERFORMED,cmd=Stop] on button1
23811,java.awt.event.WindowEvent[WINDOW_CLOSING] on frame0
23812,stop
```

Figure 1. Sample log file

In contrast to HyperCard, which defines a rigid hierarchy of GUI elements, Java programs can have an arbitrary complex structure of use interface elements defined in numerous source files. Therefore our implementation currently requires that calls to Observer's methods be manually inserted into the target Java programs. The Observer has been implemented in Java as a class whose main method records an event of user's interaction with one of the GUI elements (AWTEvent). This class also provides, among other methods, methods to start and stop a recording as well as a method to record non-standard events as strings.

The most recent version of the Java Development Kit (JDK 1.1) introduced an improved event model where the handling of events is delegated to "listener" classes. This model simplifies the inclusion of calls to Observer's methods in target applications. Figure 2 illustrates how a "listener" class was enhanced to record an event. (Note that the Observer's methods are "static" to simplify their usage across numerous classes.

```
class Command implements ActionListener {
  ...
  public void actionPerformed (ActionEvent event) {
    Observer.record (event);
    ...
  }
```

Figure 2. Sample code segment that calls Observer's recording method

A more "object-oriented" way to implement the Observer would be to offer a subclass of each of the JDK's "adapter" classes and require target programs to use these new classes. We rejected this variant as too error-prone and cumbersome as it duplicates numerous standard API (application programmer interface) classes and therefore increases the loading time of WWW applets. The most elegant option – the direct monitoring of the event queue – cannot be used for WWW applets because of security restrictions imposed on applets.

The event log files created by the Observer typically contain sequences of elements that can be combined without sacrificing the correctness of the experiment's outcome. For instance, a sequence of keystrokes within a short time span will usually be interpreted as a single action of typing a string. Combining such event sequences can reduce the size of the log files and is performed after the completion of the experimental session.

4. Analyzer

The analysis tools provided by the Experimenter's Workbench are intended to give the experimenter a variety of ways to analyze and view the recorded data. One simple tool summarizes the data collected from a group of subjects. It can be used, for instance, to define categories of events, to count the events in each category and to produce preformated data files that can be imported into conventional statistical packages for further analysis.

Analysis of the atomic low-level events is often inadequate for experiments that involve more complex problem solving tasks. Therefore we envision an "Event Editor" that allows the experimenter to structure a sequential log file into a hierarchy of higher-level events. The tool will work like an inverse outliner: a sequence of events can be grouped and represented by a new, more abstract event. For instance, a new "reorder" event can be created to encompass the subsequent activation of "Cut" and "Paste" menu items. A sequence of "reorder" actions can be in turn subsumed under a "rephrase sentence" operation. Like an outliner, the Event Editor will be able to collapse and expand the higher-level events to offer views of the user's interactions at varying levels of abstraction.

The Analyzer's tools also help in visualizing the individual user's performance. The user's interaction can be, for instance, represented as a time-event diagram. Figure 3 shows the diagram generated from the log file in Figure 1. The events are arranged on the y-axis in the order of their appearance in the log file from the "begin" of the session at the bottom to the "end" of the session at the top. Notice that the events in the center of the diagram are very close together. This often occurs when the subject quickly corrects a mistake or pauses to plan the next step. The "logarithmic" scaling of the time spans in between time events (Figure 4) which stretches the short intervals while preserving the relative differences helps in viewing such event clusters. The "equidistant" option shown in Figure 5 is useful when only the sequence of the events is important. The experimenter can also select a time interval to view an interesting portion of the session as illustrated in Figure 6.

Since the initial order of the events on the event axis may differ among the subjects, the experimenter can rearrange them by dragging their labels. A label can be also dragged out of the diagram to exclude an unwanted event from the visualization.

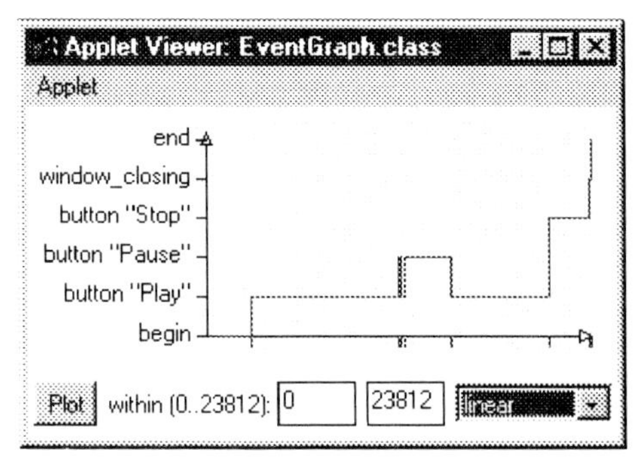

Figure 3. Time-Event diagram

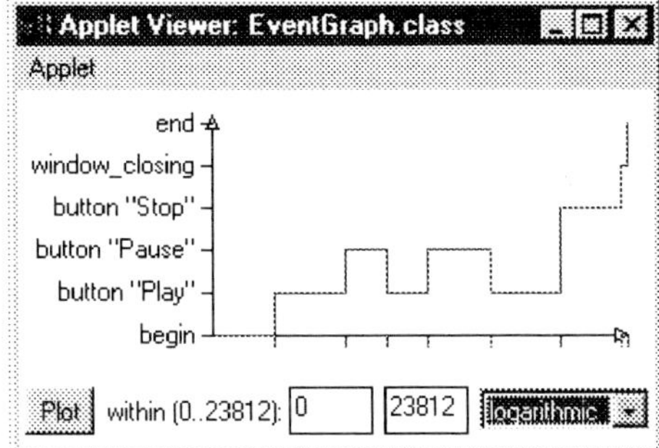

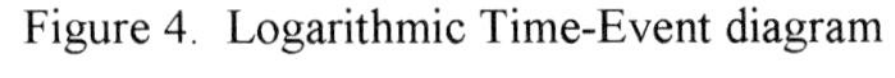

Figure 4. Logarithmic Time-Event diagram

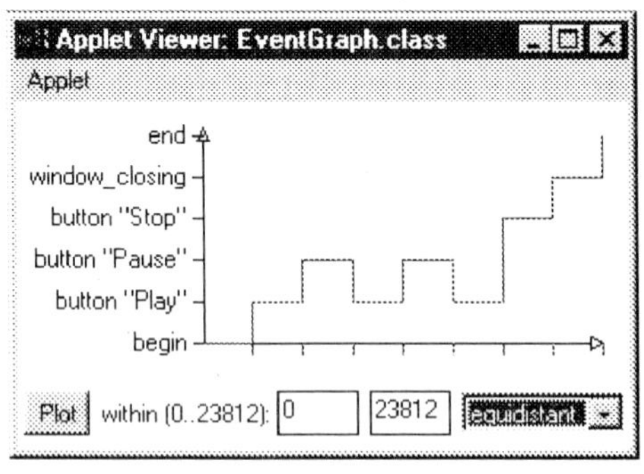

Figure 5. Equidistant Time-Event diagram

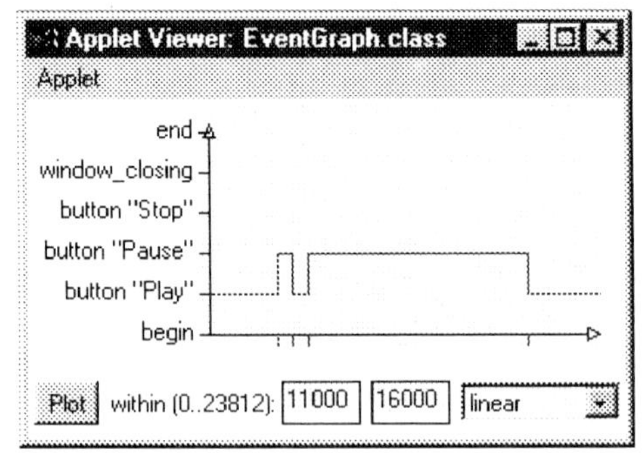

Figure 6. Zooming in Time-Event diagram

5. Future Enhancements and Conclusion

As the Administrator is essentially a GUI editor, we first intend to provide only the main features in its Java version. We plan, however, to offer an automatic parsing of Java source programs to include calls to the Observer's methods as well as log file in form of serialized time/event objects to permit direct replay of user's interactions. Our emphasis will be on the design and implementation of further visualization tools, such as 2-dimensional event and site maps, which can be animated to capture the progress to the experimental session.

Since the Observer can be incorporated into arbitrary Java applet the Experimenter's Workbench now supports distributed administration of controlled experiments, where the recording and postprocessing of log files can be performed on the client's computer and transmitted to the WWW server. With some additional effort the experimenter could call the Observer's methods from within JavaScript to record the users' interaction with a WWW site and determine their browsing activities.

References

1. Crosby, M. and Stelovsky, J., Towards Computer-Supported Testing: A Multimedia Environment for Evaluation of Distributed Multimedia User Interfaces, Proceedings of the IASTED/ ISMM International Conference, Palo Alto, CA, pp. 173 - 175, 1995.

A System for Exploring Information Spaces

Masanori Sugimoto*, Norio Katayama, and Atsuhiro Takasu
Research and Development Department, National Center for Science Information Systems,
3-29-1, Otsuka, Bunkyo-ku, Tokyo, 112, Japan

In this paper we describe a system called COSPEX (COnceptual SPace EXplorer) for assisting users in their exploration of information spaces. By using it interactively, users can effectively discover necessary information and new information which is useful for their intelligent activities.

1. Introduction

Owing to the recent development of computer network, a great number of information systems have been constructed all over the world. They are distributedly and autonomously managed, and information spaces composed by each of them have become more tremendous and complicated. It has become a difficult task to discover necessary information effectively through traditional information systems. However, by making highly use of them we can find new information which can be useful triggers for our intelligent activities. In this paper we propose a system called COSPEX (COnceptual SPace EXplorer) which assists users in their exploration of information spaces. Its features are that it provides users with an universal visual query interface for multiple databases and that it can visualize the semantic relations in a database. COSPEX proposes a new method for enhancing the interactions between information systems on the Internet and users. COSPEX is now applied to a query interface for digital libraries [1].

2. Configuration of COSPEX

The overview of COSPEX is shown in Figure 1. COSPEX is available on a WWW browser which can execute Java applets. It is composed of three different modules, Visual Query Editor (VQE) [2], Conceptual Space Visualizer (CSV) [3], and Private Digital Library Manager (PDLM). Figure 2 shows VQE. When users access a certain database, its schema information is automatically loaded and visualized on the query interface. Users can drag and drop icons (*society*, *journal*, etc.) and construct a query expression in the form of network structure as shown in Figure 2. The visual query expression in this figure means "There exists an *article*, it is written by an *author*, and the author has a name, 'Yoneda Tomohiro' ." This expression is converted into a logic-based query expression and issued to the database server. Figure 3 shows CSV. When users want to know the

*Currently a visiting researcher in Department of Computer Science, University of Colorado at Boulder, CO, USA.

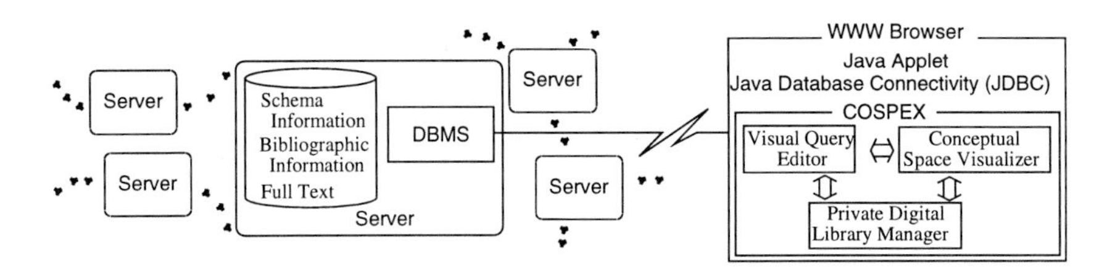

Figure 1. Overview of COSPEX

semantic relations among objects in a database, CSV visualizes object-attribute relations in a two-dimensional metric space. In this figure the relations among articles (objects) and their keywords (attributes) are visualized simultaneously and users can know that, for example, "J.Dorn@Expert S.." and "E.J.Carr..@Hypermed.." have a relation with each other by the term "knowledge" and so on. When CSV configures a space, it first counts the frequency of each attribute in each object (in this case, term frequency in each article) and uses an automatic indexing algorithm [5] in order to create attribute vectors of objects (in this case, keyword vectors of articles). They are analyzed by a statistical method called dual scaling method [4] to visualize the relation among objects and attributes. Retrieved information by using VQE and CSV is stored and managed in PDLM.

3. Usage and Effects of COSPEX

Figure 4 shows an example of a search scenario with COSPEX. Suppose a user wants to know the research on "agent." Firstly he/she searches *articles* by the term "agent" (Figure 4(a)). However, too many articles are usually retrieved when he/she accesses a large scale information system like a digital library. If he/she tries to configure a visualization space in order to know which articles he/she really needs, it may be difficult because too many objects (in this case, articles) and attributes (in this case, keywords) are visualized on the space. A user changes his/her query target and searches *journals* which include the term "agent" (Figure 4(b)), and visualizes the relations among objects ("ACM TOD" etc.) as shown in Figure 4(c). Then he/she can find a new search term ("cooperation") from the space and reflects it to his/her query expression (Figure 4(d)). Again a user visualizes the relations among objects (Figure 4(e)) and finds journals which seem to include his/her necessary articles. He/she reconstructs his/her query expression by specifying journal titles ("IEEE KDE" etc.) to search *articles* (Figure 4(f)) and the relations among articles are visualized. A user can reconfigure a space by selecting objects interactively (Figure 4(g)) in order to know the detailed relations among them (Figure 4(h)). By using COSPEX like this scenario, users can finally find their necessary or new information.

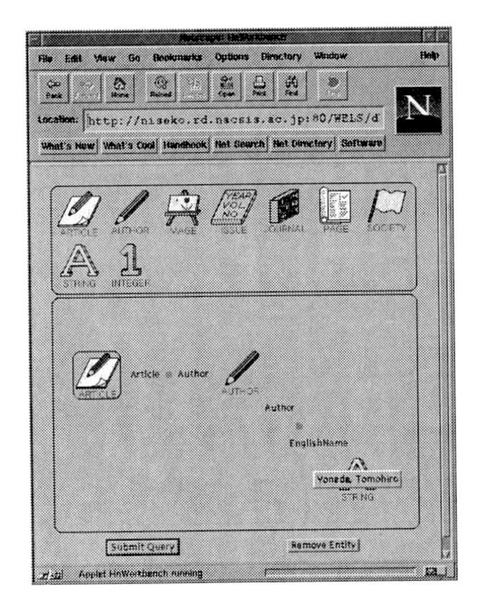

Figure 2. An example of VQE

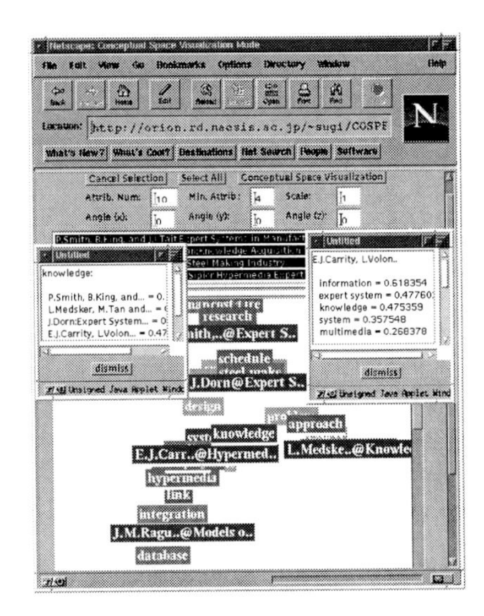

Figure 3. An example of CSV

4. Discussions

The effects of COSPEX are summarized as follows:

- Users are not puzzled with the difference between the schema of each database. They can access various databases on the Internet and easily construct visual query expressions with icons.
- By visualizing the relations among objects and attributes simultaneously in a metric space, users can effectively discover necessary information and new information which they haven't noticed so far. For example, users can find new articles which they would like to read, or new keywords which are triggers for their research activities.
- When too many objects are retrieved and it is difficult for users to know the relations among them by visualization, they can interactively change their query target, for example from *article* to *journal*.

By using VQE and CSV complementarily and interchangeably as shown in Figure 4, users can effectively explore information spaces.

5. Conclusions

In this paper a system called COSPEX for exploring information spaces was described. We have several plans to extend its functions and evaluate it through user studies.

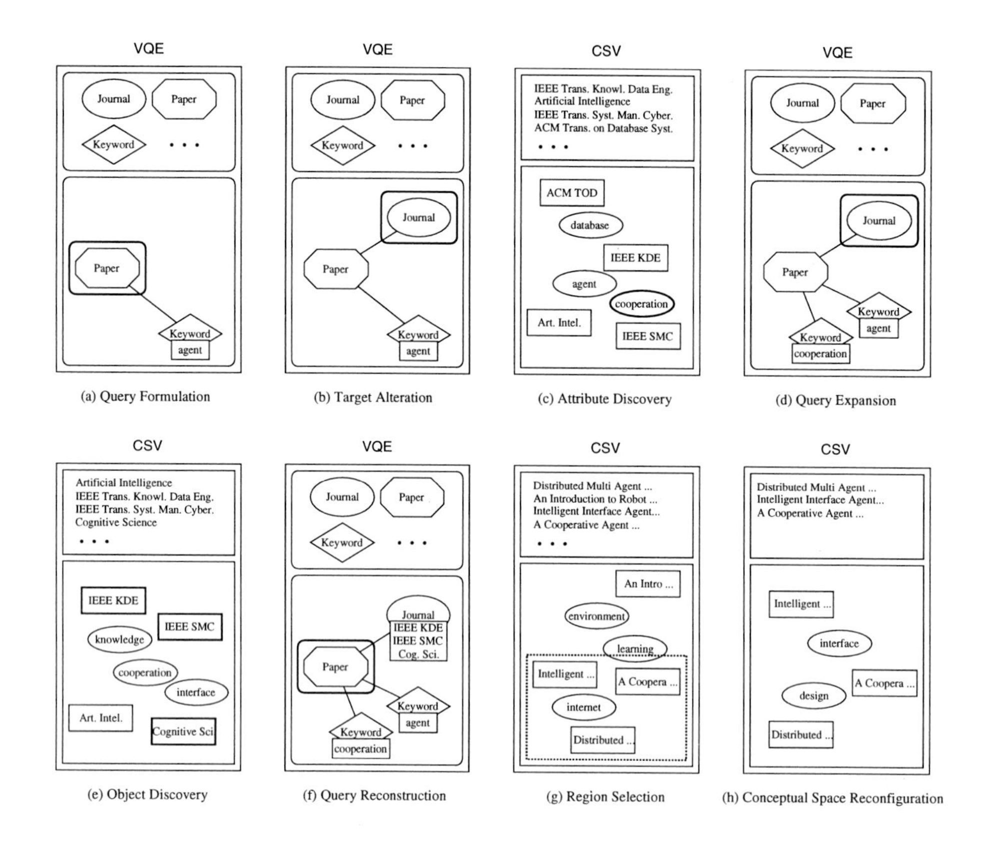

(a) Query Formulation (b) Target Alteration (c) Attribute Discovery (d) Query Expansion

(e) Object Discovery (f) Query Reconstruction (g) Region Selection (h) Conceptual Space Reconfiguration

Figure 4. An example of a search scenario with COSPEX

REFERENCES

[1.] B. Schatz and H. Chen, Digital Library Initiatives, *Computer*, vol. 29, no. 5, (1996).

[2.] N. Katayama, M. Sugimoto, and J. Adachi, A universal query interface for heterogeneous distributed digital libraries, In *Proc. of the 7th Int. Workshop on Database and Expert System Applications (DEXA'96)*, pp. 332 – 339, (1996).

[3.] M. Sugimoto, K. Hori, and S. Ohsuga. A Document Retrieval System for Assisting Creative Research, In *Proc of the 3rd Int. Conf. on Document Analysis and Recognition (ICDAR'95)*, pp. 167 – 170, (1995).

[4.] S. Nishisato, *Analysis of Categorical Data: Dual Scaling and Its Application*, University of Toronto Press, 1980.

[5.] G. Salton, *Dynamic Information and Library Processing*, Prentice-Hall, 1975.

Evaluating the Usability of Application Tools in Transferring Multimedia Information across ATM Communication Networks within The Fashion Industry in Europe.

Garry Patterson & Roisin Donnelly

Faculty of Informatics, University of Ulster, Jordanstown, Co. Antrim, BT37 0QB, Northern Ireland.

Abstract
This paper discusses the implications for the usability and evaluation of user trials as a result of the completion of the Fashion-Net project. The Fashion-Net concept provides an opportunity to illustrate the benefits that may be derived from applying the latest technological developments into working working practices of the personnel involved in the industry. It explores the packaging together of a number of separate applications to form one complete system thereby facilitating the exchange of information within the fashion industry. One of the aims of the Fashion-Net/Temin project was to use the latest advances in telecommunications to transfer information e.g. forecasting/trend data, sketches, photographs and video clips between European countries. The method of transferring information was using ATM, the most recent development in Telecommunications Technology.

1 Introduction

The rationale for the project was to support the Fashion and Design industries in geographically remote areas across Europe. More specific objectives were firstly, to establish a series of trials which consist of usage scenarios and to run these over a period as long as possible; secondly, to enhance the capabilities of the Fashion-Net infrastructure through the introduction of new multimedia oriented collaborative applications; and thirdly, by increasing the size and scope of the trials, through additional end-users and trial sites. In addition, it was aimed to run several interactions of the scenario trials over an extended period in order to establish firstly, the minimum bandwidth requirements and network usage patterns; secondly, the impact of the network on HCI and usability; thirdly, to establish the network requirements of distributed databases, co-operative design tools and videoconference packages and the resultant HCI and usability issues; and finally, to further enhance and verify the results of the cost/benefits analysis.

Furthermore, it was important to verify and enhance models of the organisational impact of network services on the fashion design domain; to develop and apply methods for establishing the business cost/benefits of introducing the technological innovations proved in the project; to evaluate and specify the cost benefits associated with implementing the proposed Broadband services throughout the European Fashion Industry, and/or within the Fashion-Net Common Interest Group. Specific objectives concerning the application tools were as follows. For FINS, it was to test an online database (currently CD-ROM) providing sourcing material for designers and to develop a Business Plan of a future Pan-European Multimedia Database for Fashion, Textile and Apparel industries. For Scribble, it was to trial and evaluate joint sketching and annotation application with audio-conferencing support, enabling designers separated over distance, to see and edit each others designs and develop a Business Plan of a future Pan-European CSCW service for Fashion, Textile and Apparel industries, in the context of an integrated CSCW/Database/Videoconferencing toolkit. The objectives for ShowMe were to test a video conferencing tool to support and aid co-operative working.

2 Fashion Design CSCW

Fashion design utilises diverse media in a wide variety of tasks. Designers, for example, create images of designs that are discussed and modified during meetings with other designers, managers, buyers, and clients. Designers receive inspiration for their designs and obtain design materials and components from a variety of sources. A typical design scenario might involve an initial search for materials and inspiration, followed by design and design appraisal. Frequently, these activities are realised collaboratively with design colleagues or mediated through staged group meetings. The three applications would support all these kinds of task and interactions between participants separated by distance [1].

As part of the usability engineering process, the designers were asked how they would normally perform each of the main tasks - i.e. sourcing information, sketching and communicating with colleagues. These questions revealed that none of the designers used computers as extensively as the project required [2]. Sourcing was normally conducted through attendance at shows searching and browsing through journals. There was some evidence that the Internet was starting to be used for tasks of this nature. Sketching of designs was normally done alone, with pen and sketchpad. Final draft ideas were shown to colleagues as story boards. Scribble goes beyond this, requiring designers to work together from the initial/concept stages of design. Communication between designers and clients was mediated through conventional methods such as telephone, visits and faxes. Within the Fashion-Net toolkit (Scribble, ShowMe and FINS) all interfacing could be achieved from one workstation. This represents a radical departure from current work practices [1, 2].

3 User Engineering Methodology

All trials took place between the sites with European ATM network conditions. The users at each site were four freelance designers. Each group undertook the three aforementioned design scenarios, under three different bandwidth conditions (2, 4, 8 Mbsec). Show-Me and Scribble ran as shared applications, FINS ran separately on a user machine. Each session lasted for approximately one hour. Restricting the bandwidth creates a bottleneck of cells during complicated activities. These are rejected or passed on later. The lower the bandwidth the more cells are rejected, especially during complicated tasks such as joint drawing and image transfer. The scenarios had been constructed to ensure that all tasks (e.g. talking, drawing, image transfer) occurred under each bandwidth [3]. The relationship between the tasks and bandwidth activity could then be determined.

It was hypothesised that the applications would run better over the higher, unrestricted bandwidth conditions, and that users would notice and be able to quantify this difference in their answers to questionnaires. The following measurements were taken preceding, during and anteceding the actual trial. The Usability Engineering techniques used were:-

1. The measurement of pro-computer attitudinal scores were completed before and after participation in each trial [4]. If the applications performed well it was hypothesised that pre and post attitudinal scores would remain constant throughout the trials.
2. Usability questionnaires relating to each application were completed after each of the trials, in order to ascertain whether there had been any noticeable effect on certain facets of system usability which could be attributed to bandwidth differences (e.g. speed, response rate, video and audio quality) [2].
3. All sessions were video recorded and subsequently subjected to a task and breakdown analysis [5].

4 User Engineering Results

Following the trials the key issues of evaluating the usability of the applications. The evaluation methodology selected plays a key role in the development of future trials within the fashion industry. Co-operative evaluation techniques were adopted and focused on four issues. Firstly, the usability of the applications, in terms of Human Computer Interaction (HCI) and co-operative functionality, through the use of questionnaires and feedback. Secondly statistical analysis of the quantitative and qualitative data relating to the cost benefit assessment of issues such as time saving and co-operative flexibility. Thirdly, usage patterns of the applications across the network and the derivation of specification details and marketing arguments in relation to multimedia teleservices and finally selective demonstrations and interviews with members of the fashion industry to ascertain the potential for multimedia services [3].

4.1 Application Bandwidth Results

The evaluation of Scribble and FINS running under optimum conditions, highlighted a number of areas where system improvements could be made. These results themselves will again be fed back to the system developers. In terms of the analysis of bandwidth requirements, an application which rates poorly in terms of usability when run over optimum conditions will probably fair even worse when the bandwidth is restricted. Regarding usability over restricted bandwidths, the teleconferencing system (ShowMe) and Scribble were rated least favourably in the lower bandwidth conditions (1 and 2Mbsec). Although FINS ran independently of bandwidth condition, images had to be exported from it into Scribble. This was difficult for subjects and sometimes resulted in system crashes (especially on the lower bandwidth conditions). The perceived usability of Scribble and FINS therefore co-varied. The mean scores that each application achieved, under respective bandwidths are discussed in [1]. The highest score attainable was 7. As FINS was run separately on the machines, it should not have been effected by bandwidth configuration. The scores, although disappointingly low, would indicate that usability is better at 4 and 8Mbsec.

4.2 Pro-Computer Attitudinal Scores

The pro attitudinal computer scores dropped slightly during the trials, regardless of the order of bandwidth presentation. The post trials scores were usually more negative than the pre trial ones. This was especially true of the 1 and 2 Mbsec bandwidth conditions. A summary of the results from the questionnaires are illustrated in [1], and show an underlying trend that the users' overall attitude to computers dropped furthest in the 2 and 4 Mbsec conditions. The highest score attainable was 140.

4.3 Breakdown Analysis Results

For the breakdown analysis, a representative subject pair was selected from each experimental condition. It was hypothesised that firstly, as the bandwidth is decreased, the number of breakdowns attributable to the system will increase; secondly, that as users were trained in the use of the systems, the 'user stuck' figure should remain fairly stable throughout the trials and thirdly, the nature of the speech breakdowns would change as a result of the bandwidth configuration.

Hypothesis 1 was upheld. More system breakdowns occurred in the lower bandwidths. They were also of greater duration (18 minutes as opposed to 90 seconds). For example in 1Mbsec condition - despite attempts by the system administrators - ShowMe remained unusable with poor sound quality, bad echo, subjects could only hear each other when they shouted through the door! In 2Mbsec bandwidth condition, the trial ended with both machines crashing. The users spent more time on Scribble, but had to wait 5 minutes for images to be transferred, during which time they were unsure as to whether they had performed the operation correctly or whether the system had crashed. In the higher bandwidths subjects were able to spend more time concentrating in the task, and seemed more confident overall.

Hypothesis 2 was upheld. Users experienced the same sort of problems on all the conditions, despite having been trained in the use of the systems e.g. they found navigating in FINS difficult, could not remember how to cut and paste images, or how to draw and delete with

Scribble. These problems were exacerbated by the lack of help facilities in both systems. These confirmed/exemplified issues raised in the usability evaluation.

In Hypothesis 3 the number of speech breakdowns increased with increased bandwidth. This could be attributable to the following reasons: firstly, in the higher bandwidths the trials went on longer; secondly, it was easier for subjects to engage in conversation in the higher bandwidth conditions; thirdly, there was a qualitative as well as quantitative change in the nature of the speech breakdowns, they became more task oriented, rather than just asking for repetitions.

4.4 Cost Benefit Assessment

The cost benefit assessment showed that major issues in the introduction of this technology would be inter and intra-organisational benefits; resulting from increased communication, concern over security and confidentiality (which is a major issue in the Fashion Industry), retraining and recognition of new skills.

5 Conclusion

The Trials have successfully demonstrated the potential benefits of delivering multimedia services into the European Fashion and Textile Industries. Each of the applications have been successfully integrated into the Fashion Design Scenarios and from these important information such as usage patterns, minimum system and network requirements have been identified. This information can now be used by network designers and operators, equipment and software vendors, in order to help provide optimum take-up of broadband services within the European Fashion, Textile, and related Industries. The trials planned within the project were completed with success, both at a local and international level. From the Technical point of view, all the work carried out during the trials provided a very important insight about the usage of network and transport protocols like TCP/IP over a transfer medium as ATM. Some very important conclusions about the performance of such protocols over ATM were drawn up. This information will be useful for further experiments in multimedia distributed environments over ATM. These are summarised on the World Wide Web [6].

From the Usability Engineering perspective, all applications provided feedback to the intervening users and will be directly used in current (FINS) and future commercial software tools (Scribble), as well as other European funded initiatives (Ten-telecom, Telematics and ESPRIT). The project concludes that due to the effort to date in the development of both a technological and evaluation methodology, they envisage a future in successful Pan-European commercial Tele-service and Teleco-operation for Fashion, Design and Marketing Intelligence in Europe.

References

1. Patterson, G., & Donnelly, R., 1997: 'Usability Engineering Within Broadband Telecommuncations for Tele-Services and Teleco-operation in the European Fashion Industry', *Proceedings of The 1997 International Conference on Industry, Engineering and Management Systems*, Florida, March.
2. Fashion-Net/TEMIN, 1996, *Fashion-Net/TEMIN B3004 Report*; Trans-European Network Integrated Broadband Communications, Evaluation Report to EU, Brussels.
3. Patterson, G., & Murphy, M., 1996: 'Multimedia within the Fashion Industry in Europe', *Interface to Real and Virtual Worlds, Informatique '96*, 161-169.
4. Badagliacco, J.M., 1990, Gender and Race Differences in Computing Attitudes and Experience, *Social Science Computer Review*, 8, 1, 42-63.
5. Urquijo, S.P., Scrivener, S.A.R. & Palmen, H., 1993: 'The Use of Breakdown Analysis in Synchronous CSCW System Design', *Proceedings of The Third European Conference on CSCW*, Milan, September, 281-293.
6. Scrivener, S: *http://dougal.derby.ac.uk/fnet* , 1996.

Re-engineering a Complex Network Interface

M. Lucas[a], J.F. Meech[b] and C. Purcell[c*]

[a]Motorola, Mahon Industrial Estate, Blackrock, Cork, Ireland

[b]University of the West of England, Frenchay Campus, Coldharbour Lane, Bristol, United Kingdom

[c]Motorola, 16 Euroway, Blagrove, Swindon, United Kingdom

1. INTRODUCTION

The GSM Cellular Network is a second-generation digital cellular telephone network which supports a range of both voice and data services. The Motorola GSM Operations and Maintenance Centre (OMC) provides cellular network management capabilities for Motorola radio base station equipment in a GSM cellular network through a workstation based graphical user interface.

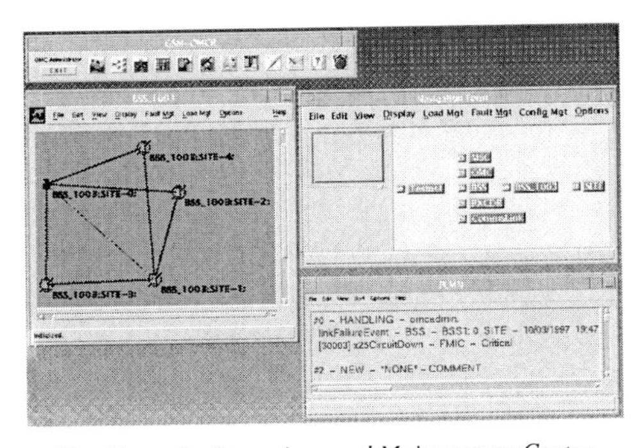

The Motorola Operations and Maintenance Centre

The GSM market is highly competitive and one of the key considerations when choosing an equipment supplier is that the OMC must be easy to use. Feedback from customers and internal field support representatives identified opportunities for improvement of system usability. As a result, Motorola launched a major effort focusing on improving the usability of the Motorola OMC beginning with an investigation centred around on-site studies of usage, using interviews and questionnaires to collect data.

* Partly funded by the Kingdom Department of Trade and Industry through the Teaching Company Scheme.

2. INVESTIGATIVE TECHNIQUES

2.1. Customer Site Visits

The decision to visit the users at their workplace was a change from previous requirements gathering activities, which had been limited to discussions with engineering staff and management rather than the end-users. The study goal was to observe and gain an understanding of end users performing real tasks in their normal work environment as well as the OMCs context of use within the customer organisation. This would allow interface improvement to be driven towards supporting actual users work on a real network.

The customer sites included in the study were chosen to provide a diverse range of customer operations in both the technical and cultural domains. Therefore a mixture of public and private cellular network operators were visited in Africa, Asia and Europe.

A site visit plan was devised which outlined the goals of the visit, the methodologies to be used, the data gathering focus and visit logistics. During the initial set of visits, it was found that the plan had to be modified based on experience.

As an example, one of the goals was to obtain a detailed step by step description of each routine task. For this reason, each interview was video-taped, so the task descriptions could be recovered from video analysis. However, this level of detail proved impossible to obtain as the interviews were controlled by the end-user rather than the interviewer. Also, a number of tasks would impact network operation and could not be performed without detailed on-site planning and customer consent.

2.2. Contextual Interviews

A methodology was required which would make the customer site visit more user-centred than those conducted in the past. The contextual inquiry technique [2] was chosen as it provides a method for focusing customer interviews on the areas of interest while allowing the user to lead the discussion.

The first stage of conducting a contextual interview is to hold focus setting meetings with interested members of the development team to determine interview topics. These meetings proved to be quite difficult to manage since the contextual inquiry technique was new to both the usability team and the developers. Initially, a brainstorming session was held, followed by an attempt to construct an affinity diagram of the brainstorm topics. This proved to be unsuccessful since the developers were skeptical of the affinity diagram approach, and were unwilling to attempt it. Eventually a set of focus topics were arrived at through discussion. The overall focus was to be on the routine tasks that the users performed on a daily, weekly or monthly basis. Other focus topics included the context of the OMC within the customer organisation, users backgrounds and the users work environment.

Two types of interviews were planned prior to the site visit. The first interview was with the operations manager for the cellular network system and the second would be with the users themselves. The purpose of the operations manager interview was to gain an overview of the overall customer organisation, the network management groups tasks and its deliverables. An area of particular interest was t he information flow between the network management group and the rest of the customer organisation. During the interview, a work flow diagram was constructed to illustrate this information. This initial meeting proved to be a rich source of background information on the customer organisation which helped to prepare for the user interviews. Notes were taken during the meeting using pen and paper.

After the operations management interview, user interviews were held. The user interviews were recorded on video-tape and an audio recording taken as backup. Paper notes were also taken. The interviews were conducted at the users workplace and started by explaining the focus topics and making the user feel at ease. The users were then asked to show how they worked with the system. It was found that the value of this interview was dependent on the individual users, with most being very forthcoming and providing an excellent overview of their work. However, on one site, single user interviews proved difficult to arrange and a multi-user round table discussion was held instead. There was a marked deference to the opinions of the team leader during this discussion.

These sessions provided a large amount of operational data which could never have been discovered in laboratory situations. For example, many operators work in conditions where constant telephone interruptions are common and multiple systems are used. Another interesting discovery was that advanced users had developed tools to enhance the systems functionality.

The contextual inquiry technique often proved difficult to conduct. Whereas most software applications such as word processors are driven by the user, the OMC is primarily driven by events on the network which the users must react to. During some inquiries there would be very little happening with the network but in others network events and telephone interruptions left the users with little time to explain what they were doing.

2.3. Usability Questionnaires

The Software Usability Measurement Inventory (SUMI) [1] is an established psychometric questionnaire which is simple to administer and provides a benchmark of system usability against a large number of computer applications (mainly PC office automation tools). It has a well-defined analytical technique associated with it which has been refined over a period of time. It has also been translated into a number of different languages.

A prerequisite for the use of SUMI is to perform a "context of use analysis." This analysis results in an additional short companion questionnaire designed to assess user training and experience in both the system problem domain and computers in general.

A system-specific questionnaire was also developed which was aimed to obtain detailed data on various aspects of the user interface such as interface artefacts, task frequency and customer likes, dislikes and suggestions.

Most users completed the SUMI questionnaire in less than 15 minutes. By asking the users to complete the questionnaire while the usability team was on site, a high rate of return was achieved. The OMC questionnaire was left behind for completion since it takes longer to complete. A correspondingly lower rate of return was achieved for this questionnaire.

The OMC questionnaire underwent some modification during the customer visits as parts of it proved to be cumbersome to answer. The part of the questionnaire which underwent the most change was where detailed responses on common tasks were required such as levels of stress felt during a task. In the end this part of the questionnaire was focused only on task frequency.

Analysis of the SUMI and OMC questionnaire responses in some ways corroborated the feedback which had been previously given by customers. The main areas of concern identified were task efficiency, system feedback and user control. We were surprised to find that overall system usability was judged to be better than we expected given the nature of some feedback from the field.

3. CONCLUSIONS AND FINDINGS

In general, the study was considered successful by both Motorola and the customer. A new perspective of the overall system was gained by focusing on end users rather than equipment specifiers and managers. This information is being used to drive future systems design.

A key to this success was the use of a site visit plan and the setting of focus topics. It was found that the plan evolved to some extent following experience and may actually have to be adapted on site during the visit. Experience with focus setting meetings also showed that you should not expect unqualified acceptance of new methods from the development community.

The contextual inquiry proved an excellent technique for user centred data gathering. However, since the interview is user driven it may prove difficult for the interviewer to cover all focus topics in a restricted time. The use of video proved of limited use as it provided little extra information over paper notes given the intensive effort required for analysis. It was found that even short visits provided large amounts of data, making exhaustive analysis difficult in a schedule driven production environment.

The SUMI questionnaire allowed successful benchmarking of system usability. The OMC specific questionnaire gave an indication of task frequency and allowed requests and suggestions to be gathered. However, at first it proved difficult to administer but evolved into a more usability format over time.

Both operations managers and end users appeared to be very pleased to be included in the study. The operations manager meetings were extremely useful for gaining an understanding of the systems "indirect" users and its context within the customer organisation. For end users this was the first time their opinions had been sought.

4. REFERENCES

[1] Kirakowski, J. *The Use of Questionnaire Methods for Usability Assessment*. Available at http://www.ucc.ie/hfrg/sumi/sumiapp.html.

[2] Whiteside, J., Bennett, J., Holtzblatt, K. *Usability Engineering: Our Experience and Evolution*. In Handbook of Human Computer Interaction, ed. Martin Helander, North Holland Press, 1988.

Web Mediator: Providing Social Interaction on the WWW

Akira Ishihara

Industrial Electronics & Systems Laboratory, Mitsubishi Electric Corporation,
1-1, Tsukaguchi-Honmachi 8-Chome, Amagasaki City, Hyogo, 661, Japan

This paper describes a technique to provide Web-based applications with a general mechanism for social interaction between persons. The Web Mediator framework adds functions supporting user awareness and cooperative work to applications running on the WWW. It enables multiple users using the same proxy to share their task information. Furthermore, it provides the proper application program for accessing their shared task information according to the context of their work.

1 WEB-BASED APPLICATION

With the explosive growth of the WWW and the appearance of Java and ActiveX, the demand for utilizing the WWW as the application platform is increasing. For example, in industrial systems, there are demands for remote monitoring of plant operations or remote maintenance of plant equipment via the WWW. The advantage of these applications is that they are built on top of Web components, such as a Web server, the HTTP protocol, and Web browsers. Their platform neutrality and centralization of software maintenance make them attractive. A key characteristic of Web-based applications is that multiple users use them either synchronously or asynchronously. The problem is how to manage information like "who worked on this?", "What and where is the result of another's work", and "How can I take over another's work".

This paper describes a technique to provide Web-based applications with a general mechanism for social interaction between persons. The Web Mediator framework adds functions supporting user awareness and cooperative work to applications running on the WWW. It enables multiple users using the same proxy to share and automatically update their task information. Furthermore, it supplies the proper application program for accessing their shared task information according to the context of their work. At first, we will describe the internal architecture of Web Mediator. Second, we will focus on the method for automatically providing the proper access program. Finally, an example application of group asynchronous browsing will be shown.

2 WEB MEDIATOR FRAMEWORK

Figure 1 shows the internal architecture and the WWW transactions of the Web Mediator. Web Mediator Core (WMC) is a WWW proxy built on top of the Java native API. When users access arbitrary Web documents through WMC, work information in Work Information Data Base(WIDB) is updated. When loading a Web document from the HTTPD, WMC composes the original document and what is generated from the WIDB into a new Web document. This composition is called "*Document Synthesis*" [1]. What we are providing is not the contents of the WIDB, but mechanisms for recording work information

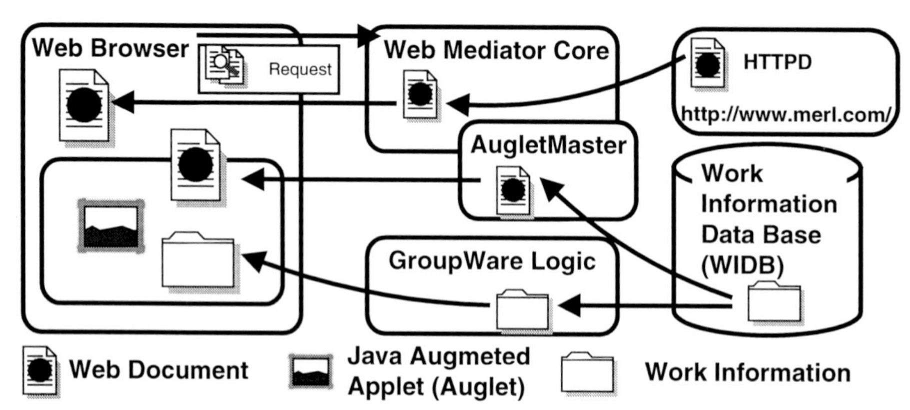

Figure 1. Web Mediator architecture

and allowing multiple users to share it on the WWW. The *AugletMaster* is a key component for this. It is a plug-in component of the Web Mediator Core. It decides how to update the WIDB from user's HTTP requests and what kinds of information to generate and merge into the original Web document. The Web Mediator architecture provides the mechanisms for activating the proper AugletMaster corresponding to the user's HTTP request and the context of their work. Web Mediator also provides a toolkit for defining a new AugletMaster (see section 3).

What is generated from the WIDB and synthesized with the original Web document may include any general Web contents, such as HTML, VRML and Java applets. In particular, a Java applet can be the user interface for the GroupWare Logic in Figure 1. In that case, it is called an *Auglet* (Java *Aug*mented App*let*) in this paper. The user can edit the information in the WIDB through the Auglet interface and GroupWare Logic. GroupWare Logic may be regarded as an embedded system under the WWW. Web Mediator is not only the framework for "*Document Synthesis*", but also the flexible communication server to be run directly on the *embedded system* [2].

3 MANAGING USER'S CONTEXT

Web Mediator can change its behavior according to the context of the user's work. In this section, we will describe the mechanism for activating the corresponding AugletMaster. In the current version of Web Mediator, the context management facility dynamically changes the activated AugletMaster as the following:

- According to user profiles: User profiles are managed in the WIDB so that they are editable through the Auglet. *(Binding with attributes of user profiles)*
- According to workgroup profiles: Users belong to some on-line workgroups in Web Mediator. *(Binding with workgroups)*
- According to URL requested by users: *(Binding with URL)*

The following is the sample code for a new AugletMaster class, which outputs the HTML applet tag to the Web Document requested by the user.

```
public class DefaultAugletMaster extends AugletMaster {
  public String service() {
    /* Here is the method for updating Work Information Data Base */
    return "<Applet Code = \"SimpleAuglet.class\">"; } }
```

The following is the example of binding this AugletMaster with the workgroup named "wg". By this method, any Web document which members of "wg" access will have the interface provided by SimpleAuglet.

```
public class WebMediatorCore extends WWWProxy {
  public static void main(String args[]) {
    WebMediatorCore wmc  = new WebMediatorCore();
    wmc.bind( wmc.getWorkGroup("wg"), new DefaultAugletMaster() );
    wmc.startServer(); } }
```

4 EXAMPLE APPLICATION

Our example application concerns group asynchronous browsing [3]. There exist few tools for organizing Web documents. Consider a tool for sharing the local hotlist of a workgroup, which consists of documents recommended by members. We want it to allow members to edit it visually. In this section, we will explain the functions for recommending a Web document and editing the shared hotlist. All the developer has to do is to implement the following two Auglets and one AugletMaster class using Web Mediator Toolkit. A *RecommendationAuglet* will be inserted in arbitrary Web documents accessed by members of the workgroup(see the bottom gray part of the bottom right browser's window in Figure 2.). By clicking the "Add URL" button, the Web document www.merl.com will be recorded in the shared hotlist. A *RecAugetMaster* is bound to the workgroup(strictly speaking, the requests from members of the workgroup) and will output the applet tag for loading this auglet. By clicking the "Session Edit" button (above the "Add URL" button), the *HotListOrganizer* auglet will appear (the bigger browser's window in Figure 2). It visualizes the shared hotlist from the WIDB. It is a kind of GUI builder implemented as a Java applet.

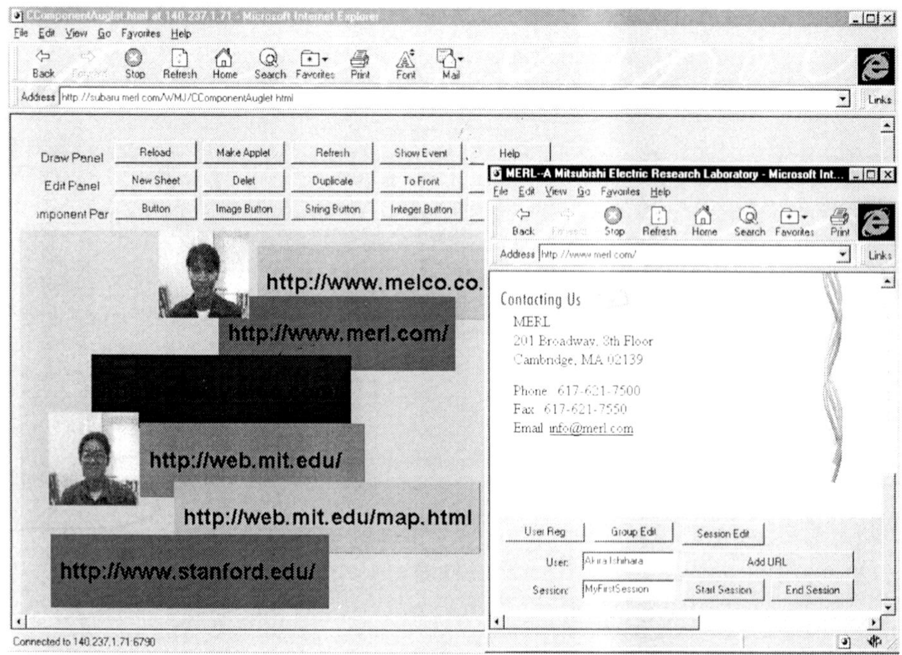

Figure 2. Example of group asynchronous browsing.

5 CONCLUSION

We have demonstrated the feasibility of an application-independent framework for providing Web-based applications with a facility supporting user-awareness and cooperative work. In particular, we have focused on the method for changing the application's behavior according to the context of user's work.

REFERENCES

1. M.Roscheisen, C.Mogensen, and T.Winograd, Beyond Browsing: Shared Comments, SOAPs, Trails, and On-line Communities, WWW95.
2. M.Bathelt, U.Gall, B.Hindel, and C.Kurzke, Accessing Embedded Systems via WWW: The ProWeb Toolset, WWW97.
3. K.Wittenburg, D.Das, W.Hill, and L.Stead, Group Asyncronous Browsing on the World Wide Web, WWW95.

From system evaluation to service redesign: a case study

P. Marti[a]., A. Rizzo[a], S. Bagnara[ab], P. Lomagistro[a], L. Tanzini[a]

[a] Multimedia Laboratory, University of Siena,
Via del Giglio 14, 53100, Siena, Italy

[b] Institute of Psychology, National Research Council
Viale Marx 14, 00137, Rome, Italy

In this work we detail a case study in which user-centred design [1] and contextual inquiry [2] were used to develop and evolve a system for accessing the multimedia database of RAI, the Italian state broadcasting television. The database stores a very large volume of audio-visual documents including programs daily broadcasted by the RAI channels; it is maintained and accessed by a very large and heterogeneous community of use.

The significance of the work is the emphasis on activity analysis, work context and community of use in the design of an information system. Contextual inquiry was an important methodological tool to reach this objective. The contextual inquiry approach emphasises the need for designers to understand users work process and the context in which this process occurs in addition to the more traditional information about the capabilities of the technology and the characteristics of the users. Our work details the impact that user-centred techniques and contextual inquiry had on the final design, in particular in terms of integration of artifacts in the work environment and service redesign.

1. THE PROBLEM

How to facilitate information access and indexing in large volume of on-line multimedia DB? How to organise an information service involving a large and heterogeneous community of use? The answers rely on the design process adopted; in our case, user-centred design techniques and contextual inquiry profoundly modified the initial requirements, bringing to re-design the entire work process. In fact, the initial commitment for our work was the usability evaluation of a new multimedia system to support information documentation and information access. A beta version of the system was installed to be tested by the RAI employees, and quite soon it became so unpopular to be refused by the existing community of use. What was wrong with the new computer application? Which were the reasons for its negative impact?

The community of users who work around the RAI database is mainly composed of people who document and store information into the database, and people who access the information with different purposes. The two groups overlap. The processes of documentation and information access to the RAI database are actually fragmented in different units working as individual actors of the process. The view that the documentation task and information access are independent activities of separate processes is mirrored in the organisation of the service. Different teams are in charge of information documentation and retrieval, they work in physically separated buildings, with no feedback between them. This way of considering information access and documentation as separated activities is even mirrored by the systems used during the activities: different environments with no integration. The audio-visual material is documented and stored in the database by RAI operators, and it is used in many fields by professionals like journalists and directors, by students and professors, by magistrates, by people who appeared on the television. Operators responsible for information documentation use many different artifacts during their job: television, newspapers, dictionaries, geographical atlases, on-line databases of the most important press agencies. They furthermore develop their own artifacts to support the activity, like hand-written annotations, quick access guides, abbreviated checklists, photos of actors or politicians postered in the office. All these artifacts used during the activity are actually poorly integrated in the working environment, causing waste of time and inaccuracies in the documentation task.

Users searching for information in the database are often required to quickly access the information relevant to the specific tasks they are performing. This is the case of television journalists who have very short time to prepare news, often no more than one hour including information retrieval, screening, editing and montage. The procedures are even longer if these users have no direct access to the RAI database: they have to request the material to the RAI Users Office, where search assistants receive the request, perform the search, and send one or more video-tapes back to the journalist.

Users who have direct access to the database face problems of getting lost in the information space, loosing information, spending a lot of time without succeeding. Often, if they find interesting documents that are not directly related to the current task, they have to memorise where to find them for future reference without any support from the system.

2. ANALYSIS OF THE CONTEXT OF USE: THE METHODOLOGICAL APPROACH

Different data-gathering approaches and analysis methods were used, all these addressing the users requirements, the context of work and the community of use. Ethnographic fieldwork was crucial to understanding the work process. It was combined with the use of a varied set of data collection techniques including interviews, observations, analysis of historical materials, artifacts walkthroughs [3], and development of use scenario [4]. The methodological approach focused on broad patterns of activity rather than narrow episodic fragments that often fail to reveal the overall direction and importance of an activity.

We conducted an ethnographic study to examine how people behave in the context of their activity, which objects they handle, which artifacts mediate their activity, their communication needs inside the work process. The users in our study were fifteen RAI employees whose jobs involved documenting and storing information into the database, or performing search on behalf of external customers. Other informants of our study were external users like journalists, directors or assistants who have direct access to the database. We interviewed and observed users at work. We asked each of them to describe the entire documentation-access process, allowing the conversation to flow naturally rather than strictly following the list of pre-determined questions.

2.1. Findings

The main results of our study was that both information documentation and information access are viewed as part of a single process by the community of practice. The problem with the new computer applications was that they supported single, episodic fragments of tasks, leaving most of the activity without support. Since this problem was recognised as the major flaw in the actual work organisation, the workers did not want to spend other resources to learn a new computer application that would have been only another object to control. The local benefits that the new application offered were not sufficient to justify the effort to pass from the old to the new system. Our study raised the need to design an artifact able to mediate the activity of different actors belonging to a community of practice, within the same work process.

3. THE SERVICE REDESIGN

We proposed a solution in which the overall process was redesigned according to the results of our study. Participatory prototyping [5] was used during the design process: we developed with the users a series of intermediate prototypes which were evaluated and incrementally refined. Two integrated prototypes were developed: i) a prototype supporting the documentation task, with on-line multimedia aids; ii) an Intranet prototype which provides user-centred information retrieval and navigation.

The first prototype was intended for the users in charge of the information documentation. We provided a support in customising the new application according to their current needs and habits. This solution provides:

1. Integration and customisation of artifacts in the work context:
   i) Geographical database and multimedia keywords thesaurus which can be customised with annotation, images or videos. These are very simple but effective tools which support the operator in solving frustrating situation like the difficulty of writing foreign names, or of providing the exact geographical reference of a place.
   ii) On-line access to the main press agencies databases;
   iii) Video and images acquisition in the documentation forms. Indeed the richer is the documentation the easier is its retrieval and exploitation. The operators who document information are well aware of the limits of the textual descriptions for information retrieval. The integration of video and images allows the users to screen them having a preview on what is actually contained in the original video tapes.
   iv) Television and video recorder accessible via computer.
2. Structured presentation of information which are relevant for the documentation task. This avoids tedious and repetitive tasks of filling in standard forms, and avoids to omit relevant information.

The second prototype was intended for different typologies of users who search information in the database. These include: RAI employees devoted to the customers assistance, or those who have direct access to the database like journalists, directors or assistants. Different searching facilities are offered more close to the search strategies of the different typologies of users rather than close to the structure of the database. The Intranet prototype provides:

1. Combination of logical hierarchical navigation with flexible searching and hypermedia links.
2. Screening facilities of videos during the search. The videos can be ordered and bought via e-mail.

3. Use of the Intranet to feedback experiences and suggestions to other users.

Both prototypes were evaluated by users and human factor experts [6]. Heuristic evaluation [7] and walkthrough analysis were used for testing and the evaluation results were fedback to the designers to fix usability problems. The first results were encouraging. Testing showed that even people who had never used an on-line service successfully navigated the information space and enjoyed the system. In particular the Intranet was viewed as a support to a cooperative process where the actors can take advantage from the experience of other people. Furthermore the integration of different artifacts used during the documentation task and the structured presentation of information reduced the workload of the operators, driving the attention on a more accurate and rich documentation.

4. CONCLUSIONS

The importance of evaluating computer application within their context of use is relevant to both the individual level and the group or organisational level [8]. The experience described in our case study demonstrates the benefits of ethnographic field studies combined with user-centred design techniques. This methodological approach allows to analyse the way in which artifacts mediate human activities, and their role within the entire work process. The analysis of the context of use reveals how the assimilation of new technologies causes new tasks to emerge (the so-called task-artifact cycle, according to Carroll et al. [9]) and which are the new needs and requirements to satisfy. Our methodological approach highlights that the design of interactive system needs to be grounded to the users practices and the work context in order to properly exploit their knowledge and capabilities.

REFERENCES

1. D. Norman and S.W. Draper (eds.), User Centered System design: New perspectives on Human-Computer Interaction, Lawrence Erlbaum Associates, Hillsdale NJ, 1986.
2. K. Holtzblatt and S. Jones Contextual Inquiry: Principles and Practices . In Participatory Design: Principles and Practices, Lawrence Erlbaum, Hillsdale NJ, 1993.
3. K. and J. Strandgaard Pedersern Workplace Cultures: Looking at Artifacts, Symbols, and Practices . In J. Greenbaum, K. Morten, Design at Work, Lawrence Erlbaum Ass., Hillsdale, 1991.
4. J.M. Carroll, Scenario-based Design, Wiley & Sons, New York, 1995.
5. K. Bødker and K. Grønbæk Design In Action: From Prototyping by Demonstration to Cooperative Prototyping . In J. Greenbaum, K. Morten, Design at Work, Lauwrence Erlbaum Ass., Hillsdale, 1991.
6. D.M. Levi and F.G. Conrad, A Heuristic Evaluation of a World Wide Web Prototype Interactions, vol. III, n.4, 1996.
7. G. Lindgaard, Usability testing and system evaluation, Chapman & Hall. London, 1994.
8. B.A. Nardi (ed.), Context and Consciousness. Activity Theory in Human Computer Interaction, The MIT Press, Cambridge, MA, 1996.
9. J.M. Carroll, W. Kellogg, and M. Rosson The task-artifact cycle . In Designing Interaction: Psychology at the Human-Computer Interface. Cambridge University Press, Cambridge, 1991.

Book based view and scroll based view: Which is suited for HTML viewers?

Kaori Ueno, Kenya Suzuki, Hideaki Ozawa

NTT Human Interface Laboratories
1-1 Hikari-no-oka, Yokosuka, Kanagawa, 239 Japan

Abstract. The usage environment of the traditional WWW viewer is unfamiliar to novice computer users. This paper proposes a WWW viewer that simplifies operations. The viewer has book like interface whose causal relation is comprehensive even for novice users. In a cognitive experiment, reading speed is compared between the proposed viewer and the traditional WWW viewer and the effectiveness of the proposed viewer is confirmed.

1. INTRODUCTION

Recently more printed information is being made electronic and the chance to see electronic information is increasing even for the general public. Information that is being continually updated is being distributed to the public via the WWW and even novice computer users are becoming common users. Electronic manuals and other texts are also being distributed to users via the WWW. It is thought that these tendencies will strengthen in the future.

Traditional WWW viewers were not intended to be used by novice users, misoperation is all too common. This distracts the user and prevents the users from concentrating on the information. Misoperation can be reduced by displaying the information on the computer display in a form more comprehensible to the users and providing more intuitive controls. Various methods of making displayed information more comprehensible have been considered.

The main idea is making a more direct relationship between the control and the response. In this vein we developed the book metaphor, a hypermedia database system with a book-like user interface[1]. This paper proposes that the book metaphor be used in a WWW data viewer and examines the effectiveness of the book metaphor as a WWW data viewer.

2. Model of reading

2.1 Information and reading

Most computer users sometimes have trouble when they read information on the WWW. For instance, the user can not easily return to the original location after moving within a long document, and the user finds it difficult to understand position within the document. The current WWW has links to large documents that have fixed contents such as stories, manuals, catalogs, and encyclopedias. We think that there are many WWW documents that are similar to paper books. The reading style of WWW should not be so different from that of traditional books. There are several different reading styles according to the kinds of contents. For instance, stories need to be read through from the beginning to the end. Procedure instructions often need to be read through in the same manner, but sometimes we concentrate on one part of the instruction. Catalogs are usually read selectively. Dictionaries and encyclopedia are read selectively and sometimes we move around among related parts (Table 1).

The information on each WWW site is usually divided into several items connected by hyper links. We think that it is important to make information on the WWW as easy to read as traditional paper documents.

Table 1. Information and operation

operation	information			
	story	proce-dure	cata-log	cyclo-pedia
turn pages from beginning to end	○	○	△	
directly open		○	○	○
immediately prior page		○	○	
refer to related items		○	○	○

Table 2. Operation and function

operation	function			
	scroll	hyper link	user's record	text search
turn pages from beginning to end	○	○		
directly open		○		○
immediately prior page			○	
refer to related items		○		

2.2 Reading style and operation

Reading style can be classified into two groups, reading through and reading selectively as stated in the previous section.
We examined the relation reading style and operations of a WWW data viewer as shown in Table 2. Looking at the relation between reading style and operation, we noticed a problem. When information is read through, it is necessary to use two operations appropriately. When moving to another file, hyper links are used. On the other hand, when moving in the same file, scrolling is used. Users find it is inconvenient to have two kinds of operations to achieve the purpose of reading the document. Accordingly, we developed another interface that solves this problem. The proposed interface has only one kind of operation; that is turning pages. We borrowed the advantage of an old medium, books, which have comprehensive causal relation between what happens on the display and the operation performed. We propose a book-like interface be used as a WWW data viewer. Many electronic books have been proposed already [2-5] and other methods also exist as book-type WWW data viewers [6]. We think the following functions are needed in the book-like interface:
Paging. WWW Data is separated into book page units so that users can remember where specific data is located in the file by memorizing the page number and the layout of the page.
Turning page. Users easily go forward and backward in the document.
Thickness. Displaying thickness so that users know the amount of total and remaining (unread) information.
Tag operation. Users directly open the book at the desired chapter.
Bookmarks. Creating user's own tags. By selecting bookmarks, the target page is opened. Users return the page at random.
Table of contents. Users go directly to the selected contents page.

3. USABILITY EVALUATION OF BOOK METAPHOR VIEWER

3.1 The first experiment

3.1.1 Method

Seven ordinary computer users were instructed to browse two sets of HTML information on the WWW using the book metaphor viewer and a traditional WWW viewer as quickly as possible (Figure 1). The information set contained text and image data and had a 2-level structure, the table of contents and the data pages.

3.1.2 Results and discussion

Completion Time. Participants browsed much more quickly with the book metaphor viewer than with the traditional WWW viewer ($p < 0.05$) as shown in Figure 2. Participants' fundamental behavior with the traditional WWW viewer was: selecting an item on the table of contents

(a) Book metaphor viewer

(b) Traditional viewer

Figure 1. Display of HTML information

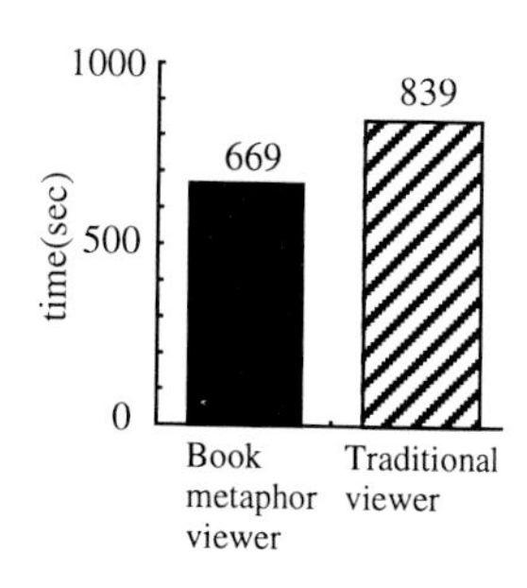

Figure 2. Completion Time

page, (automatically jumping,) seeing the item page related to the table of contents, and jumping back to the contents page. On the other hand, with the book metaphor viewer, they simply turned the pages. Because participants did not need to return to the table of contents page nor select any item, completion time was reduced. This was the reason for the difference of completion time between the two.

Operation. When the user wants to read long HTML data displayed in a scrolling field (common in the traditional WWW viewer), she cannot estimate the start point of the information because the scroll bar provides no clue as to the segmentation of the information reading area. On the other hand, the book metaphor separates long HTML data into book pages, so the reading area's location in the long data is clear to the user. The number of times of the user turned the pages backward while browsing with the book metaphor viewer was significantly fewer than backward scrolling operations with the traditional WWW viewer ($p < 0.05$). Inefficient operation, participants returning uselessly to the same page again, was less than that when using the traditional WWW viewer.

3.2 The second experiment

3.2.1 Method

It was examined whether the book metaphor viewer was effective when reading only necessity. Five ordinary computer users were instructed to look for a specified article as quickly as possible in the sets of HTML information on the WWW using the book metaphor viewer and a traditional WWW viewer. The information was part of the contents from the first experiment.

3.2.2 Results and discussion

When the participants looked for a target article, it was understood that there were two types of reading, according to having interviewed participants and observing their behavior. One was reading to understand the contents, the other was only of seeing the characters without understanding the contents. The former will be called the careful reading type and the latter be called the speed reading type. Completion time according to the reading type is as shown in Figure 3. Each type of participants accomplished the tasks much more quickly with the book metaphor viewer than with the traditional WWW viewer ($p < 0.01$, $p < 0.05$).

Careful reading type. The participants using the careful reading behavior were chiefly influenced by the difference of the operation of book metaphor viewer and the traditional WWW viewer. The amount of movement within the page was not coincident with the amount of movement of seeing position by eyes. Therefore, where the article was not understood or some lines were skipped, the participants often returned and read the same lines again.

Speed reading type. The participants of the speed reading type were chiefly influenced by the

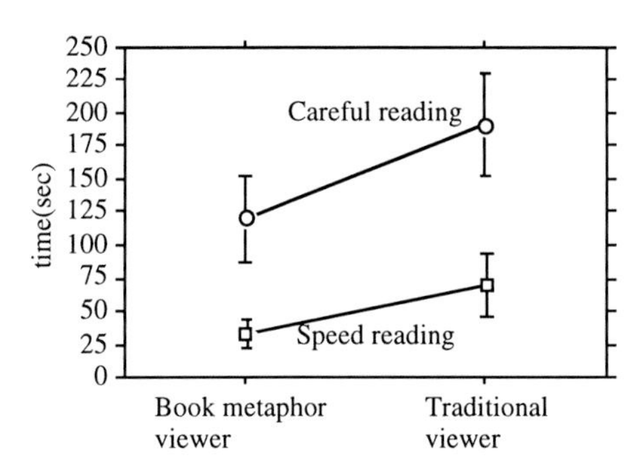

Figure 3. Completion time

difference in the layout of the book metaphor viewer and the traditional WWW viewer. The boundary increases because information is blocked and is divided into a left page and a right page when information is displayed on the book metaphor viewer. On the other hand, there is no block division of information on each page in the general type viewer. It seemed that the participants could not easily find the target items because there was no pauses on long pages.

4. CONCLUSION

We have proposed a book-like WWW viewer, and evaluated the effectiveness of the proposed viewer comparing with traditional viewer in the cognitive experiments.

In the experiment where the participants read through from the beginning to the end, they read more quickly a site of WWW data than the traditional viewer. We think that the reason is because the book metaphor viewer has a concept of page units so that only one kind of operation; that is turning pages, is necessary to move around in the file. On the other hand, with traditional viewers, it is necessary to appropriately use two operations: scroll and hyper link.

Also in the experiment where the participants read WWW data searching the target items by their eyes, they read more quickly a site of WWW data than the traditional viewer. It is presumed that this is because marks (= page breaks) in the proposed viewer, which separate data into blocks of a page, worked as a clue to find a target item.

ACKNOWLEDGMENTS

The authors would like to thank Dr. Yukio Tokunaga, and the members of Human Interface Laboratories for providing several useful comments.

REFERENCES

[1] Ogawa, K. et al., Usability Analysis of Design Guideline Database in Human-Computer Interface Design, in the Proc. of the Human Factors Soc. 36th Annual Meeting (1992) 433-437.

[2] Miyazawa, M. et al., An Electronic Book: AptBook, in the Proc. of Human-Computer Interaction - INTERACT '90, (1990) 513-519.

[3] Remde, J. R. et al, SuperBook: An automatic tool for information exploration - hypertext?, in the Proc. of ACM Hypertext '87 Conference (1987) 175-188.

[4] Weyer, S. A., Searching for Information in a Dynamic Book, Xerox T. R. SCG-82-1, 1982.

[5] Yankelovich, N. et al, Reading and writing the electronic book, Computer 18 (10), (1985) 15-30.

[6] Card, S. K. et al, The WebBook and the Web Forager: An Information Workspace for the World-Wide Web, in the Proc. of CHI 96, ACM (1996) 111-117.

Web pages: designing with rhetorical techniques

C. Dormann

Center for Tele-Information, Technical University of Denmark, Building 371, 2800 Lyngby, Denmark, Email:claire@cti.dtu.dk.

The design of web documents is approached from the perspective of visual rhetoric. The concern here is to make authors aware of rhetorical figures. The rhetorical figures are first introduced and subsequently, their application to the design of web pages is discussed. Rhetorical figures are techniques that are employed to encode meaning, find original solutions to design problems or to enhance documents. Rhetorical tools can be used as exploration tools that open choices and assist designers in creative tasks.

1. INTRODUCTION

When creating web pages, we should consider that without the visual impact of shape, colour and contrast, pages do not motivate viewers to investigate their contents. To design web pages, there are few guidelines regarding composition and visual representation, which will lead to effective visual communication.

Rhetoric is concerned with the modification of the viewer's conception and attitude toward the object of communication. The ancient Greeks developed rhetoric to discover the mean of persuasion, a method for informing and appealing to audiences. Rhetoric operates on the basis of logical and aesthetic modes to affect interaction in both an emotional and in a rational way [1]. The promise of rhetoric is that there exists a system for identifying the most effective form of expression in any given case.

The study of rhetoric has provided tools i.e. rhetorical figures that will assist web authors in their design task. Rhetoricians employ these tools for various purposes: to create emphasis, stimulate viewers' interest, guide viewers through the most significant information, enable them to compare and contrast, or persuade viewers to take action. As the many devices catalogued by rhetoricians since Antiquity have remained largely unacknowledged, the thrust of this investigation is to make authors aware of rhetorical figures. This paper aims to contribute to a richer and more systematic understanding of these techniques in relation with design tasks.

2. RHETORICAL FIGURES

Within the classical system of rhetoric, figures are usually divided into two groups, schemes and tropes. Tropes involve a deviation from the ordinary and principal signification of a word while schemes involve a deviation from the ordinary pattern or arrangement of words [2]. Tropes can be defined as semantic figures and schemes as syntactic figures.

2.1. Tropes

Metaphor is defined as one thing described as something else. In user interface design, the traditional physical media is a natural starting point for metaphoric representations. Examples include libraries with doors, help desk rooms, collections and shelves or the city of knowledge with gates, streets, buildings and landmarks [3].

Visual pun is the use of symbols to suggest two or more meanings or different associations. On the home page of an astrological site, a list of links has been arranged on dark-blue background to form a moon croissant. This is a typical example of a literal pun mixing text and image [4]. The visual language or thematic content of a page can be reinforced by using this technique. In a web page for Sam' bagels, different types of bagels are listed. In the list, the bullet points are replaced by images of bagels [5].

Personification is defined as a comparison whereby human qualities are assigned to an inanimate object, for example, by adding legs, arms or a face to an object.

Hyperbole is exaggeration for emphasis, used to make an object more prominent for example by enhancing its size. Animators and cartoonists have mastered the art of highlighting characteristics by placing in sharpest focus those features which best capture the personality or the emotion.

Metonymy is the substitution of terms suggesting an actual relationship. The substitution can be of a causal, spatial or chronological nature. For Copenhagen Council Services [6], the Town Hall has been chosen as the icon background and the different services are characterised by an additional element: a ball for sports, a tree for conservation, a medical case for health, etc. Designers employ rhetorical figures such as metonymy to encode meaning firstly because of the lack of inherent 'natural' references or secondly because computer concepts are too complex or abstract to be shown directly [7].

2.2. Schemes

Numerous rhetorical figures are classified under schemes. We are only going to discuss here, the figures of repetition and contrast.

Among the figures of contrast, antithesis and oxymoron can be found. *Antithesis* is the juxtaposition of contrasting ideas, often in parallel structure. *Oxymoron* is the yoking of two terms which are ordinarily contradictory. To explain the use of colour, an opposition of a black and white, and coloured image is created. Contrast has been found to be among those factors that invite and arrest attention.

The figures of repetition include *anaphora, polyopton and anadiplosis. Anaphora is* the duplication of a word or a group of words, it is the most elementary figure of repetition. *Polyopton* is the repetition of words derived from the same root. *Anadiplosis* is the repetition of the last word of one clause at the beginning of the following clause.

The effect of repetition can be established using virtually any design element. The most simple form consists in repeating the same element, like in a stack of videos, or a series of cars for a link on automobile. Variations are obtained by changing the element colour or size, and by varying the object position or point of view. Different object models (e.g. cars) can also be given. These figures convey an idea of profusion or multitude.

The repetition of elements throughout the web pages reinforces the nature of the product while enhancing the continuity of the document as a whole. The powerful human tendency to perceive regularity in the display leaves the designer with a wide latitude for choosing an element whose repetition facilitates communication while providing the comforting familiarity of a well-designed site [8].

3. DESIGNING WEB DOCUMENTS

The application of rhetorical figures is discussed further by taking as a design example the city metaphor. Rhetorical figures have only been presented in relation with static components of web pages however they are also applied to the development of animation and video [9].

3.1. The city metaphor

The city environment has formed the basis for a novel spatial user interface metaphor which has proved a popular organisational principle in the world wide web. At least two different groups of applications can be distinguished: sites offering goods and services (e.g. information city or commercial village) and sites offering interactive adventures or social games (e.g. MOO). The representation style varies from two-dimensional geometrical representations to very sophisticated virtual worlds.

A typical commercial Danish site guides users through a range of virtual shops giving access to a variety of products. In the home page, consumers find an image map representing a village square [10]. By looking more closely at the image map, we can see that characteristics of the shop content have been employed in the design of individual buildings. The Motorola (e.g. phone company) building has antenna on its roof and its top looks like a telephone receiver. With its bright and gay colours and style, the design is evocative of a miniature toy set. One building has round windows and an elongated red porch, reminding us of a friendly face.

This metaphor could be exploited further in the design of web documents. Other city features can be used to convey information. The village infrastructure could parallel the infrastructure of the web site, by clicking on roads, users would access an information map or by clicking on signposts a table of content, etc.

Users need predictability and structure, with clear functional and graphics continuity in web documents. The city metaphor can act as an anchor for the document design. Then other rhetorical techniques can be applied to the design of interface elements and the document's visual content. Working from the original metaphor, many other nuances or relations are expressed with these techniques. Such an approach exists in other media: many filmic metaphors are supported by underlying metonymies [11].

Thus to create continuity across a document, a variety of visual elements is repeated. Navigation buttons (e.g. house, open-door for the home page) and indices are developed. The Sun tutorial on Java uses small visual cues i.e. a hiker to distinguish between a linear or hierarchical path. Similarly, a pedestrian could signal internal links and a hot air balloon external links. Multiple illustrations can also be created e.g. caricatures of city leaders for the Copenhagen site.

3.2. Discussion

When using a metaphor, designers can explain a tool by referring to a similar process. Metaphors speak to the imagination, help visualisation, incarnate, and specify what the designers think cannot be understood otherwise. By applying hyperbole (e.g. increasing the size of the cursor), important information is made more visible and given greater presence. By concentrating on the essential features of the object (e.g. action or emotion) the message becomes more powerful. Puns surprise and entertain. With personification, all the techniques of visual comedy can be applied. An advertising agency combined Monty pitoresque humour in a site for direct debit. The result is a site that both educates the visitor and makes them smile upon remembering their experience of the site. When using these techniques, we can give viewers more attractive and effective designs.

4. CONCLUSION

Because viewers are under no compulsion to read a home page, or to continue visiting a site, an important function of rhetoric techniques is to motivate potential viewers and persuade them to stay as viewers. By providing techniques with which to produce a range of explanations, authors can choose the most appropriate one for a particular problem or audience. Rhetorical tools are also used as exploration tools that open choices and can assist the designers in the creative task. The task of finding appropriate solutions to design problems would be easier if a grammar for a visual rhetoric language was established. This paper is the first step towards defining such grammar.

REFERENCES

1. H. Ehses, Representing Macbeth: A Case Study in Visual Rhetorics, in Design Discourse, ed., V. Margolin, 187-199, Chicago University Press, London, 1989.
2. E. Corbett, Classical Rhetoric for the Modern Student, Oxford University Press, New York, 1971.
3. B. Scherdermann, Designing Information-Abundant Websites: Issues and Recommendations, to be published in International Journal of Human-Computer Studies, July 1997, <http://kmi.open.ac.uk/~simonb/ijhcs-www/>.
4. Mystic GinG (1996), http://www.killersites.com/2-sites/mystic/index.html.
5. P. Lynch, Yale C/AIM WWW Style Manual, 1995, <http://info.med.yale.edu/calm/stylemanual/M_I_2A.html>.
6. Københavns Kommune (1997), <http:// www.kbhbase.copenhagencity.dk/>.
7. W. Horton, The Icon Book, John Wiley, New York, 1994.
8. K. Mullet and D. Sano, Designing Visual Interfaces, Prentice Hall, California, 1995.
9. C. Dormann, Designing On-line Animated Help for Multimedia Applications, in Multimedia, Hypermedia and Virtual Reality,ed., N. Streitz and P. Brusilovski, 73-84, Springer-Verlag, London, 1996.
10. Zapcity (1997), <http://www.zapcity.dk/>.
11. C. Metz, Psychonalysis and Cinema the Imaginary Signifier, Macmillan, London, 1983.

Usability Issues in Web Site Design

Nigel Bevan

National Physical Laboratory, Usability Services, Teddington, Middx, TW11 0LW, UK
email: Nigel.Bevan@npl.co.uk

Why are so many web sites frustratingly slow and complicated to use? The reasons include:

- Organisations often produce web sites with a content and structure which mirrors the internal concerns of the organisation rather than the needs of the users of the site.
- Web sites frequently contain material which would be appropriate in a printed form, but needs to be adapted for presentation on the web.
- Producing web pages is apparently so easy that it may not be subject to the same quality criteria that are used for other forms of publishing.

In short, web sites provide a unique opportunity for inexperienced information providers to create a new generation of difficult to use systems! Successful web development requires the combined skills of domain expertise, HTML, graphic design and web usability.

A web site will not meet the needs of the organisation providing the site unless it meets the needs of the intended users, and provides "quality in use"[1]. To implement a web site which users find effective, efficient and satisfying requires a user centred design process[2]. This paper describes a process which integrates existing empirical evidence and guidelines for web site design into a user-centred process which is consistent with ISO 13407[3]. Due to limitations of space, the reader is referred to the references for the rationale for the individual design guidelines.

It is essential to first define the business and usability objectives, and to specify the intended contexts of use[4]. These should drive an iterative process of design and evaluation, starting with partial mock-ups and moving to functional prototypes. Continued usability requires subsequent management and maintenance.

PLANNING

Define the business objectives of the site (provider requirements)

- What is the purpose of the site? This could include disseminating information, positioning in the market, advertising services, demonstrating competency, or providing intranet services.
- Who do you want to visit the site: what are the important user categories and what are their goals?
- What type of pages and information will attract users and meet their needs? e.g. hierarchically structured information, a database, download of software/files, incentives to explore the site.
- What are the quality and usability goals which can be evaluated? e.g. to demonstrate superiority of the organisation to the competition, appropriateness of web site to user's needs, professionalism of web site, percentage of users who can find the information they need, ease with which users can locate information, number of accesses to key pages, percentage of users visiting the site who access key pages.
- What is the budget for achieving these goals for different parts of the site?

Identify responsibilities for achieving quality and usability objectives, and estimate the resources and budget for these activities.

Specify in detail the intended contexts of use (user requirements)

- Who are the important users?
- What is their purpose for accessing the site?
- How frequently will they visit the site?
- What experience and expertise do they have? What nationality are they?
- What type of information are they looking for?
- How will they want to use the information: read it on the screen, print it or download it?
- What type of browsers will they use? How fast will their communication links be?
- How large a screen/window will they use with how many colours?

Define key scenarios of use

- Describe specific examples of people accessing the site, and what they want to achieve. These will help prioritise design, and should be the focus for evaluation.
- Also identify any niche markets and interests which can be supported by the site without major additional investment (e.g. specialised information, access by users with special needs).

SITE STRUCTURE AND CONTENT

- Structure information so that it is meaningful to the user. The structure should make sense to the user, and will often differ from the structure used internally by the data provider.
- What information content does the user need at what level of detail? Use terminology familiar to the user.
- Interview users to establish the users' terminology and how they categorise information.
- Produce a card (or post it note) for each anticipated page for the site[5], and use card sorting techniques to design an appropriate structure[6].

SUPPORT NAVIGATION

Help users find their way[7]

- Show users where they are.
- Use a consistent page layout.
- Minimise the need to scroll while navigating.
- The easiest to navigate pages have a high density of self-explanatory text links[8].
- Try to make sure users can get to useful information in no more than four clicks.
- Provide links to contents, map, index and home on each page; for large sites include search[9].
- Include navigational buttons at both the top and bottom of the page.
- Use meaningful URLs and page titles. URLs should be exclusively lower case.
- Plan that any page could be the first page for users reaching the site from a search engine.

Tell users what to expect[10]

- Avoid concise menus: explain what each link contains.
- Provide a site map or overview.
- Distinguish between a contents list for a page, links to other pages, and links to other sites.
- Do not change default link colours and style, otherwise users will not recognise the links.
- Give sizes of files which can be downloaded.

Highlight important links

- The wording of links embedded in text should help users scan the contents of a page, and give prominence to links to key pages. (Highlight the topic - do not use "click here"!)
- To keep users on your site, differentiate between on-site and off-site links.

PAGE DESIGN

Design an effective home page

- This should establish the site identity and give a clear overview of the content.
- It should fit on one screen, as many users will not bother to scroll the home page[11].

Design for efficiency[12]

- Graphics add interest but are slow to load and can impede navigation.
- Use the minimum number of colours to reduce the size of graphics.
- Use the ALT tag to describe graphics, as many users do wait for graphics to load.
- Use small images, use interlaced images, repeat images where possible[13],[14].

Make text easy to read

- Never use flashing or animation, as users find this very distracting.
- Avoid patterned backgrounds, as these make text difficult to read.

Support different browser environments

- Use a maximum 640 pixel width, or 560 pixels for pages to be printed in portrait mode.
- Avoid frames - where possible use tables[15].
- Test that your pages format correctly using the required browsers and platforms.

Support visually impaired users with text-only browsers

- Use a logical hierarchy of headings and use ALT tags which describe the function of images[16].

EVALUATION METHODS

Expert inspection

- Using a checklist to inspect pages for conformance with house style (consistency of layout) and with recommendations such as those in this paper.

Early mock-ups

- Early in design evaluate a partial mock up of the site with representative users performing representative tasks. Use first drafts of screens, either on-line or as colour prints[17].

Functional prototypes

- Produce a working version of a representative part of the site, taking account of the design principles and evaluation feedback.
- Evaluate the working version with representative users performing representative tasks[18].

MANAGEMENT AND MAINTENANCE

Ensure that new pages meet the quality and usability requirements

- What skills will be required of page developers?

- What will be the criteria for approval of new pages? Is some automated checking possible?

Maintenance

Plan and review the site structure as it grows, to make sure it still meets user needs.

- Monitor feedback from users.
- Monitor the words used when searching the site.
- Monitor where people first arrive on the site, and support these pages as entry points.
- Check for broken links (e.g. using SiteMill).
- Compare your site to other comparable sites as web browsers and web design evolve.

As it is unlikely to be economic to test the usability of every page, it is important to establish a sound structure and style guide within which new pages can be developed, and for page developers to be aware of the business objectives and intended contexts of use.

ACKNOWLEDGEMENTS

This paper was developed with the support of the EC INUSE[19] and RESPECT[20] projects, and draws on information from the web sites listed below, and the results of the CHI'97 Usability Testing of World Wide Web Sites Workshop[21].

REFERENCES

A version of this paper with live links can be found at http://www.npl.co.uk/npl/cise/us.

[1] Bevan N (1995) Usability is quality of use. In: Anzai & Ogawa (eds) Proc. 6th International Conference on Human Computer Interaction, July 1995. Elsevier.

[2] Bevan N and Azuma M (1997) Quality in use: Incorporating human factors into the software engineering lifecycle. In: Proceedings of the Third International Symposium and Forum on Software Engineering Standards (ISESS'1997) (in publication).

[3] ISO DIS 13407 (1997) Human centred design process for interactive systems.

[4] ISO DIS 9241-11 (1997) Guidance on usability.

[5] http://www.sun.com/sun-on-net/uidesign/cardsort.html

[6] http://info.med.yale.edu/caim/stylemanual/M_I_3.HTML

[7] http://www.cybertech.apple.com/HI/web_design/find.html

[8] UIETips 2/14/97. Jared Spool, User Interface Engineering (uie@uie.com)

[9] http://www6.nttlabs.com/HyperNews/get/PAPER180.html

[10] http://www.cybertech.apple.com/HI/web_design/tell.html

[11] http://www.sun.com/sun-on-net/uidesign/pagedesign.html

[12] http://www.useit.com/alertbox/9703a.html

[13] http://www.pantos.org/atw/35247.html

[14] http://www.pantos.org/atw/35273.html

[15] http://www.useit.com/alertbox/9612.html

[16] http://www.useit.com/alertbox/9610.html

[17] http://www.sun.com/sun-on-net/uidesign/papertest.html

[18] http://www.sun.com/sun-on-net/uidesign/screentest.html

[19] http://www.npl.co.uk/inuse

[20] http://www.npl.co.uk/respect

[21] http://www.acm.org/sigchi/webhci/chi97testing/index.htm

Integrated User Interfaces for the Home Environment

Jürgen Ziegler and Joachim Machate
Fraunhofer IAO, D-70569 Stuttgart, Germany
{Juergen.Ziegler, Joachim.Machate}@iao.fhg.de

In this paper, we describe the development of a hand-held personal home assistant capable of controlling a wide range of electronic home devices. A demonstrator of the system was developed in the European project TIDE HEPHAISTOS (Home Environment Private Help AssISTant fOr elderly and diSabled). Its multimodal user interface is based on a coloured high resolution touch screen extended with speech input/output. The development process is focused on taking into account requirements of elderly people and people disabilities. The usability of the personal assistant was evaluated in a series of user tests with subjects from these particular user groups.

1 INTRODUCTION

It is a common observation that we are increasingly surrounded by a wide variety of interactive electronic products, not only in our professional but more and more in our private environments. Interacting with electronic products is becoming an essential skill which is needed nearly everywhere, starting from ticket sales machines and automatic teller machines to the panels of washing machines or many other kinds of electronic home equipment. In the home environment, especially the various types of audio-video entertainment devices as well as home automation and safety products are spreading rapidly.

Unfortunately, the user interfaces of these different interactive products vary vastly and often are not intuitive ones. Supplier-specific designs differ from each other and are often not even consistent within the same product line. This situation is aggravated by the fact that devices from different functional domains, say a VCR compared to a heating control system, often follow very different interaction principles. It is not surprising that users of such diverse technologies and interfaces are often overwhelmed by the complexity and inconsistency of the user interfaces they are supposed to operate in their home environment. This usability barrier is even more serious for elderly people or people with disabilities who, on the one hand, may have specific cognitive or physical impairments, and, on the other hand, may be even more dependent on the functionality provided by such devices. Supporting accessibility and usability of interactive technologies for a range of users as broad as possible, therefore, becomes a goal of highest relevance, not only from a social but also from a commercial perspective.

By the middle of the next century the proportion of elderly people (people aged 50 plus) will be bigger than the proportion of younger people and by the year 2020 the proportion of people older than 65 years will reach 20% of the European population [5]. Moreover, and often underestimated, the population of people with special needs amounts to 13% of the overall European population [2]. In order to use modern home equipment, this particular group often relies on the availability of specifically adapted control devices.

The project HEPHAISTOS, carried out in the framework of the European TIDE research programme, developed a hand-held personal assistant that puts its potential user in the position to control a wide range of electronic home products through a single consistent and easy-to-use interface. The interface was based on multimodal interaction with a touch screen and speech input and output. In this paper, we will describe the design approach used and some of the main issues and findings which resulted from the usability tests and which informed the design of the final system.

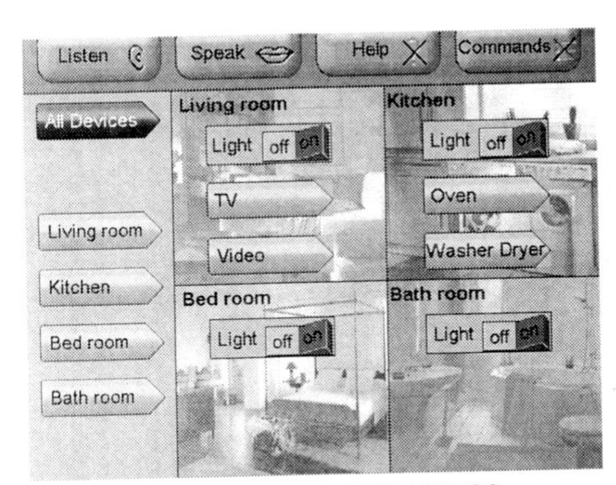

Figure 1: Main screen of HEPHAISTOS

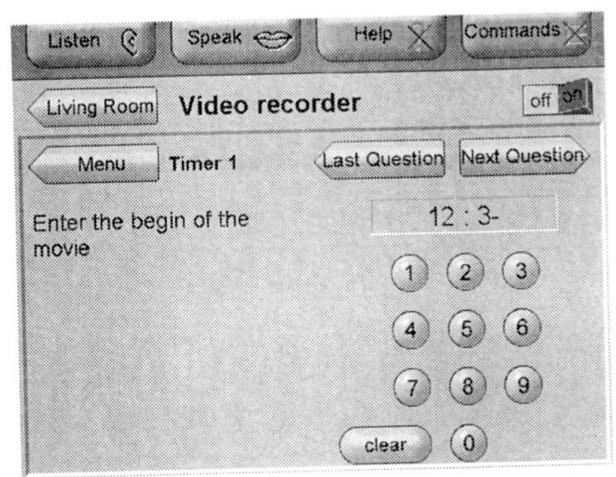

Figure 2. Time setting at the VCR panel of HEPHAISTOS

2 THE HEPHAISTOS APPROACH

The project used an iterative development with several cycles of usability testing. This process was accompanied by the development of a styleguide based on an earlier project (ESPRIT-FACE: Familiarity Achieved through Common User Interface Elements). The styleguide provided a set of basic interaction objects which had been optimized by usability testing, e.g. use of scales for parameter setting, single and multiple selection methods, or setting of clock and date [4]. The interaction techniques provided by the styleguide are designed for consistency across different functionalities and are applicable across a wide range of input/output facilities. More complex or conceptual issues, like the selection of appropriate metaphors or the design of dialog structures for systems with complex functionality, were initially not addressed but turned out to be pertinent to designing for the particular user groups the project focused upon.

The development process was set up as an iterative cycle running through four phases, namely analysis, design, simulation, and evaluation. Techniques for GUI development and design methods for customer electronics user interfaces as described in [1,6,8] were adapted for the conceptual design of the multimodal interface. In the following sections, we describe some of the main design issues and the results of the iterative user testing with respect to these issues. User testing was performed in several sessions with elderly users in Germany and users with different disabilities in Greece.

2.1 Metaphor-based Navigation

In order to control a wide range of interactive devices in a home environment, the user should have a good overview of the state of the different devices and should be able to access the different functions quickly. Therefore, an interface structure, which accommodates these requirements, is needed. Such interface structures are typically connected with a specific metaphor, which should provide a consistent visual framework and support the user with respect to transparency and familiarity of the interface. As a central *navigation* mechanism, three different interface metaphors suitable for home environments were tested in the user trials:

- A house metaphor presented an overview of the different rooms of the building without showing the different devices located in these rooms. Device control functions were only presented on the next level.
- In the 3D house metaphor, rooms were presented in a three-dimensional fashion with navigation functions presented as doors.
- The browser metaphor showed the different rooms in different screen areas in which the main devices located in these rooms were directly represented and could be accessed through buttons (Figure 1). While it is not possible to show all potential devices on the top level, the user gets a quick overview of the state of the most important devices.

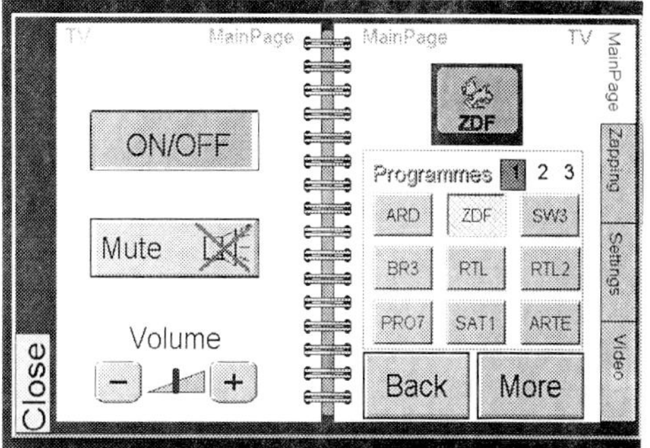

Figure 3: 2nd iteration: book metaphor

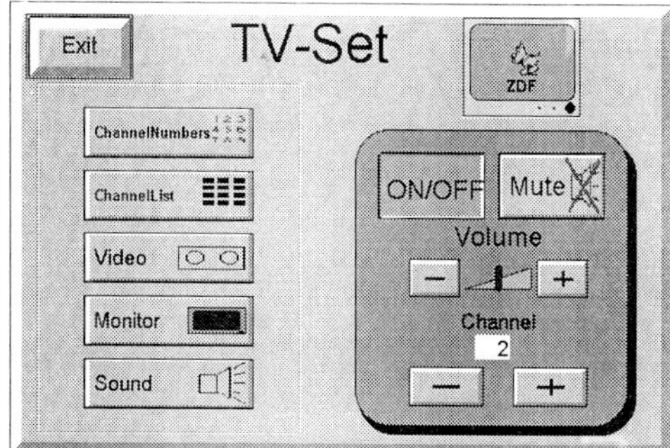

Figure 4: 2nd iteration: „magic" remote control

The user tests showed a clear preference for the browser metaphor which also resulted in less errors and problems. This finding is consistent with the results from research on menu techniques, where shallow menus are usually preferable to deep menu hierarchies [cf. 7]. In addition, direct access to often needed functions improves overall performance in using the device. While it is usually not possible to represent all devices directly on the top-level, a reasonable trade-off can be achieved by moving important or frequently needed devices to this level. It must be noted that, due to the use of a touch screen, the minimum size needed for each button leads to further restrictions with respect to the number of devices that can be shown on a single screen.

For operating the individual devices, like setting the timer on an oven or switching channels on the TV, again, two metaphors were tested against each other, a book metaphor with one page for each coherent set of functions and a gadget we called „magic remote control", which had the appearance of a standard remote control, but with an additional area showing varying information (Figures 3 and 4). In this comparison, the remote control metaphors showed to be superior in terms of user performance and preference. Due to time and cost constraints, usability testing of such complex interfaces cannot take into account all potential variations of the metaphors analysed. Nevertheless, the user test showed a clear pattern of preference even over several iterations, in which the different metaphors were improved with respect to the findings of the previous cycle.

2.2 The Assistant Concept

The more complex functions, especially those related to setting timers or searching for certain positions on a video tape, turned out to remain a source of errors for the elderly users, particularly if many different operations needed in the context were presented in parallel on the screen. For this reason, the concept of an assistant was developed, which guides the user through a sequence of interaction steps, such as programming the start and end times on the VCR. The assistant presents one interaction step at a time on the screen (see Figure 2). While this principle increases the number of interaction steps, it turned out to be very useful in order to enable the elderly users to achieve their goal.

Several design considerations resulted from the user tests: Initially, we presented each interaction step with a confirmation action, such as an OK-Button on each screen. Very often, however, users believed to have finished their task once a setting or value had been entered on the screen, and forgot to press OK. We, therefore, decided to introduce an auto-confirmation once the last entry action was performed, and to have that value confirmed explicitly in the next step, which took the form of a question. We also introduced an animated change of the assistant screen area contents, as users often did not recognize that the contents had already changed after their last action.

3 USER TRIALS

The project established two test sites. The National Institute for Rehabilitation of the Handicapped in Athens, Greece carried out trials with people impaired by speech, reading, upper and lower limb co-ordination or movement, and restricted concentration span. The University of Stuttgart IAT, Germany, carried out trials with elderly people aged from 64 to 84 years. A description of the first results can be found in [3]. In general, the interface concepts proved to be useful for better accessibility and usability of the electronic devices. With support through the assistant concept developed, the majority of elderly users tested were able to perform VCR programming tasks without previous instruction. Speech input was used and preferred by some users. Users tended to develop clear preferences with respect to the input mode and tended to stick to that mode. Interestingly, in the tests with disabled users there was no clear correlation between type of impairment and preferred mode. Actually, some of the users with speech problems rather preferred speech as they found it satisfying that the machine responded to their voice commands.

4 FUTURE DEVELOPMENTS

Whereas the simulations used for a first set of user trials consisted of a software solution solely, the final HEPHAISTOS demonstrator constitutes a touch sensitive palm top remotely controlling real devices which were connected via a home bus. A washing machine, an oven, a TV, a VCR and a lamp were chosen as representatives for the various electronic home equipment found in an average household. In a follow-up project called HOME, we are trying to overcome some of the current limitations of the demonstrator and to introduce new interaction modes. As major enhancements of HEPHAISTOS, HOME will integrate gesture recognition and free speech, which means that it will no longer be necessary to have a microphone installed in the assistant. Gestural commands will be analysed both in a isolated manner as well as in connection with speech commands, leading to a truly multimodal interaction.

5 ACKNOWLEDGMENTS

The authors wish to thank the members of the HEPHAISTOS consortium for their valuable contributions. The design of graphical elements was done by Ruth Diessl.

6 REFERENCES

1. Beck, A., Janssen, C., Weisbecker, A. and Ziegler, J. Integrating Object-Oriented and Graphical User Interface Design. Proceedings of the SE/HCI Workshop, Sorrento, Italy, May 16-17, 1994.
2. Botella, V. and Waldmeyer, M.T.A. Identification and grouping of Special Needs of PSN-Elderly with respect to User Interface Design. Deliverable 2, TIDE 1004 HEPHAISTOS, 1995
3. Burmester, M. and Fähnrich, K.P. HEPHAISTOS - A Multimodal Help Assistant for the Home Environment. In A.F. Özok and G. Salvendy (eds.): Advances in Applied Ergonomics, Proceedings of the 1st Int. Conf. On Applied Ergonomics (ICAE '96), Istanbul, Turkey: West Lafayette, 238-243, 1996.
4. Burmester, M. and Machate, J. "Common User Access" for Electronic Home Devices or 20 Ways to Set the Clock? In R. Oppermann, S. Bagnara and D. Benyon (eds.): ECCE7 Seventh European Conference on Cognitive Ergonomics. Human-Computer Interaction: From Individuals to Groups in Work, Leisure, and Everyday Life. Proceedings. GMD-Studien Nr. 233. Sankt Augustin, Germany, 97-110, 1994..
5. Dall, J.L.C. The Demography of Europe. In H. Bouma and J.A.M. Graafmans (eds.): Gerontechnology, Amsterdam: IOS Press, 31-38, 1992.
6. Görner, C. Vorgehenssystematik zum Prototyping graphisch-interaktiver Audio/Video Schnittstellen. IPA-IAO Forschung und Praxis 194, Berlin: Springer, 1994, in German..
7. Norman, K.L. (1991): The Psychology of Menu Selection - Designing Control at the Human-Computer Interface. Norwood, N.J.: Ablex.
8. Sanz, M., Gómez, E.J., del Pozo, F. e.a. Metodología de Diseno de Interfaces de Usatio Gráficas. Jornadas Técnicas del Proyecto TEMA. Telefónica Investigación y Desarrollo, Madrid, 1994, in Spanish.

Interacting with Electronic Healthcare Records in the Information Society

I. Iakovidis

European Commission, DG XIII, Telematics Applications Programme, Brussels, Belgium

This paper is concerned with technological and societal factors affecting the acceptability and usefulness of Electronic Healthcare Records. Recent efforts in Europe and the US are reviewed with the view to identify some of the key factors that relate to the technological and human-centred issues involved.

1. INTRODUCTION AND BACKGROUND

The European Union is committed to the development of Information and Communication Technologies (ICT), through a series of Research & Development (R&D) Programmes. At the forefront of the developments of new multimedia services and applications is the R&D Telematics Applications Programme (TAP) of the European Commission (EC) which supports and promotes the research, development and demonstration of telematics applications and services in areas such as healthcare, disabled and elderly, transport, education and training et al. [1]. The development of telematics applications and services provides an element of increased competitiveness for enterprises and opens up new perspectives for both work organisation and job creation. The diffusion of ICT at all levels of economic and social life is thus gradually transforming our society into an "Information Society".

The general preference in the United States, until recently, was for the term "information super-highways", implying a technology-based appreciation of current and future developments. By contrast, "Information Society" reflects European concerns about broader social and organisational changes and the quality of our lives, which will be delivered by the new information technologies. Consequently, it is concerned with the issue of *access for all* and easy use of the new services, to avoid the division of society between those who can access and exploit the new technologies and those who cannot. Putting people at the centre is a priority for the European Union as expressed in the "Europe's way to the information society: An Action Plan" (see http:\\www.ispo.cec.be). In practice, the EC intends to address social questions, to protect consumer interests and to improve the quality of public services.

This paper focuses on a specific issue relating to healthcare, namely, the issue of acceptability and use of electronic healthcare records (EHR) as the necessary step towards implementing the concept of *shared care*. This patient-centred shared care builds on health telematics networks and services, linking hospitals, laboratories, pharmacies, primary care and social centres offering to individuals a "virtual healthcare centre" with a single point of entry. The information shared by all care providers is patient health-related information which is comprehensible, reliable and confidential. In other words, the enabling factor of the patient-centred shared care is the availability of electronic healthcare records that are accessible, secure and highly usable in the European multilingual environment. The development and promotion of use of such EHR is one of the objectives of the Health sector of TAP.

2. ELECTRONIC HEALTHCARE RECORD

The experts in the field of medical informatics and telematics have been trying for many years to define the ideal EHR on both sides of the Atlantic. In 1991, the Institute of Medicine in USA published a report called "The Computer-Based Patient Record: An Essential Technology for Health Care" [2], describing the requirements of EHR, and making recommendations for the future. In the same year in Europe, the requirements of an EHR were formulated in the work-programme of European R&D Programme called AIM-Advanced Informatics in Medicine (the predecessor of the current Telematics Applications for Health sector of TAP). Further recommendations were agreed in the AIM/CEN Workshop on Medical Record in 1993 [3].

Here, we only outline the major categories of criteria that are emerging as accepted by experts world-wide. The EHR should: a) above all support the patient care and contribute to improvement of the quality of care; b) support and increase the efficiency of the healthcare professionals, as well as the non-care users such as managers and policy makers; c) provide the basis for medical research and education; and d) ensure confidentiality at all times. Today, neither the computer scanned forms in isolated PCs, nor the current hospital information systems satisfy the above criteria, despite the fact that many of them are called electronic health records or computer-based patient records. A definition of different levels of functionality of EHR is given by the Medical Record Institute (see http:\\ www.medrecinst.com). Here we refer to EHR as the one satisfying the above four criteria.

It is not difficult to imagine several direct benefits of an implemented EHR, including: rapid access to relevant data, improved communication among the users, flexible summary generation, reminders and warnings, decision support and guidelines, and many others. Among indirect benefits, one can mention billing, comprehensive data for quality assessment, research and planning, improved scheduling of resources, fewer repetition of lab tests, etc. [4].

So why despite so many benefits are there not many systems installed? There are many reasons, the main ones being categorised as (i) technology, (ii) standards, (iii) lack of leadership and vision, and (iv) legal obstacles including the security and confidentiality issues and most importantly, (v) user acceptability [5].

Concerning the technology, the concept of "virtual" or distributed EHR with data available at the point or origin of collection has been only possible in recent years thanks to the new technologies of distributed databases and the new communication possibilities (e.g. intranet and internet technologies). Some standards concerning the architecture and exchange format are appearing and being discussed within standardisation bodies, such as CEN TC 251 in Europe and ANSI-HISB in USA. The work on standardisation of vocabulary and terminology is a necessary part of communication and usability of EHR, in order to ensure the same meaning of the content to all users. This is being tackled through national and international projects such as the GALEN project [6] (see also http:\\www.ehto.be). A leadership that will foster the environment for wide implementation of EHR is needed at all levels, from hospital to regional and to international levels. This leadership is provided in Europe through the EC supported project PROREC [6], which promotes the use of EHR and establishes a network of permanent centres in Member States. In US, there is the Computer-Based Patient Record Institute with a similar mandate (see http:\\www.cpri.org).

The issue of acceptability together with the issues of security and confidentiality are the major hurdles on the way to widespread use of EHR. For example, in The Netherlands the penetration of computers is very high in general practice (80% of GPs have a computer), but the use of automated patient records is considerably less (only 40% use some form of EHR).

3. HUMAN-COMPUTER INTERACTION WITH ELECTRONIC HEALTHCARE RECORDS

The acceptability of EHR can be affected by a number of technology- and human-related factors. The main challenges from the technological point of view refer to the storage, maintenance and retrieval of multimedia information in different technological platforms and heterogeneous database systems, that may be geographically distributed. Also, the functionality that should be supported raises further research challenges, such as real-time content-based indexing and retrieval of multimedia information, development and maintenance of large active heterogeneous databases, development of high-speed telecommunications networks, etc.

One of the most important factors affecting the overall acceptability and usefulness of EHR is the user interface. The design and development of user interfaces for EHR is particularly challenging, due to the following characteristics of EHR usage:

- *diverse (groups of) users*, with diverse and changing requirements, i.e. with different background, education, training, skills, perspectives, requirements, preferences, etc; moreover, the same user may have changing requirements over time, e.g. a doctor utilising EHR either for diagnosis or for treatment;
- *changing environments*, including hospitals, health care centres, mobile care and first aid units, etc, either in rural or urban areas and patient homes;
- *different usage patterns*, e.g. diagnosis, treatment, information provision, research; personal or group collaborative work.

The above raise a series of challenges from the Human-Computer Interaction (HCI) perspective, related to capturing and input of data in EHR, as well as the presentation of the recorded data in a variety of forms, media and output systems, etc. In particular, specific technological areas that need to be addressed concern input and output devices (e.g. pen-based input, speech input), 2D and 3D interaction techniques, intuitive interface metaphors, mobile systems, multimodal interfaces, tailorable and adaptable interfaces, more natural access procedures (e.g. speech interfaces), computer-supported cooperative work (CSCW), intelligent interfaces, user identification procedures (smart cards), and user interfaces for mobile and nomadic services.

In this context, it is critical to ensure *accessibility* and *high-quality of interaction* for *all* potential (groups of) users, e.g. doctors, nurses, administrators, individuals, irrespective of the location of querying, the location and type of data storage [7], or the terminal configuration (e.g. low speed text-based, multi-modal graphical system). User interface adaptability and tailorability are, therefore, of great significance, in order to ensure that the user interface is universally accessible and meets the diverse individual, or group requirements and preferences in different systems and contexts of use.

Usability and high quality of interaction for *all* potential (groups of) users is also of great significance, since most of the users may not be experts in computer-based environments. Usability has been defined in several ways in HCI, and has been related to several attributes of the user interface, including ease of learning, consistency, efficiency, reliability, recoverability, safety, etc. In this case, particular emphasis should be given to speed of interaction, ease of use, accuracy, and flexibility, which are critical in different usage contexts, and may be difficult to satisfy simultaneously, as most of them have conflicting requirements.

Adaptivity and intelligent interface design techniques may be exploited in this direction, to enhance the quality of interaction, by detecting emerging and changing patterns of use, and dynamically modifying the user interface accordingly. For example, the information content could be filtered according to whether it is used for first aid care (where a minimal overview may be adequate), or during surgery operation (where specific details may be desired), or for research purposes (where accuracy may be the critical factor).

4. CONCLUSIONS

A usable electronic health record should reflect the needs and respect the right of access and confidentiality for all the users. This increasing demand for access and the multidimensional use of electronic healthcare records is contributing to world-wide confusion regarding what is an electronic healthcare record, and presents a major obstacle to its wider implementation and world-wide acceptability. More attention to human-computer interaction issues should be given, aiming to improve the acceptability and widespread use of EHR. Accessible and highly usable interaction with EHR will then realise one of the visions, namely the patient-centred shared care, in the context of the emerging Information Society.

REFERENCES

1. European Commission, DG XIII, Telematics Applications Programme (1994-1998), *Work-Programme*, 15 December 1994.
2. R.S. Dick, E.B. Steens eds., The Computer-Based Patient Record: An Essential Technology for Health Care, Institute of Medicine, National Academy Press, 1991.
3. Commission of the European Communities DG XIII, AIM -CEN Workshop on the Medical Record, Volume I-II, 1993.
4. K. Renner, Electronic medical Records in the Outpatient setting: Return-on-Investment Analysis, Medical Practice Management, May-June, 1996.
5. I. Iakovidis, The Research and Development Activities of the European Commission in the area of Electronic Healthcare Record, TEPR'96 proceedings, 110-117 (1996).
6. European Commission, Telematics Applications Programme, Guide to the 95-96 Projects.
7. K. Aisaka, K. Tsutsui, Y. Murakami, H. Ban, A. Hashizume, Y. Oka and S. Ishikawa, User Interface Design and Evaluation for Electronic Medical Record System, MEDINFO '95 Proceedings, 781-784 (1995).

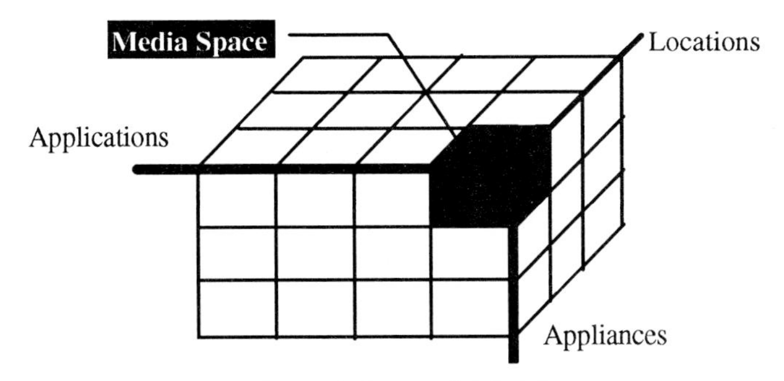

Figure 1. Research Model

3. FIELD TRIAL

What is needed is definitely not another atheoretical trial stemming from a simplistic marketing response to an engineering innovation. - J. Bryant and C. Love (1996, p:112)

Whereas planned affordances have been specifically designed for, emergent affordances exist as potentialities which may be novel and unforeseen. For example, in the University of Toronto's CAVECAT project, Mantei et al. (1991) report that an interface designed initially to allow desktop conferencing participants to view each other during the conference (the *planned* affordance) was spontaneously reappropriated by users to monitor their office for telephone calls, security purposes, and the like, when they were not in (an *emergent* affordance).

Emergent affordances, by definition, are not predictable apriori, thus rendering necessary an empirical study in order to validate the original research model. That is, hypotheses derived from the original model can only be based on an analysis of the planned affordances. The extent to which such predictions are valid can only be estimated by submitting the model to an empirical test, allowing for the emergence of unanticipated appropriations. In the context of our research, such empirical test will allow us to determine which appliance individuals indeed prefer to use to access which content, and in which location within their home.

Moreover, because emergent affordances are highly context dependent, such empirical study must take place in a natural setting which is a representative microcosm of the information superhighway of the future. Finally, because unanticipated appropriations evolve and diffuse into stable patterns of behavior over time, such study must also be longitudinal. Indeed, only a longitudinal study will allow for the identification of lasting emergent affordances which, in essence, redefine the user actions traditionally associated with a given media space. In the context of the current research, this empirical test will be provided by the infrastructure developed as part of the Intercom Ontario field trial (see Durlak and Thomassin Singh (1995) for a description of the broadband community network developed in this trial).

4. CONCLUDING REMARKS

This research constitutes a concrete step towards filling up the theoretical void which surrounds our current understanding of the social impacts of media. In particular, it develops a research model which examines the various potentialities built into a broadband digital community network which is designed to deliver several content applications over several information appliances, in a variety of locations within the home--our vision of what the information superhighway of the future should be. The knowledge gained from an empirical test of this model should allow designers, developers, providers, and users of broadband networks alike to predict which of the many potentialities enabled by such networks will *actually* be used by individuals and communities, how they will be used, and, perhaps more importantly, why--thereby providing for a well-informed management of what appears to be our inevitable transition into a digitally connected society.

REFERENCES

Abel, M.J. (1990). Experiences in an Exploratory Distributed Organization. In Galegher, Kraut, and Egido (Eds.), *Intellectual Teamwork: Social and Technological Foundations of Cooperative Work.* New Jersey: Lawrence Erlbaum Associates, 489-510.

Buxton, W. and T. Moran (September 1990). EuroPARC's Integrated Interactive Intermedia Facility (IIIF): Early Experiences. In *Proceedings of the IFIP WG8.4 Conference on Multi-User Interfaces and Applications*. Heraklion, Crete.

Bryant, J. and C. Love (1996). Entertainment as the Driver of New Information Technology. In Dholakia, Mundorf, and Dholakia (Eds.), *New Infotainment Technologies in the Home: Demand-Side Perspectives.* New Jersey: Lawrence Erlbaum Associates, 91-114.

Dutton, W.H. (1995). Driving into the Future of Communications? Check the Rear View Mirror. In Emmott, S.J. (Ed.), *Information Superhighways: Multimedia Users and Futures.* London: Academic Press.

Couch, C.J. (1996). *Information Technologies and Social Orders.* New York: Aldine de Gruyter.

Gaver (1992). The Affordances of Media Spaces for Collaboration. In *Proceedings of the ACM 1992 Conference on Computer-Supported Cooperative Work.* October 31-November 4, Toronto, Canada.

Gergen, K. (1991). *Entropy: A New World View.* New York: Bantam Books.

Gore, A. (1994). Plugged into the World's Knowledge. *Financial Times*, September 19.

Heath, C., P. Luff and A. Sellen (1995). From Video-Mediated Communication to Technologies for Collaboration: Reconfiguring Media Space. In Emmott, S.J. (Ed.), *Information Superhighways: Multimedia Users and Futures.* London: Academic Press.

Mantei, M.M., R.M. Baecker, A.J. Sellen, W. Buxton, and T. Milligan (1991). Experiences in the Use of a Media Space. In *Proceedings of CHI '91.* New Orleans, 203-208.

Meyrowitz, J. (1985). *No Sense of Place: The Impact of Electronic Media on Social Behavior.* New York: Oxford University Press.

Norman, D (1992). *Turn Signals Are the Facial Expressions of Automobiles.* Reading, MA: Addison-Wesley.

Rifkin, J. (1981). *Chaos: Making a New Science.* New York, NY: Penguin Books.

J. Durlak and D. Thomassin Singh (1995). Intercom Ontario: A User-Centered Field Trial. In *Proceedings of the 2nd International Workshop on Community Networking: Integrated Multimedia Services to the Home.* June 20-22, Princeton, NJ.

Estimating user interests from their navigation patterns in the WWW

Jesus Favela and Armando Carreon

{favela, jcarreon}@cicese.mx
Computer Science Department, CICESE Research Center, Ensenada, Mexico*

1. INTRODUCTION

The design of tools that help users navigate the Web must take into consideration the user's navigation patterns. The usability of search engines in the Web can be enhanced if we study the user's mental models while interacting with the browser. The objective of the research presented in this document is to identify the user commands on a browser that give some indication as to whether the documents being browsed are of interest to the user. Our final objective is to build a search agent that will use this information to build implicit queries by guessing the user information needs from their interest in Web documents as expressed by their navigation patterns. We have named this information retrieval strategy Guided Information Exploration [1].

2. GUIDED INFORMATION EXPLORATION IN THE WWW

Figure 1 shows the cognitive tasks that an information seeker is involved with during the process of accessing information. First, the user needs to realize that he needs information, and that this need can be satisfied. Second, he approximates his information needs in the form of a question or query. Third, the user interprets the results of the retrieval presented by the system. After these results are interpreted and evaluated, if the user is not satisfied with this information, he can either modify the query to fine-tune the search, or he can use this information to reformulate his information requirements. Declarative query languages such as SQL have simplified the formulation of queries. The later problem, that of modifying information needs in the light of new information, is what information exploration is aimed at. In the extreme case, when the formulation and subsequent changes to the information needs are not done by the user directly, but by the system (based on implicit information obtained by the monitoring of the user actions) is what we call guided information exploration. Since the user is not explicitly asking for information we say that the system is retrieving unsolicited information relevant to the problem at hand.

An important aspect for the implementation of Guided Information Exploration in the WWW is the determination of user interests from the user's interaction with documents presented in the browser.

---

* This work was partially supported by CONACYT grant No.CO24-A.

3. METHODOLOGY

To identify user interests from their navigation patterns we applied three techniques to a group of users:

1. Questionnaires. Used to determine the navigation strategies followed by the users when confronted to different use scenarios, as well as to know what options of the browser were used and how. The questionnaire was answered by 20 users.
2. Observation of navigation sessions. Users were observed while working on everyday tasks without knowing why they were being observed. 10 users were observed during 5 different navigation sessions for a period of 25 to 35 minutes. The idea of monitoring the users in 5 different sessions was to determine their recurrent navigation patterns.
3. Monitoring navigation sessions using an extension of a Web browser that was explicitly written for this purpose. From this information we related user interests to viewing time normalized by page size.

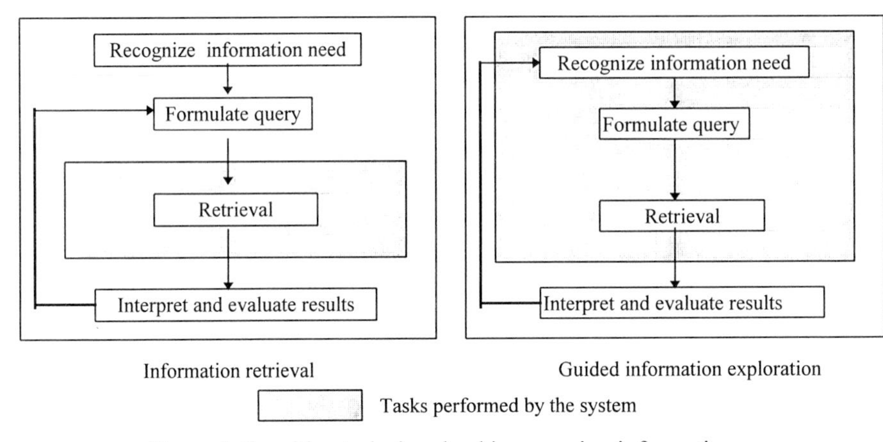

Figure 1 Cognitive tasks involved in accessing information

4. ANALYSIS OF RESULTS

An interesting finding of our observations is that most users are not aware of most of the navigation tools provided by their browsers. The command most often used was the "back" button, to return to the previous page and link selection. The browser's interface favors depth-first search as users focus on the links embedded in a page while the "back" button, used for breadth-first search, is a secondary navigation tool.

From the user navigation patterns indentified in our study we distinguished the three types of users found in [2]:

1. "Serendipitous user." Is the user that has no specific purpose while navigating the Web.
2. "General-purpose user." Is the user who consults different information sources which have high probability of having information of interest to him.
3. "Direct user." Is the one who knows where to find the information he is looking for.

It is common for users to follow links from a base document and then loose track of the place where they started browsing or even forget what they where looking for in the first place. It is also common to follow links with little or no relation with the information they were looking for, because these links satisfy other interests of the user.

We observed that most of the time users do not read the complete Web documents, but rather follow links that are contained in these documents. In average users follow between 2 and 3 links per document. This behavior however gives little indication of user interest in the document. The number of links followed in a document is thus a poor parameter to determine user interests in the anchor document.

From the analysis of our observations we determined 5 actions that give an indication that the document that is being visualized is of interest to the user. Listed in order of relevance, these user actions are:

1. Time taken to visualize the document, normalized by document size.
2. Add an URL to the "bookmark"
3. Print a document
4. Save a document
5. Send e-mail to an address contained in a document.

5. A WEB SEARCH ENGINE USING IMPLICIT QUERIES

The results of this study are being used to implement an intelligent agent to help users retrieve information relevant to the agent's best guess as to what the user information interests are. The agent formulates hypotheses of user interest by measuring the user's interest in the document he visited.

The architecture of our agent is made of the following three main modules (Figure 2):

- The first module measures the importance of a given page in the WWW to the user information interests. The agent creates a knowledge-base with a profile of each user using all the pages that the user has visited, the time it has spend with each page, and some of the most important actions performed. There are two key processes to create this knowledge base: the analysis of the information contained in the page and the monitoring of its manipulation by the user. Only the textual content is currently analyzed using the TfIDf information retrieval algorithm [3], and a weighting factor determined by the position of the text within the page and its enclosure in an HTML tag. The monitoring of the user's navigation patterns is used to estimate the importance of the information contained in that page to the user. For this purpose, we are considering the amount of time a user spends with the page, the number of times it accesses the page, explicit feedback given by the user (optional) to indicate his/her interest in the page and the keywords used in explicit search.
- The second module is responsible for automatically formulating a query to be sent to an WWW search engine. The query is made with the keywords with highest ranking from the agents knowledge base, and it is constantly being updated as the user traverses the Web.
- The third module is the user interface which allows the user to manipulate the documents that are being proposed by the system as being relevant. Through the interface the user can set query formation parameters, such as, the weight given to the different user actions or how much of the user's navigation history will be used to estimate his/her interests.

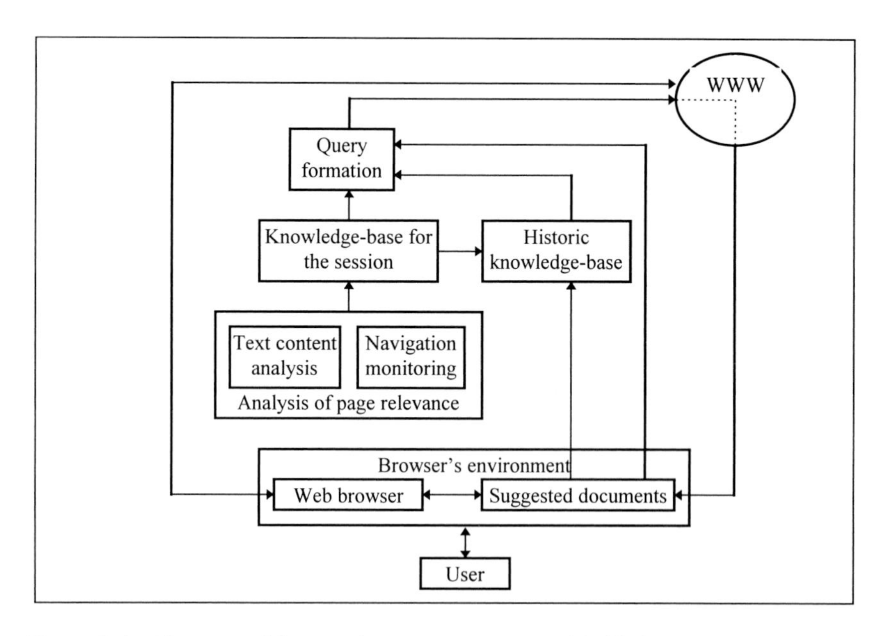

Figure 2 Architecture of the search agent to support guided information exploration

6. CONCLUSIONS AND FUTURE WORK

Given the continuous growth of the WWW, it is important to continue the development of tools that help users navigate and find information relevant to their information needs. Understanding the navigation patterns of Web users can make an important contribution to this end. In contrast with other studies of browsing strategies [4], our objective is to identify user actions that could help us estimate user interest in the information being browsed. The results of this study are being used to implement an agent that can create hypothesis of user information interests and suggest information relevant to these interests.

We are currently evaluating the effectiveness of our agent using implicit feedback of user interests and a combination of implicit and explicit user feedback [5].

REFERENCES

1. J. Favela, Capture and Dissemination of Specialized Knowledge in Network Organizations, *J. Organizational Computing*, vol. 7, (1997).
2. J, Cove and B.C. Walsh, Online Text Retrieval Via Browsing, Information Processing and Management, 24(1) (1988).
3. G. Salton, Automatic Text Processing, Addison-Wesley, 1989.
4. L. Catledge and J. Pitkow, Characterizing Browsing Strategies in the World Wide Web, Computer Networks and ISDN Systems, 27(6), (1995).
5. Hidekazu and T. Kamba, Learning Personal Preferences on Online Newspaper Articles from User Behaviors, Proc. of the 6th Conf. on the World Wide Web, (1997) 291.

Page Design Guidelines for Improving World Wide Web Navigation

Néstor J. Rodríguez, José A. Borges and Israel Morales

Center for Computing Research and Development
University of Puerto Rico, Mayaguez, P. R. 00681
nestor@exodo.upr.clu.edu

Abstract

Navigation is one of the most critical aspects of browsing pages in the World Wide Web. Users spend a significant amount of time moving from page to page in search of the desired information. The poor usability of Web pages and lack of good navigation tools wastes user's time, increases the Internet traffic, and contributes to the user's frustration. This research studies and evaluates three page design practices that can potentially improve the navigation on a Web site. These practices are: the selection of appropriate link names, the adoption of an index of links, and the use of navigation buttons.

1. INTRODUCTION

One of the most critical aspects of browsing pages in the WWW is navigation. Due to the way most Web sites are structured, users spend a significant amount of time moving from page to page in search for information. The poor usability of many Web pages is one of the major causes of delays, waste of time, Internet traffic, and user's frustration while navigating the Web. Inappropriate and misleading link names, disorganized pages, and other usability problems, contribute to the disorientation of users that become easily "lost in hyperspace".

Several alternatives have been researched and proposed to help with the problem of navigation in the Web. Wood, et. al. [1] suggested a method that provides a 3D graph of the place in which the user was located to allow an easier retracing of previously visited pages. Mukherjea, et. al. [2] proposed a method that consisted of showing the context of nodes in a path to provide users extra information about the topics of nodes. Other researchers [3,4] have proposed a multimodal interface that combines the use of speech and mouse. These projects use speech to overcome the sequential nature of present browsers and to expand the scope of selection of a hyperlink which is restricted by the screen area in the pointer based interaction.

From our heuristics evaluations of Web pages [5,6] we identified three page design practices that could potentially improve navigation on a Web site. These practices are: the selection of appropriate link names, the adoption of and index of links to pages or topics, and the use of navigation buttons. In the following sections we discuss how these practices can improve navigation and how to integrate them in the page design process.

2. SELECTING APPROPRIATE NAMES FOR LINKS

The selection of names for links is a very critical aspect in the design of Web pages. In most cases the name of a link is the only hint that a user has about the content of the page referenced by the link. If the name is misinterpreted, the user could end up reaching a useless page. This situation results in a waste of time for the user and causes an unnecessary traffic on the Internet. Thus, it is important that this problem be minimized in the design of Web pages.

To get a better understanding of the problem of link naming we decided to conduct a study to determine how well can people guess the content of a page from the link name. Fifty links randomly selected from ten commercial home pages (also randomly selected) were used for the study. Sixteen people participated in the study. Each person was provided with a hard copy of the ten home pages with the selected links identified with a number. They were asked to write what they thought was the content of the page pointed by each of the fifty links. Each response was classified according to the clue that the link name provided about the content of the page it pointed to. The following four categories were used to classify the responses:

none - the link name did not provide an idea of the content of the page
wrong - the link name provided the wrong idea about the content of the page
fair - the link name provided some idea about the content of the page
good - the link name provided a good idea about the content of the page

The category assigned to each response was determined by a consensus of the subjective opinions of three experimenters. The experimenters' opinions were based on the analysis of the content of the fifty pages pointed by the links selected for the study.

The results of the study indicate that in approximately 65% of the cases the names assigned to a link by page designers provide a good idea or some idea about the content of the pages the links point to. However, a significant number of links do not provide a good hint. In approximately 25% of the cases the link names suggest a wrong idea about the content of a page. Following these kind of links is bound to result in a waste of time. We also found that in approximately 9% of the cases the link names do not provide a clue about the content of the page. These kind of link names could discourage exploration or could result in a waste of time if followed.

An analysis of link names that provided poor clues revealed that in most cases the link name selected by the page designer did not reflect the content of the page it pointed to. In other cases the names were too general and did not provide enough information for a person to get at least a fair guess of the page content. Some of the names were just not understood.

Based on the previous analysis we developed the following guidelines for naming links:

- Link names should reflect the content of the page they point to.
- Avoid names that are too general if you can be more specific. For example, it is more meaningful to use:
  Crosswords and Comics instead of *Recreation*,
  Deadlines instead of *Important Stuff.*
- Avoid names that are not commonly understood such as technical words and terms with in-house or local meaning. For example, it is more meaningful to use:
  Industrial Affiliates Program instead of *IAP*,
  Computer and Communications Technologies instead of *Telematics.*

To determine if these guidelines could help page designers to generate more meaningful names we conducted a usability test. A group of ten people were asked to use the guidelines to provide names for links to six Web pages. Another group of ten people were asked to provide names for links of the same Web pages, without using the guidelines. The link names provided by each participant for each page were classified according to the clue that they provide about the page they pointed to (*good, fair, wrong, none*).

The results of the study indicate that in approximately 62% of the cases, the participants that used the guidelines were able to come up with a link name that provided a good or fair clue about the content of the page it pointed to. However, the participants that did not use the guidelines were only able to come up with a good or fair name in approximately 45% of the cases. Thus, use of the above mentioned guidelines could help page designer to select names for links that provide a good or fair clue of the content of the page they point to.

3. USING INDEX PAGES

The availability of an index of pages or topics of a Web site could be very useful because it allows a quick access to the pages of the site without much navigation. From our usability studies of guidelines for designing WWW pages [5] we noticed that once the users discover an index page they would frequently return to that page while searching for information on the site.

To determine whether the adoption of an index page can improve the navigation on a Web site we conducted an experiment with two groups of Web users. Each group was asked to perform 15 tasks on a version of a Web site. The tasks required the users to look for specific information from different pages of the Web site. One of the groups performed the tasks on the original version of the site and the other group on a modified version of the site. The modified version of the site included an alphabetical index page. This page consisted of a linear list of alphabetically ordered groups of links to different pages of the site. The header of the page featured the letters of the alphabet as links to an anchor at the beginning of the corresponding group. The header was made static to prevent it from getting out of sight while scrolling the list. A link to this index page was provided in the home page of the site and in the primary pages of the site (the pages pointed by the links of the home page). To prevent any bias towards the use of the index page the users of the modified version were not told about the existence of the index page. All the tasks began on the home page of the site. The time to complete the tasks (excluding page download time) was measured.

The results of the experiment are presented on table 1. The times presented in this table correspond to the average time it took the users to complete the task on each site version. The results indicate that on average the users of the site with index page were able to complete most of the tasks in a shorter amount of time than the users of the site without index page. In eleven out of fifteen tasks the average completion time of the users of the site with index page was lower than the average completion time of the users of the site without index page. The tasks for which the users of the site without index page completed in a shorter average time were four of the first five tasks of the experiment. During these first five tasks many of the users of the site with index page were not aware of the existence of the index page and thus, could not benefit from it. After the fifth task most of the users were aware of the index page and used it frequently to speedup the navigation on the site. Thus, a fair comparison of the two versions of the site should be based on the results of the last ten tasks. It is evident from the results of the last ten tasks that the adoption of an alphabetical index page can improve the navigation on the WWW because it helps the users find information faster.

Table 1. Comparison of Average Times to Complete Search Tasks on Web Sites with and without Index Pages.

	Task														
	a	b	c	d	e	f	g	h	i	j	k	l	m	n	o
Site with index page (average time in sec.)	52	27	46	13	15	63	64	19	24	17	18	27	16	18	23
Site without index page) (average time in sec.)	12	16	41	25	12	107	99	91	31	24	27	118	115	179	47
Time Ratio (Site B/Site A)	.2	.6	.9	1.9	.8	1.7	1.6	4.8	1.3	1.4	1.5	4.4	7.2	9.9	2.0

We found additional evidence in support of the incorporation of an index page. There were 23 cases in which a user of the site without index page could not complete a task. This contrasts with only two cases on the site with an index page. These results suggest that the incorporation of an index page results in a higher degree of success in the search of information on a Web site.

4. NAVIGATION BUTTONS

The adoption of navigation buttons in Web pages has the potential for improving navigation on a Web site because they facilitate a quick move to key pages of the site. Based on the results of the study with index pages it is evident that an *index* button can be very useful as a shortcut to reach an index page.

The study of index pages also revealed that the users tend to revisit the home page of the site by consecutive activations of the *back* button of the browser. This navigation method is ineffective if the back button must be activated more than once because the user would be revisiting pages that serve no purpose in their search for information. A faster way to get to the home page can be achieved by providing a button or link in each of the pages of the site that link to the home page. By pressing this button the users will get immediately back to the home page of the site without having to revisit unnecessary pages.

Other buttons such as a *predecessor* button (one that links to the page that is the logical parent of the current page) and a *top* button (one that links to the top of the current page) could be useful. However, we have not conducted studies that can support their usability.

5. CONCLUSIONS

The majority of the link names found in Web pages provide a good or fair clue of the content of the page they point to. However, there is a significant number of link names that do not. These names can mislead the users to access pages that are not of his/her interest. This problem can be improved if page designers follow simple guidelines for naming links such as the ones proposed in section 2. These and other guidelines for designing Web pages are described in more detail in the Web site *Guidelines for Designing Usable WWW Pages* [7].

Our studies with index pages demonstrated that the navigation time can be reduced by adopting and index page on each Web site and providing a link to this page in most of the pages of the site. An index page consisting of a linear list of alphabetically order links and a static header with the letters of the alphabet have proven to be very useful.

The adoption of an *index* button and a button to reach the home page of a site has a good potential for improving navigation. However, more experiments need to be conducted to study the effectiveness of these and other navigation buttons.

REFERENCES

1 Wood, A., Drew, N., Beale, R., and Hendley, B., HyperSpace: Web Browsing with Visualization. *Proceedings of The Third International WWW Conference*, Darmstadt Germany, April 1995.

2 Mukherjea, S., Foley, J., and Hudson, S., Visualizing Complex Hypermedia Networks through Multiple Hierarchical Views. *Proceedings of CHI'95*, pp. 331-337, Denver CO, May 1995.

3 Novick, D., and House, D., Spoken Language Access to Multimedia (SLAM): A Multimodal Interface to the World Wide Web. *Proceedings of Multimedia 95 Conference*, San Francisco CA, March 1995.

4 Jimenez, J., Speech-Based WWW Browsing, Master Thesis, Dept. of Electrical and Computer Engineering, University of Puerto Rico, Mayaguez, Puerto Rico, July 1996.

5 Borges, J. A., Morales, I. and Rodríguez, N. J. Guidelines for designing usable world wide web pages. *Conference Companion of the Proceedings of CHI'96*, pp. 277-278, Vancouver, Canada, 1996.

6 Borges, J. A., Morales, I. and Rodríguez, N. J. Page Design Guidelines Developed through Usability Testing. In C. Forsythe, E. Grose and J. Ratner (Eds.), *Human Factors and Web Development*. Lawrence Erlbaum Associates, Inc., 1997.

7 Rodriguez, N. J. (1996). Guidelines for designing usable WWW pages. <http://exodo.upr.clu.edu/rumhp/hci/wwwdesign.html>.

Virtual Reality

MARTI: Man-machine Animation Real-Time Interface: The Illusion of Life

Christian Martyn Jones and Satnam Singh Dlay

Department of Electrical and Electronic Engineering, Merz Court, University of Newcastle, Newcastle upon Tyne, NE1 7RU, United Kingdom

The research presents MARTI (Man-machine Animation Real-Time Interface) which combines novel studies in the field of speech recognition, facial modelling, and computer animation. The system uses vocal sound-tracks of human speakers to provide lip synchronisation of computer graphic or animatronic facial models, to realise the first natural interface and animation system capable of high performance for real-users and real-world applications.

1. INTRODUCTION

MARTI is designed as the complete human computer interface and special effect animation toolbox. Previous research has considered the extraction of control signals from human facial motion using image recognition systems, however the analysis requires that white dots are used to highlight specific key articulation points. In addition researchers have considered analysis of the acoustic soundtrack. However these systems are limited by the accuracy of their speech recognition and the requirement that they are trained on the single 'puppeteers' voice. Furthermore they do not provide timing and rhythm information to allow accurate synchronisation of the facial motion to the original speech. MARTI overcomes these limitations, allowing automatic lip synchronisation without the normal constraints of head-sets, reflectors, or other positional hardware, and uses only a single input source: speech. All the requirements of the communication and animation film industries can be met by the system including providing friendly life-like interfacing to the latest technologies and the mapping of human speech to non-human and atypical images.

The research concept is equally applicable to medical applications and may be used to further studies in the fields of speech therapy, the psychology of speech, and medical linguistics. Computer animation, including the computer game industry, 'virtual worlds' and architectural design tools, and the film industry, will benefit from the system, with the elimination of current requirements of head-sets and puppeteers to control facial motion.

MARTI uses our latest research in a number of engineering disciplines to achieve a new level of realism for animated speech [1-2]. We are working with the latest hybrid connectionist/hidden Markov model speech recognition system (STRUT [3]) and the DARPA acoustic-phonetic continuous speech database (TIMIT) to provide us with highly accurate phone recognition and timing for speaker independent continuous speech, and with knowledge from the animation industry and the physiology of human speech we have developed facial models and an accurate automated animation toolbox.

2. HUMAN LIP SYNCHRONISATION

The acoustics of speech and the articulation are obviously related. MARTI addresses speech from a phonetic view-point and recognises elements that are acoustical distinct. These fundamental sounds of speech can then be mapped into facial positions to achieve automated lip synchronisation.

The study of American English combines a theoretical understanding of the articulation of American speech and a practical consideration of conversational speech. A number of American subjects were asked to recite all 60 TIMIT phonemes and example sentences from the test data set whilst their mouth positions were videoed. From these studies we were able to suggest the minimum number of distinct mouth positions, *visemes*, used in everyday speech, Table 1. We then use wire-frame parameterized models of the tongue, lips, teeth, and jaw and match these articulation characteristics to the visemes, or control servos on the animatronic system to correspond to the facial images of the human speaker. Furthermore, our recognition stage provides timing information for each viseme in the transcription allowing us to accurately synchronise the facial model to the original speech sound-track.

The speech recognition 'front-end' achieves accuracy in excess of 58 % for speaker independent, continuous speech recognition, without any 'dictionary look-up' and 'grammatical checking' (which limits the user to sensical words and sentences) and provides 60 element phonetic transcriptions with time-signatures. The hybrid system consisted of a three layered MLP comprised of 234 input neurons (feature data), 1000 'hidden' neurons, and 61 output classifications representing the 61 phonemes. Without constraints in input speech this performance represents the highest level of speech recognition to date [4].

The speech recognition transcription returned is mapped into visemes to be modelled for graphical output. Unfortunately the recognised transcription contains inaccuracies caused by incorrectly inserted, deleted, and substituted phonemes. However, any phoneme that is incorrectly recognised as another but is in the same viseme group as the other will not affect the visual performance of the system. Furthermore, if either of the neighbouring phones of the inserted or deleted error appear in the same viseme group as the error then the output model will not be affected and the visual performance improves. Although our recognition transcriptions now contains only 26 viseme elements as opposed to the original 60 phonemes it has been shown that these groups accurately describe the 'visual' articulation of American speech. The affect of viseme groupings on insertion, deletion and substitution errors has been analysed numerically and shows an overall performance increase for the speech to lip synchronisation of 8.5% to nearly 67%, Table 2.

3. ANIMATION LIP SYNCHRONISATION

The animation industry portrays the illusion of life by the motion of non-human characters, including facial expressions and lip synchronisation. Current controls of the facial motion originate from hand puppetry and allow performers to manipulate the character by multiple analogue joystick controls. The relationship between the human motion and the output face can be complex and requires the puppeteer to be highly skilled. Hence the industry would benefit from the elimination of complex control systems with the development of automated lip synchronisation.

distinguished between search strategies which exploit the dynamic, 3-D graphical nature of the interface by navigation around the existing document space and dragging existing POIs, and more textually oriented search strategies, like typing in new keywords. Finally we considered users' subjective assessments of the system. Fifteen graduate students from the University of Nottingham took part in the study; all were computer literate, although none had ever used 3-D information visualisation tools before. Seven study tasks were designed to test interface knowledge and ability to conduct semantically driven spatial analysis of the document space with only limited training. For each task, a target document was specified and users' selections compared to the target. The tasks ranged from highly specified to less specified searches and the degree to which participants were given freedom to select their own keywords for a search. Selections were evaluated in terms of time to make a selection and accuracy of that selection, determined by considering the overall relevance rating assigned to the selected document by VR-VIBE. The methods used were established through detailed analysis of the videotaped activity protocols.

Results from this study suggested that novice users with little training were able to select documents relevant to several POIs using spatial position and item features (size and brightness) as cues. Overall, less time for selection and better accuracy was achieved on the tasks with more POIs, but less document icons displayed. Analysis of the video tapes suggested that participants who scored the highest accuracy scores frequently used POI drags, exploiting the dynamic nature of the representation to disambiguate the spatial location of documents and thus their relative and absolute relevance; this was particularly the case when many POIs had been defined. Here, the POI drag facility was used to create symmetrical, geometric shapes, e.g. a pentagon-based pyramid and a diamond with a square cross section. Participants then navigated through the 3-D space to locate the large/bright document icons around the middle of the created structures. Notably, tasks which required the specification of many keywords resulted in the least relevant selections. This suggests that whilst browsing an existing set of documents is facilitated by the visual display, specifying the appropriate keywords for the search, not surprisingly, still presents difficulties.

A second study directly addressed the interaction between interface style and task type, to address whether VR-VIBE provides relatively more success on such tasks than more traditional, text-based web, keyword interface interfaces. Again, tasks varied from highly specified searches to less specified browse-search-and-select tasks, where users were able to specify their own keywords. Twenty-six graduate students from the University of Nottingham took part in the study; subjects were randomly allocated to use the web based interface or VR-VIBE in completing the experimental tasks. Data collected were time to find an item and subjective assessments of VR-VIBE. Fifteen tasks of three different types were presented. Tasks of type 1 (TT1) were simple search tasks. For example, participants were asked to locate a specific document, given the author and title. Tasks of type 2 (TT2) were more open-ended. Participants were asked to locate the 3 most relevant documents to a particular research area e.g. groupware. Finally, tasks of type 3 (TT3) were highly underspecified and included exclusionary information. For example, participants were asked to find 5 documents which address the use of multimedia tools for network information retrieval but were told to ignore any documents pertaining to UNIX based systems. The mean time (minutes) to retrieve and record items for the different task types (TT1, TT2 and TT3) are shown in Table 1.

Table 1
Mean times to select items by group and task type

Interface type	Task type 1	Task type 2	Task type 3
WEB	2.81 (sd = 0.91)	4.54 (sd = 1.59)	3.04 (sd = 1.33)
VR-VIBE	4.00 (sd = 1.98)	4.40 (sd = 1.55)	1.64 (sd = 1.29)

The preliminary results suggested that the web-based interface proved faster for simple, highly specified queries (TT1), that there was no difference for slightly more complex queries (TT2), but that for queries which required both browsing and searching and which involved exclusionary information VR-VIBE proved quicker (TT3). Further analyses are currently being carried out on these data. As before, participants' evaluations of VR-VIBE were generally positive.

4. CONCLUSIONS AND SUMMARY

In conclusion, our results suggest that such 3-d visualisation systems may out-perform traditional text-based searches for complex, under-specified browse and search activities. However, for simple, single item queries where the query is highly specified, visualisation of database items appear to offer little advantage. Further studies concentrating on multi user searches and use over longer periods of time are currently being planned.

REFERENCES

1. Kaiser, M.K. (1993) Introduction to Knowing, in Pictorial Communication in Virtual and Real Environments, 2nd Edition, edited by Stephen R. Ellis. Taylor and Francis.
2. Shu, N.C. (1988) *Visual Programming*. Van Nostrand Reinhold Company, Inc., New York, USA.
3. Chalmers, M. and Chitson, P. (1992) Bead: Explorations in Information Visualisation, in Proceedings of SIGIRU92, ACM Press, pp330-337, June 1992.
4. Furnas, G.W., Landauer, T.K., Gomex, L.M. and Dumais, S.T. (1987) The Vocabulary Problem in Human System Communication" *Communications of the ACM,* Vol. 30, No. 11, November 1987 pp 964-971.
5. Rennison, E. (1995) Personalised Galaxies of Information, in Proceedings of CHI'95, ACM Press, pp31-32, May 7-11, 1995.
6. Benford, S. Snowdon, D. Greenhalgh, C., Ingram, R., Know, I. and Brown, C. (1995) VR-VIBE A Virtual Environment for Co-Operative Information Retrieval, Proceedings of EuroGraphics T95, Maastricht, Holland.

Display of conversation partners using 3D-facial model in the multiparty video communication system

Toshiya MURAI, Hirokazu KATO, Seiji INOKUCHI

Osaka University, Department of Systems Engineering,
1-3 Machikaneyama-cho, Toyonaka 560, JAPAN

ABSTRACT

In the traditional video communication system, the user often feels that all users in monitor are looking at him and they can not determine that who is speaking to whom. So we have already proposed the video communication system in which we can see that who speaks to whom. In this paper, to develop the system, we propose the method of the display of conversation partners by changing direction of the facial image in the system. To display the facial direction, we use a 3D facial model. The texture of the user's front view is mapped onto the front part of this facial model. So we can change the user's facial direction by rotating the facial model.

1.INTRODUCTION

In the traditional video communication system, the user often feels that all users in monitor are looking at him and they can not determine that who is speaking to whom[1]. Because all users are respectively look at their monitor or video camera. Hence, when we begin to converse, in this system, the speaker often tells his name and his listeners' name to his listner. Sometimes we can't communicate smoothly for this reason. This problem is shown in figure 1.

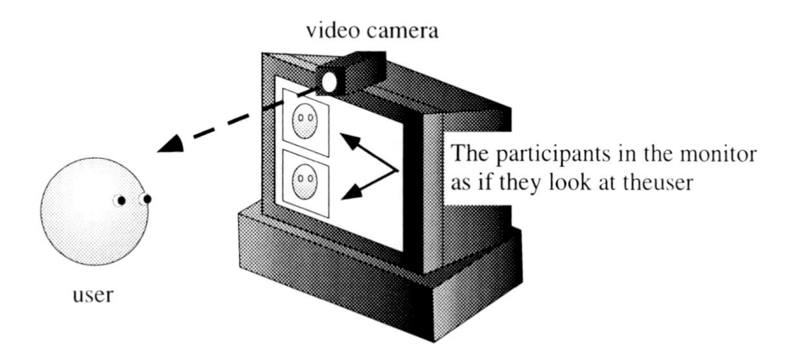

Figure 1. Problem in the video communication.

And, when users need to confer with others or need the adjustment in the video commmunication, if the user can not discern that who is visually attending to whom, the communication will not progress smoothly. So we consider that the video communication system needs to display the conversation partners, that is, the speakers and listners.

We would like to develop the video communication system in which we can see that who speaks to whom. We have already conducted evaluation experiment of the multiparty video communication system with the ability to discern speakers and listeners[2,3]. The above-mentioned output interface of conversation partners is the display of the figure of the arrow. It is already confirmed that this display of the arrow is giving good results in comparison with the display of the conversation partners' name[2].

In this paper, we describes the method of the display of conversation partners by changing direction of the facial image in video communication system. It is considered that the direction of face and gaze plays a great role in judgement of his conversation partners. If we can discern the conversation partners by their facial direction in the video communication as well as face-to-face communication, this display of conversation partners will become a good output interface.

Additionaly, in the traditional video communication system, the video camera is usually attached onto the monitor. Hence the user feels that the other users in the monitor look down. In this case, they are not often able to communicate smoothly. If the user can look as if the other users in the monitor are looking straight at him, the eye contact will be realized in this system.

In this paper, we describe method of the display of conversation partners between other users, and the display of the conversation partner with eye contact by changing direction of the user's facial image.

2. CHANGING DIRECTION OF THE FACIAL IMAGE USING 3D FACIAL MODEL

To change the facial direction, we use a 3D facial model. The texture of the user's front view is mapped onto the front part of this facial model. We change the user's facial direction by rotating the facial model.

For this method, robust tracking of head motion from video is necessary[4]. To reduce this

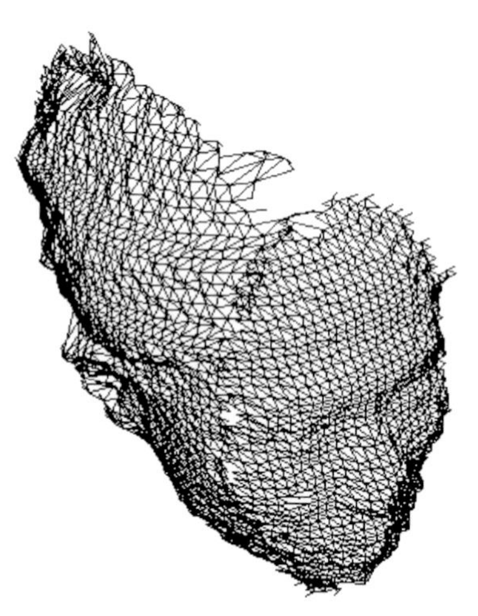

Figure 2. Facial model.

tracking cost of correspondence between the facial image and the model, we use the magnetic sensor. The sensor can detect the three translation and rotation parameters of itself in real-time, and attached to user's head. This is shown in figure 1. The face model is made by measuring user's 3D facial shape. The model has 3D object coordinate data. Since the sensor can get 3D object coordinates, we can fit the facial model to the real facial position by translating and rotating this model.

Next, we map the facial image onto the model. The camara parameter is necessary in this process. The camara parameters transforms 3D coordinates of object's point into 2D coordinates of object's image point. They are calibrated in advance. Using this parameters, we can fit 2D transformed coordinates of the model to facial image. So we can detect the color information of each pixel correspoded to 3D coordinates of facial model. Then, the model with color texture is displayed from the proper viewpoint. This experimental facial model is shown in figure 2.

3. DISPLAYOF CONVERSATION PARTNERS BETWEEN OTHER USERS

Rotating the user's facial model, we create the side view image of the user.

If we display the two user's face image looking to each other by rotating respective models arround vertical axis, the display of conversation partners between others will be realized.

But the more the model is rotated, the more difficult we create the side view image of the user, because the camera is only able to get the front view image of the user. So we map the real side view image of the user onto the side part of the model in advance, and use this side image as a side view. This image is not the real-time image. But it is considered that the static side view imge will not influence the video communication very much, because the real side view image does not change as much as the front view image. In this paper, the model is rotated at an angle of 45 degrees to display the side view image and to reduce that influence at the same time. Figure 3 is a front view image used for texture mapping, and figure 3 is an exsample of side view image created by rotating the facial model at an angle of 45 degrees around vertical axis. This side view image is transferred to the other workstation.

Figure 3. front view image used for texture mapping.

Figure 4. side view image created by rotating facial model.

4. DISPLAY OF THECONVERSATION PARTNER WITH EYE CONTACT

Assuming that user looks at his monitor, in that case, the user's image looks as if he looks down. So, rotating the user's facial model, we create the looking up view image of the user though the video camera is attached onto the monitor. We try to display the user's face image looking as if he looks straight at conversation partner by rotating the models. If possible the display of conversation partner with eyecontact will be realized. Figure 5 is an exsample of looking up view image created by rotating the model around horizonal axis. This looking up view image is transferred to the other workstation.

Figure 5. looking up view image created by rotating facial model.

5. CONCLUSION

In this paper, we propose the method of the display of conversation partners by changing direction of the facial image in the video communication system. To display the facial direction, we use a 3D facial model mapped the texture of the user's facial image onto. And we made the prototype that can display conversation partners between other users, and display the partner with eye contact. And we are going to evaluate this system for next work.

REFERENCES

1. Atsushi Fukuoka, Satoru Watanabe, Thineo Ktsuyama: "Evaluation of Communication in Teleconferencing,"Progress in Human Interface, 1994, pp.177-182. (In Japanese)

2. Toshiya Murai, Hirokazu Kato, Seiji Inokuchi: "Discremination of Conversers in TV Conference Systems," Proceedings of Human Interface, News and Report, 1995, pp177-182. (In Japanese)

3. Toshiya Murai, Hirokazu Kato, Seiji Inokuchi: " A Study on a Tereconferencing System using Conversation Partners Information," Proceedings of Twelfth Symposium on Human Interface, 1996, pp.183-188. (In Japanese)

4. Sumith Basu, Irfan Essa, Alex Pentkand: "Motion Regularization for Model-Based Head Tracking. Proceedings of ICPR'96, pp611-616.

3D Interactive Information Visualization: Guidelines from experience and analysis of applications

Richard Brath
Visible Decisions Inc., 200 Front St. W. #2203, Toronto, Canada, rbrath@vdi.com

1. BASIS for these GUIDELINES

Effective design of GUI interfaces is aided through the development and use of guidelines. Similarly, the effectiveness of 3D information visualizations will be aided through the development, testing and application of guidelines.

1.1 Expert Knowledge

The guidelines presented here are the result of the expert knowledge accumulated over the development of 50+ information visualizations[1,2] since 1993. Various techniques have been tried for communicating information in 3D visual representations drawing from knowledge in the fields of graphic design, user interface guidelines, end-user consultation, psychology and neurophysiology.

1.2. Significant Extensions to Guidelines

No comprehensive set of guidelines exist for the field of 3D information visualization. These guidelines serve as a starting point for 3D information visualization builders and researchers. These guidelines go beyond case studies and guidelines in graphic design and user interfaces to contradict some intuitive extensions of the 2D guidelines.

1.3. Iterative Methodology

Most of these visualizations have been developed for specific clients to solve specific business needs. During the development process the applications were re-evaluated with clients and modified. Additionally, some visualizations have been constructed to be generalized re-usable frameworks, or as small test applications. Over time, similarities between successful visualizations have been noted as well as recurring difficulties. These observations have led to these guidelines:

2. GUIDELINES for INTERACTION and VISUALIZATION

These guidelines are not complete - and will continue to change as more research is done on information visualization. As guidelines, they are not a guarantee to an effective visualization - poor quality data, poor mapping between data and visual attributes, or inappropriateness of visualization to the user's requirements, will all result in an unsuccessful visualization which guidelines can not rescue. Further, some goals may be better solved by breaking one or more of the guidelines - this implies that the goal is known:.

2.1 Know the goal

Before an effective information visualization can be created using any of the guidelines below, a real goal must exist. This means:

a. a problem has been identified
b. quality data exists which supports the problem
c. there exists a correlation between the supporting data and the goal
d. the goal can be found by human interpretation of the information

Without a clear identification of the goal, one can create an interesting visualization with little value It is easier to create a successful visualization for a clearly identified specific goal than it is for a weak or general goal.

2.2 Visualization

Visualization is the graphical representation of the underlying data. While an infinite number of mappings of data to representation are feasible, only a finite number will be effective. Thus:

2.2.1. Use an organizational device the user already knows:

The visual representation could look like anything - an extruded map, a star field, blobs, etc. A representation which maps closely to how the user already thinks of the data results in a visualization which is more easily accepted than one which is not. For example, a number of our visualizations include "organizational charts" to represent hierarchical data since most corporate users are familiar with organizational charts. We have often used maps, grids, time series and rooms to organize information.

Richard Saul Wurman[3] describes 5 graphical organization devices:

a. Location e.g. maps, rooms, floor plans.
b. Alphabet e.g. sorts, ranks, alphabets and arbitrary orderings based on a key.
c. Time e.g. time series, trend curves, surface plots.
d. Category e.g. collections, inventories, distributions, histograms.
e. Hierarchy i.e. multiple levels of summarization, potentially used on any of the above.

In addition, we also use:

f. Graph i.e. a set of objects (vertices) and interrelationships (edges), e.g. networks.
g. Scatterplot e.g. scatterplots, star fields, clouds, blobs.

We often use more than one of these within the same visualization, e.g. a room with an organizational chart and time series charts on two walls and a map on the floor.

2.2.2 Use small multiples:

From Edward Tufte[4], small multiples result in many tiny graphical objects which can be analyzed individually or compared across each other as a group. For example, a single small time series chart could represent sales at a store, and many of these charts could be located on a map. The user can look at one time series in detail (zoomed in close), or compare the trend across many time series simultaneously (zoomed out).

2.2.3. Include legend, scale and annotation:

Visualizations require learning. Legends, scales and textual annotations provide an immediate graphical reference which can be used in context to understand the information represented. Without these, printouts are not useable by someone unfamiliar with the visualization, and the visualization requires more effort in training.

2.2.4. Provide a reference context:

Objects floating in space are difficult to locate visually, e.g. 2 visually adjacent points in a 3D scatterplot have an unknown depth separation. But, if all objects refer to a common reference, e.g. a ground plane, then the visual identification of objects relative to each other is possible.

2.2.5. Use different visual dimensions differently:

Use different visual dimensions (e.g. size, location, orientation, form, color, transparency, motion) differently for:

a. Different types of data (e.g. enumerated type, continuous, discrete, text, images). For example, an enumerated data type with more than 10 enumerations does not map well to color (Can you name 20 unique colors?) but could but ealsiy map to unique locations.
b. Different connotations of data. Some data measures may cognitively map well to visual attributes, for example a data measure "size" maps well to the visual dimension scale, the measure "time" maps well to motion or "price" maps well to a vertical scale.

2.2.6. Avoid dis-information:

Perspective, computer graphics hardware, etc., can create visual artifacts which can be misinterpretted as significant. Two common occurences are:

a. Moire: many parallel lines or long thin bars in 3D can create moire patterns on computer graphic screens. Moire distracts attention from the information to the artifact. Use different geometric forms or interactions (e.g. filtering, scaling) to overcome moire.

b. Alignment and overlap: objects can align or overlap when viewed from a particular viewpoint and confuse the user into seeing a single object instead of 2 unique smaller objects. Use objects with borders or depth to create visual boundaries for each object.

2.2.7. Redundancy is good:

A minimalist visualization (where one data attribute maps to only one visual attribute) may be difficult to comprehend. Minimalism may result in overlap, may not map the data to a connotative representation, or may result in ambiguities between adjacent objects (all described above). Thus a more effective mapping may contain redundant geometry but result in an easier to use visualization (contradicting Tufte's[4] principle of minimizing the non-data ink to data ink ratio).

2.2.8. Visualization complexity:

The complexity and density of the information in the visualization is dependent on the:

a. User: executives who will use a visualization 10 minutes every day need a simple, less dense visualization than an analyst who may use the same visualization for a week.

b. Data Type: homogenous data can be more densely represented than highly multidimensional loosely correlated data.

Our typical information visualizations display 2,500 data points and range from 500 data points to 100,000 data points displayed simultaneously on the screen. This screen may display only a portion of a much larger data set - interaction reveals the patterns in the larger data set:

2.3. Interaction

Interaction is a key differentiator between a chart and a visualization. Interaction permits the user to manipulate the visualization to find and identify patterns visually while a chart is merely a static mapping of data to a representation.

2.3.1 Do not rely on interaction for comprehending data within a single scene:

Visualizations which rely on interaction to be comprehended cannot be printed out, published, etc. Dependence on interaction limits the audience to a small subset who can execute the interaction. Information visualization often needs to be disseminated to a wide audience - even if the application was orginally designed for a few select select users.

For example, a 3D scatterplot requires rotation of the scene so that a user can differentiate between points which are near and far. As a result, any printout of a 3D scatterplot is of limited value. We have tried various other cues (brightness, color, size) to illustrate depth, but in general we have found 3D scatterplots ineffective.

2.3.2. Analyze large data sets:

Interaction permits the user to work with a larger data set than can be presented on the screen at once. By drilling-down, animating, changing axes or adjusting the data model, the user can explore a data space orders of magnitude larger than can be assembled into a singular 3D scene.

Interaction also permits the user to remove data from the display. By slicing, filtering, zooming and querying the data the user can quickly narrow a search through the information.

2.3.3. Adjustable viewpoint::

Changing the viewpoint permits the user to:

a. move around occlusions, such as large objects infront of small objects; and,

b. disambiguate uncertainties in the scene, such as an alignment of two different lines, or two objects of the same shading overlapping from a particular viewpoint.

2.3.4. Easy navigation:

Many 3D systems (and VRML browsers) suffer from poor 3D scene navigation. An information visualization navigation model should:

a. keep the scene always on screen. It should never let the user manipulate the scene so that the user is looking at empty space.
b. use a steady-cam model. It should not permit the scene to roll - i.e. the horizon should remain horizontal. People do not experience roll for most tasks. Thus, flight simulator interfaces are inappropriate.
c. provide consistent interaction. Virtual trackball interfaces (e.g. OpenInventor and many VRML browsers) rotate the scene differently depending on the location of the initial mouse down event.
d. provide fast feedback. A user can more quickly orient a scene if provided with a subset of or proxy for the scene than if the user must wait until the entire scene redraws.

2.3.5. Drill-down to underlying data:

Users often require the underlying data behind the visual representation. This is typically used for either:

a. identification. e.g. the user needs data in its original format (numeric, text, image, etc)
b. verification. e.g. the user can validate the correctness of the data and verify his/her cognitive mapping of the data to the visualization.

For example, brushing the cursor across a visual object, which then displays the data for that object, provides immediate data and reinforces the mapping between data and representation.

2.3.6. Aid interpretation:

Interactions such as drill-down and slicing permit the user to verify or modify his/her understanding of the information. This interaction can be a learning device. Brushing as described above is a good example.

2.3.7 Permit unforeseen combinations and permutations:

There are many ways of analysing information. An interactive system will permit the analyst to change various data properties on the fly to derive new data sets and find new patterns. We have seen complex interactions to:

a. model data e.g. summaries, differences, averages, ratios, distributions, simulations.
b. manage states e.g. analysis of real-time data during operation vs after hours.

Thus, the visualization also requires interactions to adjust properties of the visual representation to accomodate unforeseen representations.

3. SUMMARY

Information visualization permits users to interactively explore large amounts of data. Poorly designed information visualizations will not reveal insight. Effective visualizations will. These guidelines can aid developers designing information visualizations and serve as reference for further research.

REFERENCES

1. W. Wright, "Information Animation - Applications in the Capital Markets", IEEE/ACM First Information Visualization Conference, October, 1995.
2. W. Wright, "Information Visualization - The Fourth Dimension", Database Programming and Design, April 1997.
3 R.S. Wurman, Information Architects, Graphis Press, Zurich, Switzerland, 1996.
4. E. Tufte, The Visual Display of Quantitative Information, Graphics Press, Cheshire, CT., 1984

Presence and side effects: complementary or contradictory?

John R. Wilson, Sarah Nichols and Clovissa Haldane

Virtual Reality Applications Research Team, Dept. of Manufacturing Engineering & Operations Management, Univ. of Nottingham, Nottingham, NG7 2RD, UK

1. INTRODUCTION

A sense of presence and side effects (particularly sickness) related to virtual environments presented in Virtual Reality systems (VR/VE) have in common that they are not easy to unambiguously define, either fundamentally or operationally. They also both appear to be key factors for potential widespread take-up of VR/VE applications.

1.1 Presence

Presence (the feeling of 'being there') can be seen as the defining characteristic of VR/VE, one of several such characteristics, merely an epiphenomenon of VR/VE or even as a desirable outcome of the experience of VE participation. Barfield et al (1995) distinguish virtual presence from real world presence and telepresence as the extent to which participants believe they are somewhere different to their actual physical location whilst they are actually experiencing a computer generated simulation. For a concept so critical to the nature and potential value of VR/VE there have until recently been few controlled experiments of presence and influences on it (see Presence, 1996). This is probably due to the multiplicity of issues involved, the difficulty of manipulating some of the relevant variables, and the absence of agreed methodology and measures; how would people define their feeling of presence in the actual world never mind an artificial one? In philosophy or art for instance, presence is often defined in terms of other concepts such as realism or perspective.

1.2 Side effects including sickness

Reports of potentially harmful or other side-effects for participants in VEs, during or post immersion, have appeared with increasing frequency in the popular media, although the number of published scientifically supportable studies is still small. Nonetheless, some respected authorities report that VR use can be difficult, disorienting, uncomfortable and nauseogenic. The important questions are how serious are these effects and what are their consequences for participant well-being and performance? Of all side effects the most widely reported and possibly most connected with presence are symptoms akin to those of motion or simulator sickness.

1.3 Relationship between presence and sickness?

There are two main possible scenarios when considering any link between presence and sickness. First it may be that a greater degree of presence provided by a greater belief in the visually simulated motion provokes a greater degree of conflict between the visual and proprioceptive senses, and so more simulator sickness symptoms are provoked. Alternatively, presence may not play a causal role in the production of these symptoms, and any symptoms experienced may in fact distract the user and thus lower the sense of presence. Alternative, less supported, connections between presence and sickness are that they are both correlated - positively or negatively - with a third factor or variable, that changes in one do initiate changes (up or down) in the other but via an intervening factor, and that there is no connection between them and any apparent correlation is spurious.

2. VRISE PROGRAMME

We have recently completed the VRISE (Virtual Reality Induced Sickness and Effects) Programme for the UK Health and Safety Executive, where interest was not only in side effects experienced by participants but also in any consequences for their subsequent work performance. For instance, if a production manager examining a new plant layout feels a little nauseous, but this wears off after five minutes, then health and safety at work may not be an issue. If, however, a maintenance engineer experiences nausea whilst rehearsing a repair task, and if this is accompanied by actual or perceived instability, then they may be a hazard to themselves or others if they immediately work on the real machine.

The programme of work consisted of twelve main experiments with a total of 223 subjects, experiencing immersion in one of three HMD VR systems for between 20 minutes and two hours. Measurement of effects was carried out through physiological monitoring, self report of symptoms and other experiences, postural assessment and visual, physical and psychomotor performance tests (see Nichols et al, submitted; Wilson, 1996). An array of symptoms and effects have been identified, some similar to those found with other types of simulators and in transportation, but the aetiology is sufficiently different to justify the new term VRISE. Approximately 80% of subjects across all experiments reported some increase in symptoms, of sickness, akin to simulator or motion sickness, but with some constituent experiences different from either of the other two better established phenomena, for most these were mild and short-lived but for 5% so severe that they had to end their participation. Visual testing revealed appreciable symptom and physiological changes (Howarth, 1997). Ergonomics assessments reveal considerable problems with current generation equipment fit and discomfort (Nichols, 1997). Evidence for effects of VR use on subsequent task performance is equivocal but there was reported increased difficulty. Amongst a variety of physiological tests (in addition to those of visual physiology) several showed evidence of significant change post-immersion, including postural instability (using relatively sophisticated measurement techniques), heart rate levels and variability, and urine and salivary-cortisol composition (Cobb, 1997; Ramsey, 1997).

Many of the data, especially related to individual differences, are being re-analysed. For now our overall conclusion is that there are undoubtedly some unwanted effects occuring but probably not to an extent to cause great concern; there is just a nagging doubt however that something potentially serious may be happening!

3. FIRST PRESENCE/SICKNESS EXPERIMENT

For the purposes of this paper two aspects of the VRISE programme are of interest. First, many of the technical (hardware and software), individual and task/environment factors influencing presence are probably the same as those influencing experience of side effects (for instance lags - registration or update, HMD optics and design, display and interaction fidelity); the same underlying theories may also be central to each (for instance, and obviously, sensory conflict theory). Secondly, "presence" has been used consistently as a multi-dimensional measure throughout the work. Data are generally still under analysis, especially looking for individual differences. One experiment specifically set out to examine the relationship between sickness and presence. This experiment used 20 subjects on a Division Provision PV100 VPX with Dvisor headset. Presence measurement was through the presence questionnaire developed by Witmer & Singer (1996).

The "interface" subscale of the presence measure showed a significant negative correlation with the change in symptoms of oculomotor disturbance, disorientation and the total symptom level, obtained from the Simulator Sickness Questionnaire. This implies that those subjects who experienced a higher level of these sickness symptoms had less of the component of presence represented by the interface subscale. (There was, however, no significant correlation between sickness reports and the other presence sub-scales or the total measure.) Furthermore, total presence and some of the sub-scales were also positively correlated with "enjoyment" and negatively with reported difficulty.

4. SECOND PRESENCE/SICKNESS EXPERIMENT

Subsequently an experiment outside the VRISE Programme was established to manipulate technical variables and assess outcomes in terms of presence and side effects. 24 subjects participated in a "duck shooting" VE with Superscape VRT on a P133 and Virtual I/O i-glasses as relevant. There were two independent variables, HMD/Desktop as a within subjects variable and Auditory/Silent as a between subject variable. Participation in each session was for up to 10 minutes. Three measures of presence were used: questionnaire items taken from a variety of sources; awareness of a manipulated level of background music as assessed by retrospective questioning and concurrent observation; and 'startle' effects of a single 'explosion' within the VE, assessed by observation and subsequent questioning. In addition sickness symptoms were collected via a Short Symptom Checklist (see Nichols et al, 1997), susceptibility profiles through adaptation of various scales and performance at the virtual task by automatic calculation. At the time of writing the experimental phase in this work has just been completed. First analyses indicate: 1) significantly higher ratings of key items of 'presence' (e.g. sense of 'being there') for the HMD condition; 2) significantly higher reports of symptoms on the SSC for HMD wearing; 3) auditory cues seemed to add nothing extra to the sense of presence; 4) significant positive correlation between presence and sickness for HMD.

CONCLUSIONS

Tests for whether human factors research into VR/VE is justifiable include whether an issue is significant for development of VR/VE and its safe and effective use (Wilson, 1997). Since it is a critical component of what distinguishes VR/VE from, say, solid modelling systems, a better understanding of presence, when it is useful and what in a VE will enhance it will be of great value. The importance of side effects will vary according to whether they are primary/secondary, positive/negative, during/post immersion and direct/indirect. Possible connections between presence and side effects (sickness) may aid investigation into the virtual experience, and thus better VEs generally. Witmer et al (1996) found negative correlation - explained as sickness detracting from presence or as presence leading to less awareness (and thus reporting) of sickness. Our own work lends some support to this, but also to the importance of 'presence' as a measure indicating enjoyment. Also, ergonomics and equipment problems, as well as sickness, detract from VE enjoyment, value and, to an extent, presence.

REFERENCES

Barfield, W., Sheridan, T., Zeltzer, D. and Slater, M., Presence and Performance within virtual environments. In: Virtual Environments and Advanced Interface Design (eds: W. Barfield and T. Furness III). Oxford University Press, 473-513. 1995.

Cobb, S.V.G., Measurement of postural stability before and after immersion in a virtual environment. To appear in Applied Ergonomics, **28**. 1997.

Howarth, P.A., Occulomotor changes within virtual environments. To appear in Applied Ergonomics, 28. 1997.

Nichols, S.C., Cobb, S.V.G. and Wilson, J.R., Effects of participating in virtual environments: Towards an experimental methodology. Submitted to Presence: Teleoperators and Virtual Environments,1997.

Nichols, S.C., Physical ergonomics issues for Virtual Reality system and virtual environment development. To appear in Applied Ergonomics, **28**. 1997.

Presence, Special issue Presence: Teleoperators and Virtual Environments, **5**/4. 1996.

Ramsey, A.D., Changes in salivary cortisol, heart rate and self report during longer immersions in interactive virtual environments. To appear in Psychosomatic Medicine, 1997.

Singer, M.J. and Witmer, B.G., Presence measures for virtual environments: background and scoring instructions. Working paper from the US Army Research Institute, 1995.

Wilson, J.R. Effects of Participating in Virtual Environments: A Review of Current Knowledge. Safety Science, 23, 1, 39-51, 1996.

Wilson, J.R.,Virtual environments & ergonomics: needs & opportunities. Ergonomics,40, 1997

Witmer, B.G., Bailey, J.H. and Knerr, B.W., Virtual spaces and real world places: transfer of route knowledge. International Journal of Human-Computer Studies, 45, 413-428, 1996.

Circumventing Side Effects of Immersive Virtual Environments

P. DiZio and J. R. Lackner

Ashton Graybiel Spatial Orientation Laboratory and Volen Center for Complex Systems, Brandeis University, Waltham, MA 02245

1. INTRODUCTION

Making head movements during exposure to some forms of passive transport or when the visual feedback about one's movements is distorted has long been recognized as disorienting and nauseogenic (1, 2). Some of the most clear and dramatic results of space flight research came from the Skylab M-131 experiment in which astronauts experienced extreme disorientation and severe motion sickness while making head movements in a rotating chair pre-flight in a 1 g terrestrial force background but were free of symptoms and dizziness when first tested in 0 g on the sixth day in orbit (3). Understanding side effects of rotation is important because NASA is contemplating using a rotating space vehicle to generate artificial gravity for long duration space missions. The rotating vehicle is a gravitoinertial interface device to create a "virtual gravity" environment. Such a virtual environment may be adequate while the user is immobile but deviates from the normal terrestrial environment when voluntary movements, especially head movements, are made (4). Motion sickness, disorientation and motoric side effects are elicited when exposure to it begins, they abate during continued exposure if a sufficient number of movements are made on a suitable schedule (5) and there is a recurrence of motion sickness accompanied by negative orientational and motoric aftereffects upon return to the normal environment (6). Understanding the self-calibration mechanisms responsible for adaptive reduction of side-effects, negative aftereffects and reemergence of motion sickness requires a) analysis and measurement of the physical environment), b) consideration of specific sensory and effector systems and c) specialized sensorimotor subsystems such as the vestibuloocular reflex and cervical motor control.

We have applied this paradigm to understanding side effects of another important virtual environment interface - the helmet-mounted visual display (HMVD). A wide field of view (FOV) HMVD can immerse the user in a virtual environment. Scenes can be presented that generate a sense of self-motion and are nauseogenic even for an immobile user. Traditional visual flight simulators can do this too. We have been interested in problems uniquely related to HMVDs that arise when scenes are presented simulating a stationary world. Such scenes are not disruptive to an immobile observer but can become extremely nauseogenic and disorienting if head movements are made. Limitations in the displays themselves or computational difficulties in image generation make these situations analogous to, for example, moving the head while wearing prism spectacles which distort the visual feedback contingent on the vestibuloocular reflex.

The specific objectives of the research were a) to document motion sickness severity and other side effects in an immersive operational virtual environment utilizing a wide FOV HMVD as well as aftereffects when the HMVD is removed, b) to measure critical characteristics of the virtual environment system and c) to evaluate etiological factors in motion sickness and other side effects.

2. METHODS AND RESULTS

2.1 Apparatus and virtual environment

The HMVD we used was a LEEP Cyberface II with a nominal (see below) FOV 138° wide by 110° high, with aproximately 40° of binocular overlap. To achieve this FOV with standard liquid crystal display units (4 inch square, color, resolution=479 horizontal x 234 vertical), a lens system is used to create a non-uniform projection of the displayed images ontc the eye. The LEEP projection involves a radial expansion that results in minimal loss of resolution in the center of the field. A Polhemus 3SPACE FASTRAK was used to track position of the HMVD. Images were generated by a Silicon Graphics ONYX Reality Engine 2 with the multi-channel video output option. The scene presented was a harbor channel from the perspective of an observer in the sail of a stationary, surfaced submarine. The open ocean was straight ahead of the boat and shorelines to the right, left and rear. Maritime scenery included buoys, other surface craft and wave motion; the land scenery included buildings, trees, and hills. The normal graphics pipeline of the ONYX could not be utilized because of the LEEP projection which had to be computed with the main processor. The contents (number of polygons) of the virtual world were adjusted so that the frame rate for update of the visual image never fell below 30 Hz.

2.2. System characteristics

The only motion present in the displays resulted from voluntary head movements because the viewpoint was stationary. Tracking delays combined with the 30 Hz frame rate resulted in a characteristic delay between head movement and update of the visual display. Our first goal was to measure this delay. We measured the delay between motion onset of the HMVD and of a high contrast contour of the harbor scene. The ONYX received its information about HMVD motion from the Polhemus but we independently measured it with a zero-latency mechanical tracker; visual motion was registered by a photodiode placed directly on the liquid crystal display. The results indicated a minimal system delay of 67 ms between HMVD motion and update of the scene. Effective FOV was measured by adjusting an object in the virtual environment until it disappeared at the edges of the HMVD. This method indicated that the FOV was 126° wide by 74° high. The helmet and its counterpoise worn by the subject weighed 2.44 kg.

3. CYBERSICKNESS SPECTRAL PROFILES

The cybersickness spectral profile in Figure 1 was derived from an analysis of the symptom profiles from eight different VE experiments. We had carried out, in collaboration with three university laboratories, four experiments (see Table 1: 1, 2, 4 & 5) using three different VE helmet mounted systems. Additionally, we had access to the data from four other experiments (two federal laboratories [3, 7] and two other university laboratories [6, 8]). Table 1 shows the pertinent details from the eight experiments involving more than 400 subjects. It may be seen that most of the experiments lasted 30 minutes, an extended period of time for purposes of immersion, but not as long as the average simulator exposure time.

Figure 2 shows the eight VE systems according to their spectral profiles (i.e., Nausea, Oculomotor, and Disorientation). Through an analysis of these profiles two distinctly different patterns can be seen. The more common (first 5 VEs) show less Oculomotor symptoms than Nausea and Disorientation symptoms. We will refer to this as VE type A. The second group (last 3 VEs), referred to as VE type B, all by the same manufacturer, show relatively less Nausea but are otherwise consistent with the type A VE. That is, all VE systems appear to exhibit a significant amount of Disorientation and lesser Oculomotor symptoms. Type A VE show significantly more Nausea (the D>N>O profile) than type B (a D>O>N profile).

Table 1
Eight Experiments Using Virtual Environments in Helmet Mounted Displays

Exp.#	Laboratory	VE System	Duration	Program	N
1	Univ. Cent FL (Orlando) (Kennedy, Stanney, Dunlap, & Jones, 1996)	Kaiser E/O Vim 500	30	WorldToolKit	34
2	Univ. Cent FL (Orlando) (Kolasinski, 1996)	i*glasses!	20	Ascent	40
3	Army Personnel Research Establishment (Farnborough, UK) (Regan & Price, 1994)	Provision 200	20	Demo software	146
4	Murray St.Univ. (Murray, KY) (Kennedy, Jones, Stanney, Ritter, & Drexler, 1996)	i*glasses!	40	Ascent	37
5	Univ. of Idaho (Moscow, ID) (Rich & Braun, 1996)	CyberMaxx 180	40	Heretic	23
6	Univ. of Houston (Bliss et al., In preparation)	Virtual Research VR-4	~20*	VrTool	55
7	U.S. Army Research Institute (Orlando, FL) (Lampton et al., 1994)	Virtual Research	20**	WorldToolKit	57
8	George Mason Univ. (Fairfax, VA) (Salzman, Dede, & Loftin, 1995)	Virtual Research EyeGen-3	75***	VrTool	39

* Time dependent upon task completion; ** Repeated exposures used; *** Total time including breaks

4. INCIDENCE LEVELS OF SICKNESS

With this understanding of the types of discomfort users of VE systems experience, next it is important to consider how pervasive a problem cybersickness presents. The incidence levels in early VE studies indicate that only 5%-10% of users report no symptoms. This is in comparison to 30%-40% asymptomatic with flight simulators. While this difference in rate of occurrence could be a manifestation of under-reporting (i.e., those in the flight simulator studies were primarily male pilots), the results suggest that adverse symptoms could be rampant among VE system users. It is important to note that the symptoms experienced by VE users can be as minimal as a slight headache, to in very rare cases an emetic response.

5. INTENSITY LEVELS OF SICKNESS

Figure 3 shows the average Total Score across the eight VE devices from Table 1. The average score over all the devices is approximately 29 on the Total Score scale and the range was broad; from 19-55. In comparison, the average from eight U.S. Army and U.S. Navy/Marine Corps helicopter simulators (which ranged from 7 to 20), is approximately 12 (see Figure 3). This indicates substantially lower severity in helicopter flight trainers as compared to VE systems. Further, the space sickness Total Score (see Figure 3), which is based on the reports of 85 astronauts upon post flight, is between the average for simulator and VE exposures. It would appear that VE users report more sickness than military flight simulators and astronauts during space travel.

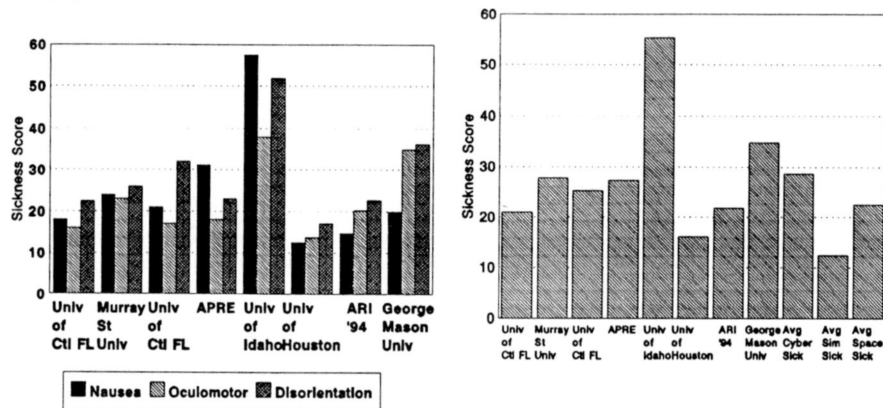

Figure 2. Spectral Profiles of Cybersickness

Figure 3. Total Score of Sickness

ACKNOWLEDGMENTS

This research was supported by National Aeronautics and Space Administration contracts 9-19482 and 9-19453 and National Science Foundation Grant DMI-9561266 and IRI-9624968. The opinions expressed here are the authors' and should not be construed to be endorsement by the sponsoring agencies.

REFERENCES

Available on request

Design concept based on real-virtual-intelligent user interface and its software architecture

S. Tano[a], Y. Namba[b], H. Sakao[b], T. Tomita[b] and H. Aoshima[b]

[a]University of Electro-Communications, Chofu 182, JAPAN
[b]Systems Development Laboratory, Hitachi, Ltd., 1099 Ohzenji, Asao, Kawasaki 215 JAPAN

1. Introduction

Recently, the multi media technology and the network technology enable us freely to access various multi media information spread in the world. However, people is tied up in front of a small CRT screen, such as 19 inches or 21 inches, and the input modalities to the computer are also limited to a keyboard and a mouse.

In this paper, we propose the new design concept of the next generation user interface, called RVI-concept, which is an acronym of the real virtual intelligent user interface concept. To show that our RVI-concept is effective and feasible, we developed a system, called RVI-desk, which provides with the intelligent user interface where a user can communicate with the computing environment through multi modalities, such as a voice, a handwriting, a pen gesture, a printed material, a touch screen and a keyboard.

2. New Design Concept "RVI(Real, Virtual and Intelligent)-Concept"

We have analyzed current leading design concepts and divided them into the following three categories.

(i) Virtual world-centered UI: Basic idea is to realize a virtual desk or a virtual office environment in a computer by the virtual reality technology and the computer graphics technology[4].

(ii) Real word-centered UI: It is a completely opposite way of thinking to augment a real world by a computer power[5].

(iii) Intelligent UI: Intelligent processing, i.e. problem solving and learning, is usually applied in the industrial world to solve complex problems. Recently, the application software becomes too complex for the users to understand the full functions. So the shift toward the intelligent UI is inevitable[6].

As explained above, the leading concepts are categorized into three by our analysis. Fig.1 illustrates the traditional user interfaces categorized by three axes which correspond to the three leading concepts. For example, since the graphical user interface is a sort of the virtual world-centered UI, "GUI" in fig.1 is located on the axis of the virtual world-centered UI. Since the level of reality is primitive, it is located near a bottom point.

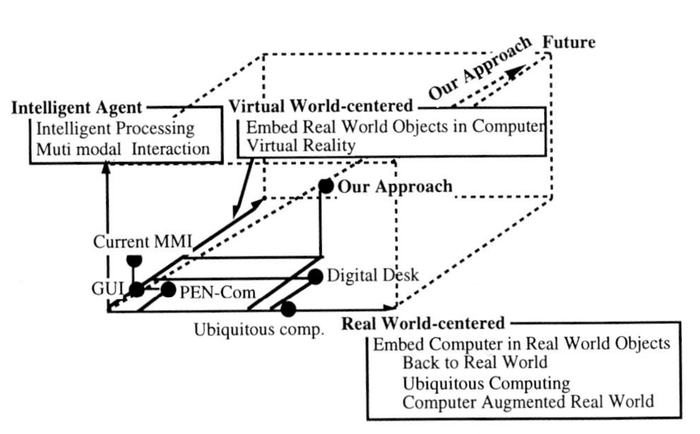

Fig. 1 Fusion of Three Exclusive Concepts

Although the pen computer bases on GUI, it takes a step toward the real world-centered UI. So "Pen-Com" is in middle of axes of the virtual world-centered UI and the real world-centered UI. Similarly, although the multi modal interface(MMI) bases on GUI, it takes a step toward the intelligent UI. So "current-MMI" is in middle of axes of the virtual world-centered UI and the intelligent UI.

It is almost impossible to achieve the ultimate goal of these three concepts. For example, the ultimate goal of the virtual world-centered UI is to produce the virtual world which is exactly the same to the real world.

So the key idea is the combination of these three conflicting concept at the adequate level. At the first glance, the three leading concepts look quit different, but look carefully, and you can see that they compensate each other. Moreover they can be synergetically combined. Our new concept "Fusion of Three Exclusive Concepts" is indicated by a thick dotted line in the middle of the virtual world-centered UI, the real world-centered UI and the intelligent UI. This idea is so simple that it is applicable to any fields. We call it RVI-concept [1, 2].

3. Experimental System "RVI-DESK" - Application to Desk Environment -

For the feasibility study, the RVI-Concept has been applied to the design of the desk environment. After analyzing the problems of the conventional PC-based desk environment, RVI-Desk was designed based on our new concepts. Fig.2 illustrates the appearance of current RVI-Desk.

Important design issues based on RVI-concept can be characterized by the following two question.

(1) How are the real word-centered UI and the virtual word-centered UI merged?

We invented the following three special devices to answer the question.

(i) InformationViewer

InformationViewer is a kind of a pen computer. It substitutes both the real paper in the real world and the CRT monitor in the PC-based desk environment. As it is intended as a real paper, it supports the following usage:

- A user can put the additional InformationViewers on the desk whenever the user wants.
- Several InformationViewers can act as one big paper when they are put side by side.
- InformationViewers can deal with the multi modal input and output.

(ii) EnvironmentViewer

EnvironmentViewer is a CRT monitor with touch sensor. Usually, it displays the desk environment which is quit similar to fig.2. At a glance, it looks like a desk top displayed in a conventional PC. However, it displays three qualitatively deferent objects;

- real objects,
- virtual objects, and
- real & virtual objects.

In this case, "virtual" means to exist only in a computer and "real & virtual" means to exist both in a computer and a real world. For example, on EnvironmentViewer, you can see a real paper (i.e. a real object), a virtual folder (i.e. a virtual object) which does not exist on the real desk but exists in a computer, and the InformationViewer whose hardware exists in the real

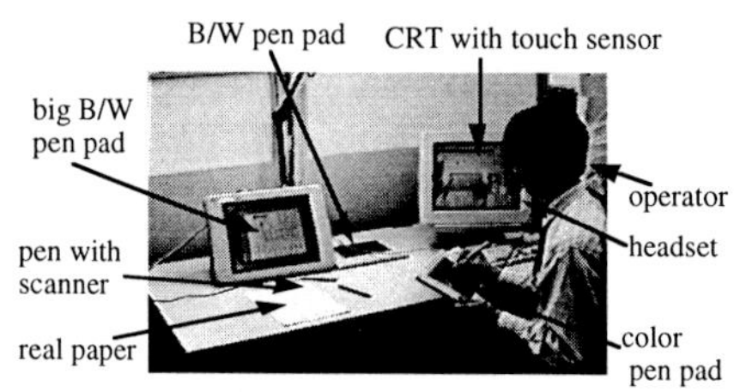

Fig. 2 Photograph of current RVI-Desk

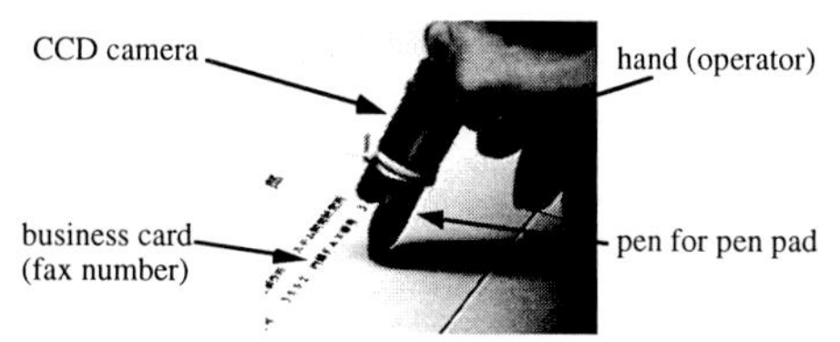

Fig. 3 Photograph of an experimental pen with scanner

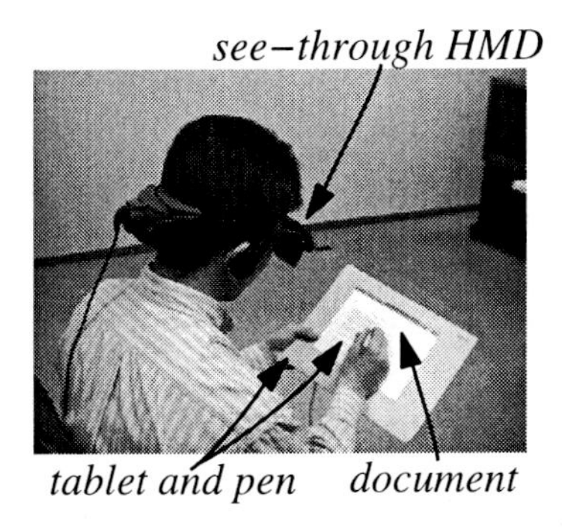

Figure 3: Simplified augmented reality system

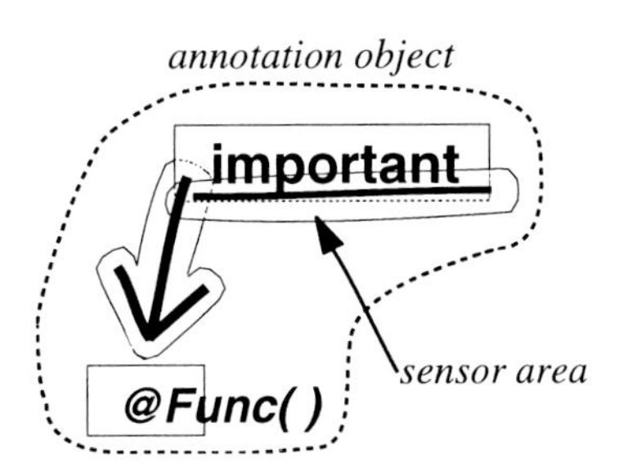

Figure 4: Example of annotation object

user manipulates the annotations using the tablet while seeing the superimposed scene through the HMD.

There is an issue concerning document identification in this configuration. To solve the issue, we printed out the documents with the id numbers. When the user inputs the id number of the document, the annotation sheet reads the corresponding annotations of the document.

3. ANNOTATION OBJECT

The annotation sheet has some annotation objects. The annotation object is a unit of annotations. It is managed by its position on the sheet. Figure 4 shows an example of the annotation objects. The annotation object consists of one of the following elements.

- Text element such as comments.
- Graphics element such as lines, arrows, or insertion marks.
- Image element such as digital pictures.
- Functional element.

The text element, the graphics element and the image element are the visual elements which are directly displayed on the sheet. The functional element is different from the other elements. When it is activated, it executes its process and displays linked visual elements as a result. It makes it possible to create hypertext like links among the annotation objects. Therefore, the annotation object could activate other annotation object. It could be used as a hidden function which does not yield any visual effects.

Each element has a sensor area. When the user's focus enters the area, the element is activated, and so is the annotation object. The sensor area is automatically generated when the element is created. It is located around the element as shown in the figure 4.

4. INTERACTION WITH ELECTRONIC ANNOTATION SHEET

There are two modes, annotating mode and browsing mode, in the annotation sheet. The user switches these modes explicitly. In this section, we describe the interface in each mode.

Annotating Mode In annotating mode, the user creates new annotations. The user registers a series of the annotation elements as an annotation object. The file linker and the English-Japanese dictionary are examples of functional elements. The file linker links from an area on the sheet to an another annotation object. The English-Japanese dictionary shows the meaning of the word in Japanese when the focus enters the sensor area of the English word.

Browsing Mode In browsing mode, the user browses previously created annotations. There are two ways of browsing annotation objects on the sheet. They are:

- moving the cursor to the focus area
- selecting a set of annotation objects explicitly

The former is based on the sensor areas of the annotation objects. By moving the focus in the document, the user can browse the related information. This method is effective when displaying an annotation object which is strongly related to its position on the document. The later is effective when the user wants to browse a set of all the annotation objects at once. It is sometimes necessary to make each individual set of annotation objects active or inactive. For example, if many people write comments on the same document, it is convenient to see an individual set of comments separately and when the user selects a set of comments, all of the comments written by the same person are displayed at once. Furthermore, combinations of these methods are possible.

5. CONCLUSIONS

We proposed the idea of the annotation sheet and its interface in this paper. We described the annotation objects, which are added electronically to the document, and the method of managing them. We showed that the system is useful to store documents and annotations in the unified manner. Using the elements of annotation objects, paper documents could be used as effective communication media. In the current system, the linkages between the paper documents and the annotations are made manually by typing the numbers printed on the top of the documents. We will make them automatic using a bar code reader system in the future. We found that HMD is not suitable for superimposing because the calibration and focusing of the display are difficult in real time. We think that a transparent flat display placed on the paper document is one of the alternatives for this purpose.

REFERENCES

1. Eric A. Bier, Maureen C. Stone, Ken Pier, William Buxton and Tony D. Derose. Toolglass and Magic Lenses:The See-Through Interface. Proceedings of SIGGRAPH'93, *Computer Graphics* Annual Conference Series, ACM, 1993, pp.73-80.
2. Walter Johnson, Herbert Jellinek, Leigh Klotz Jr, Ramana Rao and Stuart Card. Bridging the Paper and Electronic Worlds:The Paper User Interface. Proceedings of INTERCHI'93, ACM, 1993, pp.507-512.
3. Pierre Wellner. Interacting with Paper on the DigitalDesk. *Communication of the ACM*, Vol.36, No.7(1993), pp.87-96.

A Virtual Office Environment for Supporting Informal Communications and Securing Personal Space

S.Honda,T.Kimura,H.Tomioka,T.Oosawa,K.Okada and Y.Matsushita

Faculty of Science and Technology, Keio University,
3-14-1, Hiyoshi,Kohoku-ku, Yokohama 223 JAPAN

1.INTRODUCTION

Recent technological advances enable us to have a home office environment. The disadvantage of home office includes problems such as a burden from lesser communication and an excludance from society. To avoid these problems, the home office system requires the support of informal communication.

Therefore, we have realized home office system which enables the home office workers to go to an office in 3D virtual space and work with his/her colleagues. The system provides a virtual shared room as a space for working and communicating. Also, the awareness of other members were provided as a trigger for the communication[1].

Next, giving an environment which workers can plunge into the work is closely related to work efficiency. This is a very important factor to an office environment. For example, it might bother workers, if too much awareness information had been given[2]. In this system, we have defined "Awareness Space" as an aim which coexists informal communication and work space that workers can concentrate.

In this paper, we explain the features of our work, which realizes "Awareness Space" to avoid a tradeoff between supporting informal communication and securing personal space, and which provides the feeling of the ownself presence at virtual office by using "Around View" and "Sound Effect".

2.REALIZATION OF A VIRTUAL OFFICE ENVIRONMENT

2.1.The Way of Transmitting Awareness in Detail

We usually feel other members' awareness such as actions and gaze. For this reason, we are able to feel the atmosphere or the tension of the real workplace. This also triggers the occurrence of informal communication. In general, simple audio and video transmission have used to provide other's awareness, but compared to the real world office, these are not enough to transmit other members' awareness in detail.

To realize members' awareness which is similar to the real office, we provide interfaces called "Around View" and "Sound Effect".

Around View Former method of viewing 3D virtual workspace could not specify what the next person was doing[3]. But in real world, we are able to recognize if the person next to you sits down or stands up without looking. This is because the field of view of human being extends more than 180 degrees.

If the interface can view more than 180 degrees, the user has no need of turning his virtual body to communicate with next person. This looks very strange from other virtual members, since human usually communicate by looking each others' face. Human can clearly recognize the colors and the shapes only about 60 degrees in front. We cannot clearly see the angles more than that, but are adopted more to the movement of objects.

Therefore we have implemented the interface called "Around View" which is based on the field of view of real human. To be more detail, it has a field of view of 180 degrees. But only the middle 60 degrees is shown clearly, and the rest is shown vague gradually as it moves away from the middle.

Sound Effects In the real world, we can not only see but also can hear the awareness information. The examples of the awareness information of sound are divided in two types. One type is the sound made by human itself, such as the voice or the sneeze and the other is the sound made by the movement of human, such as footsteps or opening or closing of the door. If we think of this virtual office system, the former could be heard from audio communication, but the latter could not be heard. So, we have provided the sound effect which corresponds to each member's action. The type of sound effect are as follows: the sound of footsteps, chair makes, door makes and the noise of the office.

By hearing these sounds(excludes office noise), the user can sense other members' action. And the office noise were added to give a reality that the user really works in office.

2.2.Awareness Space and The Degree of Concentration

In good Japanese offices, you should have your own space to work on provided that you can feel the sorroundings. But the tradeoff between the facility of communication and keeping one's personal space from other members' awareness information exists.

To avoid the unlimited flow of other members' awareness information, we have defined the notion of "Awareness Space", and solved the tradeoff explained above.

"Awareness Space" is defined as the area which the user can sense the others' awareness information, circled centering the user in 3D virtual space. The user can receive the sound information, the movement information from the sound effect, or the realtime video information of the member who had entered the user's awareness space.

When the user is concentrating on his/her project, there is no need of grasping the surrounding information. Giving too much information will lead to the bothering of user's work. In case of working in a project with plural numbers, even when not working together, some factors such as chatting, knowing other members' mood, or feeling the team's atmosphere are thought to be important.

Therefore, we have developed the idea of "Degree of Concentration" ,about which we talk the details in the later section, to change the awareness space.

In our system, the awareness space has 3 levels as follows

NARROW The state of concentration and not wanting to bothered by others. Only the very selected awareness information, such as the colleagues next to the user or the one who walks behind the user, could be transmitted.

NORMAL The state of normal working condition. The awareness information of comparably close colleagues could be transmitted.

WIDE The state of trying to feel the atmosphere of the entire office. All of the awareness information could be transmitted.

2.3.The Degree of Concentration Detected by The Other Members

The member who is concentrating might not be able to answer others for he/she holds lesser information of surrounding. Therefore, there is a need of knowing other member's degree of concentration.

This system utilizes "headphone metaphor" to solve this problem by putting the headphone to the member who has the awareness space level of NARROW. Now, the others are able to recognize that this member is concentrating. To talk with the member with the headphone, others have to walk closer to him/her.

3.DETERMINING THE DEGREE OF CONCENTRATION

3.1.Elements to Determine The Degreee of Concentration

To detect one's degree of concentration from the usual computer work, we applied two elements to determine the degree of concentration: the frequency of key typing and chair movement. To derive the determining algorithm, we have experimented as below.

3.2.Experiment

Each of the 25 subjects was sat down in laboratory's workstation at a time to work on the menus such as programming, mailing, netscaping, or playing game. While they are working on the menu, we took the measure of the frequency of typing and moving of the chair. And to know whether the subject is concentrating or not, we observed the brain wave[4] and took the video of the subject.

In this system, the degree of concentration was provided 9 levels, when it is between 1 and 3, his/her awareness space becomes WIDE, between 4 and 6, becomes NORMAL, between 7 and 9, becomes NARROW. In this way, the awareness space has continuity and doesn't change so frequently.

4.IMPLEMENTATION

Each Member who participates in a virtual office is provided with images and audio of other members which are defined according to relationships of their positions. In this system, user's body is represented in CG and the face picture is attached. Requiring face images of various directions, we use both the continuous video images and the still pictures; the system usually uses the still picture taken from 12 directions and it changes to the continuous video images when members face each other.

Figure 1: The Office View

Figure1 shows the office view of this system and the member sitting diagonally right are putting headphone(which indicates he is concentrating).

5.Conclusion

In this paper, we describe a system that provides a “work-at-home” environment based on a virtual shared room built on a 3D graphics workstation. We realize “Awareness Space” on the system to avoid a tradeoff between providing facility of informal communication and keeping one’s workspace from others’ awareness information. Also, this system provides the feeling of the ownself presence at virtual office by using “Around View” and “Sound Effect”. We are looking forward to giving evaluation of the system.

References

[1] Paul Dourish, Sara Bly, “Portholes: Supporting Awareness in a Distributed work Group”, ACM CHI’92, 1992.

[2] Scott E. and Ian Smith, “Techniques for Addressing Fundamental Privacy and Disruption Tradeoffs in Awareness Support Systems”, ACM CSCW’96, 1996.

[3] Chris Greenhalgh and Steven Benford, “MASSIVE: A Collaborative Virtual Environment for Teleconferencing”, ACM Transactions on Computer-Human Interaction, Vol.2, No.3 Sep. 1995.

[4] http://www.opendoor.com/pagoda/IBVA/

Getting A Grip: Touch Feedback with Binary Stimulators

A. Murray[a], K.Shimoga[b], R. Klatzky[c], and P. Khosla[a]

[a]Institute for Complex Engineered Systems, and Department of Electrical and Computer Engineering; [b]Robotics Institute; [c]Department of Psychology

Carnegie Mellon University
Pittsburgh, PA 15213

ABSTRACT

Human-Machine Interface (HMI) applications that involve manipulation of a remote or virtual object, whether it's a surgical tool or hazardous waste, cannot reach their fullest potential without haptic feedback. We set out to demonstrate quantitatively what seemed apparent intuitively: namely, the objective utility of touch feedback for a pick-and-place telemanipulation task. Our initial results were surprising: we found that there were no noticeable performance benefits as measured by number of errors and mean gripping force when using *binary* tactile feedback with a good visual view of the testbed. In this paper, we summarize our experiments and our interpretation of these results.

1. INTRODUCTION

Our sense of touch allows us to manipulate objects effortlessly everyday; we often realize the amazing sensitivity and effectiveness of this forgotten sense when we wear work gloves and fumble about while trying to manipulate small objects. Several human factor [1], [2] and physiological [3] studies have documented that when touch sensations are inhibited by thick gloves (e.g., diving or space suit gloves) or blocked altogether, a person's proficiency at lifting and manipulating objects is compromised even when a direct view of the object is provided.

A virtual reality or telepresence system without touch feedback is synonymous to having a person wear work gloves while performing manipulation tasks. We believe that adding high-fidelity tactile feedback to such systems will enhance an operator's ability to manipulate remote or virtual objects. However, at this date, it is infeasible to build a tactile feedback system that provides realistic sensations, since available technologies are incapable of stimulating the human tactile sensory system in a transparent manner. Shimoga [5], and Kaczmarek and Bach-y-Rita [4] discuss the requirements for a touch display and devices that have been built.

Our desire to demonstrate the utility of touch feedback in a telemanipulation environment, even with a simple binary device, led us to perform a series of pick-and-place experiments.

We were surprised by our initial results: we found no significant improvement in performance as measured by task completion time, number of errors, mean gripping force, and maximum gripping force when using binary touch feedback in the presence good visual feedback of the testbed. The dominance of visual information over touch feedback was more extensive than we had anticipated. In this paper, we provide an overview of our experiment and our interpretation of these results.

2. TELEMANIPULATION TESTBED

The telemanipulation testbed used in our experiment consisted of a four-fingered dextrous robotic hand (Utah/MIT hand) that was mounted to the wrist of a six degree-of-freedom robotic manipulator (Puma 500 manipulator), and a sensorized glove worn by the user to track the user's hand position (Polhemus Tracker™) and finger motions (CyberGlove™). In this configuration, the robotic hand mimicked the motions of the user's hand. We attached force sensitive resistor (FSR) sensors to the fingertips of the robotic hand to convey touch information to the user's fingertips via binary tactile stimulators (or tactors) made of shape-memory alloy (SMA) actuators (from TiNi Alloy Company). The on/off state of the tactors was determined from the forces measured by the FSRs. When the measured contact force was above a specified threshold value, the tactor turned on and pulsed against the operator's fingertip at a rate of 4 Hz. A picture of the testbed may be found at "*http://www.cs.cmu.edu/~amurray*".

3. PICK-AND-PLACE EXPERIMENT

Using this testbed, we conducted a study to examine how the actuation threshold for the binary SMA tactors influenced the human operator's ability to perform telemanipulation tasks as assessed by objective measures of time-to-completion, task success or failure, mean gripping force and maximum gripping force. Four subjects performed pick-and-place telemanipulation tasks using a thumb-index fingertip pinch grasp with three blocks of different sizes and weights. The force threshold used for the onset of tactile feedback was set to one of four possible values: no touch feedback; a low threshold condition that turned on the touch feedback early in the grasp; a medium threshold condition that turned on the touch feedback just as the grip force reached the minimum force necessary to lift the block (slip threshold); and a high threshold that turned on the feedback later in the grasp.

To prevent subjects from over-gripping the objects, a maximum grip force threshold was used to disable the robot hand, if exceeded. Dropping the block and overgripping the block were considered errors that caused termination of the trial.

A total of 120 trials was conducted with each subject: a 30-trial set for each of the four threshold conditions. Within each 30-trial set, 10 trials were conducted with each of the three blocks, in a random order. The order of the threshold conditions was varied across subjects.

The subjects had a direct view of the telemanipulation testbed and they could look at a video monitor, which provided a shoulder-angle view that was orthogonal to the direct view of the testbed. The subjects were instructed to use the touch feedback signals and to concentrate more on completing the task successfully than completing the task quickly. Each subject trained on the system for 30 minutes, or until comfortable with the task and with operating in the telemanipulation testbed.

Hybridization of classical documentary techniques and techniques of three dimensional representation in space : dynamical constitution of virtual scenes.

Juliette Bacon / Xavier Soinard
juliette@cln46fw.der.edf.fr, xavier.soinard@der.edfgdf.fr

Electricité de France, Direction des Etudes et Recherches
1 avenue du Général de Gaulle, 92140 Clamart, France

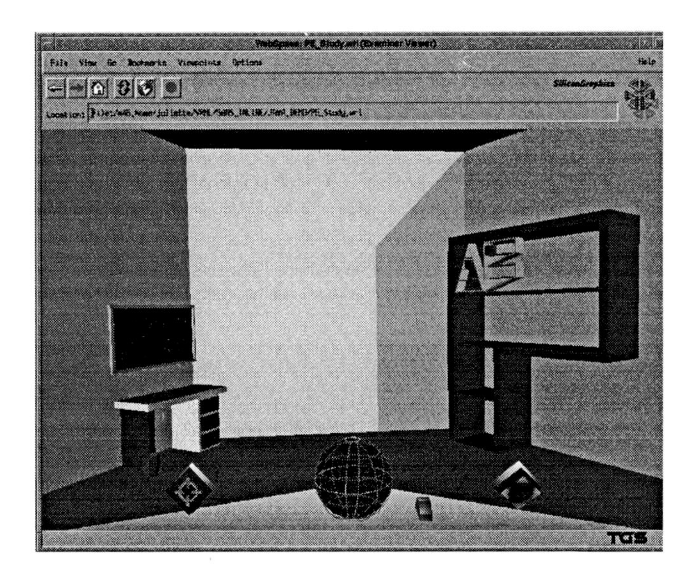

Abstract

The search and consultation of an information system comprising large amounts of data led us to investigate solutions involving the three dimensional representation of data. The conventional model of the virtual library metaphor seemed to us inadequate. What we have tried to do is enrich it by transposing it into the visual space of a computerized classification technique.

Keywords

Virtual, digital library, VRML, interface 3D, intranet, classification methods, dynamic queries visualization.

Introduction

The aim of the Electronic Library project is to create an electronic fund of structured multimedia information at the DER (Research end Development Division) of EDF (Electricité de France), and demonstrate how each agent can use it at his/her workstation.

The following are needed for this purpose:

- An object-oriented method of analysis for the specification of document models and the Electronic Library system,
- An object data base for the storage of a large quantity of closely connected multimedia information,
- Information structuring standards such as SGML (Standard Generalized Markup Language) and Hytime (Hypertext/Time-based structuring language),
- Multimedia tools (images, video, sound),
- 2D and 3D man-machine interfaces (based on HTML, JAVASCRIPT, VRML).

The IPN department (has developed the HyO_2 object data base modules, the multimedia server AthenaMuse2 (in association with the Massachusetts Institute of Technology) and search interfaces: navigation/visualization.

The data base software chosen for the prototype of the Electronic Library system is O_2 (by O_2 Technology).

Since the user of the Electronic Library is confronted with a large amount of information, various types of access are proposed:

- Textual search based on the full-text search engine (Topic/Verity and O_2 Web),
- Navigation based on the cartographic projection of data (2D in HTML, Tcl/Tk and Java),
- Synthesis view using the diminishing wall principle (3D in Open Inventor converted into VRML (Virtual Reality Modeling Language)),
- Dynamic virtual library (VRML).

The subject of this paper deals in particular with the hybridization of classical documentary techniques and three dimensional representation in space developed in the exploration module of the virtual library.

Method

Classically, the presentation of information is in the form of a list sorted according to different criteria. In our precise case, the pertinent criteria are: dates, titles, membership of an organizational structure, etc. In a classical metaphor of a virtual library, the user is confronted with similar representations of shelves containing documents of little visual significance. In

Awareness Technology: Experiments with Abstract Representation

Elin Rønby Pedersen and Tomas Sokoler

Dept. of Computer Science, Communications and Educational Research, Roskilde University, P.O.Box 260, DK-4000 Roskilde, Denmark

This paper reports on experience in designing technology for awareness, based on the AROMA project. The approach taken in AROMA is to extract essential data about activity at one site and find subtle methods of representing this information at some other site. The method of choice is *abstraction*: we abstract data from the situation, and we use abstract representation of such data. More specifically we have been experimenting with a range of abstraction in order to find a proper balance between showing too much and too little: the representation should respect concerns for privacy and perceiving the representation should not require too much attention.

1. INTRODUCTION

Human perception allows us to be aware of complex situations without apparently doing anything explicit to maintain the knowledge; we talk about peripheral awareness. Designing for peripheral awareness is a new research field; it leverage from existing communication and collaboration research, but distinguish itself by emphasizing certain automatic processes of human perception that were not previously in focus in HCI and CSCW design efforts. Awareness technology is basically an attempt to mediate and bring closer events that was otherwise outside our sensory range, i.e., through the use of communication and collaboration technology we can be aware of events that happens in remote locations as well as events that do not "really happen" except inside a computing network.

The AROMA project [2] is developing awareness technology through a minimalist approach: rather than showing exactly what is happening remotely, we are trying to show just enough for the people to make sense of.

2. AWARENESS

Awareness has been a topic of interest for some years in CSCW, e.g. for the purpose of alerting someone to a request for communication or a system malfunctioning, or just to provide the remote site with a richer picture. Recently researchers have begun looking into automatic processes of perception, such as peripheral awareness. A fascinating thing about this kind of awareness is that, although we do not seem to notice our use of it, a wealth of information is nevertheless perceived in this manner and made useful in the coordination of our everyday actions. The recent interest in awareness technology has also revealed some inherent problems, among others, the potential violation of privacy and the risk of overloading or confusing our focus of attention.

In the context of AROMA we began to speculate on the problems of privacy and of attention as a scarce resource, and it struck us that thoughtful use of abstraction might prove a solution to both problems. We talk about abstraction in the sense of "taking away" and "going symbolic".

3. ABSTRACTION AS TAKING AWAY

The abstraction that "takes away" stuff from the original source will do so in order to make fewer aspects stand out more clearly. As we remove stuff, we increase the burden of "reading" the data. Fortunately, people are amazingly apt at filling out what is missing.

Usually the removal of large parts of the original signal will be complemented by the addition of new synthetic components; we *re-embody* the compact signals. In the re-embodiment we can try to add material similar to what we took away, and that might have let us regain some of the readability that was lost. But it would have been at the expense of increased attention load: according to our intuition as well as some preliminary findings during use studies, the more naturalistic the moving imagery is, the harder it is *not* to attend to it. An example of this approach is the use of avatars.

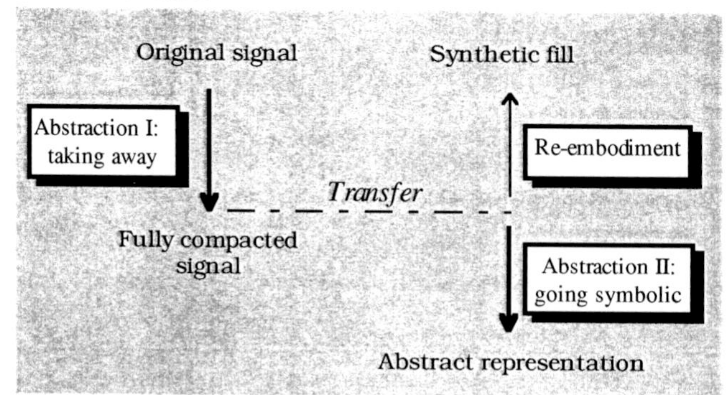

Figure 1: Two approaches to using the compacted data: re-embodiment, the movement *down-transfer-up*, versus further abstraction through remapping, the movement *down-transfer-down*. The latter is being used in AROMA.

4. ABSTRACTION AS GOING SYMBOLIC

The term "abstraction" is not only used in the sense of "taking away"; it is as often used to denote a transformation to a symbolic level. Rather than attempting the "re-embodiment", we are representing what we have left in a way that has symbolic relations to what we started out with. An example would be the display of the current sound level as the color depth (as in the "floorplan" interface, see next section) or as the number of new birds being added to blue sky (as in the "common view" interface described later).

This second kind of abstraction principle is closely related to a phenomenon we call *remapping*. The idea in remapping is to further utilize some of the separations already made, for instance; e.g., we do not *have* to "display" sound information as sound signals, they can just as well be shown as visible as well – this is *media remapping*. Such remappings seem *very* powerful in the usage domain: things that may be perceived as noise and disruption in one media, may nevertheless be informative, and by moving the data to another media we may get the benefits without the nuisance.

5. DEVELOPMENT OF AROMA INTERFACES

During the development of AROMA we have made radical changes to the user interface. This section will briefly account for the major stages and provide the rationales for our choices. We are using a general technique that combines a subtle, steady flow of motion ("self-motion") with the display of events as slightly more "live" yet smooth animations. We also attempts to embrace "calmness" as a functional and aesthetic design criterion, as suggested in [5].

5.1 Entirely abstract imagery

We worked for a while on direct representation of very simple activity data: we captured the visible activity by calculating the frame difference between each pair of consecutive video frames, and we sampled the audible activity by gauging the audio volume through a microphone. We chose one of the simplest abstract animations - a bouncing ball - and had its speed be driven by the visible activity while its size changed according to the audio level.

All kinds of activity were transformed into a visual display, despite the original capture medium. Such remapping across media was enabled by the radical abstraction applied on the capture site. We later learned more about utilizing this complete detachment of activity data from its originating media.

Use studies were less favorable: the display seemed too abstract; users found it un-pleasing, mostly because it was neither funny, pretty, nor interesting. The display showed the current level of activity, but users pointed out that they really wanted to have more information available, for instance some representation of the most immediate history. They suggested additions such as having the "present" leave a trace that would faint as time went by.

Example 1: Bouncing ball
Theme: Abstract imagery
Effects: Sound: size ; Movements: speed; History: shadows of balls

5.2 Non-computer devices

Having detached the capture media from the display media, as it happened in the case of the bouncing ball, made it rather obvious to try out other remappings, not only within the traditional output devices of the computer but also into a more ubiquitous environment.

Staying within the visual domain we built a merry-go-round, a kind of desktop sculpture (example 2). The sculpture had poles with colorful flags attached; with increased activity at the remote location, we would see the sculpture rotate faster and the flags would wave more and more vividly. Subsequently we left the standard computer domain and built a heating/cooling element into a wrist rest; a typist using the wrist rest would feel a comfortable warmth as long as the activity at the remote location was reasonably high, whereas a slight (yet not uncomfortable) decrease to a cooler temperature would happen when activity subsided (example 3).

Example 2: Merry-go-round
Theme: Non-computer object
Effects: Sound+movements: rotation speed

Example 3: Wrist pad
Theme: Non-computer object
Effects: Sound+movements: temp. of the surface

5.3 Further exploration of visual display

Most recently we have developed two visual representations: the floorplan and the common view. Both representations are based on our assumptions about the importance of subtlety, steady flow of motion and smoothness in the visual appearance.

Example 4: Floor plan
Theme: Using a conventional abstraction
Effects: Sound: floor color; Movements: speed of blob : size of blob

Using the conventional abstraction of the architectural floorplan, we are showing the general location and activity level of people in a house (example 4). The current sound level is seen as changes in the color depth of the room, while movements are mapped into the speed of a bouncing green blob. Some elements of history are also displayed: the amount of time spent in a given room is shown by increasing the blob size. Rooms in the floorplan are colored according to their level of accessibility. Our experience with this representation is that the direct mapping of location oversteps the limits of privacy, but also that the idea of showing accessibility seems promising. Thus we came up with an alternative representation showing zones of accessibility, each zone being depicted as color-coded ellipses not directly related to the physical space is currently being explored.

Most recently we have used a mixture of naturalistic and abstract imagery (example 5). A photo serves as background; it is overlaid with animated objects representing remote activ-

Example 5: Common views / outdoor scenery
Theme: Mixture of naturalistic and abstract imagery
Effects: Sound: number of birds "born" at the right edge; Movements: liveliness and position of balloon; Age: size of balloon; History: in audio and movement

ity. Sound is seen as a number of white birds traveling from right to left (sound level maps into number of birds). Movements are mapped into the liveliness and position of an animated balloon. History is represented by the total number of birds traveling across the picture (sound history) and the balloon size (time spent by person in room). Furthermore, the background picture is chosen from a selection of 12, each depicting the view outside the window at the remote site at different times of the day. By selecting the picture closest to the time of the day at the remote site, this display is particularly useful for people connecting between different time zones.

6. FUTURE RESEARCH DIRECTIONS

6.1 Screen savers

Many people have commented on the obvious resemblance between AROMA and screen savers. Contrary to AROMA, most screensavers are using entirely non-semantic imagery. But despite this difference, we may learn from their use of moving imagery, e.g., good screensavers may be bad awareness indications, and those screensavers that become annoying after a while, may actually be useful in awareness. AROMA displays could of course also be made into screensavers providing informative animations.

6.2 Integrated awareness in the primary user interfaces

In today's computer systems the user has to operate a multitude of programs - each assuming it has the user's undivided attention. Such assumptions are growing increasingly more problematic - even within the desktop metaphor - not to speak of what happens in more ubiquitous computing environments. Although the objects of awareness in the AROMA design were other human beings and their social activity, nothing precludes that our design could be extended to support awareness of processes and objects in the computer or in a computational network.

Recently we have started wondering if the term "periphery" may be more misleading than helpful. It may be a better idea to think of awareness as potentially rising from all places and directions, not only from the fringes of our vision but also from more subtle elements such as the color depth and the texture. Think of all the little feedback mechanisms that are embedded in a standard desktop user interface; such elements are extremely informative but need better design: they are too many, too un-intuitive, and too arbitrarily scattered around in the visual field.

6.3 "Things That Think"

Some strikingly elegant designs have been developed recently that brings the computation out of the computer, into the environment, without cluttering our attention space. It started with Weiser's paper on ubiquitous computing [4] and seems to have gained momentum by the launch of "Things That Think" program at the MIT MediaLab.

As described previously, we built a couple more fancy output devices: the carousel sculpture and the wrist rest with temperature control, but most of the AROMA design and development so far have focused on mapping whatever input signal and derivative of it into a visual display method. Very close to our work are recent designs involving feathers, scented oils [3], and colorful light emitted through rippling water [1]. We think awareness technology is begging for more of this kind of "out-of-the-computer thinking".

7. REFERENCES

1. Ishii, H.& B. Ulmer. Tangible Bits: Towards Seamless Interfaces between People, Bits, and Atoms. In *Proc. CHI'97*. ACM, 1997
2. Pedersen, E.R. & T. Sokoler. AROMA: Abstract Representation Of presence supporting Mutual Awareness. In *Proc. CHI'97*. ACM, 1997
3. Strong, R. & W. Gaver. Feather, Scent, and Shaker: Supporting Simple Intimacy. In *Proc. CSCW'96*, ACM, 1996
4. Weiser, M. The Computer of the Twenty-First Century. In *Scientific American, 10*, 1991.
5. Weiser, M, and J.S. Brown. Designing Calm Technology, *PowerGrid Journal, v 1.01* (July 1996), http://powergrid.electriciti.com/1.01

Manipulation Aid for Two-handed 3-D Designing within a Shared Virtual Environment

Kiyoshi Kiyokawa[a], Haruo Takemura[a] and Naokazu Yokoya[a]

[a] Graduate School of Information Science, Nara Institute of Science and Technology, 8916-5 Takayama, Ikoma, Nara, 630-01, Japan

This paper describes a case study on software aid for manipulating multiple virtual objects. First, two types of manipulation aid are introduced: *discrete placement constraints* and *collision avoidance*, and then control policies of these methods are proposed. Next, the effectiveness of our methods is examined through an empirical study. Experimental results show that the methods are efficient in both decreasing the time required for virtual object assembly and improving the manipulation feeling.

1. INTRODUCTION

Designing 3-D objects within a virtual environment rather than with 2-D projection has a number of advantages. Real-time head-tracked stereoscopic view improves the understanding of shape and spatial relationships. Spatial direct manipulation improves the accessibility to 3-D objects in intuitive and quick way.

We have been developing VLEGO II, a shared virtual environment for collaborative two-handed 3-D modeling. VLEGO II is implemented for one or two SGI graphics workstation(s). As shown in Figure 1, a user views a virtual workspace stereoscopically through a head mounted display or a liquid crystal shuttered glasses with head-tracking facility. He or she holds a pair of 3-D input devices. Each device has a 3-D magnetic tracker 3SPACE (Polhemus) and four feather touch switches on it. These devices are used for manipulating two 3-D cursors that manipulate virtual objects. VLEGO II enables two participants to construct 3-D objects by properly assembling

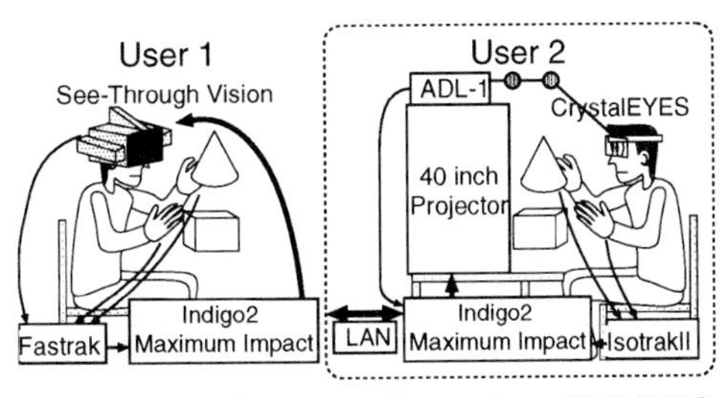

Figure 1. Hardware configuration of VLEGO II.

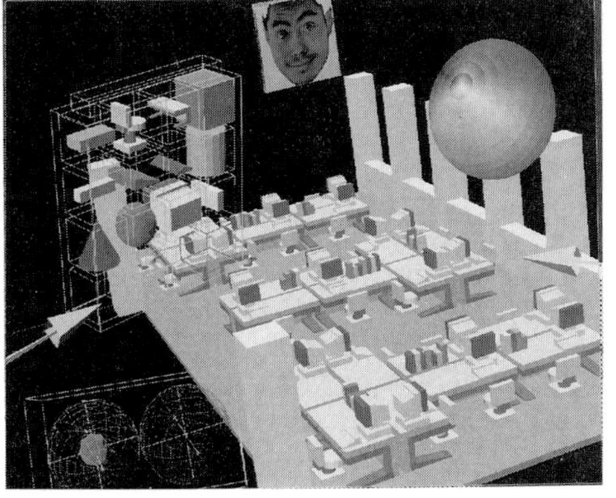

Figure 2. VLEGO II workspace.

texture-mapped primitives that have a few kinds of simple shapes like real toy blocks. Figure 2 illustrates a VLEGO II workspace.

Our VLEGO II supports two-handed and collaborative virtual object manipulation, and users have to true up the edges and corners of virtual objects to design a favorite shape. However, users are prone to be awkward in manipulating virtual objects due to lack of force-feedback, computational delay, restricted spatial resolution of input/output devices. So, in order to make the interface feasible without bulky force-feedback devices, certain practical methods for virtual object manipulation aid are required [1]. Thus, VLEGO II should provide software aid for cooperative manipulation of multiple virtual objects, which includes single-user two-handed manipulation and multiple-user collaboration.

In the following sections, we first describe a few types of manipulation aid for cooperative manipulation of multiple virtual objects employed in VLEGO II, and then an empirical study for examining the effectiveness of the methods is described.

2. SOFTWARE AID FOR VIRTUAL OBJECT MANIPULATION

2.1. Manipulation aid for single virtual object

In order to facilitate putting virtual objects in precise positions, e.g., truing up the edges and corners of virtual objects, simple and comprehensible constraints are imposed on the movement of virtual objects based on their geometric characteristics. Two types of visual constraints are explained below [2].

Discrete placement constraints

In principle, each virtual object is bound to four degree-of-freedom with discrete placement constraints (4DOFC) as shown in Figure 3. Being bound to 4DOFC, the location of the object is restricted in discrete positions at intervals of 1cm and its orientation is restricted at 0, 90, 180 and 270 degrees of horizontal rotation.

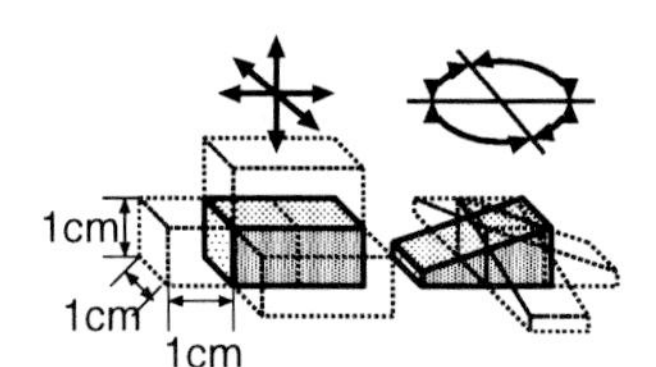

Figure 3. Discrete placement constraints.

Collision avoidance

In consequence of the 4DOFC process, the collision among objects can be easily detected using their bounding boxes. Collision avoidance is performed by moving a picked object, which was located in previous rendering cycle, towards its corresponding 3-D cursor, which is located in current rendering cycle, in X, Y and Z directions of the 4DOFC in turn so that no collision occurs.

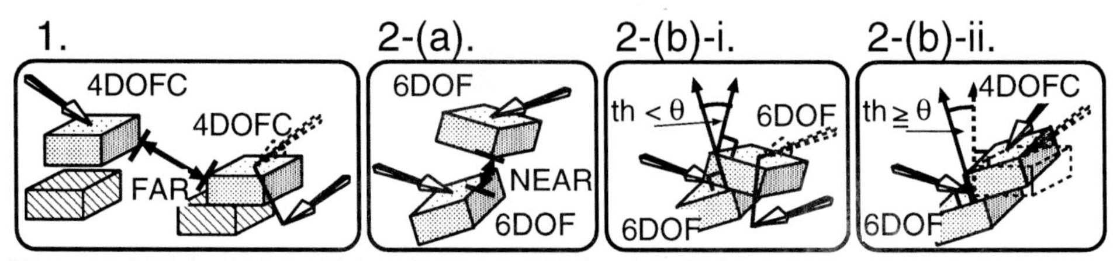

Figure 4. Manipulation aid for aligning multiple virtual objects.

2.2. Manipulation aid for multiple virtual objects

In order to provide a natural and efficient way for truing up cooperatively manipulated multiple virtual objects, we propose the following algorithm to control manipulation aid mentioned above.

1. If multiple manipulated objects are distant from each other, each object is bound to 4DOFC based on global coordinate system. The location of the constrained object is so adjusted as to avoid the collision with non-manipulated objects (Figure 4, 1.)
2. Otherwise, the object picked earliest among manipulated objects is selected as a *reference object*, which is bound to 6DOF. Other nearby objects are called *constrained objects*.
   (a) If a constrained object does not collide with the reference object, it is bound to 6DOF (Figure 4, 2-(a).)
   (b) Otherwise,
      i. If the angle θ between colliding surfaces is larger than a threshold th, in addition that the constrained object is bound to 6DOF, its location is so adjusted as to avoid the collision (Figure 4, 2-(b)-i.)
      ii. Otherwise, the constrained object is bound to 4DOFC based on the reference object to facilitate truing up each other (Figure 4, 2-(b)-ii.)

3. EXPERIMENT FOR EXAMINING THE EFFECTIVENESS OF THE PROPOSED METHOD

This section briefly describes an experiment for examining the effectiveness of the manipulation aid proposed above. A previous experiment has already demonstrated that the effectiveness of manipulation aid is generally similar for both two-handed and collaborative manipulation[3]. Hence, this experiment is performed only with two-handed manipulation.

3.1. Experimental Setup

Figure 5 shows the configuration of blocks in the experiment. There are four possible initial postures for each block, an initial layout out of 16 patterns is selected in turn for each trial. Subjects are asked to assemble two virtual blocks O_A and O_B into a box using two 3-D cursors A and B with two hands under the following eight conditions.

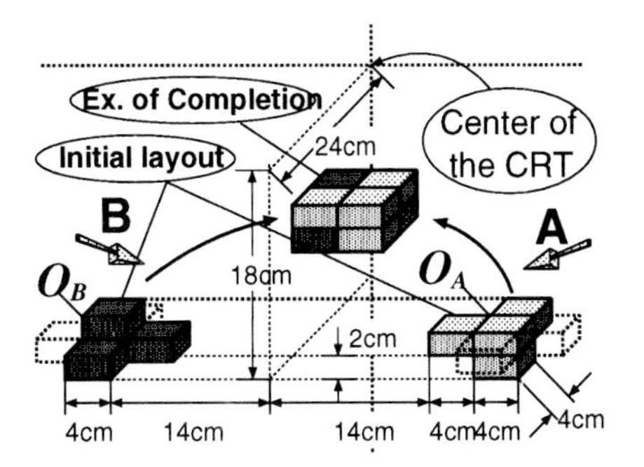

Figure 5. Configuration of blocks in the experiment.

N: No manipulation aid is provided.
C: Only collision avoidance is provided.
P_{th}: Proposed methods are provided. The thresholds th are 3, 5, 10, 15, 20 and 180 degrees.

Table 1
Average and analysis of variance of the performance of an assembling task in the experiment

Condition	N	C	P					
Threshold th (degrees)	–	–	3	5	10	15	20	180
Task completion time (sec.)	14.8 >	8.3 >	6.2 >	5.3 =	4.6 =	4.6 =	4.7 =	4.4
Error rate (%)	–	–	37.5	17.2	7.8	2.3	0	0
Subjective evaluation (1 – 7)	1.0	2.1	2.6	3.3	5.3	6.4	6.7	6.4

The following three types of data are recorded or interviewed.

- *Task completion time* which is the time required for assembly in each trial.
- *Error rate* which is the percentage of the number of trials, in which the angle θ was larger than th when one of the blocks is released.
- *Subjective evaluation* about manipulation feeling in 1 (very bad) to 7 (very good).

3.2. Experimental Results

Eight students of our laboratory performed the task, each is a novice for virtual object manipulation and performed 256 trials, i.e., 32 for each manipulation condition. Table 1 shows the average of experimental results and the analysis of variance for task completion time ($\alpha = 0.05$). We can observe that the proposed methods reduce the task completion time by 60% to 70%, and substantially improve the score in subjective evaluation. Generally speaking, the smaller the threshold th is, the more it gets natural for a user to rotate virtual objects. However the error rate and subjective evaluation are getting worse drastically in $th < 15$. Hence, it can be said that the most reasonable value of th is about 15 degrees.

4. CONCLUSION

Manipulation aid for two-handed and collaborative virtual object manipulation was proposed. The empirical study for examining the proposed methods demonstrated that appropriate visual constraints substantially reduce the time required for virtual object alignment and improve the manipulation feeling.

For enhancing virtual collaboration, future works include developing and evaluating other supporting methods, e.g., providing voice channel, showing participants' viewing positions by polygonal faces with real-time video captured texture.

REFERENCES

1. Kitamura, Y., Yee, A. and Kishino, F., Virtual Object Manipulation Using Dynamically Selected Constraints with Real-Time Collision Detection, Proc. ACM Sympo. on Virtual Reality Software and Technology (VRST '96), pp.173–181, (1996).
2. Kiyokawa, K., Takemura, H., Katayama, Y., Iwasa, H. and Yokoya, N., VLEGO: A Simple Two-handed Modeling Environment Based on Toy Blocks, Proc. ACM Sympo. on Virtual Reality Software and Technology (VRST '96), pp.27–34, (1996).
3. Kiyokawa, K., Takemura, H. and Yokoya, N., An Empirical Study on Two-Handed / Collaborative Virtual Assembly, Proc. The Third International Display Workshops (IDW '96) Vol.2, pp.477–480, (1996).

REQUIREMENTS AND DESIGN OF THE INFORMATION POD INTERFACE

M. R. Stytz, Ph.D.; S. B. Banks, Ph.D.; J. J. Kesterman; J. J. Rohrer; and J. C. Vanderburgh

Virtual Environments Laboratory
Human-Computer Interaction Laboratory
Air Force Institute of Technology
mstytz, sbanks@afit.af.mil

1. INTRODUCTION

As the scale, complexity, unpredictability, and realism of virtual environments have grown with the degree of user immersion and presence in the environment, there has been a concomitant increase in the amount of information available in a virtual environment and the number of activities a user might want to perform. However, users of current virtual environments typically have few tools available to assist them in accomplishing meaningful work. Currently, users are limited to manipulating items, moving within a limited range, and viewing the activity of other participants in the virtual environment. However, these are not sufficient capabilities for accomplishing useful work within a complex virtual environment. If virtual environments are to be useful, tools must be developed that reduce the user's cognitive workload and enhance mission performance while exploring, analyzing, operating, and working within the virtual environment. One aspect of providing this required assistance is to provide an interface that is rich with controls, does not obscure the display of the environment, and allows the user to remain immersed.

The exemplar immersive virtual environment interface would enable the user to perform a wide variety of useful work and would not intrude into the user's observations of the virtual environment. The interface would give the user convenient access to virtual environment display parameters, analysis reports, conferencing and collaboration capabilities, intelligent agents, motion and orientation controls, recording devices, and situation awareness aids. The user could move across large spaces. The interface controls would be familiar so that the user could build upon past interface experience to perform tasks. To address these needs we developed the Information Pod, an interface approach that satisfies these requirements and effectively supports the user within large and complex virtual environments. The Pod allows us to provide controls that support user movement, observation, decision making, and situation awareness. The final design for the appearance components of the Pod's interface is the result of experimentation with several hardware input devices and interface presentation elements.

The motivation for the work reported here is the Synthetic BattleBridge (SBB) and Solar System Modeler (SM) virtual environment projects[1,2,3,4]. These projects use distributed virtual environment technology[5]. The Information Pod has been incorporated into these applications and can be incorporated into any immersive virtual environment application.

Our work builds upon the concepts developed by many others. Robinett and Holloway[7] describe a system that permits the user to fly through a virtual world, scale it, and grab objects.

The system described by Venolia[8] centers around an experimental interface that uses a 3D mouse and 3D interaction techniques. Bolt[9] describes a multimodel user interface that employs both direct manipulation and language. Ware[10] presents a series of techniques for manipulating objects in a virtual environment. The Zashiki-Warashi system[11] is a 3D direct manipulation interface that allows users to directly manipulate furniture and lighting to design a virtual room and uses a beam cursor to select objects for direct manipulation. Figueiredo[12] reports on work to enable a user to directly manipulate objects in a virtual environment.

In this paper we describe the requirements and design criteria for the Information Pod. Section 2 contains a discussion of the requirements and design for the Pod and a description of the components of the interface and the interface's user paradigm. Section 3 presents illustrative results for the Pod in supporting virtual environment interaction within three virtual environment projects. We also discuss areas for future work in the Pod design and for virtual environment user interaction support.

2. REQUIREMENTS AND DESIGN FOR THE POD INTERFACE

The goal for our immersive virtual environment interface is to support a visually rich environment in which information manipulation and display is facilitated by a variety of mutually supportive, natural modes of interaction. The user interface must support user viewpoint movement in any direction and around any axis and control the display of the components of the environment. The interface must also provide attachment controls so that the user can tether to other participants. Other required capabilities are to support calculation of a viewpoint automatically based on some condition, timing, actor distribution, or important event. The interface must provide tools that allow the user to remotely explore the virtual environment, collaborate in performing work, and develop situational awareness for the environment. In addition, the interface must provide the user with the ability to control the autonomous agents and the information summary capabilities that they provide. The interface must be usable and acceptable when using a head-mounted display or when using a CRT for viewing the virtual environment. The interface must also be easy to modify. These needs led us to the Information Pod paradigm.

The Pod is a user "capsule" within the virtual world that supports immersed user operation within a virtual environment. The Pod replaces a desktop metaphor for the interface with one of a control console or the bridge of a ship. The Pod virtually encases the user with controls and with displays mounted upon 2D panels in 3D space, with each panel devoted to a specific set of tasks. Tasks are assigned to panels and subpanels by factoring the tasks to be performed by the affinity of the tasks. The controls and displays change contents depending upon the user's focus of attention and the task to be accomplished. The Pod capitalizes upon the user's natural inclinations, so if the user wants to see what is to the right in the virtual environment, the user just needs to look to the right. The user performs large-scale movement and reorientation in the virtual environment by moving the Pod. Fine viewpoint adjustments are accomplished by head movement alone. As a result of this design, the Pod decouples the user position and orientation relative to the controls and displays from user position and orientation regarding the virtual environment.

The 2D panels in 3D space allow the user to control the Pod, receive and display information, and change the characteristics of the virtual environment. Each panel can contain subpanels, with each subpanel in-turn containing associatively grouped controls and displays.

The panel placement design selected for the Pod allows the designer to place different functional categories in different locations and then factor the specific functions onto additional panels co-located (or stacked) under the primary panel locations. Sub-tasks are identified on the main panel and their associated subpanel is activated using a button on the main panel. This strategy for panel placement allows subpanels to be added to the application without cluttering the Pod's display environment.

The Pod is designed so that the viewer selects the panel to be activated and only one panel is active at any time. Panel activation occurs when the user looks at the panel. If the viewer does not want to activate the panel in the field of view, a button allows the user to hide the panel. Because user interaction is completely panel-based, buttons are the only means for interaction with the virtual environment. The Pod's buttons operate as described in Table 1.

BUTTON TYPE	FUNCTION
single-fire	one signal to the button's function regardless of duration of selection
continuous-fire	updates the button's functionality the entire time of selection
toggle	changes the button's functional state each time the button is depressed

Table 1: Types and Operation of Control Buttons in the Pod

The final issue addressed was button selection. Our initial requirements for the selection device were that it be easy to use, not be a significant cause of user fatigue, and use typical selection gestures. Because the Pod's panels are within arm's reach, a dataglove seemed to be an ideal candidate. The dataglove approach was unworkable due to noise and jitter, so we experimented with a variety of devices before settling on a mouse selection device.

4. RESULTS AND CONCLUSIONS

The Synthetic BattleBridge immerses a user within a large-scale virtual military battlespace environment involving air and ground vehicles.

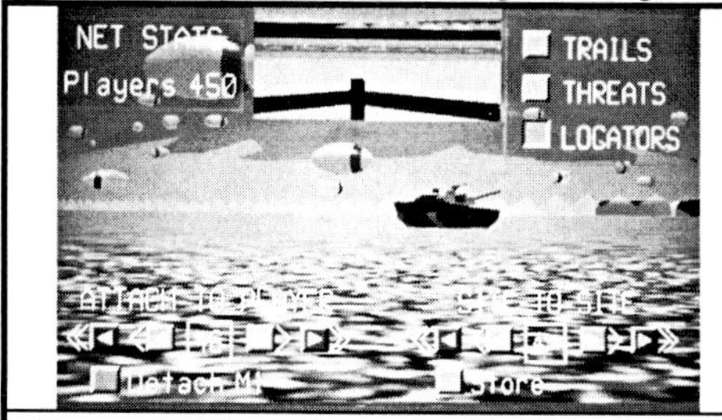

Figure 1: The Information Pod's Left Panel in the Synthetic BattleBridge

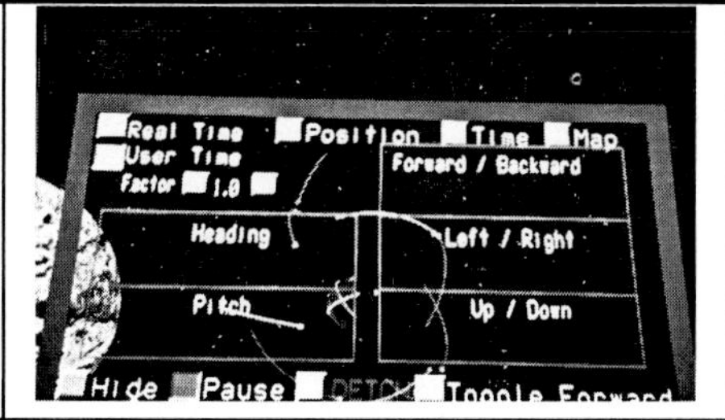

Figure 2: The Information Pod's Center Panel in the Solar System Modeler

Figure 1 presents a view of the left panel in the Synthetic BattleBridge. The Solar System Modeler immerses a user within a space-based large-scale virtual environment involving planets, the Sun, Moon, and man-made satellites. Figure 2 presents a close-up view the center panel in the Solar System Modeler, showing the subpanels for motion and orientation control as well as some Solar System Modeler specific controls.

The research reported in this paper arose from the conviction that virtual environments provide a potentially revolutionary means for humans to interact with each other and with computers. However, to achieve this potential, techniques that allow users to accomplish a wide variety of work and communication within a virtual environment must be developed. In this article, we described our own initial efforts toward this potential as realized in the Pod. These efforts centered upon the specification of a software architecture for the interface, an interface paradigm, and the interface elements. The Pod is an interface designed to support immersed user operation within a virtual environment; however, it does not provide a direct manipulation capability for objects within the environment. The Information Pod does allow the user to manipulate components of the virtual environment using an interface paradigm and interface widgets of our own design. The interface allows us to provide controls for precise user movement, observation, decision making, and environment contents. The final design for the appearance components and interaction affordances for the Pod is the result of experimentation with a variety of hardware input devices and of the presentation elements.

REFERENCES

1. Stytz, M. R.; Hobbs, B.; Kunz, A.; Soltz, B.; and Wilson, K. "Portraying and Understanding Large-Scale Distributed Virtual Environments: Experience and Tentative Conclusions," *Presence*, 4(2), 146-168, Spring 1995.
2. Stytz, M. R.; Amburn, P.; Lawlis, P. A.; and Shomper, K. "Virtual Environments Research in the Air Force Institute of Technology Virtual Environments, 3D Medical Imaging, and Computer Graphics Laboratory," *Presence*, 4(4), 417-430, Fall 1995.
3. Stytz, M. R.; Block, E.; Soltz, B.; and Wilson, K. "The Synthetic BattleBridge: A Tool for Large-Scale VEs," *IEEE Computer Graphics and Applications*, 16(1), 16-26, Jan. 1996.
4. Stytz, M. R. and Kunz, A. "A Distributed Virtual Environment for Satellite Orbital Modeling and Near-Earth Space Environment Simulation and Portrayal," *Simulation*, 67(1), 7 - 20 July 1996.
5. Stytz, M. R. "Distributed Virtual Environments," *IEEE Computer Graphics and Applications*, 16(3), 19 - 31, May 1996.
6. Robinett, W. and Holloway, R. "Implementation of Fying, Scaling, and Grabing in Virtual Worlds," *Proceedings of the 1992 Symposium on Interactive 3D Graphics*, 189 - 192, Cambridge, Massachusetts, 29 March - 1 April 1992.
7. Venolia, D. "Facile 3D Direct Manipulation," *INTERCHI '93 Conference Proceedings: Conference on Human Factors in Computing Systems, INTERACT '93 and CHI'93*, 31 - 36, Amsterdam, The Netherlands, 24 - 29 April, 1993.
8. Bolt, R. A. and Herranz, E. "Two-Handed Gesture in Multimodal Natural Dialog," *Proceedings of UIST '92, The Fifth Annual ACM SIGGRAPH and SIGCHI Symposium on User Interface Software and Technology*, 7-14, Monterey, California, 15 - 18 Nov. 1992.
9. Ware, C.; Arthur, K.; and Booth, K. S. "Fish Tank Virtual Reality," *INTERCHI '93 Conference Proceedings: Conference on Human Factors in Computing Systems, INTERACT '93 and CHI'93*, Amsterdam, The Netherlands, 37-42, 24 - 29 April 1993.
10. Yoshimura, T.; Nakamura, Y.; and Sugiura, M. "3D Direct Manipulation Interface: Development of the Zashiki-Warashi System," *Computers & Graphics*, 18(2), 201 - 207, March/April 1994.
11. Figueiredo, M.; Bohm, K.; and Teixeira, J. "Advanced Interaction Techniques in Virtual Environments," *Computers & Graphics*, 17(6), 655-661, Nov./Dec. 1993.

The human-machine interface challenges of using virtual environment (VE) displays aboard centrifuge devices

B.D. Lawson, A.H. Rupert, F.E. Guedry, J.D. Grissett, and A.M. Mead.

Spatial Orientation Systems Department, Naval Aerospace Medical Research Laboratory, Pensacola, FL. 32508-1046. blawson@namrl.navy.mil

1. CENTRIFUGE-BASED VE: FLIGHT TRAINING FOR THE 21st CENTURY?

While VE displays composed of synthetic visual and auditory inputs to a stationary user can often accomplish the user's goal, fully immersive "virtual reality" will remain a virtuality until realistic simulations of self-motion are incorporated. The perception of acceleration involves the integration of vision, audition, touch, kinesthesia, and especially the vestibular senses, which evolved specifically to transduce acceleration. Incorporating these modalities into a VE display by basing it aboard a centrifuge would expand the flight maneuvers that could be demonstrated to aviator trainees and would aid in mishap reconstruction. However, integrating virtual displays with centrifuge flight demonstrators will present significant human-machine interface challenges. The operating characteristics of the vestibular system can cause troublesome effects during certain combinations of centrifuge motion, subject movement, visual field motion, and visual scanning. These effects include: misperceptions of self- orientation, decreased ability to read visual displays, motion sickness, and Sopite Syndrome. When these effects do not occur in "real" flight, they should be minimized in the training simulations.

2. INTRODUCING HELPFUL EARTH-REFERENCED INFORMATION

In a centrifuge-based VE, a head movement in an axis that is not parallel to the central axis of centrifuge rotation will often produce Coriolis Cross-Coupling (CCC) effects, which include perception of tilt or tumbling and feelings of motion sickness. The angular velocities experienced during real flight are usually too small to generate appreciable CCC effects[1]. Certain vestibular or visual inputs can reduce the CCC effect; e.g., the presence of an acceleration stimulus to the horizontal semicircular canals diminishes the disturbing effect of an earthward head movement during body rotation while seated upright[2]. Similarly, earthward head movement after prolonged rotation is less disorienting if it is preceded by viewing an earth-fixed visual reference[3]. We have extended these observations as follows: a) the aforementioned visual reference tends to be helpful for several gaze strategies and field-of-view widths[4], b) the visual reference will be most helpful when the plane of stimulation of the vestibular system prior to the head movement is in yaw[5], c) the helpful effect of an earth reference can also be achieved via the tactile and kinesthetic modalities[6], d) systematically manipulating the CCC stimulus produces the expected increase in disorientation and sickness[7], and e) we may be able to account for the perception of CCC over time and in directions of felt motion not commonly described in the literature[8].

3. INTEGRATING VISUAL DISPLAYS WITH CENTRIFUGE MOTION

Most whole-body motions will elicit a vestibulo-ocular reflex (VOR) that has the effect of keeping the eyes steady in space as the head or body moves. This natural coordination of visual and vestibular function is disrupted by a centrifuge-based VE when the user attempts to read a head-fixed display during rotation. In this situation, interpreting the display requires visual suppression of the VOR. Those who design VE simulations involving centrifuge motion should understand the conditions in which a viewer's ability to suppress his VOR is enhanced[9,10]. If documented VOR suppression effects are observed among VE users, we may be able to introduce peripheral visual stimuli that enhance the user's ability to interpret a head-fixed display during whole-body acceleration. For example, VOR suppression and visual interpretation are superior when visual background movement is "concordant" with the vestibular signal[11]. It is also likely that VOR suppression will be better as the visual background velocity matches the rate of turn signaled by the vestibular input[12]. However, since visual suppression of the VOR is sickening, we will need to enhance visual suppression without causing greater discomfort.

4. DEALING WITH "CYBER-SICKNESS" AND THE SOPITE SYNDROME

While we should optimize centrifuge-based VE devices so they are less acutely disturbing, we should also be alert to subtle and chronic problems that occur after prolonged or repeated exposure to mild stimuli. Studies of sickness in flight simulators reveal that about 25% of trainees report symptoms of "simulator sickness" that last more than an hour after training, and some report effects lasting six hours or longer[13,14]. Effects that persist after the cessation of stimulation have been observed in laboratory studies involving visually induced motion sickness[15], and adaptation to a rotating environment[16]. Even after individuals are adapted to an unusual sensory environment, they may suffer from the Sopite Syndrome[17], which is characterized by drowsiness and mood changes. Recent studies suggest that the syndrome occurs frequently enough to warrant further study[18] and may not be limited to conditions of long term or repeated stimulation[19]. Such side effects probably occur routinely in untrained observers without being fully recognized and attributed to the causal stimulus [17,20]. Moreover, some individuals who consider themselves immune to motion sickness exhibit Sopite symptoms[21]. Waiting for trainees to adapt to a VE is not the answer, because they may adopt sensory-motor strategies that are inappropriate during real flight[22]. We must determine which conditions allow for isomorphic simulations without initiating inappropriate reflexes. We should also develop simulator profiles and training schedules to avoid prolonged aftereffects and we should devise tests to identify susceptible individuals.

5. CONCLUSIONS

The side effects likely to occur during centrifuge-based VE flight simulation can be reduced under certain circumstances. Designers of centrifuge-based VE simulations should evaluate strategies to minimize unwanted effects while optimizing the ability to interpret head-fixed displays during rotation. Such information will be important to the design of 21st century VE systems. For example, display time-lag can be diminished by reducing display complexity peripheral to the direction of gaze of the user, but certain centrifuge profiles will cause gaze direction to vary due to the VOR rather than the user's intended focus of visual attention. Systems that adjust scene complexity or allow device control inputs based on gaze responses will require careful human factors engineering before integration with centrifuges. Otherwise, faithful simulations are unlikely, and symptoms of motion sickness will be widespread.

REFERENCES

1. Gilson RD, Guedry FE, Hixson WC, Niven HJ. Observations on perceived changes in aircraft attitude attending head movements made in a 2-g bank and turn. Aerosp Med 1973;44:90-92.
2. Guedry FE, Benson AJ. Coriolis cross-coupling effects: Disorienting and nauseogenic or not? Aviat Space Environ Med 1978;49(1):29-35.
3. Guedry FE. Visual counteraction of nauseogenic and disorienting effects of some whole-body motions: A proposed mechanism. Aviat Space Environ Med 1978;49(1):36-41.
4. Rasey H, Lawson BD, Anderson AM. Efficacy of a visual reference in diminishing Coriolis cross-coupling effects in a variety of conditions. Proceedings of the 66th Annual Meeting of the Aerospace Medical Association, Anaheim CA 1995 May 7-11.
5. Lawson BD, Guedry FE, Rupert AH, Anderson AM. Spatial disorientation induced by head movement during whole-body rotation: Further tests of a predictive model of human performance. Proceedings of the Advisory Group for Aerospace Research and Development Symposium: "Virtual Interfaces -- Research and Applications," Lisbon Portugal 1993 October 18-22.
6. Lawson BD, Guedry FE, Rupert AH, Anderson AM, Tielemans WCM. Multimodal influences on perception of spatial orientation during unusual vestibular stimulation. Proceedings of the Inaugural Meeting of the Cognitive Neuroscience Society, San Francisco CA 1994 March 27-29.
7. Lawson BL, Tielemans WCM, Rameckers FHJI, Guedry FE. Disorientation during multiaxis rotation is influenced by the resultant angular impulse vector. Proceedings of

the 24th Meeting of the Society for Neuroscience, Miami Beach FL 1994 November 13-18.
8. Grissett JD, Lawson BD. Modeling human perceptions of an unusual vestibular stimulus. Proceedings of the 2nd Annual Meeting of the Cognitive Neuroscience Society, San Francisco CA 1995 March 26-28.
9. Barnes GR, Benson AJ, Prior ARJ. Visual-vestibular interaction in the control of eye movement. Aviat Space Environ Med 1978;49:557-64.
10. Lawson BD, Mead AM, Rupert AH. Effect of visual task conditions on visual performance & motion sickness during whole-body oscillation while viewing a head-fixed display. Aviat Space Environ Med 1995 May;66(5):480.
11. Guedry FE, Lentz JM, Jell RM. Visual-vestibular interactions: I. Influence of peripheral vision on suppression of the vestibulo-ocular reflex and visual acuity. Aviat Space Environ Med 1979;50(3):205-12.
12. Guedry FE, Lentz JM, Jell RM, Norman JW. Visual-vestibular interactions: The directional component of visual background movement. Aviat Space Environ Med 1981;52(5):304-9.
13. Baltzley DR, Berbaum KS, Lilienthal MG, Gower DW. The time course of postflight simulator sickness symptoms. Aviat Space Environ Med 1989;60(11):1043-8.
14. Ungs TJ. Simulator induced syndrome in coast guard aviators. Aviat Space Environ Med 1988;59:267-72.
15. Teixeira RA, Lackner JR. Optokinetic motion sickness: Attenuation of visually induced apparent self rotation by passive head movements. Aviat Space & Environ Med 1979;50:264-266.
16. Graybiel A, Kennedy RS, Knoblock EC, Guedry FE, Mertz W, McLead ME, Colehour JR, Miller EF, Fregly AR. Effects of exposure to a rotating environment (10 RPM) on four aviators for a period of twelve days. Aerosp Med 1965;36(8):733-54.
17. Graybiel A, Knepton J. Sopite syndrome: A sometimes sole manifestation of motion sickness. Aviat Space Environ Med 1976;47(8):873-82.
18. Lawson BD, Mead AM, Apple A. The sopite syndrome revisited: Drowsiness and mood changes in student aviators. Proceedings of the 12th Annual Meeting of the Man in Space Symposium: "The Future of Humans in Space," International Academy of Astronautics, Washington DC 1997 June 8-13.
19. Askins K, Mead AM, Lawson BD, Bratley MC. Sopite syndrome study I: Isolated sopite symptoms detected post hoc from a preliminary open-ended survey of subjective responses to a short duration visual-vestibular stimulus. Proceedings of the 67th Annual Meeting of the Aerospace Medical Association, Chicago IL 1997.
20. Bratley MC, Lawson BD, Mead AM. Sopite syndrome study II: Further evidence of sopite syndrome among aviation students indoctrinated aboard the multi station disorientation demonstrator (MSDD). Proceedings of the 67th Annual Meeting of the Aerospace Medical Association, Chicago IL 1997.
21. Mead AM, Lawson BD. Sopite syndrome case report I: Motion-induced drowsiness and mood changes in an individual with no other motion sickness symptoms -- a case of a pure sopite syndrome? Proceedings of the 67th Annual Meeting of the Aerospace Medical Association, Chicago IL 1997.
22. Calkins DS, Reschke MF, Kennedy RS, Dunlop WP. Reliability of provocative tests of motion sickness susceptibility. Aviat Space Environ Med 1987:(58)Suppl9:50-54.

A real world oriented interface technique using computer vision

Takeshi Ohashi, Takaichi Yoshida and Toshiaki Ejima [a]

[a]Department of Artificial Intelligence, Kyushu Institute of Technology
Kawazu 680-4, Iizuka, Fukuoka 820, JAPAN

Augmented reality(AR) focuses on the interaction in dealing with the real world. Science frontiers has investigated studying the human and augmented real world interactions but did not consider much intraction using the AR system. We propose a communication technique which will use by humans and a virtusl world model to share a complex world model constructed from some real world models. For instance, we built a virtual billiard system for players who are in different environment. The prototype system shows that sharing a complex world model to players whose interface and environment are different made them play altogether.

1. INTRODUCTION

In human computer interaction research, augmented reality(AR) is considered as a new interface paradigm. The main theme of AR is how to deal with the real world on interactive system. For example, *the digitaldesk*(P. Wellner 1993[1]) superimposes information on a real desk and *Navicam*(J. Rekimoto 1995[2]) augments almost everything that surround the user. Most studies paid attention to investigate interactions between a human and the augmented real world, but did not much consider studying the interaction using the AR system. Of course, some researchers are investigating about collaborative sharing a virtual world, but are less considering to import objects from the real world to the virtual world. K.H. Ahlers et al. proposed *a distributed Augmented Reality*(K.H. Ahlers 1995[3]). In their approach, users could share a virtual environment whose databases are separated. It is hard to make interactions between humans in the virtual world.

We propose a real world oriented interface which will be used by humans and a virtual world model to share a complex world model constructed from some real world models. Here, humanbeings will enter the complex world which includes some shadows of the objects in the real world and virtual objects generated by a computer. The following sections shows our general idea and a case study where we apply our idea to a billiard game system.

2. A REAL WORLD ORIENTED INTERFACE

We assume that there are two humanbeings who are working together or competing. Each exists in the real world and shares a virtual world generated by a computer. If each wants to use a separate AR system, the system should be able to handle two real world and one virtual world and maintain consistency. To solve the problem, a new

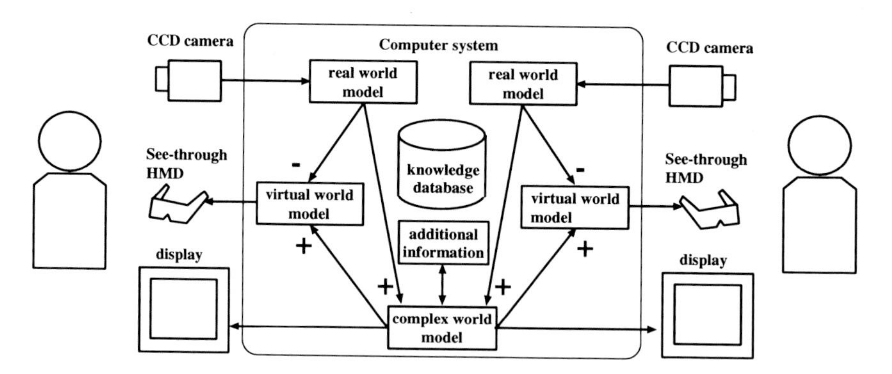

Figure 1. Communication model in sharing a complex world.

framework to treat real and virtual world models is needed. Figure 1 shows the concept of our strategy. In Figure 1, there are two people and their activity is supported by a computer system. The computer system observes the environment around them by CCD cameras and displays information for augmentation to the head mounted displays(HMD) or monitor screens.

The knowledge database has information of all the components in the system. It has the descriptions about the shapes and colors of the components in the total world of the system. Examples are the following: *Shape and size*: 3D/2D shape, ex. the radius of a sphere, *Position*: location and direction in the complex world model, *Color*: color information to be extracted by image processing, *Relation*: Physical touch or encapsulation condition.

Figure 1 shows some world models which are constructed based on a common knowledge database. A real world model is constructed by extracting and measuring the materials in the database. In the process, the system picks up an object in the database and tries to find out a similar object in the real world by image processing. Because the shape and color of these objects are already known, the image processing task might be simple. The position of the objects found and measured and registered in the real world model. Some objects are observed in some real world models.

These real world models are integrated in the complex world model. If an object has multiple registration in real world models, one of them is selected by priority in the real world models. If objects are not registered in any real world model but are in the database, then they are virtual objects generated by a computer. The complex world model is the total world model and is shared for human interface.

A virtual world model is constructed for each person. This will show the person the difference between the complex world model and real world model, and this will be used to augment his viewing information.

2.1. Case study: Billiard game system

To evaluate our proposed technique, we made a virtual billiard system for players who are in different environments. Figure 2 shows the concept of our approach as applied to a billiard game. The left player uses an AR interface and the right player uses a traditional graphical user interface(GUI).

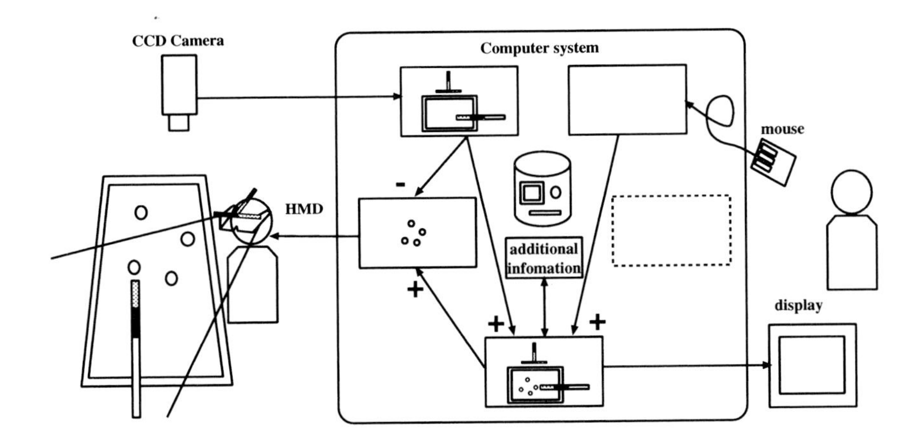

Figure 2. AR system applied to a billiard game.

In the billiard world, there are two real world for the players, and a complex world which represents the game table, some balls, cues and players. The table and balls are in simple shape and color, and so they are easy to store in the database and to extract from a camera mounted on the table. The cue is also in simple shape but the most part of it is overshadowed by the player. To measure the cue tip coordinates and position, the tip part is marked with two certain lengths of two colors. Finding the player in real time is difficult but the system needs his eye sight only. Then we mark the head mounted display that the player wears. The mark is in T shape and colored like a cue. By these, objects will be easy to extract by simple image processing which simply depend on its color. In the billiard game, each player plays alternately except during bunking. If the bunking rule changes to alternate playing, the most number of players is one and its real world model is primary. In this system, the player can play using all the real objects to interact with all virtual objects.

For example, he can choose to play using a table, a cue and virtual balls. He could see real objects overlaid with the virtual balls and additional information, eg., the probable balls' trajectories. He could not experienced the shot feeling but he could hear sounds when balls hit each other.

We built a prototype system on a personal computer with a video capture card and a CCD camera. The system is written on FreeBSD, XFree86, xforms and Mesa that is an OpenGL APL compatible 3D graphic libraries. Figure 3 shows the top view of a toy billiard game table. The CCD camera lense is directed downward. Figure 4 shows the player's view.

3. DISCUSSION

The prototype system validates our approach to provide an environment for players who are in different environment to play a game. On the other hand, two problems where observed. One is that when both players use real balls, they should move their balls by hand to maintain consistency with the complex world model. This is not convenient

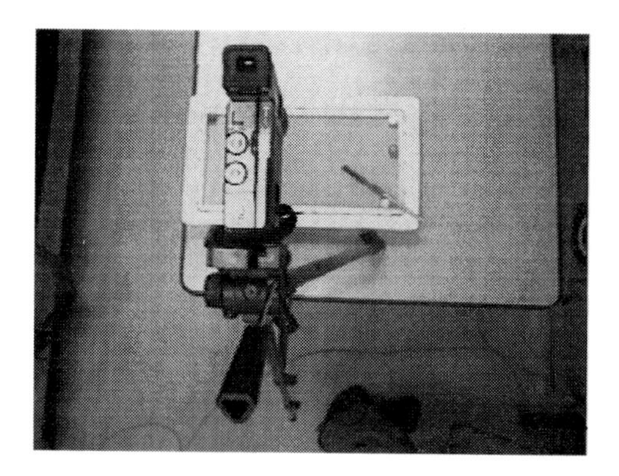

Figure 3. Overview of a toy billiard game table and a CCD camera.

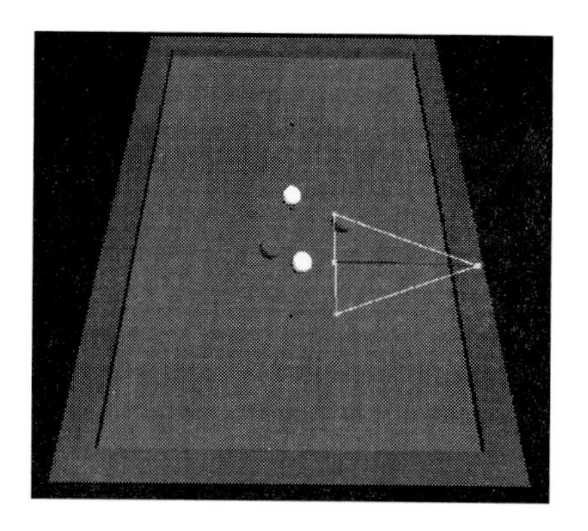

Figure 4. A computer graphics image on the simulator.

for users but may be acceptabe for carom billiard where there are only few balls. This problem is minor compared to the benefits of realizing a game played remotely using real objects. The other problem is when a real cue ball and virtual balls are used. The player can experience the shot feeling and his work is only to move the cue ball. In the complex world, when the cue ball hits other balls the real cue ball fowards the edge of the table, but in the virtual world, the cue ball should be directed away from hitting the edge of the table. These makes the big difference between the typical real world and the real world as used in complex world model. This case is similar to the first problem which we consider minor compared to the benefits.

4. CONCLUSION

We have proposed a real world oriented interface use to share a complex world model. We applied our idea to a billiard game and built a prototype system. The prototype system shows that sharing a complex world model between players whose interface and environment are different can make them play a game. We pointed out not only the difficulty of dealing with real and virtual objects but also important of achieving a remote communication.

REFERENCES

1. P. Wellner, "Interacting with paper on the digitaldesk," Communication of the ACM, Vol.36, No.7, pp.87-96, 1993
2. J. Rekimoto, "Augmented Interaction: Interacting with the real world through a computer," Symbiosis of Human and Artifact, 20B, Y. Anzai etal (editors), Elisevier Science B.V., pp.255–260, 1995
3. K.H. Ahlers et al., "Distributed Augmented Reality for Collaborative Design Applications," Proceedings of EUROGRAPHICS'95, pp. C-3–C-14, 1995

HCI Issues about Immersive Modeling

K. Coninx, F. Van Reeth, E. Flerackers

Expertise Centre for Digital Media (EDM), Limburg University Centre (LUC), Wetenschapspark 2, B-3590 Diepenbeek, Belgium

Abstract

The primary objective of our research is to investigate how immersion can be introduced in the modeling process in such a way that the modeller's performance is enhanced. Examples of modeling systems are CAD applications, industrial product design, architectural design and the realization of virtual worlds to be used in virtual reality applications. To investigate HCI issues about immersive modeling, a framework for a modeling application has been realized. A hybrid 2D/3D user interface is used to interact with the immersive modeling application. 2D interfacing elements such as menus and dialog boxes are presented to the user. Particular interaction scenarios benefit from 3D spatial input and 3D perception. In this paper we elaborate upon human-computer interaction principles and interaction metaphors that play a role when designing an immersive modeling system.

1. INTRODUCTION

1.1. Immersive modeling

The emergence of more and more virtual reality applications results in a need for virtual world models. The design of these virtual world models is usually realized by means of through-the-window modeling (using conventional input devices and workstation displays). We decided to make use of virtual reality techniques to support the design process. Intuitively, we believe that immersing the designer in the design space facilitates a number of modeling tasks. An immersive modeling framework has been realized to investigate this statement.

Immersive modeling offers a combination of technologies that can improve the interaction for design purposes. Head tracking can be used as a paradigm for viewpoint specification: the view of the user depends on the position and orientation of his head. Stereoscopic display exploits binocular parallax as a depth cue. 3D input devices such as gloves provide the user with additional degrees of freedom for spatial interaction.

Introduction of immersion in the modeling process seems promising. On the other hand, working in an immersive environment has its own drawbacks, such as the difficulty to precisely control the input device for accurate work in three dimensions. A number of HCI issues about immersive modeling, amongst others the design of a user interface, still have to be investigated.

1.2. HCI in the context of virtual reality

Human-computer interaction (HCI) is the research domain dealing with effective and efficient interaction between people and the computer systems that support their task. Virtual reality (VR) applications try to realize natural and intuitive user interfaces. Immersion is used to maximize the involvement of the user in his task. How can the goals of virtual reality and HCI be combined? VR techniques can be applied to achieve effective and efficient systems, but the question remains to what rules and guidelines user interfaces for virtual environments must obey. D. Norman presents key principles for design of everyday things that apply as well to user interfaces (visibility, affordances, mappings, feedback and constraints) [1]. However, a set of general guidelines does not exist. The complexity of VR systems and the absence of a unified interaction paradigm for virtual environments makes the design of the user interface a hard job. In practice, the designers base their work on former VR realizations and on experiences from 2D Graphical User Interfaces. In this paper we illustrate particular design decisions, without going into detail about related work concerning VR interface metaphors.

1.3. Universal interaction tasks

An interactive virtual environment usually has to define interaction scenario's for the following tasks: (1) command issuance and use of interfacing elements such as menus and dialog boxes, (2) navigation, (3) object selection and (4) object manipulation. Due to its dynamic nature (being a virtual world that is constantly shaped by the user) the immersive modeling system is an excellent research environment for these interaction scenario's. We continue by describing our solutions for the universal interaction tasks in the immersive modeling system.

In the research experiments a BOOM (Binocular Omni-Orientational Monitor) is used for immersive viewing. The input device is a Pinch Glove, this is a glove with fingertip pads to sense contact between the thumb and any of the four fingers.

2. INTERFACING ELEMENTS

A number of approaches are seen in present-day virtual environments with regard to dialogue style interaction, many of which give menus and dialog boxes their own 3D virtual representation. We decided to integrate 2D interaction techniques in 3D virtual environments for particular tasks for the following reasons:

* 2D menus/buttons are efficient for command issuance, and the third dimension would not provide additional information (looking to the side of a menu is useless)
* 2D control panels (dialog boxes) are efficient tools to visualize and manipulate textual and numerical information, under the condition that special attention is given to the entry of text and numbers (the keyboard is not accessible in immersive systems)
* The user can make an appeal on his skills in using traditional 2D GUIs (Graphical User Interfaces) to interact with these widgets.
* A vast amount of standard interfacing objects comes at the developer's disposal.
* There is no need to have virtual 3D representations of the interfacing elements in the same 3D space as the application objects. Keeping the interfacing elements "on the display surface" instead of in the virtual world ensures their accessibility.

A widget cloning technique has been implemented to make X (Motif) widgets usable for stereoscopic viewing e.g. through a BOOM (Binocular Omni-Orientational Monitor). The main idea in this implementation is to clone the interface widgets to provide each eye with its own copy of the interface widgets. Details about the implementation can be found in [2]. The user interacts with the interface widgets by means of the Pinch Glove. Therefore the Pinch Glove's position tracking and fingertip contact data are transformed into 2D mouse events.

2D interfacing elements are used in the immersive modeling system for all interfacing elements such as menus and dialog boxes.

3. NAVIGATION

Probably the most common task of all in virtual environments is that of navigating through the environment. Navigation can be decomposed in two subtasks that can be carried out in one or in several steps: (1) changing the user's position in the environment and (2) changing the user's viewpoint on the environment.

Regarding navigation a flight metaphor is used for general exploration of the virtual design space. Head-tracking provides a natural mapping (as defined in [1]) between head movements of the user and a changed viewpoint in the environment. Targeted navigation exploits the hierarchical scene/object representation to implement more direct access to objects than through a flight metaphor. Figure 1 shows a sample scenario for targeted navigation. In order to prevent disorientation, the viewpoint's orientation remains the same.

4. OBJECT SELECTION

The chosen object selection technique should be consistent with the "click-to-select" equivalent used in the interaction with the 2D interfacing elements. A ray-casting metaphor that is not transparent in the user interface is applied. This metaphor enables remote selection.

5. OBJECT MANIPULATION

The approach to object manipulation is based on the principle of the hybrid user interface. Object manipulation scenario's switch between 3D interaction, and interaction with 2D interfacing elements. 3D spatial input is used for specific tasks that benefit from specifying simultaneously multiple degrees of freedom. Examples are object placement (position and orientation), scaling, extrusion, orientation of reference axes etc. Several techniques are used to facilitate accurate work in 3D. As an extension of the object selection metaphor, remote manipulation is possible. 2D interfacing elements are used for command issuance by means of menus, for input of numerical parameters etc.

6. CONCLUSION

In this paper we have presented HCI issues about immersive modeling by describing our approach to universal interaction tasks. The basic concept is a hybrid 2D/3D user interface. Besides for the creation of virtual worlds, the proposed modeling framework can be extended and specialized for other design tasks and computer animation. More details concerning the modeling functions can be found in [3]. The interaction techniques and metaphors that are used in the immersive modeling system can be useful in other immersive applications as well.

7. ACKNOWLEDGMENTS

Part of the work presented in this paper is subsidized by the Flemish Government and EFRO (European Fund for Regional Development). We also acknowledge the work of B.

Rassaerts and D. Nouls who were instrumental in the programming work for the research experiments.

REFERENCES

1. Norman D., 1988, The psychology of everyday things, BasicBooks
2. Coninx K., Van Reeth F., Flerackers E., 1996, 2D Human-Computer Interaction techniques in Immersive Virtual Environments, *Proc. Compugraphics'96*, p. 163-171.
3. Coninx K., Van Reeth F., Flerackers E., 1997, A hybrid 2D/3D user interface for immersive object modeling, *Proceedings of Computer Graphics International '97*.

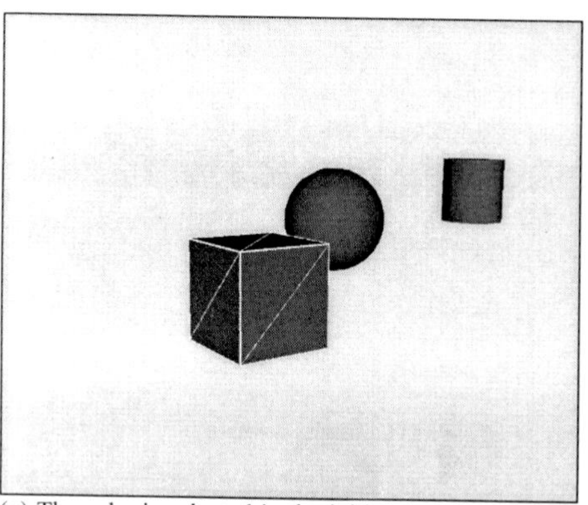

(a) The cube is selected in the initial situation, before the targeted navigation scenario

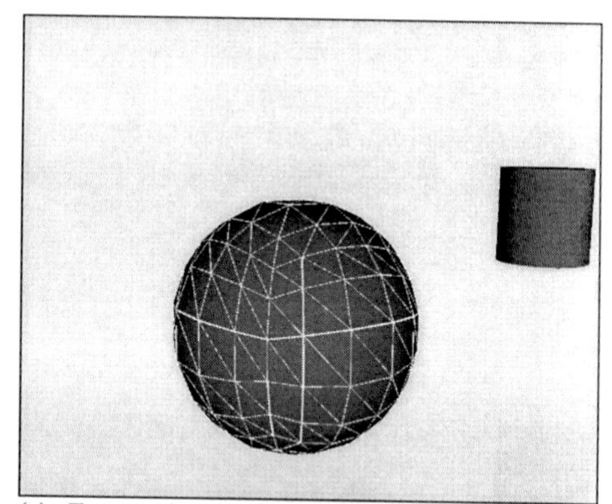

(c) Target situation: the sphere is selected and centered in the field of view after the navigation

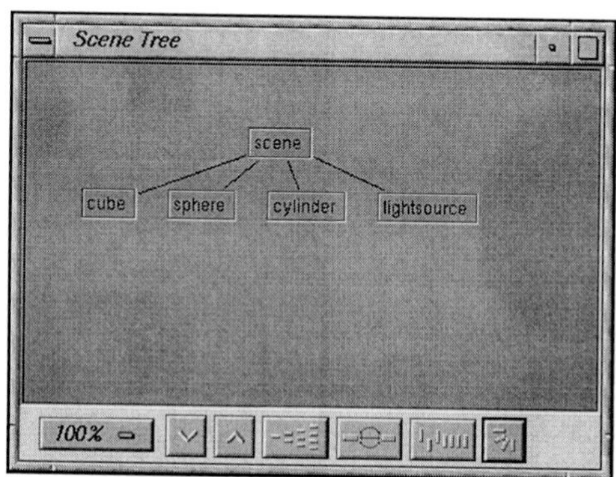

(b) Scene representation used to indicate the target object

Figure 1: The steps in a targeted navigation scenario. (a) Depicts the initial situation. The cube, shown in wireframe representation, is the selected object. In order to center the sphere in the field of view, the user pops up a 2D window showing the hierarchical scene/object representation. (b) The node with the name of the targeted object, sphere, is selected. (c) Shows the target situation with the sphere as the selected object in the center of the field of view.

Assisting Remote Instruction Using Copied Reality

Hideaki Kuzuoka[a] and Tatsuhiro Nakada[b]

[a]Institute of Engineering Mechanics, University of Tsukuba,
1-1-1 Tennoudai, Tsukuba, Ibaraki 305, Japan

[b]NEC Cooperation

For a remote instructor to give instruction to an operator on how to operate a machinery, it is better for an instructor to have the same machinery as an operator's. It is not common, however, that both of them have the same machinery. If a copy of an operator's machinery is created in a virtual environment an instructor will be able to give instructions using it. In this paper, a simplified system is introduced and the result of a preliminary experiment is described.

1. INTRODUCTION

The authors have been studying on remote instruction support systems which mainly use audio-video communication links [1]. With our systems, a remote instructor can give instructions on how to operate machinery to an operator. From our experiences, we have noticed that if an instructor has exactly the same machinery as an operator, he/she can give instruction efficiently and correctly by showing the actual operation to his/her machinery. It is not common, however, that an instructor and an operator have exactly the same one.

On the other hand, there are some research which constructs solid models from observations of real objects [2]. Using this technology, an instructor can instantly make a copy of an operator's machinery in a virtual reality (VR) environment and he/she can use it to give instruction to an operator.

2. TAECHING BY SHOWING IN THE COPIED REALITY ENVIRONMENT

Figure 1 shows the conceptual image of a system. An operator and an instructor sit face to face and there are two tables between them. The same machinery are placed on both tables. An instructor teaches how to operate a machinery by showing the actual operation on an instructor's machinery.

It is not common, however, that an instructor always have the same machinery that an operator have. Thus, we decided to create a copy of an operator's machinery in a VR environment. Since a copied machinery is in VR environment, an instructor can manipulate the machinery as he/she intended. An instructor can also make another copy

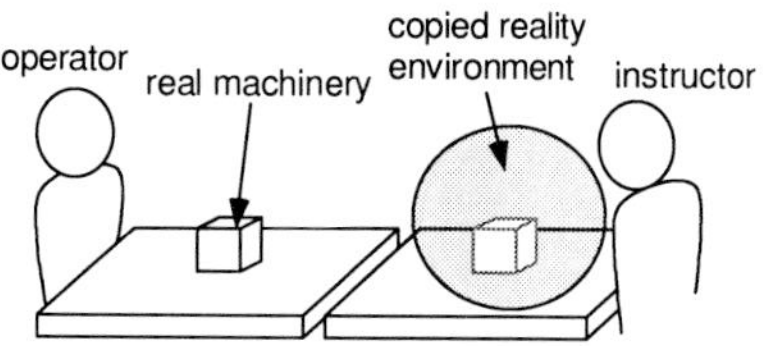

Figure 1. The conceptual image of a copied reality system.

of a machinery as many as he/she wants. Therefore, the instructor's environment is not just a copy of the operator's environment, but it has a characteristics of a virtual reality. In this paper, we call such an environment as copied reality environment.

Although we think this system is effective mostly for instruction on three-dimensional (3-D) operations of a machinery, the technology to create a solid model from observations of real objects will not be widely available for a few more years. Thus, as a beginning stage of our research, we decided to deal only with two-dimensional (2-D) operations of 2-D objects.

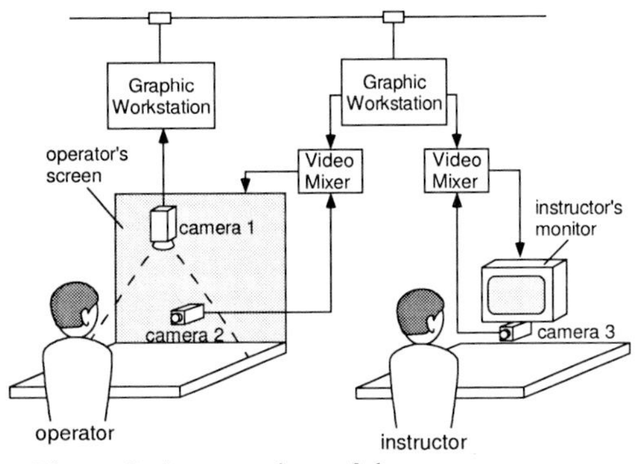

Figure 2. An overview of the system.

3. SYSTEM ARCHITECTURE

3.1. Overview of the system

Figure 2 shows the overview of the system. The camera 1 is placed above an operator's desk. The image from the camera 1 is processed by the workstation, then shapes and positions of objects on the desk are recognized. The current system recognizes objects only when an instructor pushes the recognize button of the user interface. Recognized objects are reproduced as a computer graphics. Figure 3-(a) is a reproduced image from an instructor's point of view and figure 4-(a) is a reproduced image from an operator's point of view.

Every time an operator operates a real machinery, an instructor need to confirm if the

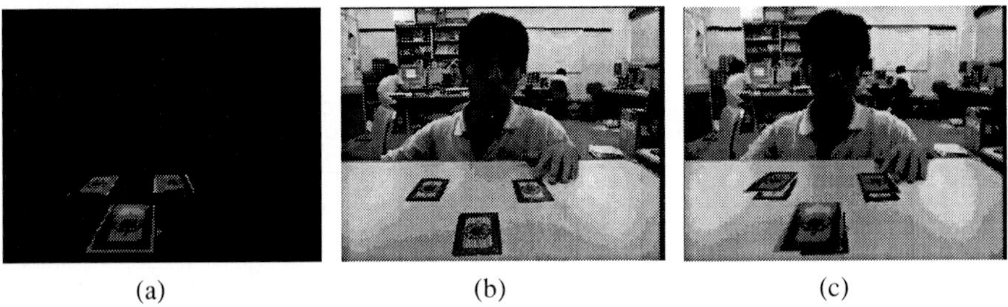
(a) (b) (c)

Figure 3. Synthesizing the copied reality environment for an instructor.

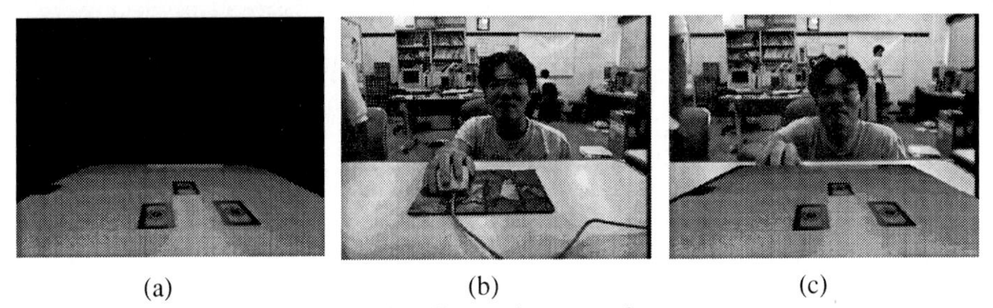
(a) (b) (c)

Figure 4. Synthesizing the copied reality environment for an operator.

operation is fulfilled correctly. For this purpose, the image of the copied objects is overlapped on the image of the real environment. Although it makes difference from the conceptual image (Fig. 1), an instructor can easily compare the current state of objects this way. For instance, the camera 2 takes the image of both the upper half of an operator's body and real objects on the desk (Fig. 3-(b)). Then, Fig. 3-(a) is superimposed on Fig. 3-(b) such as Fig. 3-(c) and shown on the instructor's display. In case of an image for an operator, the camera 3 takes the upper half of an instructor's body (Fig. 4-(b)). Then, Fig. 4-(a) is superimposed on Fig. 4-(b) such as the Fig. 4-(c) and shown on the operator's screen. Not only the work area but also face images are synthesized because we think that it help them to realize another person's viewpoint.

3.2. Instructor's manipulation in copied reality

An instructor can choose one of the copied objects and he/she can move or turn it. The current system enables these manipulation using an ordinary mouse with three buttons. In the copied reality environment, hand shaped mouse cursor moves according to the motion of the instructor's mouse. When the mouse cursor overlaps one of the objects, the object is highlighted to show that it is selected. Then an instructor can grasp a selected object by clicking a left button. A selected object rotates clockwise 90 degrees every time a middle button is pressed. When the right button is pressed, a copy of a selected object is created.

3.3. Operator's manipulation

An operator manipulates objects according to an instructor's operation in the copied environment. It is not very easy, however, to make correspondence of position of an object in VR and a real object. In order to alleviate this problem, a copy of an operator's hand and objects are superimposed on the image Fig. 4-(c) (Superimposed hand and objects are not shown in the figure). Due to the limitation of the computational performance of the workstation, however, a hand is shown like a gray shadow (Fig. 5).

4. PRELIMINARY EXPERIMENTS

4.1. Settings

Preliminary experiments were conducted to get implications to improve the current system. In the experiment, the instructor gave instructions on placing some cards into predefined arrangements. During the sessions, the instructor had to show position and orientation to put each card.

There were two cases of communication settings. In case 1 the image from the camera 2 was shown both on the operator's screen and the instructor's display. The instructor could draw lines which was superimposed on the image. This case was similar to an ordinary desktop conferencing system. In case 2 copied reality system was used. 4 pairs of subjects served as

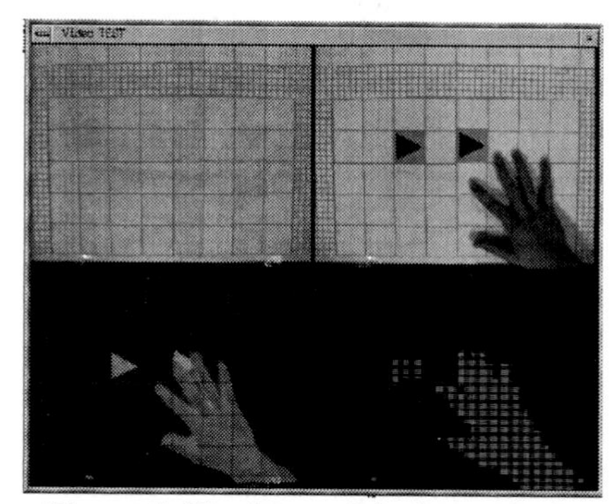

Figure 5. The lower right image is the created shadow image of an operator's hand and objects. This image is created by subtracting the original desktop image (upper left) from the current desktop image (upper right).

the instructor and the operator. Each pair conducted the remote instruction both in case 1 and case 2. Before the experiment, each pair had about 5 minutes practice time to get use to the user interface.

Table 1. The average time to complete the task and the average number of errors. Standard deviation is shown in the parenthesis.

	time (sec)	errors
case 1	202 (43.2)	2.5 (1.3)
case 2	250 (46.0)	0.0

4.2. Results

Table 1 shows the average time to complete the task and the average number of errors in each case. The number of errors means that the number of misunderstood operations by the operators. As the result shows, case 2 tended to take more time to complete the task compared to case 1. From the observation, we noticed that the instructor had to spend relatively longer time to manipulate the virtual objects. In other words, it was faster to use superimposed drawings to show place and orientation of each card.

The number of errors shows, however, that there were no errors in case 2. In case 1, since the image of the objects that the operators saw on the display was taken from the opposite point of view from his/hers, it was easy especially for the operators to misunderstand the instructions. In case 1 there were no errors since the instructor could just show the actual operation using copied objects. From the interview to the operators, one of the operators said that the image of the instructor's face helped him to understand from which direction the instructor was observing the environment. We think this understanding is one of the reasons that reduced the number of errors in case 2.

From these results, we noticed that the copied reality system is effective to reduce communication errors. The user interface to manipulate copied objects, however, should be improved so that the objects can be manipulated much faster.

5. RELATED WORKS

ATR have been developing the networked reality environment in which not only objects but also human faces and bodies are reconstructed [3]. In this system, everything is done in the virtual space. Thus it cannot be used when it is necessary to give instruction on real object.

Double DigitalDesk [4] may be more effective for remote instruction for 2-D task. However, it cannot deal with 3-D objects.

6. CONCLUSIONS

From the preliminary experiments, we found some advantages of the copied reality system. In order to confirm the effect of the system, however, experiments on more complex task is necessary. Furthermore, the system needs to be improved so that it can deal with 3-D objects

REFERENCES

1. Kuzuoka, T. Kosuge and M. Tanaka, GestureCam: A Video Communication System for Sympathetic Remote Collaboration, Proc. of CSCW'94, pp. 35--43 (1994).
2. VASC Virtual Reality Home Page, http://www.cs.cmu.edu/afs/cs/project/vision/www/VR/vr.html.
3. ATR MICRL Dept 1 Research Topics, http://www.mic.atr.co.jp/organization/dept1/research.html.
4. Pierre Wellner, The DigitalDesk: Supporting Computer-based Interaction with Paper Documents, Proc. Of Imagina '93, pp. 110-119 (1993).

Texture display for tactile sensation

Y. Ikei, K. Wakamatsu and S. Fukuda

Department of Production and Information Systems, Tokyo Metropolitan Institute of Technology, 6-6 Asahigaoka, Hino, Tokyo 191, Japan

This paper describes a texture display and its rendering technique which produce tactile stimulation on the user's fingertip raising virtual texture sensations of simulated object surfaces. The texture display makes use of a vibrating pin array as a window through which the user touches a virtual object. The control of the display is based on the image data obtained as a photograph of the object surface. Texture sensations were synthesized satisfactorily provided that the image data was adequately modified so that the vibratory stimulation might replicate the touch sensation with much similarity to the real object.

1. INTRODUCTION

There are two approaches to present a sensation of touch with an object to the user of a virtual environment: tactile and force feedbacks. The tactile feedback is intended to stimulate the cutaneous sensation during the contact exists between user's skin and the surface of a physical object. On the other hand, the force feedback is designated to stimulate the deep sensation that originates from the forces imparted on the limbs and fingers as a result of the weight or the stiffness of a manipulated object. The texture display we developed conveys the information of object surface properties to the user's fingertip by vibratory stimulation as he/she is exploring the virtual surface.

The vibratory stimulus as a means to provide information has been utilized since the 1970's in the context of sensory substitution. The reading aid for the blind, Optacon was equipped with a vibratory pin array, and it was designed for displaying printed characters to the finger [1]. With the similar device, we have discussed the method to present the surface texture sensation arising from the variation of physical properties on the surface by controlled vibrations [2, 3].

2. TEXTURE DISPLAY SYSTEM

2.1. Display hardware

The texture display system consists of a display box, power circuits, and a control computer. The pin (contactor) array window on which the user fixes his/her index finger puff is at the top plate of the display box (Figure 1). A piano-wire pin 0.5 mm in diameter is arranged in a 5 x 10 matrix of a 2 mm pitch, and it is driven by piled piezoelectric actuators. The amplitude of the actuator is augmented by a mechanical lever to yield a sufficient amount (about 60 microns at maximum) of vibratory stimulus. The display box is attached to a mouse which counts the movement of the box on a table. The user explores within the 2-dimensional plane holding the box with his/her finger contacted to the pin matrix. The tactile stimulus changes along with the finger (the box) position in the texture data which provides the intensity map of the surface.

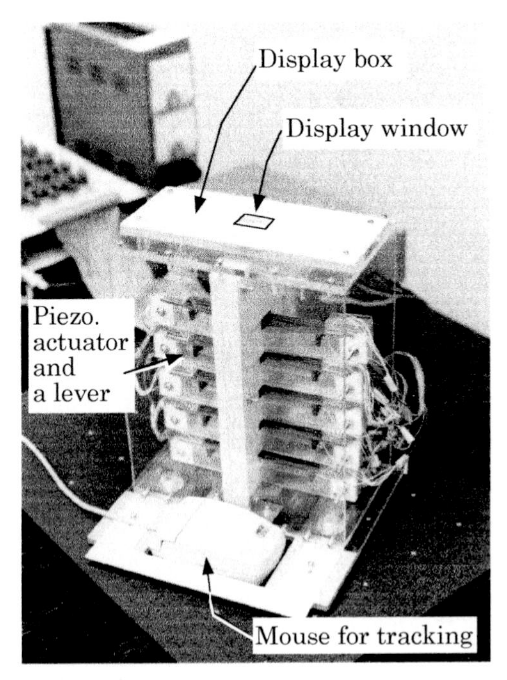

Figure 1. Texture display

2.2. Scaling

The frequency of vibration is 250 Hz around which the sensitivity of cutaneous sensation takes the maximum [4]. The experiment of the sensation intensity scaling was performed by the JND (Just-Noticeable-Difference) method which compiled the difference threshold of vibratory stimulus at controlled amplitudes. Five experienced subjects executed the experiment. As the averaged result, ten levels including no stimulus were obtained defining the sensation intensity scale of this display.

3. IMAGE BASED RENDERING OF A TACTILE TEXTURE

3.1. Production of the data for texture presentation

The accurate distribution of sensation intensity as observed by the user on the object surface is almost impossible to measure because the tactile sensation results from the integration of complicated processes at numerous mechanoreceptors of five kinds in the skin and/or subsequent perception procedures in the central nerve system. In addition, the surface state of objects referred to as *texture* consists of variations in various properties, therefore the data of the intensity distribution is just defined as no other than the approximation in any cases. Among them, using the photograph image (black and white) of the surface has a great advantage of easy data acquisition, although it requires a certain modification of intensity distribution and imposes some conditions on properties of target objects.

The photo-image has to be taken as it best approximates the tactile intensity observed while the object is touched by the finger. Therefore in general, the brightness at the point within the surface where tactile stimulus is stronger is to be higher than other points. This condition is generally met with not so much difficulty if the surface does not have intense color changes and the lighting for the object is properly arranged. However the obtained image needs a certain modification before it is used as a stimulus distribution.

The process to prepare the image data for the display is described as follows:

1. Take a picture of the target surface by a digital camera with low-angle lighting from both sides to make the brightness distribution reflect the surface height map.
2. Transfer the image to a file on a personal computer.
3. Convert the color image to a gray scale, then adjust brightness and contrast manually so that the peak of the histogram is moved near center and the width of it covers almost all the brightness range.
4. Apply a specific filter to the image to enhance the similarity of tactile impression to the real surface.
5. Pick up the value of the image data at each display pin position and reduce it linearly to ten levels to drive the pin during the exploration of the user.

Based on the images obtained through this process, five wall paper samples were displayed. The wall papers were made of synthetic resin with similar uneven surfaces. The specific filters to enhance the similarity of the presentation were selected through the experiment in which the subjects compared the touch feel of six image data (including original) processed by different filters to that of the real sample. The subjects selected the most preferable data, or the filter, for that real sample. The original images (brightness and contrast manually adjusted) of the wall papers are shown in Figure 2, A to E (with 0 meaning original), and the images selected by three experienced subjects in Figure 3. The size of the image is about 55 x 41 mm and 320 x 240 pixels. The filters used are numbered as follows: 1) Average filter, 2) Prewitt x-derivative filter, 3) Prewitt y-derivative filter, 4) Laplacian filter, 5) Contrast enhancement filter of 3 x 3 digital operators.

The selected filters were not a single kind but peculiar to the individual image,

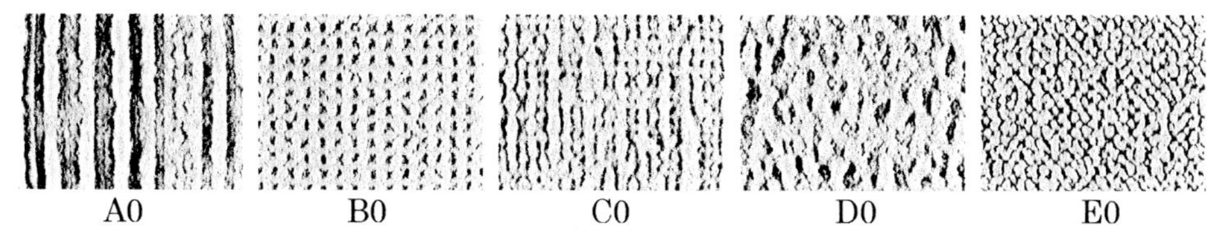

Figure 2. Original image of wall paper A - E

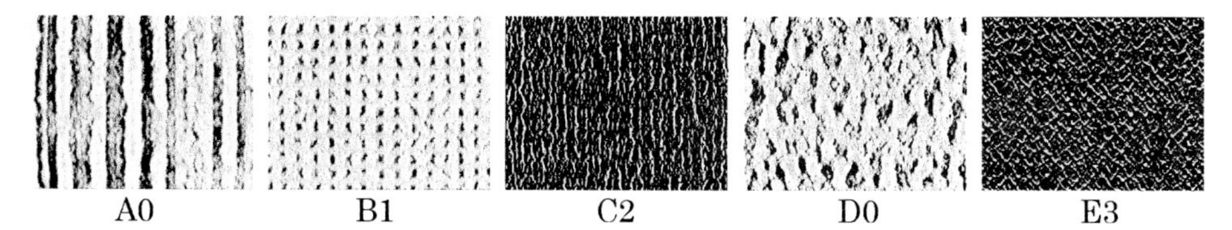

Figure 3. Selected images

which seemed to be determined in terms of both the shape of the ruggedness of the surface and the state of its image. For example, the sample C had more sharp and fine protrusions than the others, and the basic feature of vertical lines, so a derivative filter for x direction contributed to enhance the display fidelity. As for the sample E, a y-derivative filter was selected because the real sample E was observed more rugged when it was rubbed in the vertical direction than in the horizontal. However the original image had the opposite property as it was lighted from the side, horizontally. Thus, the optimal filters for the image differed with the samples.

3.2. Presentation based on the height equivalent images

The variation in optimal filters selected in the previous section originated from the lack of accurate information in the image about the shape of the surface. However, the sample with a particular shape do produce the image equivalent to the shape, or the height map. Of course the height map image itself is obtainable by some measurement apparatus. In the both cases, the images equivalent to the height map are expected to accept a common filter for modification at least for the similar surface materials, with which the optimal data are obtained.

We investigated a histogram modifying filter by which the histogram of the height equivalent image was reshaped to match a given template histogram. The template histogram was such that it divided the intensity range into three sections and each section had a constant frequency. Nearly fifty templates were designed first, and they were applied to a simple image of a semi-sphere array. Then they were reduced carefully to five of distinctive features by the visual inspection of result images, then by their histograms, and finally by tactile difference. Their features had two categories of contrast enhancement and brightness reduction. The real texture samples of three basic shape feature, a semi-sphere array, a cylindrical bar array and a prismatic edge array, were examined with the filters. The result suggested that strong brightness reduction was effective for the prismatic array and weak brightness reduction for the semi-sphere and the cylinder arrays.

4. CONCLUSION

The tactile texture sensations were successfully rendered to the fingertip based on the modified image data. Although the modification depended on the feature of the target surfaces, two modifying filters were identified for two specific shape groups. The future work includes the exploration of filters for more shapes.

REFERENCES

1. J. G. Linvill and J. C. Bliss, Proc. IEEE, 54 (1966) 40.
2. Y. Ikei, S. Fukuda, Trans. Info. Process. Soc. Japan, 37(3) (1996) 345.
3. Y. Ikei, K. Wakamatsu, S. Fukuda, Proc. VRAIS '97, IEEE, (1997) 199.
4. R. T. Verrillo, A. J. Fraioli, and R. L. Smith, Perception & Psychophysics, 6(6A) (1969) 366.

Real-time numerical simulation in haptic environment

T. Ogi[a], M. Hirose[b], H. Watanabe[b] and N. Kakehi[b]

[a]Intelligent Modeling Laboratory, University of Tokyo
2-11-16 Yayoi, Bunkyo-ku Tokyo 113, Japan

[b]Engineering Research Institute, Faculty of Engineering, University of Tokyo
2-11-16 Yayoi, Bunkyo-ku Tokyo 113, Japan

1. INTRODUCTION

When we manipulate and/or examine objects, our sense of touch plays an important role. In physical simulation such as in deformation analysis of structural dynamics, display of data based on haptic sensation is thought to be an effective method. For example, in this method when force is applied to an object, the deformation of the object is visualized in real time and the change in the reaction force is represented using a force feedback device.

In achievement of this kind of "interaction", it is necessary both to calculate the deformation of the object and control the force feedback device in real time. In a virtual environment, a simple spring model based on Hooke's law is often implemented to simulate the elastic deformation of objects[1]. However, in this study, numerical simulation using a FEM (finite element method) is introduced to simulate various phenomena more realistically. In this case, the FEM program should be implemented independent of the force feedback control program in a distributed architecture because it requires extensive calculation.

In this paper, the framework of a haptic environment in which the results of deformation analysis using a FEM is represented in real-time is described. In addition, the construction of an elastic world in which objects can be deformed using the FE model is discussed.

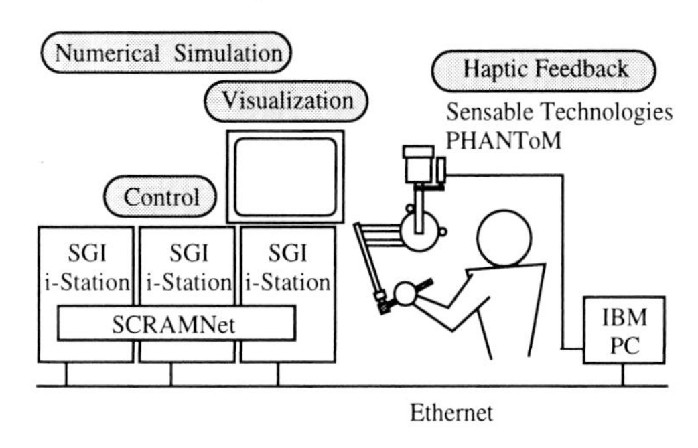

Figure 1. System architecture of haptic environment for deformation analysis

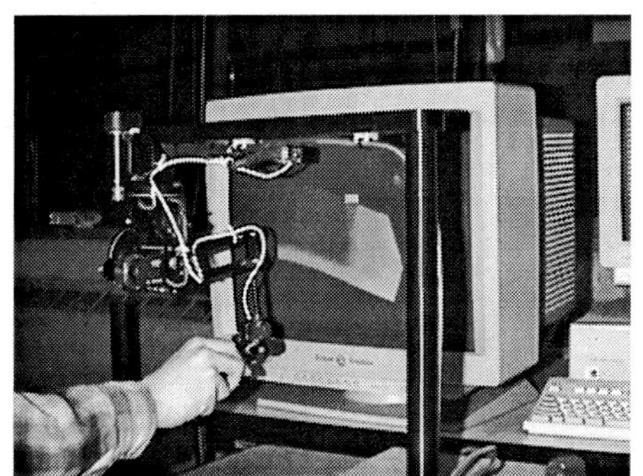

Figure 2. Appearance of haptic environment using PHANToM

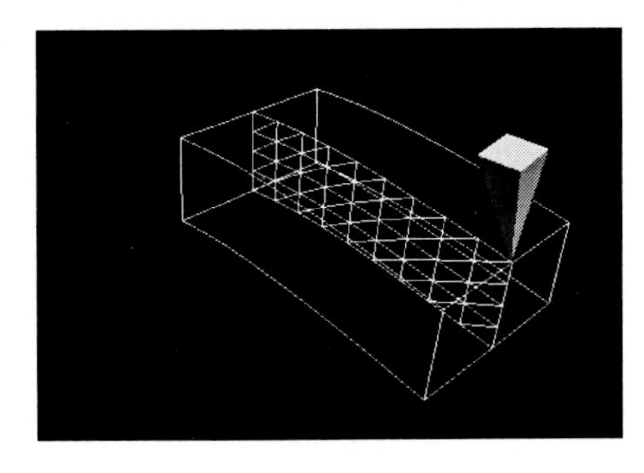

Figure 3. FEM data for the deformation analysis of the elastic cantilever

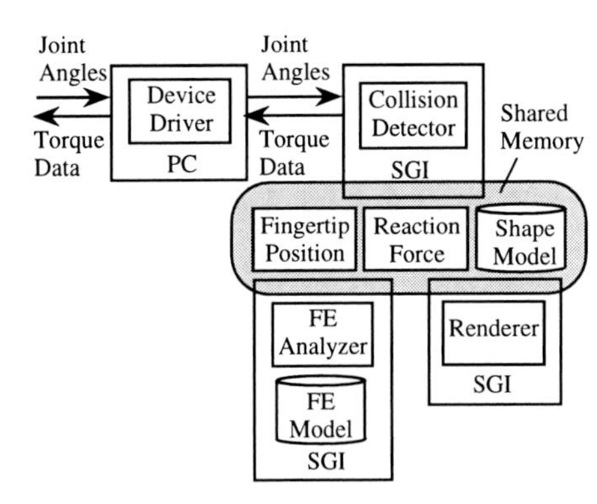

Figure 4. Distributed software construction for the force feedback

2. SYSTEM ARCHITECTURE

Figure 1 illustrates the system architecture of the haptic environment used in this study. PHANToM (SensAble Technologies Co.) controlled by an IBM PC (Pentium 133MHz Linux) is used as the haptic feedback device[2]. As for the computer environment, three workstations (SGI i-Station, R4400 200Hz) that have 1 Mbytes of shared memory (Systran SCRAMNet) are used for the distributed computation such as numerical calculation and visualization. The SGI i-Stations and the IBM PC are connected via an Ethernet. Figure 2 shows the appearance of the haptic environment.

3. HAPTIC REPRESENTATION OF FINITE ELEMENT ANALYSIS

As an example of the FE analysis, the deformation of the elastic cantilever shown in Figure 3 is considered. The task is to enable the user to sense the reaction force of the deformed cantilever using the force feedback as well as to touch and feel the shape. The FE model for the deformation analysis includes two-dimensional 10 x 5 triangle elements, and is analyzed as a linear problem.

In general, force feedback needs higher frequency control than visual feedback[3]. Therefore, in the design of the simulation program, the following points have to be taken into consideration.

- Distribution of the simulation tasks such as deformation analysis, force feedback control, and visualization.
- Reduction of the communication load between the distributed processes.

Figure 4 illustrates the construction of the simulation software developed in this study. This software consists of four processes; a device driver, a collision detector, a FE analyzer and a renderer. The device driver works on the PC and communicates with the collision detector via a TCP/IP socket. The collision detector, the FE analyzer and the renderer work on the SGIs and communicate with each other via shared memory. The data of the shape model, the fingertip position and the calculated reaction force are read into the shared memory and are transferred at high speed from one process to the others.

The device driver measures the joint angles of the PHANToM arm, and sends them to the

Table 1. Control Bandwidth of the Force Feedback Loop

	Collision Detection	Deformation Analysis
Device Driver	1,139 Hz	1,139 Hz
Collision Detector	1,139 Hz	1,139 Hz
FE Analyzer	-	30.9 Hz
Renderer	41.8 Hz	41.8 Hz
Total	1,139 Hz	26.0 Hz

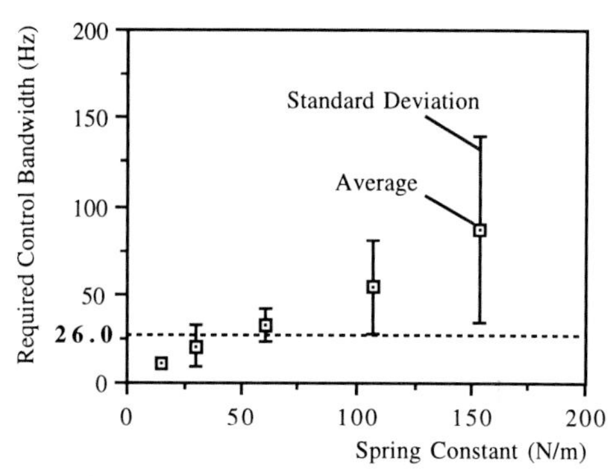

Figure 5. Required control bandwidth to represent elastic objects

collision detector. The collision detector calculates the fingertip position from the joint angles and detects whether the user's fingertip touches the shape model. When force is applied to the object, the FE analyzer calculates the deformation and the reaction force based on a FE model. The calculated reaction force is translated into torque data which are sent to the device driver. The shape model in the shared memory is modified by the deformation data and is visualized by the renderer.

Table 1 shows the performance characteristics of this program. The force feedback rate was 1139 Hz for the collision detection without deformation, and 26 Hz for the deformation analysis. In both cases, the bandwidth of the torque control was 1139 Hz.

4. CONTROL BANDWIDTH AND HUMAN PERCEPTION

The question of whether this performance is sufficient to represent the deformation of objects should be discussed. In general, the required bandwidth of the force feedback depends on the elasticity of the displayed objects and the velocity of the finger's motion. If the user applies the force to the solid objects quickly, the force feedback needs high frequency control. In this study, the required bandwidth to represent the elastic object was evaluated experimentally.

The subjects were asked to push a virtual spring of various spring constant changing the control bandwidth of the force feedback, and to indicate the minimum bandwidth at which they could feel smooth deformation. Figure 5 shows the results of the experiments for seven subjects. The large standard deviation values are thought to be caused by differences in finger velocity. From these results, we conclude that a force feedback bandwidth of 26 Hz is sufficient to represent elastic objects with spring constants of up to 40 N/m.

5. VIRTUAL WORLD BASED ON FINITE ELEMENT MODEL

In the above-described method, haptic feedback is used to represent the results of FE analysis. This method can also be used to construct a haptic virtual world based on FE analysis. In this world, when force is applied to the object, it is deformed based on the FEM calculation. This kind of world can be constructed by definition of a FE model as well as a shape model for each object.

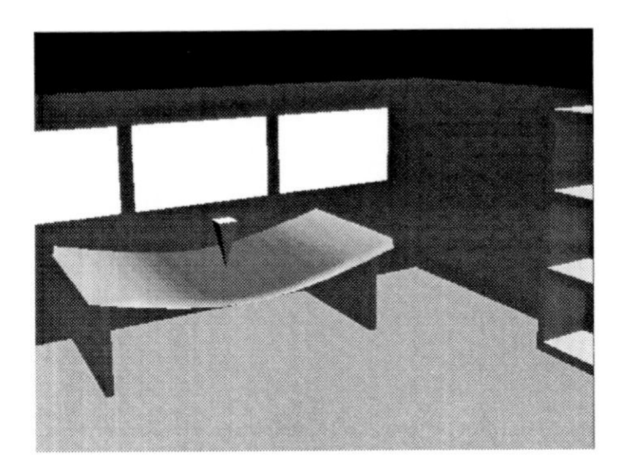

Figure 6. Virtual world based on finite element analysis

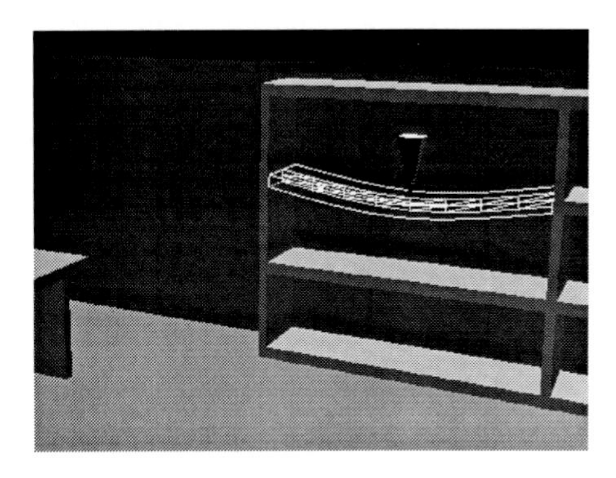

Figure 7. Finite element model used for deformation in the virtual world

Figure 6 and 7 show an example of the virtual world based on FE analysis. This world consists of several objects such as a desk and a bookshelf which have both shape and FE models. Though the shape model is constructed using rough polygons, the FE model is defined using detailed mesh data that would be used in deformation analysis. In this world, the user is able to walk through the room and touch the shapes of the objects using the force feedback device. When force is applied to the objects, the FEM data for the selected object are loaded and the deformation analysis is performed. This simulation program is developed using the framework shown in Figure 4.

6. CONCLUSIONS

In this paper, a framework for haptic representation for deformation analysis using a FEM was described. The results of the performance test indicate that in the case of a small problem, the deformation can be represented using force feedback in real time. In addition, the construction of a virtual world in which objects can be deformed using a FE model was discussed. Future work will include use of a parallel supercomputer and a high-performance network to simulate more complex and larger-scale problems.

ACKNOWLEDGEMENT

This work was supported by the Information-technology Promotion Agency, Japan.

REFERENCES

1. H. Iwata, "Pen-based Haptic Virtual Environment", Proc. of IEEE VRAIS'93, pp.287-292, 1993.
2. T. Massie, K. Salisbury, "The PHANToM Haptic Interface: A Device for Probing Virtual Objects", ASME Winter Annual Meeting, DSC-Vol. 55-1, pp.295-300, 1994.
3. K.B. Shimoga, "A Survey of Perceptual Feedback Issues in Dexterous Telemanipulation: Part I. Finger Force Feedback", Proc. of IEEE VRAIS'93, pp.263-270, 1993.

Implementation of elastic object in virtual environment

Koichi Hirota and Toyohisa Kaneko

Department of Information and Computer Sciences, Toyohashi University of Technology
1-1 Hibarigaoka, Tempaku, Toyohashi, Aichi 441, Japan
hirota@mmip.tutics.tut.ac.jp, kaneko@mmip.tutics.tut.ac.jp

In this paper, we describe an idea of capturing models of soft and transforming objects through experimental transformations on real objects. We propose an approach to solve this problem from the viewpoint of the identification on physical parameters. We employ an elastic model and determine model parameters that minimize the transformation errors between the real object and the model using a method of descent.

1. INTRODUCTION

Implementation of soft and transforming objects in virtual environment came to be one of important technical topics. To represent the sensation of softness to users, various types of haptic devices have been developed[1]. However, before we can represent the sensation of touching on a soft object through these interface devices, we need to construct a model on the physical characteristics of the object[2]. Especially, in medical application such as surgical simulation, precise information on the softness is essential in executing operation and diagnosis, and therefore precise physical models on transformation characteristics are required.

In this paper, we will discuss an approach to acquire the model on physical characteristics through the measurement on the softness of real objects.

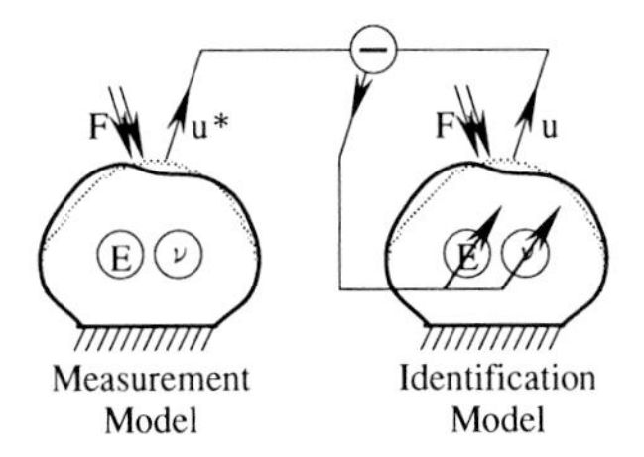

Figure 1. Identifying Physical Parameters

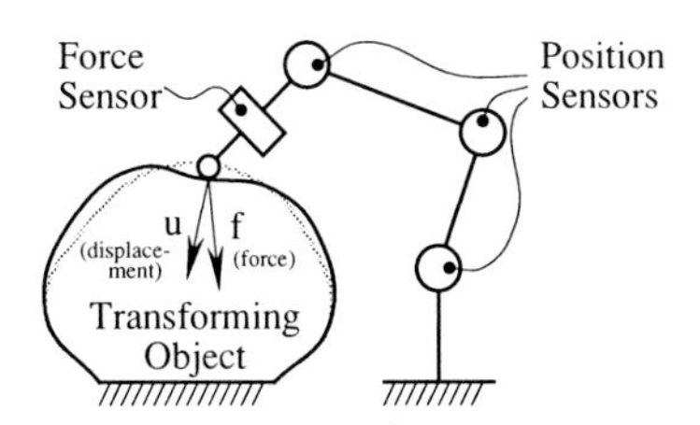

Figure 2. Measurement Approach

Table 1. Materials used in Experiments

Material	E [N/m^2]	ν [−]	
1	1.0×10^5	0.45	(Rubber)
2	1.0×10^7	0.30	(Plastic)

Table 2. Number of Components

Exp.	Meas. Model	Id. Model	Num. Meas.
1	1	1	10
2	2	2	10
3	2	64	300

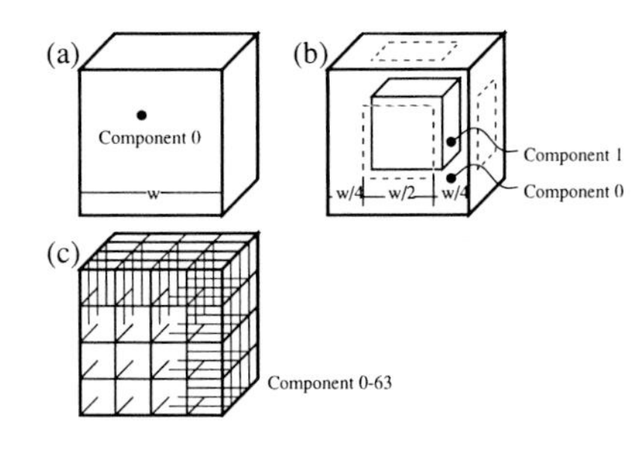

Figure 3. Structure of Components

2. IDENTIFICATION OF ELASTIC PARAMETERS

We regarded the problem of capturing soft objects as a problem of identifying physical parameters in virtual object models (Figure 1).

2.1. Physical Representation

We applied elastic representation to describe physical characteristics of virtual objects[3][4]. In the representation method, objects are described as collections of elements. We employed a tetrahedral element model, and assumed linear elasticity in the transformation. Under this assumption, the behavior of an element is defined by two parameters: Young's modulus (E) and Poisson's ratio (ν).

2.2. Measuring Method

We assumed that we can affect force at one point and measure the displacement at the same point (Figure 2). In our experiments, we simulated the measurement process in a computer by constructing a measurement model based on elastic representation, and derived pairs of force and displacement vectors in experimental transformations.

2.3. Error Function

In the identification process, we employed another model whose physical parameters can be changed arbitrarily (identification model). We defined the error function as a sum of square errors in the displacement values that were caused in the measurement and identification models under the same force conditions.

Also, we introduced an idea of component to reduce the degrees of freedom in the identification problem based on the knowledge on the distribution of materials in a object. Component is a collection of elements that consist of the same material, and therefore, shares same physical parameters with each other.

2.4. Identification Method

We determine the parameters by minimizing the error function. For the minimization, we employed the conjugate gradient method, which is one of the method of descent[5].

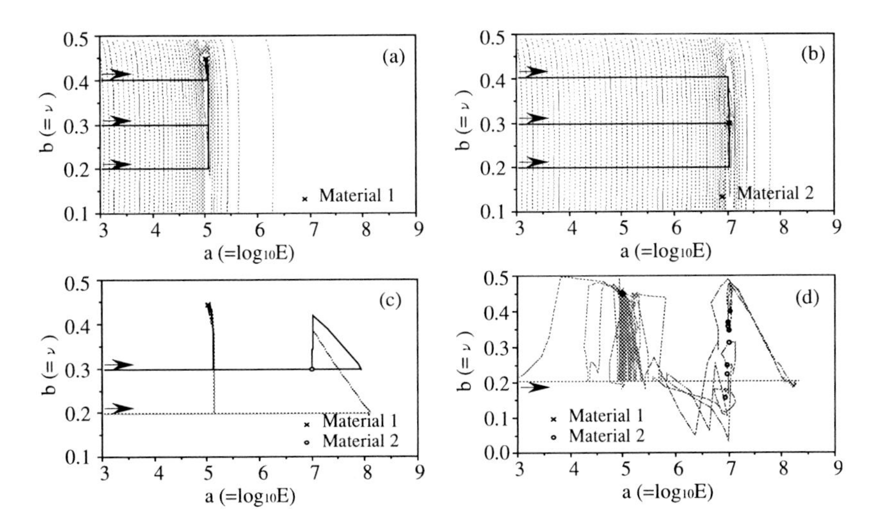

Figure 4. Identification of One-Component Model

3. EXPERIMENTS AND RESULTS

In all experiments discussed here, we employed cubic objects whose length of a side (w) is 10[cm]. Both the measurement and identification models consisted of 125 ($= 5 \times 5 \times 5$) nodes and 320 ($= 4 \times 4 \times 4 \times 5$) tetrahedral elements. We also assumed two kinds of materials, whose physical parameters were defined as listed in Table 1.

We executed three experiments according to the difference in the number of components in the measurement and identification model (Table 2, Figure 3). In the measurement of transformation, we applied the force of 10, 5 and 2.5[N] to nodes on the surface, edges and vertices respectively. The node was selected randomly and the orientation of the force was determined also randomly.

Further more, we decided to apply the descending algorithm in two steps: firstly, the descent was made in the space of parameter a, and next, in the space of a and b. We also implemented an exceptional rule that the value of b jumps to 0.25 when it exceeds the theoretical boundary of $0.0 < b < 0.5$.

Experiment 1

In this experiment, the physical parameter of a uniform object is identified under the knowledge that it is uniform. This is the most simple case of the identification problem.

In Figure 4(a)(b), we plotted the trajectories of the descent, and the contour lines of the error functions. According to this result, these error functions had good nature that they have no local minimum, and therefore, the parameters were identified correctly.

Experiment 2

In the cases where we have information on the internal structure of objects, we can reduce the number of components in the identification model by using this knowledge. In this experiment, we identified physical parameters of two-component hybrid object (i.e. material 1 outside, and material 2 inside) using two-component identification model.

We plotted the descending trajectories of parameters in two components overlaying each other in Figure 4(c). Consequently, we could obtain almost precise parameters on the material of these two components.

Experiment 3

In the case where the internal structure of components in an object is unknown, we have to estimate it from the spatial distribution of physical parameters in the object. It is a natural approach for this purpose to divide the identification model into small parts and identify the physical parameters of each part as independent variables. In this experiment, we tried to identify the parameters of the two-component measurement model by 64-component identification model.

The result of the experiment is shown in Figure 4(d). As is found in the figure, the values of ν in material 2 has variance in the estimated results. While, other parameters converged to the collect values.

4. CONCLUSION AND FUTURE WORK

In this paper, we described an idea of capturing models of soft and transforming objects through experimental transformations on real objects. Through experiments using computer simulation, we confirmed the feasibility of this approach.

In the next step of this study, we are going to apply our method to the capturing of real elastic objects. For this purpose, we need to investigate on the effects of errors caused in the measurement, non-linearity or plasticity of the material, and the affection of gravity, on the characteristics of descent that are unavoidable in the measurement on real transforming objects.

References

[1] Koichi Hirota, Michitaka Hirose, Providing Force Feedback in Virtual Environment, *IEEE CG&A*, Vol.15, No.5, pp. 22-30, 1995.

[2] K. Hirota, T. Hattori, and T. Kaneko: Implementation of Elastic Objects in Virtual Environment, *Proc. of MMJ'96*, pp. 242-248, 1993.

[3] D. Terzopoulos, J. Platt, A. Barr, and K. Fleischer: Elastically Deformable Models, *Computer Graphics*, Vol. 21, No. 4, 1987.

[4] M. Yagawa, S. Yoshimura: Computational Dynamics and CAE series 1 - Finite Element Method, Baifu-kan, 1991 (in Japanese).

[5] H. Imano, H. Yamashita: OR Library 6 - Non-linear Programming, Nikkagiren Publishing Corp, 1978 (in Japanese).

Operator Interaction with Virtual Objects: effect of system latency

S.R. Ellis[a], F.Bréant[b], B. M. Menges[c], Richard H. Jacoby[d] and B. D. Adelstein[e]

[a]Flight Systems and Human Factors Division, NASA Ames Research Center, Moffett Field, CA, 94035, USA

[b]MASI Laboratory, University of Paris VI, France

[c]San José State University Foundation, San José, CA, USA

[d]Sterling Software, Palo Alto, CA USA

[e]University of California, Berkeley, CA , USA

A see-through head-mounted visual display was used to present computer generated, space-stabilized, nearby wire-like virtual objects to 14 subjects. The visual requirements of their experimental tasks were similar to those needed for visually-guided manual assembly of aircraft wire harnesses. An experiment examined the precision with which operators can manually move ring-shaped virtual objects over virtual paths as a function of required precision (Figure 1), path complexity, and system response latency. Tasks with placement precision better than 1.8 cm will require system latency of less than 50 msec for asymptotic performance.

1. INTRODUCTION

Interactive 3D computer graphics displays have attracted considerable interest as human interfaces for a wide variety of scientific, industrial, medical, educational and entertainment applications. But the real world has such high detail that even current high powered graphics workstation have difficulty rendering it with low latency, i.e., < 30 msec, at the frame rates required for high fidelity simulation & natural interaction, i.e., > 60 Hz [1].

But since the real work is in the real world, one alternative is to let the world render itself and to overlay geometrically conformal graphics on it for specific purposes. This approach is similar to that taken by aircraft Heads-Up-Displays or by the see-through displays for wire harness assembly and inspection work at Boeing Computer Services (BCS) and McClellen AFB (e.g. [2]). But even such displays which overlay computer generated virtual objects on the real world have difficulty achieving precise dynamic registration due to latency in the image generation [3]. Accordingly, representative manipulative tasks need to be studied to determine the dynamic performance requirement needed by human operators to productively use these displays. The following experiment attempts to determine such requirements through the study of a 3D tracing task in which an operator attempts to pass a virtual ring over a virtual path without making contact [4,5].

2. METHODS

The entire display system, called an electronic haploscope, (Figure 2), is built around a rigid head-mounted carbon fiber frame worn by a freely moving, tethered subject. The configuration used weighs 1.26 kg, provides a bright (65 cd/m^2), high resolution (< 5'), stereoscopic see-through display with 21° binocular field of view [4]. Each channel of the display can be bore-sight-aligned to allow users to distinguish stereoscopic disparities of several millimeters [4, 6].

2.1. Stimuli

A set of 30 distinct paths were constructed for this experiment (See [6] for details). To obtain a fair random selection of the paths across categories, series of lists containing only 1 occurrence of each path were generated for each block of conditions. All paths were 76 cm long with a square cross-section of 5 mm and were of two general categories: smooth and angular (Figure 3). Subsets of the paths were randomly selected for each subject.

Large and small rings (ID/OD: 5.08/9.65 cm and 1.78/3.30 cm) were used as low precision and high precision cursors to be passed along the paths. These rings were defined by meshes with 300 facets and were positioned in 6 degrees of freedom (DOF) just in front of the subjects' hands. Figure 2 show a large ring being passed over a smooth path.

The subjects head and hand positions were tracked with the Polhemus FasTrak electromagnetic tracking system which was positioned so that its transmitter was within about 1 meter of the head and hand position sensor. Successful alignment of the rings was possible to within 1 cm of the intended position.

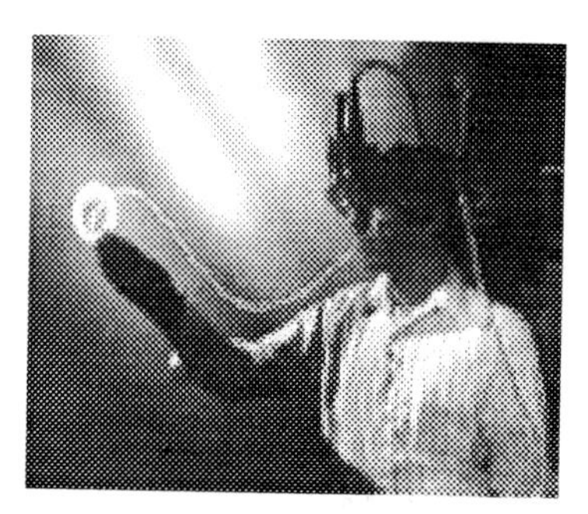

Figure 1.

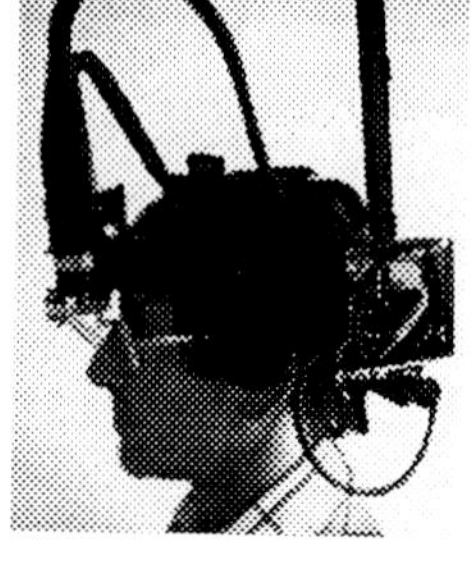

Figure 2

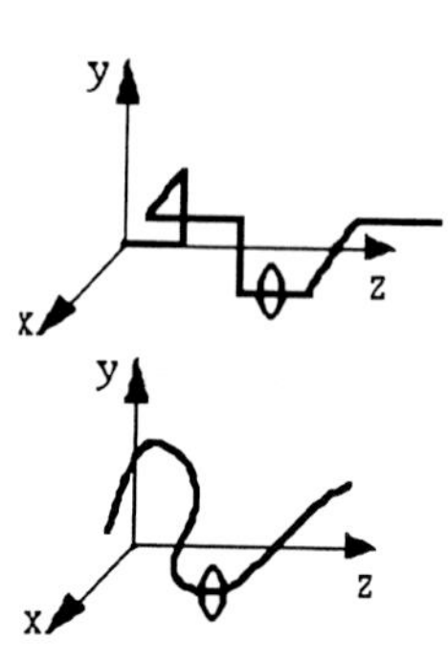

Figure 3

For the simple 3D imagery used the computer could stabilize the 45 Hz stereoscopic graphics update rate with a measured full system latency of 48 msec. This fast simulation update and low latency response was possible due to a number of hardware and software enhancements described in [7]. Interaction dynamics involving forces were not simulated. Contact between the virtual ring and the virtual path was algorithmically detected and signaled by an audible "beep" and a "blink" of the simulated room lights.

2.2. Task

After alignment and calibration of the display, subjects were given several minutes of informal practice moving the virtual ring along the virtual path without touching it. The task was self-paced but subjects were told to complete the block of paths as quickly as comfortably possible. Pilot studies showed that subjects did not trade-off completion time for improved accuracy. Thus, the frequency of collisions (errors) between the ring and the path was highly correlated with path completion time for all subjects.

2.3. Subjects and Design

Ten subjects selected from the paid subject pool and four laboratory personnel participated in the experiment. All demonstrated at least 1 arcmin stereo resolution as measured with the B&L

Orthorater. Seven used the large ring (22-47 yrs) and seven different subjects (19-40 yrs) used the small ring. Three different angular paths and three different smooth paths were generated and crossed with the five different latency conditions of 50, 100, 200, 300, and 500 msec. This produced blocks of 30 conditions for tracing that were randomly presented as three blocks to provide a total of 90 paths for tracing. Each specific path type and latency condition was collected into a block of conditions that was internally randomized. Performance was regarded as asymptotic when the differences between successive blocks were no longer statistically significant. Analysis, in fact, was restricted to the third block. If they wished, subjects could take short breaks between blocks.

3. RESULTS

The absence of a speed/accuracy trade-off was verified by individually correlating each subject's tracing completion time and their number of collisions measured by number of collisions. All subjects had positive, statistically significant pearson correlations between completion time and number of collisions across the 90 paths. They ranged between 0.492 and 0.940.

The time to complete each path and number of collisions were subjected to ANOVA but since observation indicated marked inequality in the within-group cell variances with the variance roughly proportional to the means, the data were transformed by log(x) or log(x+1) respectively to equalize variances for statistical analysis. All statistical analyses of the collision data were based on the transformed data, but the graphs below reflect the untransformed data.

Only two main effects, path type and latency, show that the experimental conditions significantly affected completion time. The smooth path took an average of 14.4 seconds for completion while the angular path took 23.2 sec. ($F(1,12) = 111.2$, $p < .001$). The effect of latency on completion time ($F(4,48) = 66.16$ $p < 0.001$) is plotted in Figure 4. The standard errors plotted in the figure are based on N = 14 subjects. Interestingly, the effect of latency is almost perfectly linear. No interactions involving time as a dependent measure were statistically significant.

Analysis of the experimental effects on the number of collisions also showed reliable effects. Main effects of Ring size ($F(1,12)=112.7$, $p < 0.001$), Path complexity ($F(1,12)=46.2$, $p < 0.001$) and Latency ($F(4,48)=31.8$, $p < 0.001$) as well as the Path x Ring x Latency interaction ($F(4,48)=3.8$, $p < 0.009$) were statistically significant (Figure 5).

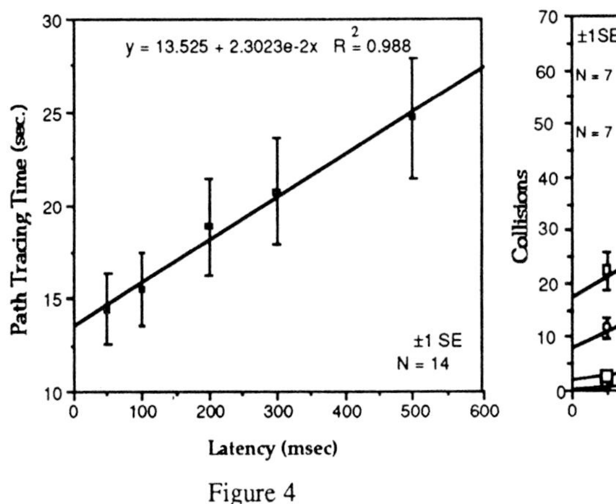

Figure 4

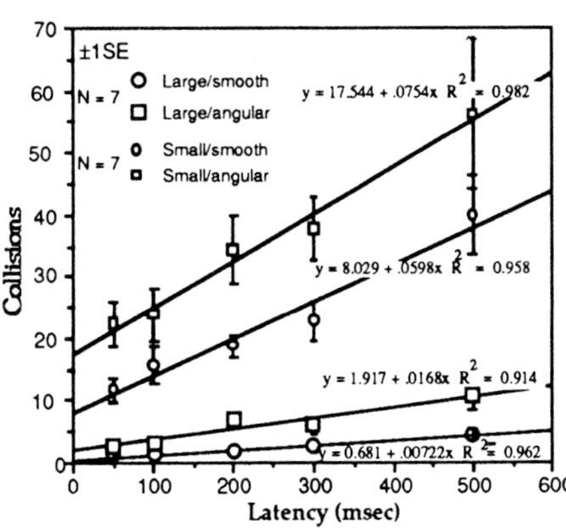

Figure 5

4. DISCUSSION

The absence of a speed-accuracy trade-off suggests that the subjects were not able to maintain a constant level of performance across the various task conditions probably indicating that the task is not as over learned as a classic Fitts tapping task. The absence of speed/accuracy trade off could be due to the increase in the number of control movements required for a given task as a system response latency is increased. This behavior is typical during tracking with low inertia cursors [8]. Increased latency, which also would increase the likelihood of contact between the ring and path by making it difficult for the subjects to avoid contact errors, would thus introduce a correlation between time to completion and number of errors (contacts).

The increased tracing time associated with the angular trajectories is an expected result. These paths required a larger number of discrete movements than the smooth paths which could be traced out with single, smooth movements resembling hand-waving.

Furthermore, as is well know from classical tracking literature, introduction of time lags increases task completion time [9] and the current experiment confirms this effect. Experiments with 3 DOF control tasks incorporating delays in the range used in the present experiment also show task completion time to be linearly dependent upon response latency e.g. [10] .

The tracing characteristics of most practical significance in the current data is shown by the 3-way interaction between Path type, Ring size, and Latency in Figure 8. This effect shows that as the complexity and the required precision of the tracing task increases, overall performance become increasingly sensitive to system latency. This increased sensitivity is expected since the greater precision will require higher tracking gain making the integrated human-machine system prone to instabilities which can avoided by slowing performance down.

Since many possible applications of virtual object displays will require manipulation precision greater than those used in the present experiment, it is evident that practical display system use will need to have latencies less than the minimum 48 msec latency we achieved used. This implication is consistent with results reported in Poulton that even 40 msec of latency can measurably degrade tracking performance [8].

REFERENCES

1. Barrette, R., D., R., Kruk, R., Kurtz, D., Marshall, S., Williams, T., Weissman, P., Antos, S. (1990) Flight simulation advanced wide FOV helmet mounted infinity display," AFHRL-TR-89-36, Air Force Human Resources Lab., Williams AFB, AZ.
2. Janin, A.L., Mizell, D.W., & Caudell, T.P. (1993) Calibration of head-mounted displays for augmented reality applications, Proc. of IEEE VRAIS '93, Seattle, WA. pp. 246-255.
3. Azuma, R. (1993). Tracking requirements for augmented reality. Comm. of the ACM, Vol. 36, No. 7, 50-51.
4. Ellis, S. R. and Menges, B. M. (1997) Judged distance to virtual objects in the near visual field *Presence, 6,* (in the press).
5. Poston, T. & Serra, L.(1996) Dexterous Virtual Work, Comm. of the ACM, 29, 5, 37-45.
6. Ellis, S.R., Bréant, F., Menges, B. M., Jacoby, R. H. & Adelstein, B. D. (1997) Factors influencing operator interaction with virtual objects viewed via head-mounted see-through displays.Proc. VRAIS '97, Alburquerque, N. M. IEEE: New York. pp. 138-145.
7. Jacoby, R., Adelstein, B.D. & Ellis, S.R. (1996) Improved temporal response in virtual environment hardware and software. IS&T/SPIE Proceedings, Conference. 2653B, Session 2653-39, pp. 271-284.
8. Poulton, E.C. (1974) Tracking skill and manual control, Academic, N.Y. 199-206.
9. Sheridan, T. B. & Ferrell, W.R. (1963) Remote manipulative control with transmission delay. IEEE Trans. on Human Factors in Elect., HFE, *4*, 1, 25-29.
10. Liu, A., Tharp, G., French, L., Lai, S., & Stark, L.. (1993) Some of what one needs to know about head-mounted displays to improve teleoperator performance. IEEE Trans. on Robotics and Automation, 9, 3, 638-648.

Experiences with User Interactions in a CAVE™-Like Projection Environment

Roland Blach, Andreas Simon, Oliver Riedel

Department for Virtual Reality, Fraunhofer Institute for Industrial Engineering, Nobelstrasse 12 D-70569 Stuttgart, Germany,
email: roland.blach@iao.fhg.de, andreas.simon@iao.fhg.de, oliver.riedel@iao.fhg.de

Abstract
Multi wall stereo projection (MWSP) systems are an emerging display paradigm, promising a new quality in 3d-real-time interactions. Not much is known about the ergonomics of these systems. This paper describes two experiments of user perception and interaction to obtain a better understanding of user interactions with existing projection technology. The first task is the estimation of absolute geometrical dimensions of simple objects. The second task is grabbing simple objects of different sizes. For both experiments quantitative data was collected as a measure of interaction quality. In order to classify MWSPs, these tasks were compared to other display devices and compared to physical reality. Due to the limited number of participants these experiments are considered as case-studies only.

1. Introduction

Virtual Environments (VE) are getting more and more mature. New interaction and display devices appearing on the market claim to fulfill promises of the early days of virtual reality: natural and intuitive user interaction with computer generated environments. MWSP systems [1] as a special form of display environment recently became manageable because of new and powerful computer hardware technology available. They present the user with an immersive quasi-holographic visual impression in adequate resolution. The user is surrounded by the projection environment seeing generated virtual space and the true physical space including his own body. This is an overlay of virtual space with real space as it is performed by augmented reality systems [2]. A very important fact is the awareness of the users own body and therefore the existence of a natural frame of reference. This geometrical reference brings back the need for exact matching of true physical and virtual space. This is obviously necessary if the user problem domain is evaluation of proportions like in architecture, virtual product design and ergonomics. But even in abstract data spaces one needs the exact control of the geometrical overlay in order to be able to design effective direct manipulation 3d-interfaces like 3d-menus. These preliminary experiments lead to a better understanding of properties and limitations of synthetic 3d-interaction environments. The long-term research goal is the design of adequate and usable 3d-interfaces. Following simple tasks are investigated in this study:

- Estimation of absolute geometrical dimensions in VEs.
- Grabbing simple objects in a 3d-environment

Ergonomics of different stereo projection systems were researched by [3][4] which are preliminary examinations as well. We would like to add new results to the research area of user interaction in 3d viewing environments.

2. Experimental Setup

A generalized formulation of the projection equation was used to obtain a normalized geometrical appearance for the virtual environment. All computer generated scenes have the same calculated dimension regardless of the projection system geometry as described in [5]. In order to correctly normalize the geometry of different display environments various physical dimensions of the projection system have to be known. This includes the dimensions of the projection plane, the interocular distance and the position and orientation of the viewpoint relative to the projection planes. Position and orientation of the viewpoint has to be collected in real time, i.e. between 10 and 30 times per second. Measurement of the viewpoint location should be done with a sufficiently fast and accurate tracking system, while all other data can be obtained off-line.

As image generation system the same graphics workstation is used throughout all experiments. The display devices that were used are:

- A four wall stereo projection (4WSP) system of 3x3x3 m with a resolution of 960x960 pixel
- A stereoscopic dual monitor device (BOOM) with a resolution of 1280x1024 pixel
- A stereoscopic CRT Head Mounted Display (HMD) with a resolution of 1280x1024 pixel

As tracking device for the HMD and the 4WSP an electromagnetic system is used. The BOOM system has a mechanical position and orientation tracking. As interaction device in the grabbing task an 18-degree of freedom tracked dataglove is used.

The main goal in these first experiments is not to obtain 100% valid statistical data but to get some hints what might be worth investigating further. Therefore we chose 10 people, with proven stereoscopic sight ability, with different experience with VEs. They had to perform the defined tasks without prior training.

3. Description of the Experiments

3.1 Experiment 1: Estimation

Two different environments were presented where the user had to estimate the size of cuboids and spheres, which are laid out in a 3d-scene:

- Scene 1: tiled ground plane and floating objects
- Scene 2: simple room with a chair, a table, walls and daylight conditions with objects which are laid out in the room

The size of the objects was in the range of 5x5x5 cm to 100x100x100 cm. Both scenes exist as a computer generated environment as well as physical model. The scenes were very simple to obtain a 'fair' comparison between the physical and the computer generated scenes (images of the scenes can be found at http://vr.iao.fhg.de/vr/information/Publications/hci/exp/).

Participants had to estimate the size of the objects (length width and depth) from a fixed distance but they were able to change the head orientation. The user had to perform the estimation task with ten objects to in both scenes. All these tests were done with the following display configurations: 4WSP, HMD, BOOM and a real physical model.

The estimates were collected and evaluated with simple statistical methods. To fade out the inability of some people to tell distances in metrical units, they had to show the expected size with their hands. This distance was used as measure of quality of the estimation.

With this experiment we can gain some insight if computer generated environments are useful for the evaluation of 3d-scenes and if a MWSP is a major improvement.

3.2 Experiment 2: Grab Task

Scene 1 of Experiment 1 was presented to the participant. Cuboids of different size have to be grabbed between thumb and index finger in one continuous movement. If the user believes that he has successfully grabbed the object, he should turn the object. The objects were located within natural reach. Their size was between 3x3x3 cm and 15x15x15 cm. It was possible to move in the scene within the range of tracking (ca. 3x3 m) only without the use of an additional navigation device. The participants had to grab and turn ten different objects. The error between the measured distance between thumb and index finger in the moment of rotating and the siye of the virtual object is considered as a measure of quality.

These experiments were performed with HMD and the 4WSP only. Performing this experiment with the BOOM was considered pointless, because the user has to hold the viewing device with both hands. A physical world setup was excluded because of the presence of force feedback making a comparison to computer generated environments difficult.

To identify the influence of tracking errors in the case of the 4WSP projection environment we compared two setups: direct and indirect interaction. In the direct interaction mode the position of the real hand was matched exactly with the virtual space, such that a virtual representation should not be necessary. The indirect interaction mode shows a virtual model of the hand in a distance of 30 cm in front of the user. The participant controlled this representation with the dataglove.

The main concern of this experiment is gaining some insight into the ability of users to perform grabbing tasks in VEs. The task should be as natural as possible in the sense of distance and size of objects. We did not use gesture recognition to obtain the information if the user has actually grabbed an object. Instead we used a collision detection algorithm to determine contact between fingertips and virtual objects.

4. Results

Results can be obtained in form of tables and diagrams at http://vr.iao.fhg.de/vr/information/Publications/hci/exp/ as well as significant protocols of the experiments.

5. Conclusion

Two experiments were performed to evaluate user perception and interaction with 3d VEs in MWSP systems compared to more established viewing devices. The first task was the estimation of absolute (perceived) size under different viewing conditions. The second task was grabbing and holding of virtual objects. The ability to estimate size as established by the first experiment is an essential part of the interaction in this task. We consider these experiments to be case-studies with their results representing a preliminary investigation in a technology driven field.

We observed a strong relationship between tracking accuracy and satisfying completion of interaction tasks, making it a crucial factor for good 3d-user interfaces. Much care has to be taken for calibration and error correction of electromagnetic tracking systems. This is less obvious for the size estimation task, but it is the dominating factor in the grabbing task. We hope that more reliable tracking systems will be available soon to improve direct 3d-user interaction. While users like the comfort and freedom MWSP environments provide [4], the lack of adequate tracking systems significantly impairs their interaction performance.

The dataglove with electromagnetic tracking was used to measure the exact location of the fingertips. Contact between the fingertips and the virtual object determines the status of the interaction. This interpretation of a grabbing motion is fairly close to the natural process of handling an object. It is worth noting that customary use of the device is for pure gesture control, circumventing the need for exact position measurement. While the dataglove is certainly not ideal for precise and pleasing interaction control it is nonetheless the only proven turn-key device for real-time tracking of a users hand and fingers.

6. Future Work

For the design of truly interactive user interfaces it will be necessary to investigate interactions with dynamic environments. We will therefore adapt experiments of estimation and spatial interaction to more dynamic scenarios. The ability to perceive and correctly estimate the velocity of a moving object is next on our list of open questions. We hope to derive guidelines for designing real interactive 3d-interfaces for different projection system environments.

References

[1] Cruz-Neira C., Sandin, D.J., De Fanti T.: Surround-Screen Projection-Based Virtual Reality: The Design and Implementation of the CAVE. Annual Conferences Series ACM SIGGRAPH 1993 pp. 135-142

[2] Azuma R.: A Survey of Augmented Reality. SIGGRAPH 95 Course Notes #9 (Developing Advanced Virtual Reality Applications)

[3] Deisinger J., Riedel O: Ergonomic Issues of Virtual Reality Systems: Head Mounted Displays. Virtual reality Word 1996. Conference proceedings, München 1996

[4] Deisinger J., Cruz-Neira C., Riedel O., Symanzik J.: The Effect of Different Viewing Devices for the Sense of Presence. To be published in Proceedings of the HCI 1997

[5] Landauer J. et al: Toward Next Generation Virtual Reality Systems. To be published in Proceedings of the IEEE Multimedia Conference 1997

3D-Rendering and the Eye: Aspects for Efficient Rendering through Physiological and Anatomical Parameters of the Eye

Oliver H. Riedel

Department for Virtual Reality, Fraunhofer-Institute for Industrial Engineering, Nobelstrasse 12, D-70569 Stuttgart, Germany, email: oliver.riedel@iao.fhg.de

Abstract
There are different algorithms for efficient rendering of 3D scenes which pay attention to technical aspects of the used computer hardware. Tuning all aspects of the hardware is one way of fast and good quality rendering especially for virtual environments. Another way is take some of the physiological and anatomical parameters of the eye in account. This could result a better renderer on the same machines as before or a same renderer on cheaper machines. This paper describes some of those aspects and shows some new approaches and their results for the 3D real-time rendering process. The basis of this paper is a research work on enhancement of a real-time rendering algorithms that was conducted for the virtual reality kernel Lightning developed by the VIS-Laboratory at the Fraunhofer-Institute for Industrial Engineering. The new rendering algorithms have proved their usability during several industrial projects.

1. Human Factors of the State of the Art 3D Rendering

For most of the computer graphics applications image quality and speed are the most important factors. Therefore research and development aims towards an optimal teamwork of components from hard- and software. Also the direct relationship of the scene complexity and the quality and/or the speed of the rendering is important. In the field of 3D real-time rendering this problems raises due to the demands from the aspects of fast response to the users interaction and the necessity to render the scene twice. There are immensely algorithms known which improve the use of the hardware but only a small amount pays attention to the parameters of the eye. Some of those are:

- *Representation of Geometry*: Projection of geometry, hierarchical and nonhierarchical representation (BSP, CSG), visibility algorithms, levels of detail
- *Models for color perception*: Colors models for additive, subtractive and special color composing, interpolation, dithering in the color-space
- *Shading algorithms*: Textures, transparency, shadows, reflections, ray-tracing, radiosity
- *Filtering in the 2D*: Antialiasing, multisampling, motion-bluring, dithering etc.

A further description of most of this algorithms can be found in [8] and [2].

2. Physiological and Anatomical Aspects of the Human Eye for 3D Real-time Rendering

The spatial vision of the human is based on numerous information which are collected by the eye and it's apparatus before being send to the brain. Some of the information can be recognized by one eye but must need two functional eyes. The parameters of the eye which come under evaluation in this work are both from the anatomy and the physiological (focus, adaptation, field of view (FOV), movement of the eyes, color perception) side of the eye. Not all of those parameters can be adopted to the 3D Real-time Rendering. We tried to match values from literature concerning the human eye [3], [4], [5], [6] with the limits of the available computer hard- and software and put together a first list of requirements for a new 3D real-time renderer which are possible to realize:

- *Improve of the FOV and the view volume*: The FOV has a irregular shape and can be simplified to a circle or an ellipse, but most of the displays devices have a rectangular form. The binocular overlap and the used parallaxes are other important parameter which should fit to the eye. According to the Federal Aviation Administration and other literature we recommend that the FOV is described by an inlaying rectangle with an overlap of about 50% for normal applications and with 100% for applications with visual demands. The parallaxes should be positive and about the same value as the interpupillar distance.
- *Level of detail (LOD) switching as a function of the focus*: The criteria for the switching between to levels of detail should be on the one hand a function of the maximum focus of the eye and the linear distance between each object and the eye. On the other hand one can use the fact that the focus is a radial symmetrical function with the center in the fovea centralis. Combining those two facts the LOD-switching becomes a combination of the radial and linear distribution of the focus.
- *Refresh- and framerate according to the temporal resolution of the eye*: An unstable framerate confuse the user, can cause simulator sickness and lead to unintended learning effects. Therefore the used hardware should support a refreshrate up to 85Hz which is the actual standard for computer working placed. To achieve a stable framerate the renderer should use a load balancing mechanism with smooth transitions between different famerates. A framerate is a direct function of the complexity of the displayed scene. The load balancing is normally implemented in combination with LOD.
- *Perception of movements*: The calculation of the speed of both, the user and the objects, and the fact that the eye recognizes only angle velocities up to 400°/s gives another change for efficient rendering. This can be implemented in the LOD as a function of the relative velocity between the object and the eye, too.
- *Perception of color*: The perception is a function of the vision, too. In the peripheral view the sensitiveness is lower as in the near of the foveal center. In addition the perception of color also depends on the wavelength of the displayed color. However, the average hardware supports a color depth of 8bit per red, green and blue which is fairly enough. The graphics-hardware is not able to vary the color resolution over the display space. That leads for the perception of color to the single demand of using 24bit color space.

According to this list we implemented and tested some enhancements for the 3D real-time renderer Lightning, described in [7]: Exact mathematical description and use of model for the vision, predictive tracking for reducing the total system lag, extended algorithms for LOD, correction of optical errors from HMD´s lenses. Following the new LOD algorithms will be explained as an example for the new algorithms because most of the mentioned requirements can be implemented in the LOD-algorithms.

3. New LOD-Algorithms as an Example for Efficient Rendering

Based on the standard algorithms of the rendering library IRIS Performer some enhancements were defined and implemented:

- Automatic computing of the ideal switching point based on the visus of man (ideal linear switching). It is a function of each bounding box, the distance and the individual visus.
- Switching between the LOD states based on the real location of the object on the retina (radial switching). It is a function of each bounding box, the number of available LOD states and the location of the object on the retina. the Location on the retina can be substituted by the angle between the line of sight and the line trough the eye point and the pivot point of the object.
- Switching based on the angular velocity of the rendered objects (dynamic switching). It is a function angular velocity of the object.

All the new algorithms use special tracking technology for the measurement of the location and orientation of the user and its eyes. Another prerequisite is the exact computation of the line of sight, the distance between the user and all relevant objects and the bounding-volumes of the objects. All new algorithms can be combined with each other or with traditional LOD methods. The combination could be a multiplication or a weighted sum. A hysterese should be added to the algorithm for the final decision of the rendered LOD state of each object to avoid an oscillating between two states.

Derived from the parameters of the eye the following table gives an example for typical switching points for objects with 5 states. φ_{obj} is the angle covered by the object, δ_{obj} the angular location of the object on the retina and ω_{obj} the angular velocity of the object. The recommended values are derived from the literature. As an example for the differences between the different LOD algorithms figure 1 shows a principle diagram of the linear and radial switching process.

LOD state	Ideal LOD	Radial LOD	Dynamic LOD	Grade of Detail
1	$\varphi_{obj}>16°$	$\delta_{obj}<15°$	$\omega_{obj}<90°/s$	All details of the object visible
2	$8°<\varphi_{obj}<16°$	$15°<\delta_{obj}<30°$	$90°/s<\omega_{obj}<180°/s$	Do without small details
3	$4°<\varphi_{obj}<8°$	$30°<\delta_{obj}<45°$	$180°/s<\omega_{obj}<270°/s$	Render still bigger details
4	$2°<\varphi_{obj}<4°$	$45°<\delta_{obj}<60°$	$270°/s<\omega_{obj}<400°/s$	Render only outline
5	$\varphi_{obj}<2°$	$\delta_{obj}>60°$	$\omega_{obj}>400°/s$	Replacement by geoprimitives

Table 1: Values for switching between LOD states for different algorithms

4. First Results and further development

For the evaluation of the new algorithms we measured the speed improvements of the renderer and the subjective enhancement of the visual quality. The new mathematical view models were matched with the parameters of the eye with the new model. The predictive tracking was evaluated by error calculation. The quality of the 2D filters was measured by a differential comparison of the images. Table 2 shows a part of the measurements for the new LOD algorithms. The values are the time needed for rendering in percent based on the no LOD as 100%. Four different types of hardware were used for the testing.

As an examples for further developments the LOD problems give some good examples for further development. Future enhancements based on anatomical and physiological parameters of the eye can be: LOD switching based on the color space, LOD switching based on the brightness, LOD methods based on the total system delay. In all other fields of this research future trends are identified and subject of the next period of work.

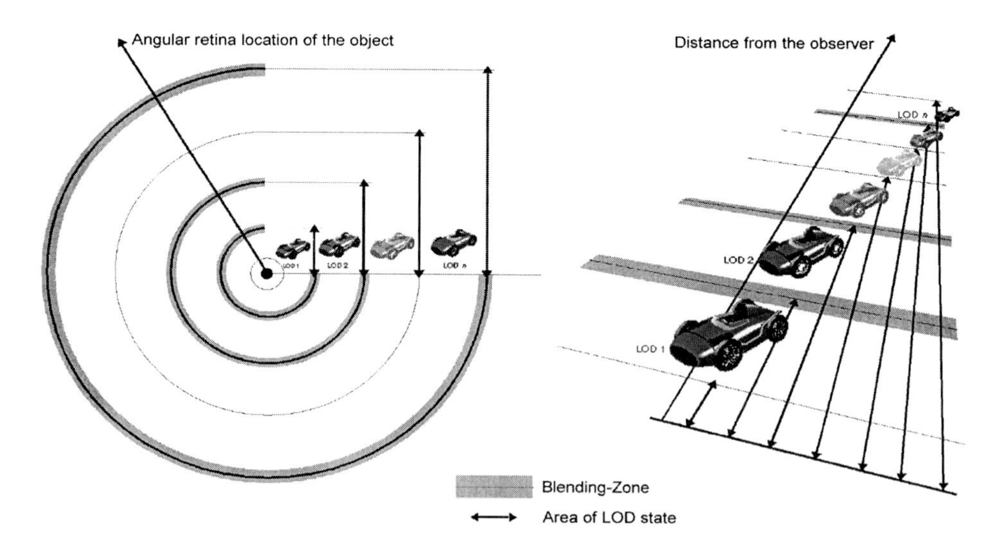

Figure 1: Radial vs. linear switching of LOD states

	Type of LOD					Combination	
	No	Old	Ideal	Radial	Dynamic	Max.	Add.
1	100%	43%	39%	41%	72%	8%	7%
2	100%	37%	32%	36%	69%	18%	15%
3	100%	26%	31%	47%	89%	23%	19%
4	100%	39%	37%	49%	78%	22%	18%

Table 2: Results for the rendering time from the new LOD algorithms

For further references about this work and all results from the measurement of the performance enhancements please refer to http://vr.iao.fhg.de/vr/information/Publications.

References

[1] Bryson, S.; Fisher, S.: Defining, modeling and measuring system lag in virtual environments. In: Conf. on Stereoscopic Displays and Appl.. SPIE, Vol. 1256, p. 98-109

[2] Clay, S.R.: Optimization for Real-Time Entertainment Applications on Graphics Workstations. In: Designing Real-Time 3D Graphics for Entertainment, Course Note 33 of the SIGGRAPH ´96, New Orleans. New York: ACM, p. 77-114.

[3] Davson, H.: Physiology of the Eye. 5th Ed. Houndmills: The Macmillian Press Ltd., 1990.

[4] Kern, P.; Riedel, O.: Ergonomic Issues of Head Mounted Displays. In: Proc. Of the 11th Annual Int. Occupational Ergonomics and Safety Conf., 1996.

[5] Kulikowski, J.J. et al (Eds.): Limits of Vision. Vision and Visual Dysfunction, Vol. 5. Houndmills: The Macmillan Press Ltd., 1991.

[6] Kennedy, R.S.; Fowlkes, J.E.: What does it mean when we say that simulator sickness is polygenic and polysymptomaic. In: Proc. of the 1990 Image Conf., 1990, p. 44-45.

[7] Landauer, J. et al: Toward Next Generation Virtual Reality Systems. In: Proc. of the IEEE Multimedia Conf. 97, 1997, in print.

[8] Foley, J. D., et al.: Computer Graphics - Principles and Practice. 2nd ed. Reading: Addison-Wesley, 1990.

Barriers to Industrial Application of Virtual Environments

John R. Wilson

Virtual Reality Applications Research Team, Dept. of Manufacturing Engineering & Operations Management, University of Nottingham, Nottingham, NG7 2RD, UK

1. INTRODUCTION - ANY 'REAL' APPLICATIONS?

In a research council funded programme of work in 1994-95, examining potential industrial application for virtual environments (VEs), we collaborated directly with about a dozen UK companies and surveyed, in varying degrees of depth, over 250 more. Despite considerable enthusiasm for the potential of the technology (in planning, design, training, layouts) all we could find were internal "marketing" demonstrations or else exemplar VEs built by outside research or consultant teams (Wilson et al, 1996). Based on evidence collected some 12-18 months later the same story is reported in a UK government report on the Virtual Reality market (Stone, 1996). Across the world there are few if any working applications of virtual environments built within VR systems (VR/VE) which have replaced another technology or else are doing something new.

One possible exception to this dearth of ***real*** working applications of VR/VE in manufacturing is in training. The nature of the technology and its capabilities for visualisation, learner involvement and interaction give some of the reasons for this. Also, the nature of the training function means it is ready for VR/VE before other applications. Both the activity and related technology can be 'stand-alone' within a company, reducing interfacing problems. Desktop VR is quite adequate, even preferred for widespread distributed training and so technical or user problems with HMDs are not an issue. Training also does not need high photorealism and display fidelity, and in fact may gain from a degree of simplification.

In spite of the use in training, we have to ask hard questions about the very limited applications to date. Is it because of technical capabilities, interface elements, user attitudes or behaviour, task environment appropriateness or cost? Whatever the reasons, will difficulties be reduced as new generations of systems and ideas for appropriate use emerge or is VR/VE flawed in concept and delivery for manufacturing use?

2. BARRIERS TO TAKE-UP OF VR

If VR is to establish itself as a major technology in industrial, educational, medical and public service use, then it has a number of barriers to overcome. These are summarised as: technical - trade-offs and systems integration difficulties; practical - poor evidence of added value from evaluated working demonstrations; and user - usability and potential side effects.

It is not simple to render realtime, recognisable 3D representations, which update synchronously with participants' inputs or movements in the virtual environment. Today and into the foreseeable future, VR application must address a number of well-known trade-offs in planning and building VEs, including:

- update rate .v. photorealism or complexity
- detail and immersion .v. field of view
- extra peripheral vision .v. headset weight
- depth cues through stereoscopic 3D .v. cost and possible side effects
- better resolution and colour contrast of CRTs .v. greater weight/size
- total immersion (for 'presence') .v. possible side effects or discomfort

Whilst these barriers are real ones they are not insurmountable, given likely VR technical developments of the next few years and through careful selection of appropriate applications now. Potential industrial users raise the issue of efficient data input/output more frequently than any other so this may be the key to mass industrial use, and currently is the focus of urgent development. The paucity of salient working demonstrations of VEs, and poor current evidence of added value is summed up by one manufacturing engineer ".... I can't even begin to discuss use of VR sensibly with them [company engineers and decision makers] unless they sit with it, try it, use some worlds they can relate to ....". The needs for evaluated demonstrations of applications are related to the need for a structured methodology to guide application identification, building and evaluation against operational criteria. The third set of barriers, to do with usability, will be the providence of hci specialists amongst others. Some means of assessing the usability of VR technology is required, embracing utility, usability, likeability and cost, and adding system integratability or organisational usability. This last relates to the integration of VR/VE with other technologies and to how VR changes existing company strategies and groupings.

Many traditional usability criteria will not apply directly to VR/VE, and those that do will require redefining. Relevant experimental work is emerging to address this. Requirements for usability of VR/VE include:

- support use of, and switching between, egocentric and exocentric views
- allow representation of participant, as visible 'object' or invisible presence
- support the users' navigation and orientation within the VE
- minimise effect of spatial distortions, for instance of scale
- provide depth perception and perspective cues
- minimise effect of temporal resolution, lags from sensor update or frame rate
- support compromise between HMD field of view and visual acuity of scene
- provide interface support to harness the possible natural semantics of dialogue
- improve the ergonomics of equipment and input devices
- produce guidance on procedures and training for appropriate system use
- understand mechanisms, predisposing factors and consequences of any side effects

This last is a particular concern of the UK Health & Safety Executive who have funded us to investigate any consequences for health and safety at work. A proportion of participants in some systems might display effects of nausea, or occular discomfort, visual performance change or disorientation, but if these effects are minimal, wear off very quickly or do not

Issues for Integrating Virtual Teaming Into the Organization

Vincent G. Duffy

Department of Industrial Engineering and Engineering Management, Hong Kong University of Science and Technology, Clear Water Bay, Kowloon, Hong Kong, vduffy@uxmail.ust.hk

1. INTRODUCTION

What is needed is the establishment of techniques that will allow more flexible organizational structures for companies that want to have design and manufacturing located in different sites. This would allow more local companies to take advantage of opportunities provided by increased development of manufacturing capabilities in China. Companies based in Hong Kong can not afford to have the effectiveness of integrated product development efforts compromised by distance between the team members. The competitive advantage in using integrated product and process development, reduced time to market, can still be realized for remotely located teams using the appropriate technology with the proper organizational and human conditions.

The goal of this research is to assess Virtual Teaming Technology for its ability to create or enable a sense of presence and offset the effects of distance on effectiveness of remote integrated product and process development efforts. Team members are observed during task completion in order to determine the human, organizational and technology factors and differences in patterns of communication for successful and unsuccessful teams. The proposed model suggests that synchronous, remote technologies allow for a greater feeling of presence by the users and should yield similar results to co-located work. This would minimize the need for proximity and could provide for more flexible and effective product development efforts as well as reduced time to market.

2. BACKGROUND

The proposed approach incorporates the Virtual Teaming Technology within the human-computer interaction model of 'groupware in the time/space matrix' presented by Dix [1] and a proposed model for integrating technology in the workplace. The model builds on the previous findings for effectiveness in concurrent engineering which considers the critical human, organization and technology aspects. A relationship is expected between proximity and technology, based on feeling of presence provided by that technology. It is anticipated that effectiveness of real industrial teams will also depend on the pre-established level of quality of communication as well. If shown to be true, these results would support the notion that concurrent product development, which is likely to see increased use of remotely located teams, should not suffer adverse effects.

2.1 A Trend Toward Global Efforts in Product Development

The trend that has been shown, a shift of location of the manufacturing function of many companies from Hong Kong to China, is changing the way in which products are developed in this region. It is imperative that the differences between the more and less effective teams are better understood, particularly due to the added complexity of coordinating product development efforts with team members working in separate locations.

When technologies such as this are implemented without concern for the organizational and people factors, full benefits of the technology are not realized [2]. As well, it has been shown that the likelihood of communication is found to be higher for teams which are located at closer distance to one another and the communication patterns have been shown to be related to effectiveness [3,4].

2.2 A Sense of Presence: Video Conferencing and CSCW

The ability to communicate using video, by itself, would be considered video conferencing. However, when combined with the ability to do work in this environment it could be considered computer supported cooperative work (CSCW) [5]. This cooperative work in the virtual space, is virtual reality rather than simply multi-media based on the definition of virtual reality provided by Wann and Mon-Williams [6]. This definition as well as most definitions of virtual reality would include some description regarding the users having some sense of presence and interactivity. Using video, this virtual teaming environment would provide for the interactivity part and, as was found by Watts, et. al. [7], a strong sense of presence. This implies that the video and shared workspace environment or Virtual Teaming should also create a strong sense of presence.

2.3 Effective Integration of New Technology in the Workplace

With the advent of new technologies which allow for video communication and computer supported cooperative work, it is important to consider the potential benefits of the use of those technologies. However, lags are different and further studies are desirable. A review of literature indicates that there is a lack of experimental studies on the effects of *finger movement-related lag*.

3. LAG COMPENSATION BY IMAGE DEFLECTION AND PREDICTION ALGORITHMS

3.1 Principle of image deflection

With a virtual reality simulator, *head movement-related lags* can cause image position errors during head movements (Bryson and Fisher, 1990). So and Griffin (1992) proposed an image deflection technique to compensate for the lag-induced image position error. Figure 2 illustrates the view of a 'virtual switching panel' at time t=0s and time t=Ts during a rotational head movement. With a '*head movement-related lag*' of Ts, an image position error of θ degrees would occur at t=Ts. This image error (θ) is the difference between the head pointing angles at t=Ts and t=0s. Knowing this error (θ), images can be displaced appropriately back to the correct pixel positions by a raster scan system. Detailed explanation of the benefits and short-comings of the image deflection technique can be found in So and Griffin (1992, 1996). Image deflection has been shown to be effective in compensating lags introduced by *rotational* head movements.

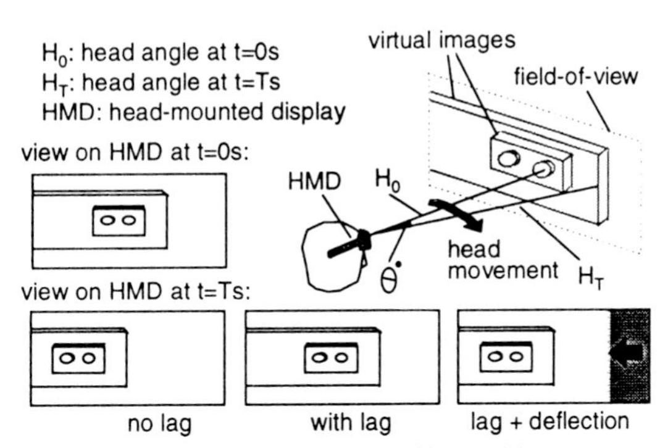

Figure 2 Illustration of the effects of lag and image deflection on images viewed on a HMD during head movement.

3.2 Principle of prediction algorithms

Lag compensation by prediction assumes that an up-to-date head, hand, or finger position can be predicted from the appropriate past (or delayed) positions. The use of prediction algorithms in virtual reality systems has been the subjects of many simulation studies (e.g. Non-linear prediction algorithms: List, 1983; Adaptive algorithms: Albrecht, 1989). So and Griffin (1996) reported an experimental study in which head tracking performance with a virtual reality system of 40 ms lags or more was significantly improved by a real-time phase lead filter algorithm.

4. BENEFITS OF LAG COMPENSATION IN VIRTUAL TRAINING APPLICATIONS

Image deflection could compensate for the image error introduced by lags related to rotational head movement. Also, depending on the lag durations and the frequency ranges of head, hand, and finger movements, a set of optimized predictive algorithms could be obtained to compensate for the *head, hand, and finger movement-related lags*. A review of literature indicated that there is a lack of experimental studies in this area.

5. DISCUSSIONS AND RECOMMENDATIONS

Time lags occur in a virtual training system. These lags can be divided into three categories: head movement-related; hand movement-related; and finger movement-related. Both head and hand movement lags have been shown to degrade object handling task performance in virtual training applications. The effects of *head and hand movement-related lags* on task performance have also been found to be different. Although the degrading effects of lags have been reported in the literature, studies concerning the mechanism by which lags affect the performance of an object handling task has not been found.

At the Hong Kong University of Science and Technology, a research programme investigating the effects of time lags on object handling tasks has been launched. Both the individual effects of lags and combined effects of individual lags will be modelled and studied. The aim is to identify an optimized lag compensation method for virtual training applications.

REFERENCES

Albrecht R.E (1989) An adaptive digital filter to predict pilot and head look direction for helmet-mounted displays. Unpublished M.Sc. Thesis, University of Dayton, Ohio, USA.

Bryson, S. and Fisher, S.S. (1990) Defining, modelling, and measuring system lag in virtual environments. SPIE vol. 1256. Conference on stereoscopic display and applications, Santa Clara, 1990.

Chorafas, D.N. and Steinmann, H. (1995) Virtual reality: practical applications in business and industry. Prentice Hall. ISBN 0131856383.

Kawara, T., Ohmi, M. and Yoshizawa, T (1996) Effects on visual functions during tasks of object handling in virtual environment with a head mounted display. Ergonomics, Vol.39, No.11, pp. 1370-1380.

Kenyon, R.V. and Afenya, M.B. (1995) Training in virtual and real environments. Annals of Biomedical Engineering. Vol.23, pp. 445-455.

Kozak J.J., Hancock, P.A., Arthus, E.J. and Chrysler, S.T. (1993) Transfer of training from virtual reality. Ergonomics, Vol.36, No.7, pp.777-784.

Lin, F., Hon, C.L. and Su, C.J. (1996) A virtual reality-based training system for CNC milling machine operations. Annual Journal of IIE (HK), pp.13-16.

List, U.H. (1983) Non-linear prediction of head movements for helmet-mounted display. AFHRL Technical paper 83-45. Operations training division, Williams AFB, AZ 85224, USA.

So, R.H.Y. and Griffin, M.J. (1992) Compensating lags in head-coupled displays using head position prediction and image deflection. J. Aircraft. 29(6): 1064-1068.

So, R.H.Y. and Griffin, M.J. (1995) Effects of lags on human operator transfer functions with head-coupled systems. Aviation, Space and Environmental Medicine, Vol.66, No.6, June 1995, pp. 550-556.

So, R.H.Y. and Griffin, M.J. (1996) Experimental studies of the use of phase lead filters to compensate lags in head-coupled visual displays. IEEE Transactions on Systems, Person, and Cybernetics. Vol. 26, No. 7, July 1996.

Data Input Devices and Application in Virtual Reality

Hanqiu Sun

Dept. of Computer Science & Engineering, The Chinese University of Hong Kong
E-mail: hanqiu@cse.cuhk.edu.hk

Virtual reality has emerged as an evolving technology that integrates 3D input/output devices to multi-discipline applications. Within the 3D virtual world of a working environment, human operators can perform various tasks by manipulating graphical models. This paper presents the up-to-date VR technology by describing Glove-based input devices and applications using the devices, including the structure of hand, its multi-degrees interface, the current state of posture and gesture recognition, and precise-work applications.

1 VR and Data Input Devices

Virtual reality (VR) is a newly emerged three-dimensional environment. The sense of a third dimension is created if a pair of stereo pictures is displayed separately for two eyes on a wearable helmet or head-mounted unit. Alternatively, pictures can be projected onto the surrounding surfaces, such as the ceiling, floor, and walls, of an enclosed space to create the impression of an immersive environment. On the other hand, the three-dimensional interactive devices introduced in virtual reality extend the conventional control space to three or higher dimensions. Data glove devices [1] are one of the new VR devices, equipped with sensors to track both static hand shapes and dynamic hand movements. Data glove devices have the advantages in creating simpler human-computer interfaces than most conventional input devices for their highly coupled degrees of freedom, familiar sign language, and body reference coordinates.

2 Glove-based Input

Object handling with hands involves different hand shapes (or postures) and motions (or gestures) [2]. Postures distinguish the relative finger positions and orientations that hand can form at a time. Postures are recognized at discrete times and used to trigger individual signals or commands. Gestures, on the other hand, form another type of interaction that inputs the continuous feedback of hand movements to a control process. This type of interface interprets the dynamic transition from one hand shape to another over time. After hand shapes and motions are detected from data gloves in a device coordinate system, their coordinates need to be mapped to an intermediate room coordinate system in which the device is used first and then to the coordinate system that stands for the virtual environment where applications are developed.

2.1 Hand Structure

Hand itself is an articulated tree structure with multiple degrees of freedom (DOFs). There are 17 active joints in the hand, together providing 23 degrees of freedom, and 6 degrees of freedom in the free motion of the palm. The distinguishing characteristic of whole-hand input is that the interface is transparent to the user, unlike other input devices whose movements are manipulated by the hand (e.g., mouse, joystick, trackball). The whole-hand input naturally adopts the most intuitive and flexible hand functions people use in daily life. It performs complex control problems such as grasping directly in the task space. The user may not be aware of using the device while making familiar signals or signs in the task space.

2.2 Hand Interface

The whole-hand input interface has three novel properties: intuitive use of the hand, smooth transition of control modes, and reduced interface complexity. The hand interface is natural in a sense that it is not a device manipulated by hand but a direct hand input device. The free poses and motion of hand directly map to the 3D task space. Since familiar hand skills are used for performing the task, less user training time is required for using the interface. Also, the hand is capable of quickly and smoothly switching functions. Hand shapes and motion can be interpreted to various states that switch among multiple tasks. Such a capability reduces the space and concentration required for using multiple devices. Another capability of the hand is the natural coordination between fingers. Highly coordinated work can be fairly easily produced using this interface. By the way it is structured, the hand can easily perform multi-degree tasks in a single device interface.

Hand interface is formed by both postures and gestures. The recognition of both can be difficult due to two reasons: the complexity of hand structure and the lack of accurate measure. Both the sample patterns stored in the database and the ones to be recognized may be imprecise due to mechanical noise or human factor. Even the same person cannot repeat a posture exactly. It must be slightly different from the original one. Gesture recognition faces extra difficulties such as time-varying input signals and gesture segmentation.

2.3 Hand Recognition

A detailed analysis on some approaches to posture and gesture recognition is given by Watson [3], which include template matching, neural network, statistical classification, and discontinuity matching. These methods have their own pros and cons. The first approach is simple and efficient, but fails to provide accurate results. Neural network is highly tolerant of noise and incomplete data. However, this approach has a number of disadvantages. The training phase is too long, and a large set of labeled examples is needed. Manual tuning of parameters is required and there is no obvious rule which guides the tuning and design of the network. Adding new records may require re-training or even re-design of the network architecture. The problems of statistical classification are similar to the previous method. Discontinuity matching requires a lot of computation and thus cannot be integrated with VR applications, which are usually very CPU intensive.

More recent approaches include model-based analysis for hand postures by Lee [4] and Hidden Markov Model (HMM) for gesture recognition by Liang [5] and Nam [6]. The first one uses an iterative improvement approach and it takes a long time to recognize a posture. The success of HMM in handwriting and speech recognition has drawn researchers' attention recently. HMM is able to extract and recognize dynamic pattern. However, similar to the neural network approaches, the system design and training algorithm may not be obvious and manual tuning is needed. Another approach using fuzzy logic concept for hand posture recognition is presented by Tsang and Sun [7], whose primary goal is to improve the tolerance of noise or imprecise data in an efficient way.

3 Applications

While developing, VR technology has been widely used in many application areas, including interactive games, computer animation, industrial design, scientific visualization, robotics simulation, remote-control manipulators. The common framework of these applications is to transport and immerse users into a virtual space, in which they can view and manipulate the surrounding objects with the movements of the hand(s) and body. The diversity of these applications can be classified to two streams: One creates virtual environments which are viewed using the HMD-based (head-mounted display) systems for walkthrough and entertainment applications [8]. The other alternatives to HMD-based systems are being explored for precision work and extended use. These include industrial design [9], robotics simulation [10, 11], and surgical planning [12]. In these applications, the mapping of high-volume, multidimensional data into meaningful displays is important. The interactions with these displays are often too small, too abstract, and too complicated to be otherwise produced by the conventional 2D control interfaces. For instance, Poston and Serra [12] describe the virtual workbench which offers the virtual imaging tools for precise work of medical data at various levels.

Virtual environments allow production engineers to specify, verify, and optimize manufacturing tasks without touching any physical objects. Conventional on-line teaching pendants that program robots for assembly tasks are replaced by human-computer interactions in an off-line programming style. An operator can use such a system to insert, for example, a bushing into a roller simply by grabbing the bushing's graphics icon with a mouse, dragging it across the 2-D window, and placing it inside the graphical model of the roller. Possible paths of moving the bushing freely within a virtual workcell can then be produced with artificial reasoning and animated with robot simulation.

Similarly derivable are different assembly sequences that accomplish a given production process. In front of a computer monitor, the operator is able to check these simulated assembly operations to verify their feasibility and to improve the efficiency of an assembly plan. He/she works in the virtual workcell to detect potential collisions and to perform accurate cycle-time analysis. Furthermore, once the process is optimized, the programs can be downloaded directly to the machines and robots on the shopfloor. Programming in virtual environments [13] therefore offers the benefits of higher-quality products, lower costs, fewer prototypes, optimal efficiency, minimal risks, reduced time-to-market, and fewer problems when production begins.

Even through the whole-hand input interfaces have been applied in industrial and medical applications, there are still many issues and problems that need to be resolved, including constraints on degrees of freedom, sensory feedback, gesture recognition, and two-handed input, before whole-hand input becomes a generally useful tool.

References

[1] D. Sturman and D. Zeltzer, A Survey of Glove-based Input, *IEEE Computer Graphics and Applications*, 14(1); 30-39, 1994.

[2] H. Sun, Hand-guided Scene Modeling, *Virtual Reality and Its Applications*, Academic Press Ltd, 41-52, May 1994.

[3] R. Watson, A Survey of Gesture Recognition Techniques, *Technical Report TCD-CS-93-11*, Department of Computer Science, Trinity College, July 1993.

[4] J. Lee and T. L. Kunii, Model-based Analysis of Hand Posture, *IEEE Computer Graphics and Application*, 77-86, September 1995.

[5] R. H. Liang and M. Ouhyoung, A Sign Language Recognition System Using Hidden Markov Model and Context Sensitive Search, In *ACM Symposium on Virtual Reality Software and Technology*, 59-66, July 1996.

[6] Y. Nam and K. Wohn, Recognition of Space-Time Hand-Gestures Using Hidden Markov Model, In *ACM Symposium on Virtual Reality Software and Technology*, 51-58, July 1996.

[7] K. H. Tsang and H. Sun, An Efficient Posture Recognition Method Using Fuzzy Logic, submitted to *ACM Symposium on VRST '97*, April 1997.

[8] Chris Longhurst, Event-driven Visual Effects in Flight Simulation Virtual Environments, *Virtual Reality Applications*, Academic Press, 231-244, 1995.

[9] C. Wang and D. Cannon, Virtual-Reality-Based Point-and-Direct Robotic Inspection in Manufacturing, *IEEE Transactions on Robotics and Automation*, 12(4): 516-530, August 1996.

[10] H. Ogata and T. Takahashi, Robotic Assembly Operation Teaching in a Virtual Environment, *IEEE Transactions on Robotics and Automation*, 10(3): 391-399, 1994.

[11] H. Sun, X. Yuan, Y. Gu, CAD-Oriented Robot Programming in a Virtual Environment, To appear in *Virtual Reality Journal of Research, Development and Applications*, December 1997.

[12] T. Poston and L. Serra, Dextrous Virtual Work, *Communications of the ACM* Virtual Reality 39(5): 37-45, May 1996.

[13] Gilad Lederer, Making Virtual Manufacturing a Reality, *Industrial Robot*, 22(4): 16-17, 1995.

Quantification of Human Performance in Extreme Environments

Corinna E. Lathan,* and. Dava J. Newman
Department of Aeronautics and Astronautics, The Massachusetts Institute of Technology, Cambridge, MA.

1. MENTAL WORKLOAD AND PERFORMANCE EXPERIMENT (MWPE)

The Mental Workload and Performance Experiment (MWPE) was developed in the mid-80's to look at human performance during spaceflight. MWPE flew on the International Microgravity Laboratory-1 Space Shuttle mission in January of 1992 with four astronaut subjects [1]. Human performance has many aspects that are difficult to measure and distinguish from one another. Space Station operations, however, will put particular emphasis on astronauts' interaction with the Station's many computer control systems. The MWPE experiment was therefore designed to focus on motor and cognitive skills associated with such interactions, specifically computer cursor control and short-term memory. Though narrowly focused, the experiment serves as a prototype for further investigations to pursue broader, multidimensional measures of in-space performance. The MWPE performance assessment test is based on the *Fittsberg* task, a combination of Fitts' Law and Sternberg tasks, that combines tests of short-term memory and motor control [2-4].

2. MEMORY PROCESSES AND MOTOR CONTROL (MEMO)

MWPE was enhanced for the Canadian Astronaut Program Space Unit Life Simulation (CAPSULS) mission and was renamed the MEmory processes and MOtor control experiment (MEMO). This mission studied four Canadian astronauts during seven days of isolation at the Defense and Civil Institute of Environmental Engineering, Toronto, Canada. The CAPSULS 7-day isolation mission offered an ideal opportunity to collect human performance data for individuals in extreme isolation to compliment the data collected in space on the IML-1 mission using a similar protocol and ground hardware. The CAPSULS experiment duplicated the space flight workload and conditions of isolation without the physiological changes due to exposure to microgravity. The experiment then evaluated operator performance on the same short-term memory and fine motor control tasks that were performed for MWPE. MEMO then examined human performance when a sensorimotor transformation was deliberately induced in order to evaluate the limitations of different human operator control strategies. Specifically, our subjects wore left-right reversing prism goggles for approximately one-third of the trials.

---

* *Currently in the Department of Mechanical Engineering, Biomedical Engineering Program, The Catholic University of America, Washington, DC.*

3. METHODS

For both the MEMO and MWPE task, the four subjects used either a position-control device (trackball) or a rate-control device (joystick) to perform the experiment. A Grid 1530 microcomputer was used to present the experimental paradigm and collect the data. Finally, only during the MEMO experiment, the subjects repeated the task wearing left-right reversing prisms to induce a sensorimotor transformation. In other words, while wearing the reversing prisms, when the subject moved the joystick to the left, the cursor was seen to move to the right.

We used the "Fittsberg" experimental paradigm [3] that provides independent control and measurement of two tasks: response selection and response execution, where the former represents a cognitive task and the latter, a neuromuscular task. The selection of a response is based upon the Sternberg memory search task [4] that requires the subject to determine if a displayed item is a member of a previously memorized set. Fitts' paradigm [5] was developed to examine the control and accuracy of movement and was used here to measure response execution. Subjects were required to manually acquire a target of a certain size and distance away from an initial cursor position as quickly and as accurately as possible. The Fittsberg paradigm is illustrated in Figure 1.

From the time the targets appear to the time it takes for the subject to identify the letter is the reaction time (RT) and is a measure of short-term memory. From the time the subject starts to move the cursor on the screen via the computer device to the time he or she reaches the target is the movement time (MT) which is a classical measure of motor control. For each memory set, 8 test stimuli were presented, while 12 memory sets were presented for each device.

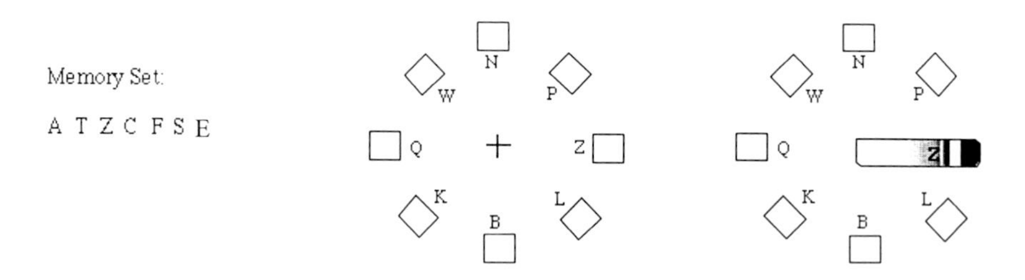

Figure 1: The Fittsberg Paradigm A) The subject was presented with a memory set consisting of 1-7 letters which they were asked to memorize. The subject pressed return on the keyboard to indicate the end of the memorization time. B) The subject was then immediately presented with a test stimulus with a cursor in the center. Only one of the letters from the memory set is presented, the letter Z in this case. C) As soon as the subject spotted the letter, the subject moved the cursor to that location. Once the location is reached, a new test stimulus appears immediately.

KEYWORD INDEX